P9-AFY-848

Frontiers in Astronomy

Readings from

SCIENTIFIC AMERICAN

Frontiers in Astronomy

With Introductions by

OWEN GINGERICH

Harvard University

and

Smithsonian Astrophysical Observatory

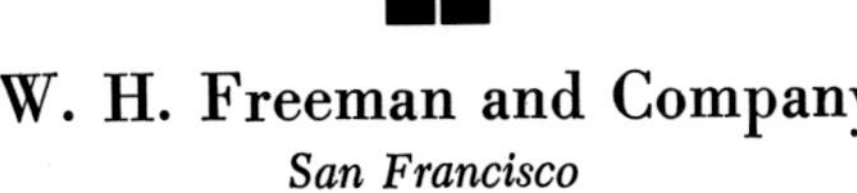

W. H. Freeman and Company

San Francisco

Seven of the SCIENTIFIC AMERICAN articles
in *Frontiers in Astronomy* (articles 1, 7, 14, 15, 28, 29 and 30)
are available as separate Offprints.
For a complete list of
approximately 700 articles now available as Offprints,
write to W. H. Freeman and Company,
660 Market Street, San Francisco, California 94104.

Printed in the United States of America

Library of Congress Catalog Card Number: 71–129925

International Standard Book Number:
0–7167–0948–1 (cloth); 0–7167–0947–3 (paper)

Preface

These are exciting and fast moving times for astronomy. At least four remarkable discoveries of the 1960's were essentially unanticipated when the decade began: the quasars, the pulsars, the 3° background radiation, and the intense infrared sources with the associated anomalous OH radio radiation.

This collection of articles from the *Scientific American* documents these and other recent developments in astronomy. Half the articles appeared since 1964; some describe developments so new that they are scarcely mentioned in textbooks. Others were published as early as 1956; with their earlier vantage point, they are now partly historical in character, yet for the most part these selections are still quite up to date. Witness the 1957 article on the Crab Nebula, in which Jan Oort draws particular attention to the star that has since turned out to be the first known optical pulsar (and at present, the only one).

In a sense, all the articles are historical. In Grant Athay's "The Solar Chromosphere" we are left hanging in breathless anticipation: will the 1962 eclipse expedition to New Guinea find clear skies? (*Note:* There was a dramatic clearing two minutes before totality.) Or will the rocket flight proposed by Herbert Friedman in "X-Ray Astronomy" blast off to a successful mission? And even such a recent an article as "Pulsars" by Antony Hewish is probably as incomplete as any in the collection. This is the small price we pay for articles that at some rather recent moment carried the latest astronomical news, often from a leading investigator just flushed with success.

My introductions to the eight sections have attempted to fill in some of the more recent developments in each area, for example, the success of Friedman's rocket flight. The introductions are admittedly sketchy, and they, too, will go out of date, but at least they clarify developments through the spring of 1970. Furthermore, I have added over sixty new references to the bibliography, especially for the older articles, and these sources will allow the reader to pursue the topics introduced here.

The most agonizing task in preparing this collection of readings from the *Scientific American* was neither the writing of the introductions nor the selection of the 33 representative articles, but rather the rejection of an even larger number of other contenders from the 1960's and at least 30 more from the 1950's. The high quality and broad coverage of the *Scientific American* has made the selection both difficult and ambiguous—nearly everyone with whom I have discussed this project has cited one or more articles not included in the final choice. More than a few interesting articles were eliminated because they seemed too specialized for a short general book. Nevertheless, at least one got in: Robert Richardson's "The Discovery of Icarus," which seemed just too readable not to be included. Like the *Scientific American* itself, this volume has been designed for the interested nonspecialist. Hence readability was an important criterion in the selection.

If the 1970's are filled with as many unexpected astronomical discoveries as the 1960's, there will soon be exciting new articles demanding to be read. Nevertheless, I hope that this collection, which includes writings by some of the most distinguished astronomers of our times, will continue to provide a background and perspective for what is to come.

Cambridge, Massachusetts
March, 1970

OWEN GINGERICH

Contents

VI Galaxies

VII The New Astronomy

VIII Cosmology

NOTE ON CROSS-REFERENCES

Cross-references within the articles are of three kinds. A reference to an article included in this book is noted by the title of the article and the page on which it begins; a reference to an article that is available as an offprint but is not included here is noted by the article's title and offprint number; a reference to a SCIENTIFIC AMERICAN article that is not available as an offprint is noted by the title of the article and the month and year of its publication.

Frontiers in Astronomy

I

The Earth and Moon

I

The Earth and Moon

INTRODUCTION

The choice of the first article in the collection, geologist Patrick Hurley's "The Confirmation of Continental Drift," may surprise many astronomers who rarely think of the ground beneath their feet as an astronomical object. Yet the crustal rocks provide information about the age, composition, and internal structure of the earth that is of inestimable value in deriving the evolutionary pattern of the solar system. And the fossil record contained within these crustal rocks not only assures us that the solar energy output has remained fairly constant for at least 5×10^9 years but gives mute testimony on the lengthening of the earth day through geological epochs.

In the course of describing the new evidence for continental drift (which has brought for geologists as sudden and exciting a revolution as the resolution of the nature and distances of spiral galaxies gave to astronomers in the period from 1918 to 1924), Hurley touches on many points of interest to planetary astronomers: the repeated reversal of the terrestrial magnetic field, the methods of dating ancient rocks, the elastic and convective nature of the earth's mantle. (A curious footnote: Alfred L. Wegener, one of the originators of the continental drift theory early in this century, began his scholarly career as an historian of astronomy, analyzing the 13th-century Alfonsine Tables of planetary motions!)

Incidentally, astronomy students unfamiliar with the geologist's timetable ought to memorize the order of the five geological periods at the left of the diagram on page 6 before tackling Hurley's article.

Perhaps the claim made in the opening paragraph that crustal rocks provide evidence of internal structure requires further elucidation. The average density for terrestrial surface rocks is 2.8 grams per cubic centimeter (g/cm^3), but the mean density of the earth (as derived from the strength of its surface gravity) is 5.5 g/cm^3. Hence we deduce the presence of a high density core, plus the possibility of substantial geochemical fractionation in the earth's interior. This now stands in distinct contrast to the moon, the only other astronomical object for which this comparison can be made. Lunar rocks retrieved by the Apollo 11 and 12 explorations have a surprisingly high density range of 3.1 to 3.5 g/cm^3, not appreciably different from the moon's mean density of 3.3 g/cm^3.

Now that men have set foot on the moon, most of the numerous *Scientific American* articles on our earth's nearest neighbor are too outdated for inclusion in this collection. The article on the Lunar Orbiters, however, has much of the same timelessness as Galileo's *Sidereus Nuncius,* with its pioneering telescopic charts of the moon. The magnificent pictures of Tsiolkovsky, Mare Orientale, and even the track of the lunar rock that rolled down the crater Vitello—part of the first comprehensive look at the entire moon—will remain a triumph of startling freshness even after dozens of manned landings on moon. And the text of the article by the Boeing engineers Levin, Viele, and Eldrenkamp will remind us of the ingenious technical achievement that has recreated here on earth the 1,950 photographs taken and how lost in space in the five (out of five) successful Orbiter spacecrafts.

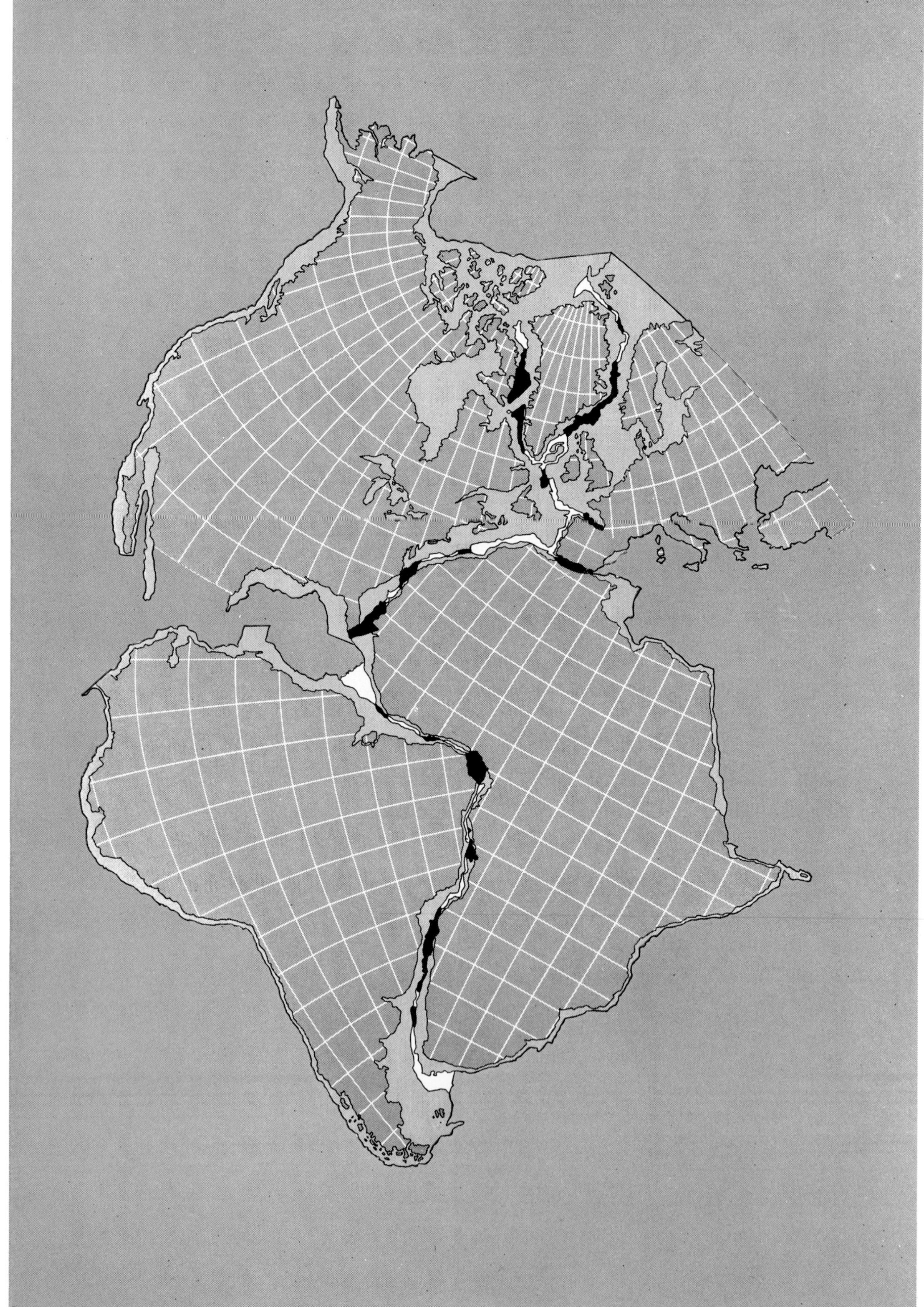

1

The Confirmation of Continental Drift

PATRICK M. HURLEY

April 1968

As recently as five years ago the hypothesis that the continents had drifted apart was regarded with considerable skepticism, particularly among American investigators. Since then, as a result of a variety of new findings, the hypothesis has gained so much support that its critics may now be said to be on the defensive. The slow acceptance of what is actually a very old idea provides a good example of the intensive scrutiny to which scientific theories are subjected, particularly in the earth sciences, where the evidence is often conflicting and where experimental demonstrations are usually not possible.

As long ago as 1620 Francis Bacon discussed the possibility that the Western Hemisphere had once been joined to Europe and Africa. In 1668 P. Placet wrote an imaginative memoir titled *La corruption du grand et du petit monde, où il est montré que devant le déluge, l'Amérique n'était point séparée des autres parties du monde* ("The corruption of the great and little world, where it is shown that before the deluge, America was not separated from the other parts of the world"). Some 200 years later Antonio Snider was struck by the similarities between American and European fossil plants of the Carboniferous period (about 300 million years ago) and proposed that all the continents were once part of a single land mass. His work of 1858 was called *La Création et Ses Mystères Dévoilés* ("The Creation and Its Mysteries Revealed").

By the end of the 19th century geology had come seriously into the discussion. At that time the Austrian geologist Eduard Suess had noted such a close correspondence of geological formations in the lands of the Southern Hemisphere that he fitted them into a single continent he called Gondwanaland. (The name comes from Gondwana, a key geological province in east central India.) In 1908 F. B. Taylor of the U.S. and in 1910 Alfred L. Wegener of Germany independently suggested mechanisms that could account for large lateral displacements of the earth's crust and thus show how continents might be driven apart. Wegener's work became the center of a debate that has lasted to the present day.

Wegener advanced a remarkable number of detailed correlations, drawn from geology and paleontology, indicating a common historical record on the two sides of the Atlantic Ocean. He proposed that all the continents were joined in a single vast land mass before the start of the Mesozoic era (about 200 million years ago). Wegener called this supercontinent Pangaea. Today the evidence favors the concept of two large land masses: Gondwanaland in the Southern Hemisphere and Laurasia in the Northern.

In the Southern Hemisphere an additional correlation was found in a succession of glaciations that took place in the Permian and Carboniferous periods. These glaciations left a distinctive record in the southern parts of South America, Africa, Australia, in peninsular India and Madagascar and, as has been discovered recently, in Antarctica. The evidence of glaciations is compelling. Beds of tillite—old, consolidated glacial rubble—have been studied in known glaciated regions and are unquestioned evidence of the action of deep ice cover. In addition many of the tillites rest on typically glaciated surfaces of hard crystalline rock, planed flat and grooved by the rock-filled ice moving over them.

This kind of evidence has been found throughout the Southern Hemisphere. In all regions the tillites are found not only in the same geological periods but also in a sequence of horizontal beds bearing fossils of identical plant species. This sequence, including the geological periods from the Devonian to the Triassic, is called the Gondwana succession. The best correlations are apparent in the Permocarboniferous beds, where two distinctive plant genera, *Glossopteris* and *Gangamopteris,* reached their peak of development. These plants were so abundant that they gave rise to the Carboniferous coal measures, which are commonly interbedded in the Gondwana succession [*see top illustration on pages 6 and 7*].

The South African geologist Alex L. du Toit and others have sought out and mapped these Gondwana sequences so diligently that today they provide the strongest evidence not only that these continental areas were joined in the past but also that they once wandered over or close to the South Pole. It is inconceivable that the complex speciation of the Gondwana plants could have evolved in the separate land masses we see today. It takes only a narrow strip of water, a few tens of miles wide at the most, to stop the spread of a diversified plant regime. The Gondwana land mass was apparently a single unit until the Mesozoic era, when it broke into separate parts. Thereafter evolution proceeded on divergent paths, leading to the biological diversity we observe today on the different continental units.

Wegener and Du Toit published their work in the 1920's and 1930's. The de-

FIT OF CONTINENTS (*opposite page*) was optimized and error-tested on a computer by Sir Edward Bullard, J. E. Everett and A. G. Smith of the University of Cambridge. Over most of the boundary the average mismatch is no more than a degree. The fit was made along the continental slope (*light gray*) at the 500-fathom contour line. The regions where land masses, including the shelf, overlap are black; gaps are white.

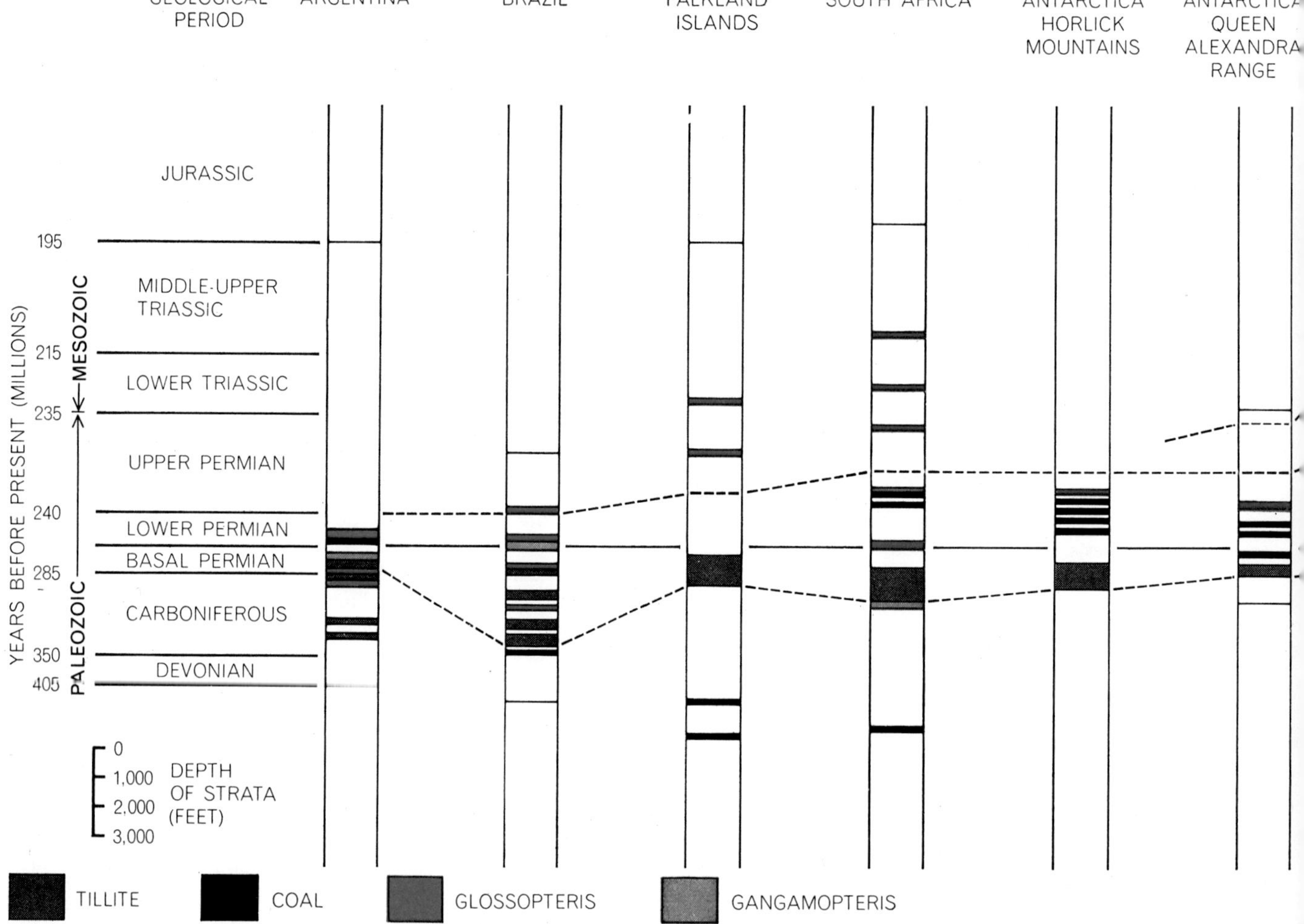

GONDWANA SUCCESSION is the name given to a late Paleozoic succession of land deposits found in South America, Africa, Antarctica, India and Australia. The succession contains beds of tillite (glacial rubble), coal deposits and a diversity of plants arranged in such a way that perhaps 200 million years ago the different areas must have been a single land mass known as Gondwanaland, or at

bate for and against drift became polarized largely between geologists of the Southern Hemisphere and the leaders of geophysical thought in the Western Hemisphere. Eminent geophysicists such as Sir Harold Jeffreys of the University of Cambridge voiced strong opposition to the hypothesis on the grounds that the earth's crust and its underlying mantle were too rigid to permit such large motions, considering the limited energy thought to be available.

Not all felt this way, however. In the late 1930's the Dutch geophysicist F. A. Vening Meinesz proposed that thermal convection in the earth's mantle could provide the mechanism. His ideas were supported by his gravity surveys over the deep-sea trenches and the adjacent island arcs of the western Pacific. The results implied that some force was maintaining the irregular shape of the earth's surface against its natural tendency to flatten out. Presumably the force was somehow related to thermal convection. Arthur Holmes of the University of Edinburgh added his weight to the argument in favor of the hypothesis, and he was followed by S. W. Carey of Tasmania, Sir Edward Bullard and S. K. Runcorn of Britain, L. C. King of South Africa, J. Tuzo Wilson of Canada and others [see "Continental Drift," by J. T. Wilson; SCIENTIFIC AMERICAN Offprint 868]. The historical and dynamical characteristics of the earth now engaged the attention of many more geophysicists, and today the interplay of all branches of geology and geophysics generates the excitement of a new frontier area.

Continents and Oceans

Although the general nature of the earth's crust is familiar to most readers of *Scientific American,* it is worth reviewing and summarizing some of its major features while asking: How do these features look in the context of continental drift? The earth's topography has two principal levels: the level of the continental surface and the level of the oceanic plains. The elevations in between represent only a small fraction of the earth's total surface area. What maintains these levels? Left alone for billions of years, they should reach equilibrium at an average elevation below the present sea level, so that the earth would be covered with water. Instead we see sharp continental edges, new mountain belts, deep trenches in the oceans—in short, a topography that appears to have been regularly rejuvenated.

The continental areas are a mosaic of blocks that are roughly 1,000 kilometers across and have ages ranging from about 3,000 million years to a few tens of millions. In Africa there appear to be several ancient nuclear areas, or cratons, surrounded by belts of younger rocks. Most of the younger belts have an age of 600 million years or less, contrasting sharply with an age of 2,000 million to 3,000 million years for the cratons.

A closer look at the younger belts tells us that although much of the material is

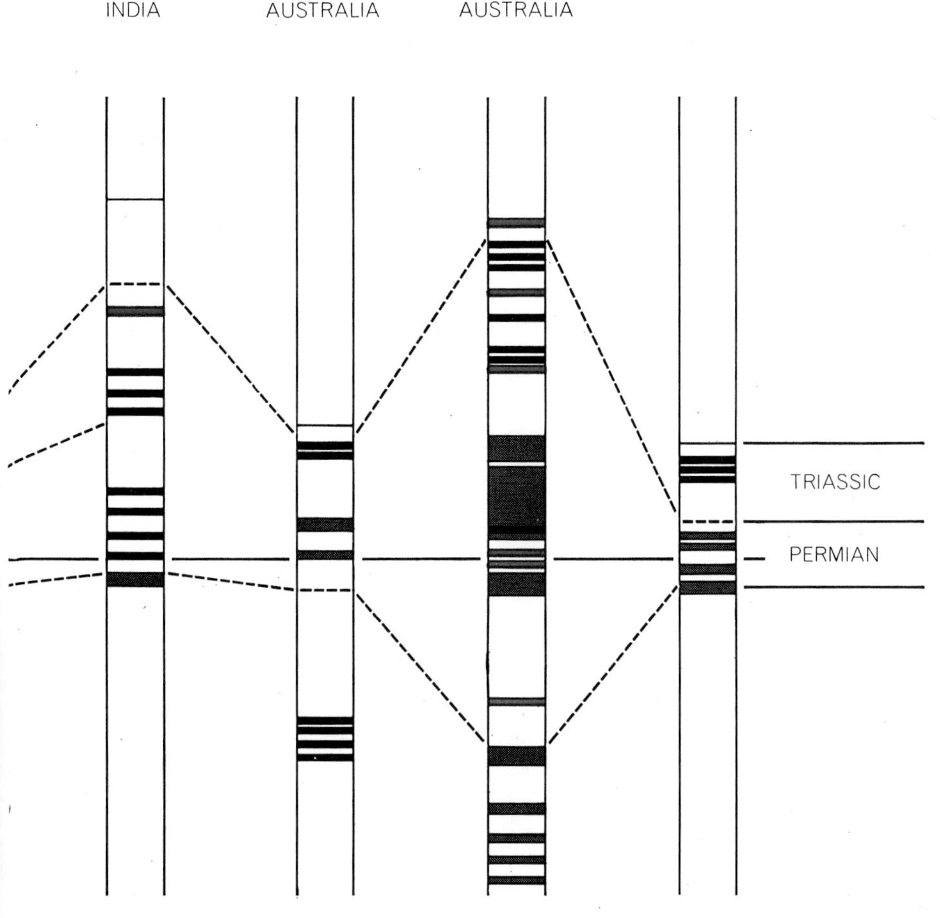

the very least a closely associated mass connected by land bridges. Only two of several major plant genera are plotted here: *Glossopteris* and *Gangamopteris*. The depths of the various deposits have been arbitrarily aligned between the lower and the basal Permian.

apparently new, there are large blocks that have the same age as the cratons. It looks as if the earth's surface has been warped and folded around the ancient continental masses, catching up segments of the crust and intruding younger igneous rocks into the folds. In some places the ancient material has been altered beyond recognition, but elsewhere it has been left fairly undisturbed and its antiquity can be determined by radioactive-dating methods. These composite belts are termed zones of rejuvenation. When they are eroded down to sea level, all we see, as far as topography is concerned, is another part of the continental platform. Geological mapping, however, reveals the belt structure clearly. A closer look at the cratons shows us that they too have the structure of preexisting mountain belts that have been carved into segments, with the younger material always cutting across the older structural pattern.

We see the process in action today. Our young mountain belts have not been eroded to sea level but show high elevations that are clearly apparent; we do not need geological surveys to observe them. It is only when we see the global distribution of these mountain belts on land areas, together with the distribution of rifts and their associated ridges under the oceans, that we begin to perceive the possibility that vast motions of the earth's surface may be their cause [*see illustration on next two pages*].

The earth is also encircled by belts of geological activity in the form of volcanoes, earthquakes and high heat flow, and observable motions in the form of folded rocks and the large displacements known as faults. In recent years the direction of displacements that are not observable on the surface has been deduced by the study of seismic waves arriving at various points on the earth's surface from earthquakes. It is now possible to tell the direction of slippage in the zones of rupture within the solid rocks of the earth's near-surface regions, so that the directions of the forces can be obtained.

If one looks at a map such as the one on the next two pages, one is immediately struck by the large scale and systematic distribution of these lines of geologic activity. Some of the systems are coherent over distances of several thousand kilometers. This immediately suggests the large-scale motion of material in the earth's interior. It does not, however, necessarily imply motions extending a similar distance into the interior. It is

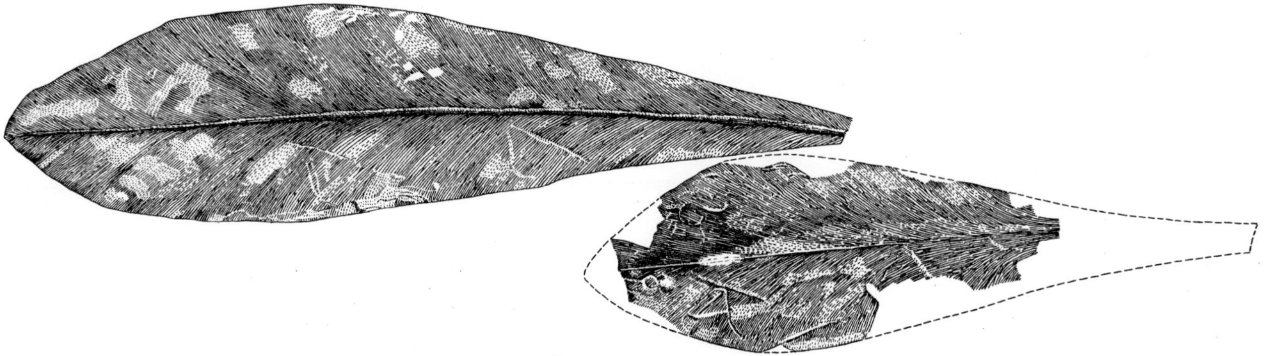

TYPICAL GONDWANA FLORA are *Glossopteris communis* (*leaf at left*) and *Gangamopteris cyclopteroides* (*right*), two species of fern that are identified in the Gondwana succession illustrated at the top of these two pages. The fossils from which these drawings were made were uncovered in the central part of Antarctica in 1961–1962 by William E. Long of Alaska Methodist University.

possible to have sheets of rigid material supporting stresses and fracturing over great distances if the underlying material is less rigid.

The topography of the ocean floors has been rapidly revealed in the past two decades by the sonic depth recorder. The principal systems of ridges and faults have been mapped in considerable detail by such oceanographers as Bruce C. Heezen and Maurice Ewing of Columbia University and H. W. Menard of the Scripps Institution of Oceanography. The layers of sediment on the sea floor have also been explored by such methods as setting explosive charges in the water and recording the echoes. It became a great puzzle how in the total span of the earth's history only a thin veneer of sediment had been laid down. The deposition rate measured today would extend the process of sedimentation back to about Cretaceous times, or 100 to 200 million years, compared with a continental and oceanic history that goes back at least 3,000 million years. How could three-quarters of the earth's surface be wiped clean of sediment in the last 5 percent of terrestrial time? Furthermore, why were all the oceanic islands and submerged volcanoes so young? The new oceanographic investigations were presenting questions that were awesome to contemplate.

In the early 1960's Harry H. Hess of Princeton University and Robert S. Dietz of the U.S. Coast and Geodetic Survey independently proposed that the oceanic ridge and rift systems were created by rising currents of material which then spread outward to form new ocean floors. On this basis the ocean floors would be rejuvenated, sweeping along with them the layer of sedimentary material. If such a mechanism were at work, no part of the ocean basins would be truly ancient. Although this radical hypothesis had much in its favor, it appeared farfetched to most.

Tracking the Shifting Poles

During this time a group of physicists and geophysicists were studying the directions of magnetism "frozen" into rocks in the hope of tracing the history of the earth's magnetic field. When an iron-bearing rock is formed, either by crystallization from a melt or by precipitation from an aqueous solution, it is

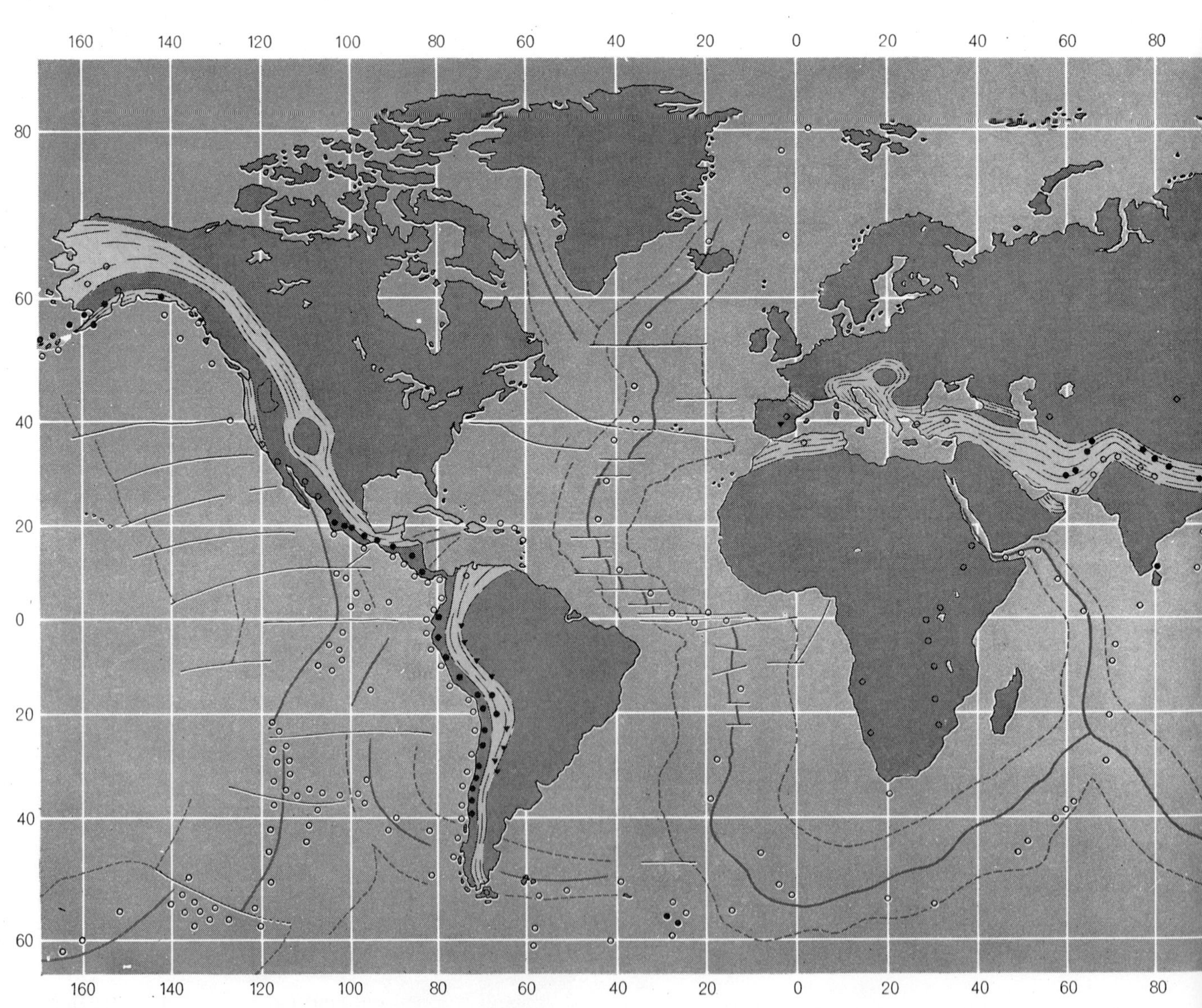

WORLDWIDE GEOLOGICAL PATTERNS provide evidence that the major land masses have been driven apart by a slow convection process that carries material upward from the mantle below the earth's crust. The dark-colored lines identify the crests of oceanic ridges that are now believed to coincide with upwelling regions. These ridges are crossed by large transcurrent fracture zones. The broken lines show the approximate limits of the oceanic rises. The light gray areas identify the worldwide pattern of recent mountain belts, island arcs, deep trenches, earthquakes and volcanism that apparently mark the downwelling of crustal material. The

slightly magnetized in the direction of the earth's magnetic field. Unless this magnetism is disturbed by reheating or physical distortion it is retained as a permanent record of the direction and polarity of the earth's magnetic field at the time the rock was formed. By measuring the magnetism in rocks of all ages from different continents, it has been possible to reconstruct the position of the magnetic pole in the past history of the earth. Great impetus was given to this study by P. M. S. Blackett and Runcorn, who with others soon found that the position of the pole followed a path going backward in time that was different for each continent [*see top illustration on next page*].

The interpretation of this effect was that the continents had moved with respect to the present position of the magnetic pole, and that since the paths were different for each land mass, they had moved independently. Because it was unlikely that the magnetic pole had wandered very far from the axis of the earth's rotation, or that the axis of rotation had changed position with respect to the principal mass of the earth, it was concluded that the continents had moved over the surface of the earth. Moreover, since the shift in latitude of the southern continents was generally southward going backward in time, the motions were in accord with the older evidence pointing toward a Gondwanaland in the south-polar regions. In short, the magnetic evidence supported not only the notion of continental drift but also the general locations from which the continents had moved within the appropriate time span.

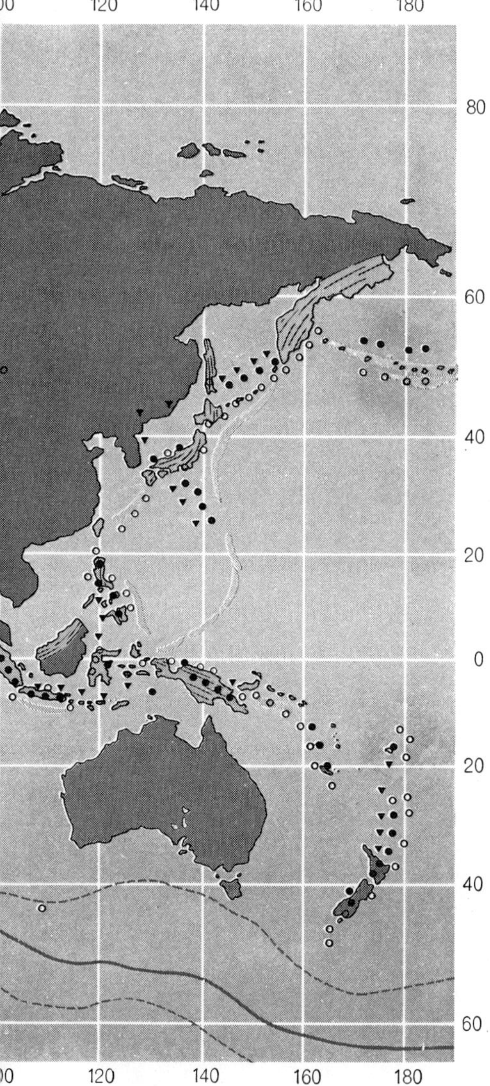

downwelling seems to coincide with the occurrence of deep earthquakes (*triangles*) and earthquakes of intermediate depth (*solid dots*). Upwelling zones seem to coincide only with shallow earthquakes (*open dots*).

This was still not enough to sway the preponderance of American scientific opinion. Finally, at the annual meeting of the Geological Society of America in San Francisco in 1966, came the blows that broke the back of the opposition. Several papers put forward startling new evidence that related the concepts of ocean-floor spreading and continental drift, the cause of the oceanic-ridge and fault systems and the direction and time scale of the drift motions. In addition, the development of new mechanisms explaining displacement along faults brought into agreement some of the formerly contradictory seismic evidence.

In the study of rock magnetism it was observed that the earth's magnetic field not only had changed direction in the past but also had reversed frequently. In order to study how frequently and when the reversals occurred three workers in the U.S. Geological Survey—Allan Cox, G. Brent Dalrymple and Richard R. Doell—carefully measured the magnetism in samples of basaltic rocks that they dated by determining the amount of argon 40 in the rocks formed by the decay of radioactive potassium 40. They noted a distinct pattern of reversals over some 3.6 million years [see "Reversals of the Earth's Magnetic Field," by Allan Cox, G. Brent Dalrymple and Richard R. Doell; SCIENTIFIC AMERICAN, February, 1967]. Their finding was soon confirmed when Neil D. Opdyke and James D. Hays of Columbia University found the same pattern in going downward into older layers in oceanic sediments. It was thus established that the polarity of the magnetic field had universally reversed at certain fixed times in the past.

Meanwhile an odd pattern of magnetism in the rocks of the ocean floors had been detected by Ronald G. Mason and Arthur D. Raff of the Scripps Institution of Oceanography. Using a shipborne magnetometer, they found that huge areas of the ocean floor were magnetized in a stripelike pattern. Putting together these patterns, the discovery of magnetic reversals and Hess's idea that the oceanic ridges and rifts were the site of rising and spreading material, F. J. Vine, now at Princeton, and D. H. Matthews of the University of Cambridge proposed that the hypothesis of the continuous creation of new ocean floors might be tested by examining the magnetic pattern on both sides of an oceanic ridge. The extraordinary discovery that the pattern was symmetrical with the ridge was demonstrated by Vine and Tuzo Wilson, who studied the two sides of a ridge next to Vancouver Island.

The history of the magnetic field going back into the past was laid out horizontally in the magnetism of the rocks of the sea floor going away from the ridge in both directions. It appeared that new hot material was rising from the rift in the center of the ridge and becoming magnetized in the direction of the earth's field as it cooled; it then moved outward, carrying with it the history of magnetic reversals. Since the dates of the reversals were known, the distance to each reversed formation gave the rate of spreading of the ocean floor [*see bottom illustration on next page*].

This important piece of work was quickly followed up by James R. Heirtzler, W. C. Pitman, G. O. Dickson and Xavier Le Pichon of Columbia, who have now shown that the ridges of the Pacific, Atlantic and Indian oceans all exhibit similar patterns. In fact, these workers have detected recognizable points in the history of magnetic reversals back about 80 million years, or in the Cretaceous period, and have drawn isochron lines, or lines of equal age, over huge strips of the ocean floors. Hence it is now possible to date the ocean floors and perceive the direction and rate of their lateral motion simply by conducting a magnetic survey over them. The implications for the study of drifting continents are immediately apparent.

These and other new findings do not unequivocally call for continental drift. It might be possible to have sea-floor spreading without drifting continents. Nonetheless, the directions and rates of motion for both sea-floor spreading and continental drift are entirely compatible. Above all, the principal objection to a hypothesis of continental movement has been removed.

Looking back, it is interesting to ob-

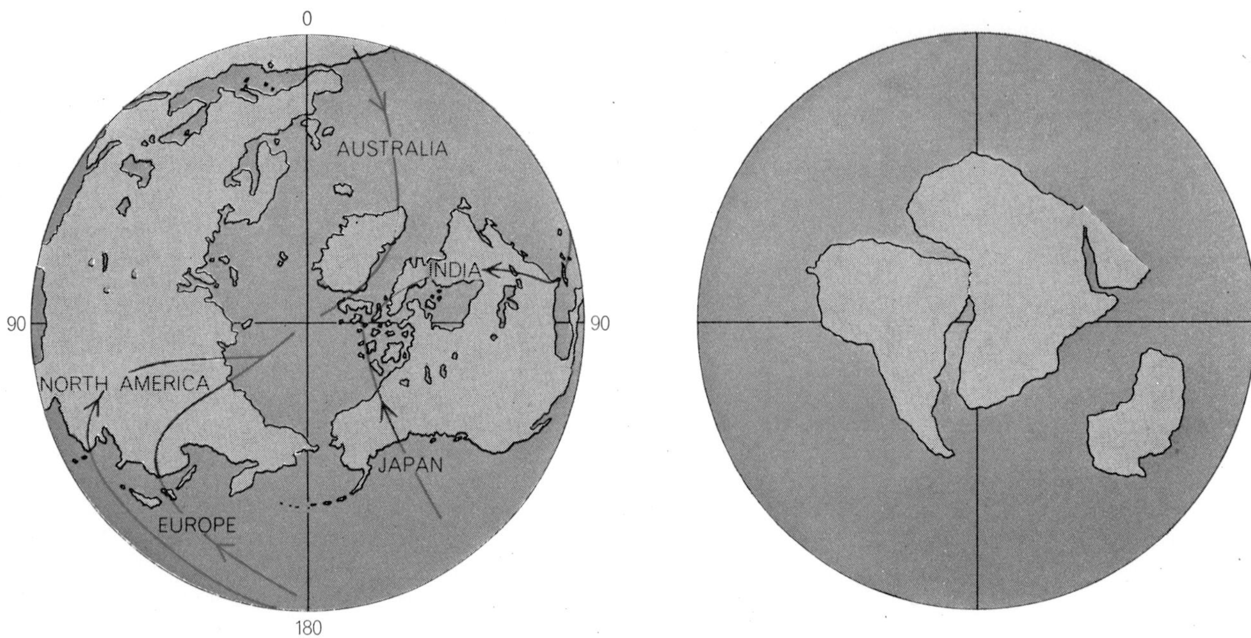

NORTH MAGNETIC POLE would appear to have wandered inexplicably during the past few hundred million years (*colored lines at left*), on the basis of "fossil" magnetism measured in rocks of various ages in various continents. The diagram is based on one by Allan Cox and Richard R. Doell of the U.S. Geological Survey. The pole could hardly have followed so many different tracks simultaneously; evidently it was the continents that wandered. K. M. Creer of the University of Newcastle upon Tyne found that the tracks could be brought together if South America, Africa and Australia were grouped in the late Paleozoic as shown at the right.

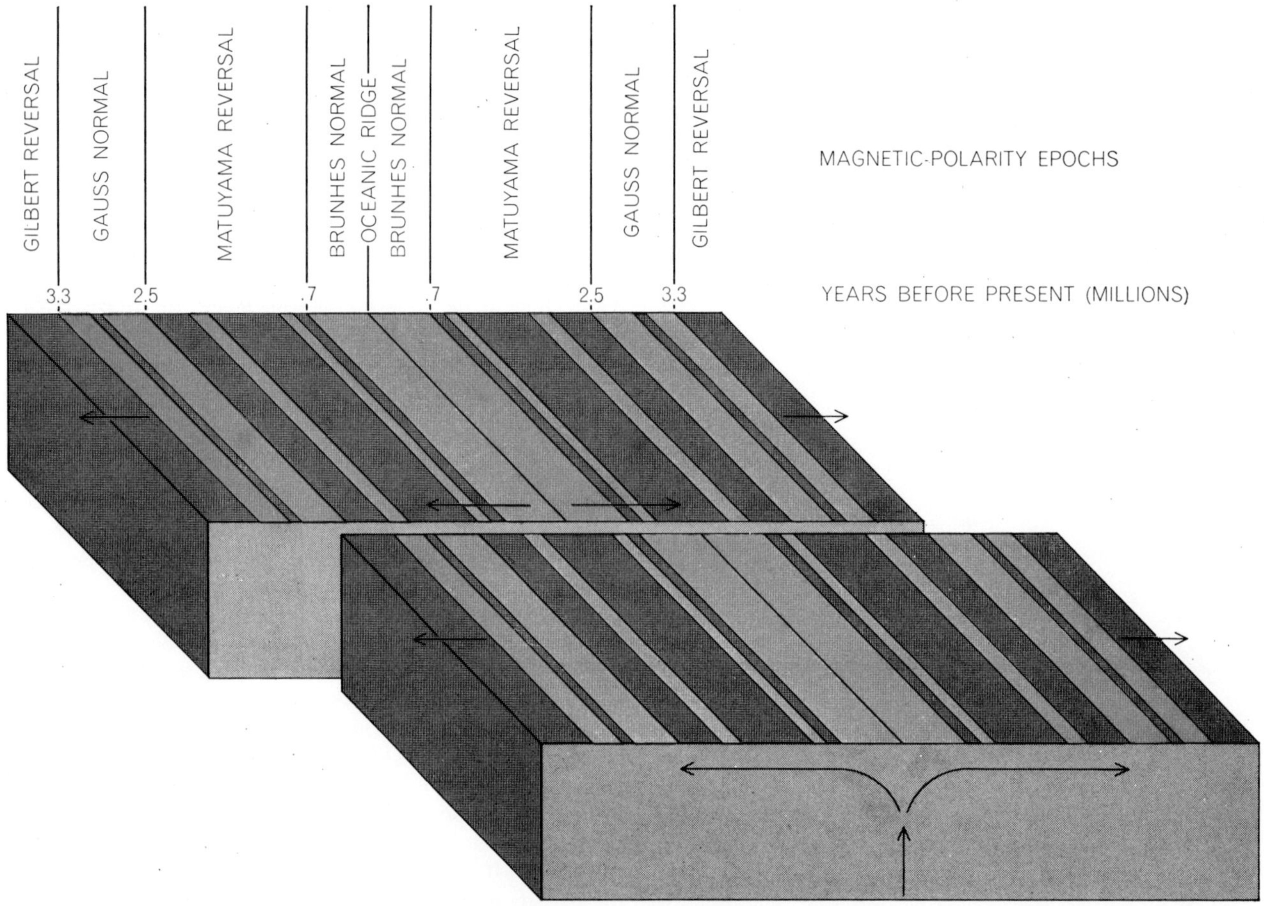

GILBERT REVERSAL
GAUSS NORMAL
MATUYAMA REVERSAL
BRUNHES NORMAL
OCEANIC RIDGE
BRUNHES NORMAL
MATUYAMA REVERSAL
GAUSS NORMAL
GILBERT REVERSAL

MAGNETIC-POLARITY EPOCHS

3.3 2.5 .7 .7 2.5 3.3

YEARS BEFORE PRESENT (MILLIONS)

EVIDENCE FOR SEA-FLOOR SPREADING has been obtained by determining the polarity of fossil magnetism in rocks lying on both sides of oceanic ridges. In the diagram rocks of normal, or present-day, polarity are shown in color; rocks of reversed polarity are in gray. The displacement of the two blocks represents a transcurrent fracture zone. The symmetry suggests that the rocks welled up in a molten or semimolten state and gradually moved outward. The diagram is based on studies by a number of workers.

serve how each new piece of evidence presented in the past was met by counterevidence. Wegener's reconstruction, for example, was countered by numerous geologists who took exception to his detailed arguments. The arguments for the Permocarboniferous Gondwana glaciations were countered by Daniel I. Axelrod of the University of California at Los Angeles and others in this country. They contended that most species of fossil plant tend to be restricted to zones of latitude that hold for the continents in their present position, a fact that is hard to reconcile with the presumed pattern of glaciation. The idea that the great Gondwana land masses drifted in latitude has also been opposed by F. G. Stehli of Case Western Reserve University; his studies suggest that ancient fauna were most diverse at the Equator, and that the Equator defined in this way has not shifted.

Another Test of the Hypothesis

Any hypothesis must be tested on all points of observational fact. The balance of evidence must be strongly in its favor before it is even tentatively accepted, and it must always be able to meet the challenge of new observations and experiments. My own interest in the problem of continental drift was stimulated at a 1964 symposium in London sponsored by the Royal Society and arranged by Blackett, Bullard and Runcorn. At that time Bullard and his University of Cambridge associates J. E. Everett and A. G. Smith presented an elegant study of the geographic matching of continents on both sides of the North and South Atlantic. They had employed a computer to produce the best fit by the method of least squares. Instead of using shorelines, as had been done in earlier attempts, they followed the lead of S. W. Carey; he had chosen the central depth of the continental slope as representing the true edge of the continent.

The fit was remarkable [*see illustration on page 4*]. The average error was no greater than one degree over most of the boundary. My colleagues and I at the Massachusetts Institute of Technology now began to think of further testing the fit by comparing the sequence and age of rocks on opposite sides of the Atlantic.

Radioactive-dating techniques for determining the absolute age of rocks had reached a point where much could be learned about the age and history of both the ancient cratonic regions and the younger rejuvenated ones. For such purposes two techniques can be used in combination: the measurement of strontium 87 formed in the radioactive decay of rubidium 87 in a total sample of rock, and the measurement of argon 40 formed in the decay of potassium 40 in minerals separated from the rock. A collaborative effort was arranged between our geochronology laboratory and the University of São Paulo in Brazil (in particular with G. C. Melcher and U. Cordani of that institution). We also enlisted the aid of field geologists who had been working on the west coast of Africa (in Nigeria, the Ivory Coast, Liberia and Sierra Leone) and on the east coast of Brazil and Venezuela. The São Paulo group made the potassium-argon measurements of the Brazilian rock samples; we did the rubidium-strontium analyses on samples from all locations.

European geochronologists (notably M. Bonhomme of France and N. J. Snelling of Britain) had done pioneering work on the Precambrian geology of former French and British colonies and protectorates in West Africa. Of special interest to us at the start was the sharp boundary between the 2,000-million-year-old geological province in Ghana, the Ivory Coast and westward from these countries, and the 600-million-year-old province in Dahomey, Nigeria and east. This boundary heads in a southwesterly direction into the ocean near Accra in Ghana. If Brazil had been joined to Africa 600 million years ago, the boundary between the two provinces should enter South America close to the town of São Luís on the northeast coast of Brazil. Our first order of business was therefore to date the rocks from the vicinity of São Luís.

To our surprise and delight the ages fell into two groups: 2,000 million years on the west and 600 million years on the east of a boundary line that lay exactly where it had been predicted. Apparently a piece of the 2,000-million-year-old craton of West Africa had been left on the continent of South America.

In subsequent work on both sides we have found no incompatibilities in the age of many geological provinces on both sides of the South Atlantic [*see illustration on next page*]. Furthermore, the structural trends of the rocks also agree, at least where they are known. Minerals characteristic of individual belts of rocks are also found in juxtaposition on both sides; for example, belts of manganese, iron ore, gold and tin seem to follow a matching pattern where the coasts once joined.

Can such comparisons be made elsewhere? To some extent, yes. Unfortunately the rifting process by which a continent breaks up seems to be guided by zones of rejuvenation between cratons, as if these zones were also zones of weakness deep in the crust. It is necessary for the break to have transected the structure of the continent, cutting across age provinces, if one is to get a close refitting of the blocks. In the North Atlantic this is not the case, but the continental areas on both sides were simultaneously affected by an unmistakable oblique crossing of a Paleozoic belt of geological activity [*see illustration on page 13*]. Actually the belt covers the region of the Appalachian Mountains and the Maritime Provinces of North America, with an overlap along the coast of West Africa, and then splits into two principal belts: one extending through the British Isles and affecting the Atlantic coast of Scandinavia and Greenland and the other turning eastward into Europe. There is a superposition of at least four periods of renewed activity affecting the various parts of this complex. All four are represented on both sides of the North Atlantic, making this correlation extremely difficult to explain unless the continents were once together.

My colleagues H. W. Fairbairn and W. H. Pinson, Jr., and I, as well as other workers, have made age measurements in the northern Appalachians and Nova Scotia for many years, and we have found all four periods well represented in New England. The earliest period of activity (which Fairbairn has named Neponset) is dated about 550 million years ago; it is seen in some of the large rock bodies in eastern Massachusetts and Connecticut, in the Channel Islands off the northern coast of France, in Normandy, Scotland and Norway. The next-oldest period (the Taconic) was about 450 million years ago and is found on the western edge of New England and in parts of the British Isles. The next period, going back about 360 million years, is strongly represented in the entire span of the Appalachians and Nova Scotia (where it is called the Acadian) and in England and Norway (where it is called the Caledonian). Finally, about 250 million years ago, the activity seemed to move into southern Europe and North Africa, where it has been called the Hercynian. This activity, however, also extended into New England; much of southern Maine, eastern New Hampshire, Massachusetts and Connecticut show rocks of this age. Here the event is called the Appalachian.

Farther south the Lower Paleozoic section of the northwest coast of Africa (Senegal) appears to continue under the younger coastal sediments of Florida.

This African belt shows large rock units with ages equivalent to the Neponset, and also evidence of the younger events.

The Fitting of Antarctica

The recent extensive geological surveys in Antarctica have been highly rewarding in reconstructing Gondwanaland. Prior to the end of the Permian period the younger parts of western Antarctica were not yet formed. Only eastern Antarctica was present, including the great belts of folded rocks that form the Transantarctic Mountains. These consist of two geosynclines, or sediment-filled troughs: the inner Eopaleozoic and the outer Paleozoic [*see illustration on page 14*]. The inner belt includes late Precambrian and early Cambrian sediments, which were folded and invaded by igneous rocks during late Cambrian or early Ordovician times (about 500 million years ago). Thus the inner belt is similar in age to the widespread event in the rest of Gondwanaland. It is marked by the Cambrian fossil Archaeocyatha, an organism that formed barrier reefs. These coral-like structures are found transecting sediments in bodies known as bioherms. The outer belt, farther within western Antarctica, is a geosyncline filled with Lower Paleozoic sediments. Like the northern Appalachians, it was deformed and invaded by igneous rocks in the middle and late Paleozoic. Later it was covered with a quite representative Gondwana succession, with its glacial deposits, coal and diverse plants.

There seems to be a similar record of events in eastern Australia. The bioherms of the Cambrian Archaeocyatha are found in a belt extending northward from Adelaide and mark the edge of an early geosyncline filled with sediments including late Precambrian and Cambrian ones. Later in time, and farther to the east, great thicknesses of Silurian and Lower Devonian sediments accumulated in the Tasman trough. Compression and igneous intrusion occurred in this Tasman geosyncline mostly in the late Lower Devonian period to the middle Devonian (about 350 million years

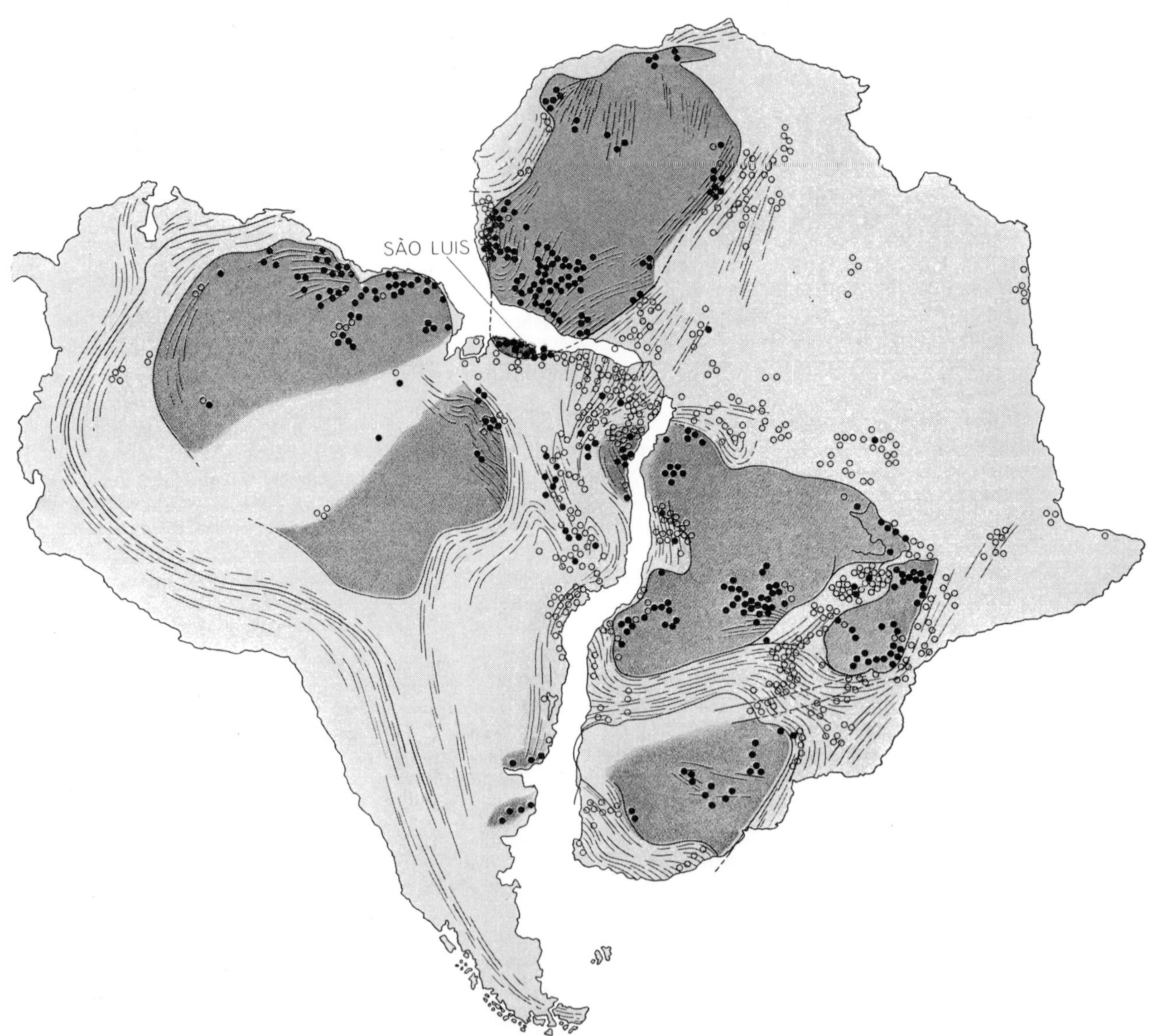

TENTATIVE MATCHING of geological provinces of the same age shows how South America and Africa presumably fitted together some 200 million years ago. Dark-colored areas represent ancient continental blocks, called cratons, that are at least 2,000 million years old. Light-colored areas are younger zones of geological activity: mostly troughs filled with sediments and volcanic rocks that were folded, compressed and intruded by hot materials, forming granites and other rock bodies. Much of this activity was 450 million to 650 million years ago, but some of it goes back 1,100 million years. The dots show the sites of rocks dated by many laboratories, including the author's at the Massachusetts Institute of Technology. Solid dots denote rocks older than 2,000 million years; open dots denote younger rocks. The region near São Luís is part of an African craton left stranded on the coast of Brazil.

ago). The later cover of sediments includes a Gondwana succession similar to the one in Antarctica.

There is also strong evidence for a juncture between Australia and India, particularly in the Permian basins of sedimentation of the two continental blocks and in Gondwana sequences of coal and plants. Limestone beds containing the same Productid shells are found in the upper layers of the sequence on both sides. A correlation also exists between the banded iron ores of Yampi Sound in northwestern Australia and the similar ores of Singhbhum in India.

The illustration on page 14 is a reconstruction of Gondwanaland based on the evidence we have discussed so far. The three land masses—Antarctica, Australia and India—have been fitted together not at their present shorelines but where the depth of the surrounding ocean reaches 1,000 meters. As can be seen, the fit of the edges is good. The detailed fit of this assemblage into the southeastern part of Africa is still debated because most of the edges lack structures that cut across them. Nevertheless, I have included the edge of Africa in the map to show how it might possibly fit on the basis of limited age data from Antarctica.

This arrangement of land masses in the late Paleozoic is extremely tentative. It is now up to the geochronologists to test each juncture more closely for correlations in geologic age, and up to the field geologists to match structure and rock type. One particularly interesting fit may be forthcoming in a study of the boundaries of shallow and deep marine glacial deposits, and of the land tillites around what appears to be the start of an oceanic basin at the time Antarctica was breaking away. This attempt to establish the former position of Antarctica, which is being made by L. A. Frakes and John C. Crowell of the University of California at Los Angeles, may set in place the key piece in the puzzle. A detailed correlation of fossil plants in Antarctica with those of the adjacent land masses, which has been undertaken by Edna Plumstead of the University of Witwatersrand, is similarly limiting the possible position of the blocks.

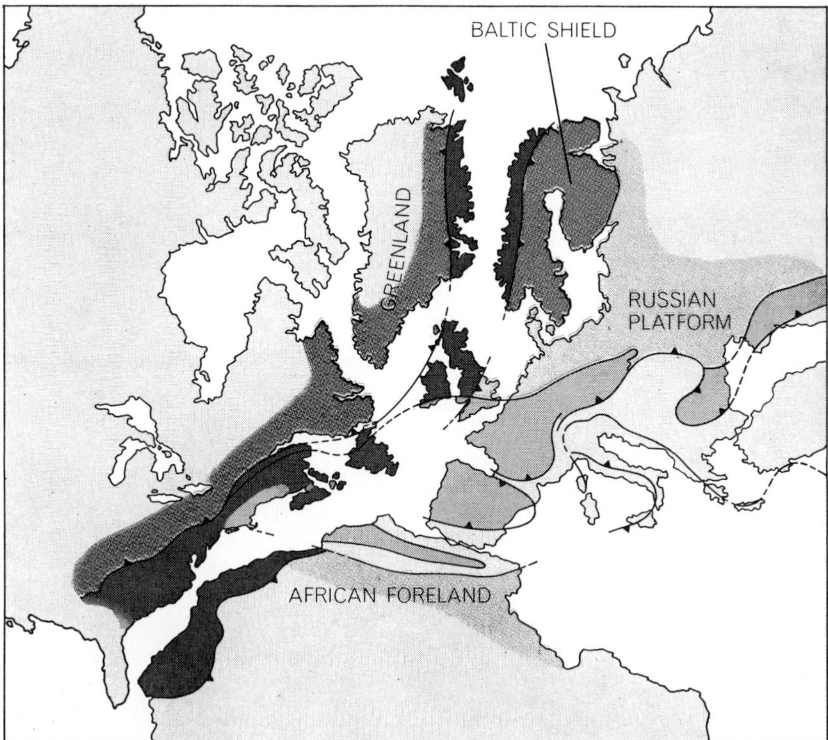

MATCHING OF NORTH ATLANTIC REGIONS is more difficult than in the South Atlantic. This tentative, pre-drift reconstruction of a portion of Laurasia depends on matching ancient belts of similar geological activity. The dark gray belt represents the formation of sediment-filled troughs and folded mountains in the early and middle Paleozoic (470 million to 350 million years ago). The medium gray belt was formed in the late Paleozoic (350 million to 200 million years ago). The latter belt overlapped the region of the former in the northern Appalachians and in southern Ireland and England, but diverged eastward in Europe. Four distinct and superimposed periods of geological activity occur on both sides of the present North Atlantic, providing strong evidence for a previous juncture.

The Age of the Atlantic

When did Gondwanaland begin to break up? One of the best pieces of evidence for the start of the opening of the South Atlantic is the age of offshore sediments along the west coast of Africa. Drilling through these sediments down to the ancient nonsedimentary rocks shows that the layer of sediments is quite young: not older than the middle Mesozoic (about 160 million years ago). If the South Atlantic had been in existence for a major part of geologic time, the continent of Africa would unquestionably have developed a large shelf of sediments along the entire length of its western margin. The continental shelf would consist of sediments dating all the way back to the time of the ancient cratons. This is not the case. It looks as though the rift started from the northern edge of western Africa in the middle Triassic and slowly opened to the south until the final separation occurred in the Cretaceous. The east coast of Africa, on the other hand, apparently started to open earlier, in the Permian.

With the acceptance of sea-floor spreading and continental drift the global problems of geology are beginning to be solved. Although the train of thought on such matters is not universally accepted in detail, it is something like the following. Continental areas appear to have greater strength, to a depth of 100 kilometers or so, than ocean basins do, so that they tend to maintain themselves as buoyant masses that are not destroyed by sinking motions. They can, however, be ruptured. Rising material pushes the surface apart; sinking material pulls the surface together and toward the region of sinking. Therefore if a sinking zone is established in an oceanic region, the continents will move toward the zone, and if a rising zone is established under a continent, the continent will split apart and the parts will move away from the zone. When the ocean floor moves toward a sinking zone in an oceanic region, it forms a deep trench bordered by volcanoes, chains of islands or elongated land masses such as the Philippines and Japan. When an ocean floor moves toward a continent, it appears to pass under the continental border, forming a great mountain chain. The mountain chain may be in part piled-up material that was already present and in part volcanic material that rose as the ocean swept its load of sediment, underlying volcanic rock and the continental shelf itself toward and under the edge of the continent. The process leads to a melting of underlying rock and to the intrusion of new volcanic material. The west coast of South America is a good example.

Another example is the thrust of India into Eurasia that formed the Himalayas. It has long been known that there was a large body of water between Africa and Eurasia and that a great thickness of sediments was deposited there at some time during the past 200 million years. This body is known as the Tethys Sea. It was located north of Arabia and extended from the former location of the Atlas Mountains to east of the Himalayas. As I have mentioned, it appears that Gondwanaland not only broke up but also moved northward, with India and Africa pushing up into Eurasia. This motion apparently caused the buckling up of sediments in the Tethys Sea, giving rise to the mountain ranges that now form a contorted chain from the western Atlas range through the Mediterranean, the western Alps, the Caucasus and the Himalayas.

The way the present mountain systems of the earth fall along great circles suggests that the motions in the earth's interior have a large-scale coherence, of the order of the dimensions of the earth itself. The prevailing explanation stems from a new lead in seismology: a zone in the earth at a depth of 100 or 200 kilcm-

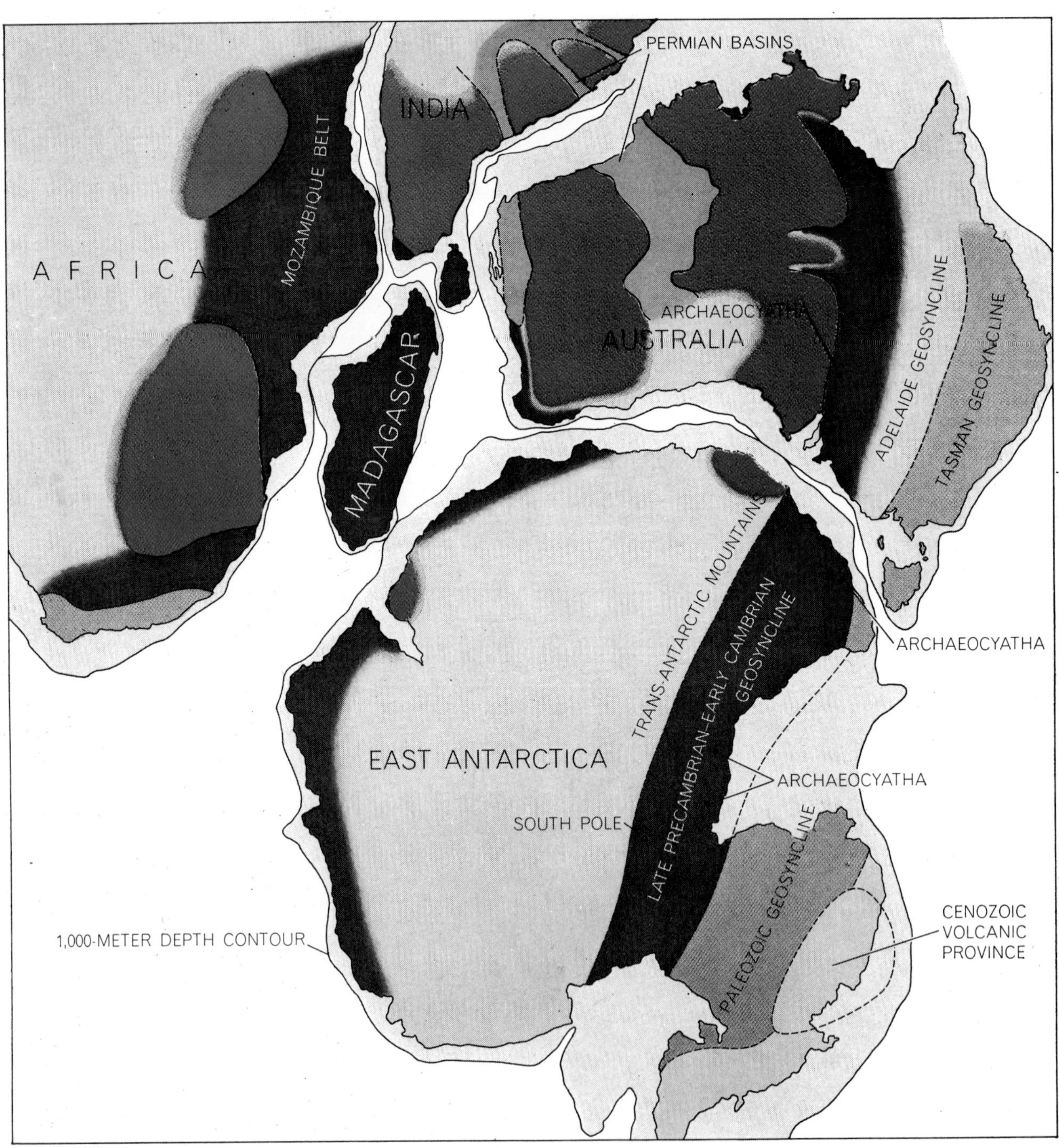

PART OF GONDWANALAND, tentatively reconstructed, brings together East Antarctica, Africa, Australia, Madagascar, India. The fit is at the 1,000-meter depth contour of the continental slope. Late Precambrian and Paleozoic geosynclines, or sediment-filled troughs, in eastern Australia are correlated in age and location with similar troughs along the Transantarctic Mountains. The deep Permian basins of northwest Australia match those of India. Glacial deposits, fauna and metal ores provide other correlations.

eters has been found to transmit seismic waves more slowly than the layers above and below it and to absorb seismic energy more strongly. This low-velocity zone is generally thought to consist of a material whose strength is reduced because a small amount of it is molten or because its temperature is approaching the melting point. The surface of the earth may therefore move around on this low-strength layer like the skin of an onion. It is believed the earth loses heat partly by conduction outward and partly by convection currents in the relatively thin layer above the weak zone. These currents, as they have been depicted by Walter M. Elsasser of Princeton and Egon Orowan of M.I.T., form rather flat convection cells.

A hypothesis that is currently popular is that the mechanism of spreading at the oceanic ridges involves the intrusion of hot material into ruptures near the surface. This material is the same as that in the low-velocity zone, lubricated by partly molten rock. A small proportion of the intruded material actually loses some of its melted fraction upward, giving rise to volcanoes and creating a thin layer (about five kilometers thick) of volcanic rock at the surface. The masses of intruded material cool as they move sideways from the central ridge, which is overlain by the thin layer of volcanic rock. This results in the observed distribution of seismic velocities at various depths, helps to explain why the flow of heat to the surface decreases with distance from the ridge and accounts for the pattern of magnetic reversals. At the sinking end of the convection cell this relatively rigid block of mantle material with its thin cover of basalt (plus a thin cover of new sediment) moves downward on an inclined plane.

It is clear where these concepts will lead. If folded mountain belts are the "bow waves" of continents plowing their way through ocean floors and ramming into other continents, we can use them to show us the relative directions of motion prior to the last great drift episode. If we look at the pre-drift Paleozoic mountain belts, such as the Appalachian belt of North America, the Hercynian of Europe and the Ural of Asia, we find that they are located *internally* in the great continental masses of Gondwanaland and Laurasia. This suggests that these pre-drift supercontinents had been formed by the inward motion of several separate blocks, which came together before they broke apart. Geologists have a new game of chess to play, using a spherical board and strange new rules.

2

The Lunar Orbiter Missions to the Moon

ELLIS LEVIN, DONALD D. VIELE,
AND LOWELL B. ELDRENKAMP

May 1968

In a corridor of the Boeing Space Center near Seattle there is a little sign that reads "5 for 5." The sign briefly summarizes the fact that all five Lunar Orbiter missions, which were primarily designed to make photographs of the moon from spacecraft in lunar orbit, were successful. The results included complete photographic coverage of the side of the moon that is visible from the earth and coverage of more than 99.5 percent of the side that cannot be seen from the earth. Most of the near side is now depicted in Lunar Orbiter photographs that provide resolution at least 10 times better than it is possible to obtain with the best earth-based telescopes. Although the photographs of the far side were made at higher altitudes than most of the photographs of the near side, the far side is now depicted in greater detail than was previously obtainable of the near side from earth-based observations.

The Lunar Orbiter series was one of three programs organized by the National Aeronautics and Space Administration in preparation for the Apollo missions in which men will land on the moon. The first of the other programs was the Ranger series, which provided experience in reaching the moon with spacecraft and also yielded the first closeup pictures of the lunar surface. The second of the other programs was the Surveyor series, in which spacecraft made "soft" landings in prospective Apollo landing areas, tested the mechanical and chemical properties of the lunar surface and made pictures of the immediate surroundings. Ranger and Surveyor spacecraft provided detailed information about small areas of the moon; the purpose of the Lunar Orbiter missions was to obtain good photographic coverage of large areas.

As defined in a contract between the Langley Research Center of NASA and the Boeing Company, which was the prime contractor for the Lunar Orbiter project, the main objective of the series was to obtain detailed photographic coverage of a number of areas along the equator on the near side of the moon that were regarded as potential landing sites for Apollo spacecraft. The project was designed to be flexible, however, and as a result it was possible to expand the objectives as the series progressed. For example, the coverage of potential Apollo sites from the near-equatorial orbits of the first three missions was good enough to warrant putting the last two spacecraft into near-polar orbits, thereby providing for virtually complete photographic coverage of the moon.

The five Lunar Orbiter spacecraft made a total of 1,950 photographs. Already the photographs have been put to several uses. NASA has used them to select five potential Apollo landing sites from some 40 candidates that had been identified from earth-based observations [*see illustration on page 25*]. The photo-

FAR SIDE of the moon appears in a photograph made through the wide-angle lens of *Lunar Orbiter III*. The spacecraft was about 900 miles above the moon and 250 miles south of the lunar equator. At the top, which is north, the photograph spans about 700 miles; the curved southern horizon stretches to within 400 miles of the moon's south pole. The conspicuous crater filled with dark material is called Tsiolkovsky.

graphs have been used to choose landing sites for several of the Surveyor spacecraft and to help interpret results from the Surveyor missions. The photographs have been used as a basis for detailed maps of the lunar surface, including far-side maps that could not have been made previously. From Orbiter photographs NASA has made models of the lunar surface for study and practice landings by astronauts. The photographs are also providing valuable information for investigators studying the processes that have shaped the surface of the moon.

In addition to the photographs, the Lunar Orbiter missions have secured several other kinds of information. Each spacecraft carried devices for detecting radiation and the impact of micrometeoroids. Observations of the orbits of the spacecraft, as they were tracked by stations of the Deep Space Network in California, Spain and Australia, yielded improved data on the moon's overall shape and gravitational field. Separate tracking by the Manned Spaceflight Network gave the Apollo tracking teams valuable training for the manned missions of the future. In another experiment *Lunar Orbiter V* was carefully maneuvered in lunar orbit to reflect sunlight from its solar panels and its underside mirrors, and the reflection was photographed with a telescope on the earth.

Design of the Missions

In planning the Lunar Orbiter missions a considerable number of objectives and constraints had to be identified and taken into account. Each spacecraft had to be launched from the Kennedy Space Center by an Atlas-Agena rocket on a course that would bring it to a predetermined point in the vicinity of the moon four days later. There its velocity would be reduced and redirected by the firing of its rocket engine so that the spacecraft would be captured by the moon's gravitational field and would achieve a suitable initial orbit of the moon. After being tracked for several days the spacecraft would be further slowed so that its perilune, or closest approach, would be reduced to about 28 miles above the lunar surface, which would be the primary altitude for photography.

Originally the plan was to make the initial orbit circular at about 575 miles above the lunar surface; the slowing maneuver would put the spacecraft into an elliptical orbit with a perilune of 28 miles and an apolune, or maximum altitude, of 575 miles. During the design

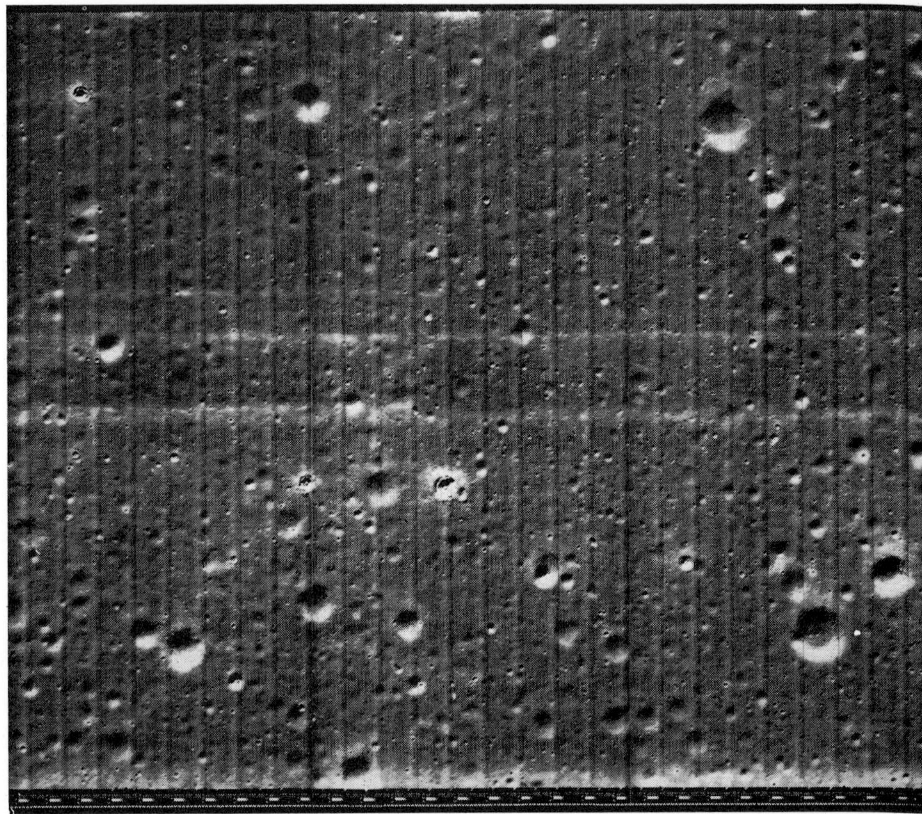

LANDING AREA of the U.S. spacecraft *Surveyor I* is included in a view obtained by the telephoto lens of *Lunar Orbiter III*. The scale is about .6 mile per inch. Vertical

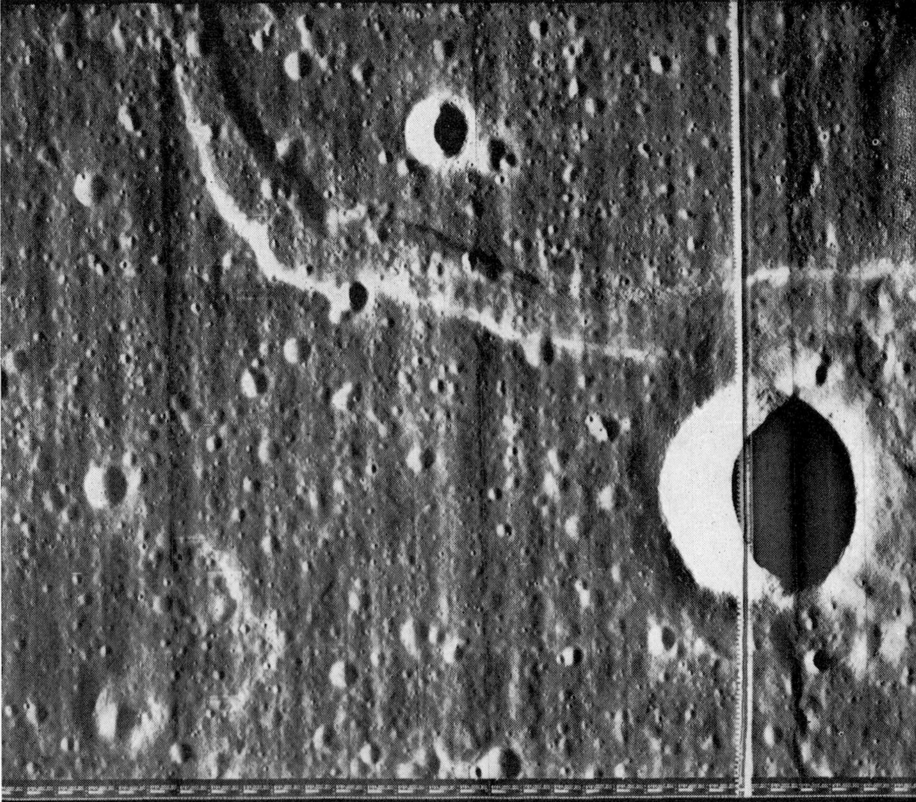

RILLE STRUCTURE in the Marius Hills region was photographed with the telephoto lens of *Lunar Orbiter V*. The scale is about 1.2 miles per inch. The spoonlike rille, or valley, that

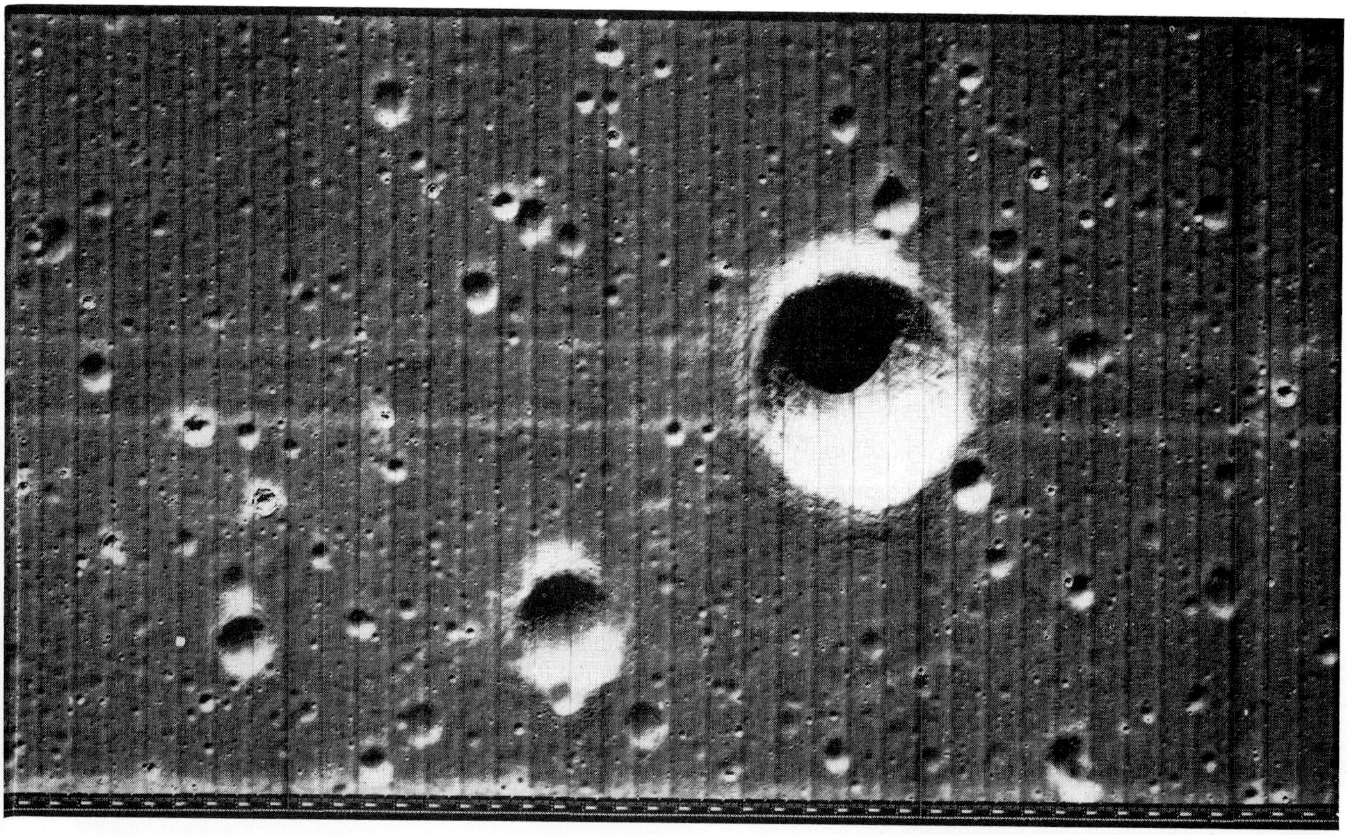

bands result from the scanning process by which the spacecraft transmitted its photographs to the earth; light horizontal bands were made in the spacecraft's photographic subsystem. A few craters appear by their distinctness to be of more recent origin than the others. At left below the photograph is an enlargement of the edge data preexposed on the film. The data include framelet numbers, gray scale and resolution bars to aid in the calibration of photographs and the control of the readout of photographs.

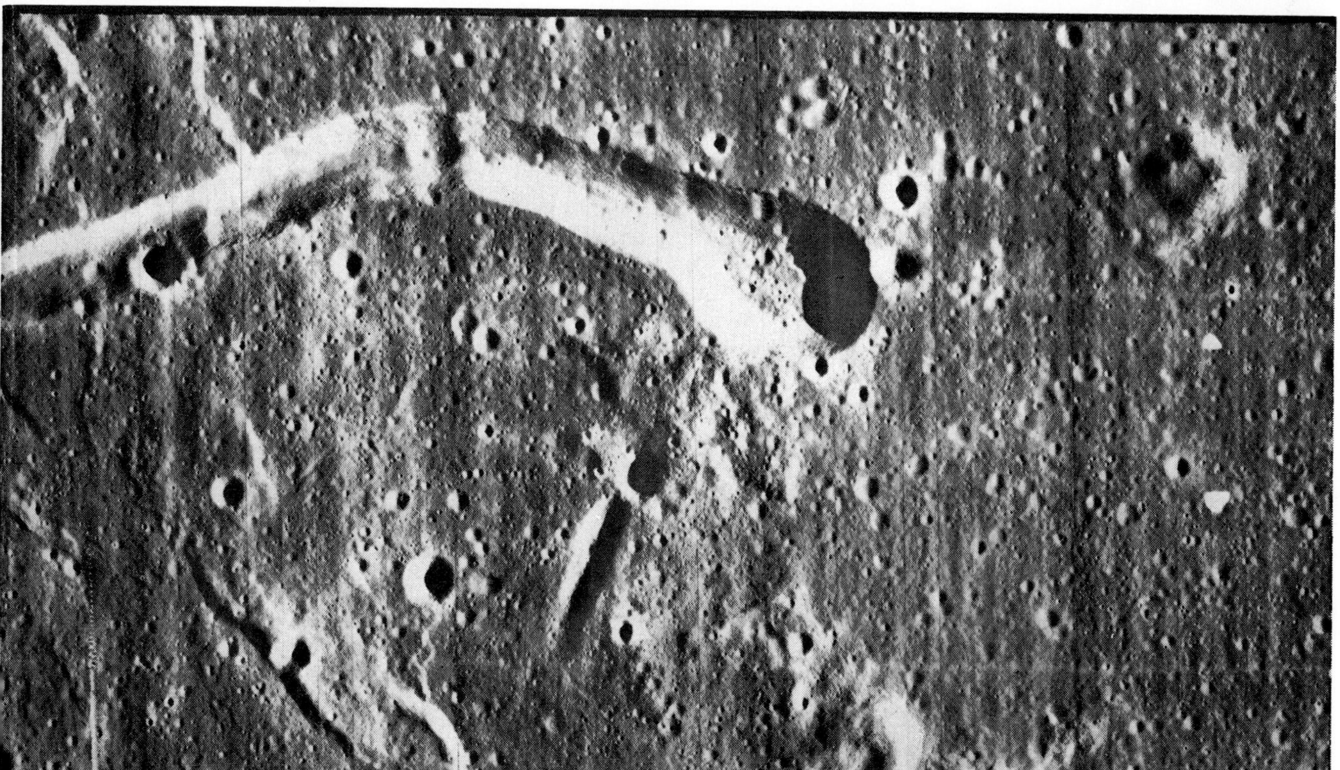

winds through most of the photograph is one of a number of unusual features in the area; among the others are the domes shown in the illustration on page 63. Region was one of several included in the mission of *Lunar Orbiter V* as being of special interest.

phase the plan was changed to make the initial orbit elliptical with a perilune of 120 miles and an apolune of 1,150 miles. This change avoided the impractical alternative of increasing the weight of the spacecraft by adding batteries; in an elliptical orbit the spacecraft would spend more time in the sun and so could make more use of solar power. The change also reduced the amount by which the velocity of the spacecraft would have to be modified in the maneuvers near the moon. Another significant change was to provide for four days for the spacecraft to travel from the earth to the moon instead of the three days taken by the Ranger and Surveyor vehicles; in this way the velocity-change requirement for the Orbiters was further reduced and there was more time to track an Orbiter and analyze its trajectory as it traveled to the moon.

The greater part of the photography would be done near perilune. The perilune altitude was chosen to achieve a balance between good photographic resolution and coverage of sufficiently large areas. Beyond that the shapes of the orbits had to be established in such a way that the location of perilune would coincide with the time when the illumination of the area being photographed was suitable for photography. On the moon the illumination would be suitable when the sun was between 10 and 30 degrees above the local horizon, so that it would be possible to detect small slopes, depressions and protuberances.

In principle the times of suitable illumination could be either shortly after sunrise or just before sunset. In actuality, since the spacecraft would be traveling from west to east across the visible face of the moon, at least on the initial missions, the time of the most suitable illu-

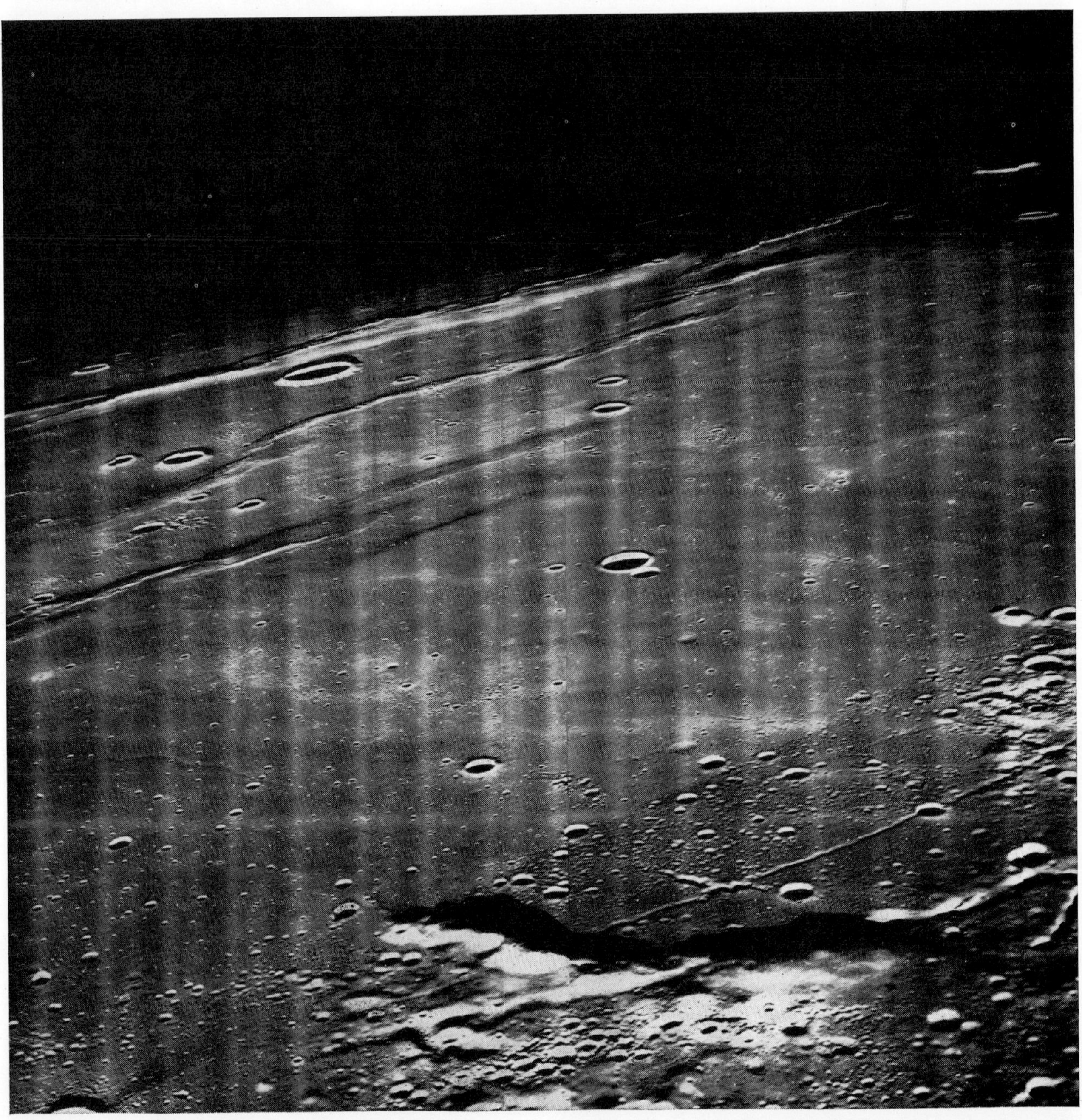

POTENTIAL LANDING SITE for manned Apollo spacecraft was photographed by *Lunar Orbiter III* in an oblique view that resembles what the astronauts would see as they approached the lunar surface. The chief mission of the Orbiter project was to obtain photographs of a number of potential landing sites for Apollo vehicles. This one is near the western edge of the Apollo zone.

mination was sunrise. Then the spacecraft would be moving into the sun after photography and so could draw for extended periods the solar power needed to process the photographs of the near side—the side of primary concern.

In order to plan for thorough photographic coverage of the equatorial band containing the potential landing sites for Apollo spacecraft, we had to give special consideration to the inclination of the orbit, that is, the angle between the plane of the orbit and the plane of the lunar equator. The inclination had to be low enough to obtain suitable overlapping of photographs from adjacent orbits and yet high enough to provide photographic coverage in good lighting over an acceptably wide range of latitudes. In the end an inclination of 12 degrees was chosen for the first two missions; for the third it was 21 degrees.

Another consideration was to achieve orbits that were sufficiently stable with respect to the moon to permit control of the spacecraft's perilune altitude for photography and to avoid having the spacecraft crash onto the moon before the transmission of its photographs to the earth could be completed. Instabilities arise not only from variations in the moon's gravitational field but also from the gravitational attraction of the earth and the sun on the spacecraft. The gravitational variations of the moon are due to the distribution of the lunar mass and to the moon's departures from a spherical shape. The variations could not be predicted accurately before the first Orbiter flight. As a result we had to make allowances for the extremes in the best available theoretical model based on observations from the earth [*see bottom illustration on page 23*]. We charted a course between the extremes and made

LUNAR DOMES near the crater Marius, which is near the horizon, were seen in detail for the first time in this oblique photograph made with the wide-angle lens of *Lunar Orbiter II*. The domes, which are from two to 10 miles in diameter and 1,000 to 1,500 feet high, are of special interest because they resemble volcanic features found on earth. Marius is about 25 miles wide and one mile deep.

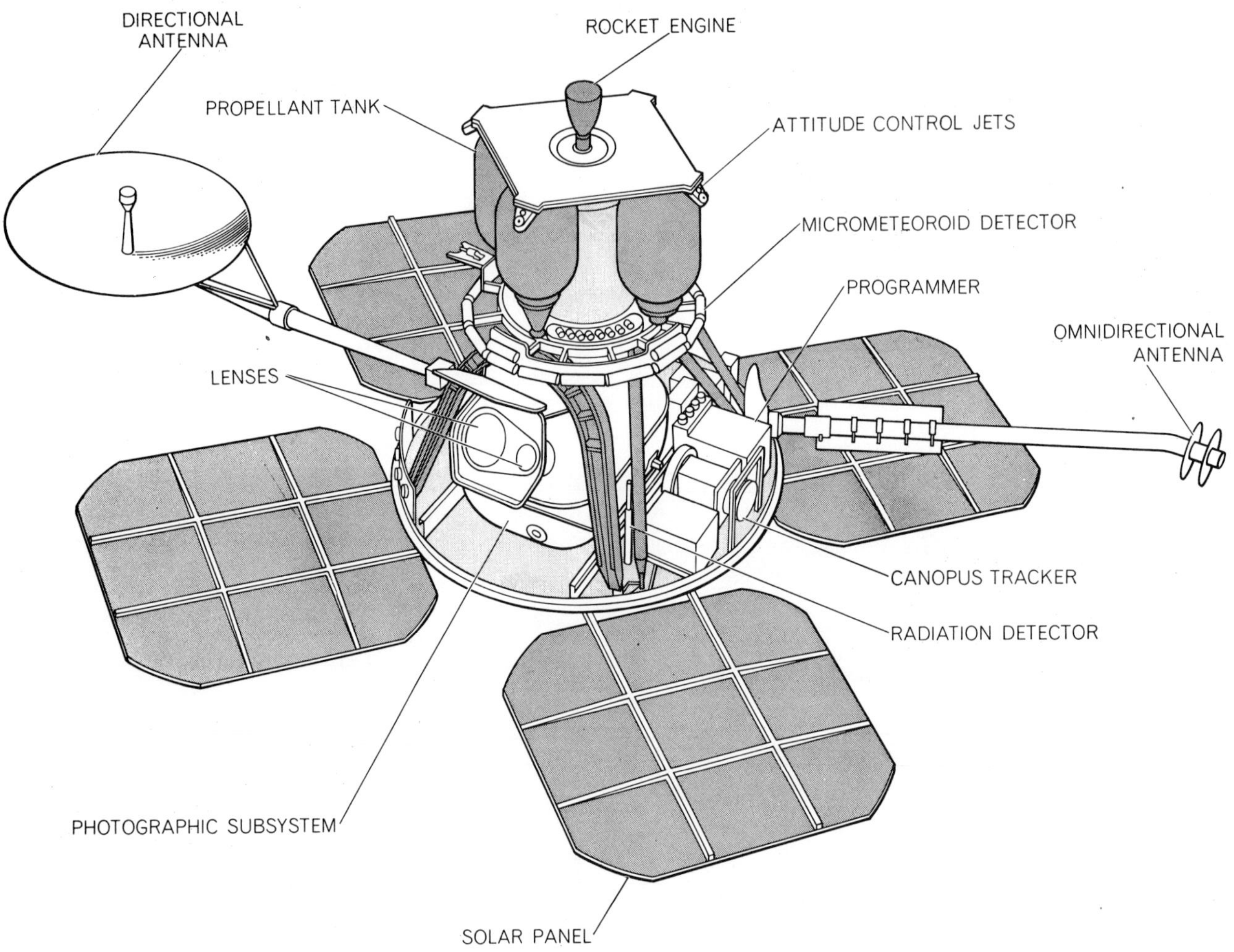

LUNAR ORBITER SPACECRAFT is depicted as it appeared while it was in operation, except that an aluminized Mylar wrapping used as a thermal barrier is not shown. Antennas and solar panels were folded close to the body of the spacecraft during launching.

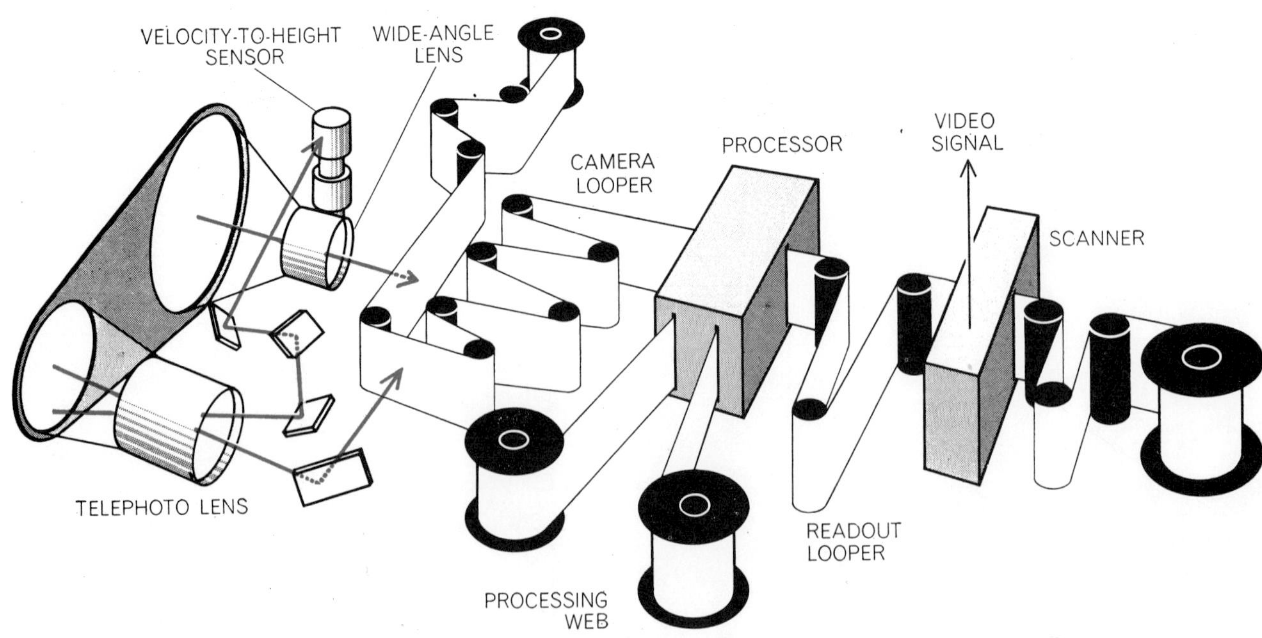

PHOTOGRAPHIC SUBSYSTEM of the Orbiter spacecraft had 260 feet of 70-millimeter film. Simultaneous exposures were made through the two lenses. Film was stored on loopers until it could be put through the processor, where it was developed. In scanning the direction of movement was reversed, so that a full readout could not be done until photography was complete and the processing web had been cut. Readout loopers made it possible to read out selected portions of film before photography was completed.

provision for changing the course of the spacecraft after it had been in orbit long enough for us to determine the combined effects on the spacecraft of the earth, the moon and the sun.

Several other matters had to be taken into account; here we can only mention them. One was the effect of radiation on the film. Another was the choice of exposure settings. Another was wisely budgeting the spacecraft's film and fuel, which entailed choosing the optimum sequence of activities for the missions. It was also necessary to observe a limitation that the capacity of the launch vehicle put on the weight of the spacecraft and to make sure that adequate tracking facilities were available.

The Spacecraft

On the launching pad a Lunar Orbiter spacecraft was a compact package five and a half feet high and five feet wide. Once in space it unfolded two antennas and four solar cells [*see top illustration on opposite page*]. The high-gain antenna was highly directional, the low-gain one nearly omnidirectional. When they were deployed, on booms that extended from opposite sides of the spacecraft, they spanned 18½ feet from tip to tip. The solar panels, spanning a little more than 12 feet when they were extended, provided the electrical energy for the operation of the spacecraft and for recharging the battery used when the spacecraft was not in the sun. Power for changes of velocity was provided by a rocket engine; changes of attitude were achieved with nitrogen-gas jets. At launch the craft weighed close to 850 pounds, including about 150 pounds for the photographic subsystem and 262 pounds of propellant for the engine.

In space an Orbiter was stabilized on three axes: pitch, yaw and roll. A sun-sensor provided the angular reference needed to control pitch and yaw; roll was controlled by a sensor that locked on the bright star Canopus. When these celestial references were not visible to the sensors, stabilization was provided by three gyroscopes.

The photographic subsystem, which was built by the Eastman Kodak Company, contained elements for making photographs, processing the film and converting the image on the developed film into electrical analogue signals that were communicated to the earth. Although the system carried a single 260-foot roll of 70-millimeter film, the handling of the film was arranged—by means of storage loopers—to provide flexibility in moving the film through the camera, the processor and the readout apparatus at different rates and at appropriate times [*see bottom illustration on opposite page*].

The camera had a wide-angle lens with a focal length of 80 millimeters and a telephoto lens with a focal length of 610 millimeters. Each had its own shutter. The shutters operated simultaneously, so that each exposure produced two photographs. At any given altitude the telephoto lens viewed about 5 percent of the terrain covered by the wide-angle lens but gave a resolution about eight times better than that provided by the wide-angle lens. For example, in vertical

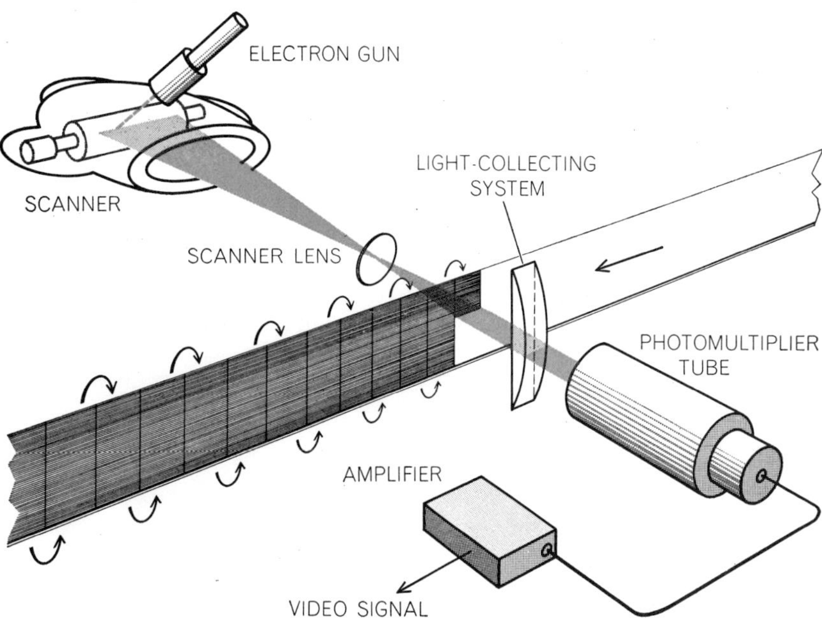

SCANNING PROCESS moved a microscopic spot of light, generated by a beam of electrons aimed at a drum coated with a phosphor, along a portion of the film. The density of the image on the film governed the strength of the light reaching the photomultiplier tube, which generated electronic signals for transmission to the earth. One complete scan read out the information on a band of film that took up .1 inch along the length of the film and 60 millimeters of its width. These are the bands one sees in photographs from the Orbiters.

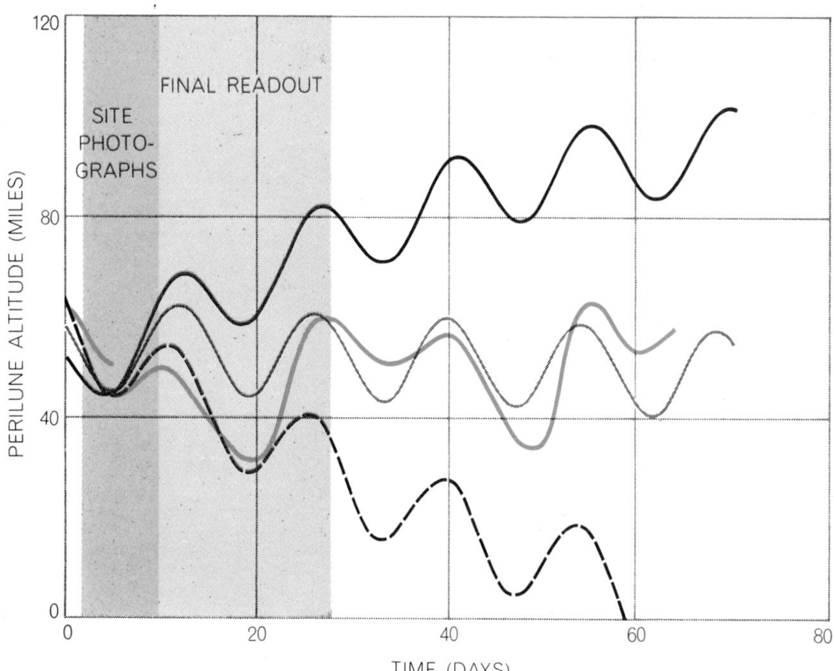

ACTUAL ALTITUDE of first Orbiter at perilune (*color*) is compared with three preflight calculations using different lunar-gravity assumptions. Earth's gravity causes 14-day variation. Perilune had to be chosen to avoid crashing prematurely or impairing photography.

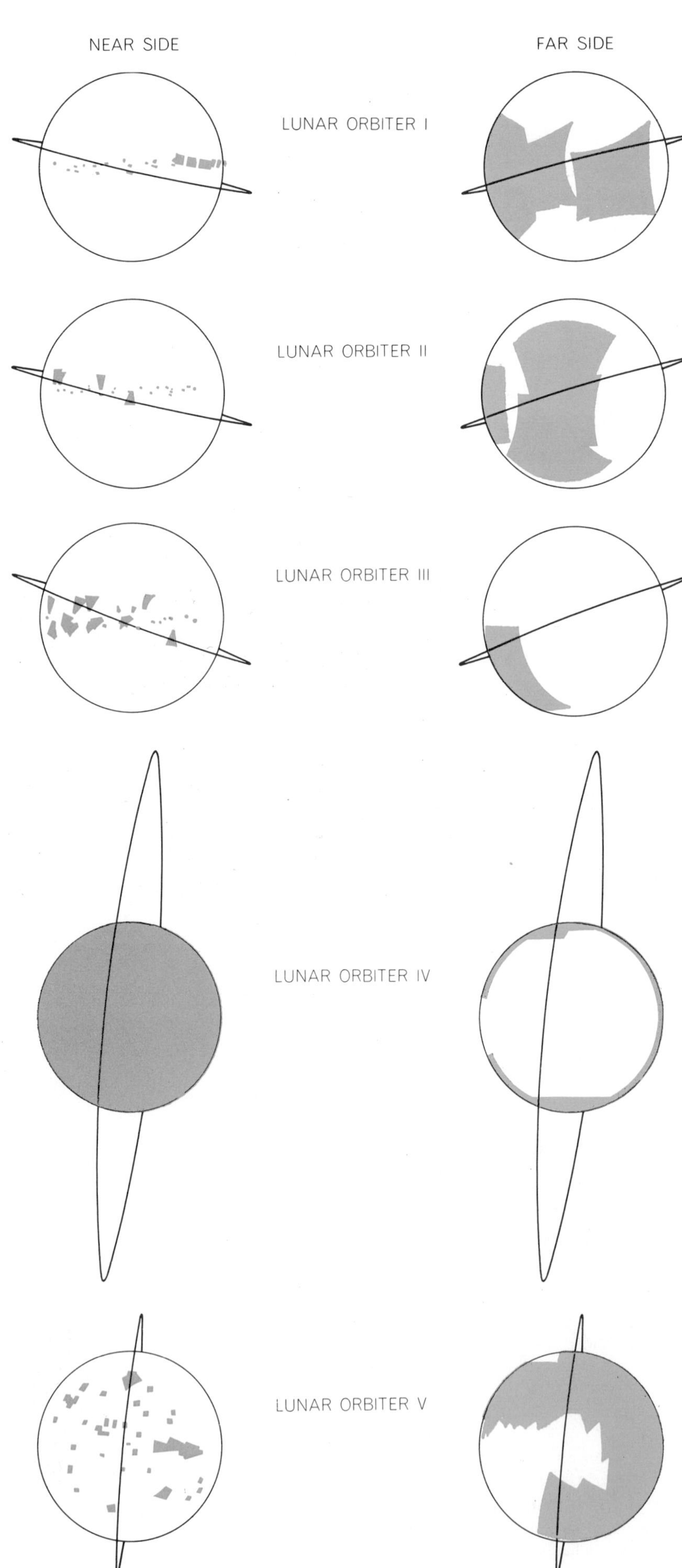

photography at an altitude of 28.5 miles the telephoto lens covered an area 2.6 by 10.2 miles at a resolution of three feet; the wide-angle lens covered an area 19.6 by 23.2 miles at a resolution of 24 feet. Coverage of large areas was obtained by overlapping photographs made successively in the direction of flight or made on succeeding orbits.

Because the spacecraft moved at a speed of about 4,300 miles per hour while making photographs near perilune and the film exposures were rather long (1/25, 1/50 or 1/100 second) the photographs would have been blurred if the camera had not had a mechanism that compensated for the motion of the image at the plane of the film. This compensation was achieved by means of a device that sensed the velocity-to-height ratio of the spacecraft and drove the film platen to follow the image along the direction of flight. The sensor determined the velocity-to-height ratio by measuring the apparent motion of a small portion of the lunar surface as viewed through the telephoto lens.

The exposed film was developed in a processor using the Bimat technique devised by Eastman. In the Bimat method the exposed film is pressed against a special processing web, or rolled strip, with a gelatin layer that has been soaked in a solution that develops and fixes the film in one step. The film was then separated from the processing web and was dried by being passed over an electrically heated drum.

The developed film was read out by an electronic scanner that worked with a rapidly moving spot of light. One transverse band of film, called a framelet, was scanned at a time. A framelet occupied .1 inch of the length of the film and 2¼ inches of the film's width [*see top illustration on preceding page*]. One band can be regarded as analogous to a page in a book; reading it required 17,000 horizontal movements by the spot of light and took 22 seconds. At the end of a scan the film was advanced .1 inch and the next band was scanned in the same way, although in the opposite vertical

SUMMARY OF MISSIONS by the five Lunar Orbiter spacecraft shows the orbit of each one and, in color, the areas each one covered photographically. The primary mission of the Orbiters was to photograph potential Apollo sites on the near side of the moon along the equator. That work was essentially completed after the first three missions, so that the last two spacecraft were put into near-polar orbits to photograph other sites and most of the moon.

direction. It took about 43 minutes to scan the 11.6 inches of film taken up by one wide-angle and one telephoto exposure.

As the spot of light passed through the image on the negative it was modulated by the density of the image: the darker areas of the film transmitted less light than the lighter areas. The intensity of the light passing through the film at any moment was sensed by a photomultiplier tube, which generated an electrical signal proportional in strength to the intensity of the light. This signal, amplified and transmitted to the earth by means of the spacecraft's high-gain directional antenna, could be used to reproduce the images that had been obtained by the camera in the spacecraft.

The communications subsystem, provided by the Radio Corporation of America, used a technique called vestigial side-band amplitude modulation in conjunction with a 10-watt amplifier and the directional antenna to pack the wide-band video information and the accompanying telemetry into the band of frequencies allotted to the Deep Space Network. When the spacecraft was not transmitting photographs, telemetry from the spacecraft and commands to it were communicated through the low-gain omnidirectional antenna.

Reconstruction of the photographs on the earth involved recording each scanned band on 35-millimeter film. When the appropriate strips of 35-millimeter film are placed side by side, a duplicate of a photograph made by the spacecraft is produced (at an enlargement of about 7.5 diameters). The strips of 35-millimeter film, each representing

APOLLO SITES chosen by the National Aeronautics and Space Administration on the basis of photographs from Orbiter and Surveyor spacecraft are shown on a photograph made with an earth-based telescope. NASA first chose eight sites and then reduced the number to five, which are represented by triangles; circles indicate eliminated sites. Eventually the five sites will be reduced to three.

one scanned band of spacecraft film, give the Lunar Orbiter photographs their characteristic striped appearance.

A complete readout could not be started until all the film had been exposed. The direction of film movement during readout had to be the reverse of the direction in which the film moved as the photographs were being made. It was not possible to run film backward through the processor while the Bimat system was still functioning, because the dried negative might have stuck to the processing film. When all the film had been exposed, however, the Bimat web could be cut on command from the earth, and then the film could be pulled backward for readout.

It seemed evident during the planning of the missions that some provision should be made for partial readout be-

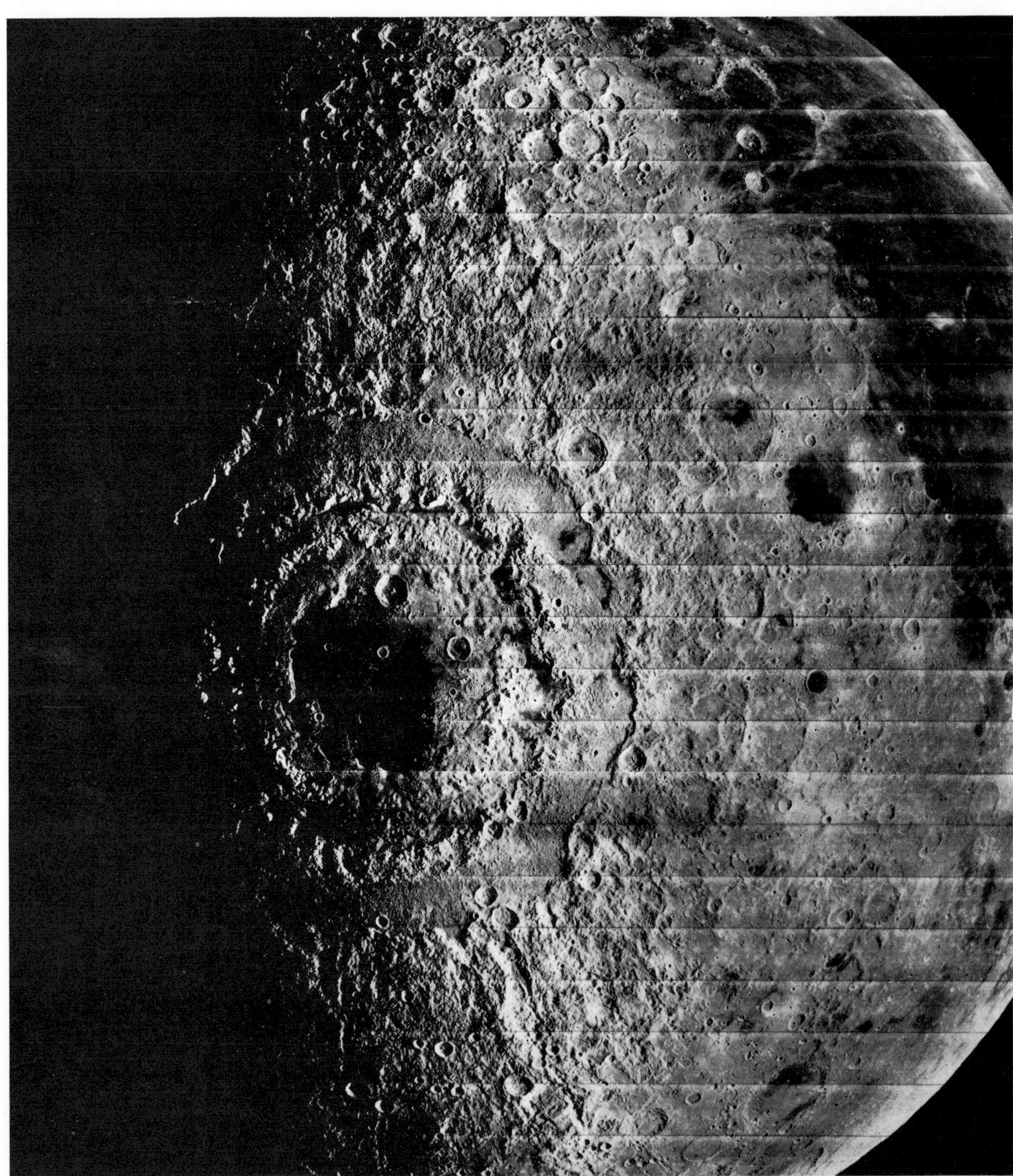

ORIENTALE BASIN is an enormous lunar feature that had not been viewed from above until *Lunar Orbiter IV* made this photograph from an altitude of 1,690 miles. The view is what one would see if it were possible to look around the lower left edge of the moon as it is seen from the earth. The Cordillera Mountains, which rim the basin on the right, can be seen from the earth, but only in profile. Across the outer ring the feature is about 600 miles in diameter. Orientale may be the most recent of the large circular lunar basins. The large dark area in the upper right part of the photograph is Oceanus Procellarum, which is visible from earth.

fore the photography was completed, not only because of the need to check on the quality of the photographs but also because there would be intense interest in what kind of scenes the cameras were recording. Accordingly a looper was built into the photographic subsystem between the processor and the scanner. A small amount of film could be read out and temporarily stored on the looper without going back into the processor. Many of the photographs that were published in newspapers and magazines during Lunar Orbiter missions were obtained with this system. Early readouts were also used for planning subsequent Surveyor and Orbiter missions.

The first of the Lunar Orbiter spacecraft was launched from Cape Kennedy on August 10, 1966. Succeeding ones were launched at intervals of approxi-

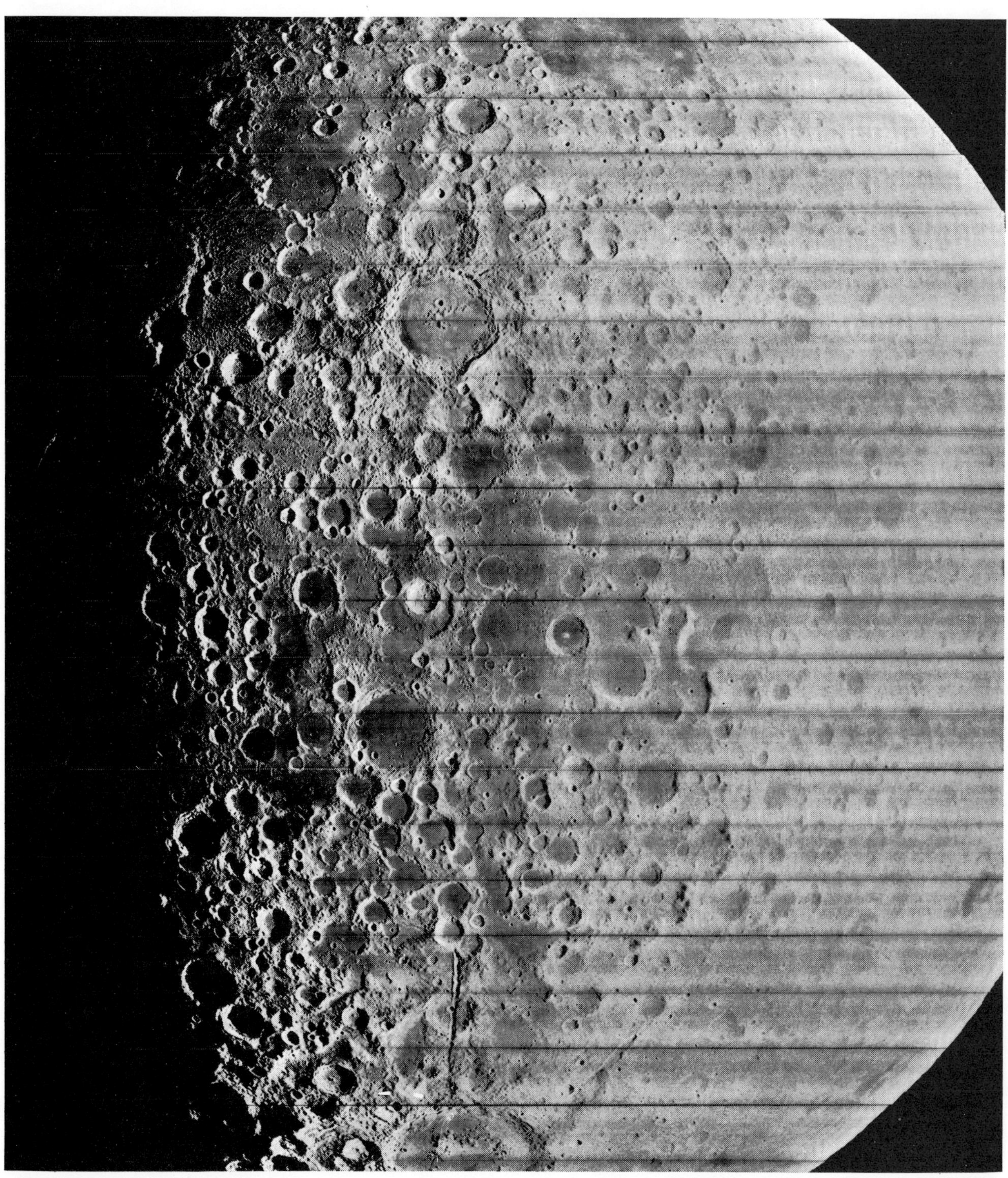

SOUTH-POLAR REGION of the moon was photographed through the wide-angle lens of *Lunar Orbiter IV* from an altitude of 1,856 miles. The view is what one would see if it were possible to look around the lower right edge of the moon as it is seen from the earth. The conspicuous trough near the bottom of the photograph is about 150 miles long. In places it is nearly five miles wide. It is also fairly young, as indicated by the fact that it cuts through several features. A still younger crater can be seen near the middle of the trough. It has been speculated that one way a trough could be made is by a rigid body that strikes the surface at a low angle.

CRATER COPERNICUS was photographed vertically through the wide-angle lens of *Lunar Orbiter V.* The crater is likely to be explored by a party from one of the Apollo missions because the slumping walls of the crater may expose a sequence of geological events. North is at left; the rim there is the one that appeared in widely reproduced oblique photographs of Copernicus from *Lunar Orbiter II.* The small dots in that area are marks resulting from the developing process in the spacecraft's photographic subsystem.

mately three months, so that the last of the Orbiters began its mission a little less than a year after the first one—on August 1, 1967. On many occasions the flexibility built into the system proved invaluable.

For example, the mission of *Lunar Orbiter I* had no advance plan for using certain frames of film involved in dealing with what was called the film-set problem: the fact that the film had to be moved after a certain amount of time, even if no photography was in progress, so that it would not be deformed by remaining too long on any of the sharp bends in the winding mechanism. As the mission proceeded it was decided to use these frames to obtain previews of sites for later missions, photographs of the far side and a picture showing the earth as it is seen from just beyond the limb of the moon.

The instrument for such flexibility was the digital programmer carried by each Lunar Orbiter. The programmer controlled more than 70 functions of the spacecraft. It could store a large program of instructions, which could be changed by command from the earth whenever the spacecraft was within reach of communications. With each successive mission more use was made of the programmer's capabilities.

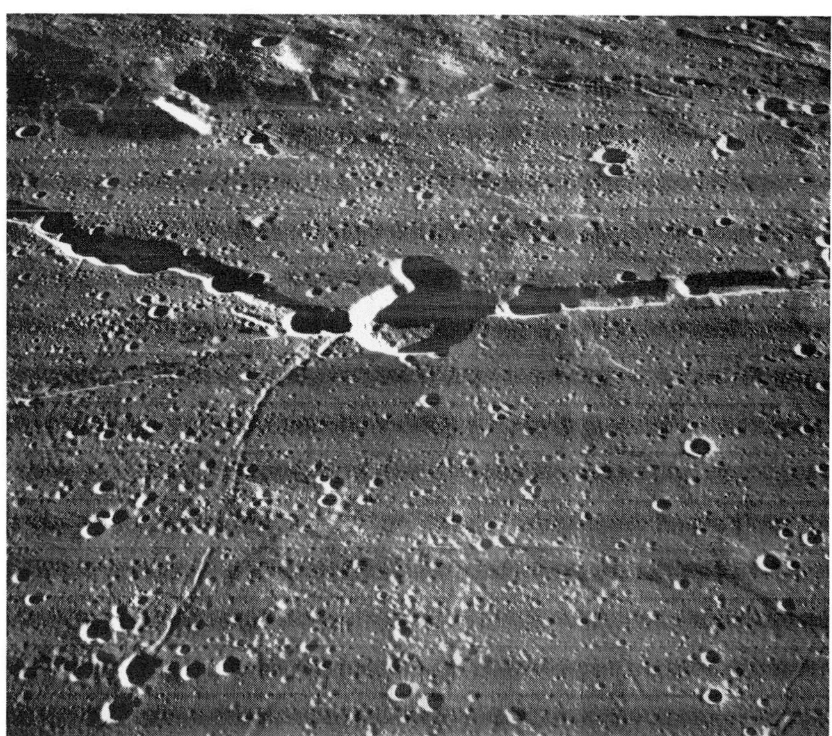

HYGINUS RILLES extend to the northwest and the east of the crater Hyginus, which is about 6½ miles in diameter and 2,600 feet deep. The oblique photograph was made by *Lunar Orbiter III*. The two large rilles appear to be associated with Hyginus and a smaller rille extending to the southwest appears to be associated with a smaller crater near Hyginus.

The Missions

Lunar Orbiter I primarily accomplished an examination of nine potential Apollo sites in the southern portion of the equatorial band in which all such sites are located. *Lunar Orbiter II* made a similar examination of sites in the northern portion of the band. In addition *Lunar Orbiter II* undertook an experiment in converging stereoscopic photography, which entails making overlapping photographs of a given site from two different orbits. The technique is particularly valuable for photographic analysis of the topography of an area. The results were so good that it was decided to add such coverage to the photography planned for the Apollo sites. *Lunar Orbiter II* also made oblique photographs of many areas of scientific interest. One such picture, widely published at the time, showed the inside of the crater Copernicus from a perspective that could not have been obtained from the earth.

The third Orbiter made a detailed examination of 10 promising Apollo sites selected from the photographs secured during the first two missions. After the third mission NASA made its selection of eight sites as being the most promising

FINE DETAIL of a terrace, a steep slope and a mare area in Oceanus Procellarum was obtained through the telephoto lens of *Lunar Orbiter III* at an altitude of about 32 miles.

LARGE OBJECTS, probably boulders, cast long shadows in an enlargement of a photograph made with the telephoto lens of *Lunar Orbiter II*. The object casting the longest shadow is about 50 feet wide at its base and stands as much as 75 feet high. The white cross is a reference mark preexposed on the film; width of cross represents about 25 feet of lunar surface.

DISLODGED ROCK in the crater Vitello has left a 250-yard trail on the lunar surface after rolling down a slope into a depression and up the other side. The rock is about 19 yards in diameter. The view is an enlargement of a photograph made with the telephoto lens of *Lunar Orbiter V*. Many rocks, including several that had rolled, were observed in Vitello.

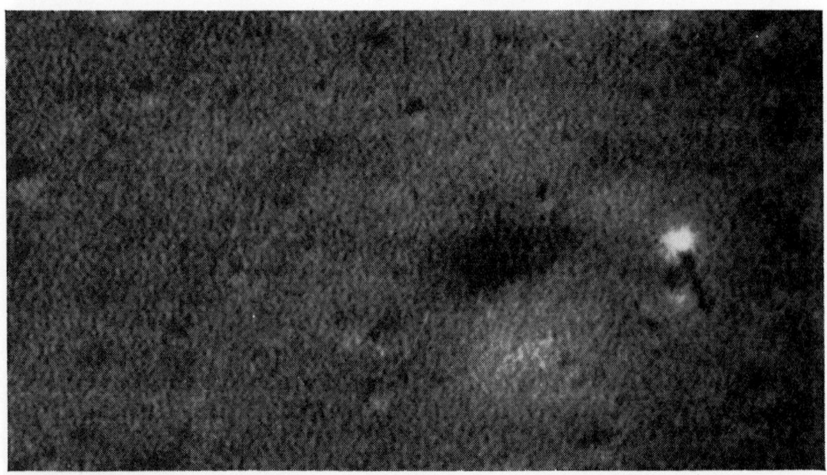

***SURVEYOR I* SPACECRAFT is identifiable by the shadow it casts. Its landing place was photographed with the telephoto lens of *Lunar Orbiter III* and its location was established by means of triangulation from objects that appear in photographs made by *Surveyor I*.**

for the Apollo missions. Since then NASA has reduced the number of sites to five; eventually it plans to focus on three sites for planning Apollo missions.

The primary task of the Lunar Orbiter program was essentially completed during the first three missions. Accordingly NASA used the last two Orbiters to photograph areas of scientific interest. *Lunar Orbiter IV* was assigned the task of photographing the entire front face of the moon, which it did at resolutions of 150 to 300 feet. It also made excellent vertical photographs of the lunar limbs.

Lunar Orbiter V photographed 51 selected areas on the near side at resolutions of six to 16 feet, virtually completed the coverage of the far side and added a photograph of what an observer on the moon would call a nearly full earth. The mission also secured stereoscopic photographs of Apollo areas not previously covered by that technique and obtained oblique photographs simulating what Apollo astronauts might see as they approach the moon.

Among the sites of particular scientific interest was the crater Aristarchus, where some earth-based observers had detected signs of what they thought might be current volcanic activity. Similarly, *Lunar Orbiter V* obtained photographs of the Marius Hills region, which has a number of domelike structures that might be indicative of past volcanic activity. Another area of interest is Copernicus, where the slumping of walls might be expected to have exposed a sequence of geological events on the moon. Such a site would be a prime target of exploration during one of the Apollo missions. *Lunar Orbiter V* made vertical photographs of Copernicus to complement the oblique photographs obtained by the second Orbiter.

All the Lunar Orbiters remained in orbit for a time after completing their photographic mission. During that period they continued gathering data from their nonphotographic experiments on radiation, micrometeoroids and lunar gravitation. They also were used for other special tests. Eventually each Orbiter except the fourth was deliberately crashed onto the lunar surface. The reason was to make sure that they would not become sources of interference with communication between the earth and other spacecraft that are sent to the moon or elsewhere in the future. Communication between the earth and *Lunar Orbiter IV* was lost in July, 1967, but computations of its orbit indicate that it crashed on the moon in October.

We are not in a position to interpret in detail the photographs from the Orbiter missions. Analysis of the lunar surface by means of the photographs is being done by NASA, other Government agencies and independent investigators. We think the photographs accompanying this article, which provide examples of both the wide-angle and the telephoto coverage obtained by the Orbiter cameras, speak for themselves in terms of photographic quality and the detail they show.

A Sampling of Photographs

It is possible to point out here several of the interesting lunar features in the photographs that either cannot be seen from the earth or are viewed from perspectives not available to earth-based observers. In a wide-angle photograph of the far side of the moon [*pages 16 and 17*] the most conspicuous feature in the rugged terrain is the crater called Tsiolkovsky, which is largely filled with the dark material characteristic of a lunar mare, or "sea."

The Orientale Basin [*page 26*] can be partly seen from the earth, but in a perspective that gives no hint of its extraordinary bull's-eye configuration or its great size. The Cordillera Mountains, which form the eastern rim of the basin, are visible from the earth, but only in profile because they appear at the left edge of the moon's visible face.

Near one of the potential Apollo landing sites is a crater [*illustration at right*] that gives the appearance of having been formed more recently than most of the others so far studied on the moon. The crater, which is about 500 feet in diameter, is in Oceanus Procellarum about 35 miles north of the lunar equator and 800 miles west of the center of the moon's visible face. The material ejected from the crater overlies all the surrounding material and appears to be much less eroded than other features.

Evidence that objects on the moon sometimes move appears in a photograph showing part of the inside of the crater Vitello [*middle of opposite page*]. A boulder visible in the photograph has rolled into and out of a small depression, leaving a distinct trail. The boulder is about 60 feet in diameter.

The kind of terrain that NASA is considering for the landing of Apollo spacecraft can be seen in a photograph [*page 20*] made with the wide-angle lens of the camera aboard *Lunar Orbiter III*. The terrain is fairly smooth. The site is 48 degrees west of the center of the visible face of the moon.

An unusual view of the rille, or valley, associated with the crater Hyginus appears [*top of page 29*] in an enlargement of part of a wide-angle photograph secured by *Lunar Orbiter III*. The feature is visible from the earth but not in this perspective. The crater itself is about 6½ miles wide and 2,600 feet deep.

Nonphotographic Results

By means of the Lunar Orbiter spacecraft it was possible to gather a considerable amount of information about the moon's gravitational field, the intensity of solar radiation in the vicinity of the moon and the frequency of micrometeoroids. These data have been summarized for us by three investigators at NASA's Langley Research Center in Virginia. William H. Michael, Jr., is principal investigator for the experiments in selenodesy, Trutz Foelsche for the radiation measurements and Charles Gurtler for the detection of micrometeoroids.

Michael reports that the analyses of data obtained from tracking Lunar Orbiter spacecraft have provided new information on the moon's mass, its density and its radius. He writes: "The new determination of the mass of the moon has been performed by a more direct procedure than that used in previous analyses, and the result is in close agreement with the value obtained from analyses of Ranger spacecraft data. Present indications... are that the material in the interior of the moon is probably of more or less uniform density, that the density increases somewhat toward the center of the moon, but that a small core of higher-density material cannot yet be completely ruled out. These indications are in contrast to somewhat questionable previous results, obtained prior to the establishment of lunar satellites, which suggested that the density of the moon was greater in the outer portions of the moon.

"The radius of the moon in the equatorial region on the near side... is about two kilometers less than that previously obtained from analyses of earth-based photographs. This result may indicate either that the mean radius of the moon is less than was previously thought or that the center of mass of the moon is slightly displaced toward the earth relative to the center of figure."

Among the conclusions reached by

RECENT CRATER is identifiable by the freshness of the material around it. Boulders as small as one yard in diameter can be seen in the pattern of rays. Sliding of the material inside the crater has caused the double-wall appearance. The crater, which is in Oceanus Procellarum, is about 165 yards wide. The third Orbiter made photograph at 31-mile altitude.

VIEW OF THE EARTH from a distance of 224,000 miles was made through the telephoto lens of *Lunar Orbiter V* last August 8. The time was noon in Saudi Arabia, which occupies most of the Arabian Peninsula (*left center*). Spacecraft was about 3,640 miles above moon.

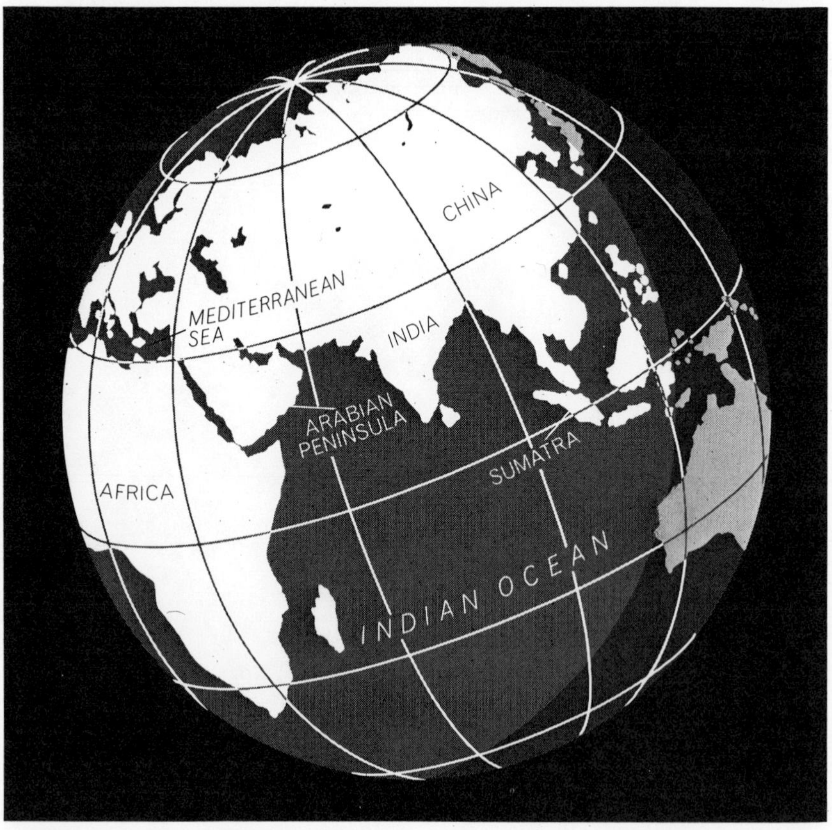

PRINCIPAL FEATURES of the earth that can be seen in the Lunar Orbiter photograph at the top of this page are identified. Some features were partly obscured by light clouds.

Foelsche and his colleagues are the following: (1) No trapped energetic particles are observed near the moon over long periods when the sun is quiet. Thus the intrinsic magnetic field of the moon is weaker than the one at the boundary of the earth's radiation belts; indeed, the moon may have no magnetic field. This finding leads to the conclusion that the moon, unlike the earth, does not have an extended liquid core in which slow fluid motions can occur. (2) Three events involving the emission of energetic particles by the sun were detected by the Lunar Orbiter spacecraft. All the events were of moderate scale and fairly low particle energy.

Gurtler reports that 18 punctures were recorded by the 100 micrometeoroid detectors carried by the Orbiter spacecraft. (There were 20 detectors on each spacecraft.) "These data," he writes, "indicate that the measured rate in the lunar environment is less than half the rate in the near-earth environment as measured by the same type of detectors on *Explorer XVI* and *Explorer XXIII*." According to Gurtler the punctures were distributed randomly around the Orbiter spacecraft, suggesting that the micrometeoroids did not come from any preferred direction.

The work of many agencies, companies and individuals was brought together on the Lunar Orbiter team. The efforts of all who participated in the program are acknowledged, even though it is not possible to name them here.

In the American program of lunar exploration attention is now focused on the Apollo project. When the men in an Apollo landing vehicle approach the moon, they will have the benefit of Lunar Orbiter photographs to help them identify their landing area and the landmarks around it. Later missions will use Lunar Orbiter photographs to select areas on the moon that seem particularly likely to reward the kind of close inspection that parties of astronauts will be able to conduct. The astronauts will also have the benefit of the nonphotographic findings from the Orbiter missions as a major contribution to the effort to make their mission to the moon a safe one.

Analysis of the photographs should yield knowledge about the processes that shaped the lunar surface. That knowledge could be expected to contribute to the understanding of the processes that formed the surface of the earth. Finally, the extensive knowledge and operational experience gained from the Orbiter missions will surely contribute to the exploration of the planets beyond the moon.

II

The Planetary System

II

The Planetary System

INTRODUCTION

The Natural History of the Heavens, although a cognate and most interesting field of inquiry, is not included in the domain of Astronomy proper. This latter treats of laws, and of motions; and as has been well said by one of the first of modern authorities, all else that we learn of the heavenly bodies, as for example, their appearance and the character of their surfaces, although, indeed, far from unworthy of attention, is not of astronomical interest."

This opinion, expressed by the distinguished astronomer B. A. Gould, scandalized the 19th-century supporters of the Dudley Observatory to whom it was addressed. It was an unexpectedly parochial view on Gould's part, for he was a collector of scientific classics, and had in his library Galileo's *Sidereus Nuncius,* wherein a discussion of the surface of the moon gave astronomical reform a major impetus. Yet in the ensuing years after Galileo reported his telescopic discoveries, studies other than celestial mechanics seemed to lead nowhere. In 1859 Gould could scarcely have guessed, for example, that historical observations would give theoreticians important clues to the stability and longevity of the Jovian surface features. Raymond Hide utilizes precisely such descriptions in his article "Jupiter's Great Red Spot."

What was missing early in 1859 was a sound technique for conducting meaningful astrophysical planetary studies. In December of the same year, however, Kirchhoff described his spectroscopic experiments and interpretation of the dark Fraunhofer lines in stellar spectra; this historic work opened the way to an understanding of the chemical composition of the universe. The attack on the chemical natures of the planets did not come easily, for the distinctive molecular bands of planetary spectra lie mostly in the comparatively inaccessible infrared. Nevertheless, what seemed irrelevantly remote in Gould's day has transformed all of astronomy beyond his wildest imagination.

Similarly, in the 1880's, when Schiaparelli sketched the dusky markings on Mercury to determine its rotational period, could he have foreseen how radar would reveal a quite different rotation? The story of this and the equally unexpected results for Venus is told by Irwin Shapiro in "Radar Observations of the Planets" in this collection—results, I should add, as unexpected for contemporary astronomers as they would have been for Schiaparelli!

Or would Percival Lowell, whose years of observations of the surface of Mars were summarized in his *Mars and Its Canals* in 1906, have imagined that astronomers would study close-up photographs of the red planet only six decades later? Two of the great pictures of Mars, obtained from Mariner 7 in the summer of 1969, are reproduced here, and a discussion of other spacecraft results is found in Von Eshleman's "The Atmospheres of Mars and Venus."

Techniques of spectroscopy and photometry, applied to the electromagnetic spectrum from radio wavelengths through the infrared and to the ultraviolet, together with the methods of deep space probes, have combined to recapture planetary exploration from the realm of science

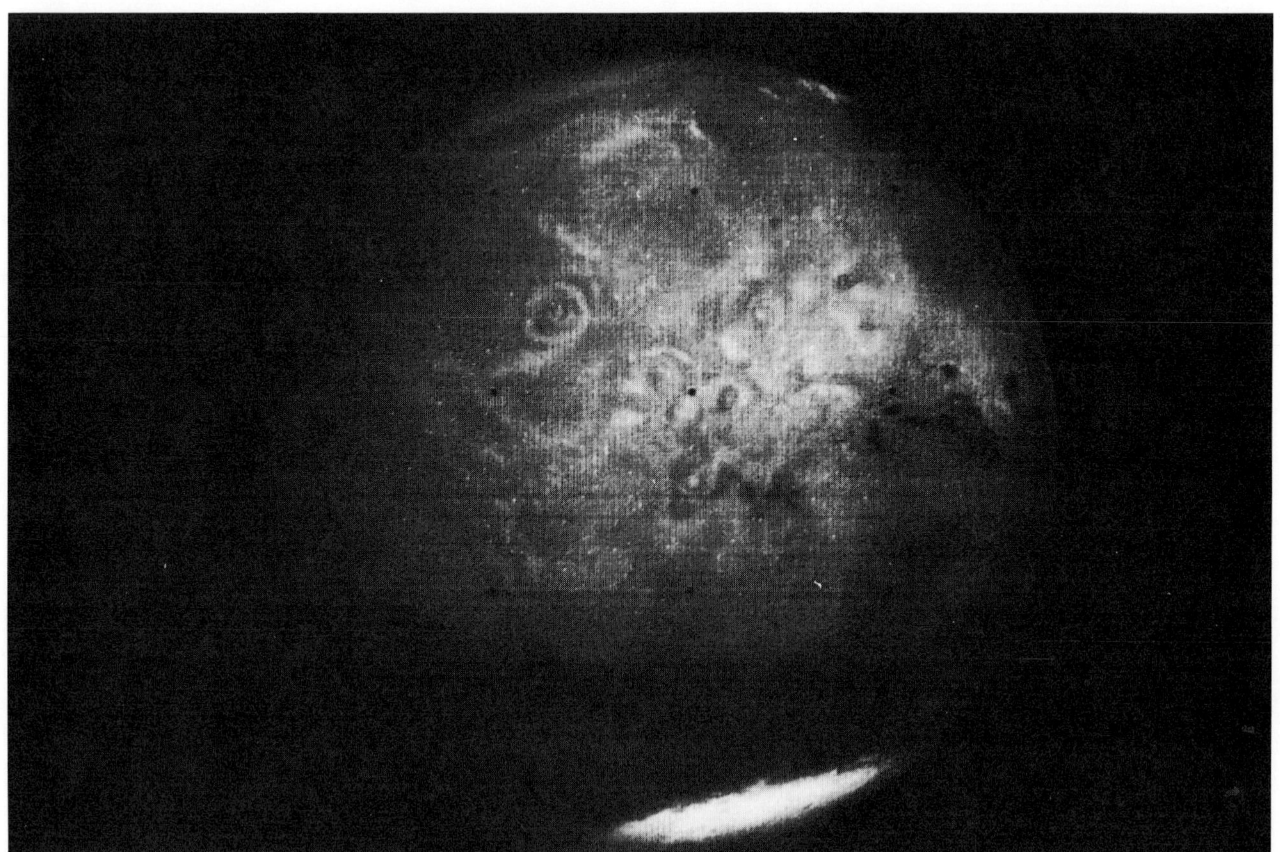

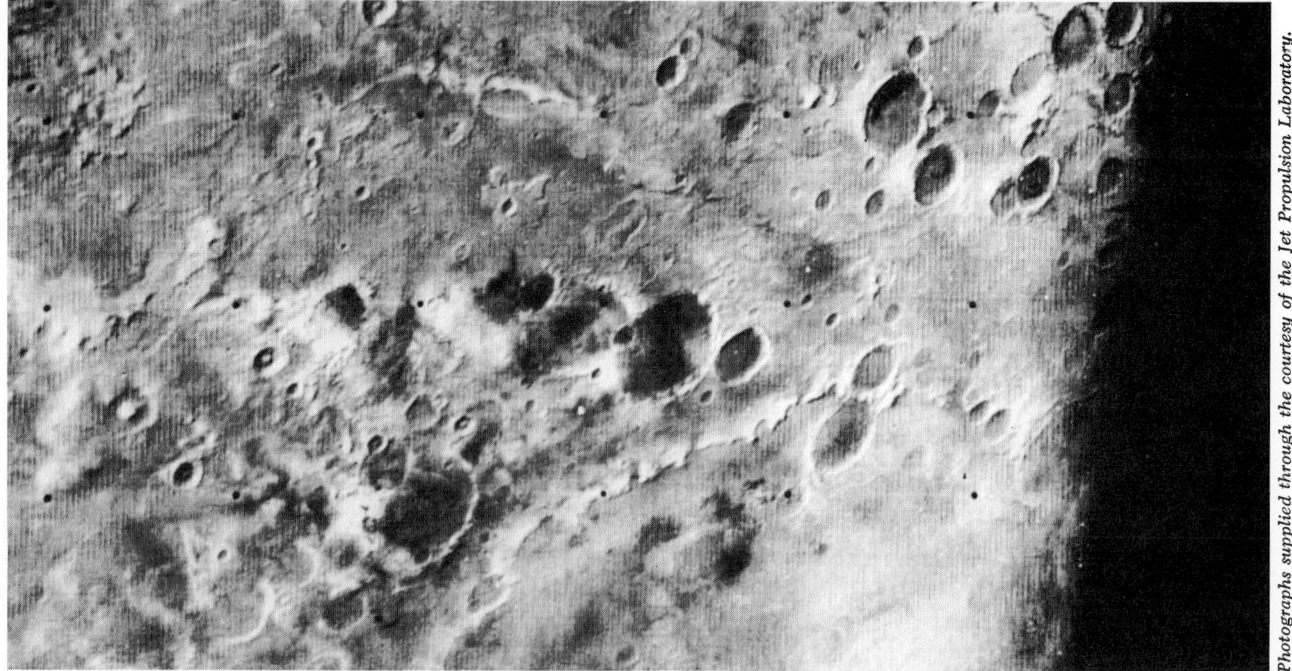

Photographs supplied through the courtesy of the Jet Propulsion Laboratory.

THE SNOWS OF MARS are shown in these two *Mariner 7* photographs. The picture at the top of this page was taken at a distance of some 290,000 miles. Typical of the far-encounter views that provide a resolution up to 10 times better than earth-based photographs, it shows that the region slightly above and to the left of the middle of the planet, long familiar to astronomers as *Nix Olympica,* is actually a giant crater some 300 miles in diameter, larger than any found on the moon. The picture also shows that the south polar cap has a serrated edge. In the close-up of the cap (*bottom*) Martian "snow" covers craters of all sizes. Because infrared instruments gave ambiguous readings for the surface temperature, the nature of the "snow" is still unresolved. The bottom view, taken by *Mariner 7* at a distance of 3,100 miles, covers an area 550 by 1,100 miles.

fiction. To these direct developments of technique must be added the psychological impact of the rejuvenated nebular hypothesis for the origin of stars and planetary systems—a hypothesis that makes the formulation of other planets a commonplace phenomenon, and makes the study of the accessible planets all the more urgent. Su-Shu Huang addresses some of these problems in a later section in his article "Life Outside the Solar System."

The oldest article in this section is my own, "The Solar System beyond Neptune." Although it antedates the results from the space exploration of the solar system, its discussion of the evolution of the planets is to a large extent as good—or as bad—as it was a decade ago. The article can perhaps be criticized for describing the origins of solar system in terms too definite, as if the solution were nearly at hand. A principal problem, the differences in composition of the planets, is stated there, together with the ages and compositions of the moon and meteorites; but a definitive evolutionary picture of the solar system seems as elusive as ever.

3

Radar Observations of the Planets

IRWIN I. SHAPIRO

July 1968

Just a decade ago planetary radar astronomy was only a glint in the eyes of a few electrical engineers and physicists. Yet by 1961 the first echoes of radar signals reflected from Venus had been detected. Radar contacts with Mercury and Mars soon followed. Although radar astronomy is still a very young science, it has already produced results whose scientific importance has far exceeded the most enthusiastic predictions. With this new technique one can, for example, investigate the spin and orbital motions of planets, their surface characteristics and the properties of the medium through which the radar signals pass. The data obtained and the theories developed are already far too extensive to be discussed adequately in a single article. I shall therefore concentrate primarily on a few of the more significant results: a new test of Einstein's general theory of relativity and, most surprising of all, discoveries concerning the rotation of Mercury and Venus.

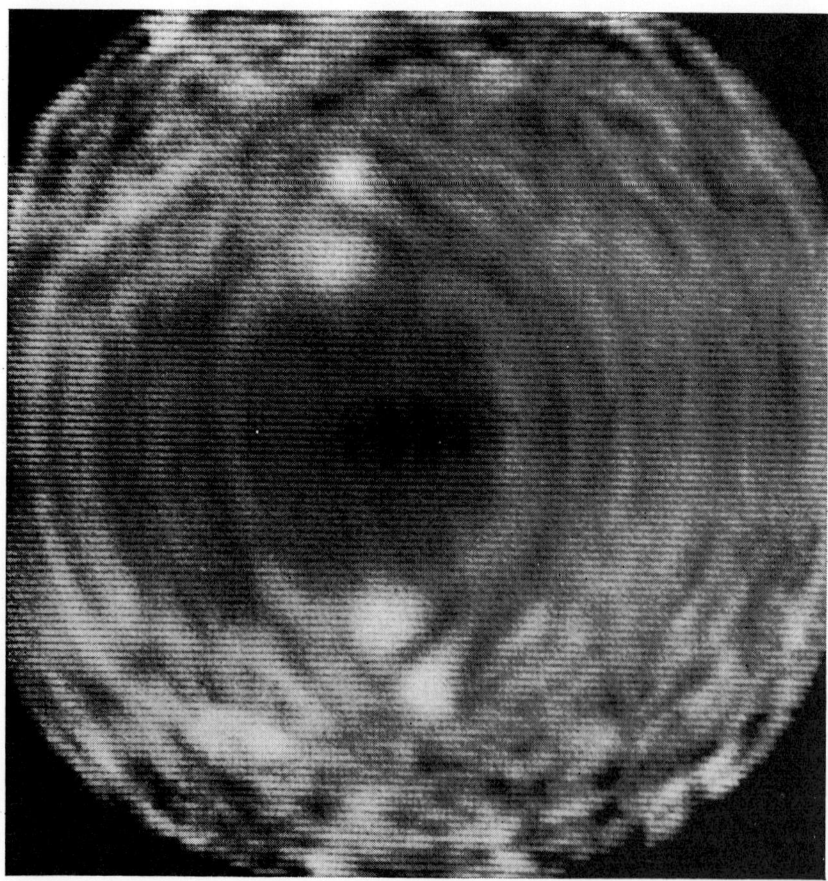

RADAR PICTURE OF VENUS was made by Raymond F. Jurgens of Cornell University with the 1,000-foot radio telescope at Arecibo in Puerto Rico. In radar mapping the power of an echo from a given area of the target is determined as a function of the time the echo is received and the shift of the echo's frequency from that of the transmitted signal. A two-fold ambiguity in plotting each point (*see text*) gives the planet a symmetrical appearance. This picture was obtained from the data shown in more direct form at the bottom of p. 46.

The basic experimental problem is to aim a carefully controlled radio signal at a distant planet, detect the echo and analyze it for the information it carries. For this to be done the energy in the echo must be sufficiently large. Since the energy density of both the transmitted and the reflected waves decreases with the inverse square of the radar-target distance, the energy of the echo received at the radar antenna varies inversely with the fourth power of that distance [*see top illustration on page 40*]. Hence there are enormous differences in the relative difficulty of detecting various objects in the solar system. The step up from the detection of echoes from the moon to the detection of echoes from Venus (at its closest approach to the earth) required about a 10-million-fold improvement in the sensitivity of radar systems. Yet only 15 years was required for this advance.

The primary instruments now being used for planetary radar astronomy are the Lincoln Laboratory's Haystack facility in Tyngsboro, Mass., operated at a frequency of 7,840 megacycles per second; Cornell University's 1,000-foot telescope at Arecibo in Puerto Rico (40 and 430 megacycles), and the Jet Propulsion Laboratory's radars at Goldstone, Calif. (2,388 megacycles). These three sites are continually jockeying for the lead in overall system sensitivity, having far outdistanced the competing facilities at Jodrell Bank in England and at the Crimean Tracking Station in the U.S.S.R., which have apparently given up the radar race.

Just how sensitive are these contending radars? As an illustration, consider the Haystack facility, which can transmit continuously up to 400 kilowatts of power and routinely detects echoes almost 27 orders of magnitude weaker. Such echoes—about 10^{-21} watt—represent far less power than would be expended by a housefly crawling up a wall at the rate of one millionth of a meter per year. These minuscule signals are extracted from the much more powerful radio noise (originating partly in the receiver) by computer processing. Because the received signals are accumulated for many hours, the quantum aspects of detection can be safely ignored.

Considering the centuries of precise optical observations of the planets, one might wonder how a few years of radar measurements could make an important contribution to the study of planetary orbits. The significance of the radar data flows from two facts. First, radar adds two new types of measurement: the echo time delay, which is related to the round-trip distance through the travel time of light, and the Doppler shift of the echo from the planet, which yields the velocity of the reflecting surface along the line of sight. Second, many of the radar measurements are of unprecedented accuracy; precision exceeding one part in 100 million is now achieved routinely in interplanetary time-delay observations.

Why should anyone want to add yet another significant digit to the expressions for the orbital positions of the planets? One reason is that in planning planetary spacecraft missions knowledge of the distance to the objective is vital. Such distances were traditionally expressed in terms of the astronomical unit (in essence the average distance between the earth and the sun). Before 1961 the relation between this unit and the kilometer was known to only about one part in 1,000. Since no planet comes closer to the earth than 40 million kilometers, .1 percent errors can be serious! The first measurements of interplanetary time delays and Doppler shifts made by Gordon H. Pettengill and others at the Lincoln Laboratory and by Richard M. Goldstein, Duane Muhleman and others at the Jet Propulsion Laboratory contributed significantly to the success of the *Mariner II* mission to Venus in 1962. Recently Michael E. Ash, William B. Smith and I, using Lincoln and Arecibo radar data, have established the light-time equivalent of the astronomical unit

210-FOOT ALUMINUM ANTENNA operated by the Jet Propulsion Laboratory at Goldstone, Calif., is now being used for planetary radar astronomy. Its size and almost perfectly paraboloidal surface make it a sensitive instrument for aiming radio signals at planets and detecting the echoes. The "feed cone" protruding through the center contains both the transmitter and the receiver when it is used for radar observations. Signals are reflected between cone and antenna by a subreflector above them (*not visible in photograph*).

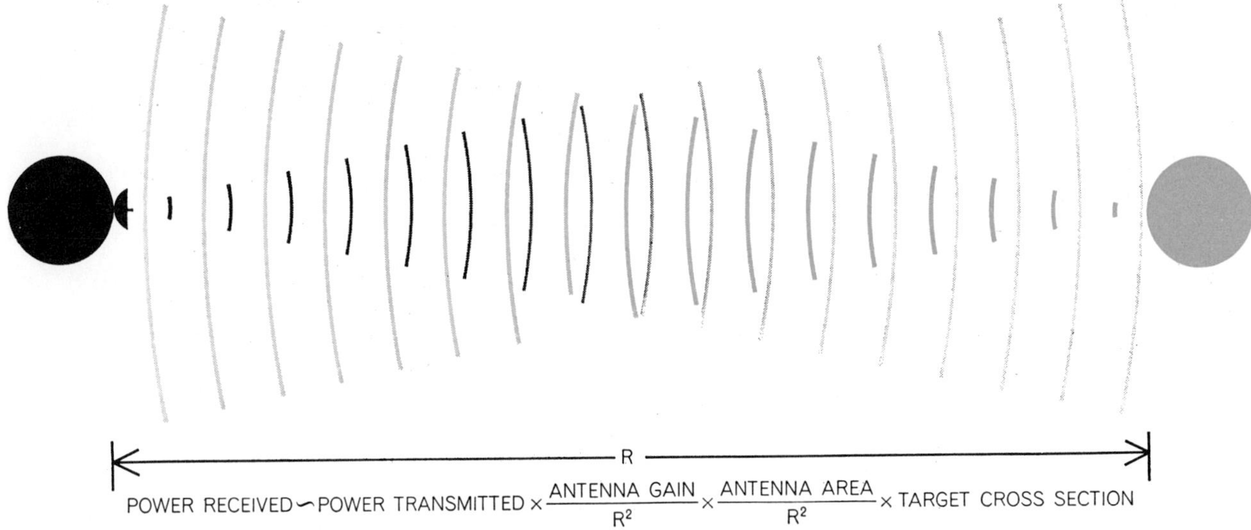

RADAR EQUATION shows how the received echo power is related inversely to the fourth power of the range (*R*). The planet (*right*) intersects a very small segment of the total spherical wave front emanating from the radar transmitter and reflects only a fraction of that power; the receiver, in turn, can collect only a small fraction of the signal (*colored wave front*) returned by the target.

to about one part in 100 million. (The kilometer equivalent remains less accurate because the speed of light is still known only to one part in a million.)

In addition to the practical value of precise interplanetary distances there is an important potential theoretical gain: The more accurately one knows the positions of planets, the more decisively one can test the physical laws that attempt to account for them. These laws are embodied, for example, in Einstein's general theory of relativity, which had been subjected to only three experimental tests when, in 1964, improvements in radar sensitivity led me to point out that a fourth test was technically feasible. According to general relativity, a light wave or radio wave passing the sun is not only deflected (the basis of one of the three classical tests of the general theory) but also slowed. (This effect is just the opposite of what one might have expected, since matter speeds up on approaching a massive body. In both cases, however, the trajectory is bent toward the attracting center.) Can such a slowing be measured by radar? Suppose we continually beam radar signals at a planet. Only when it is on the opposite side of the sun from the earth at "superior conjunction" will the signals pass close to the sun and so be significantly affected by solar gravity [*see top illustration on page 43*]. The maximum increase in delay, when the signal just grazes the sun, was predicted to be some 200 microseconds, about one part in 10 million of the total echo time delay.

The experimental problem, then, is to search for this very small difference. One should, of course, first know what delays to expect in the absence of a relativistic effect. That is, one must determine the orbits of the earth and the target planets far more accurately than was previously possible. This intermediate goal is accomplished by making a large number of accurate radar time-delay and Doppler-shift measurements and using them to determine the orbits and such necessary quantities as the masses and radii of the target planets. All calculations are carried out within a consistent (relativistic) theoretical framework. By following this procedure we can now predict planetary positions to within about 10 microseconds in terms of the travel time of light, or to within roughly 1.5 kilometers one way.

So far I have made the tacit and false assumption that the radar signal is unaffected by the medium between the earth and the target planet. The most important influence is the plasma of the solar corona, which also tends to decrease the speed of propagation of the radar signal. Fortunately this slowing can be distinguished from the corresponding effect of relativity: The plasma influence is inversely proportional to the square of the radar frequency, whereas the relativity effect should not vary with frequency. These facts present us with several experimental options, the most convenient of which is to make the measurements with waves of a frequency high enough to reduce the plasma effect to a negligible level.

A major question remains: How is it possible to measure the round-trip radar delay to within 10 microseconds considering that a planet is a huge target and that the radar beam is wider still? A key point in the answer lies in the planet's reflecting radar waves in a quasi-specular manner, that is, reflecting them much as a polished ball bearing reflects light. Most of the echo power therefore stems from reflections near the subradar point: the point where the surface of the planet

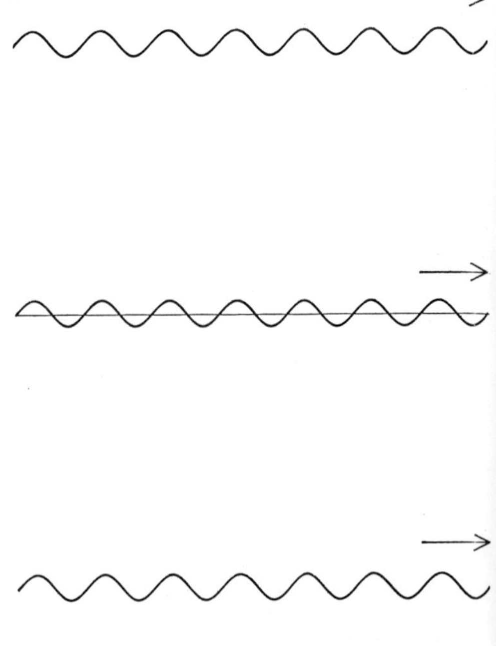

DELAY-DOPPLER MAPPING depends on the delay and frequency of the echo from each reflecting area (*a*). The signal (*black waves*) is reflected first from the subradar point (*on line from radar to center of tar-*

intersects the line from the radar site to the planet's center. Further refinement in establishing the time delay specifically to the subradar point is achieved through delay-Doppler mapping, a major tool of the radar astronomer that was first suggested by Paul E. Green, Jr., of the Lincoln Laboratory in 1959. This technique is based on the planet's being so large compared with the wavelength of the radar signal that one can consider its surface to be composed of separate elements, each reflecting independently of the others. Every surface element is therefore considered to reflect an incident wave front at a particular time delay and to impart to that reflection a particular Doppler shift.

If the transmission consists of short pulses, then the echo received at any instant will correspond to reflections from a set of points equidistant from the earth. These points define contours of equal time delay, which correspond to rings on the planet that are perpendicular to the line from the radar to the planet's center. Similarly, there are contours of equal Doppler shift. As viewed from the radar the target planet appears to have a certain rotation, the result of the relative orbital motion of the earth and the planet and of the planet's intrinsic spin. This apparent rotation causes each reflecting element to impart a characteristic Doppler shift to the single-frequency radar signal: A surface element moving toward the earth will return an echo with a frequency higher than the frequency of the transmitted signal; an element moving away from the earth will return an echo with a lower frequency. Points on the surface that impart the same Doppler shift to the incident signal define an arc on the surface in a plane parallel to the plane containing the radar site and the target's apparent rotation axis [*see illustration below*].

A combination of the contours of equal time delays and equal Doppler shifts yields a delay-Doppler map of the planetary surface, a map in which the echo power of elements on the surface is plotted against delay and frequency [*see illustration on next page*]. Such maps have an unusual property: Since the same Doppler shift is imparted to the signal by two different points on the surface, there is a twofold ambiguity for all points not lying on the apparent equator. Along this equator a Doppler contour intersects just a single segment of a delay ring. Since the Doppler and delay contours are parallel at that point, the area of intersection is greatly enlarged.

How do we use the delay-Doppler mapping principle to pinpoint the time delay and the Doppler shift that correspond to reflections from the subradar point? First we prepare a two-dimensional "template" from observations made when the target planet is nearest the earth and the echo signal is consequently strongest. This is an electronic record of the expected form of the echo: its power as a function of the delay and Doppler coordinates in relation to the subradar point. Now, in any given situation—when the planet is far away, for example—the template, correctly adjusted, shows the expected echo power with respect to the subradar point. What will be unknown are the actual values for the delay and Doppler shift corresponding to the reflection from that point. These values are obtained by cross-correlating the actual echo with the template; the values for which the cross correlation is maximum constitute the best estimates of the delay and Doppler shift undergone by the signal reflected from the subradar point.

This procedure, first suggested by Robert Price of the Lincoln Laboratory, has been applied since 1966 to planetary echoes observed at Haystack. From such measurements made near the superior conjunction of Venus in that year and during the first three superior conjunctions of Mercury in 1967, Pettengill, Ash, Melvin L. Stone, Smith, Richard P. Ingalls, Richard A. Brockelman and I concluded that the sun's gravity does decrease the speed of propagation of electromagnetic waves by about the amount

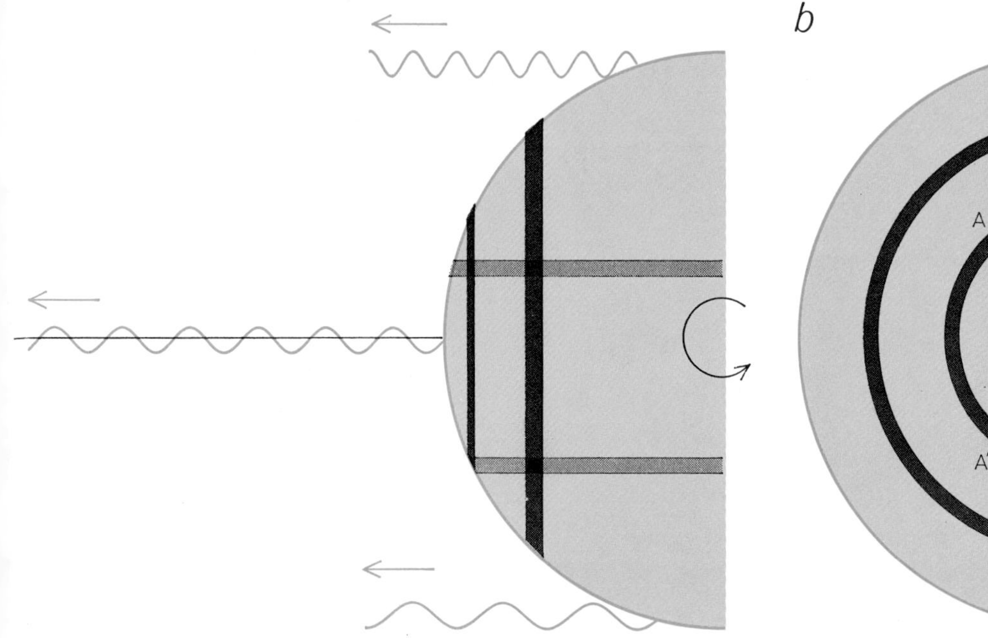

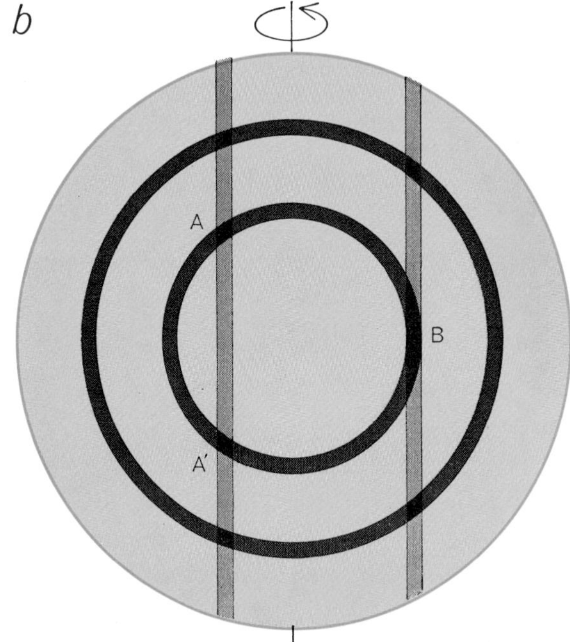

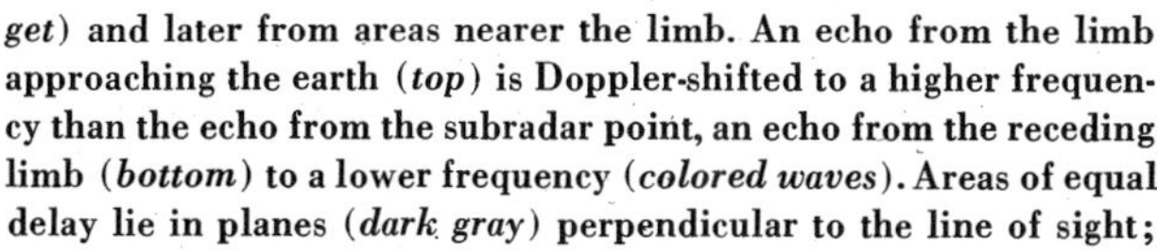

get) and later from areas nearer the limb. An echo from the limb approaching the earth (*top*) is Doppler-shifted to a higher frequency than the echo from the subradar point, an echo from the receding limb (*bottom*) to a lower frequency (*colored waves*). Areas of equal delay lie in planes (*dark gray*) perpendicular to the line of sight; areas of equal Doppler shift lie in planes (*light gray*) parallel to the axis of apparent rotation and the line of sight. On the face of the target (*b*) these planes establish delay (*rings*) and Doppler (*strips*) contours. There is a twofold ambiguity in mapping (*A, A′*) except at the apparent equator, where the intersection area is enlarged (*B*).

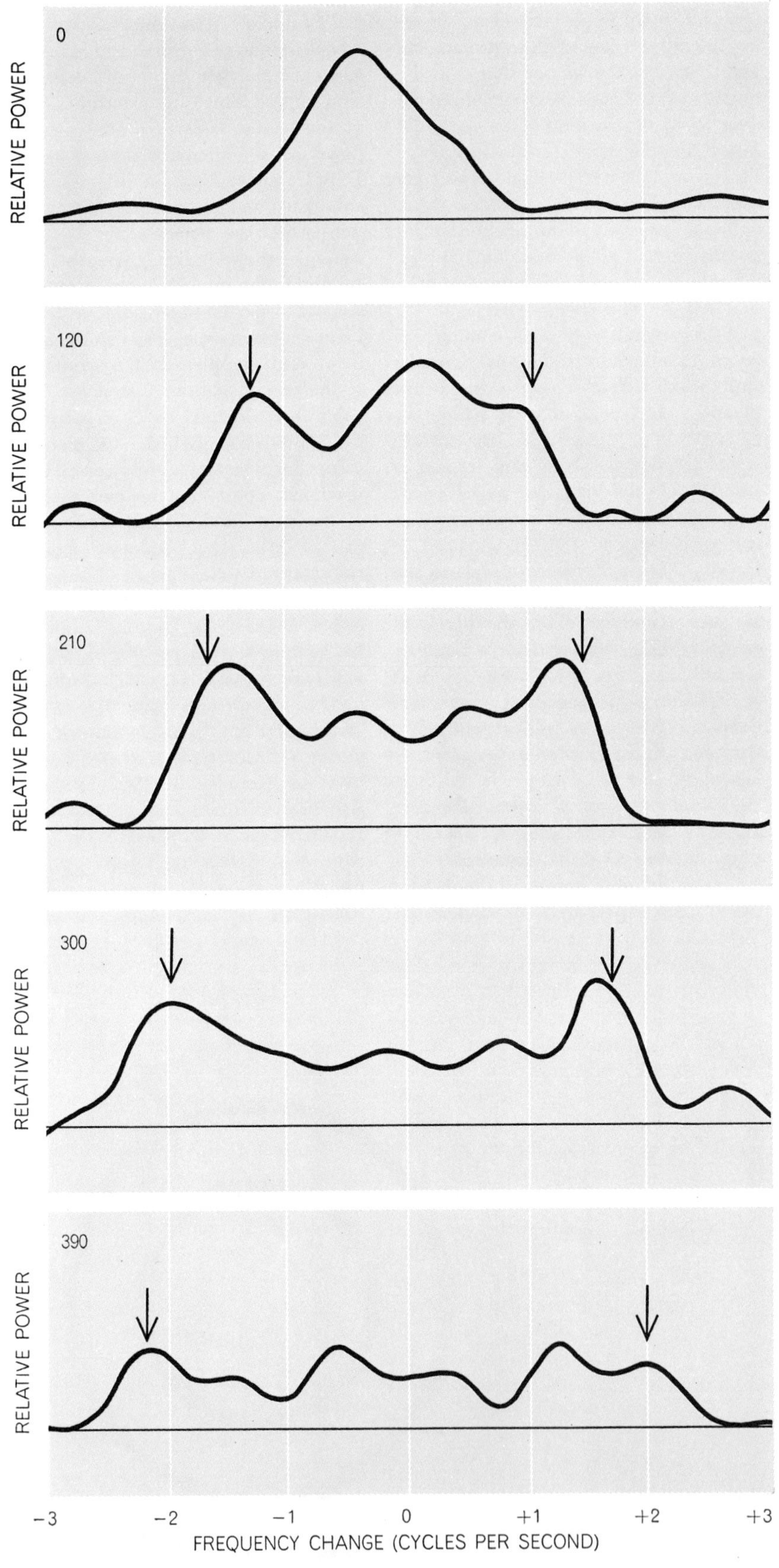

SPECTRA of echoes from Mercury provide a crude delay-Doppler map of the surface. The echo power is plotted, for returns from each of five delay contours up to 390 microseconds from the subradar point, as a function of the frequency shift in relation to the echo from the subradar point. The arrows show the positions of the edges of the spectra (the enlarged areas of intersection of contours at the apparent equator) that correspond to a rotation period for Mercury of 59 days. The spectra were obtained at the Arecibo observatory.

predicted by general relativity; the experimental uncertainty was about 20 percent [*see bottom illustration on opposite page*]. It should be possible to increase the accuracy significantly with the aid of space probes or of radio beacons placed on the surface of planets.

The same delay and Doppler measurements can also be used to improve the accuracy of the third classical test of general relativity—the measurement of the advance of the perihelion of Mercury's orbit—and to show whether the sun is oblate and whether gravitational forces change with time. Such deductions, which depend primarily on the gradual accumulation of small effects, will become increasingly precise as measurements are collected over a long period. For example, within the next five years a change in the gravitational constant of less than one part in 10 billion per year should be discernible!

The most striking discoveries made with planetary radar observations concern the rotation rates of Mercury and Venus. It may seem strange that these rates have not been well known from telescopic observations. Astronomers had indeed thought they knew the spin of Mercury, but thoughts change. Venus, on the other hand, is hidden behind the veil of its atmosphere, and optical observations of its surface have been impossible.

Early in the 19th century the German observer J. H. Schröter noted mountains rising to 20 kilometers on Mercury and made many drawings of the appearance of the planet's surface. From these the mathematician Friedrich Wilhelm Bessel deduced a rotation period of just over one day. Although some astronomers remained skeptical, many found it aesthetically appealing to think that Mercury, like Mars, had a day about as long as the earth's. Not until the late 1880's was this "fact" discredited: Giovanni Schiaparelli's extended series of observations then convinced almost everyone that Mercury was rotating much more slowly. He himself concluded that Mercury's spin was actually synchronous with its orbital motion, the planet making one rotation on its axis for each revolution about the sun. From that time until the spring of 1965 all observations—and there were literally hundreds—were interpreted as being consistent with the 88-day synchronous spin period. (One noted astronomer, in affirming this conclusion, even scoffed indignantly at earlier observers who had accepted the 24-hour period, "as if God would adhere to the aesthetic views of Man." He went on, however, to point out the aesthetically pleasing similarity between Mercury's synchronism with the sun and the moon's with the earth!) Although partial rationalizations can be given, it is nonetheless unsettling to contemplate this persistence of self-deception.

Then, in the spring of 1965, Pettengill and Rolf B. Dyce used radar observations made at Arecibo to show rather conclusively that Mercury's spin period is nearly 59 days. The basis of this radar determination was again delay-Doppler mapping: the faster the apparent rotation of the planet, the greater the difference in frequency between reflections from the two sides of each time-delay ring [*see illustration on opposite page*]. By making a large number of such measurements at different orbital positions, the various ambiguities were resolved and the speed of the planet's rotation was established.

How is this result for Mercury's spin to be interpreted? Does it represent a stable, final spin state or a stage in evolution toward a synchronous spin? Stanton J. Peale and Thomas Gold of Cornell suggested that because of the large eccentricity of Mercury's orbit the present spin state could be stabilized by the average torque exerted by the sun on the tidal bulge it raises on Mercury. Shortly thereafter Giuseppe Colombo of the Smithsonian Astrophysical Observatory noted that 59 days is almost exactly two-thirds of Mercury's orbital period of 88 days. He proposed that Mercury's spin might be locked to its orbital motion in a 3/2 spin-orbit resonance, with the planet rotating precisely one and a half times during each revolution about the sun. Colombo and I then developed a mathematical model, taking account of the tidal effects and also of the torque exerted by the sun on any permanent axial asymmetry in Mercury's mass distribution, and found that the resonance lock was possible and indeed would be stable. Peter Goldreich and Peale confirmed this, showing that the probability of capture into the 3/2 state might be about one chance in five. That is not a large probability, yet no one could be mortally offended that our one sample (Mercury) corresponded to an event with a probability of .2.

The story does not end here, however. The probability calculation was based on Mercury's having maintained its present orbital eccentricity during the process of capture into the spin-orbit resonance, but in fact the pull of other planets tends to change the eccentricity

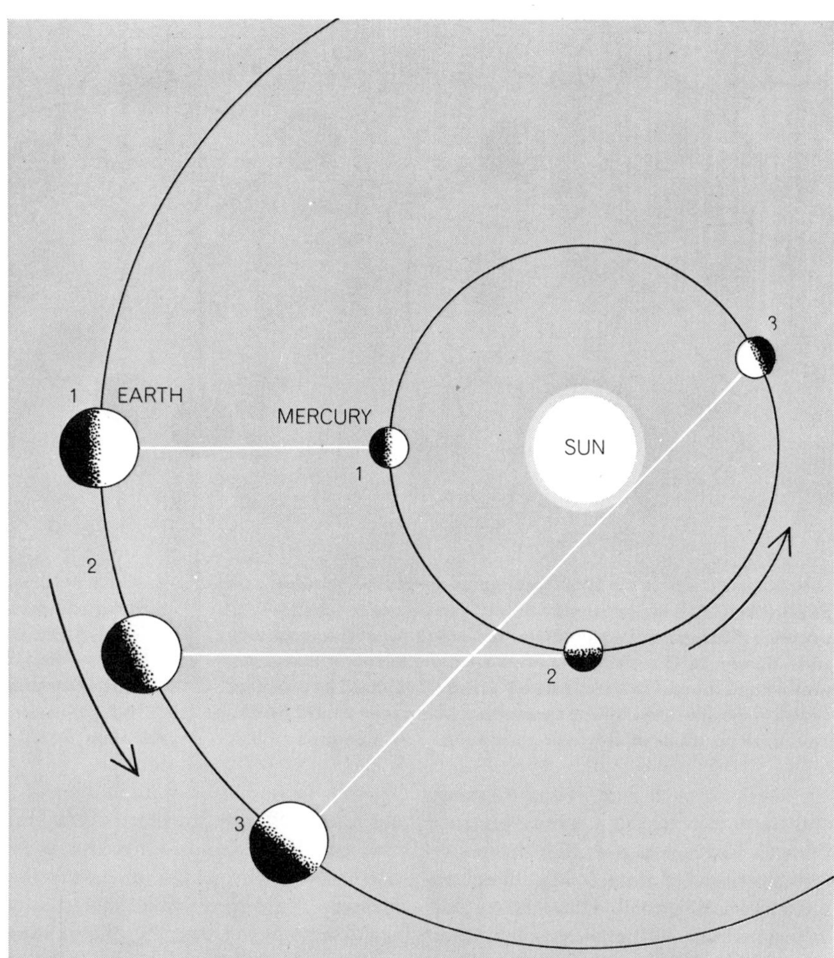

SUN'S EFFECT on radar travel time between the earth and Mercury is very small at inferior conjunction (*1*) and elongation (*2*), but near superior conjunction (*3*), as the general theory of relativity predicts, radar signals will be slowed down appreciably as they pass the sun.

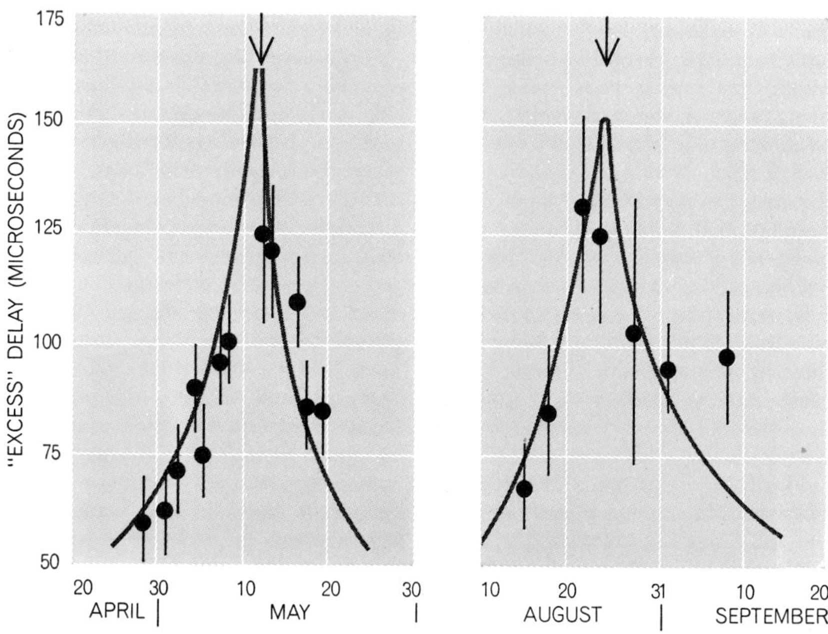

EFFECT OF SOLAR GRAVITY on earth-Mercury delays near two 1967 superior conjunctions (*arrows*) was about as predicted by general relativity. From delays computed on the basis of radar determinations of orbits, the "excess" delay caused by the sun was predicted (*gray curves*). Then actual delays were measured and "excess" delays determined (*black*).

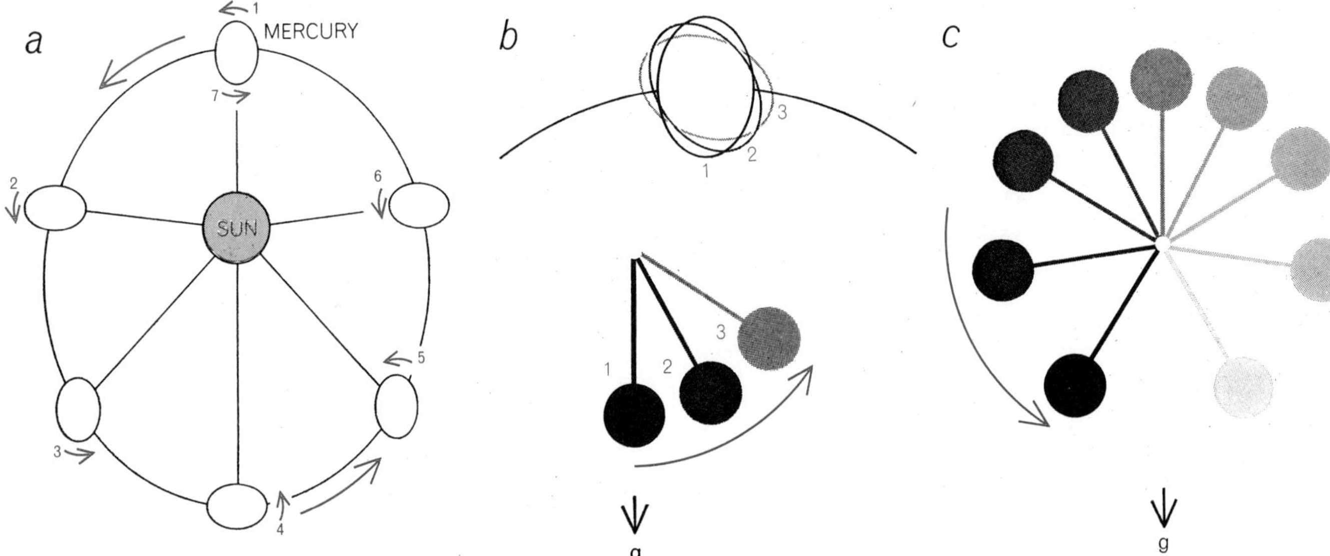

SPIN CAPTURE of Mercury by the sun can be understood by analogy with the motion of a pendulum. In the 3/2 spin-orbit resonance state Mercury will rotate one and a half times during each revolution around the sun (*a*). If it rotated slightly "too fast," its positions at successive perihelia (close approaches to the sun) would be like successive positions of a swinging pendulum (*b*). Coming to a halt at *3*, the pendulum swings back under the influence of gravity (*g*); similarly, Mercury's orientation at perihelion swings back through the action of the solar torque exerted on Mercury's egg-shaped equator. However, Mercury probably once rotated much faster than the 3/2 rate, its orientations at perihelia corresponding to a rotating pendulum (*c*). Mercury's rotation rate was undoubtedly slowed by

and thus affects the capture process. Charles Counselman of the Massachusetts Institute of Technology therefore introduced the available material on past variations in eccentricity and found that the probability of capture was reduced to about .02. That a single sample should correspond to an event with this low a probability is quite offensive to one's sensibilities. What could be wrong?

The value for the capture probability is clearly no better than the assumptions on which it is based. Two important assumptions may contain serious flaws. First, values of Mercury's orbital eccentricity have only been calculated reliably back a million years; perhaps deeper in the past the eccentricity was larger, a condition that would enhance the probability of capture. Second, Mercury may not be a rigid solid, as has been assumed; if its outer portion is coupled to an inner core by a viscous medium of some kind, the drag forces introduced could significantly raise the probability of capture. Both possibilities are now being investigated.

Finally, we must entertain the possibility that Mercury's spin may not be in the 3/2 resonance after all. The radar result established the spin period only approximately, as 59 ± 3 days. Some astronomers, studying old drawings and photographs, have pinned it down more closely, however, to within .01 day of the exact resonance value of 58.65 days. (Astronomers have tended to observe Mercury once a year, when conditions are most favorable. In that period Mercury completes about four orbits. It did not occur to the earlier observers that Mercury could meanwhile have rotated not four times but, say, six.) This optical confirmation is heartening, but past experience suggests that we should await more evidence before concluding that Mercury's spin is adhering to our current aesthetic values.

The Venus story is equally fascinating. Because of its veil—now thought to resemble smog more closely than clouds—the surface of Venus has never been seen from the earth. Before the first radar observations estimates of Venus' rotation period were therefore based on the Doppler shifts of spectroscopic lines from constituents of the planet's atmosphere and on the motion of features in the atmosphere. Even the sense of the rotation—direct (in the same direction as Venus' orbital motion) or retrograde—could not be determined. The evidence seemed to favor a slow rotation; most astronomers thought the spin was direct and synchronous, as Mercury's was then thought to be. The first Venus radar measurements in 1961 established only that the rotation was slow. Later more detailed analyses of these data by Smith and independently by Roland L. Carpenter of the Jet Propulsion Laboratory suggested that the direction might be retrograde. By the time Venus was again near the earth (in 1962) the sensitivity of the Goldstone facility had improved enough for Carpenter and Goldstein to prove that the rotation is indeed retrograde. On Venus (at least above its atmosphere) the sun rises in the west and sets in the east.

The retrograde spin is demonstrated when bandwidth data, obtained by the delay-Doppler technique, are plotted against time. The curve always dips near inferior conjunction, showing that the total angular velocity is least when the earth-Venus distance is smallest. Now, the apparent rotation caused by the relative orbital motions of the earth and Venus is inversely proportional to the distance between the planets. The dip in the curve therefore implies that the two contributions to the total angular velocity are opposing each other. Since the apparent rotation is in a direct sense (because Venus moves around the sun faster than the earth does), the intrinsic spin of Venus must be retrograde.

Since 1962 the spin period has been determined with increasing precision from bandwidth data and from following the motion of distinguishable surface features that appear on radar maps. My most recent calculations yielded a period of 243.1 ± .2 days. This number is significant because, as Goldstein noted in 1962, a retrograde rotation with a period of 243.16 days implies that Venus presents the same face to the earth at every close approach. The experimental results suggest strongly, therefore, that the rotation of Venus is being controlled not by the sun but by the earth!

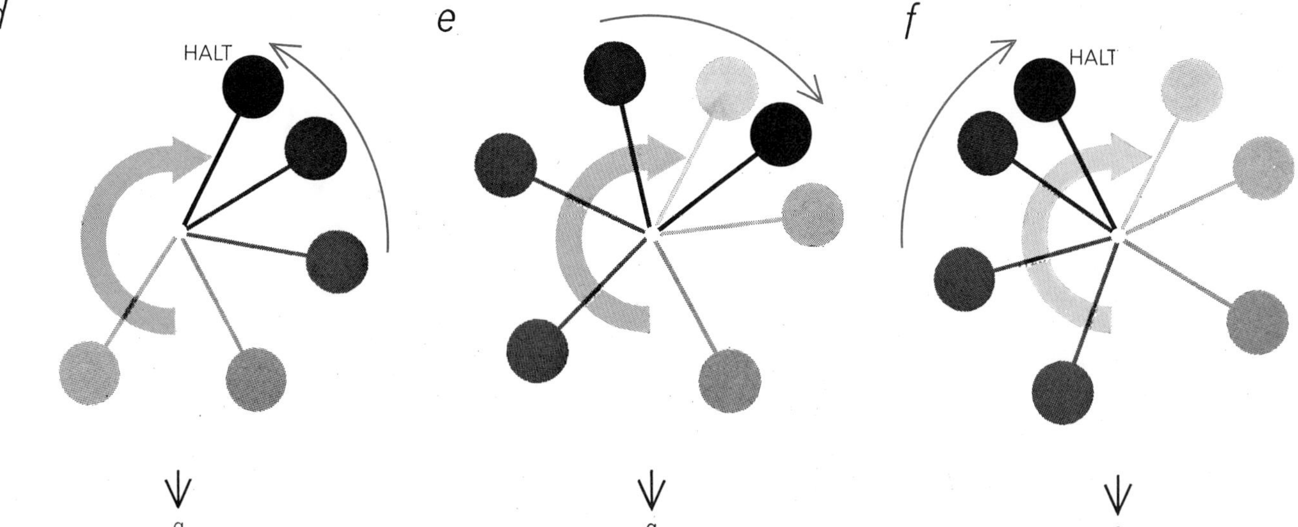

tidal friction as an analogous pendulum (*d*) is brought to a halt by an opposing torque (*broad arrow*). Further evolution can proceed by one of two paths. If the opposing torque maintains its strength, the pendulum will continue to accelerate, with the rotation now in the opposite direction (*e*); this possibility corresponds to Mercury's penetrating the resonance "barrier." If the opposing torque has a velocity-dependent component, and hence has a magnitude that depends on the direction of rotation, the pendulum may halt again (*f*), oscillate and finally come to rest. This possibility corresponds to Mercury's being trapped in the spin-orbit resonance (*a*). The outcome also depends on the initial conditions of the rotational state, which makes the capture process a probabilistic affair.

How can the retrograde motion and the control apparently exerted by the earth be explained? One might assume that Venus' primordial spin was retrograde, but data from other planets suggest that this possibility is unlikely; the initial rotation of Venus was probably rapid and direct. Solar tidal forces could have been strong enough to slow the rotation to a synchronous value, but for Venus such a state should be quite stable, and it is hard to imagine torques strong enough to break the resonance lock. One rather fanciful possibility is that Venus once collided with an asteroid with a diameter of about 200 kilometers, a cataclysm that could have reversed the planet's rotation, converting a synchronous spin into a retrograde one near the present value.

Given that Venus' spin was somehow established at a rate near resonance with the relative orbital motion of the earth and Venus, it is conceivable that the earth could acquire a controlling influence over that spin through the gravitational torque exerted on a permanent axial asymmetry in Venus' distribution of mass. The probability is vanishingly small, however, unless—as Ettore Bellomo, Colombo and I suggested—a viscous tidal force is operative. Goldreich and Peale went considerably further and assumed that Venus had a liquid core, suitably coupled to its mantle. They showed from this model that the probability of spin capture by the earth could be raised to a respectable level—perhaps to about .1. In spite of all this theoretical activity our understanding of the axial rotation of Venus is still far from complete. Yet one cannot help but be impressed at the thought that sensible inferences about a liquid core in Venus can be made from earth-based radar observations.

Radar reflections from a planet can, as one might expect, disclose properties of the planet's surface and sometimes of its atmosphere as well. Simply from the fraction of incident signal power that is reflected (more precisely, the radar cross section, which takes into account the size of the target, its surface roughness and its reflectivity), one can infer the approximate dielectric constant of the surface material in the vicinity of the subradar point and thereby obtain an important clue as to its composition. The average values of the radar cross section for Mars, Mercury and the moon are all about the same (approximately 7 percent of the geometric cross section) and seem not to vary much as the radar frequency is changed. There is, however, substantial variation over the surface, apparently greatest for Mars. For example, at Arecibo's 430-megacycle frequency the effective dielectric constant was observed to vary from about 2 to 4.5 around the 22nd parallel of north latitude. Such values suggest a material much like some of the earth's rock and soil and eliminate the possibility of appreciable expanses of water.

The cross-section values for Venus are generally higher and show less change with aspect but a marked change with frequency: the cross section was discovered to drop off sharply at high frequencies, a change that is attributed largely to the attenuation of short radar waves in Venus' atmosphere. The dielectric constant of 4.5, determined in 1961 at longer wavelengths, seemed clearly incompatible with large bodies of water, as has been verified by later ground-based and space-probe data.

Planetary topography can be studied with radar by monitoring the time delay of the echo from the subradar point as the planet rotates and thus recording fluctuations in altitude. For a rapidly spinning planet such as Mars the variations in delay caused by changing topography are readily differentiated from those caused by relative orbital motions because the periodicities of the two effects are so different—one day as compared with two years. Measurements made by Pettengill at Haystack in 1967 along the 21st parallel of north latitude on Mars disclosed height variations of about 12 kilometers, the most striking aspect of which was the regular, wavelike nature of the fluctuations [*see top illustration on next page*]. An unexpected finding was the lack of a strong correlation between elevation and either optical or radar brightness.

The determination of corresponding height variations on Mercury and Venus is greatly complicated by the similarity of their spin and orbital periods, which

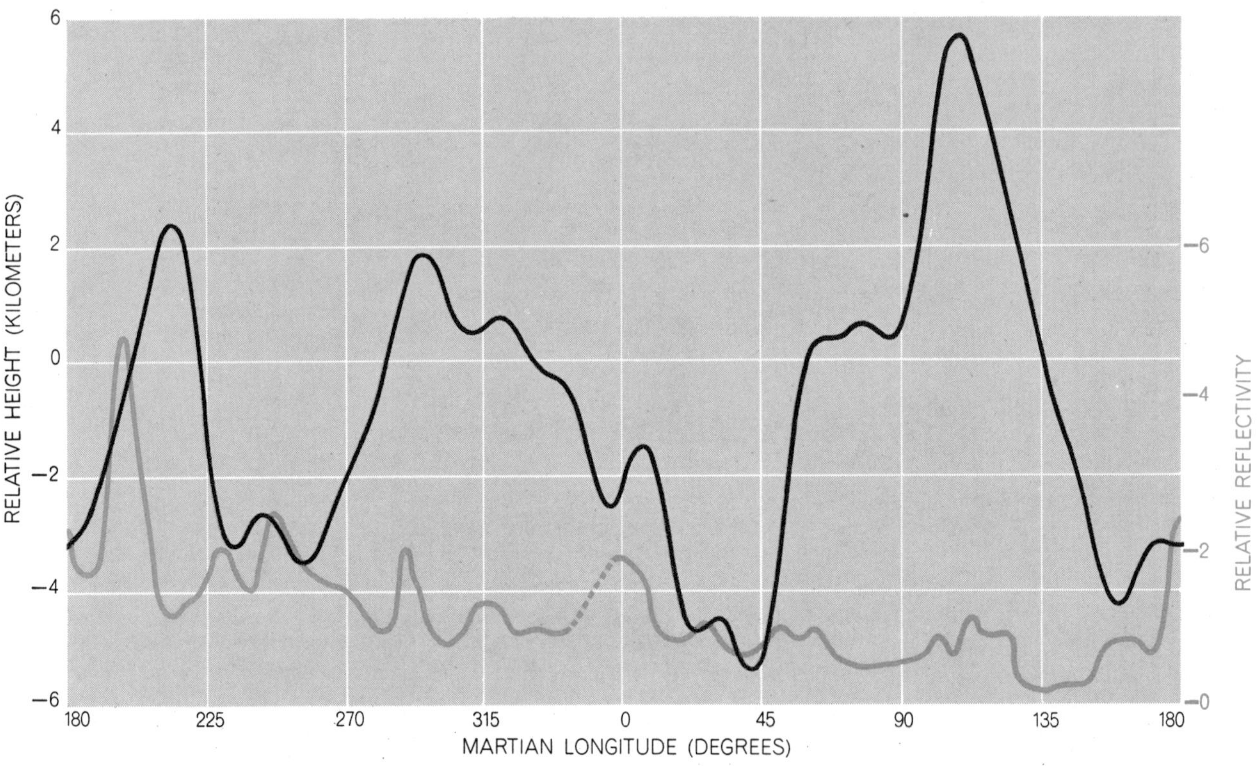

TOPOGRAPHY AND REFLECTIVITY of Mars along the 21st parallel of north latitude are based on measurements made as the planet rotated 360 degrees with respect to the earth. The altitude (*black*) does not seem to be strongly correlated with the radar reflectivity (*color*). Some of the fine structure shown is probably attributable to measurement error and may not be significant.

inhibits the separation of topographic and orbital effects. Nonetheless, the separation is possible, and preliminary analyses indicate that the height variations are far less pronounced than they are on Mars.

Using the delay-Doppler technique, one can in principle construct a picture of a planet from the signal power reflected from each surface element (or, more precisely, from each delay-Doppler "resolution cell"). Such a radar picture, which is a plot of power as a function of delay and frequency, is analogous to an optical photograph, which is a plot of intensity as a function of latitude and longitude. The major difficulty in making radar pictures is a practical one: insufficient signal strength requires integration of the power over a comparatively large area, making the delay-Doppler cell

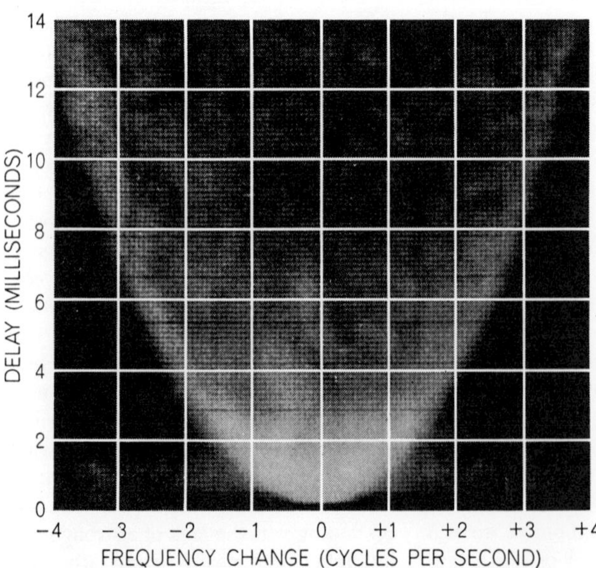

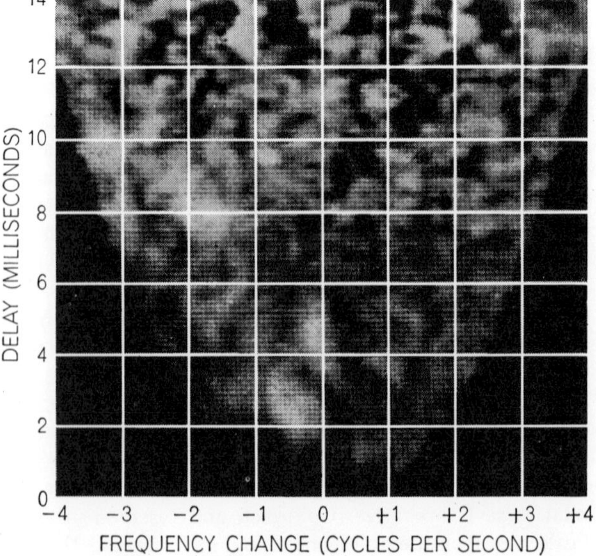

PRODUCTION of the radar picture of Venus (*see page 28*) began with the delay-Doppler map (*left*) based on August 6, 1967, observations, in which brightness is proportional to the logarithm of the power received at a given delay and frequency. The very bright area surrounds the subradar point. Its brightness is reduced and surface features are revealed when the theoretical values for a smooth, spherical surface are subtracted, leaving a representation of anomalous reflecting areas (*right*). This information is mapped into the usual surface coordinate system to prepare the final picture, which covers an area of Venus about 100 degrees in diameter.

large and sacrificing the inherently high resolution of the technique.

At present only radar pictures of the moon can be produced with high quality. These are comparable to the best earth-based optical photographs but are naturally far inferior to the Lunar Orbiter or Surveyor results. Radar pictures of the surface of planets that are not readily accessible to optical observation would of course be most valuable. Relatively crude (100- to 200-kilometer resolution) radar pictures of Venus have been prepared recently by Tor Hagfors and Alan Rogers at Haystack and by Raymond F. Jurgens at Arecibo [*see illustration on page 38*]. The echoes from Mercury are quite weak for mapping, but there are indications that the planet may turn out to have unusual surface characteristics. To achieve the goal of mapping more of the surface of these inner planets with substantial resolution, better radar systems are required.

Another approach to taking radio-wavelength pictures of planetary surfaces is to use two terminals, a large one on the earth and a second in a spacecraft orbiting the target planet. With one acting as transmitter, the other can receive both the direct rays and those reflected by the planet, thus enabling a hologram to be constructed. An attempt at applying this technique to the moon was made recently by G. L. Tyler, Von R. Eshleman and their colleagues at Stanford University using the Lunar Orbiter. The results are promising but much work must still be done to perfect the analysis.

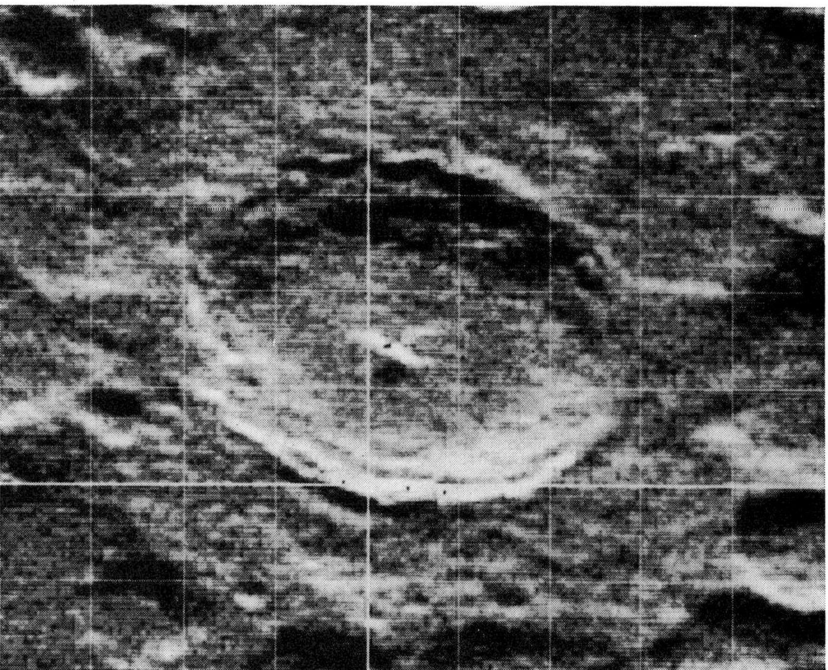

PICTURE OF MOON made with radar at the Lincoln Laboratory by Gordon H. Pettengill and others is comparable in resolution (one or two kilometers) to a ground-based optical photograph. The major feature is the crater Tycho in the moon's southern hemisphere.

Quite surprisingly earth-based radar observations have also played a crucial role in determining the surface temperature and pressure on Venus. Last October Russian workers reported that their spacecraft *Venus 4* had radioed such information from the surface. The data indicated that the temperature there was about 550 degrees Kelvin (280 degrees centigrade) and the pressure approximately 20 atmospheres. The following day *Mariner V* passed by Venus. From an analysis of the changes in Doppler shift of the signals transmitted from the U.S. probe it was possible to infer temperatures in the higher regions of the atmosphere as a function of the distance from the *Mariner*-earth ray path to the center of Venus. On the basis of a single radar measurement made by *Venus 4* of its altitude, the Russian values of temperature were related to the height above the surface. Combining the measurements from the two spacecraft led to a determination of the radius of Venus. There was only one difficulty: the radius deduced from the probe data was about 25 kilometers higher than the one Ash, Smith and I had found in 1966 from an analysis of the radar time-delay data. This year several independent studies were made of the accumulated radar data from Arecibo, Goldstone and Haystack. All confirmed the previous radar value of the radius.

Assuming that the *Mariner V* measurements and the radar radius are correct, one can extrapolate to find a surface temperature of about 700 degrees K. and a pressure of about 100 atmospheres. These values have the virtue of being able to account, through absorption, for the low radar cross section observed at high frequencies by Smith, John V. Evans and others at the Lincoln Laboratory. Consistency is then also achieved with earth-based measurements of the frequency dependence of Venus' radio-brightness temperature. The surface atmospheric properties inferred from *Venus 4* can explain neither without the aid of rather arbitrary assumptions. One is almost forced to conclude that the altitude measurement made aboard *Venus 4* was in error and that the spacecraft stopped transmitting its steady stream of remarkable data about 25 kilometers above the surface. Thus the seemingly innocuous determination of the radius of Venus from radar data will play a pivotal role in the planning of further space-probe investigations of the lower atmosphere and surface of our sister planet.

The most recent triumph of radar astronomy was the first direct contact with a minor planet, Icarus, achieved at Haystack and Goldstone in mid-June during Icarus' close approach to the earth.

What can be predicted for the future of radar astronomy? The past growth in radar capability can only be termed explosive. Since World War II overall system sensitivity has improved on the average by a factor of almost four each year. Although sustaining this growth rate permanently might prove a bit difficult, the pace could be maintained easily for at least another decade using only present technology and without spending a significant fraction of the gross national product. In fact, a radar telescope has already been designed that would be nearly 1,000 times more sensitive than Haystack. The estimated costs for the new facility are less than for a single Mariner spacecraft. This increased sensitivity would bring many more objects in the solar system—such as the Galilean satellites of Jupiter and the two tiny moons of Mars—under radar surveillance. The solar system abounds with mysteries, and new experimental information seems to solve fewer than it creates. There is little danger that future radar observations of the planets will reverse this trend.

4

The Atmospheres of Mars and Venus

VON R. ESHLEMAN

March 1969

The atmosphere of Venus is hot and dense and the atmosphere of Mars is cold and thin. Spacecraft have gathered information about the atmospheres of our nearest planetary neighbors that could not be obtained any other way. The results of these flights, together with new measurements made from the earth, mean that one can begin to speak with considerable confidence about the temperature, density and chemical composition of the atmospheres of Venus and Mars. It also becomes profitable for the first time to consider how the atmospheres of these two planets came to be so different from the atmosphere of the earth. It is now clear that the presence of life has been, and continues to be, a controlling influence on the composition of the earth's atmosphere. Conversely, the apparent absence of life on Venus and Mars may explain much about the nature of their atmospheres. In short, the experiments carried out by American and Russian planetary probes already constitute a preliminary life-detection test for our nearest neighbors beyond the moon.

Not quite four years ago (in July, 1965) the American spacecraft *Mariner IV* passed within 6,118 miles of the surface of Mars. When the spacecraft began its atmospheric measurements, it was early in the afternoon of a winter's day in a "desert" known as Electris in the planet's southern hemisphere; *Mariner IV* indicated that the temperature of the atmosphere at that point was about −113 degrees Celsius, 26 degrees below the coldest air temperature ever recorded on the earth. Nearer Mars's poles during the long winter night some of the atmospheric gas itself may freeze and fall to the surface [*see top illustration on opposite page*].

On successive days in October, 1967, the Russian space probe *Venera 4* and the American probe *Mariner V* found that the temperature of Venus' atmosphere increased nearly 10 degrees C. for each kilometer of decrease in altitude, reaching 100 degrees C.—the boiling point of water at sea level on the earth—at an altitude of about 45 kilometers. It is now apparent that neither spacecraft measured conditions all the way down to the surface, but by extrapolation it seems likely that the temperature there is 425 degrees C., or some 100 degrees above the melting point of lead.

Extrapolations of measurements from *Venera 4* and *Mariner V* also indicate that the atmospheric pressure at the surface of Venus is roughly 100 times that on the earth, comparable to the pressure at a depth of one kilometer in our oceans. The thick air on Venus would seem like a hot, dilute fluid with a tenth the density of water. Such an atmosphere would strongly bend the paths of light rays and create strange optical distortions. In contrast the pressure of the atmosphere at the surface of Mars is only about a hundredth that on the earth.

In the atmospheres of both Mars and Venus the main component is carbon dioxide. On the earth carbon dioxide is only .03 percent of the atmosphere; nitrogen accounts for 78 percent and oxygen for 21 percent. On both Mars and Venus free oxygen is certainly rare and may be virtually absent; the proportion of nitrogen in their atmospheres does not exceed 20 percent and could be much smaller.

What about water, which is so plentiful on the earth? If the earth were as hot as Venus, the oceans would evaporate, creating an atmosphere so thick with water vapor that the pressure at the surface would be about 300 times its present value. Apparently Venus has less than .7 percent water vapor in its lower atmosphere, making atmospheric and surface water 400 times less abundant on Venus than it is on the earth and possibly much less so. In Mars's atmosphere water vapor is barely detectable, but large amounts could possibly be frozen below the solid surface of the planet.

I find such comparisons fascinating and also very puzzling. Venus, the earth and Mars are within a factor of two of being the same distance from the sun and the same size [*see top illustration on next two pages*]. All three planets were doubtless formed from the same primordial source of matter at the same time. The atmospheres of all three appear to have been created after the initial supplies of light gases (hydrogen and helium) were largely lost to space. It is generally thought that these secondary atmospheres represent gases "exhaled" from the main body of the planet, with atmospheric evolution proceeding for aeons to produce the markedly different conditions we find on the three planets today.

Observations carried out by the space probes, together with new observations from the earth, have sharpened speculation about the reasons for the present differences in the atmospheres of the three planets. For example, it has recently been suggested that if all life on the earth were destroyed, the atmosphere would slowly begin to change and eventually would resemble the atmosphere of either Venus or Mars. I shall return to such speculations after first describing some of the experiments performed by the American and Russian spacecraft [*see bottom illustrations on next two pages*].

Until a few months before *Mariner IV* was launched in 1964 the plan for the mission did not include any radio measurements of the atmosphere of Mars. Although various methods of mak-

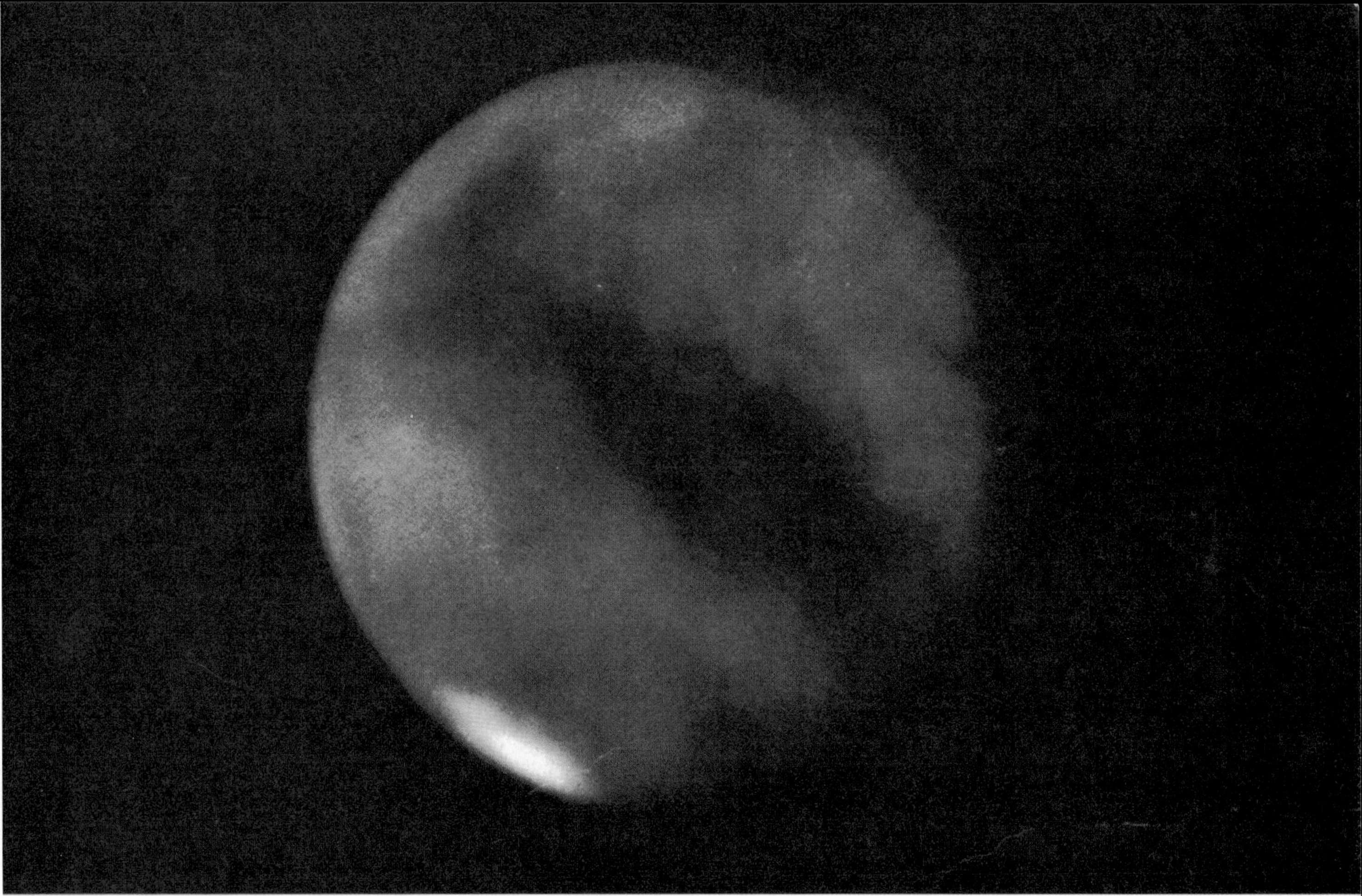

MARS (*above*) has a thin, cold atmosphere consisting chiefly of carbon dioxide, which in solid form may be the material seen in the planet's permanent polar caps. This color picture is a composite of several 16-millimeter motion-picture frames taken in July, 1956, by Robert B. Leighton of the California Institute of Technology.

VENUS (*below*) has a dense, hot atmosphere, which also consists chiefly of carbon dioxide. This view taken at the Lowell Observatory when Venus was almost directly in line between the sun and the earth shows how the planet's atmosphere scatters sunlight, creating a thin luminous arc opposite the directly lighted crescent.

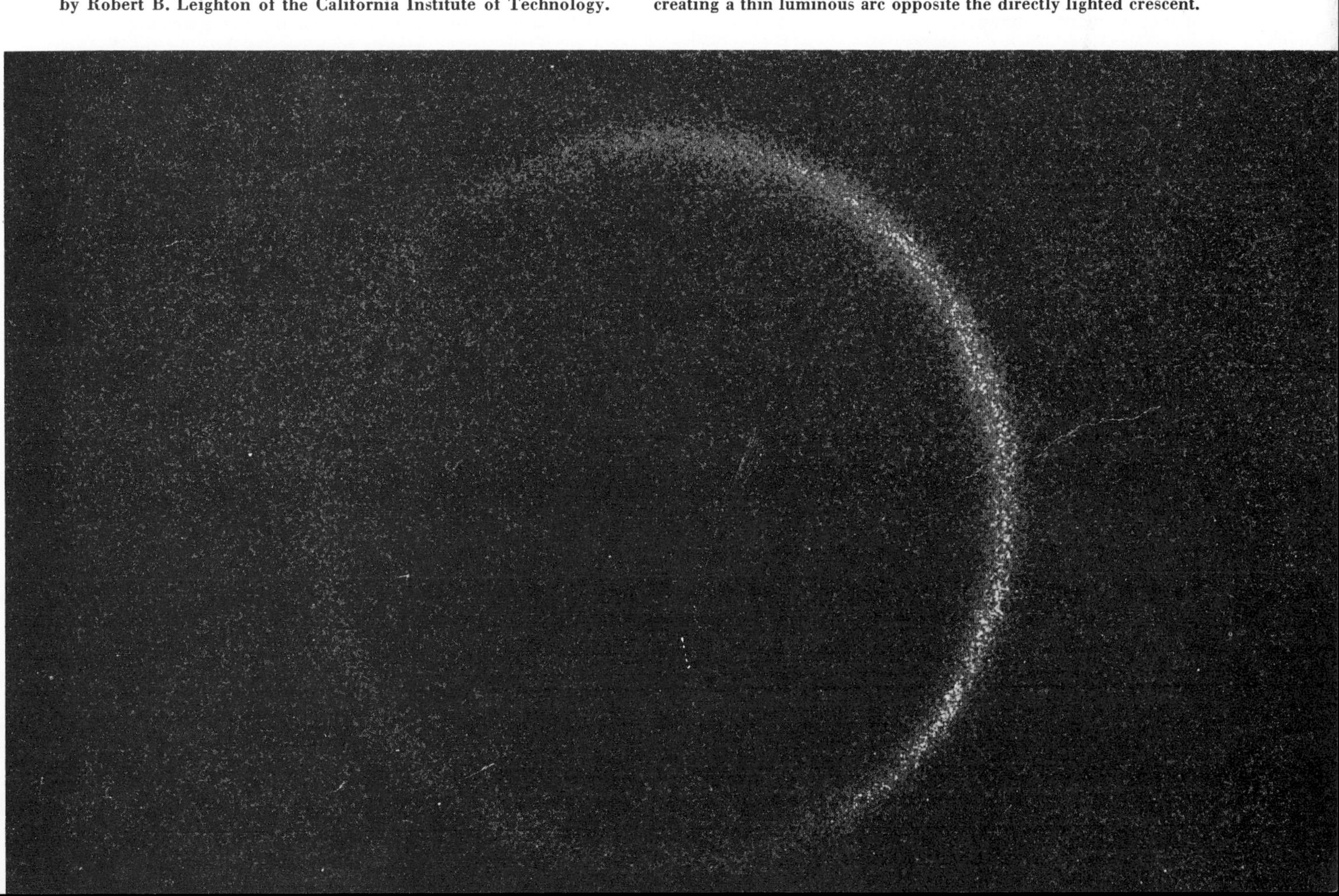

VENUS, EARTH AND MARS resemble one another much more closely than they resemble any of the other planets in the solar system. Their distance from the center of the sun and their size are plotted here on two scales: the size scale (*far right*) is arbitrarily 1,500 times the distance scale (*bottom*). On this size scale the curvature of the limb of the sun (*left*) is barely evident. The minimum

ing such measurements had been discussed since 1960, there was a fundamental problem facing the scientists and engineers concerned with the project. The best experiment would be to have the spacecraft pass behind Mars and be occulted, or eclipsed, so that the radio signals between the earth and the spacecraft would pass through Mars's atmosphere. Because the 1964–1965 flight to Mars was to be only the second interplanetary mission attempted by the U.S., the project managers were understandably reluctant to accept the occultation trajectory. They had plotted the path of the spacecraft specifically to avoid such a temporary but critically timed loss of their long, tenuous thread of radio contact with it [see "The Voyage of *Mariner IV*," by J. N. James; SCIENTIFIC AMERICAN, March, 1966].

A group from Stanford University and the Stanford Research Institute was the first to describe how planetary atmospheres and surfaces could be studied by occultation and related radio measurements. This group had proposed an experiment and instrumentation for this flight that would be particularly sensitive to Mars's ionosphere, the ionized upper part of the planet's atmosphere. Another group involved in the radio tracking of space probes first showed in detail, however, that the radio systems already designed for communications and for tracking the spacecraft could supply such precise data on signal characteristics that valuable atmospheric measurements could be made even without adding any new equipment. During the same period that the two groups were refining their predictions of the scientific potential of radio-occultation measurements, other scientists and engineers were beginning to worry that lack of knowledge about the atmosphere of Mars would make it difficult to plan for missions in which spacecraft would land on the planet. It had been thought for years that the pressure of the atmosphere at Mars's surface was about 8.5 percent of the pressure at the earth's surface, but new measurements from the earth seemed to place the value closer to 2.5 percent. This combination of events led to the 11th-hour change in plan to accept the hazard of the temporary loss of communications in order to make atmospheric measurements with these same radio links in the moments before and after occultation. The men who were assigned the responsibility for this experiment were Arvydas Kliore (the principal investigator), Dan L. Cain, Gerald S. Levy and Frank D. Drake of the Jet Propulsion Laboratory of the

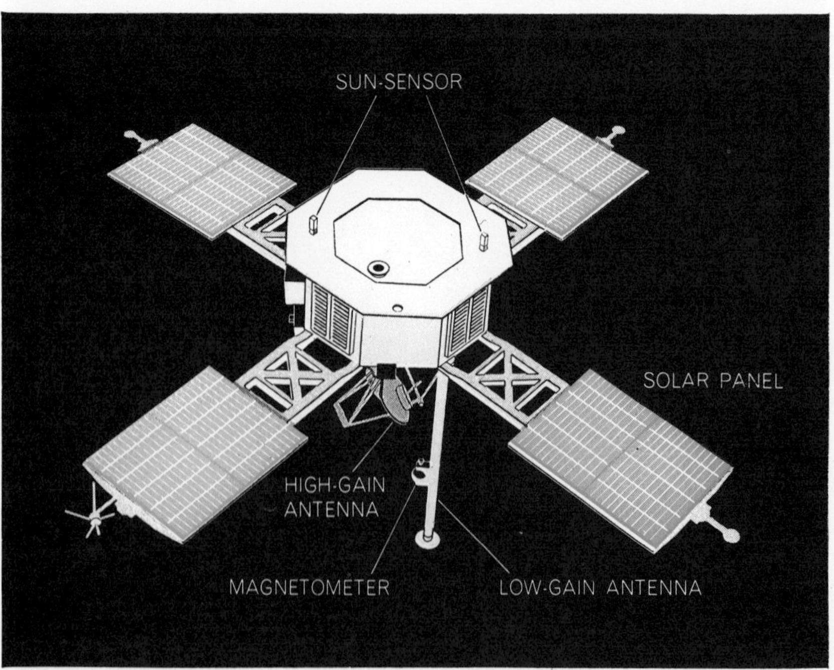

***MARINER V* approached to within 2,544 miles of the surface of Venus on October 19, 1967, after a voyage of 127 days. The 540-pound spacecraft was designed by the Jet Propulsion Laboratory of the California Institute of Technology. *Mariner IV,* which passed within 6,118 miles of Mars on July 15, 1965, after a trip of 228 days, was very similar in design.**

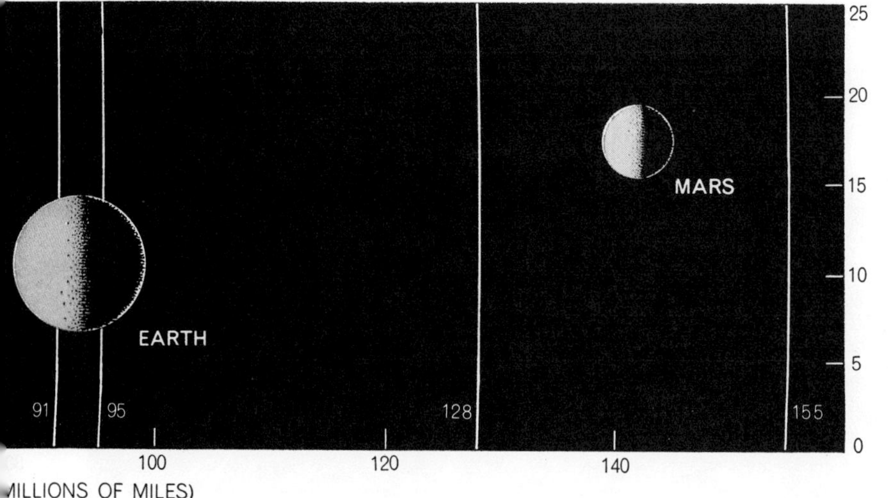

and maximum distances of the three planets as they travel in orbit around the sun are indicated by the vertical white lines. The atmospheres of the three planets, now very different, appear to have been formed after the light gases initially present escaped into space.

California Institute of Technology, and Gunnar Fjeldbo and myself of Stanford University. (Drake is now at Cornell University.)

The principle involved in the experiment is as follows. The atmosphere of Mars acts as a radio lens, causing the spacecraft to seem to depart from its actual path when it passes behind the atmosphere, just before going into occultation and just after emerging from it. The apparent distance to the spacecraft is determined by sending signals of precisely controlled frequency to the vehicle, which then retransmits the signals to the earth. At the tracking station on the earth the frequency of the returned signal is compared with what was originally transmitted, and the difference (the Doppler frequency) is precisely measured.

By counting the number of cycles difference in a second, one can determine the apparent distance (in radio wavelengths) the spacecraft has moved in a second, either toward or away from the tracking station. For example, if the frequency of the returning signal is 200,000 cycles per second lower than the outgoing signal and if the radio wavelength of the tracking signals is 13 centimeters, then in one second the spacecraft has apparently moved 13 kilometers away from the tracking station. (The distance is not 26 kilometers, as one might think, because one must divide by two to allow for the two-way path to and from the spacecraft.) If there is nothing but empty space between the spacecraft and the tracking station, the radio wavelength will remain constant at 13 centimeters between the two points, and the apparent motion can be regarded as the true motion of the spacecraft. As the signal passes through the atmosphere of Mars, however, several changes in wavelength occur. In the un-ionized lower part of the atmosphere the phase velocity of the radio waves is less than *c*, the velocity in a vacuum, so that the wavelength becomes less than 13 centimeters. In the ionized upper part of the atmosphere the phase velocity is greater than *c*, and the wavelength becomes greater than 13 centimeters. In addition the path followed by the signals is bent by both the un-ionized and the ionized part of the atmosphere [*see illustration on next page*].

These changes are small, but the accuracy of measurement is great. As *Mariner IV* first moved behind the daytime ionosphere of Mars, the ionosphere made it appear that in one second the spacecraft's motion away from the earth was as much as four centimeters less than the actual increase in distance, which was about 13 kilometers. The sum of such extra effects during the entire period *Mariner IV* was approaching occultation resulted in an apparent displacement of .75 meter toward the earth followed by an apparent displacement of about two meters away from the earth before the signal was completely blocked by the planet. On emergence from occultation the planet's neutral atmosphere made it appear that *Mariner IV* was about 2.4 meters beyond its actual position, but it "came back into focus" without showing a discernible negative shift, which would have been expected if an ionosphere were present on the night side of Mars. In order to appreciate the precision involved in these measurements one should bear in mind that the spacecraft was then about 216 million kilometers from the earth.

It may be hard to believe such small changes could be attributed to the at-

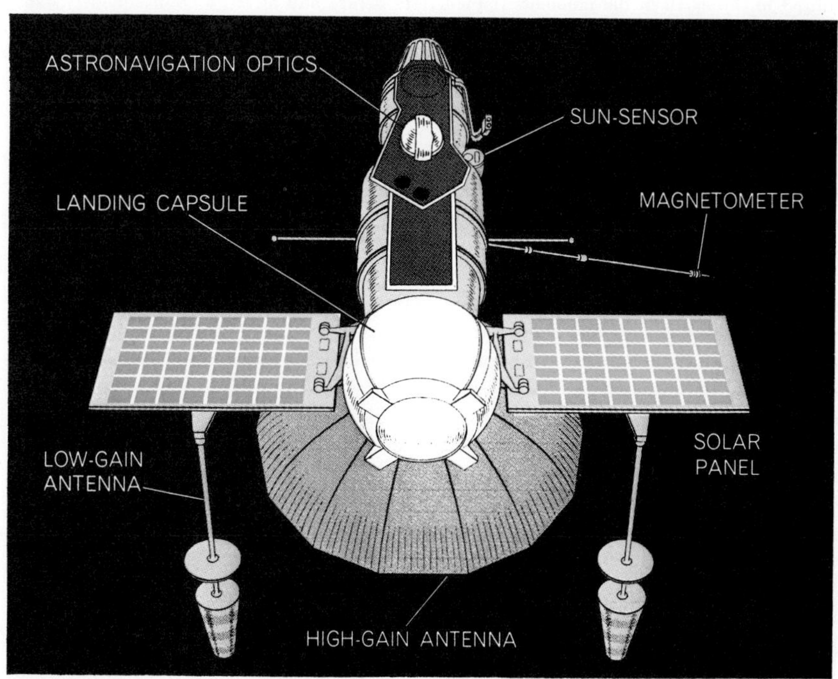

***VENERA 4*, the 2,433-pound Russian probe, reached Venus some 36 hours before *Mariner V* and ejected an instrumented capsule equipped with a parachute. The Russians reported that the capsule measured atmospheric conditions all the way to the planet's surface. There is now good evidence that the last reading was at an altitude of about 25 kilometers.**

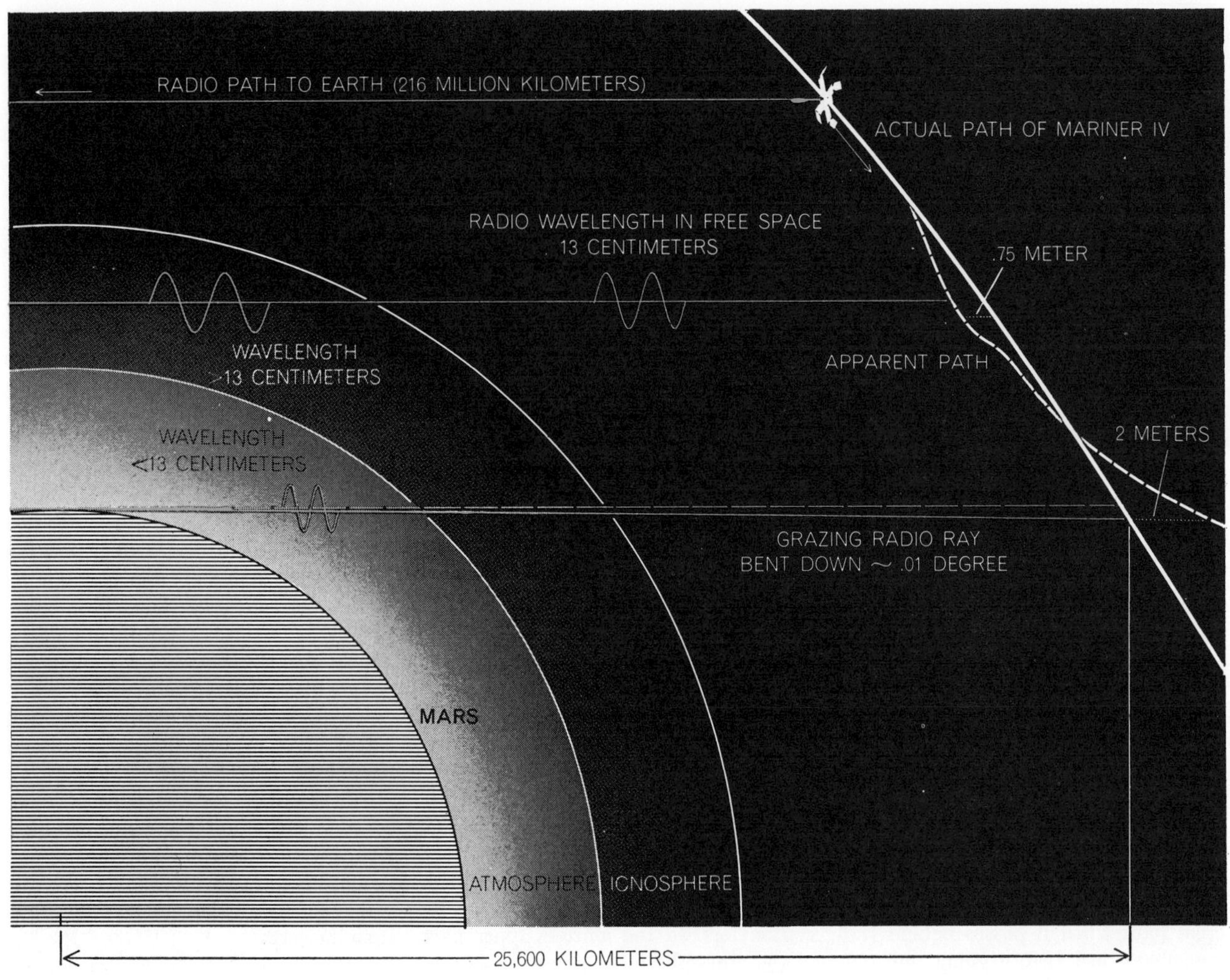

MARS'S ATMOSPHERE was studied by measuring the changes in the frequency of radio signals beamed to *Mariner IV* and retransmitted to the earth in the moments immediately before and after the signals were blocked by the planet. Signals that passed through the ionosphere, the electrically charged part of the upper atmosphere, were increased in wavelength, making it appear that the spacecraft had deviated from its actual path by about .75 meter. Subsequently signals passing through the neutral, lower part of the atmosphere were shortened in wavelength, making it appear that *Mariner IV* had deviated about two meters in the other direction.

mosphere of Mars. After all, the radio signals also had to pass through the much denser ionosphere and lower atmosphere of the earth, and the position and actual path of the spacecraft were not known down to the last meter. In regard to the first point, irregularities in the earth's ionosphere and lower atmosphere did not create a problem because the period of measurement was very short. Most of the effect of the lower atmosphere on both sides of Mars showed itself within 10 seconds, as the spacecraft flashed behind the planet. In such a brief period changes in the earth's atmosphere and in interplanetary space are almost always small enough to ignore, even though they could become important over a longer time. As for the second point, the answer is that, although the exact position and trajectory of the spacecraft are not known on an absolute basis, we do know that the path must be very nearly the path of a point mass in a gravitational field that, at a distance of several tens of thousands of kilometers from the planet, must be highly symmetrical. In the brief period that interests us the law governing motion in such a field can be fitted to the measurements to determine to what extent the law is broken by the extra effects of Mars's atmosphere.

Similar considerations apply to the radio measurements of the atmosphere of Venus by *Mariner V* in 1967, although here the distortion of the radio signals was much greater. Kliore again headed a team that used radio links to study atmospheric effects. In addition a two-frequency occultation experiment was conducted by H. Taylor Howard, Roy A. Long, Fjeldbo, myself and others from Stanford University and the Stanford Research Institute. It measured very tenuous regions in the upper atmosphere of Venus and also provided additional data on characteristics of the lower atmosphere.

One important difference between the results for the two planets arises from the fact that the dense atmosphere of Venus is super-refractive [*see illustration on opposite page*]. It is well known that light and radio rays in the earth's atmosphere are bent by changes of density with altitude; the maximum deflection due to the lower atmosphere is about half a degree when one looks toward the horizon. Thus the disk of the setting sun, which subtends an angle of half a degree, appears to be just above the horizon when in reality it has already gone below it.

Light rays are bent the same amount on Venus at an altitude of about 50 kilometers. At somewhat lower altitudes a horizontal radio ray is bent even more before it escapes into space, and at an

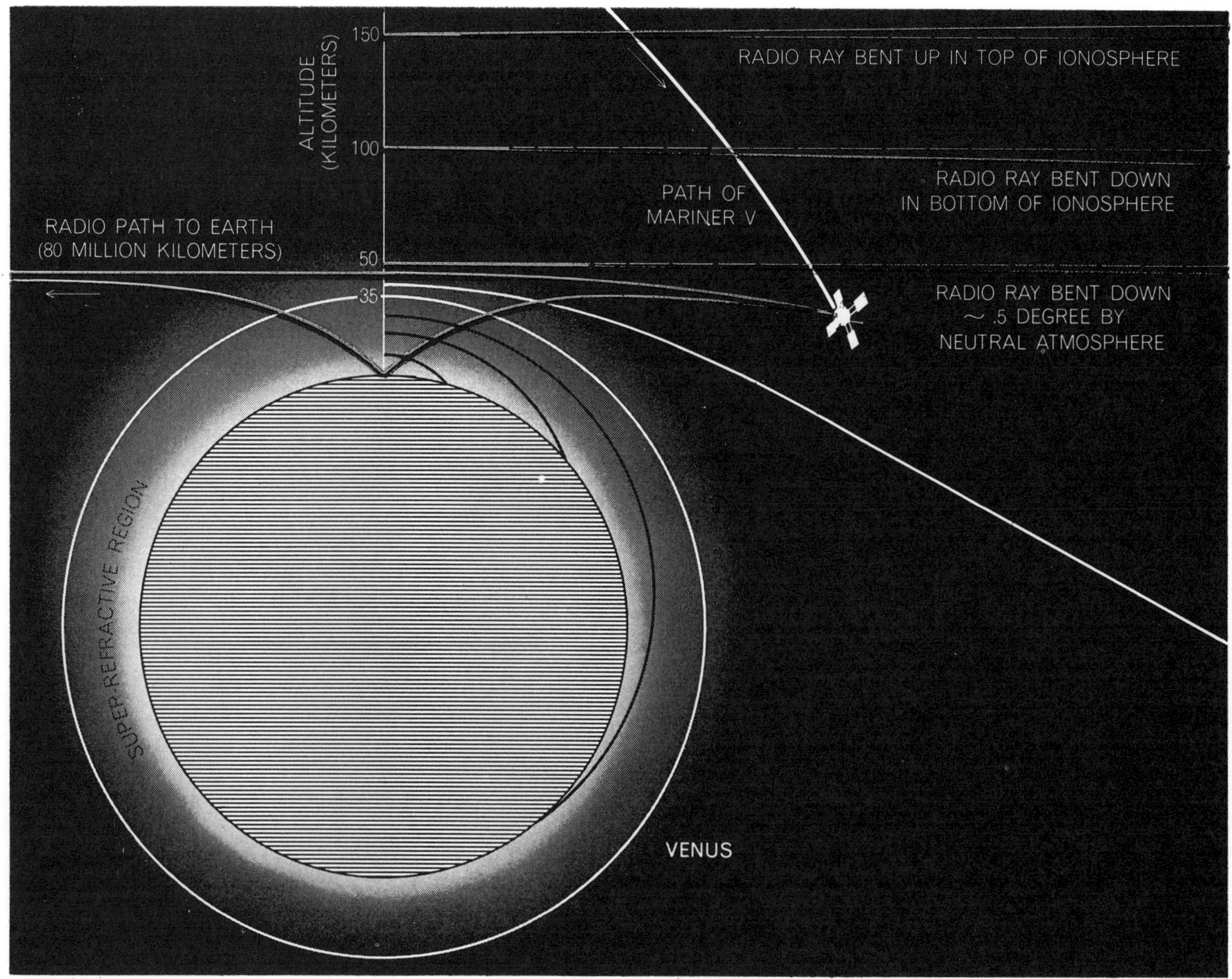

VENUS' ATMOSPHERE, studied with the aid of *Mariner V*, turned out to have a pressure about 100 times that of the earth and some 10,000 times that of Mars. Below an altitude of about 35 kilometers the atmosphere is super-refractive, which means that a ray of light would follow an arc that is sharper than the curvature of the planet. *Mariner V*'s antenna beam was tilted to help compensate for the ray curvature expected in the dense atmosphere. Tape recordings of the signals are still being searched by computer for evidence of the reflected ray that presumably bounced off the planet to demonstrate a possible way to study the super-refractive region.

altitude of about 35 kilometers the ray would curve so much that it would circle the planet and return to its starting point. Below this critical level Venus' atmosphere is super-refractive, which means that the arcs of rays bend more sharply than the arc of the surface of the planet. Hence the radio-occultation measurements of the atmosphere can be made only down to the region of super-refraction. An observer on a planet that had a transparent, super-refractive atmosphere could in principle see the entire surface of the planet spread out around him as if he were on the bottom of a bowl [*see illustration on next page*].

How can these lenslike effects on the radio waves passing through the atmospheres of Mars and Venus be interpreted in terms of density, temperature and pressure? Since the measured effects are the bending of radio rays and apparent changes in distance, the crucial property of the atmospheric gas is its refractive index. This value is inversely proportional to the phase velocity of light through the gas, and it controls ray-bending just as it does in a glass lens. Because the refractive index of a gas is very nearly unity, it is convenient to use the derived quantity called refractivity, which is defined as a million times the difference between the refractive index and unity. Of course the measured refractivity is averaged through almost tangential paths through the atmosphere. Determining the refractivity as a function of height in the atmosphere requires that the sequence of measurements be mathematically inverted. Techniques for solving this complex problem on a digital computer have been worked out by my colleague Fjeldbo. For Mars the maximum atmospheric refractivity turns out to be about four, that is, the refractive index differs from unity by only four parts in a million.

The refractivity of most un-ionized gases is equal to the number of molecules per unit volume (called the molecular number density) times a quantity that is characteristic of that gas. For a gas mixture it is the sum of such products for the individual constituents. Therefore if the constituents and their relative abundances are known, a measure of refractivity directly yields the molecular number density, which in turn is proportional to the ratio of pressure to temperature. Then the refractivity profile, or change of refractivity with altitude, is proportional to the molecular number density profile. In a mixed planetary atmosphere the number density will decrease exponentially with increasing height, falling to 37 percent of the initial value over a height difference called the scale height. The scale height, in turn, is directly pro-

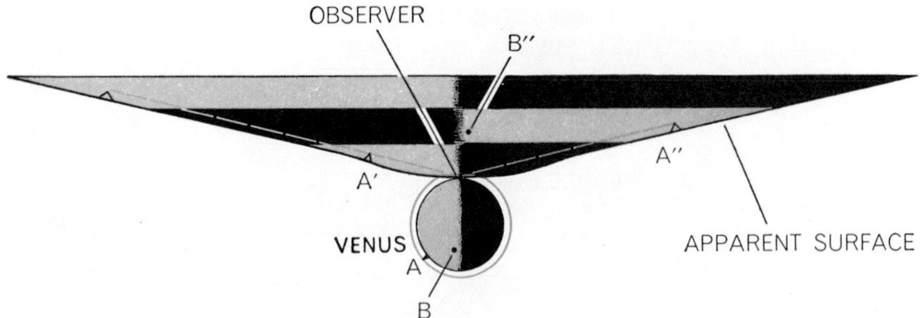

SUPER-REFRACTIVE ATMOSPHERE would persuade an observer on Venus that he was living at the bottom of a gigantic bowl, illustrated here in cross section. If the atmosphere were perfectly clear, he would see the entire surface of his planet endlessly repeated in expanding rings. Because of the bending of light rays an object at position *A* would seem to lie at *A′*, *A″* and *A‴*. Similarly, an object at *B* would appear in the second ring, at *B″*. Its appearances at *B′* and *B‴* cannot be shown because they lie in the missing front half of the bowl.

portional to the gas temperature and inversely proportional to the mean molecular mass of the mixture and the known planetary gravity. If the pressure-temperature ratio is found from the local refractivity and the temperature from the profile, then both pressure and temperature can be determined for various heights in the atmosphere.

The foregoing assumes that the atmospheric composition is known. Alternatively, if other information is available about temperature or pressure at some point in the atmosphere, then some inferences about constituents can be made, but a unique solution is not possible. Fortunately the composition of the atmospheres of Venus and Mars now appears to be known well enough so that pressure and temperature can be derived from occultation measurements with comparatively little uncertainty.

Spectroscopic observations made from the earth before the Mariner flights had indicated that carbon dioxide is a constituent of both atmospheres, but the relative abundance had not been established. For Mars the approximate amount of carbon dioxide and the total refractivity combine to indicate that there is no room for much else—carbon dioxide must be the principal constituent. New observations made from the earth are in agreement. In addition the temperature derived by assuming that Mars's atmosphere consists entirely of carbon dioxide indicates little leeway for lighter constituents; for example, if there were as much as 20 percent nitrogen, the temperature at the southern hemisphere point measured by *Mariner IV* would have been at the freezing point of carbon dioxide. In that case some of the carbon dioxide in the atmosphere should have turned to a snow of "dry ice" and fallen to the surface. Robert B. Leighton and Bruce C. Murray of the California Institute of Technology have presented strong evidence that the polar caps of Mars may consist predominantly of dry ice, with the atmosphere and the caps exchanging carbon dioxide on a seasonal basis. Inasmuch as *Mariner IV* flew over an area well away from the southern polar cap, the atmosphere it measured should have been somewhat above the freezing point of carbon dioxide.

Venera 4 directly measured the carbon dioxide content of Venus' atmosphere and found it to be 90 ± 10 percent, a value that is in general agreement with less direct measurements made with earth-based instruments and by *Mariner V*. *Venera 4* carried two instruments for measuring nitrogen, and both gave negative results; the more sensitive instrument could have detected nitrogen if as little as 2.5 percent were present. The probes to Mars and Venus have overturned the long-held belief that, by analogy with the earth, nitrogen must be the predominant atmospheric constituent on both planets.

The atmospheric-pressure profiles for Mars and Venus show that the surface pressure of Mars is roughly a hundredth of the earth's and that the surface pressure of Venus is 100 times greater than the earth's. The actual measurements at the occultation points on Mars give about .45 and .8 percent of the earth's pressure, but this difference is doubtless due to differences in the height of the particular surface features that cut off the lowest radio path. Radar studies by Gordon H. Pettengill of the Lincoln Laboratory of the Massachusetts Institute of Technology show height variations of at least 11 kilometers in Mars's topography. Hence the atmospheric pressure at the surface must vary at least by a factor of three for different locations on Mars, with the occultation measurements of pressure probably being somewhat below the average.

The pressure curves for Mars, the earth and Venus are remarkably parallel [*see illustration on opposite page*]. The slope of a pressure curve would be steeper with higher temperature, with lower gravity and with a lower mass of the molecules in the gas. The earth and Venus, with nearly the same gravity but different temperatures, still have parallel pressure lines because the average mass of the molecules in the atmosphere of Venus is greater. In comparing the earth and Mars one finds that Mars's lower gravity (only 39 percent that of the earth) is counteracted by the higher molecular mass of its atmosphere and by its lower temperature, keeping the pressure curves approximately parallel.

The temperature of the atmosphere near the surface of Mars is approximately what one would predict on the basis of its distance from the sun [*see illustration on page 56*]. If Mars, Venus and the earth were black spheres without atmospheres, they would reach equilibrium temperatures in which the amount of solar radiation they absorbed would be exactly matched by the amount of heat they radiated into space at infrared wavelengths. Temperature would depend only on distance from the sun. Venus would be 330 degrees Kelvin (degrees C. above absolute zero), the earth would be 280 degrees K. and Mars 225 degrees K. (On the Fahrenheit scale these three temperatures would be 134 degrees, 45 degrees and −54 degrees.) *Mariner IV* provided two values for the lower air temperature on Mars: 160 degrees K. and 235 degrees K. The first figure applies to a winter day at 50 degrees south latitude; the second, to a summer night at a latitude of 60 degrees north.

On Venus the temperature in the lower atmosphere far exceeds the temperature of a simple black body at Venus' distance from the sun. There is still considerable uncertainty about the temperature below an altitude of 25 kilometers. The Russians originally reported that *Venera 4* reached the surface of Venus, where it recorded a pressure of 20 atmospheres (roughly 300 pounds per square inch) and a temperature of 545 degrees K. It now appears that these figures must apply to the region about 25 kilometers above the surface. The figures for the lower atmosphere are determined from extrapolation, for which two different assumptions can be made. One is an "adiabatic" extrapolation, in which it is assumed that the measured rate of change in temperature from 57 kilometers to 25 kilometers continues all the way to the surface of the planet. This is about the maximum temperature gradient possible in the atmosphere of Venus. If the lower region were hotter than this value, the atmosphere would be unstable: the lower, hotter gas would tend to rise through the higher, cooler gas, redistributing the heat energy to re-create an adiabatic temperature gradient at a higher absolute value of temperature.

The other assumption, which I consider the less plausible, is that the rate of change in temperature does not continue all the way to the surface, so that the surface temperature is no higher than about 600 degrees K. In any event, it follows that the temperature near the surface is within 100 degrees of 700 degrees K. Venus, unlike Mars, appears to exhibit little daily or seasonal temperature change and is probably very hot even at the poles. Measurement and theory also indicate that the surface of Venus does not vary greatly in altitude, perhaps less than a few kilometers

It was known soon after the *Venera 4* and *Mariner V* measurements were made that this evidence could not be reconciled with radar determinations of the radius of Venus without a major change in interpretation of at least one of the three sets of measurements. Comparing the atmospheric measurements made by the two spacecraft establishes the connection between the *Mariner V* scale, which is distance from Venus' center of mass, and the *Venera 4* scale, which is purported to be height above the solid surface. From this comparison it follows that the radius of the planet on the *Venera 4* scale would be about 25 kilometers greater than the value determined by earth-based radar.

This contradiction was troublesome because it seemed that the radar-determined radius was accurate and that the *Venera 4* data were self-consistent. On the other hand, either could be in error; determining the radius on the basis of radar data alone is quite involved and the *Venera 4* interpretation depends entirely on a single altimeter report. Although a number of other arguments could be brought to bear on the problem, what was really needed was an independent measurement either of the radius of the planet or of the atmospheric properties near the surface.

Such a measurement (for the radius) proved to be feasible by combining the results from the radio tracking of *Mariner V* with measurements of the radar distance to the surface of Venus, which had been made simultaneously [*see illustration on page* 57]. When the spacecraft was near Venus, its trajectory was influenced primarily by the planet's center of mass; hence the radio-tracking data can be used to establish the distance from the earth to the center of Venus. A simultaneous radar echo yields the distance from the earth to the surface of Venus. The difference between these two distances is thus a relatively direct measurement of the radius of Venus. This measurement is largely independent of the complexities of the experiments conducted with radar alone, which involve the simultaneous solution for some 20 parameters associated with the orbits of the earth and Venus, along with perturbations due to other planets. The radar-echo distance to the surface of Venus and the radio-tracking data for calculating the distance to the center of Venus provided a value for the planet's radius that agreed closely with the earlier radar measurements: 6,053 ± 5 kilometers.

What could have gone wrong with *Venera 4*? Its atmospheric measurements are compatible with the radar and *Mariner V* data only if they actually began 50 kilometers above the surface and ended about 25 kilometers above it. The Russians have stated that the aim of *Venera 4* was not necessarily to sample conditions all the way to the surface but rather to penetrate deep into its atmosphere—to a level where conditions were approximately those of the actual measurements. Nevertheless, they initially concluded that *Venera 4* did reach the surface on the basis of (1) a single altimeter reading, apparently of about 25 kilometers, at the start of the spacecraft's parachute descent and (2) the subsequent elapsed time to the termination of its signal, during which time it must have fallen about 25 kilometers.

Suppose, however, that the initial altitude were 50 kilometers instead of 25. Then all the other data would be in excellent agreement, with *Venera 4* penetrating about as far as it was designed to go and giving its last report at an altitude of 25 kilometers. The logic of this explanation is consistent with an ambiguity in the interpretation of the radio-altimeter reading. It happens that certain simple systems for indicating a particular altitude (say 25 kilometers) are unable to distinguish heights that are exact multiples of the intended altitude. We do not know, however, if this kind of system was used on *Venera 4*.

The temperature of Venus is of partic-

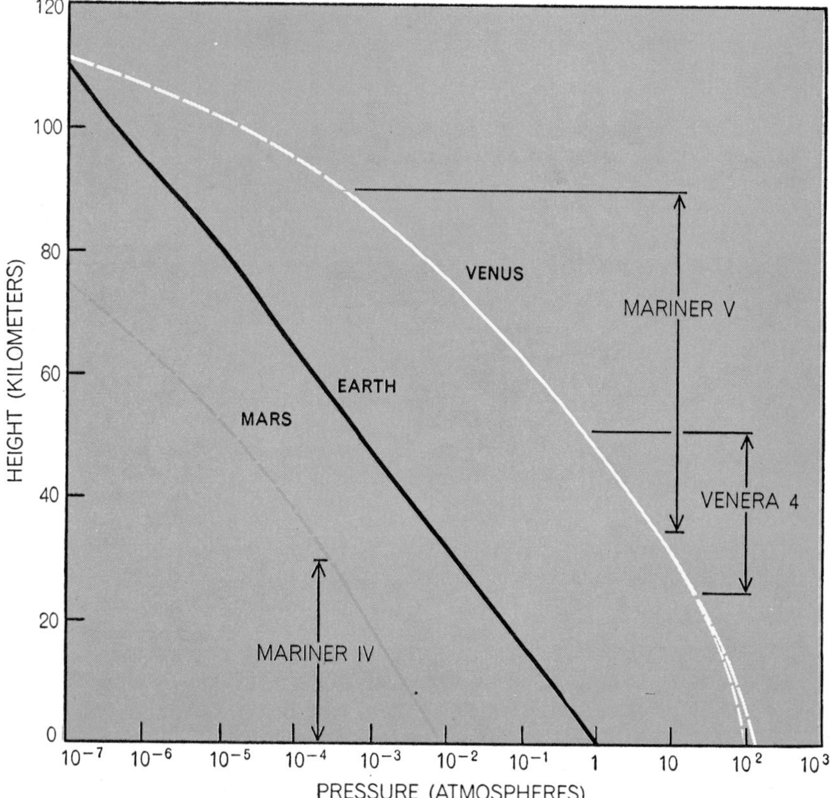

ATMOSPHERIC PRESSURE on the earth is about 100 times that on Mars and the pressure on Venus is about 100 times that on the earth. The values for Venus provided by *Venera 4* and *Mariner V* are in excellent agreement in the region between 35 and 50 kilometers, where the two craft made overlapping measurements. The broken lines are extrapolations.

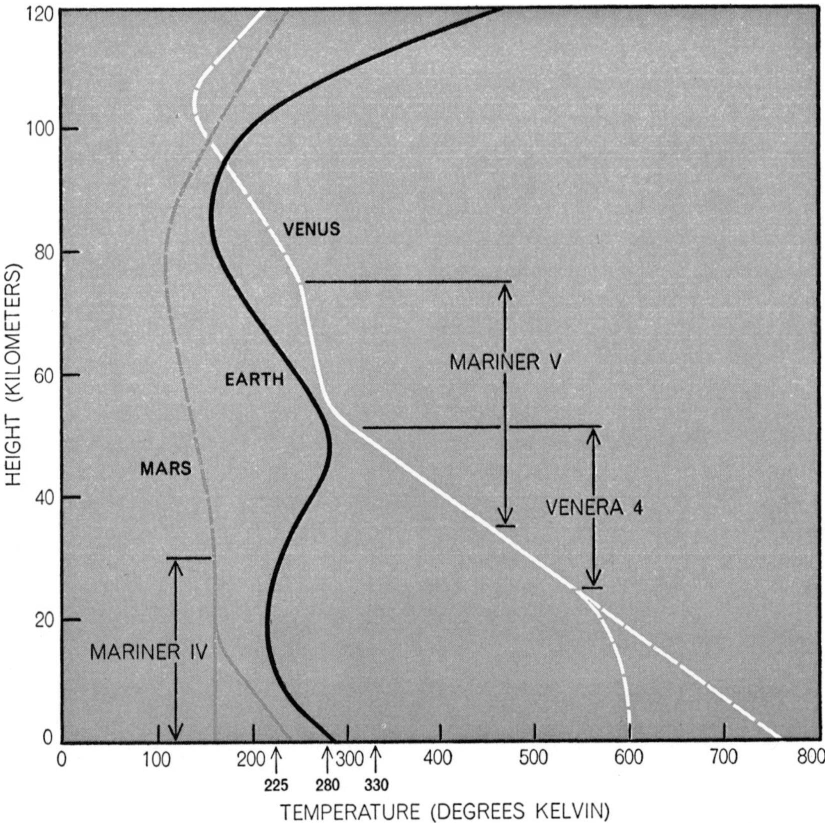

ATMOSPHERIC TEMPERATURE near the surface of Mars is about two-thirds that on the earth and only about a third that on Venus, as measured on the Kelvin, or absolute, scale. The three values at the bottom of the chart (225, 280 and 330 degrees) are the theoretical temperatures expected at the surface of the planets if they behaved as ideal black bodies in space. The two legs of the actual curve for Mars represent values at different seasons and latitudes as measured by *Mariner IV*. The two legs of the temperature curve for Venus represent different extrapolations for the lowest 25 kilometers of the atmosphere.

ular interest. Why is the planet approximately twice as hot as the black-body equilibrium temperature of 330 degrees K.? Venus rejects (by reflection) about 76 percent of the radiant energy it receives from the sun; this effect alone should make it 100 degrees cooler than the equilibrium value. It would seem that the high temperature of the planet can be explained only in terms of either an extremely strong internal source of heat or an extremely efficient trapping of the infrared radiation that otherwise would carry away the heat energy.

The hypothesis favored at present is that solar radiation is trapped by the "greenhouse" effect. Carl Sagan of Cornell University and James Pollack of the Smithsonian Astrophysical Observatory have studied this problem in detail. Their view, in brief, is that the carbon dioxide and water vapor in the atmosphere of Venus act as a highly efficient glass in the greenhouse, allowing the energy of the sun's visible light to enter and be deposited throughout the lower atmosphere, perhaps down to and including the surface. The energy that is reradiated at infrared wavelengths finds the atmosphere so nearly opaque that the temperature must rise to the point where enough heat is radiated, in spite of this opacity, to balance the incoming energy. The higher the infrared opacity is, the higher the temperature must go before this balance is reached. For the earth and Mars the amount of solar energy lost by reflection from clouds and from the surface of the planet just about balances the weak trapping of infrared in their atmospheres, but for Venus the greenhouse effect greatly predominates and is the determining factor in its temperature.

The constitution of the clouds and the amount of atmospheric water are particularly relevant to our understanding the temperature of Venus. *Venera 4* indicated that below the clouds the water content is between .1 and .7 percent, but the reliability of this measurement has been contested. Earth-based spectroscopic observations indicate a very dry atmosphere above the clouds, and there is controversy over whether or not clouds of ice particles are compatible with these observations. Among the alternatives that have been suggested are clouds of dust, carbon suboxide (C_3O_2) and compounds of mercury and chlorine. If the *Venera 4* water measurement is accepted, then it would seem that the clouds are ice crystals, with a lower boundary some 60 kilometers above the surface.

I am struck by the sudden break in the temperature profile near this height as measured by *Mariner V* on both the day and the night side of Venus. The temperature at this break is near the freezing point of liquid water, if one assumes that the atmosphere is almost entirely carbon dioxide. The break suggests to me that the clouds may be a mixture of ice and water, in which case the percentage of water vapor condensing and sublimating at the cloud level should be even higher than the limit established by *Venera 4* at lower altitudes. If raindrops formed in such a cloud, they would not fall to the surface but would vaporize at a height of about 38 kilometers. Alternatively the drops might consist not of pure water but of acid; hydrogen chloride and hydrogen fluoride have been detected in the atmosphere.

Even if one accepts the presence of extensive water clouds on Venus, it is a mystery why the planet should have so much less surface and atmospheric water than the earth. If one assumes that Venus and the earth have exhaled from their interior about the same amount of water since their formation, this means that Venus ultimately lost to space an amount of water comparable to the quantity in the oceans of the earth, while the earth lost essentially none. The upward flow of water vapor in an atmosphere is controlled by a "cold trap," an altitude of minimum temperature; when the vapor reaches this altitude, much of it may condense or freeze, and then fall. The water vapor that reaches higher altitudes is dissociated into hydrogen and oxygen by solar radiation; the free hydrogen may be lost by escape into space and the free oxygen by being bound into compounds, notably in the oxidation of surface materials (such as iron) and of atmospheric constituents (such as carbon monoxide and methane). If atoms in the atmosphere of Venus are ionized by being stripped of electrons by solar radiation, their escape from the top of the atmosphere may be strongly affected by Venus' lack of a magnetic field, which allows the charged particles of the "solar wind" to interact more directly with Venus' atmosphere than they do with the earth's. Clues to the characteristics of such interactions were obtained from three experiments on *Mariner V.* One dealt with solar-wind particles, another with magnetic fields and the third was the two-frequency occultation experiment.

Unfortunately there is disagreement on the magnitudes of the various effects. Some investigators have concluded that Venus could have lost an amount of water equal to the quantity in the earth's oceans over the past 4.5 billion years; others have concluded that it could not have done so. If the second conclusion is correct, it may be that Venus originally condensed from materials poorer in water than those that formed the earth. An interesting alternative to this assumption has recently been outlined by Peter E. Fricker and Ray T. Reynolds of the Ames Research Center of the National Aeronautics and Space Administration. They suggest that Venus still has much of its original water in solution in subsurface molten rock. Because Venus was presumably always hotter than the earth, much of the silica rock at its surface was probably molten over most of geologic time. The early characteristics of the atmosphere would therefore be determined primarily by the relative solubility of water and carbon dioxide in the silica melt. One would expect water to be much the more soluble of the two, with the result that carbon dioxide would have been driven off to form the bulk of the atmosphere. When the surface temperature cooled to the present value of about 700 degrees K., the silica would finally form a thin crust, but one so plastic that it would tend to prevent molten material from escaping to the surface. Most of the small amount of water initially in the atmosphere eventually escaped into space, but the major part of the planet's initial supply of water may remain trapped below the surface. The cooler earth, by comparison, has long had a thick, fractured crust through which molten material was able to escape to the surface. As the molten volcanic effluents cooled, essentially all their water and carbon dioxide entered the atmosphere; eventually most of the water condensed to form the oceans and the carbon dioxide became stored in carbonate rocks near the surface.

Of what use is this intense effort to understand inhospitable environments on our neighboring planets, apart from man's urge to explore the unknown? I think it is clear that we gain a new understanding of our terrestrial environment from these initial but highly informative measurements of other atmospheres. It has often been pointed out that biologists have only one planet as an example of the rules governing the origin and evolution of life, whereas the physical scientist has abundant evidence that his physics and chemistry are applicable in the most distant galaxies. Nevertheless, the origin and evolution of a planetary atmosphere such as our own are so very complex and may be so finely tuned to small changes in temperature, pressure and chemical composition that the physical rules are not sufficient to bestow full understanding. We need examples of other atmospheres to guide our thinking and to throw new light on our attempts to understand puzzling terrestrial phenomena. It has also been suggested that the space techniques developed for studying Mars and Venus could be used to answer important questions about the earth. For example, Bruce B. Lusignan of Stanford and several others have proposed that radio occultation ex-

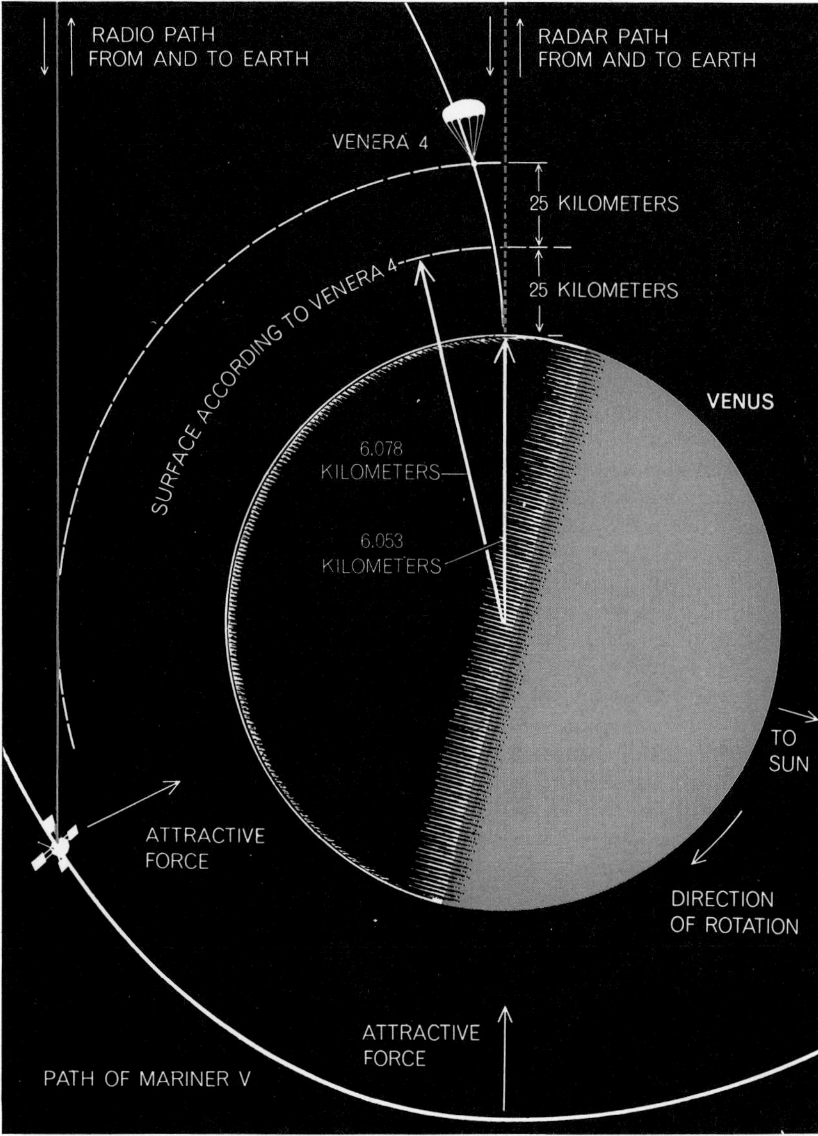

DISCREPANCY IN *VENERA 4* AND *MARINER V* DATA for atmospheric pressure can be resolved if the Russian craft's readings apply to the region between 50 and 25 kilometers rather than to the region between 25 kilometers and the surface, as the Russians reported. *Venera 4*'s data began with a single altimeter reading indicating an altitude of 25 kilometers. The final pressure reading, presumably at the surface, showed the pressure to be only about 20 times that on the earth, whereas *Mariner V*'s data show the surface pressure to be about 100 times the earth's. If *Venera 4*'s last reading was indeed at the surface, the radius of Venus would have to be some 25 kilometers greater than determined previously by radar measurements from the earth: 6,078 kilometers rather than 6,053. The text explains how simultaneous radio-range data for *Mariner V* and radar-range data for the distance to the surface of Venus confirmed the smaller value for the radius and support the hypothesis that *Venera 4*'s last readings were actually made at an altitude of 25 kilometers. *Venera 4*'s initial reading then agrees with *Mariner V*'s value for the pressure at 50 kilometers.

periments be carried out by satellites in orbit around the earth to provide atmospheric measurements over inaccessible regions so as to serve as a powerful aid in weather prediction.

In a more speculative vein, Venus poses an intriguing challenge for planetary engineering. If the thesis of Fricker and Reynolds is correct, one might be able to create oceans on Venus by finding a way to lower its temperature. Sagan has made an imaginative suggestion for achieving the cooling and at the same time producing atmospheric oxygen. He suggests that microorganisms be introduced into the cool high-level clouds, where they would break down the carbon dioxide in two steps into carbon and oxygen. As oxygen replaced carbon dioxide in the atmosphere the greenhouse effect would diminish, lowering the temperature to the point where Venus might eventually become suitable for human colonization.

The space probes to Mars and Venus have emphasized, I think, that biological processes themselves may play a more central role in the characteristics of planetary atmospheres than we have thought. James E. Lovelock and Charles E. Giffin of the Jet Propulsion Laboratory have advanced the view that measuring atmospheric characteristics of Venus and Mars, including minor constituents, would actually be a powerful experiment for the detection of extraterrestrial life. If these planets support life of any kind, it should develop to the limit set by energy sources and raw materials, with the atmosphere most likely acting as a medium for transferring the products of the life cycle. Lovelock and Giffin hold that this role would so modify the composition of the atmosphere that it would be recognizably different from a composition that had resulted simply from the distribution of available chemical compounds in the absence of life.

A clearer view of the earth's ecosystem—the totality of living organisms and their manifold interactions with their environment—could result from the pursuit of such a thesis. For example, the amount of oxygen, nitrogen, carbon dioxide, methane and nitrous oxide in the earth's atmosphere may be governed by life on the earth. The biological sources and sinks maintain a large and balanced flow of oxygen, carbon dioxide and nitrogen into and out of the atmosphere. The abiological sources and sinks of these atmospheric constituents are weaker than the biological ones by factors of 10 to a million.

In the absence of life on the earth the abiological sources and sinks would be controlling. The abiological sinks for oxygen (oxidation of compounds on the earth's surface and oxidation of atmospheric nitrogen and methane) and the abiological sinks for nitrogen (incorporation into stable nitrate compounds) are apparently stronger than the abiological sources, so that both the oxygen and the nitrogen in the atmosphere should slowly decrease to only trace levels. For carbon dioxide the strengths of abiological sources and sinks appear to be approximately equal, making it difficult to predict the steady-state concentration. If the amount of carbon dioxide in the atmosphere slowly increased, a critical level might be reached where the enhanced greenhouse effect would raise the temperature and drive more water vapor and carbon dioxide into the atmosphere; this would further enhance the greenhouse effect and further raise the temperature. This runaway greenhouse effect would end only when the earth resembled Venus in temperature and atmospheric density. If, on the other hand, the amount of carbon dioxide stabilized at a low level, the lifeless earth would have an atmosphere more like that of Mars. On this basis it can be argued that our first life-detection experiment has already been conducted. The tentative conclusion is that on Mars and Venus life is absent.

It would seem that our relative ignorance of the possible global effects of small changes in the composition and temperature of the earth's atmosphere should give us serious and immediate concern about man's use of the atmosphere as a garbage dump. In the past 100 years the percentage of atmospheric carbon dioxide has increased 15 percent, primarily owing to man's burning of fossil fuels. This same activity now uses 15 percent of the biological production of oxygen and in the future could become the major sink. Smog, jet contrails, deforestation and pollution of the earth's waters may affect the constituents and temperature of the atmosphere by only minute amounts, yet they may nonetheless have far-reaching consequences. Let us hope that continuing studies of the earth and its planetary neighbors will lead rapidly to a fuller understanding of planetary atmospheres in general. At the same time it behooves us to take strong steps to eliminate the known harmful effects of man's activities and to minimize those effects that are not yet fully understood. Whatever else we may wish to know, we hardly wish to test whether a lifeless earth would become like Mars or like Venus.

5

The Discovery of Icarus

ROBERT S. RICHARDSON

April 1965

The probability that the earth will collide with an asteroid is extremely small, but not so small that it can be totally ignored. Asteroids are irregularly shaped bodies, also called minor planets or planetoids, that range in diameter from slightly less than a mile up to about 500 miles. The *Ephemerides of Minor Planets for 1965* lists 1,651 that have been assigned a number and 1,520 with proper names. Practically all of these objects travel uneventfully around the sun in orbits lying between those of Mars and Jupiter. About a dozen, however, are known to move in elongated, comet-like orbits that periodically bring them near the orbit of the earth. We are aware of these close-approaching asteroids only through the accident of discovery. No one knows how many objects ranging in size from a few miles in diameter downward may pass near the earth each year without being noticed.

I shall recount the discovery of an asteroid about six-tenths of a mile in diameter that comes closer to the sun and the earth than any other planetary object whose orbit has been well established. It was named Icarus, after the adventuresome youth who flew too close to the sun with wings made of wax and feathers. Asteroid Icarus passes within 17 million miles of the sun, and when it was discovered accidentally in 1949, it had just passed within four million miles of the earth.

The story of Icarus' discovery illustrates how easy it is to be wholly unaware of a massive body hurtling past the earth at high speed and how difficult it can be to obtain enough observations to calculate a reliable orbit. For lack of such observations three asteroids that came even closer to the earth than Icarus—Apollo, Adonis and Hermes—have not been seen again since their original close approach. In 1932 Apollo passed within two million miles of the earth; in 1936 Adonis came within slightly less than a million miles; in 1937 Hermes came within 485,000 miles, or about twice the distance from the earth to the moon, a record that still stands today. Adonis was lost in spite of being carefully tracked for two months in 1936. In June, 1943, when it was supposed to make another close approach, I failed to find it even though I made an extensive search with the 100-inch telescope on Mount Wilson. Apollo, Adonis and Hermes are estimated to be about a mile in diameter and thus are probably a trifle larger than Icarus. These estimates are quite uncertain because they are not derived from direct measurements but depend on the amount of light assumed to be reflected from the surface of the asteroid.

In 1968 Icarus will again pass within four million miles of the earth. The prediction can be made with confidence because it is based on observations obtained in 1949, 1950, 1952, 1953, 1954 and 1957. These have been combined in a definitive orbit prepared under the supervision of Samuel Herrick of the University of California at Los Angeles; it includes the perturbing effects of all the planets except Pluto. A change of only a few degrees in the position of the descending node of Icarus' orbit, the point at which the asteroid crosses the plane of the earth's orbit from north to south, would make it possible for Icarus and the earth to be at the same place at the same time.

There is no unequivocal record that the earth has ever been hit by a body the size of Icarus or larger. It is a matter of definition whether certain large craters, such as Meteor Crater in Arizona, were produced by meteorites—the term reserved for meteoroids that strike the earth—or by small asteroids. A meteoroid can be defined operationally as an object too small to be observed in space by reflected light even with the most powerful telescope; it becomes visible only when it penetrates the earth's atmosphere and leaves a glowing trail. It has recently been estimated that an excavation the size of Meteor Crater, which is 4,000 feet across and 570 feet deep, could be produced by a 20-megaton nuclear explosion set off a few hundred feet below the earth's surface. The force of such an explosion could be duplicated by a meteoroid weighing about 180,000 tons, which, if it were made chiefly of iron, could be contained within a sphere only 110 feet in diameter. Such an object would be much smaller than any of the asteroids whose orbits are known.

Even if close surveillance revealed that a collision between the earth and Icarus were imminent sometime in the future, what could be done about it? Conceivably one might try to destroy it by intercepting it with a nuclear-armed space vehicle. A cleverer idea would be to land a rocket engine on it capable of pushing it slightly off course. Either measure would provide a challenging assignment for space technologists.

The remote possibility of collision is not the only reason for keeping Icarus under close surveillance; the asteroid could provide an important test of the general theory of relativity. One prediction of the theory is that the elliptical orbit of a body revolving around the sun undergoes a slow change in orientation: the point of perihelion and with it the major axis of the ellipse rotate slowly in the direction of the body's motion. The major axis of Mercury's orbit is rotating eastward at the rate of nine minutes 34 seconds of arc per century. Most of this displacement is caused by the gravitational effects of Venus and other planets, but it has been known for a long time that the total is 43 seconds more than can be accounted for by the Newtonian theory of gravitation alone. The discrepancy between theory and observation was finally explained by the general theory of relativity, which predicted an advance of 43 seconds per century in the perihelion of Mercury, in exact agreement with observation. The relativistic effect is more pronounced the smaller the orbit, and the greater its elongation, or degree of eccentricity. Until 1949 Mercury was the only body that furnished a definite test of the theory. Because the orbit of Icarus is more eccentric than that of Mercury it should eventually provide an even better test.

One asteroid, Eros, has already earned a place of honor in the history of astronomy by serving as a yardstick for measuring the size of the solar system. At the time of its discovery in 1898 Eros was the first asteroid known to pass inside the orbit of Mars. On the most favorable occasions it comes within 14 million miles of the earth and is large enough to be easily photographed. In 1931 Eros came within 16.2 million miles of the earth and in 1938 within 20 million miles.

The fundamental measurement for scaling the solar system is the "astronomical unit," the mean distance from the center of the earth to the center of the sun. The sun itself is too distant and too bright for accurate trigonometric measurement from two widely separated stations on the earth's surface. For a long time Venus and Mars were the only bodies suitable for such measurements, but even our nearest neighbor, Venus, does not come closer than 26 million miles. Eros comes so close, however, that it shows an easily measured displacement with respect to the stars when it is photographed simultaneously from observatories in the Northern and Southern hemispheres. Moreover, positions of Eros are accurately defined because its image on photographic plates is pointlike, resembling a star, rather than disklike, which is the case with planets. At these close approaches the motion of Eros is strongly perturbed by the earth. From an analysis of these perturbations it is possible to derive a highly accurate value of the ratio of the mass of the earth to the mass of the sun, as well as of the mass of the moon to that of the earth. These ratios provided the most accurate determination of the astronomical unit until they were superseded in recent years by radio-echo measurements that directly determine the distance from the earth to the nearby planets.

The first asteroid was discovered by Father Giuseppe Piazzi of the astronomical observatory at Palermo in Sicily, who was engaged in making an extensive star catalogue. On January 1, 1801—the first night of the 19th century—he observed through his telescope a starlike object that he had difficulty identifying with stars shown in previous catalogues. The next night the object had moved, proving that it was not a star but an object inside the solar system.

For the first two weeks the object moved slowly eastward, then came to a stop and began traveling in a retrograde, or westerly, direction. It was these loops that had led the early astronomers to believe the planets moved in complex epicycles as they revolved around the earth. Piazzi was of course aware that the back-and-forth motion was simply an illusion due to the fact that the earth is also revolving around the sun, thus regularly overtaking planets farther away from the sun and being overtaken by the planets nearer the sun.

At first Piazzi thought his new object might be a comet, but since it lacked the fuzzy outline characteristic of most comets he concluded that he had discovered a new small planet. He named it Ceres, for the guardian divinity of Sicily. After observing the asteroid for six weeks Piazzi became ill; when he returned to his telescope, Ceres was lost in the glow of the evening sky.

As it happened, the great mathematician Karl Friedrich Gauss had just worked out a simplified method for calculating orbits from only three or four observations. With Piazzi's charts he computed an orbit for Ceres, which enabled him to tell astronomers where to look for the planet. It was picked up without difficulty on December 31, 1801, exactly a year after it was first seen, within the apparent diameter of the moon of its predicted position. According to Gauss's calculations the orbit of Ceres was located between the orbits of Mars and Jupiter.

By coincidence an association of astronomers had just been formed for the purpose of looking for a small planet in just that region when they received word that Ceres had been discovered.

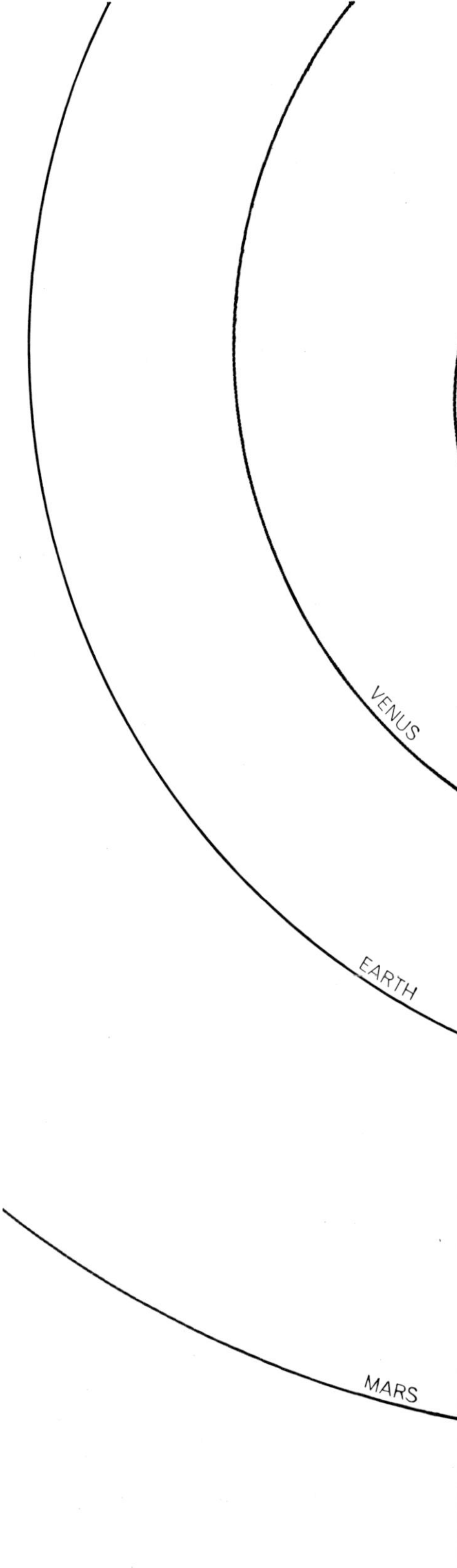

ORBIT OF ICARUS passes closer to the sun than that of any other known asteroid. At perihelion, or point of closest approach, Icarus is 17 million miles from the sun, compared with 28.6 million miles for Mercury,

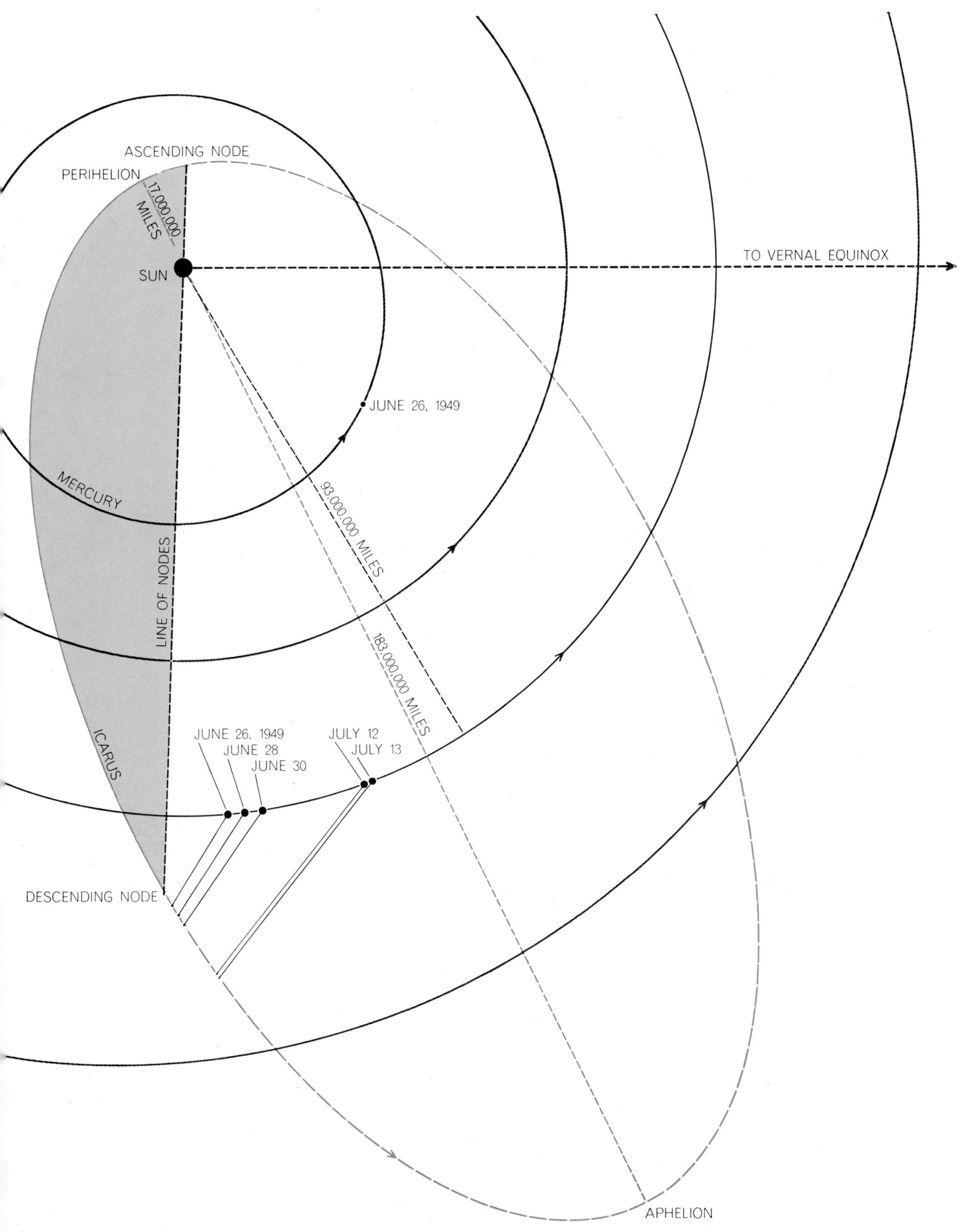

the innermost planet. At aphelion Icarus is 183 million miles from the sun. The orbit of Icarus is tilted 23 degrees to the plane of the earth's orbit; the line of nodes shows where the planes of the two orbits intersect. The part of Icarus' orbit filled in with color is above the plane of the earth's orbit. Although Icarus missed the earth by four million miles in 1949 and should miss by about the same margin in 1968 (*see illustrations on page 64*), only a small shift in its orbit would be needed to send it much closer on subsequent passages. Because it passes near Mercury as well as near the earth its orbit will change significantly with the passage of time.

DISCOVERY PHOTOGRAPH made by Walter Baade on June 26, 1949, shows the asteroid later named Icarus as a faint streak in a rich field of stars. The photograph, a 60-minute exposure, was among the first made with the then recently completed 48-inch Schmidt telescope on Palomar Mountain. This is an enlarged negative print of a section that measures about 1/2 by 3/4 inch on the original 14-by-14-inch Schmidt plate. Icarus is moving to the right. It was then about 22 million miles away, having passed within four million miles of the earth on June 15. The path of Icarus at the time of its discovery is traced on a wide-angle view of the sky below.

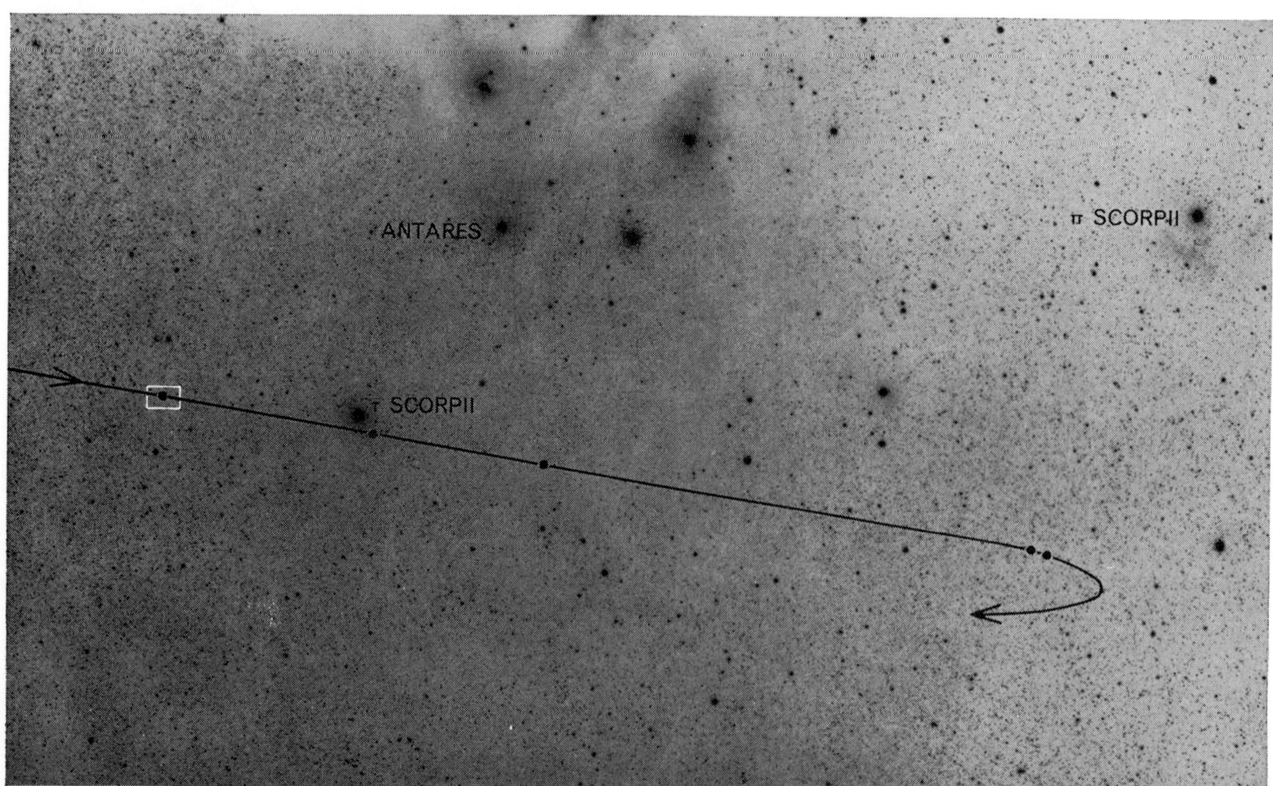

PATH OF ICARUS took it through the region of Scorpius. The three widely spaced dots show the position of Icarus on June 26, 28 and 30, when it was first photographed; it was moving southwest in rapid retrograde motion with respect to the background stars. The next two dots show its position on July 12 and 13, when it was photographed again. The rectangle outlines the region shown in the photograph at the top of the page. This wide-field picture was made by F. E. Ross of the Yerkes Observatory.

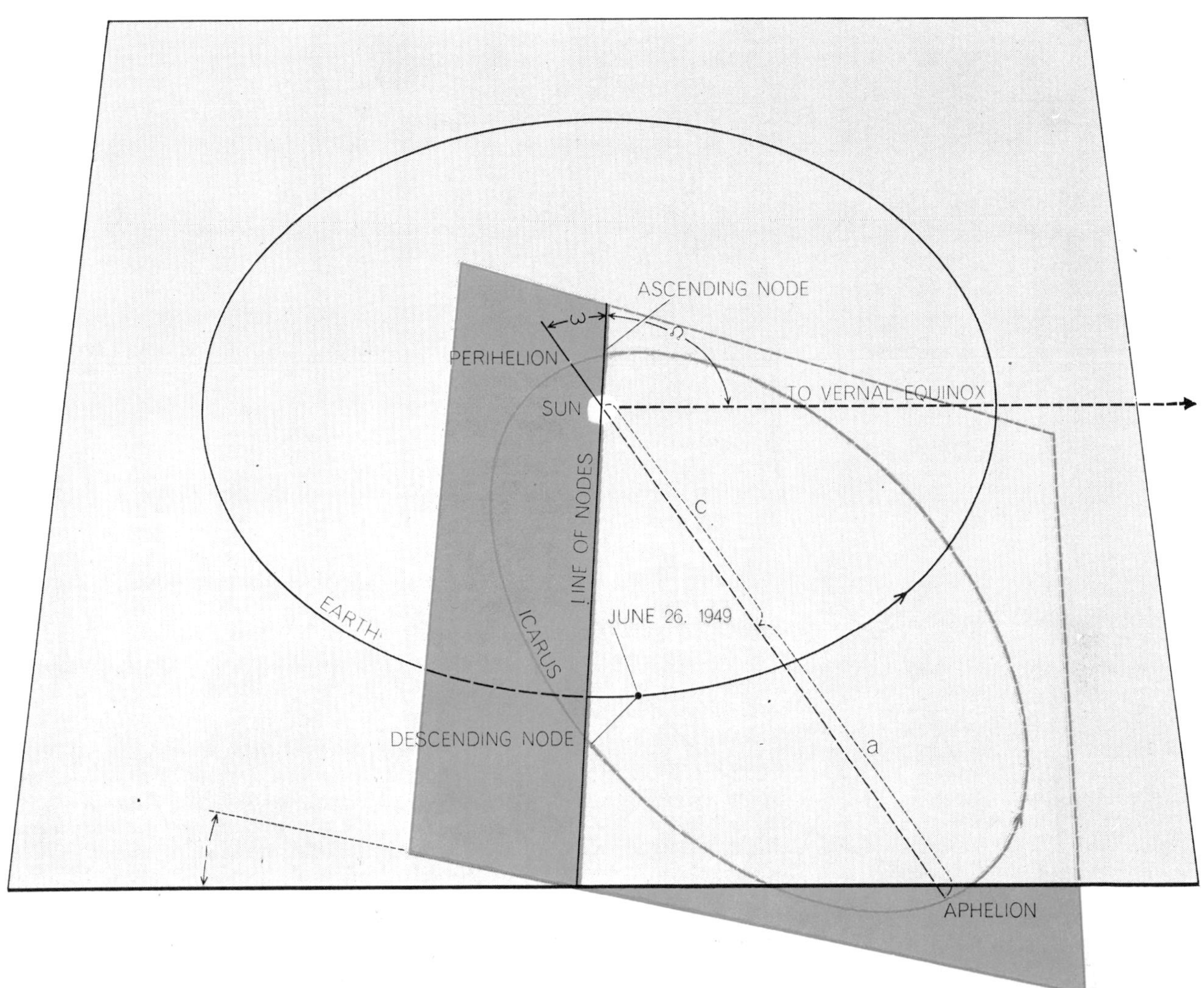

COMPUTATION OF ASTEROID ORBIT requires the determination of six elements, which can be expressed in a variety of ways. Five elements are needed to describe the size, shape and orientation of the ellipse itself; the sixth gives the asteroid's position at a particular time. The size and shape of the ellipse are given by two elements: the length of the semimajor axis, a, and the eccentricity, e, defined as c/a, where c is the distance from the sun to the center of the ellipse. A third element, i, gives the angle between the plane of the asteroid's orbit and that of the earth's. A fourth element, ☊, is the longitude of the ascending node. It is the angular distance measured eastward in the plane of the earth's orbit, or ecliptic, from the vernal equinox to the point where the asteroid crosses the ecliptic from south to north. A fifth element, ω (small omega), defines how the major axis of the ellipse is oriented in its orbital plane by giving the angle between the ascending node and the perihelion point, measured in the plane of the asteroid's orbit and in the direction of the asteroid's motion. The sixth element is usually given as the time when the asteroid passes perihelion. For Icarus a is 1.0777 astronomical units (100.2 million miles), e is .827, i is 22.979 degrees, ☊ is 87.746 degrees and ω is 30.912 degrees. The time of Icarus' first perihelion passage after its discovery was June 4, 1950.

The large gap between the orbits of Mars and Jupiter had puzzled astronomers since the time of Johannes Kepler early in the 17th century. In 1772 J. D. Titius, a professor at the University of Wittenberg, had published an empirical law of planetary distances that provided for a planet at 2.8 astronomical units; Titius' "law" would probably have been forgotten had it not come to the attention of Johann Bode, editor of the influential *Astronomisches Jahrbuch*. Bode gave it such wide publicity that it became generally known as "Bode's law."

The law can be expressed by adding .4 to the progression 0, .3, .6, 1.2, 2.4, 4.8, 9.6, 19.2 and so on; this yields a sequence in which the first eight numbers are .4, .7, 1.0, 1.6, 2.8, 5.2, 10.0 and 19.6. If these numbers are regarded as astronomical units, the first four and the last three agree quite closely with the positions of Mercury, Venus, the earth, Mars, Jupiter, Saturn and Uranus—the seven planets known at the close of the 18th century. Bode became convinced there must be a planet at the unoccupied distance of 2.8 units. The discovery of Ceres precisely at this distance just 20 years after Uranus had been discovered only .4 astronomical unit away from its predicted position seemed an amazing confirmation of Bode's "law." Faith in the relation was badly shaken in 1846 with the discovery of Neptune, which turned out to be nearly nine astronomical units from its predicted position. The law breaks down completely in the case of Pluto, which is at 39.5 astronomical units rather than the predicted 77.2 units [*see top illustration on page 66*].

Three more asteroids were discovered in rapid succession: Pallas in 1802, Juno in 1804 and Vesta in 1807. All have orbits quite close to 2.8 astronomical units. The rate of discovery was slow and sporadic until photographic methods of search were instituted in

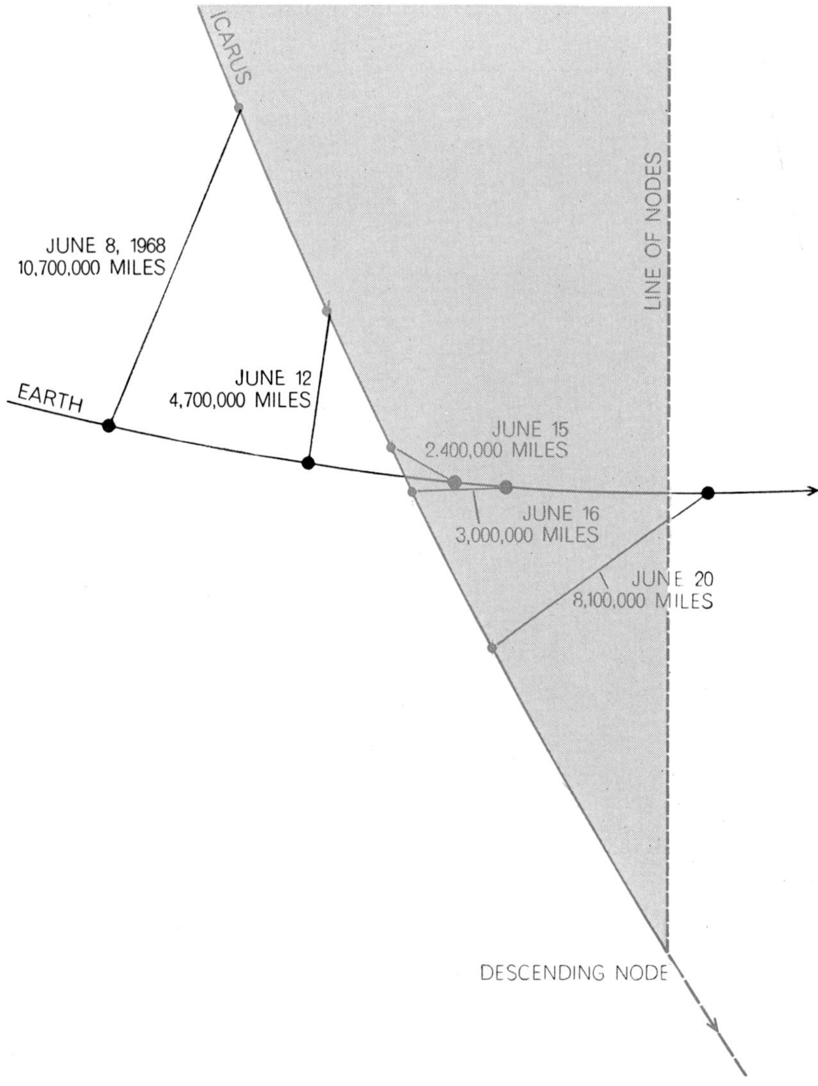

NEXT ICARUS-EARTH MEETING will be in June, 1968, when the two bodies will again pass each other at a distance of about four million miles. Here the path of Icarus is shown projected onto the plane of the earth's orbit. The distances refer to this projection and are therefore somewhat less than the actual distances computed in three dimensions. As shown in the projection below, asteroid Icarus will be about 4.2 million miles above and slightly behind the earth on June 15, when the two planetary bodies make their closest approach.

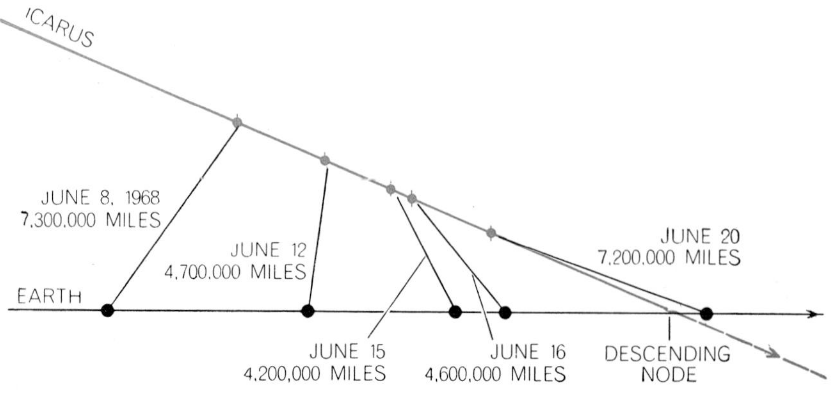

EDGE-ON VIEW during the close approach of 1968 shows the orbits of Icarus and the earth projected onto a plane perpendicular to the line of nodes. Beginning about June 10 Icarus will move into the evening sky and should then remain observable for about two weeks.

1891 by Max Wolf of the observatory at Heidelberg. In the 50 years from 1890 to 1940 the number of recognized asteroids increased from fewer than 300 to nearly 1,500.

As a result astronomers have become increasingly selective in their search. Today when an astronomer sees on one of his plates an elongated image that is presumably an asteroid, he seldom bothers to pursue it further unless the streak is an exceptionally long one, indicating an object moving close to the earth. It has been estimated that there are 55,000 asteroids bright enough to be photographed with a 100-inch telescope when they are nearest the earth.

We can be reasonably sure that there are no undiscovered asteroids more than a few miles in diameter. Ceres, the first asteroid discovered, is also the largest known, with a diameter of 480 miles. Only it and three others are large enough for their diameters to be determined by direct measurement of their disk. The other three are Pallas (304 miles), Vesta (248 miles) and Juno (118 miles). Curiously Vesta, the third largest in size, is the only asteroid that becomes visible to the unaided eye. Its visibility is attributable to the high reflectivity of its surface.

On February 22, 1906, Wolf discovered asteroid 588, later named Achilles; it was the first member of what has since become known as the Trojan group. (In his busy career Wolf discovered 582 asteroids, of which 228 received recognition—a personal record that still stands.) The Trojan asteroids, of which 14 are now recognized, revolve around the sun at about the same distance as Jupiter in approximately the same orbital plane. Nine of them precede Jupiter by about 60 degrees and the other five trail by about 60 degrees. They are of considerable interest in the theory of planetary motions, since they occupy the "Lagrangian points": positions that the French mathematician Joseph Louis Lagrange had predicted more than a century earlier would be stable for secondary bodies moving in the same plane as the primary planet if both bodies were revolving around the sun in circular orbits undisturbed by other planets. Under such ideal conditions if the primary planet and asteroid are 60 degrees apart, they will remain 60 degrees apart. The Trojans only satisfy these conditions approximately; moreover, their motions are disturbed by Saturn. Thus their positions deviate widely from the theoretical 60-degree point [*see the illustration on page 67*].

In 1920 the distance record held by the Trojan asteroids was broken when Walter Baade—using the 39-inch reflector of the Hamburg Observatory in Bergedorf, Germany—discovered asteroid 944, later named Hidalgo. Its orbit has a semimajor axis (half the longest dimension of an ellipse) of 5.8 astronomical units, and is so eccentric that at perihelion it comes within two astronomical units of the sun and at aphelion recedes to a distance of 9.6 astronomical units, which is Saturn's mean distance from the sun. The orbit of Hidalgo is inclined to that of the earth by 42.5 degrees, which is exceeded only by that of asteroid 1580 (Betulia), which has an inclination of 52.03 degrees.

Although Baade was only casually interested in asteroids, it was he who happened to discover Icarus, the asteroid that comes closer to the sun than any other. Baade joined the staff of the Mount Wilson Observatory in 1931 and retired in 1958, 10 years after it had become the Mount Wilson and Palomar Observatories. On the night of June 26, 1949, he was working on a program with the newly completed 48-inch Schmidt telescope on Palomar Mountain that involved a 60-minute exposure of a region in the constellation of Scorpius near the bright star Antares. When he examined the plate the following day, he found a streak among the stars that had undoubtedly been made by an asteroid. Moreover, it was obvious from the length of the streak that the asteroid must be rather close to the earth [*see top illustration on page 62*].

Baade decided that the object was sufficiently remarkable to warrant trying for the two additional photographs needed for computing an orbit. The direction in which an asteroid is moving is often revealed by the nature of the photographic track: the track usually tapers from dark to light in the direction of travel. This is because the emulsion is more sensitive to sky brightness after it has been exposed to the image of the passing asteroid than before. With this clue and the knowledge that the object had passed the point in the sky directly opposite the sun, Baade could be quite certain that the object was moving in a retrograde, or westerly, direction. Accordingly two nights later, on June 28, Baade made another exposure slightly to the southwest of the object's previous position. His surmise proved correct—there was the streak again. It was an easy matter now to get a third observation on the night of June 30.

When Baade returned to the Mount Wilson and Palomar office in Pasadena, he asked Seth B. Nicholson and me if we would like to compute an orbit for the new body. Astronomers develop different specialties and computing orbits did not happen to be one of Baade's. Nicholson and I still retained a keen interest in the solar system; Nicholson in particular had become an expert at tracking down minor members of the solar system and determining their orbits. As a graduate student at the Lick Observatory in 1914 he had discovered the ninth satellite of Jupiter (J IX), and later at Mount Wilson he had found J X, J XI and J XII. I had become familiar with orbit work by helping him keep track of his satellites, which were always getting lost.

Our first step in computing the orbit of Baade's object was to measure its position with respect to the star images around it on the plate. We supposed that the positions of these stars would be readily available in one of the volumes of the star catalogue called the *Carte du Ciel*. Unfortunately for us, the particular volume we needed had apparently not been published as yet. (After all, the *Carte du Ciel* project had only been started in 1887.) This meant that we had to provide our own positions for the stars around the asteroid by measuring their distances in relation to stars in adjacent zones. As a result the positions we obtained for the asteroid were not as accurate as we could have wished. By July 5 we finally had three positions established for the asteroid that were reduced to the proper form for publication. We airmailed these to the Harvard College Observatory, which in turn relayed them on "announcement cards" to other observatories.

We were now ready to begin the computation of the orbit itself. This would merely be a preliminary orbit, suitable for predicting the whereabouts of the asteroid for about a month at the most. Unless we could secure additional observations to improve the orbit, Object Baade was likely to be lost forever. Nicholson had all the formulas written down step by step in an old notebook, a relic from his student days, with warnings about the traps and pitfalls besetting the unwary computer. One of the features of the formulas, which were based on a method devised by Pierre Simon de Laplace, is that the accuracy of the orbit can be checked by seeing how closely it will reproduce the original observations. If the observed and computed positions are not in satisfactory agreement, they must be made to agree by means of a differential correction. It is my experience that most orbits usually require such forcible treatment as a matter of routine. A preliminary orbit is imperfect not because of errors in computation but because the observations themselves are imperfect and the orbit method involves approximations.

From our preliminary orbit it was evident that the object discovered by Baade was a most exceptional asteroid; hence our increasing anxiety over its escaping us. Unless more observations were available very soon, it would be hopelessly lost. But now, to add to our difficulties, the moon had moved into the Scorpius region and would reach full phase on July 10. Its light would fog a plate so badly that another photograph was temporarily out of the question. We hoped that a few days later, when the moon would be out of the way, we might be able to pick up the object on a short exposure.

The first opportunity for a photograph would be Tuesday, July 12, although the moon would still be uncomfortably close. We completed work on the orbit and had predicted positions ready by late that afternoon. We immediately telephoned them to Bruce Rule, a colleague who was scheduled to use the 48-inch Schmidt that night and who had agreed to try for some plates of Baade's asteroid.

While awaiting news from Palomar we received a disquieting letter from Leland E. Cunningham of the University of California at Berkeley. He also had been doing some figuring with our three positions on the Harvard announcement card. His results showed that they could be represented about equally well by half a dozen orbits in which the semimajor axis ranged all the way from .9 astronomical unit to infinity! In other words, our orbit as it stood was virtually indeterminate and therefore of doubtful value for predictive purposes. This indeterminacy could only be removed by securing more observations. But how could we secure more observations if we didn't know where to observe?

On July 12 and 13 Rule succeeded in photographing the region in which the asteroid should be according to our calculations. Finding the asteroid on these new plates, however, was quite a different proposition from what it had been three weeks earlier. At that time

	BODE'S FORMULA	PREDICTED DISTANCE (ASTRONOMICAL UNITS)	ACTUAL DISTANCE (ASTRONOMICAL UNITS)
MERCURY	.4 + 0	.4	.39
VENUS	.4 + .3	.7	.72
EARTH	.4 + .6	1.0	1.0
MARS	.4 + 1.2	1.6	1.52
ASTEROIDS	.4 + 2.4	2.8	2.1-3.5
JUPITER	.4 + 4.8	5.2	5.2
SATURN	.4 + 9.6	10.0	9.5
URANUS	.4 + 19.2	19.6	19.2
NEPTUNE	.4 + 38.4	38.8	30.1
PLUTO	.4 + 76.8	77.2	39.5

BODE'S "LAW" is an empirical scheme that seemed to predict the distances from the sun at which planets should lie. It worked well for the six planets known to the ancients and its prediction was satisfied when Uranus was discovered by Sir William Herschel in 1781. The predicted position at 2.8 astronomical units was subsequently filled in 1801 with the discovery of the asteroid Ceres. The law collapsed with the discovery of Neptune and Pluto.

the asteroid was moving rapidly in a retrograde direction, so that it left an easily recognizable trail among the stars. Now the asteroid had almost reached its second stationary point, which meant that its motion with respect to the stars was scarcely perceptible [*see bottom illustration on page 62*]. Furthermore, it had moved into a section of the Milky Way that was particularly rich in stars. Each of the 14-by-14-inch Schmidt plates was covered with hundreds of thousands—perhaps millions—of star images. Our job was to find among those myriads the one particular image that happened to interest us.

The situation was not quite as hopeless as it sounds. We calculated the position of the asteroid on the plate and drew a box about an inch square around it. If our orbit was any good at all, the asteroid should be somewhere within that square inch. What we were looking for was an image elongated just enough to be distinguishable from the multitude of star images around it. We spotted half a dozen hopeful-looking images on the two plates, all of which on closer scrutiny turned out to be defects in the emulsion.

Finally we put the two plates on the "blink comparator," a device that makes the star images in the two fields merge when they are viewed through a single eyepiece. The two plates are then illuminated alternately so that the two fields are seen superposed in rapid succession. Since the stars are fixed in the sky their images remain stationary. If an object has moved between the two exposures, however, its image on one plate will not quite match its image on the other; hence the image will appear to "blink" or "jump." It would seem that such a procedure should have revealed the asteroid at once, but it was not that easy. Nicholson persisted, however, and finally found two displaced images that appeared to be real. Identification was confirmed beyond doubt when the positions of the images were found to be separated by the expected amount.

The new positions enabled us to derive an improved orbit in which we could place considerable faith, although it was somewhat disappointing in one respect. In our preliminary orbit the semimajor axis of the ellipse had come out to be less than one astronomical unit. The only other bodies in the solar system with a semimajor axis less than one astronomical unit are Venus and Mercury. Now in this asteroid we thought we had a third. When the orbit was corrected, however, the semimajor axis was just a little over one unit—1.066. The best value available today is 1.0777.

Even without this distinction Baade's object is remarkable enough. It is the only asteroid known to pass inside the orbit of Mercury; it approaches to within 17.4 million miles of the sun. At aphelion it travels beyond the orbit of Mars until it reaches a point 183 million miles from the sun. Its inclination to the plane of the earth's orbit is 23 degrees, about twice that of the average asteroid.

To define an orbit completely six numerical quantities are necessary [*see illustration on page 63*]. These can be expressed in a variety of ways, depending on how they are to be used in the process of computation. Five of the elements specify the size, eccentricity and orientation of the ellipse in relation to a standard reference system. In the case of asteroids this reference system is the plane of the earth's orbit. The sixth element tells where on the ellipse the asteroid or other body can be found at a particular time, usually the time of passing the perihelion.

ASTEROID EROS, discovered in 1898 by the German astronomer Gustav Witt, was the first asteroid known to pass inside the earth's orbit. This photograph was made in 1931 by George Van Biesbroeck, using the 40-inch Yerkes refracting telescope, when Eros came to within 16.2 million miles of the earth. The asteroid is approximately 15 miles long and five miles wide.

The name of Icarus for the new asteroid was suggested independently by R. C. Cameron of the University of Indiana and G. E. Folkman of Mount Clemens, Mich. The privilege of selecting a name rested, of course, with its discoverer. Baade liked the name Icarus, and so it was adopted.

The date of the next close approach of Icarus to the earth can easily be found from the respective orbital periods. The period required for Icarus to travel once around the sun is 406 days, or 1.119 years. A little arithmetic shows that 19 revolutions of the earth

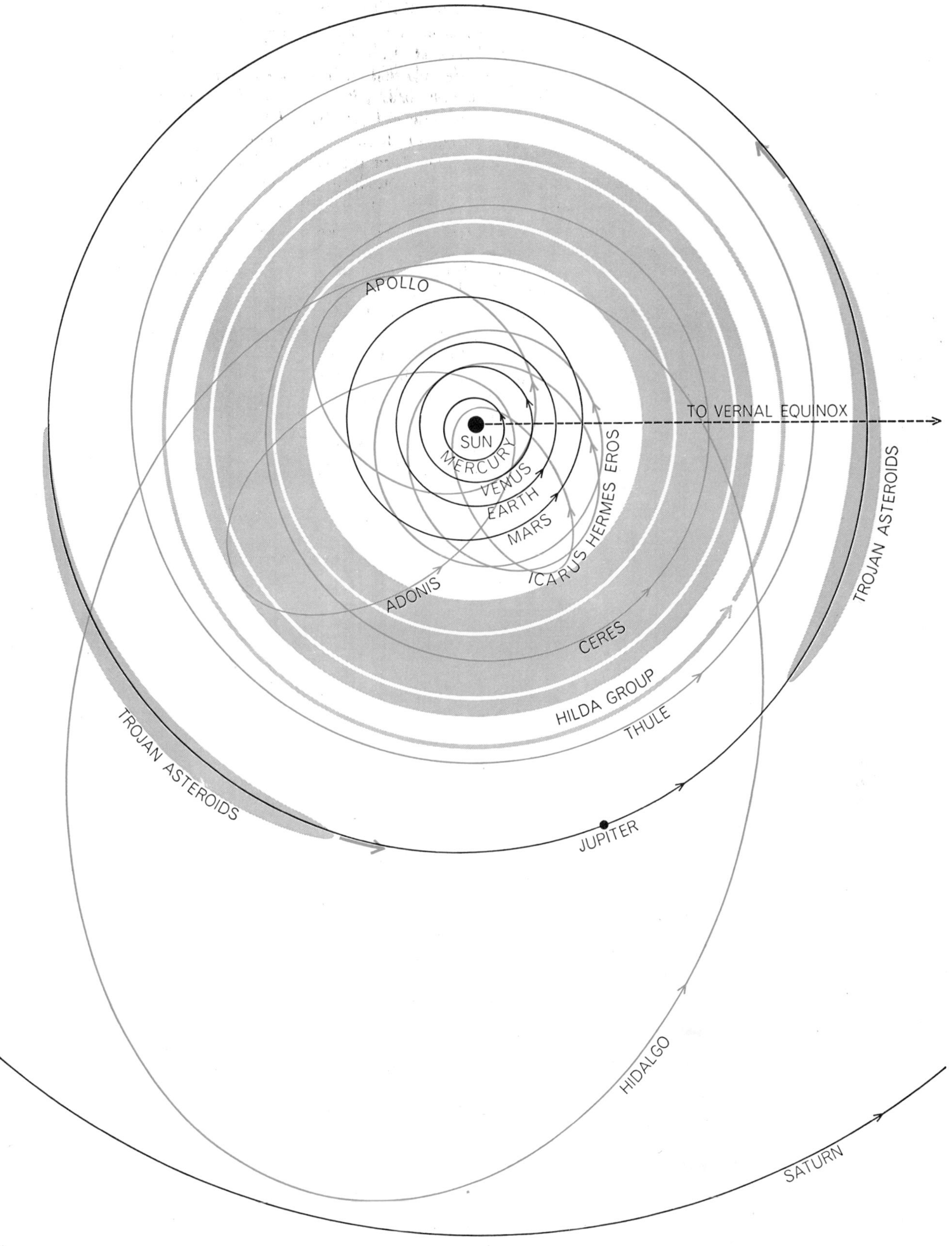

ASTEROID PATHS lie predominantly between the orbits of Mars and Jupiter where the first asteroid, Ceres, was found in 1801. The three closely spaced colored bands show the general location of more than 1,500 asteroids discovered since then. The two white separations between bands indicate relatively sparse regions where a resonance is set up between the period of an asteroid and the period of Jupiter, with the result that the asteroid is shifted into a nonresonant orbit. The 14 known Trojan asteroids are remarkable in that they precede and follow Jupiter by about 60 degrees, occupying what are called the Lagrangian points. They oscillate around these points within the two lenticular regions. Hidalgo, discovered by Baade in 1920, travels farther from the sun than any other known asteroid. Apollo, Adonis and Hermes made the closest-known approaches to the earth—within two million miles—but have been lost from sight. Eros helped to refine the value of the astronomical unit, the mean distance from the earth to the sun.

are equal approximately to 17 revolutions of Icarus ($17 \times 1.119 = 19.023$). This means that if Icarus and the earth are close together on a certain date, they will be close together again 19 years later. Since the last close approach was in 1949, the next one will occur in 1968, on June 15, when the minimum distance between the two bodies will be four million miles [*see illustrations on page 64*]. In its present orbit Icarus could come to within 3.5 million miles of the earth on another occasion.

I have left to the end any mention of the origin of asteroids, because this is still a matter of speculation. In 1802 the German astronomer H. W. M. Olbers, the discoverer of Pallas and Vesta, advanced the hypothesis that the asteroids are fragments of a former planet revolving between Mars and Jupiter that had been shattered. Perhaps the most remarkable feature of Olbers' hypothesis is its extraordinary vitality in view of the total lack of evidence to support it. If the asteroids had such a catastrophic origin, their orbits, if undisturbed, would all intersect at the point of disintegration. All attempts to locate this point, however, have proved futile for the simple reason that all trace of it would have been erased long since by the perturbing effect of Jupiter and Saturn. If the asteroids had once formed a primeval planet, its mass could hardly have exceeded that of the moon. A hypothesis that has been received with more favor is that the asteroids originated not from the disruption of a single planet but from a collision between two planets, or possibly a series of collisions among several planets.

The asteroids have demonstrated their value to astronomy in a variety of ways. They have stimulated cooperation among astronomers. They have presented astronomers with problems of great interest in celestial mechanics. The discovery of Eros enabled us to get the first accurate value for the length of the astronomical unit, as well as to improve our values of the ratio of the mass of the earth to the sun and of the moon to the earth. Eventually Icarus should furnish us with an excellent test of the general theory of relativity. In addition the asteroids may provide clues to the origin of the solar system. In time, no doubt, we shall wish to send out a space vehicle to photograph an asteroid at close range and see what one really looks like. A photograph of Ceres on the first night of the 21st century would make a fitting commemoration of Father Piazzi's discovery.

6

Jupiter's Great Red Spot

RAYMOND HIDE

February 1968

The surface markings of the planets have always had a special fascination, and no single marking has been more fascinating and puzzling than the great red spot of Jupiter. Unlike the elusive "canals" of Mars, the red spot unmistakably exists. Although it has been known to fade and change color, it has never entirely disappeared in the 130 years it has been regularly observed. The red spot appears to be embedded in the banded clouds of Jupiter's atmosphere, and its period of rotation about the axis of the planet undergoes slight but persistent fluctuations. For this reason astronomers have generally thought that the spot could not be attached to the solid planet (assuming Jupiter has a solid surface underlying its atmosphere) and that it must be a solid object rather like a huge raft floating in the atmosphere.

An alternative suggestion was put forward in 1961: I proposed that the great red spot might be visible evidence of a hydrodynamic phenomenon essentially similar to a "Taylor column." This is a more or less stagnant cylinder that can be produced in a rotating fluid by either a protuberance or a depression at the base of the fluid. Such columns, which tend to be parallel to the axis of rotation of the fluid, are named for their first investigator, Sir Geoffrey Taylor of the University of Cambridge. According to my hypothesis, the presence of even a shallow topographical feature on the surface of the solid planet, if indeed it be solid, would give rise to a pronounced disturbance in the atmosphere because the planet is rotating so rapidly (one revolution in less than 10 hours). This hypothetical disturbance could well give rise to a permanent marking, such as the red spot, at the top of the cloud layer, where it would be visible to outside observers.

To explain the fluctuations in the red spot's period of rotation one must assume that there are forces acting on the solid planet capable of causing an equivalent change in its rotation period. In other words, the fluctuations in the rotation period of the red spot are to be regarded as a true reflection of the rotation period of the planet itself. Theoretical studies suggest that the fluid regions of Jupiter—its atmosphere and its fluid core, assuming there is one—are together sufficiently massive and well agitated to create the forces needed to alter the planet's speed of rotation. At the same time the topographical feature responsible for the red spot would be kept from wandering in latitude because of the planet's great gyroscopic stability. Indeed, one of the principal objections to the "raft hypothesis" is its apparent inability to explain why the latitude of the spot has remained fixed.

Fifth in order of distance from the sun, Jupiter is a giant among the planets. Its diameter is some 138,000 kilometers (about 86,000 miles), or roughly 11 times the diameter of the earth. Jupiter circles the sun once every 11.8 years. It can be observed profitably for about 10 months out of every 13; the rest of the time it is on the far side of the sun from the earth and its apparent diameter is much reduced. At opposition, when both planets are in line with the sun on the same side of it, Jupiter is twice as bright as Sirius, the brightest star, and its apparent diameter reaches 50 seconds of arc. This corresponds to a 33rd of the diameter of the moon. Among the planets only Venus and occasionally Mars exceed Jupiter in brilliance.

Studies of various kinds indicate that the visible surface of Jupiter is made up of clouds of ammonia and ammonia crystals suspended in an atmosphere that is mainly hydrogen admixed with water and perhaps methane and helium. Other lines of evidence, particularly the fact that Jupiter's density is only 1.3 times the density of water, suggest that the main constituents of the planet are hydrogen and helium. It was once conjectured that Jupiter had a metallic core similar to the core of the earth, jacketed by a thick mantle of ice. Ideas changed in 1951 when William H. Ramsey, then at the University of Manchester, pointed out that the high pressures prevailing deep inside the planet would have the effect of converting hydrogen from its ordinary liquid or solid form into a metallic, electrically conducting form.

Theoretical models of Jupiter's structure based on Ramsey's ideas have since been constructed by a number of workers. Although these models give a fairly complete description of the distribution of density and pressure within the planet, they do not lead to predictions about such important properties as temperature, thermal and electrical conductivity, viscosity and mechanical strength. Moreover, they cannot predict whether the material at the base of the atmosphere is a solid or a liquid. Thus for all anyone knows Jupiter could be fluid throughout [*see illustration on page 73*].

In 1610 Galileo turned his primitive telescope on Jupiter and discovered the four largest of the planet's 13 satellites. Twenty years later Nicolas Zucchi and Daniel Bartoli observed and recorded the large-scale features of Jupiter's visible disk. The most prominent are a series of bright and dark bands, numbering 14 or more, that run parallel to the planet's equator. The bright bands are usually called zones, the dark ones belts [*see top illustration on page*

JUPITER THROUGH BLUE FILTER confirms the visual impression that the light reflected by the great red spot is primarily yellow to red. The equatorial zone brightened markedly in blue light between October 23, 1964 (*left*), and December 12, 1965 (*right*). Photographs on these two pages were taken at New Mexico State University Observatory with either a 12-inch or a 24-inch telescope.

72]. Because the zones and belts present a continually changing pattern it is obvious that they are cloudlike structures in a fluid atmosphere, not markings on the surface of a solid planet. Isolated dark spots and brilliant white areas frequently appear in the zones. At other times an entire belt or a large portion of it will disappear from view and reappear after weeks or months.

The great red spot was probably first observed in 1664 by Robert Hooke. That same year Giovanni Cassini made drawings of the spot and began recording its period of rotation. He found that it speeded up slightly between 1664 and 1672, when its rotation period, originally nine hours 55 minutes 59 seconds, decreased by five seconds. "Hooke's spot," as it was later called, was observed intermittently until 1713.

The next known record of the spot is a drawing made in 1831 by Heinrich Samuel Schwabe, the German apothecary who is best known for his discovery that sunspots wax and wane on a cycle of roughly 11 years. Drawings of Jupiter showing the spot were made in 1857 by William Rutter Dawes, an English clergyman, in 1870 by Alfred M. Mayer, an astronomer at Lehigh University [*see bottom illustration on page* 72] and thereafter by many other observers.

Mayer described his observation of Jupiter with a 15-centimeter refracting telescope in January, 1870, as follows: "I was struck with the beautiful definition and steady sharpness of outline of the details of [Jupiter's] disk, and especially was my attention riveted on a ruddy elliptical line lying just below the South Equatorial belt.... This form was so remarkable that I was at first distrustful of my observation; but...I perceived that the ellipse became more and more distinct as it advanced toward the centre of the disk." Mayer believed the feature he was recording had "never before been noticed."

The great red spot became so conspicuous in 1879 that it was widely publicized; it was then that the spot received its present name. In 1882 the spot began to fade. Its decline was so steady that by 1890 astronomers believed it would eventually disappear. By 1891, however, the fading had halted and was

TWO-COLOR SERIES taken on November 8 and 9, 1964, makes it possible to compare the markings on Jupiter as seen in blue light (*left*) and in red light (*middle and right*). The two photographs at the right, which were taken only 30 hours apart, show how cloudlike structures in the equatorial zone can change quite considerably in the time required for the planet to make only three rotations.

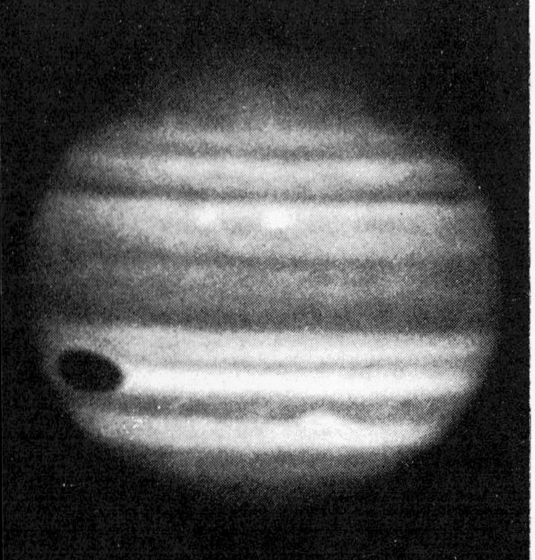

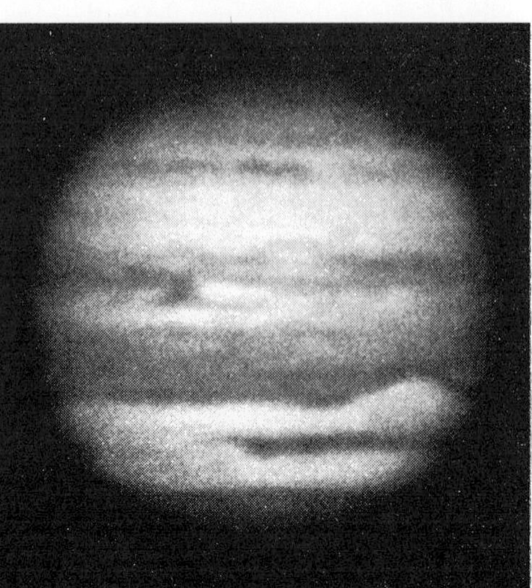

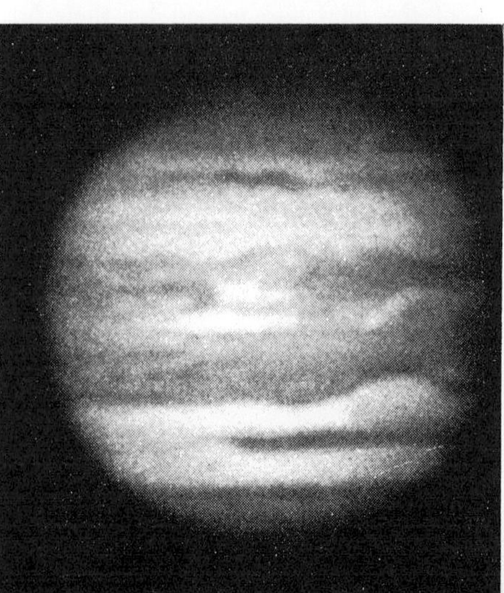

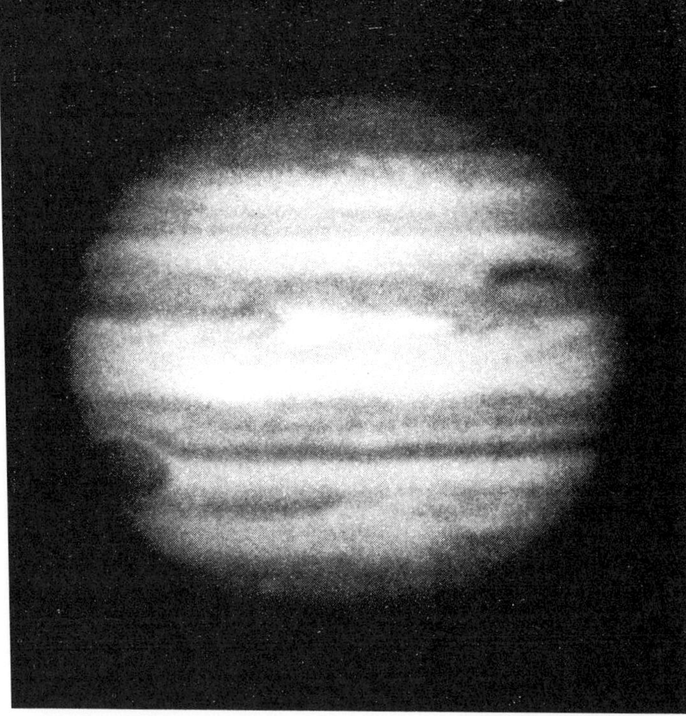
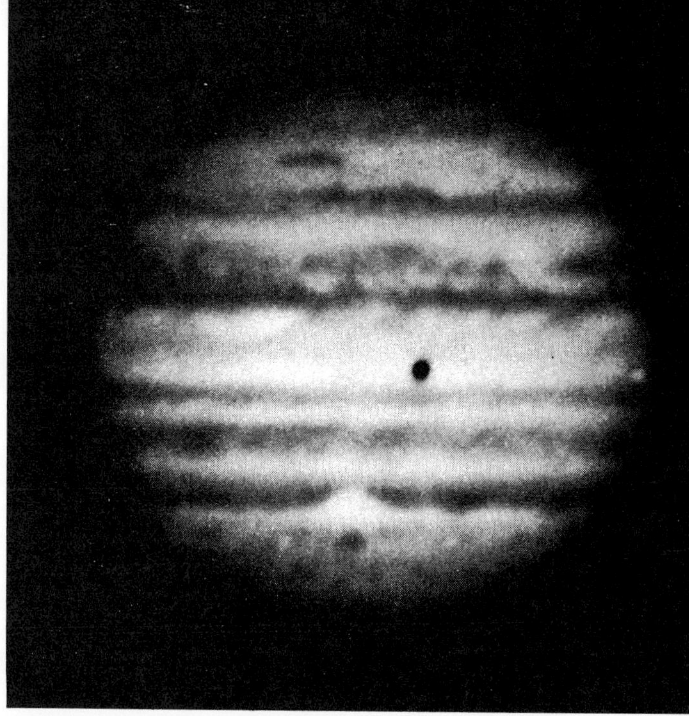

JUPITER THROUGH GREEN FILTER shows the planet with the red spot visible (*left*) and out of sight on the far side of the planet (*right*). Extensive changes in band structure took place between October 29, 1965 (*left*), and March 2, 1967 (*right*). The dark spot near the equator at right is the shadow of Jupiter's third largest satellite Io, which is a barely visible white spot near the right limb.

soon reversed. Although the spot has varied since then, it has never disappeared from view.

Any acceptable theory of the great red spot must account for its principal properties without being at variance with what is known about the planet as a whole. These properties can be listed as follows.

First, the spot has a specific size, position and form. It is an ellipse about 40,000 kilometers long and 13,000 kilometers wide, centered on about 22 degrees south latitude. The spot is roughly equal in area to the entire surface of the earth. Its shape, size and orientation undergo slight fluctuations.

Second, the spot has a rotation period that fluctuates by more than 10 seconds [*see top illustration on page 74*]. These fluctuations, when plotted as variations in longitude with respect to a mean period, show that the spot has at times advanced or fallen back as much as 500 degrees. Significant accelerations of this motion occurred in 1880, 1910, 1926 and 1936, years when the spot was also very conspicuous.

Third, the spot exhibits only slight excursions in latitude. In fact, variations in the latitude of the center of the spot have never exceeded the probable error of measurement.

Fourth, the spot is associated with a persistent indentation, known as the red spot hollow, in the southern boundary of the south equatorial belt. On no occasion have the hollow and the spot been simultaneously absent.

Fifth, the spot has interacted in a specific way with a phenomenon called the south tropical disturbance. The disturbance made its appearance in 1901 as a short dark streak in the south tropical zone some distance from the red spot. Eventually it grew in length until it stretched nearly two-thirds of the way around the planet. In 1939 the disturbance vanished, seeming to give way to three "bright ovals" that can still be seen in the belt just south of the red spot. The rotation period of the disturbance was somewhat less than that of the red spot, and the two features came in contact on nine occasions. The most remarkable behavior at these conjunctions was

THREE-COLOR SERIES was made in March, 1967. The red-light view (*left*) was taken on March 4, only 32 minutes before the green-light view (*middle*). The elliptical shape below the great red spot is one of three "bright ovals" in the south temperate belt. The blue-light view (*right*), taken 17 rotations after the green-light view, shows that the ovals travel slightly faster than the spot.

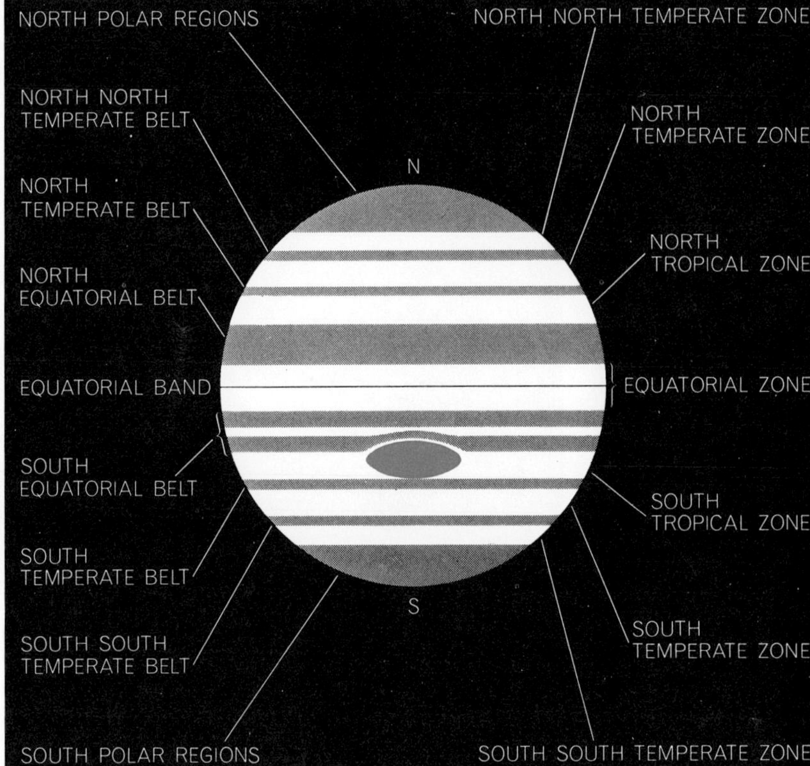

JUPITER'S BANDS are commonly called zones if they are bright and belts if they are dark. At any given time some of the bands may be indistinct or even absent. Nevertheless, most of the zones and belts shown here can be identified in the photographs on the preceding pages.

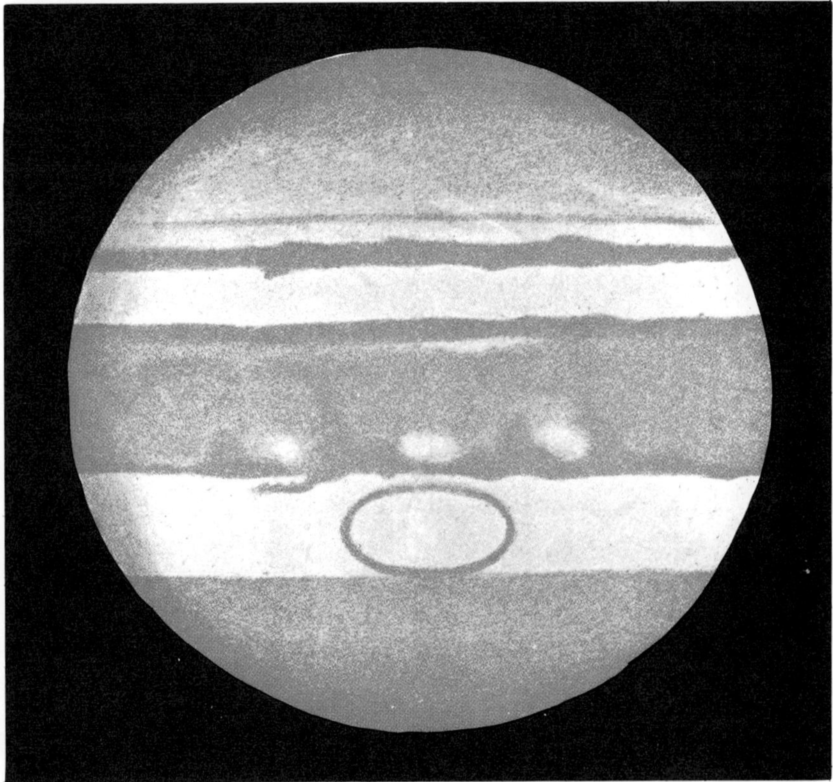

THE RED SPOT IN 1870 was drawn by Alfred M. Mayer of Lehigh University and published in a paper titled "Observations of the Planet Jupiter." A feature now believed to be the red spot was first reported in 1664 by Robert Hooke. For more than 200 years the spot had attracted so little attention that Mayer wrote in his account that the "ruddy elliptical [feature] lying just below the South Equatorial belt...has never before been noticed."

the tendency for the disturbance to skirt around the edge of the red spot at about 10 times the speed with which it approached the spot and receded from it.

Sixth, the color of the spot seems to fluctuate, although the changes are hard to measure. At its faintest the spot has been described as "full gray" and "pinkish" and at its most prominent as "brick red" and "carmine." In black-and-white photographs made through colored filters the spot usually shows up very dark when the filter is blue and is almost invisible when the filter transmits only red or infrared. In 1963, using the 200-inch telescope on Palomar Mountain, Bruce C. Murray and his colleagues at the California Institute of Technology compared the infrared emission of the spot with the emission of surrounding areas. They found that the temperature of the spot was about 127 degrees Kelvin (degrees centigrade above absolute zero), or about two degrees cooler than the adjacent regions.

Finally, a truly satisfactory theory of the great red spot should be able to explain its uniqueness. Why are there not several spots of various sizes? Until evidence to the contrary is provided, the theorist must also assume that the red spot is a permanent feature that has existed for thousands or millions of years.

The great red spot is such a weird phenomenon that few serious attempts have been made to account for it. The raft hypothesis seems to have been first proposed in 1881 by G. W. Hough in his annual report for the Dearborn Observatory. One of its strongest advocates was Bertrand M. Peek, an amateur astronomer whose book *The Planet Jupiter* contains a valuable collection of visual observations of the planet. He proposed that the raft was mainly made up of several solid forms of water that have been produced in the laboratory at high pressures. Peek suggested that the fluctuations in the red spot's rotation period might be due to slow variations in the depth at which the raft floated in the planet's atmosphere. He also proposed that as the raft rose and fell one form of ice would change into another form with a consequent absorption or release of heat that might account for the variations in the appearance of the spot.

In a recent appraisal of the raft hypothesis Wendell C. DeMarcus of the University of Kentucky and Rupert Wildt of the Yale University Observatory said: "It has proved difficult to conceive of an object able to float in a surfaceless ocean of... hydrogen gas." They assert, however, that in principle the red spot could be a solid object with the same density as the fluid surrounding it and having the same constituents but in different proportions.

When the hypothesis has been fully tested against observation and when its other consequences have been carefully examined, it may well look more attractive than it does now. For example, Robert H. Dicke of Princeton University has suggested that Jupiter's atmospheric winds, being strongly channeled in zones parallel to the equator, might inhibit any tendency for a raft to drift in latitude. If this is so, one of the strongest objections to the raft hypothesis would be removed. One would still be puzzled, however, as to why Jupiter's atmosphere contains only one such object and how it resists disruptive forces.

The Taylor-column hypothesis avoids many of the problems that beset the raft hypothesis. I should point out, however, that a raft could also give rise to a Taylor column. In what follows, therefore, I shall use "Taylor-column hypoth-

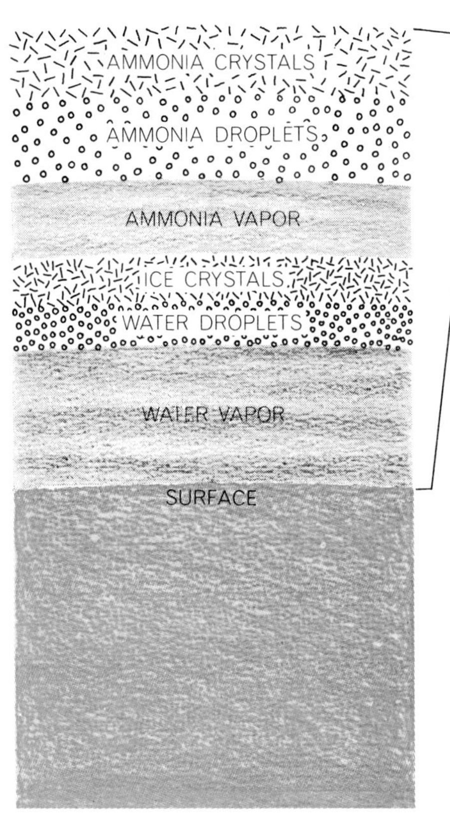

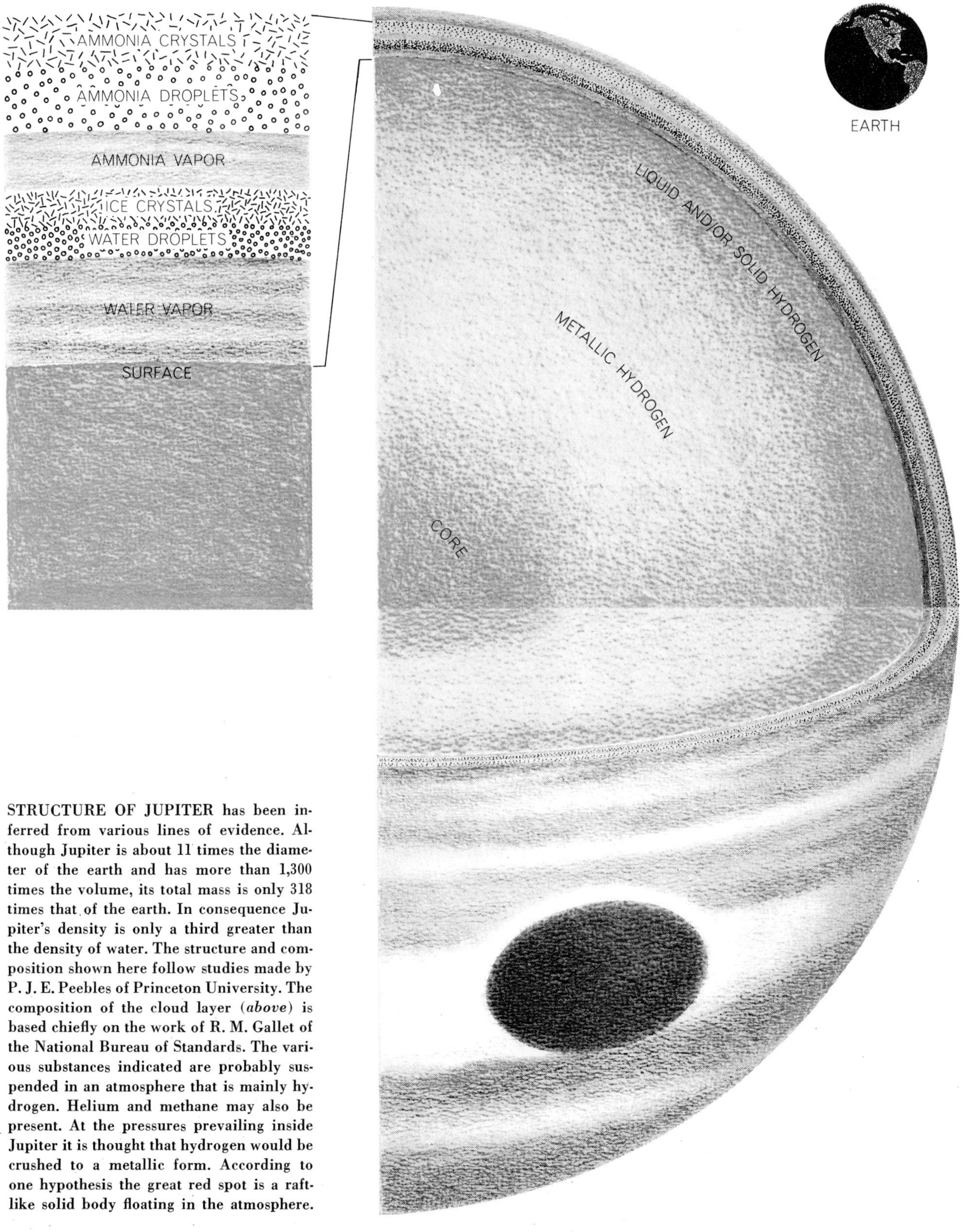

STRUCTURE OF JUPITER has been inferred from various lines of evidence. Although Jupiter is about 11 times the diameter of the earth and has more than 1,300 times the volume, its total mass is only 318 times that of the earth. In consequence Jupiter's density is only a third greater than the density of water. The structure and composition shown here follow studies made by P. J. E. Peebles of Princeton University. The composition of the cloud layer (*above*) is based chiefly on the work of R. M. Gallet of the National Bureau of Standards. The various substances indicated are probably suspended in an atmosphere that is mainly hydrogen. Helium and methane may also be present. At the pressures prevailing inside Jupiter it is thought that hydrogen would be crushed to a metallic form. According to one hypothesis the great red spot is a raft-like solid body floating in the atmosphere.

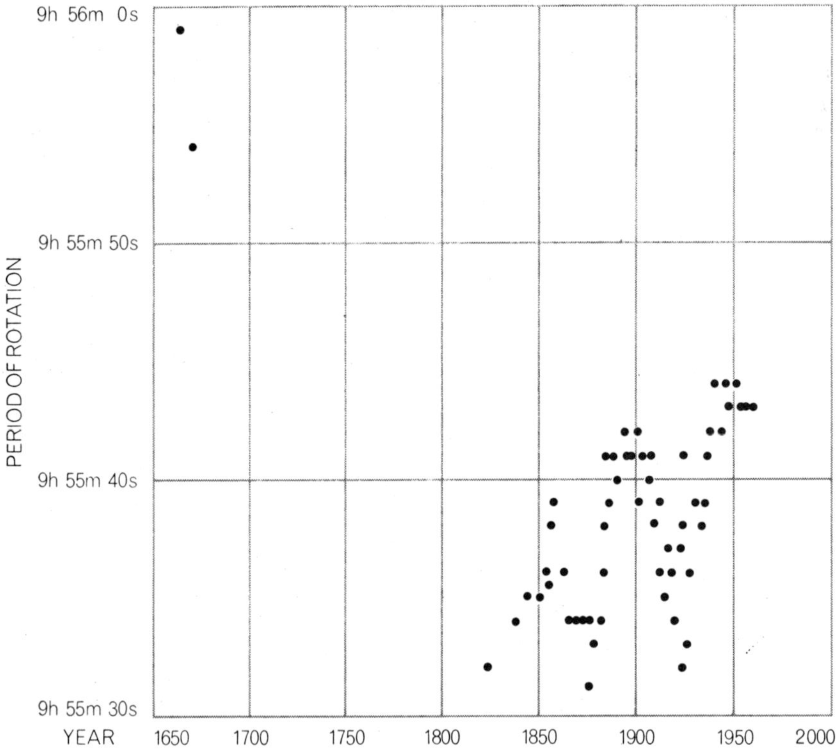

ROTATION PERIOD OF GREAT RED SPOT has varied considerably in the 135 years since it has been recorded with care. The variation is even greater if the earliest observations of Giovanni Cassini, made between 1664 and 1672, are regarded as accurate. It was long thought that a feature with such a variable period could not be connected in any way with the surface of the "solid" planet below, which one would expect to rotate with great constancy. But if the author's hypothesis is correct, the red spot is produced by a surface feature, either a raised area or a depression, and the planet's rotation rate is indeed variable.

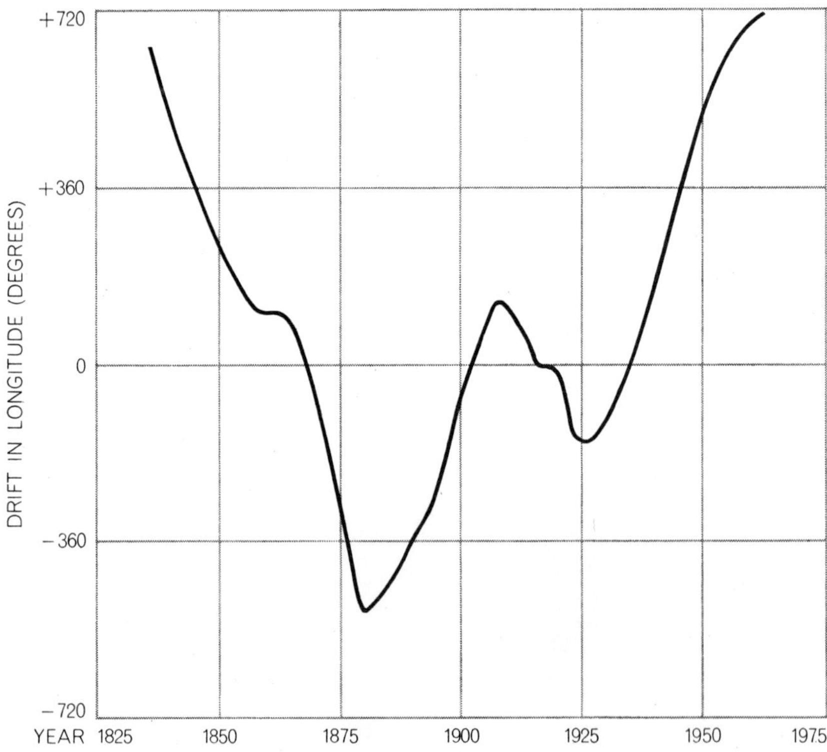

LONGITUDINAL WANDERING OF RED SPOT can be visualized by defining a mean period of rotation and plotting in degrees how much the spot advances or retreats with respect to the mean. The diagram is based on one in *The Planet Jupiter*, by Bertrand M. Peek.

esis" to mean a hydrodynamic phenomenon caused by an irregularity on the surface underlying Jupiter's atmosphere.

In the simplest conceivable fluid system, in which the fluid has uniform density and zero viscosity, the effects of rapid rotation on hydrodynamical flow can be expressed in terms of a theorem first proposed in 1916 by James Proudman of the University of Liverpool. This theorem simply states that the flow must be the same in planes perpendicular to the axis of rotation.

In 1921 Sir Geoffrey Taylor recognized, and verified by experiment, an important implication of Proudman's theorem: If a solid object were to be moved slowly through a rotating tank of fluid, the object would carry with it a relatively stagnant column of fluid aligned parallel to the axis of rotation. Taylor columns can also be created by irregularities such as bumps and corrugations at the bottom of a rigid container when the fluid is in motion with respect to the container.

In considering whether a topographical feature on the surface of Jupiter might give rise to a Taylor column it is important to know how high (or alternatively how deep) the feature must be to be effective. It turns out that the height (h) must exceed a value determined by a relation involving the total depth of the fluid (D), the horizontal dimension of the topographical feature (L), the speed of the fluid over the feature (U) and finally the angular speed of rotation of the entire system (Ω). Specifically, the value of h must exceed $D(U/2L\Omega)$, all expressed in appropriate units.

To investigate this relation, one of my students, Alan Ibbetson, and I conducted a series of laboratory experiments [*see illustrations on opposite page*]. In addition to confirming the relation, we found that some of the details of the flow patterns we observed have their counterparts in the theoretical studies of Taylor columns in fluids of zero viscosity conducted by Keith Stewartson of University College London and by Michael J. Lighthill of the Imperial College of Science and Technology. Nevertheless, certain questions concerning Taylor columns remain unanswered; both theoretical and experimental work are being continued.

The extent to which a planet's rotation influences the flow of its atmosphere depends chiefly on its size and rotation speed. The linear speed of Jupiter's surface owing to the planet's rotation is more than 25 times that of the earth. Consequently Jupiter's rotation dominates the winds in its atmosphere even more effectively than the earth's rotation dominates terrestrial winds. A measure of this "domination" is the Rossby number (R), named for the Swedish meteorologist Carl-Gustaf Rossby. This number is equal to $U/2L\Omega$, which is the factor multiplied by D in the equation given above. The smaller the Rossby number, the larger the dominance of rotation. R for large-scale motions in the earth's atmosphere is about .1. The corresponding value of R for Jupiter's atmosphere is .0002 (except near the equator, where R is .01).

Any realistic discussion of Jupiter's meteorology would have to take account of many complicating factors, such as the nonuniform density of the planet's atmosphere and the possible effects of magnetic fields. In a first approximation, however, the complicating effects are not expected to vitiate the essential idea behind the Taylor-column hypothesis. In any case, the banded appearance of Jupiter makes it clear that large-scale winds on Jupiter must blow mainly parallel to the equator along circles of latitude, showing that the rotation of the planet indeed dominates the motions of the atmosphere.

In order to estimate the height (or depth) of a topographical feature capable of creating a Taylor column in Jupiter's atmosphere, we proceed as follows. We begin with the assumption that the atmosphere is underlain by a material of such high viscosity that it flows very much more slowly than the atmosphere; this material is the "solid" planet.

For the sake of carrying out a fairly definite calculation let us make the innocuous assumption that the depth (D) of Jupiter's atmosphere is 3,000 kilometers. Since the red spot itself is about 40,000 kilometers long let us use this as the value of L, the major horizontal dimension of the topographical feature. A plausible value for the velocity (U) of the wind passing over the surface might be two meters per second, or about four miles per hour. For this value of U the Rossby number is .0004. If one inserts these values into the preceding equation, one finds that a Taylor column will be produced if the height (or depth) of the topographical feature exceeds only one kilometer. By earth standards this is a very modest dimension, but until more is known about the mechanical properties of the "solid" part of Jupiter one cannot be sure that a topographical feature one kilometer high could in fact be supported against gravitational forces nearly three times those on the earth. The Tay-

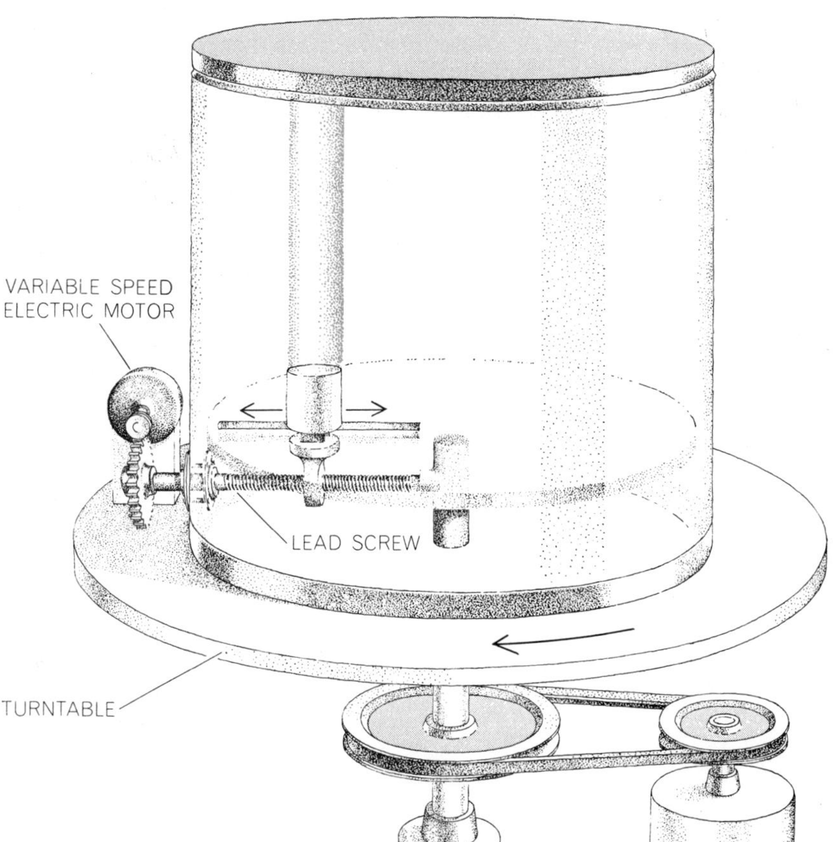

TAYLOR-COLUMN EXPERIMENT was carried out by the author and Alan Ibbetson, using the apparatus diagrammed here. A liquid, usually water, was rotated at a uniform speed in a cylindrical tank 12.2 centimeters deep and 14.5 centimeters in radius. A short cylindrical obstacle of variable height was mounted so that it could be driven slowly across the radius of the rotating tank. The author and his colleague established the conditions under which a stagnant column of liquid, a Taylor column, would form above the obstacle.

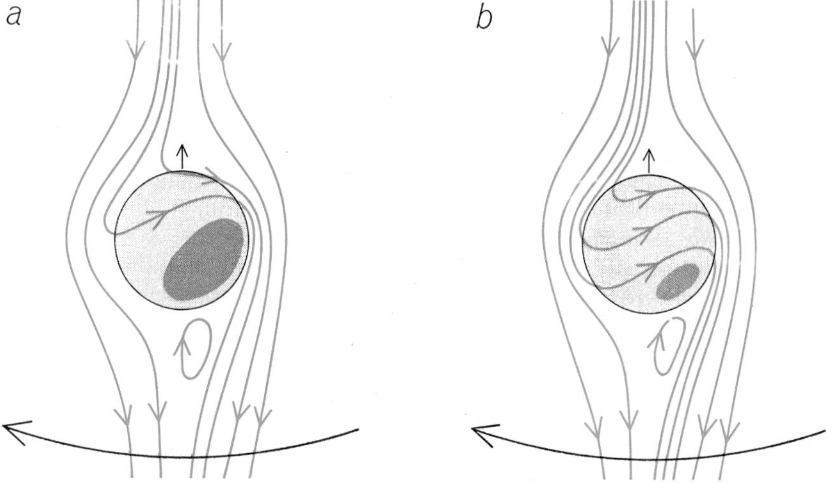

FLOW PATTERNS OVER OBSTACLE in the experimental tank were drawn by the author and Ibbetson, who observed the streaks made by a dye tracer about 10 centimeters above the obstacle, which projected about two centimeters from the base of the tank. The obstacle was moved toward the center of the tank at various speeds while the tank and liquid rotated at about 40 r.p.m. A Taylor column formed (*colored ellipse in "a"*) when the obstacle was moved at 1.2 centimeters per minute. When the rate was increased to nine centimeters per minute (*b*), flow was diverted but no pronounced Taylor column was visible.

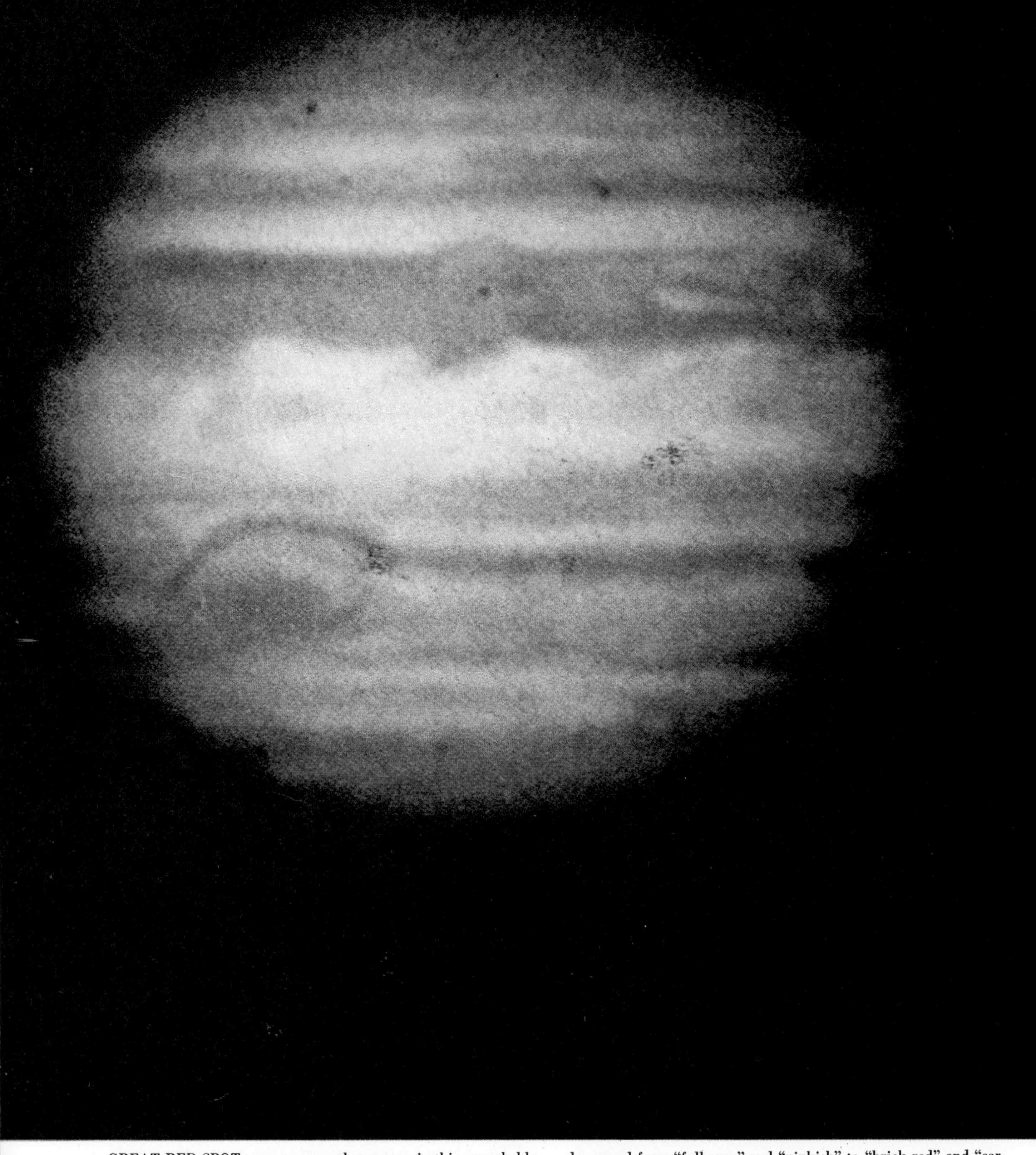

GREAT RED SPOT appears somewhat orange in this remarkably detailed photograph made at the Lunar and Planetary Laboratory of the University of Arizona. During the century or so that Jupiter's great red spot has been closely observed its color has reportedly ranged from "full gray" and "pinkish" to "brick red" and "carmine." The photograph was made on December 23, 1966, by Alika Herring and John Fountain, who used a 61-inch reflecting telescope. The exposure was one second on High Speed Ektachrome.

lor column would not necessarily rise vertically over the surface feature. In the simplest case the column would have its axis parallel to the axis of rotation of the planet [*see at right*].

Perhaps the most obvious weakness of the Taylor-column hypothesis is one it shares with the raft hypothesis: Why should there be only one such feature or object? There is, however, a lead toward an explanation of the uniqueness of the Taylor column in Jupiter's atmosphere that has no ready analogy in the case of a raft. The conditions for creating a Taylor column discussed above show that for a given wind speed a column would not form over a topographical feature of a kilometer in height if its horizontal dimensions happened to be much less than 40,000 kilometers.

If, however, Jupiter's atmospheric winds are sufficiently variable and occasionally shift from one semistable pattern to another, it should be possible for such topographical features to give rise temporarily to Taylor columns. Clark Chapman, working in my former laboratory at the Massachusetts Institute of Technology, recently made a search for such columns, using a large amount of observational data supplied mainly by Bradford A. Smith and his colleagues at New Mexico State University. He found none. However, J. H. Focas of the Meudon Observatory in France has reported seeing "false red spots" in various locations not far from the red spot itself. Conceivably these are eddies in the downstream wake produced by the Taylor column underlying the red spot.

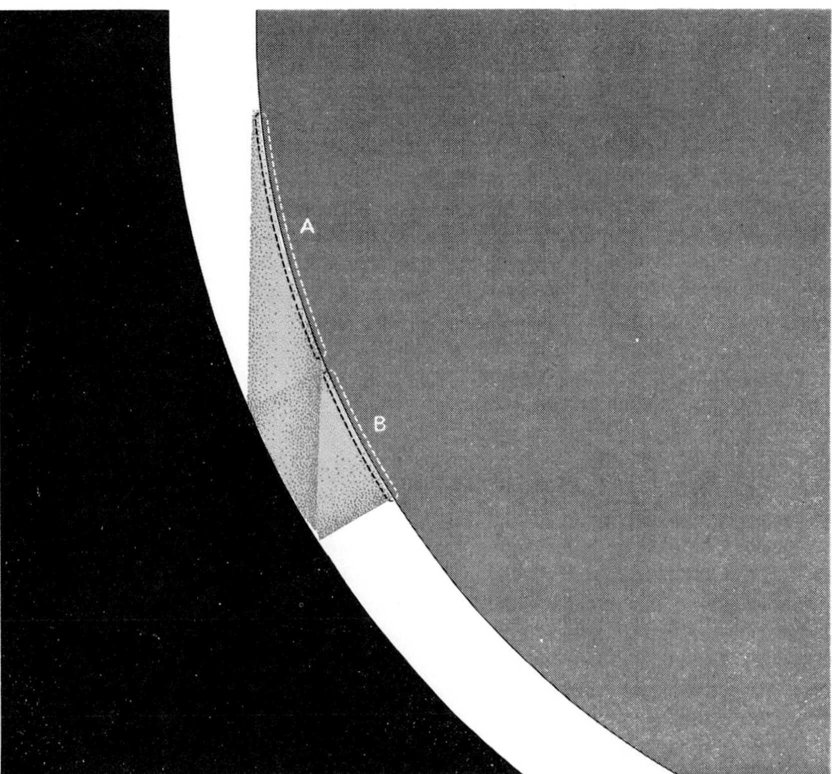

TAYLOR COLUMN ON JUPITER, which the author believes accounts for the great red spot, might assume various orientations with respect to either a plateau or a depression on the surface of the planet. In the simplest theoretical case (*A*) the axis of the column would be parallel to the planet's axis of rotation. But depending on the characteristics of the planet's atmosphere and other variables, the Taylor column might rise more or less vertically (*B*) above the surface feature. The depth of the atmosphere is unknown, hence not to scale.

The other properties of the great red spot are in principle readily accounted for by the Taylor-column hypothesis. For example, the variable rotation rate of the red spot must reflect the variable rotation rate of the main mass of Jupiter. This may seem difficult to accept, involving as it does huge changes in the planet's angular momentum, but a variable rotation rate is by no means inconceivable. Since Jupiter does not rotate as a solid body, it must be constantly agitated by some energy source or other. Various lines of evidence suggest that Jupiter may still be contracting. Although the contraction would be much too slow for verification by direct observation, it could still convert gravitational energy into internal heat energy at a significant rate (a rate comparable, in fact, with the amount of solar energy received by the planet).

If the changes in the rotational energy of Jupiter were brought about entirely by frictional processes, the associated dissipation of heat would be quite excessive. The mechanical coupling between different parts of the planet need not, however, be due only to friction. The electrical conductivity of most of Jupiter's interior and lower atmosphere should be high enough for the internal magnetic field of the planet to contribute significantly to mechanical coupling. In contrast to frictional coupling, which transforms rotational energy irreversibly into heat, magnetic coupling transforms rotational energy reversibly into magnetic energy. The magnetic field involved will be mainly of the toroidal type, with a strength probably exceeding 1,000 gauss.

Observations of the intense bursts of radio emission from Jupiter provide information about the part of the magnetic field whose lines of force pass through the surface of the planet and out into surrounding space. This part of the magnetic field (the "poloidal" part) is probably much weaker than the internal toroidal field. The significance of the external field for this discussion is that the pattern of radio emission associated with it has its own period of rotation, which differs slightly from the rotation period of the great red spot. This difference, which amounts to several seconds, is expected on the Taylor-column hypothesis because the red spot motion is the motion of the "solid" planet, whereas the radio emission is related to the motion of the fluid parts of the planet where the magnetic field originates.

It will be interesting and most important, therefore, to compare future observations of the radio period with those of the red spot period, with the expectation of learning more about the dynamics and magnetohydrodynamics of Jupiter's interior. As one investigator has remarked: "This is the first instance in astronomy [where] the distribution of angular momentum within a rotating body... manifests itself in observational effects measurable within a short time scale." Meanwhile observers will continue to find the red spot a fascinating object. Their studies can hardly fail to turn up evidence that will help to support either the raft hypothesis or the Taylor-column hypothesis. And, of course, there is always the chance that someone will have a still better idea.

7

The Solar System beyond Neptune

OWEN GINGERICH

April 1959

Early in 1979 the planet Pluto, having crossed inside the orbit of Neptune, will be proceeding on its orbit toward a perihelion that will bring it 16 million miles nearer to the sun than Neptune ever comes. Thus for two decades Neptune will resume the position it held, before the discovery of Pluto, as the outermost planet. Many astronomers think that Neptune should be reinstated now, in advance of the event; in their relatively brief acquaintance with Pluto they have begun to doubt that this object is a planet at all. Pluto's eccentric orbit is tilted at a considerable angle to the plane of the ecliptic, in which the orbits of the other planets lie. Even on its closest approach to our region of the solar system, it will shine no brighter than Triton, one of Neptune's two satellites, suggesting that it is no larger. There is suspicion that Pluto is an illegitimate offspring of Neptune, a satellite that escaped, as two man-made satellites recently did, to ply its own orbit around the sun.

The discovery of Pluto in 1930 had seemed a triumph of classical celestial mechanics. Computation from perturbations in the orbits of Neptune and Uranus had shown where it might appear in the sky. Astronomers had looked there and found it, in apparent fulfillment of the Newtonian vision of the solar system as a perpetual clockwork whose totality might ultimately be inferred from the motions of its nearer parts. What we know about Pluto now, however, leads some astronomers to declare that its discovery was an accident. Certainly its dubious status as a planet must dash the expectation that we might go on discovering new planets at ever greater distances from the sun. But as it brings one era of astronomy to a close, Pluto opens a new period of exploration at the outer edge of our solar system. It tells us to look not for planets but for other objects, including errant bodies like itself. Trans-Neptunian space holds increasing interest for what it may disclose about the origin and evolution of our no longer unchanging solar system.

It has taken a little more than a century to complete this revolution in astronomy. The civilized world was electrified in September, 1846, by the announcement from Berlin of the discovery of a new planet circling the sun in the remote space beyond Saturn and Uranus. Its existence had been predicted by the young French celestial mechanician Urbain Jean Joseph Leverrier from his study of small irregularities in the orbit of Uranus. His computations, communicated to observers in Berlin, pinpointed the new planet's location within a degree of where it was found in the constellation Aquarius. Meanwhile a parallel investigation in England had led John Couch Adams to essentially the same, but unpublished, prediction. The coincidence served to heighten the celebration of the discovery of the new planet, which was named Neptune, and encouraged astronomers to hope for a repeat performance. Leverrier, addressing the French Academy of Science, declared: "This success allows us to hope that, after 30 or 40 years of observation of the new planet, we should be able to use it in its turn for discovering the planet next in order of distance from the sun. Continuing this process, we should eventually arrive at planets which, unfortunately, because of their immense distance from the sun, would be invisible, but whose orbits may be worked out in ensuing centuries and traced with great accuracy. . . ."

By the end of the 19th century celestial mechanicians had gathered enough data to make a fresh attack on the motions of the outermost planets. W. H. Pickering of the Harvard College Observatory, working from historical and contemporary records of the orbits of Neptune and Uranus, predicted in 1919 the location of a trans-Neptunian planet near the foot of the constellation Gemini. A search of photographs made at the Mount Wilson Observatory, however, failed to show the hypothetical planet. Meanwhile, Percival Lowell of the Lowell Observatory in Flagstaff, Ariz., was launched on a parallel effort. Though a trans-Neptunian planet would necessarily disturb Neptune more than Uranus, the planet next closer to the sun, Neptune had not yet been observed for a complete revolution around the sun. Lowell accordingly based his computations on Uranus, which had been observed for two full revolutions. The disturbances in the orbit of Uranus did not, however, lead to a strong solution for the place of the unknown body. Lowell found two possible regions on opposite sides of the sky. He believed that one of these areas—an area near Pickering's predicted position—was the more likely.

Recognizing that the hypothetical planet would be faint and thus hard to find among the myriad of zodiacal stars, Lowell undertook a systematic photographic search of the indicated region of the sky. But the cameras available during his lifetime proved inadequate. Not until 1929 did the Lowell Observatory secure the necessary equipment. A young assistant, Clyde Tombaugh, then took on the task of making pairs of matched plates covering the region of the sky staked out by Lowell. The plates in each pair were taken several days apart and recorded the same 170 square degrees of sky. If the hypothetical planet were in the area covered by a pair of plates, the movement of the earth on its orbit would cause the image of the planet to appear in a different position in the second plate. Tombaugh then compared the matched plates in a "blink comparator," an instrument in which the two plates are precisely aligned and presented alternately to the observer's eye by means of a hinged mirror. Star by star Tombaugh searched the plates, looking for a pinpoint of light that would appear to jump to a new position on the second plate when he flopped the mirror.

After a year he located a pair of 15th-magnitude images that showed the motion required of a trans-Neptunian object. On March 13, 1930, the anniversary of Lowell's birth, the discovery of the ninth planet was announced to the world. Pluto revolves about the sun at a mean distance of 3,670,000,000 miles from the sun, 40 times farther out than the earth and nearly 900 million miles beyond the mean orbit of Neptune.

A reinvestigation of the Mount Wilson plates which had been made in 1919 revealed that Pluto had actually registered its existence then. It had been overlooked partly because it was unexpectedly faint. It was nearly 1,000 times fainter than Neptune, which itself would have to be about five times brighter to be just visible to the naked eye. Astronomers were perplexed. If Pluto were faint because it was small in diameter, then it would probably not be very massive. But if it had small mass, Pluto could not have perturbed Uranus or Neptune enough for its place to be found theoretically. On the other hand, if Pluto were small but massive, its density would be unreasonably high. And if Pluto were large in diameter yet faint, then its reflectivity would be unexpectedly low.

The first task was to determine Pluto's mass. In the case of Uranus and Neptune the motions of their satellites accurately reveal the masses of the mother planets. But Pluto had no apparent satellites; its mass could be determined only indirectly by its perturbations of Uranus or Neptune. In the nearly two centuries since Uranus was found and the century since the discovery of Neptune, Pluto has caused such tiny displacements in their positions that the mass of Pluto cannot be satisfactorily established from them alone. Lowell and Pickering had depended upon observations of Uranus and Neptune that had been recorded before the two objects were recognized as planets.

The French astronomer Joseph Jérôme Lefrançais de Lalande had recorded a pair of fixes on Neptune in 1795. He thought it was a star, because of its pointlike appearance, and overlooked its snail-like pace across the sky. His observations, compared with modern positions, would indicate that Pluto had produced a displacement of at least six seconds of an arc in Neptune's celestial longitude. To cause such an effect the mass of Pluto would have to be nearly equal to that of the earth. Allowance for reasonable error in Lalande's observations would make Pluto only half as massive as the earth. By comparison Uranus has a mass of 14.5 earth masses, and Neptune a mass of 17.2.

It was the small indicated mass of the new planet that prompted several astronomers to pronounce its discovery a lucky accident. The perturbations caused by such a small body would be concealed by the random errors of observation. How, then, did the two independent calculations of Lowell and Pickering converge on the same general area of the sky?

Vladimir Kourganoff, a French astrophysicist, provided one explanation after a careful re-examination of Lowell's and Pickering's work. He found that both had relied on the 18 observations that had been made of Uranus before its discovery in 1781. These prediscovery measurements show an unfortunate amount of statistical scatter. But they had indicated a seemingly extraordinary displacement of Uranus around 1710 as

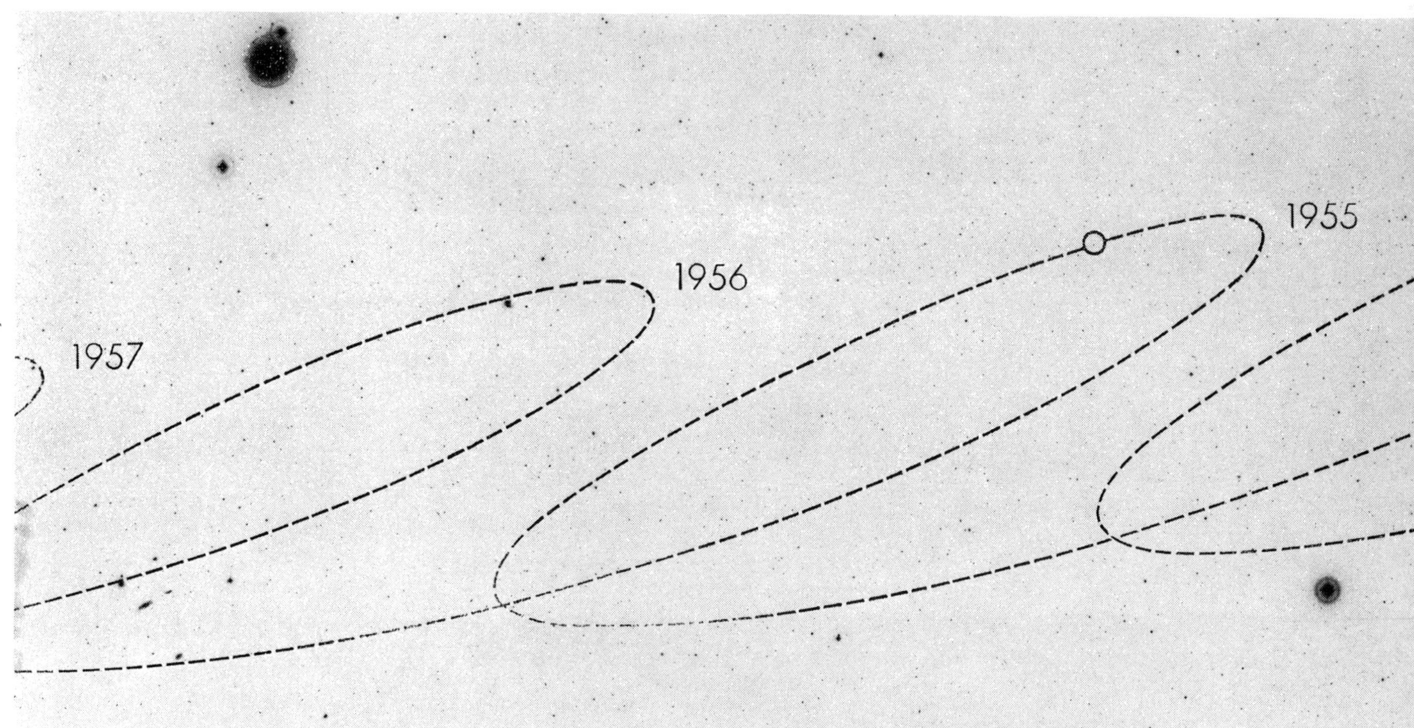

PLUTO appears as the tiny black speck within the solid circle on this photograph made in 1955 with the 48-Schmidt telescope on Palomar Mountain. The earth's own rotation about the sun makes Pluto appear to spiral slowly through the heavens (*broken line*). The picture is reproduced from a negative print. Such a print increases the contrast and visibility of objects on the plate.

the result of a particular juxtaposition with Pluto, and had apparently defined the time well enough to permit the predictions by Lowell and Pickering. The 18th-century observations do not, however, yield any reliable determination of Pluto's mass.

Unfortunately the calculation of Pluto's mass from the observed positions of the other two outer planets cannot be more exact than the observations themselves. Recently W. J. Eckert of the Watson Computing Laboratory, Dirk Brouwer of Yale University and G. M. Clemence of the U. S. Naval Observatory programmed an electronic computer to calculate the positions of the outer planets throughout the four centuries from 1653 to 2060. For this definitive study they assigned to Pluto a mass that had been derived in an earlier investigation by L. R. Wylie at the Naval Observatory. They found that this mass, slightly less than that of the earth, made possible a satisfactory agreement between the computed and observed positions of both Neptune and Uranus.

Astronomers would probably accept this value if it were not for a set of quite different observations made in March, 1950, by Gerard P. Kuiper of the Yerkes Observatory. At Palomar Mountain, with the help of Milton Humason, Kuiper used the 200-inch reflector (in one of the rare visual observations made with that instrument) to compare the size of Pluto's image with a standard set of disks. Kuiper and Humason agreed that the image had a diameter of only .23 second, corresponding to 3,600 miles, or less than half the diameter of the earth.

Now if Pluto's mass were equal to that of the earth, this determination of its

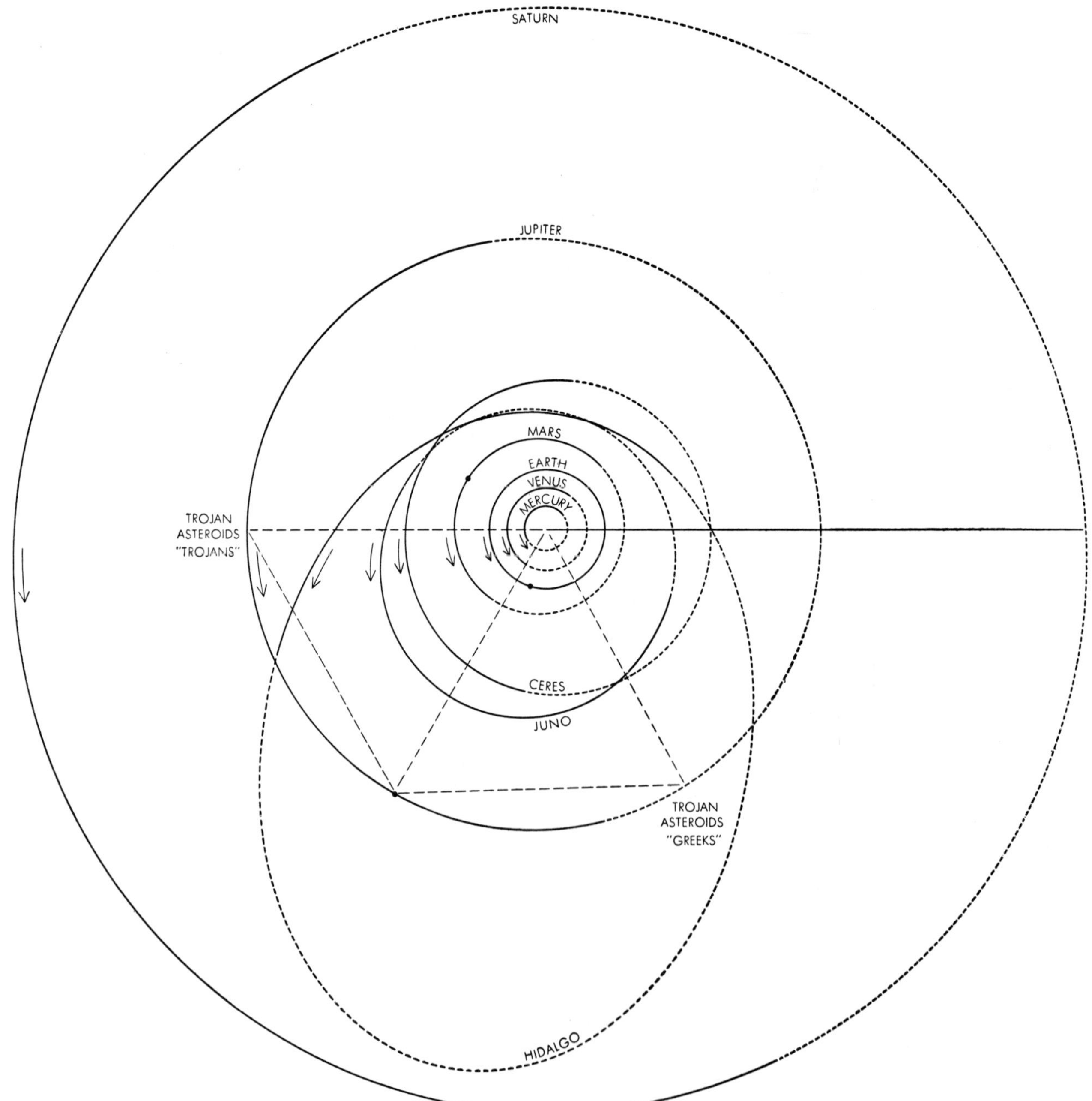

INNER SOLAR SYSTEM comprises the orderly orbits of six planets and a population of many hundreds of asteroids. Typical asteroids, such as Ceres and Juno, lie between Mars and Jupiter. The peculiar orbits of Hidalgo and the two Trojan groups suggest that these bodies are former satellites of Jupiter. The two groups of Trojan asteroids are locked by Jupiter's gravitational field in relative positions indicated by the adjacent equilateral triangles. The broken portions of orbits lie below the earth's orbital plane.

diameter would give it a density 10 times that of the earth! Such an enormous density, five times greater than that of lead, is highly unlikely in a member of our solar system. Could something be wrong with the measurement of Pluto's diameter? Dinsmore Alter of the Griffith Observatory in Los Angeles pointed out that if Pluto reflected light like a polished ball bearing, Kuiper would have observed not the disk of Pluto but the reflected image of the sun. But there is no reason to believe that Pluto's surface is abnormally smooth. An icy surface would be quickly eroded by evaporation, temperature changes and bombardment from cometary material.

Since the Kuiper measurement cannot be controverted, astronomers feel compelled to reject the rather weak determination of Pluto's mass from celestial mechanics. Granting the planet an acceptable density, the 3,600-mile diameter yields a calculated mass about .03 that of the earth. Such a tiny mass could not have produced the apparent perturbation of Uranus in 1710 on which Pickering and Lowell had relied so strongly. Whether Pluto was discovered by chance or prediction remains a major mystery. Brouwer writes off the prediction as "a fantastic coincidence."

Pluto's status as a planet is further weakened by the eccentricity of its orbit. When Pluto swings inside Neptune's orbit to perihelion, it comes more than 1.8 billion miles closer to the sun than at aphelion. Moreover, the inclination of the orbit to the plane of the ecliptic exceeds 17 degrees. All of these considerations led Issei Yamamoto of the Kyoto Observatory to speculate in 1933 that Pluto was not a true planet but an escaped Neptunian satellite. Shortly afterward calculations by R. A. Lyttle-

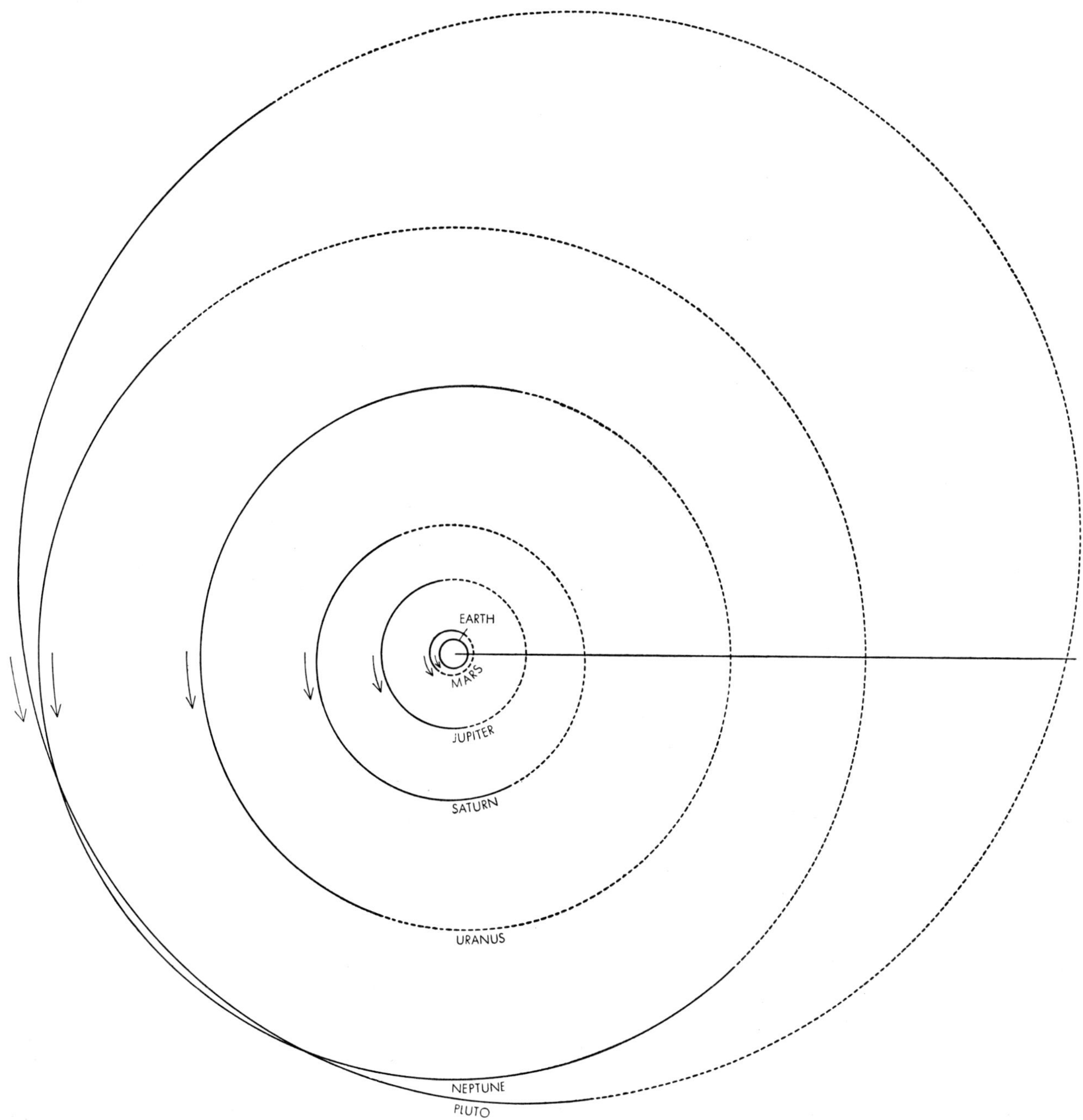

OUTER SOLAR SYSTEM comprises the three planets discovered in modern times: Uranus, Neptune and Pluto. Pluto's abnormal orbit crosses Neptune's at lower left. As in the diagram on page 80, the broken portions of orbits lie below the earth's orbital plane. However, only Pluto's orbit is significantly inclined to the orbital plane of the earth; its orbital "tilt" is about 18 degrees.

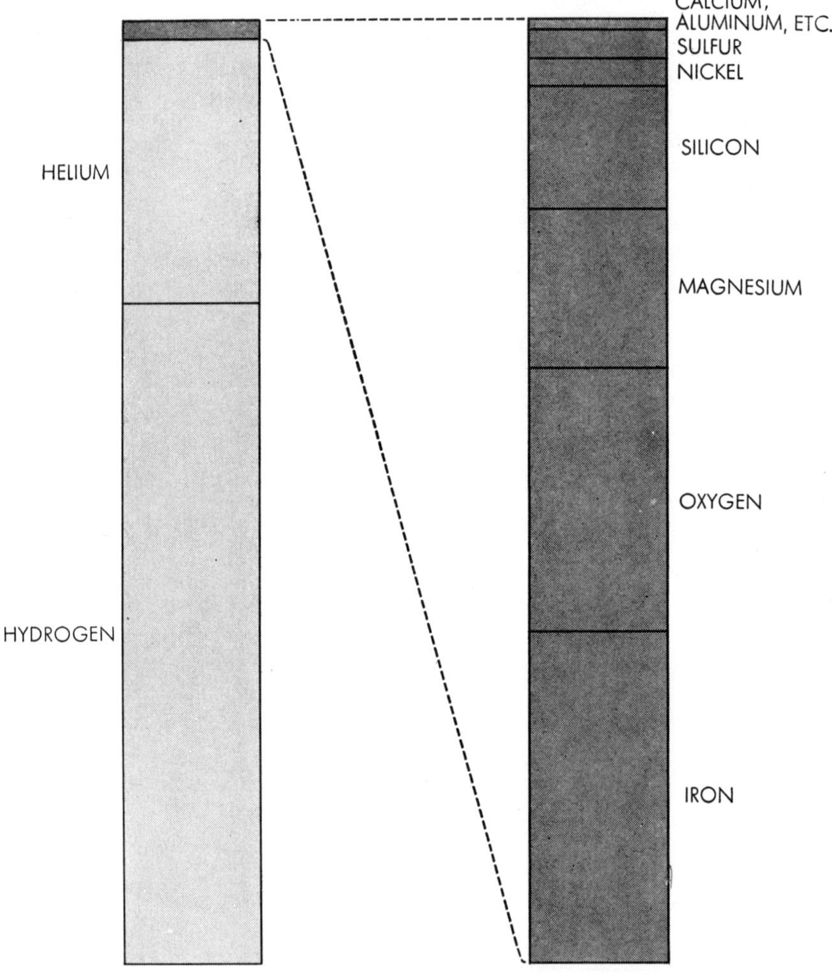

COMPOSITION OF PROTO-EARTH probably resembled the present composition of the sun (*left*): 98 per cent hydrogen and helium with heavier elements (*dark shading*) making up the remainder. Present composition of earth (*right*) includes 35 per cent iron, 28 per cent oxygen. The lighter elements have evaporated almost completely except for oxygen, most of which is bound into heavy mineral oxides. The larger planets such as Jupiter, being more massive than the earth, have retained a higher proportion of light elements.

ton of the University of Cambridge showed how a close encounter between Pluto and the Neptunian satellite Triton might have ejected Pluto from the gravitational system of the planet.

Kuiper agrees that Pluto began its existence as a satellite of Neptune but has proposed that it escaped to a planetary orbit under entirely different circumstances. In this account of its genesis, Pluto figures as a major item of evidence for an explanation of the origin of the solar system which is gaining increasing acceptance in modern astronomy. According to this view the planets and the sun began by condensing from an extended mass of dust and gas. Numerous investigators have contributed to the current revival of this nebular hypothesis, now also called the "dust-cloud" theory; Kuiper has given his protoplanet theory the fullest development.

The birth of the solar system begins with the contraction, under its own gravitation, of a cold cloud of dust and gas. Rotation of the cloud sets up centrifugal force that works against the gravitational collapse and flattens the cloud into a wheeling disk. As the cloud contracts and grows more dense, local instabilities break it up into individual self-gravitating units, with a protostar at the center and a series of protoplanets out toward the periphery.

The protoplanets formed in such a cloud do not necessarily condense into planets. Tidal attractions from the protostar may disrupt these individual clouds so they cannot contract into spheres. The ring system of Saturn gives us a scale model of what happens in this circumstance. The disruptive action of Saturn apparently prevents the material in the rings from forming an inner satellite. Saturn's rings thus represent the vestiges of a discoidal protoplanet.

Condensation can occur, however, in those protoplanets whose density exceeds a certain critical limit. Here the dust-cloud theory requires a certain elaboration in order to reconcile it with observed facts. If the present mass of the planets and their satellites in our solar system were smeared out into a dusty disk around the sun, the density of the cloud would not attain the critical condensation limit. Modern theories ingeniously avoid this difficulty by postulating that the mass of the protoplanets must have greatly exceeded the present planetary mass. As condensation proceeded, radiation from the protosun, already glowing from the release of gravitational energy, caused the surplus material (approximately 99 per cent of the original mass) to evaporate from the protoplanets and swept it out into space. This argument is supported by the observation that such dissipation of material from the atmospheres of the planets is still going on, though at a greatly reduced rate. The lightest elements, hydrogen and helium, which must have constituted the major bulk of the solar nebula and therefore of the protoplanets, have almost entirely escaped from the earth, leaving behind the tiny residue of heavier elements which now constitute our planet. The higher escape velocities of the more massive Jupiter, Saturn, Uranus and Neptune have allowed them to retain a larger proportion of light elements, giving them a lower average present density.

To assist us in visualizing the further evolution of the solar system we have another scale model at hand in Jupiter and its dozen satellites. Proto-Jupiter, the largest and most massive cloud to have broken from the solar nebula, must itself have broken down into a central body and a retinue of protosatellites. In the evaporation stage proto-Jupiter would have continually lost mass. Its outer satellites, bound more and more tenuously to the system, must eventually have slipped away. Only the five innermost satellites remained. They still resemble a miniature solar system, these "planets" being spaced at regular intervals in a common plane of revolution. In contrast, the seven tiny outer satellites of Jupiter present a scene of great irregularity. Their highly inclined orbits interloop one another, and the outer four revolve in a direction opposite that of the rest. We may deduce that these satellites once took up independent orbits around the sun. Upon later encounter with the extended gaseous atmosphere surrounding proto-Jupiter, they were captured in irregular orbits bearing little resemblance to their original paths. Other proto-Jovian satellites may have been lost completely, to become asteroids orbiting around the sun with unusual inclinations and eccentricities.

The parallel process of evolution in the case of proto-Neptune fully accounts for Pluto's remarkable orbit. Both of Neptune's present satellites represent recaptures; Triton's orbit is retrograde and Nereid's is highly eccentric. Any remaining original protosatellites would have been removed by massive proto-Triton as it spiraled in through proto-Neptune's gaseous envelope on its return journey. The inclination of Pluto's orbit suggests that it too escaped from orbit around Neptune but did not return. It is otherwise difficult to explain the interlooping of its ellipse with Neptune's orbit.

Convincing evidence for this reconstruction of Pluto's origin came in 1956, when Merle Walker and Robert Hardie at the Lowell Observatory measured variations in the light reflected by the planet. The small fluctuations in brightness repeated themselves regularly, as if dark markings were passing across the planet's disk. This cycle indicated that Pluto rotates in 6.39 days, much more slowly than any of the massive outer planets, which rotate in 10 to 15 hours. Kuiper suggests that Pluto originally revolved around Neptune once in 6.39 days, rotating on its own axis in the same period. Now, on its lonely orbit around the sun, Pluto rotates just as slowly as when it was attached to Neptune.

Has Leverrier's dream of extending the solar system step by step by means of celestial mechanics been shattered? The answer is yes. Pluto lies just at the threshold of gravitational detection. From now on we must look to accidental photographic discovery rather than celestial mechanics if we hope to add any remote members to our solar system.

Following the discovery of Pluto the Lowell Observatory continued a systematic search of a wide belt, covering three quarters of the sky, on either side of the ecliptic. Altogether 90 million star images passed under Tombaugh's examination. The survey turned up about 3,000 asteroids but found no trans-Plutonian object. The limiting magnitude of this investigation is represented by the edge of the darker colored area in the brightness-distance diagram reproduced on this page. Any planet as bright as Neptune would have been discovered out to 180 astronomical units (an astronomical unit is equal to the distance from the earth to the sun). Pluto itself lies close to the threshold; a planet the size of Pluto would have escaped the net beyond 50 astronomical units. Trans-Neptunian objects intrinsically fainter than Uranus's moon Titania, and therefore presumably smaller than 600 miles in diameter, would not have been caught.

The 48-inch Schmidt telescope on Palomar Mountain casts a considerably finer net. It could show Pluto at a distance of 150 astronomical units or a planet 200 miles in diameter at Pluto's

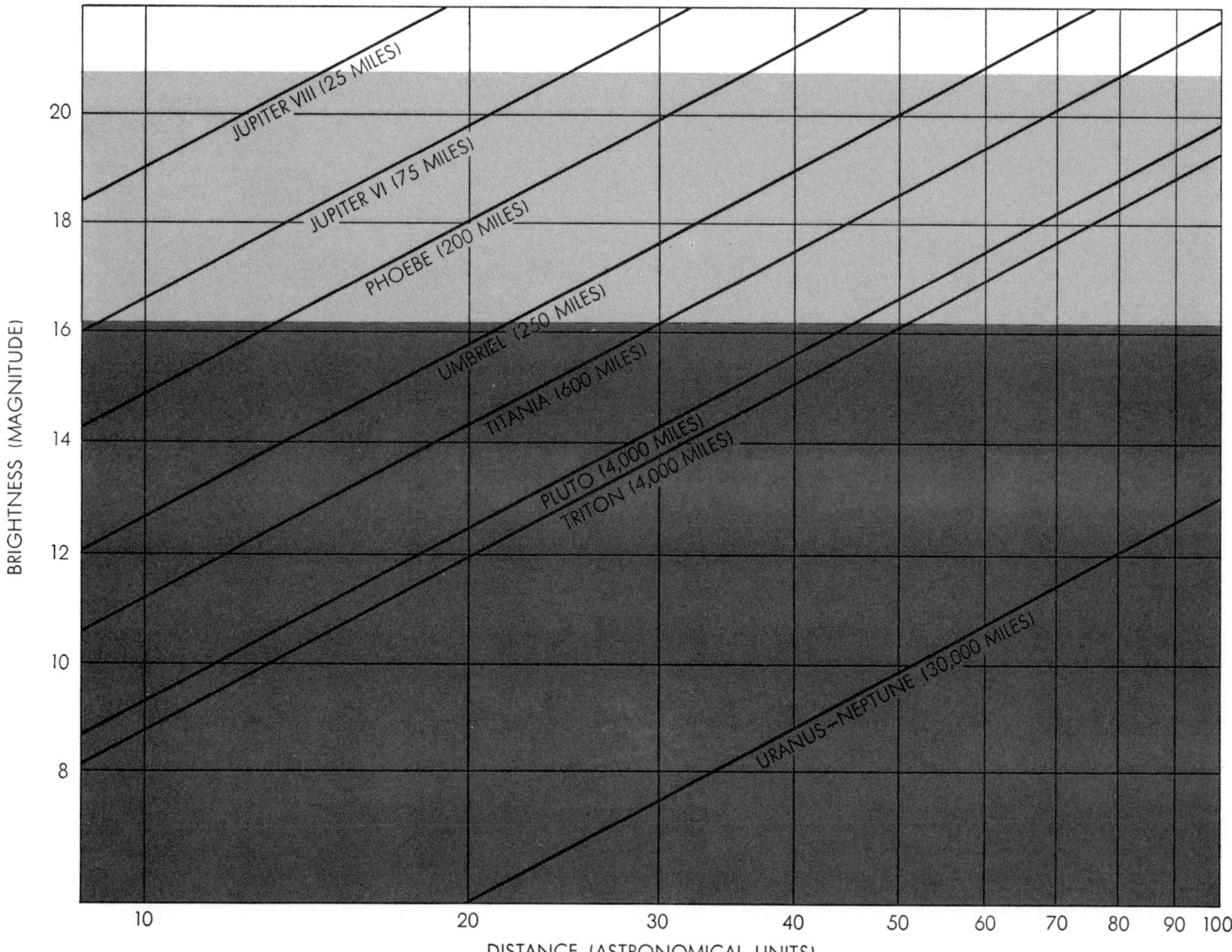

LIMITS OF OBSERVATION for objects of different sizes are shown on this brightness-distance diagram. Numbers in parentheses give approximate diameters of nine known objects; one astronomical unit equals about 93 million miles. The Lowell Observatory sky survey (*dark color*) extends to slightly beyond the 16th magnitude (increasing magnitude means decreasing brightness). The Palomar survey with the 48-inch Schmidt telescope (*light color*) reaches the 20th magnitude. Thus a planet the size of Pluto would be invisible beyond 100 astronomical units even with the Schmidt. Other considerations make it unlikely that such a planet exists.

planet, according to the protoplanet theory. Examination of the mass distribution of the outer planets supports the idea that beyond Neptune the density of the primordial dust cloud must have been too low for a protoplanet to form. But at such vast distances the gravitational force of the protosun would also have been much attenuated, reducing in turn the critical density of condensation. Smaller, less massive bodies might therefore have formed from the cloud in this region. Comets appear to have just the right size and composition, being made up of light elements and their compounds, frozen with a sprinkling of heavier stony or metallic particles. When occasional comets come out of their cosmic deep-freeze into the central regions of the solar system, they evaporate, forming gaseous atmospheres that are dissipated into space via their long and often spectacular tails. The comets appear to come from a tremendously distant swarm, suggesting that Pluto might have scattered them into their remote orbits. Other cometary bodies may still circle the sun on less eccentric orbits just beyond Neptune. But all known comets are much too small to be visible at these distances.

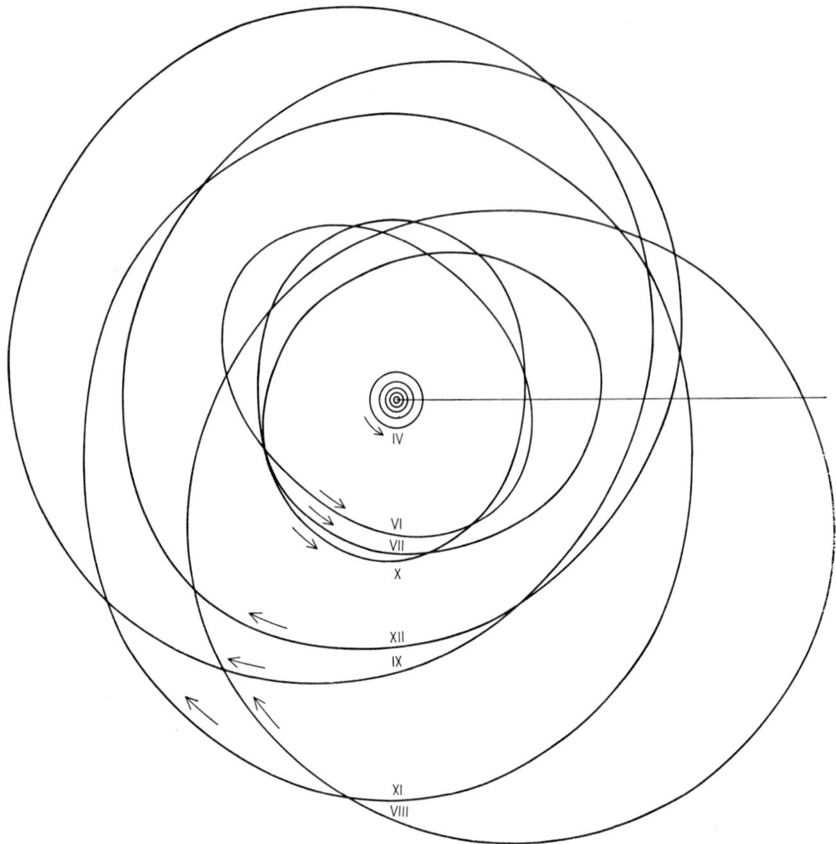

SATELLITES OF JUPITER fall into three groups. The inner satellites, V, I, II, III and IV, all revolve in the same plane and from west to east; they have presumably always been satellites. The next three, VI, VII and X, revolve in the same direction but in three different planes; VIII, IX, XI and XII also revolve in different planes, but from east to west. These two groups apparently represent "escaped" satellites that were recaptured at two different times. These Jovian satellites are numbered in the order of their discovery.

present distance. But a search down to magnitude 20.5 with the Palomar Schmidt would be a fantastic undertaking; the number of faint star images on some individual plates exceeds two million. The colored area for the Palomar Schmidt on the brightness-distance diagram represents the practical limit for the accidental discovery of a trans-Plutonian object.

What members of the solar system might be found in the unsearched range, fainter than magnitude 16 but brighter than magnitude 20? Probably no major

The escape of Pluto from Neptune, however, raises the possibility that we may yet find objects larger and more interesting than comets out on the fringes of the solar system. Three or four additional Neptunian satellites may have escaped to travel even more highly elliptical orbits around the sun. The accidental discovery of one or more such objects is no less improbable than the discovery of Pluto itself. They would add considerable substance to speculation about the origin of our solar system.

III

The Sun

III

The Sun

INTRODUCTION

To repeat a well-worn cliché, the sun is our nearest star. At a distance of 150,000,000 kilometers, its visible surface subtends an angle of about 31 minutes of arc. In the far ultraviolet the angle is slightly larger, and in some radio wavelengths the sun displays a diameter almost twice as great. If we include the tenuous outer layers as part of the sun, then the earth itself is bathed in the solar atmosphere, with the terrestrial magnetic field creating a shock front in the ionized material flowing out from the sun.

Astronomers have not been able to examine the surface disk of any other star directly, and for eclipsing stars, where indirect measurements are possible, relatively little information has been forthcoming. On the sun, however, astronomers have discovered features that would have been overlooked if the sun were at the distance of Alpha Centauri, the next closest stellar system—features that nonetheless, in exaggerated or more violent forms, can be detected in more distant stars. Among these are:

—the sun's magnetic activity, waxing and waning in an 11-year cycle, but repeating the polarity of the magnetic fields in a 22-year cycle. Several score magnetic variable stars are known, with fields ranging up to 5000 gauss. These magnetic fields are necessarily stronger than the sun's, for the general solar field of 1–2 gauss would be undetectable at stellar distances.

—solar flares, mighty short-lived outbursts of radiant energy that sometimes spray a stream of ionized particles into space. As yet still partially unpredictable, flares are intimately related to rapid magnetic changes in active regions on the solar disk. A bright flare of the size observed on the sun, although tiny compared to the sun's continuous energy output, could easily double the luminosity of an intrinsically faint M-dwarf star, and several such stars do indeed show brilliant, erratic flares of a few minutes duration.

—the chromosphere, a hot atmospheric shell surrounding the cooler photospheric layers. A testing-ground for radiation theories, the chromosphere thus far defies explanation of the detailed causes for its high temperature, although it seems related to violent convective motions of the underlying gases. The chromosphere reveals itself through the presence of emission lines in the solar spectrum (especially during eclipses, as described by Grant Athay in the article on "The Solar Chromosphere") and also by the high intensity of the continuum spectrum at extremely long and extremely short wavelengths. The existence of other stellar chromospheres can be inferred from emission lines in spectra from certain distant stars, but above-the-atmosphere instrumentation is not yet sensitive enough to find evidence for chromospheres in the infrared or ultraviolet spectra.

—the solar wind, most spectacularly revealed in the creation of comet tails and in the aurora borealis, but also disclosed by the more subtle effects also described by Eugene Parker in the article "The Solar Wind." In some red supergiant stars, stellar winds of monumental proportions continually blow atoms into space—diminishing the masses of these stars, and replenishing interstellar nebulae with processed stellar material.

Beginning with the work of Eddington and Jeans in the 1920's, astronomers have formed an increasingly detailed picture of the interior of the sun, entirely without the expectation of ever undertaking any direct observation of radiation from this inaccessible zone. One of the exciting developments of solar astronomy in the 1960's was the possibility that the solar interior might after all be observable, through corpuscular radiation of neutrinos. Though the results are still negative, John Bahcall's article "Neutrinos from the Sun" gives the interesting story of a search that may significantly affect our ideas of the construction of the sun as well as the evolution of stars.

8

The Solar Chromosphere

R. GRANT ATHAY

February 1962

On October 2, 1959, I stood on a barren hilltop at the southern tip of Fuerteventura, one of the Canary Islands off the west coast of Africa, viewing the image of the eclipsing sun in a small telescope. My task was to cry "Now!" at the instant when the moon covered the last visible portion of the brilliant disk of the sun's photosphere. That instant would bring into view the rosy-red annulus of the chromosphere and the pearly-white corona extending into space beyond it. The signal from me would set in motion a battery of six cameras and expose more than 1,000 square feet of film in the next 2½ minutes in order to record the spectra of these two elusive features of the sun. My colleagues and I, from the High Altitude Observatory of the University of Colorado and the Sacramento Peak Observatory in Sunspot, N.M., were particularly concerned to capture good images of the chromosphere. Two years of costly engineering and shop time spent in perfecting the apparatus, months devoted to preparations for transporting the delicate instruments, electric-power sources, machine tools, a 10-man crew and the necessities of life 6,000 miles from Boulder, Colo., to the remote island, and over a month of instrument-checking, rehearsing and waiting—were to yield their harvest in those 150 seconds.

Seconds before totality, however, the sun disappeared behind high, broken clouds, and I had to give my command at the computed time of totality rather than by observation. The cameras whirred, and the clouds did thin for a precious few seconds, giving us a faint view of the corona and some bright prominences. Five minutes after the eclipse the clouds moved away from the sun. We had not observed the chromosphere at any time.

Our failure was made the more bitter by the still vivid memory of a similar fiasco the year before at Danger Island in the South Pacific and by the knowledge that seven other expeditions in the Canaries had failed along with us. Yet when this issue of SCIENTIFIC AMERICAN comes from the press, we shall be off again, this time to Lae in New Guinea. In July, 1963, we shall send another eclipse expedition into northern Canada, beyond the Arctic Circle. Whether it is cloudy or fair at these eclipses, we shall make still more journeys to other remote sites (eclipses are seldom so accommodating as to pass over established observatories). A sane person may reasonably wonder why we continue such costly and difficult expeditions in the face of a high probability of failure.

The answer is this: The chromosphere, which can be viewed in the required detail only during a total eclipse, holds clues to many physical processes in the sun itself, in interplanetary space and in the upper atmosphere of the earth. Of greatest interest is the nature and the source of the nonradiative energy that wells up out of the sun and creates the chromosphere and the corona in the first place. This energy is absorbed primarily in the chromosphere and is there transformed into visible light and other forms of electromagnetic radiation. Without the nonradiative energy and its mysterious transformation in the chromo-

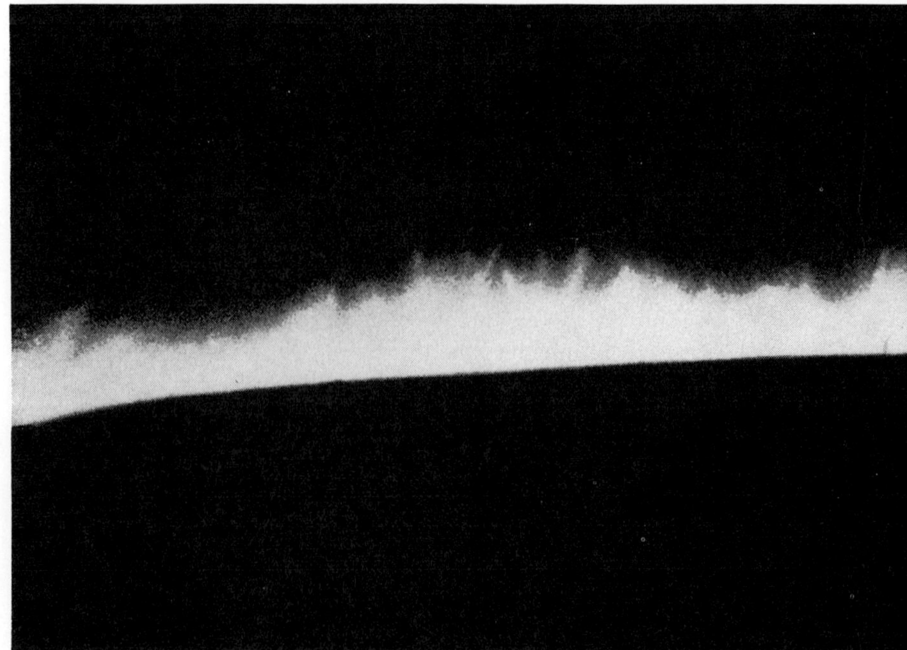

MIDDLE AND UPPER CHROMOSPHERE appear in this photograph, made by R. B. Dunn with the 16-inch coronagraph at the Sacramento Peak Observatory in Sunspot, N.M. The coronagraph is a telescope that artificially eclipses the bright disk of the sun. A highly

sphere the solar atmosphere would simply grow cooler and thinner in direct proportion to the distance from the sun. Instead, the temperature rises from 4,300 degrees Kelvin (degrees centigrade above absolute zero) at the surface of the photosphere to more than 6,000 degrees in the lower chromosphere. It climbs to approximately 50,000 degrees in the middle chromosphere and then takes a huge leap to nearly a million degrees in the corona.

The chromosphere, under excitation by the nonradiative energy, emits intense ultraviolet radiation. In the upper atmosphere of the earth this radiation sets up the electrically charged layer known as the ionosphere, which reflects radio waves and makes possible long-distance radio communication. The chromosphere is also the birthplace of the great solar flares. Energetic particles reaching the earth from these flares generate auroras and, on occasion, disrupt the ionosphere and the radio communications that depend on it. The geyser-like projections called spicules spring up continuously from the chromosphere, feeding into the corona the protons that eventually escape into interplanetary space to make up the "solar wind." Finally, certain atoms in the chromosphere absorb specific wavelengths of radiation from the photosphere, producing the strongest of the many absorption (dark) lines that cross the otherwise continuous spectrum of the photosphere.

So little is known about the chromosphere that one incomplete set of observations of its spectrum made by our first expedition, to the Sudan in 1952, lays much of the basis for the present conception of its physics. In contrast to the spectrum of the photosphere, the spectrum of the chromosphere consists of bright, narrow emission lines separated by broad, dark regions. The light at each wavelength represents the radiative output of excited atoms or of ionized atoms (that is, atoms with one or more electrons stripped from them) of one or another element in the chromosphere, such as hydrogen, helium, oxygen, calcium, iron and titanium. The chromosphere gets its name from the red color of one of the stronger emission lines of hydrogen. At the top of the chromosphere the spectrum changes abruptly into the continuum of colors, punctuated by a few faint emission (bright) lines, that characterizes the corona.

During an eclipse the moon moves across the face of the sun at some 200 miles per second; this means that the most interesting portion of the chromosphere, its lower 600 miles, shows for only three or four seconds. The upper chromosphere, which can extend for as little as 1,000 miles to as much as 10,000 miles above the photosphere, is seen for less than a minute. In the span of a century the lower chromosphere is visible for a total of less than 10 minutes. An astronomer who observes 10 eclipses—far more than the average—may expect to observe this region of the chromosphere for no more than a minute in his lifetime, and that only if he is luckier than any other astronomer ever born.

Apart from total eclipses, it is possible to make a few, somewhat compromised observations of parts of the chromosphere. The entire disk of the sun may be photographed, for example, in the monochromatic light that comes from the center of one of the chromospheric absorption lines (since the light is less intense than that of the surrounding portions of the spectrum, the line still appears to be dark). Such pictures contain a wealth of detail, but there is as yet no way to extract very much useful information about the chromosphere from them. They are made today in order to detect solar flares and other events on the sun. The study of the chromosphere would be a by-product of this effort, but one that could well prove to be more useful in the long run than the detection of unusual solar activity.

The highly active upper part of the chromosphere shows up to advantage in the coronagraph, a telescope that produces artificial eclipses, primarily for studies of the corona. The lower chromosphere lies too close to the photosphere to be visible in the coronagraph, but the middle region, which extends from 600 miles to about 2,000 miles,

monochromatic filter excluded almost all light except the red color emitted by excited atoms of hydrogen at a wavelength of 6,563 angstrom units. Top of chromosphere consists of spicules, geyser-like columns each 500 miles wide, that shoot up at 20 miles per second from lower chromosphere. Spicules inject protons into the corona, from which they move outward as the "solar wind."

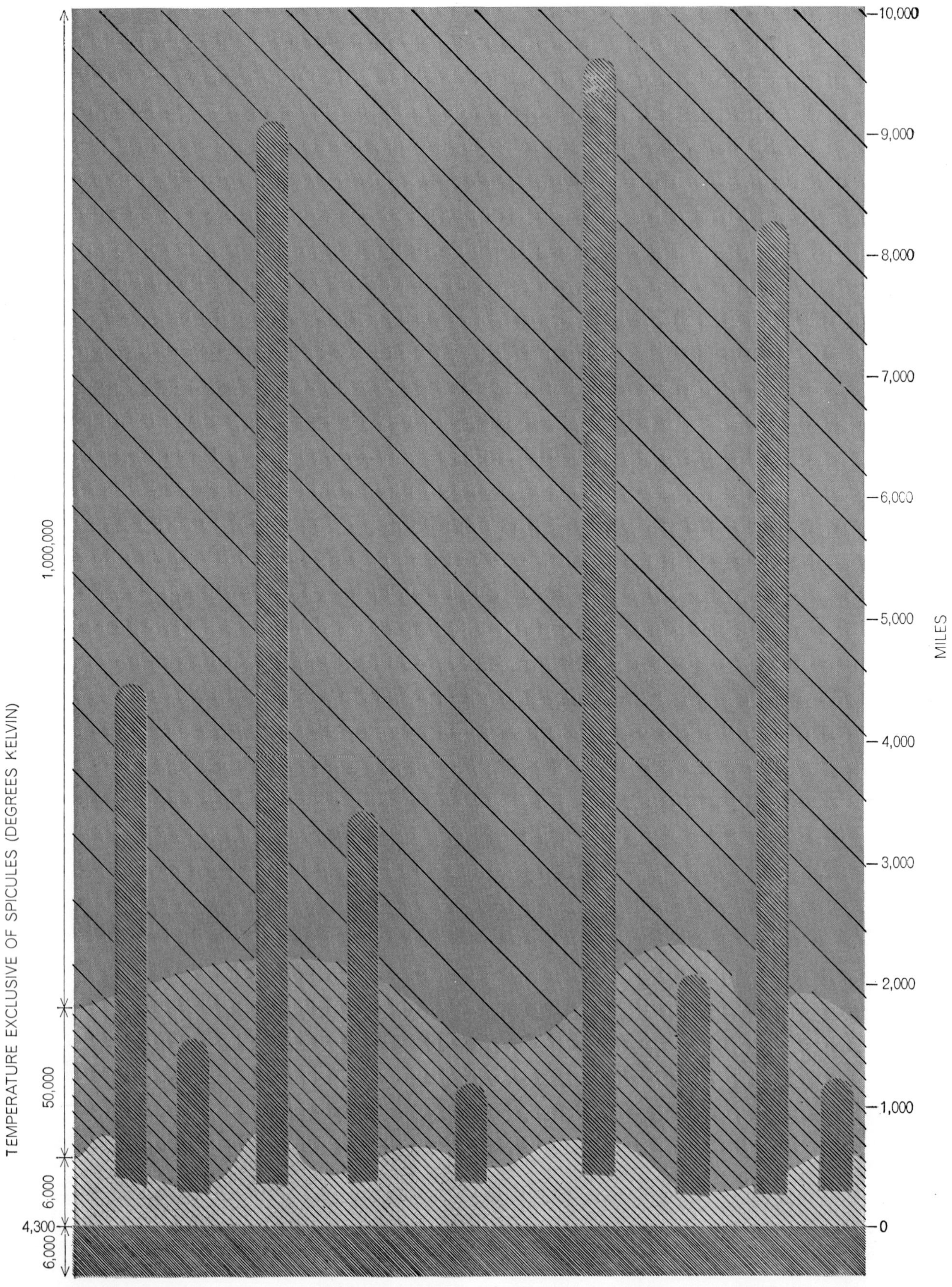

SOLAR CHROMOSPHERE, seen here in a highly schematic diagram, consists of three layers, decreasing in density (*hatching*) and increasing in temperature by sudden jumps (*color*). Upper photosphere is at bottom. Lower chromosphere averages 6,000 degrees Kelvin, middle chromosphere 50,000 degrees and corona, between spicules, nearly a million degrees. The spicules show evidence of two temperature ranges throughout their length, about 10,000 degrees and near 50,000 degrees, possibly because a hot envelope encases a cooler core. This complex temperature structure is not shown. Spicules make up third, or top, layer of the chromosphere.

can be seen. Above this height the chromosphere assumes an extremely rough appearance: it bristles with spicules that well up from the lower chromosphere and fall back or disappear into the corona at a height of about 10,000 miles. An early observer, seeing the spicules, likened the glowing chromosphere to "a burning prairie."

Radio astronomy provides a fourth way to observe the chromosphere: radio noise of wavelengths shorter than 10 centimeters originates in the region. As yet radio telescopes cannot resolve the disk of the sun in sufficient detail to provide the kind of data needed. Eclipses improve the radio definition but so far few radio observations have been made during eclipses.

For the present most of what is known or postulated about the chromosphere comes from analysis of its optical spectrum. As in all spectroscopy, the wavelength of an emission line provides reliable identification of the atoms that are the source of the radiant energy of that point in the spectrum. The absolute intensity of the spectral line can be taken, although with much less reliability, as an index to the number of atoms present; in other words, to the density of the chromosphere in the region observed. This deduction is heavily qualified, as will be seen, by other variables. With corresponding qualifications and uncertainty, both the width and the relative intensity of an emission line can be taken as indicators of the temperature.

The scantiness and uncertainty of the data can be judged by the wide swings in interpretation that have engaged students of the chromosphere in recurrent controversy over the past half-century. Early observers were impressed by what they found to be, from measurement of the intensity of the spectral lines, the unexpectedly high density of the chromosphere. The density of any atmosphere tends to decrease exponentially with altitude. In the upper photosphere, which like the rest of the sun is gaseous, the density falls off by a factor of 2.7 (the base for natural logarithms) for each 60-mile increase in altitude; the 60-mile unit is called the scale height. Measurement of the intensity of the hydrogen lines in the spectrum of the chromosphere gave the startling scale-height value of 600 miles; that is, the 2.7 decline in density in the chromosphere was found to occur over a depth of 600 miles, compared with 60 miles in the photosphere immediately below.

Astronomers had to postulate some "support mechanism" capable of holding up the chromosphere. In 1929 W. H. McCrea of the University of London suggested that the chromosphere was in violent, turbulent motion, which produced enough pressure to force the atmosphere outward. Observations made during a total eclipse in 1941, however, led R. O. Redman of the University of Cambridge to propose another mechanism. After measuring the width of many chromospheric spectral lines he concluded that the temperature in the lower chromosphere was 35,000 degrees Kelvin. Since the scale height of a quiescent atmosphere is directly proportional to its temperature, 35,000 degrees K. was clearly enough to produce a large scale height in the chromosphere without the help of turbulence or any other support mechanism. Now the question was: What produces the high temperature?

The most effective challenge to Redman's results came in 1954, when Redman himself and Z. Suemoto of the Tokyo Astronomical Observatory showed that the 1941 work had failed to take into account factors other than temperature that may broaden spectral lines. As a result they concluded that the temperature of the lower chromosphere is only 5,000 to 10,000 degrees K. The support problem was now to the fore again,

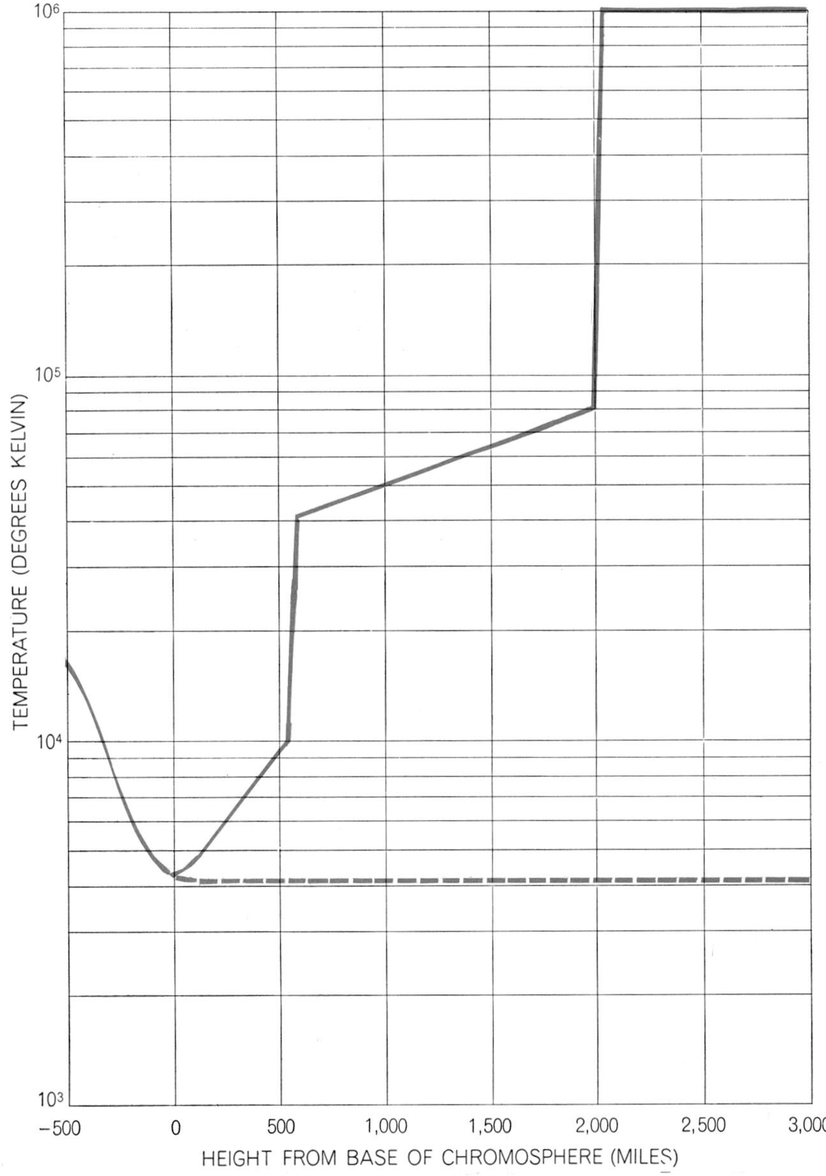

TEMPERATURE STRUCTURE of chromosphere is represented by solid line. The jumps are caused by chromospheric response to nonradiative energy emitted by the sun. If energy flowed out of the sun by radiation alone, the temperature would decline with distance but so slowly that, in the distance shown, the drop would not be apparent (*broken line*).

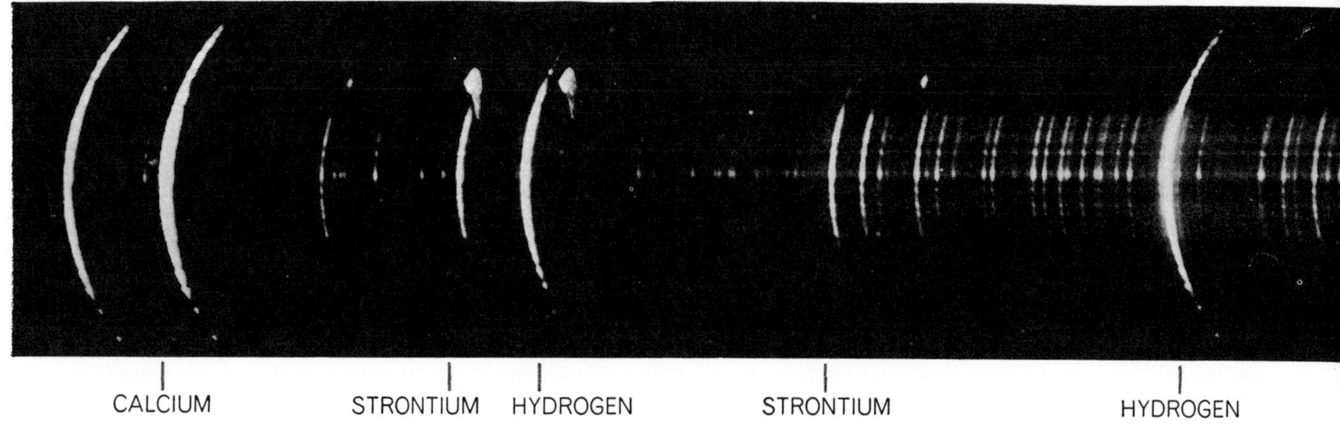

SPECTRUM OF CHROMOSPHERE consists of many bright emission lines separated by dark areas. This is the violet-to-blue portion of the spectrum, photographed at Khartoum in the Sudan by the 1952 eclipse expedition of the High Altitude Observatory of the

but Redman's suggestion of a high temperature was nevertheless to lead eventually to new concepts in chromospheric physics.

Shortly after this a largely empirical analysis of our data from the 1952 eclipse unexpectedly revealed a serious flaw in the simple reasoning that for so long had assigned a scale height of 600 miles to the lower chromosphere. Our analysis included an allowance for the changes in the intensity of the hydrogen lines that would result from ionization. Since it is the excited electron that emits radiation as it gives up surplus energy, hydrogen ceases to radiate when it has been ionized, that is, stripped of its single electron. Accordingly we derived a new density scale height of only about 100 miles. The same calculations gave a sharp rise in temperature from about 5,000 degrees at the base of the chromosphere to about 6,000 degrees at a height of 600 miles, which would ac-

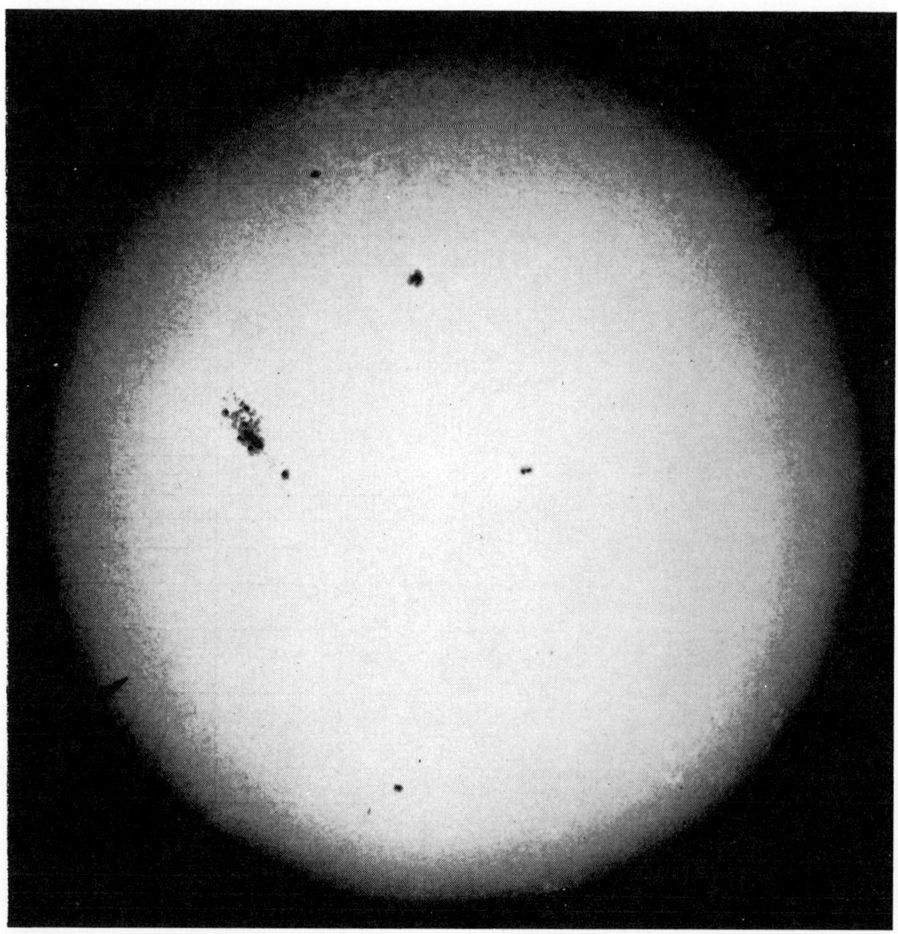

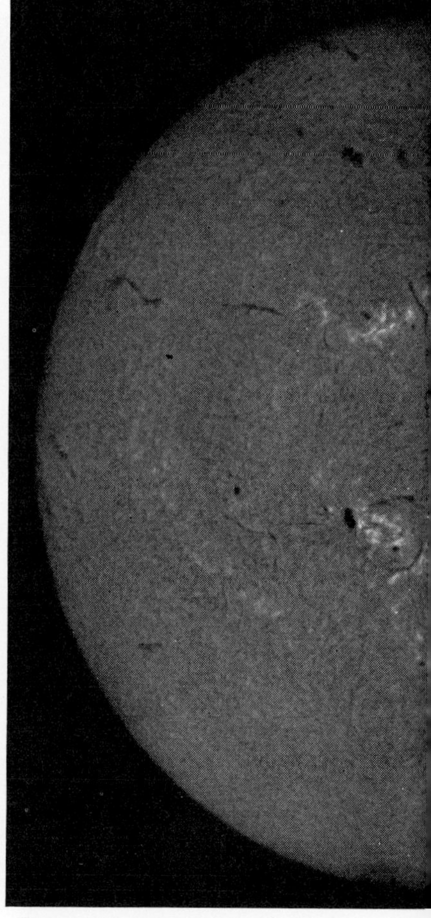

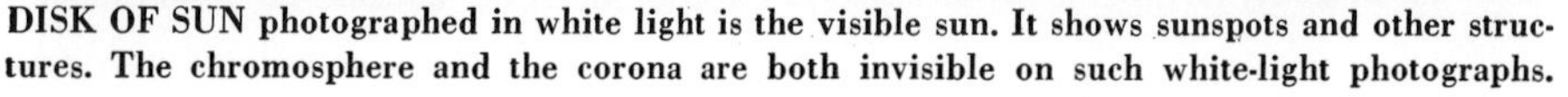

DISK OF SUN photographed in white light is the visible sun. It shows sunspots and other structures. The chromosphere and the corona are both invisible on such white-light photographs.

CHROMOSPHERIC DISK shows plainly in photograph made by red

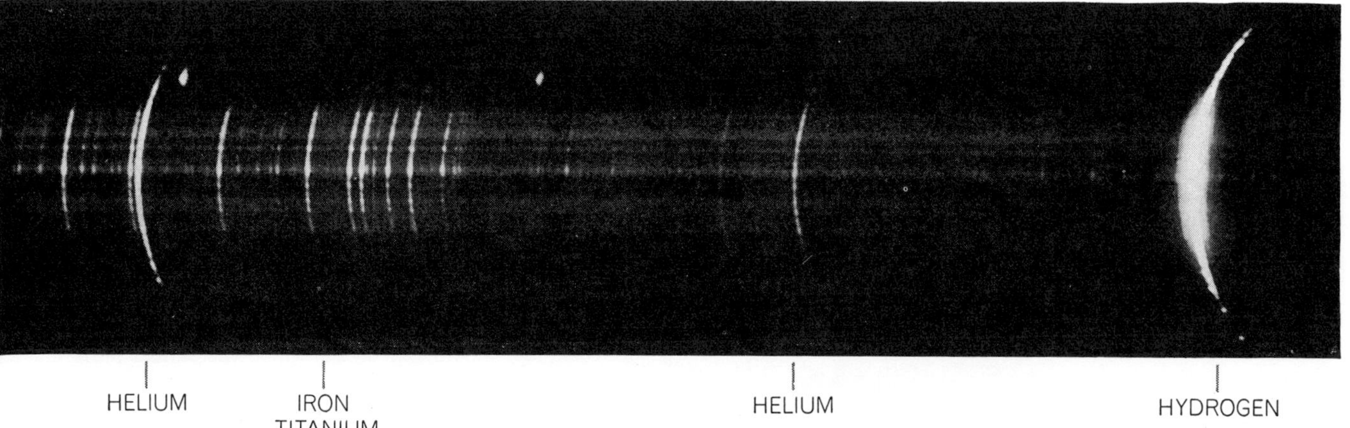

University of Colorado. The two strong calcium lines at far left appear on the spectrum of the photosphere as the heaviest of many absorption (*dark*) lines. Spectra such as this, available only at total eclipse, are the major source of data on the chromosphere.

count adequately for the new scale height without any other support mechanism.

Thus the question of support vanished once again, this time, we feel, permanently. Another question was raised, however, by the rise in temperature. Our initial results were later refined by Stuart R. Pottasch of the Institute for Advanced Study in Princeton, N.J., and Richard N. Thomas of the National Bureau of Standards, who deduced a temperature at the bottom of 4,500 degrees, which rose to 7,500 degrees at 600 miles. This sharp rise poses the main problem of the lower chromosphere. The temperature change, incidentally, is matched by a constant increase in the density scale height.

Above the 600-mile level, in the middle and upper chromosphere, two other factors enter the problem: the structure and movement of the spicules, and the state of the gas within and between the

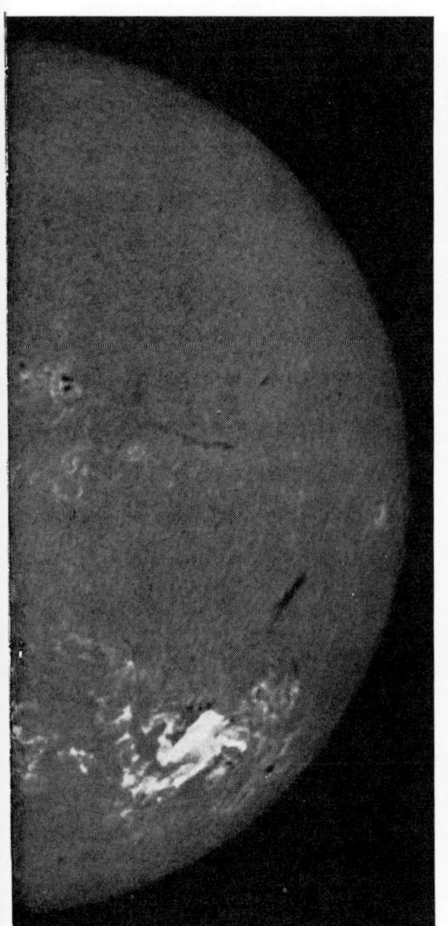

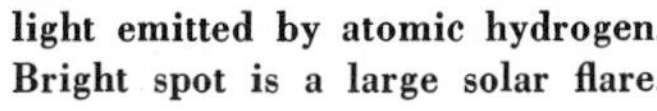

light emitted by atomic hydrogen. Bright spot is a large solar flare.

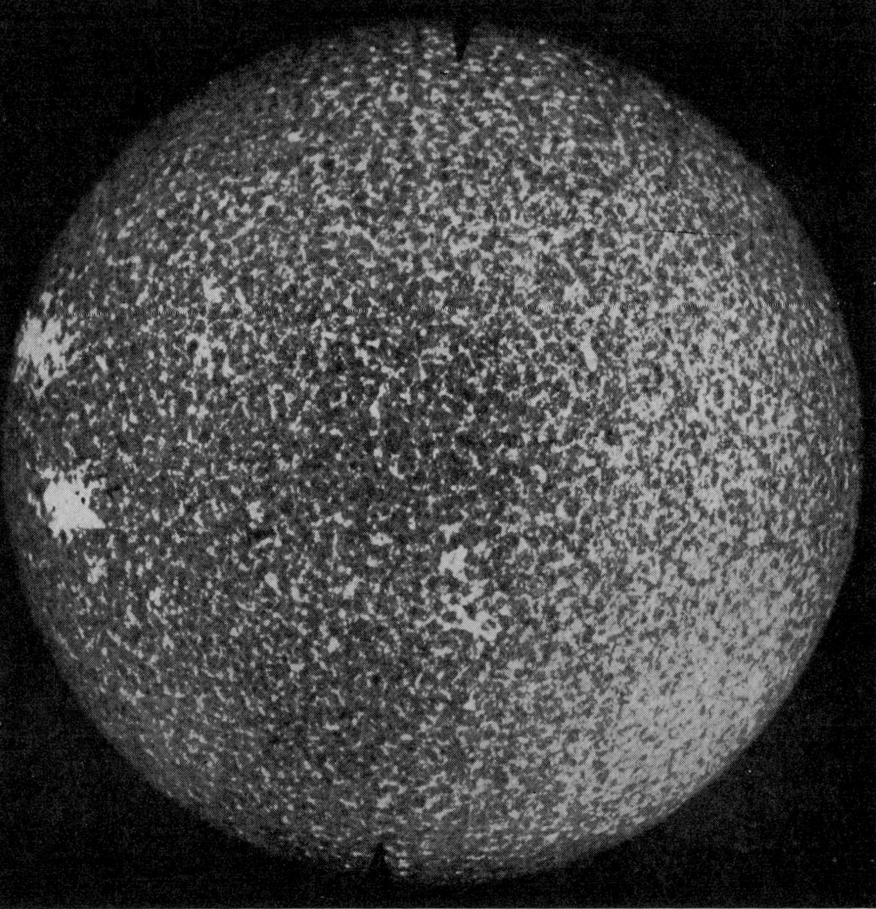

CALCIUM EMISSION along violet line at far left in spectrum at top of these two pages produces another detailed view of chromosphere. The bright patches at left surround regions of sunspots.

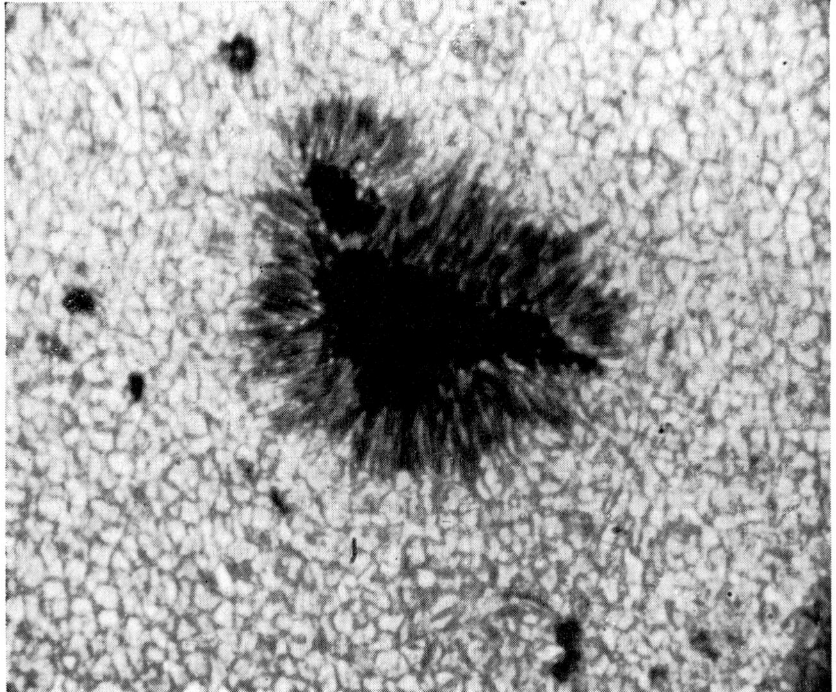

SURFACE OF PHOTOSPHERE is highly granulated. One sunspot appears in this photograph, made with a 12-inch telescope carried to 80,000 feet by an unmanned balloon. Unusual detail appears because balloon was outside most of earth's atmosphere. Photograph is from Project Stratoscope I of Princeton University, sponsored by Office of Naval Research, National Science Foundation and National Aeronautics and Space Administration.

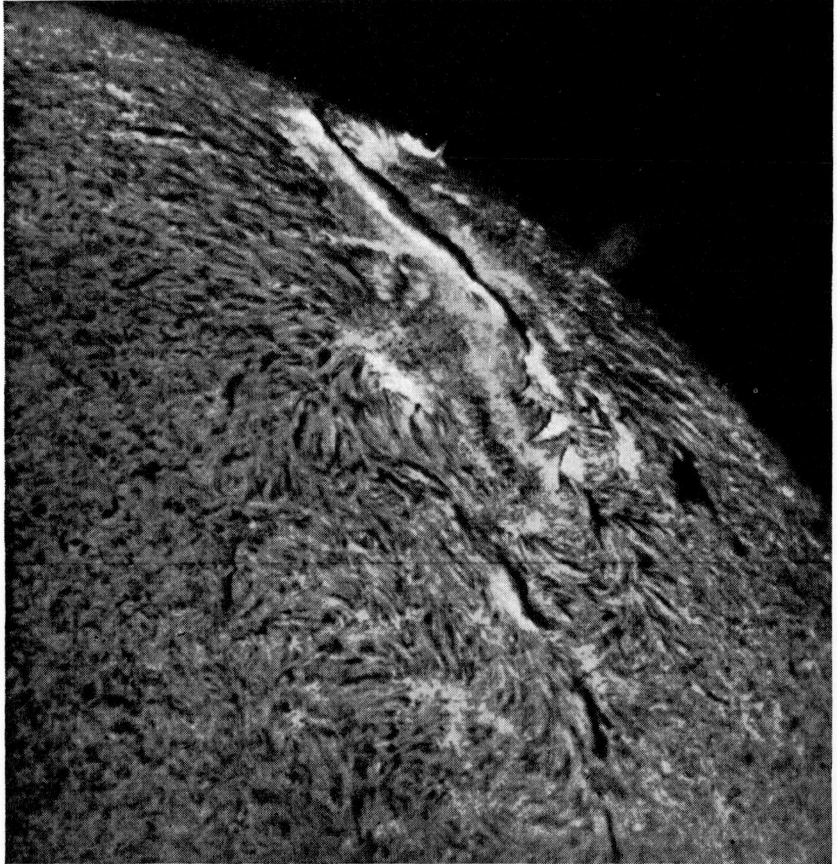

GRANULAR STRUCTURE characterizes chromosphere in this photograph made by the red light of atomic hydrogen. Large features are over sunspot. This photograph, from Mount Wilson Observatory, shows a far larger portion of solar disk than photograph at top of page.

spicules. It is here that the geometry of the chromosphere becomes extremely important.

Oddly enough, Redman's mistaken announcement of a 35,000-degree temperature in the lower chromosphere led the way to present understanding of the chromosphere. Thomas and R. G. Giovanelli of the Commonwealth Scientific and Industrial Research Organization in Australia, seeking to reconcile the high temperature with seemingly contradictory data, were led to suggest the abandonment of the whole theoretical basis of chromospheric studies—the idea that the chromosphere is in a state of thermodynamic equilibrium. This ideal state normally prevails in the laboratory, where the temperature of a gas can be determined by measuring the intensity of radiation at one or two points of the spectrum. On the basis of such laboratory studies, the German physicist G. R. Kirchhoff and others as long ago as 1860 deduced very simple concepts for explaining the properties of spectra produced by atoms in a gas. Under equilibrium conditions each microscopic process in the gas is exactly balanced by its inverse process: for every atom that absorbs a photon of light, another atom of the same kind radiates an identical photon; for each atom that gains energy in a collision, another loses exactly as much energy in an identical collision. So long as thermodynamic equilibrium prevails—or is assumed—all processes balance, and there is no need to know such details as how often one or another event takes place or which processes are the most important.

The solar physicist quite naturally applied this simplifying assumption to the sun. In the case of the photosphere it has given many valuable and undoubtedly correct results. In the early 1940's, however, it became apparent that the extremely hot, tenuous gas of the solar corona could not exist in the state of thermodynamic equilibrium. More recently investigators have found themselves compelled to set the concept aside in coping with the problems of the chromosphere. In fact, the idea of equilibrium has been shown to be invalid for the formation of some of the spectral absorption lines produced by atoms within the photosphere. Still other absorption lines that were attributed to the photosphere are now known to originate in the chromosphere without the blessings of thermodynamic equilibrium.

The abandonment of this concept is not a trivial matter for the theoretician or for the observer. According to Kirchhoff's law, one could assume that a darker region of the sun is at a lower temperature than a brighter region. Studies have now shown that a cloud of hydrogen at a temperature of, say, 50,000 degrees K. can show up as a dark feature when viewed in the line spectrum against a brighter background arising from hydrogen at 10,000 degrees. On the other hand, a dark cloud can also be cooler than the brighter background. Similarly, a bright cloud may be hotter or cooler than its darker background. Only detailed scrutiny of the spectrum can reveal the true situation. By allowing for departures from thermodynamic equilibrium, however, one can begin to form some idea of events on the microscopic scale and can determine which processes dominate the scene.

From such analysis of the spectra recorded in 1952 we have been able to piece together otherwise contradictory data into a consistent description of the chromosphere that is completely different from earlier ideas. It is at greater altitudes in the chromosphere that these departures become more significant. Beginning at about 600 miles the chromosphere separates into relatively cool, high-density columns, which are the spicules, and hot, low-density regions between them. As the coronagraph so clearly reveals, spicules shoot out at about 20 miles per second, attaining a height of 8,000 to 10,000 miles before they fade from view or fall back to the sun. Each spicule is about 500 miles in diameter (the width of the state of Colorado) and endures only five or 10 minutes. A given point on the solar surface undergoes spicule upheaval about once in 24 hours, and at any one time 100,000 spicules are passing through some stage of their evolution over the surface of the sun. It is true, as McCrea suggested, that the middle and upper chromosphere is in a continuous state of agitation. But the motions bear little resemblance to the usual concept of

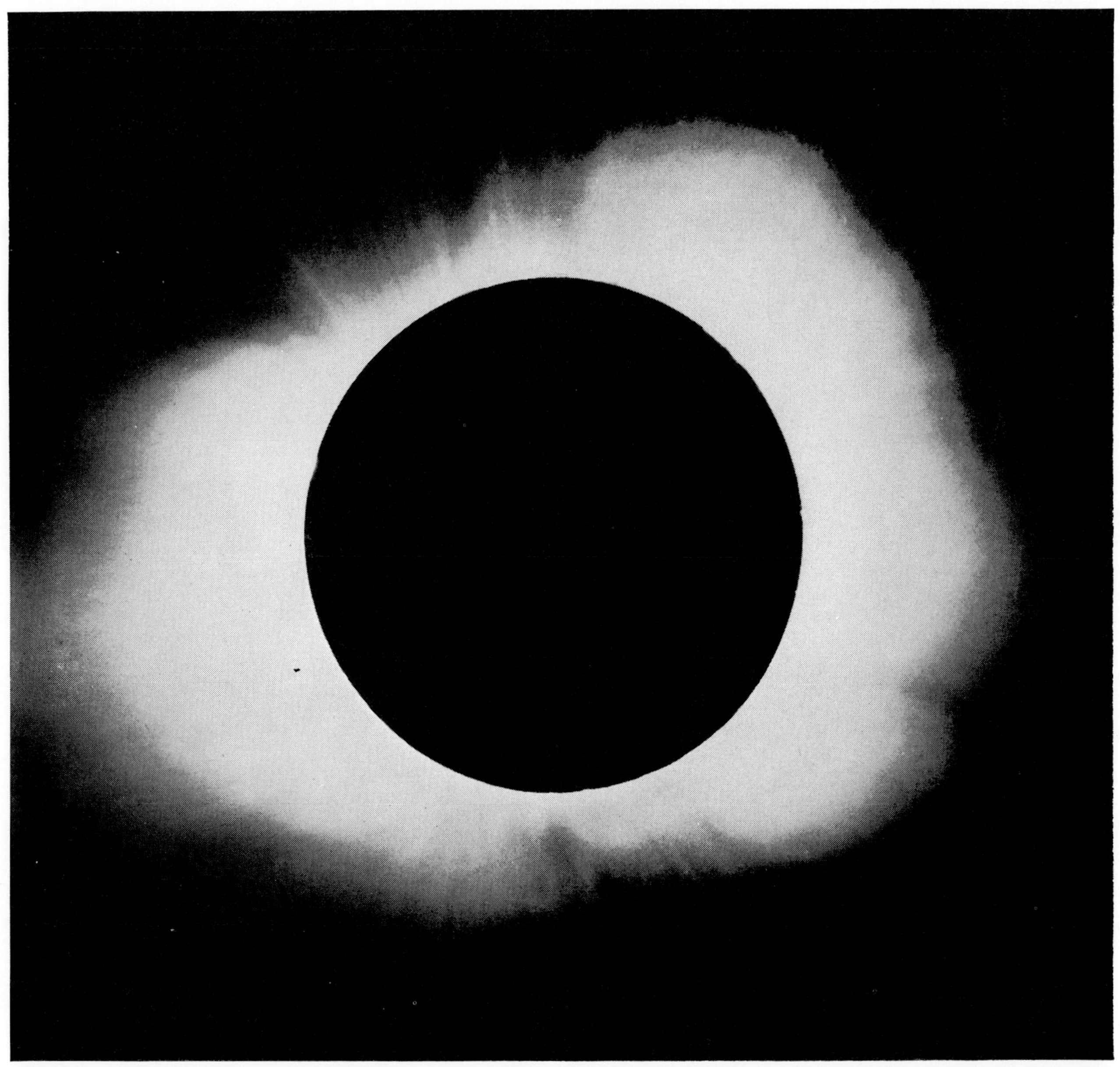

SOLAR CORONA, the tenuous outer atmosphere that lies beyond the chromosphere, is pearly-white. Although some of its features can be viewed with a coronagraph, it is best seen, as here, during the natural total eclipse of the sun that occurs about once a year.

turbulence; they are essentially well ordered. My colleagues and I feel intuitively that the spicules must somehow result from convection starting in the hydrogen convection zone that lies beneath the photosphere. They may, however, arise from some basic instability in the lower chromosphere that we do not yet understand.

Without doubt spicules occupy an important place in the over-all economy of the sun. Enough protons move outward across the 3,000-mile level of the chromosphere, for example, to replace the entire corona in less than three hours. This is more than a thousand times the observed flow of protons out of the sun into interplanetary space in the solar wind. If even a small fraction of the spicule flux continues to move upward, the spicules play a decisive role in the chromosphere-corona complex. Actually, so far as we can tell from observations, about half of the spicules move upward with no evident return flow. The flux of energy due to the spicules exceeds the energy radiated by the corona in the ultraviolet and X-ray regions of the spectrum, and it just about equals the flux of ultraviolet radiation from the chromosphere that is represented by the strong Lyman-alpha lines of hydrogen and ionized helium. This amounts to about a ten-thousandth of the sun's total energy output—a small fraction but a far from negligible quantity in absolute terms.

Apparently the spicules have an exceedingly complicated temperature structure. There is strong evidence for two discrete ranges of temperature—either near 10,000 degrees or near 50,000 degrees—rather than for a continuous variation. The two temperature ranges appear to occur at every height in a spicule, as if the spicule had a cool core encased in a hot envelope, or vice versa. It is no exaggeration to say that the explanation of the origin and subsequent history of spicules is one of the most challenging problems of solar physics.

The relatively quiescent regions between the spicules, which make up most of the upper and middle chromosphere, present still other problems. The low density and high temperature of the interspicular spaces with respect to the spicules make them difficult to observe. From a height of 600 miles up to about 2,000 miles much of the radiation from un-ionized helium atoms and from ionized helium atoms seems to originate in the region between the spicules, where the temperature is about 50,000 degrees. At 2,000 miles the spectrum of the interspicular spaces throughout the visual and photographic regions suddenly takes on all the characteristics of the corona at a temperature of a million degrees. At such heights the ultraviolet portion of the spectrum displays strong lines of oxygen, nitrogen and carbon atoms that have been stripped of several of their outer electrons. These lines probably originate within the interspicular spaces, but little more can be said about them at present. In any case, it seems ap-

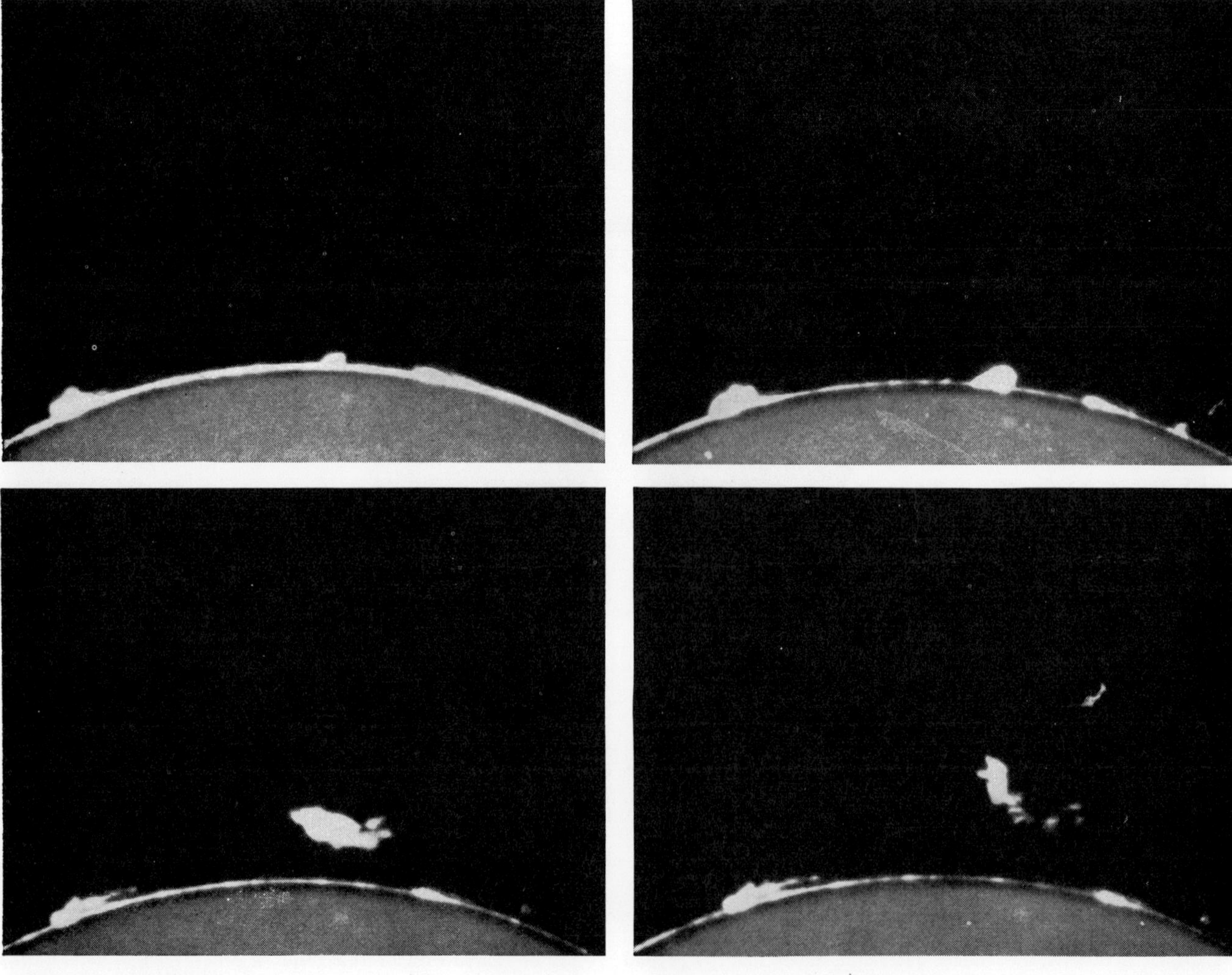

EXPLOSIVE FLARE, which erupted on November 20, 1960, in the chromosphere on the limb of the sun, ejected a huge mass of incandescent gas into space. Sequence was recorded in red light by a High Altitude Observatory coronagraph at Climax, Colo. Second

parent that the helium emission from between the spicules disappears at the 2,000-mile level.

The curious steplike temperature gradient of the chromosphere and the lower corona, with sudden increases from about 10,000 degrees to 50,000 degrees and then to a million degrees, demands an explanation. Thomas and I have proposed a theory that relates the temperature plateaus to the chemical composition of the solar atmosphere as observed in the spectra.

We assume that the initial rise in temperature in the chromosphere requires a nonradiative source of energy, perhaps sound waves originating in the hydrogen convection zone below the photosphere. Such a source of energy would square the account left unbalanced in the energy-conservation ledger by the strong outflow of radiant energy. We also assume that the chromosphere and the corona absorb the nonradiative energy and re-emit it in the form of radiation. These conditions permit only certain ranges of temperature in the solar atmosphere. For each small increase in temperature up to 10,000 degrees, it can be shown that un-ionized hydrogen will radiate proportionately more of the inflowing nonradiative energy and so keep the inflow and outflow of energy about equal. Above 10,000 degrees, however, any increase in temperature makes the hydrogen radiate less energy. The temperature must therefore rise until some other element in the solar gases is able to dispose of the energy. The next possibility is ionized helium at temperatures above about 40,000 degrees. Helium will continue to radiate more and more energy until the temperature reaches about 80,000 degrees. This is the temperature approximately 2,000 miles above the photosphere. There a new regime must take over.

The next possible absorber and radiator of energy appears to be heavy elements that have lost about 10 electrons. The observed temperature of a million degrees in the corona will strip that many electrons from an atom. According to theoretical calculations, the amount of energy an atom radiates increases as the fourth power of the electric charge on its nucleus. A loss of 10 electrons thus increases the ability of an atom to radiate by a factor of 10^4. Such a high degree of ionization is required because the corona possesses only a few atoms of the heavy elements; the deficiency in numbers must be made up by an ability to radiate more intensely. It can be seen that the relative numbers of atoms of various elements, as shown by the spectrograph, becomes a crucial factor in explaining the temperature structure also indicated by the spectrograph.

If our interpretation of the chromospheric structure holds up, then the major unsolved problems are the source and nature of the nonradiative energy that creates and powers the chromosphere, and the cause of spicule upheaval. The two problems are probably closely related.

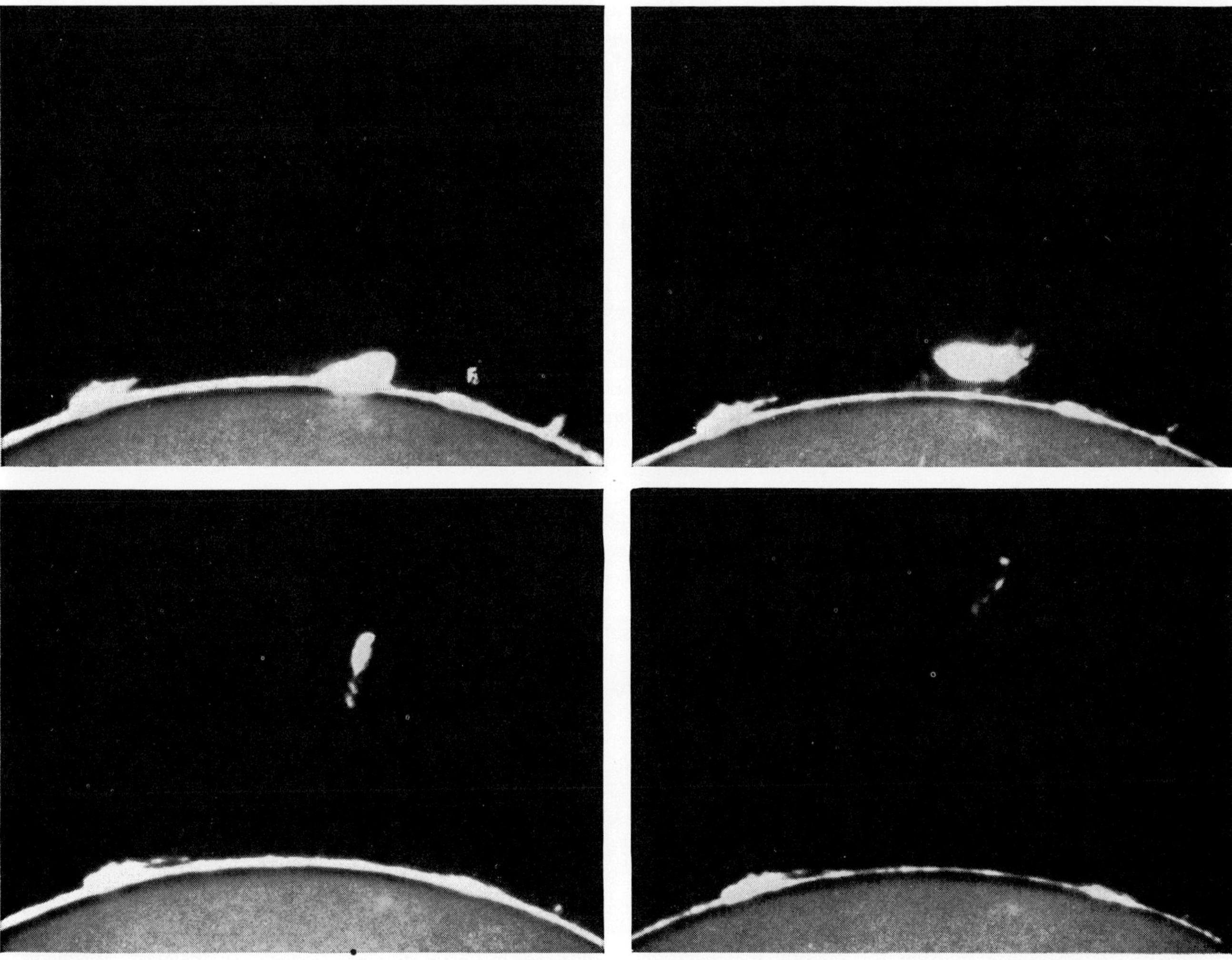

photograph was made 10 minutes after the first. Remainder were made 20, 25, 27, 28, 30 and 32 minutes respectively after start. The region of the sun shown is more than 340,000 miles across. Particles from such solar flares can disrupt communications on the earth.

9

The Solar Wind

E. N. PARKER

April 1964

A swift wind of hydrogen blows continuously through the solar system. Emanating from the sun, it speeds past the earth at 400 kilometers per second (about 900,000 miles per hour) and rushes on past the planets into interstellar space. Like a broom, it sweeps up the gases evaporated from planets and comets, fine particles of meteoritic dust, and even cosmic rays. It is responsible for the outer portions of the Van Allen radiation belts around the earth, for auroras in the earth's atmosphere and for terrestrial magnetic storms. It may even play a part in shaping the general pattern of the earth's weather.

The existence of this solar wind, which had long been suspected, has now been verified by space vehicles. They have measured its velocity and density. And studies of another kind have unraveled the mystery of its origin and given us an understanding of its effects.

The realization of the solar wind's existence came only gradually, over a period of several decades. The first explicit assertion that something besides light was coming to the earth from the sun was made in 1896 by the Norwegian physicist Olaf K. Birkeland. He suggested that the aurora borealis might be caused by electrically charged "corpuscular rays" shot from the sun and "sucked in" by the earth's magnetic field near the poles. He was led to this suggestion by the fact that the aurora looked very much like the electric discharge in the then newly invented tubes generating streams of charged particles ("cathode rays").

Birkeland's idea was taken up by the Norwegian mathematician Carl Størmer, who went on to calculate the paths that streams of charged particles from the sun should follow when they entered the earth's magnetic field. As it happened, his theoretical scheme of looped and spiral paths did look like patterns seen in the aurora, but this resemblance turned out to be a coincidence; nothing else in his theory worked. To this day there is still no complete theory explaining how the solar wind produces the aurora, although some interesting ideas are beginning to develop. The fact remains, however, that Birkeland and Størmer were on the right track and started an important new line of thinking by calling attention to the possibility of charged particles coming from the sun.

Magnetic Storms

Further evidence for the sun's emission of particles came from another phenomenon (and many years later). This had to do with the magnetic storms that are associated with the disruption of radio, telephone and telegraph communication. The storms are evidently caused by fluctuations in the earth's magnetic field. Because they usually came a couple of days after a flare on the surface of the sun, they were at first attributed to a burst of ultraviolet radiation from the flare or some similar cause. Then the British geophysicist Sydney Chapman surmised that corpuscular emissions from the sun offered a more reasonable explanation. In the 1930's he and V. C. A. Ferraro carried out a series of calculations and demonstrated that a cloud of ions ejected from the sun, traveling at 1,000 or 2,000 kilometers per second, would reach the earth in a day or two and ruffle the earth's magnetic field as it passed. Their theoretical picture of such a disturbance of the field so closely resembled the actual fluctuations during a magnetic storm that Chapman's idea was widely accepted.

The third manifestation of the solar corpuscles was noted in the late 1940's. This time they came up in connection with fluctuations in the bombardment of the earth by cosmic rays. Scott E. Forbush of the Carnegie Institution of Washington had discovered that the intensity of cosmic radiation reaching the earth was low during the height of solar activity in the sunspot cycle and often fell abruptly during a magnetic storm. In other words, the more active the sun, the smaller the number of cosmic ray particles impinging on the earth. It was at first supposed that this effect must be due to solar-caused changes in the earth's magnetic field and atmosphere that deflected the cosmic particles away from the earth. But the University of Chicago physicist John A. Simpson, who began to keep track of cosmic ray variations with a neutron monitor he had just developed, soon found that the fluctuations were much greater than had been supposed. They could not be produced merely by changes on the earth; they must reflect a rise and fall of cosmic ray intensity in solar-system space as a whole.

Apparently something in the sun's radiation tended to impede the flow of cosmic rays into the solar system, and this obstruction increased when the sun was particularly active. What might the impeding agent be? The general mechanism is to be found somewhere in the magnetohydrodynamic theory of the Swedish physicist Hannes Alfvén [see "Electricity in Space," by Hannes

"SOLAR-WIND SOCK," the tail of a comet, always points away from the sun, blown back by the high-speed stream of hydrogen in space. Comet Mrkos (*opposite page*) was photographed in August, 1957, with a five-inch camera on Palomar Mountain. Irregularities in the comet tail are probably caused by turbulence in the solar wind.

Alfvén; SCIENTIFIC AMERICAN, May, 1952]. He had pointed out that an ionized gas in motion must carry a magnetic field with it. This being so, it was suggested by Philip Morrison of Cornell University and by others that a stream of charged corpuscles from the sun, carrying a magnetic field, would tend to sweep cosmic ray particles out of the solar system, and the effect would be strongest when the solar radiation was most intense. Such a theory would explain the cosmic ray fluctuations.

At about the same time there emerged a fourth and decisive line of evidence for the corpuscular radiation from the sun. For centuries it has been known that the tails of comets always point away from the sun. No matter where a comet may be in its orbit through the solar system, its head is always toward the sun and its gaseous tail streams away. Why is this so? In modern times the almost universally accepted theory has been that it is the pressure of sunlight, pushing the extremely tenuous matter of the comet, that drives the tail in the opposite direction. But in the 1950's Ludwig F. Biermann of the University of Göttingen showed that the pressure of the sun's light was not nearly sufficient to account for the violence with which a comet's gases are blown away from the head. He suggested instead that the only solar radiation that could account for the pushing away of the comet's tail was a stream of actual particles. He pointed out that such radiation from the sun would also account for the existence of excited, light-emitting ions seen in comet tails [see "The Tails of Comets," by Ludwig F. Biermann and Rhea Lüst; SCIENTIFIC AMERICAN, October, 1958].

Biermann's discovery conveyed something else that had considerable bearing on the question of how this corpuscular radiation from the sun originated. Speculation up to that time had centered on two possibilities: that the corpuscles were sent out in bursts by solar flares (which were known to emit very energetic protons, or hydrogen nuclei) or were projected in beams from sunspots (by some unknown electromagnetic acceleration process). But Biermann's evidence now made it plain that the corpuscular radiation could not be coming merely in bursts or isolated beams. The comet tails showed that the radiation was blowing continuously in all directions outward from the sun. The comet tails were in effect interplanetary "wind socks" demonstrating the existence of a steadily blowing, space-filling radiation. The streaming of the particles might intensify when the sun became particularly active, but it was present all the time, with or without sunspots or flares.

The Corona and the Wind

So it seemed that the flow of corpuscles must stem from something that went on all the time all over the sun's

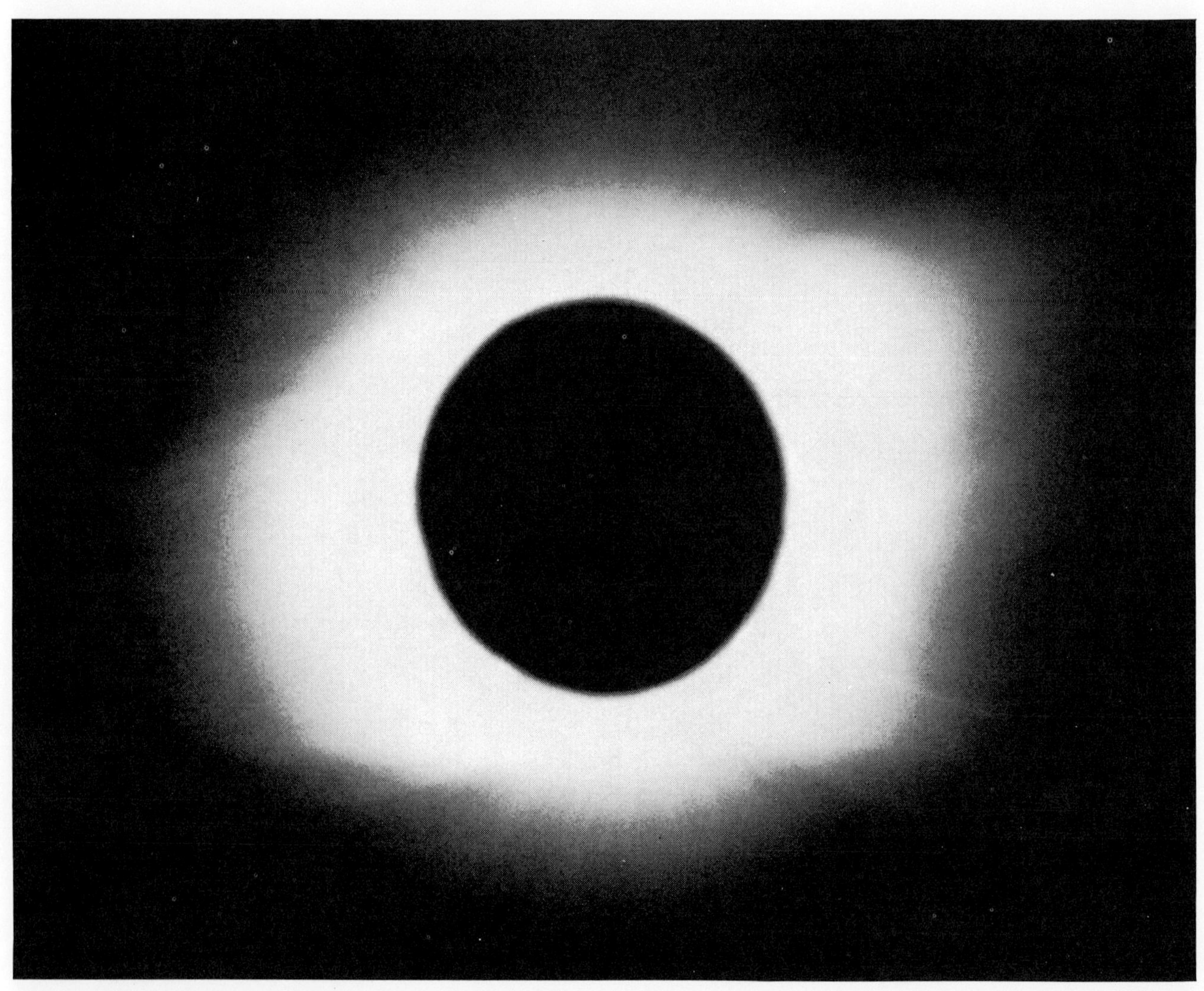

SOLAR CORONA, the source of the solar wind, was in a fairly quiet state during the eclipse of July 20, 1963. The photograph at the left was made at Talkeetna, Alaska, by an expedition from the High Altitude Observatory of Boulder, Colo. That at the

surface. The sun was continuously shooting a thin hail of projectiles in all directions out into space. By what process could it do such a thing? A suggestion of a possible answer came one afternoon in 1957 when I was visiting Chapman at the laboratories of the High Altitude Observatory in Boulder, Colo., where he was then working.

Chapman was studying the sun's corona, from the standpoint of whether or not it might be responsible for heating the outer regions of the earth's atmosphere. Soundings of the upper atmosphere had brought out the curious fact that it got hotter, rather than colder, with increasing altitude. This suggested that the upper air was heated by hot gases in outer space. Chapman suspected that these hot gases might be maintained by the solar corona.

The corona is the tenuous outer atmosphere of the sun. It is very thin indeed: even close to the sun it contains only about 100 million to a billion hydrogen atoms per cubic centimeter, a density only a hundred-billionth that of the air we breathe. The temperature of the corona, however, as measured by the velocity of its atoms, is extremely high: about a million degrees centigrade near the sun. Because of its high temperature the coronal gas is completely ionized and therefore consists of separate protons and electrons.

Because the corona is so tenuous it is not self-luminous, in spite of its high temperature. It is visible, however, by virtue of the fact that its atoms scatter the light from the sun's luminous photosphere, just as grains of dust in the earth's atmosphere become visible by scattering sunlight. When the brilliant light of the sun itself is dimmed by an eclipse, the white corona can be seen stretching far out from the hidden solar disk. Photographs of the corona show that it extends for millions of miles from the sun, and were it not for interfering haze and light in the sky we could probably see its fainter reaches extending many times farther than that.

Now, Chapman knew from his pioneering theoretical studies of the properties of ionized gases that a tenuous ionized gas at a million degrees must have an extraordinary ability to conduct heat. According to his calculations, the heat flow through ionized gas increases at a rate almost equal to the fourth power of the increase in temperature. At a million-degree temperature this means a great deal of heat flow. Chapman calculated that, if the corona extended as far as the earth's orbit, its temperature that far from the sun would be about 200,000 degrees, owing to its high conduction of heat. This was a very interesting figure as support for his theory that the corona might heat the earth's upper atmosphere.

But Chapman had made another discovery that impressed me even more as he told me his results in our talk that afternoon. He had gone on to make

right was made 115 minutes later by another team from Boulder on Cadillac Mountain in Maine. Careful measurements show motions of solar plumes: either lateral displacements or "virtual" motions due to appearance and disappearance of adjacent plumes.

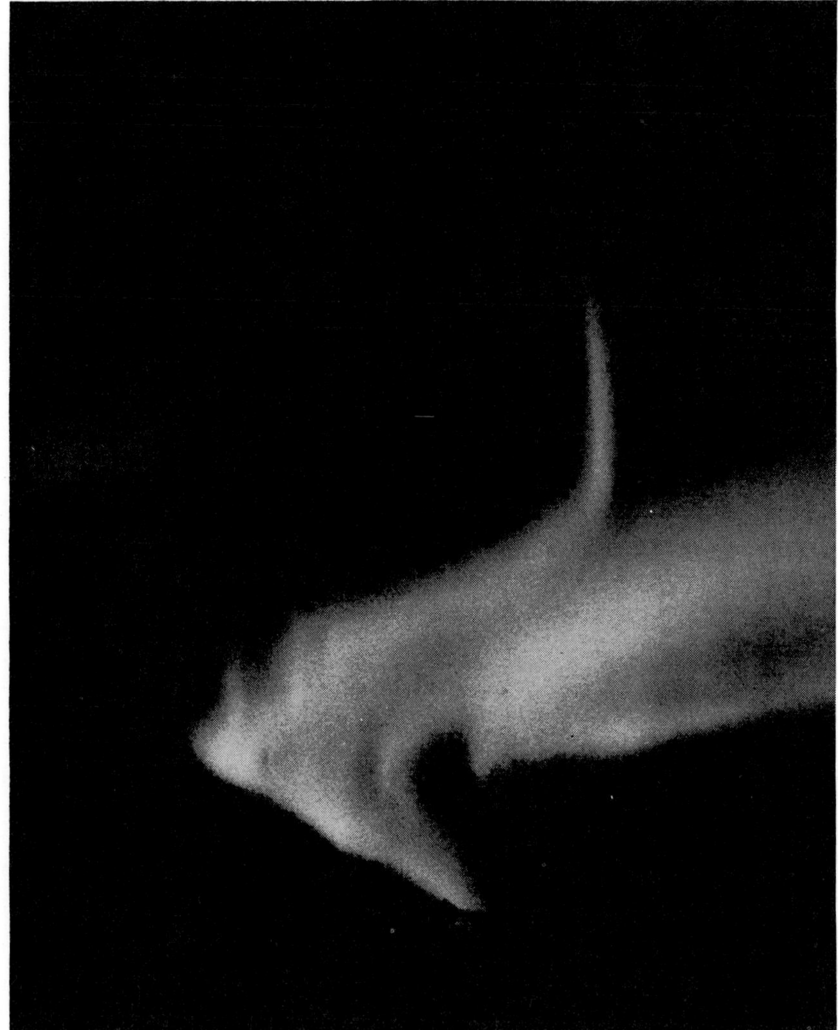

AURORAL DISPLAY, which results from interaction of earth's atmosphere and solar wind, was photographed in Alaska by Victor P. Hessler of the University of Alaska. Lines of force of the earth's field funnel solar particles into the atmosphere at high latitudes.

some calculations to determine if the corona did reach to the earth. For these he used the equation of the barometric law, which states the obvious fact that in an atmosphere the pressure at any given height must be just sufficient to support the weight of the portion of the atmosphere above it. (If it were not, the atmosphere would collapse.) Starting from the known density of the corona near the sun (which can be ascertained approximately), he was able to estimate its density at the earth's distance. This turned out to be roughly 100 to 1,000 hydrogen atoms per cubic centimeter. In other words, the corona, although it was highly tenuous at this distance, did reach all the way from the sun to the earth and beyond!

It was a startling idea: The earth in its orbit around the sun moves within the sun's hot corona. The corona is not a limited blanket enveloping the sun the way our atmosphere envelops the earth; on the contrary, the corona fills the whole solar system.

The Corona in Motion

It took a while for Chapman's statement to sink in. When it did, I recalled Biermann's description, during a visit to Chicago, of the corpuscular radiation that blows comet tails away from the sun. Now there were apparently two bodies of solar vapor to think about: the steady corona and the stream of particles flowing out from the sun at high speed. This, however, was impossible. In a magnetic field one stream of charged particles cannot pass freely through another, and it was known that solar-system space was filled with magnetic fields. Therefore the corona and the solar stream could not be separate entities. They must be one and the same. The corona, behaving like a static atmosphere near the sun, must become a high-velocity stream farther out in space. How could this come about?

I examined the mathematics of the barometric law in more detail and saw that, in the absence of a large inward pressure from outside the solar system, the high-temperature corona must flow away from the sun. To find the nature of this flow I then applied the hydrodynamic equations for the flow of a gas. These nonlinear equations are so complex that it was out of the question at the time to find a general solution covering all possible assumptions; I settled for a simple case that approximately represented what Chapman had calculated, namely, that the temperature of the corona remains high for a distance of several million kilometers from the sun and then drops to a lower figure. This made the mathematics relatively straightforward.

The mathematical solution of the equations produced a result that must be considered surprising in view of the traditional idea that the corona is a static atmosphere. It showed that with increasing distance from the sun the corona tends to expand. At first the expansion is slow, but as the distance increases, the pressure within the corona gradually overcomes the weight of the overlying gas and rapid expansion takes over. At 10 million kilometers (some six million miles) from the sun the corona is expanding at a speed of several hundred kilometers per second—faster than the speed of sound. At that point it must be considered a supersonic wind rather than the sun's atmosphere. It continues to accelerate and reaches velocities several times the speed of sound as it moves out of the sun's gravitational field [*see top illustration on page 106*].

The application of the equations showed that away from the sun the erstwhile corona *must* expand rapidly and become a high-velocity stream. I have called it the "solar wind" because this seems to me now a more accurate description of the phenomenon than the older pictures of a static "atmosphere" or a bullet-like "corpuscular radiation." Biermann's comet tails are truly "solar-wind socks," signaling the direction and strength of the corona's expansion.

The Heat of the Wind

The corona's expansion arises from the fact that its temperature at the sun is of the order of a million degrees. What makes it so hot? We know that the temperature at the sun's photosphere is only about 6,000 degrees, and one would suppose that the nonincandescent corona outside it should be cooler. But about 15 years ago Martin Schwarzschild of Princeton University and Biermann independently presented a now accepted explanation of the paradox of the corona's high temperature. The corona is so tenuous that it takes very little heat to raise its temperature. Schwarzschild and Biermann suggested that the churning motions of the gas at the surface of the sun generate low-frequency waves that provide enough energy to heat the corona to a million degrees. The action is somewhat akin to that of the boy scout who, although his body temperature is only 37 degrees C., produces enough heat to start a fire by rubbing two sticks of wood together until they reach a temperature of several hundred degrees.

Our theoretical calculations cannot give us really accurate numbers for the speed and density of the solar wind, because this would require precise knowledge of the temperature and density of the corona near the sun—for which we have only rough estimates. But if we

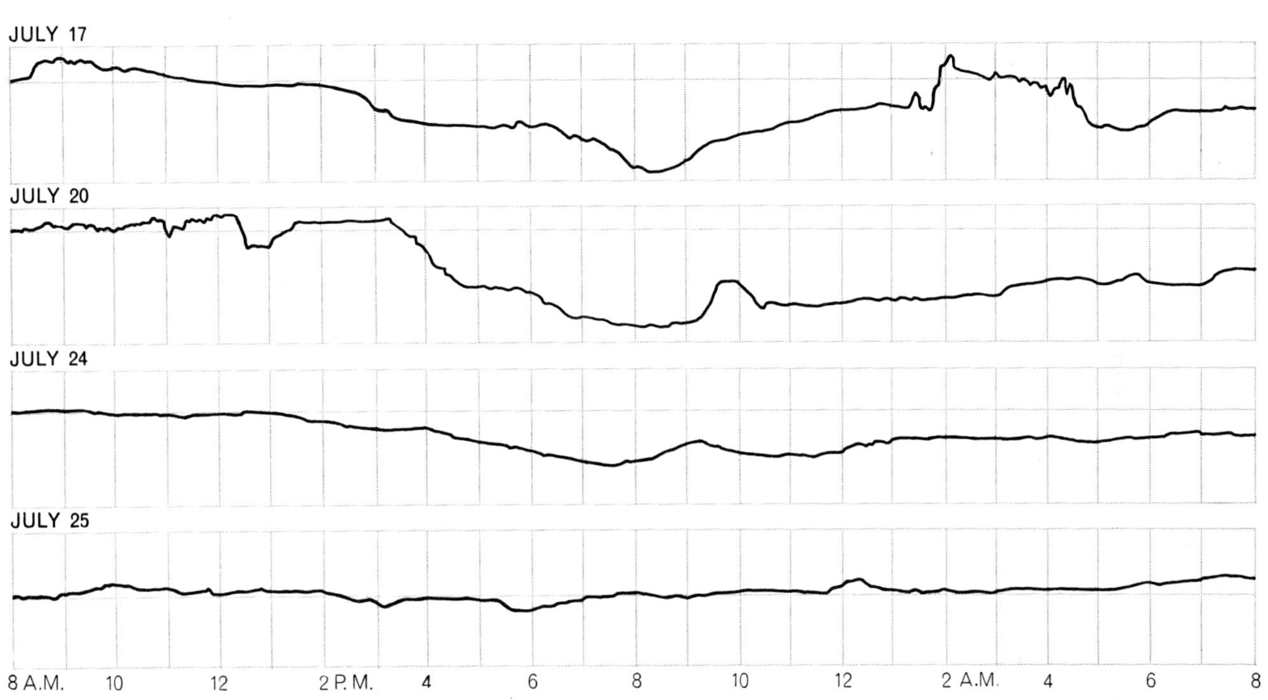

EARTH'S MAGNETIC FIELD fluctuates in response to changes in the solar wind as it blows past. These records of variations in strength of the horizontal component of the earth's field cover four full days during July, 1961. They were made in Honolulu.

COSMIC RAY BOMBARDMENT of earth decreases as solar activity and wind increase during the 11-year "sunspot cycle." Black curve shows changes in cosmic ray intensity compared with maximum in 1954. Colored curve is a plot of solar activity as indicated by the number of sunspots. Spots are another manifestation of processes on the sun that cause fluctuations in solar wind.

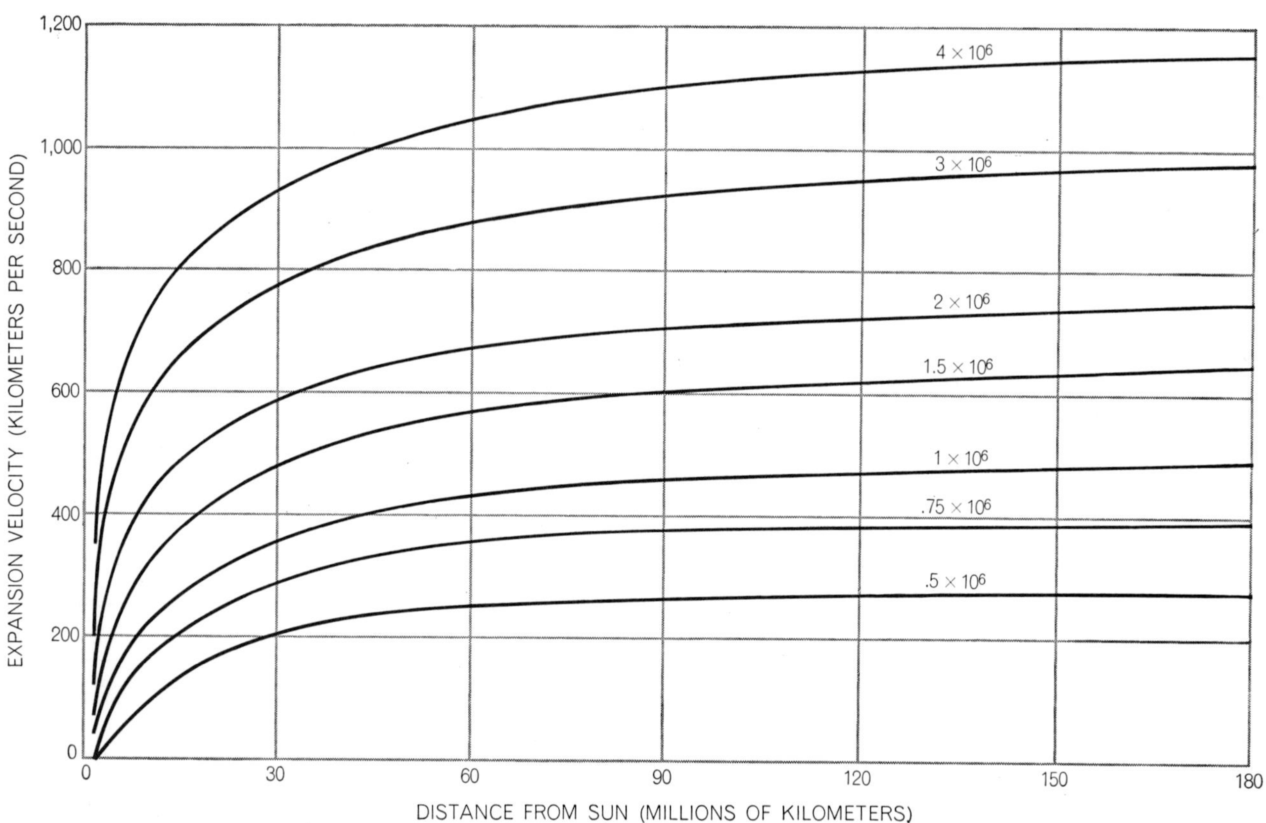

RATE OF EXPANSION of the solar wind into space depends in part on temperature of the corona. Temperatures are given on each curve in degrees absolute, ranging from 500,000 (*bottom*) to four million degrees. Orbit of earth is at 150 million kilometers.

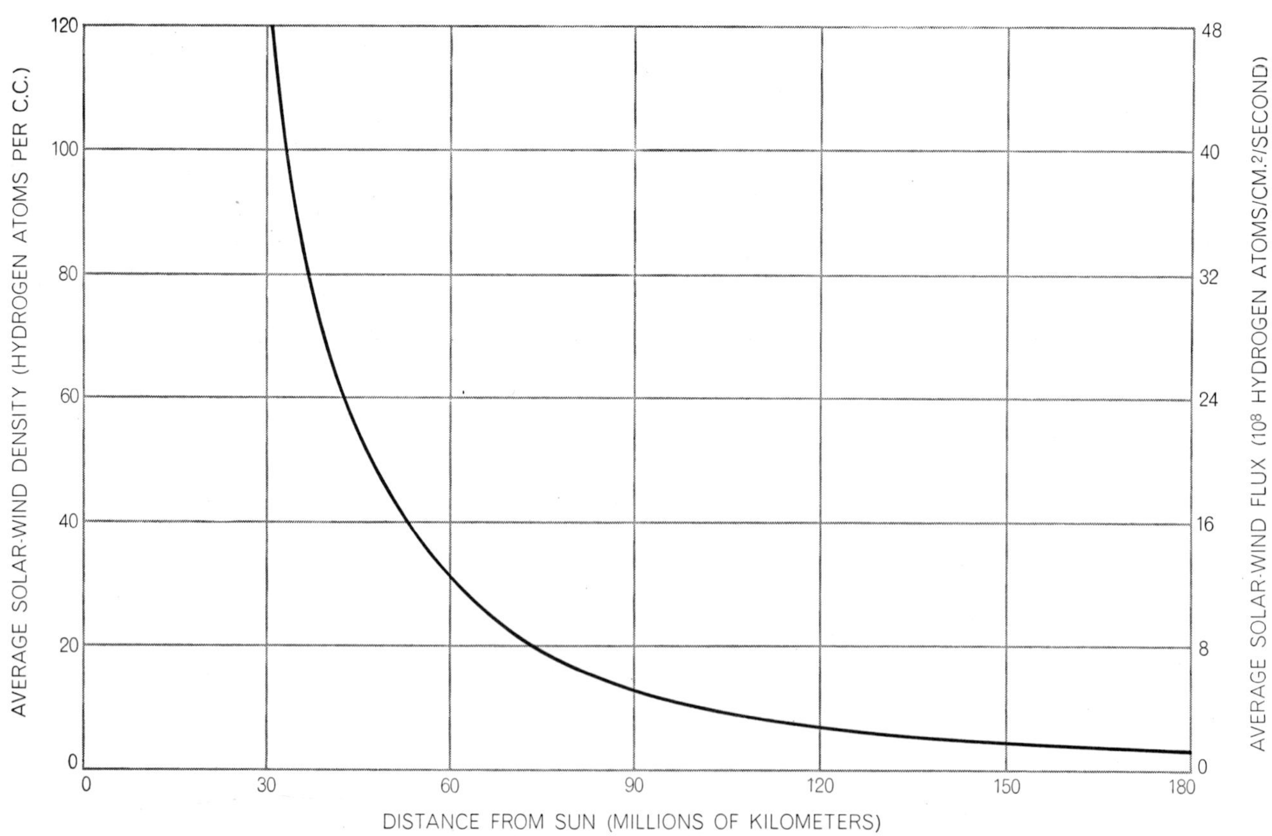

DENSITY AND FLUX of solar wind as function of distance from the sun are plotted from calculations. Scales are at left and right. The flux of the wind is defined as the number of hydrogen atoms passing through an area one centimeter square in one second.

assume that the temperature at the base of the corona is a million degrees, we can draw the following approximate picture of the rise and progress of the solar wind. At the bottom of the corona the gas is almost stationary (in cosmic terms): it moves away from the sun's surface at the rate of only a few hundred meters per second. As it moves on it is replaced by more gas rising from below the photosphere. The coronal gas streaming slowly away is accelerated very gradually: it takes about five days and about a million kilometers of travel to get really under way. Thereafter it speeds up to hundreds of kilometers per second, and in four more days it has spanned the 93 million miles to the earth. The gas we see at the bottom of the corona on a Sunday will be passing us about Tuesday of the following week. Two weeks after this gas zooms by us it will pass Jupiter.

The Magnetism of the Wind

The solar wind carries a magnetic field along with it because the gas is ionized. (It remains ionized all the way out through the solar system, even though its temperature may drop to a low level; the gas is so tenuous that the separated protons and electrons have only a small probability of combining.) What is the nature of this magnetic field? Presumably its source is the general magnetic field of the sun. The corona cannot carry away the sun's concentrated local fields associated with sunspots and active regions, because these are strong enough to prevent the portions of the corona in their vicinity from streaming away at all. The sun's general field amounts to one or two gauss. (The earth's is about half a gauss.)

If the sun did not rotate (as it does once every 25 days), the solar wind would draw its general magnetic field straight out into space, so that the lines of force would stretch radially from the sun and a compass in solar-system space would always point straight toward or away from the sun. The sun's rotation, however, imposes on this radial field a circular field, with the result that the field carried by the solar wind takes a spiral form [*see illustrations on this page*].

The strength of a radial magnetic field, like that of gravity and light, weakens at a rate proportional to the square of the increase in distance from the source. It can be calculated, therefore, that at our distance from the sun

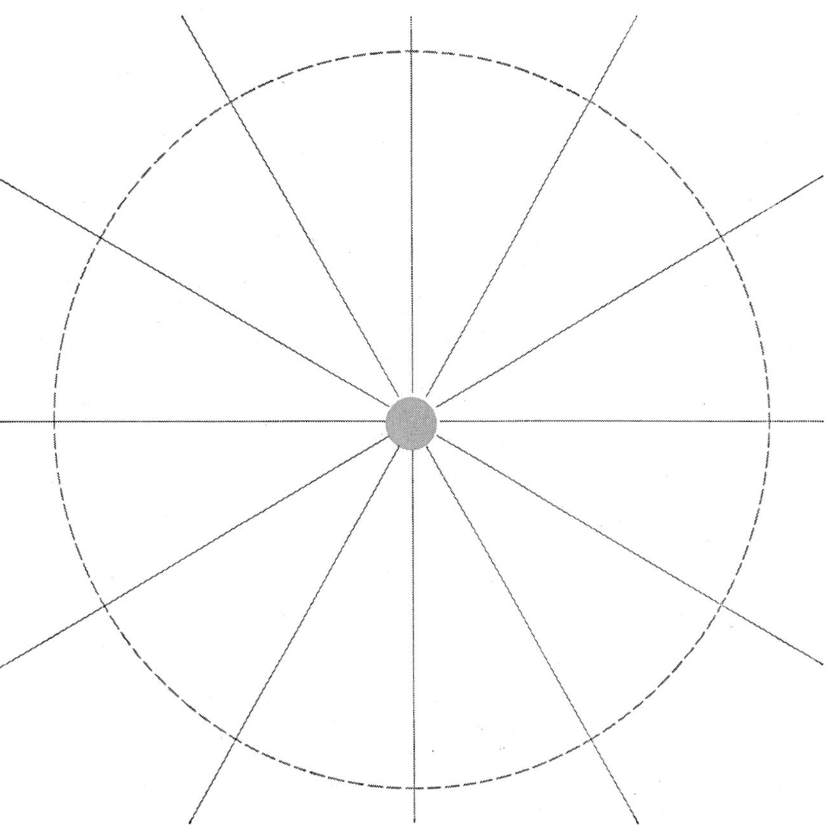

MAGNETIC LINES OF FORCE associated with the solar wind would appear as shown above if the sun did not rotate. The lines are in the equatorial plane of the sun. The broken circle marks the orbit of the earth, which is one astronomical unit from the sun.

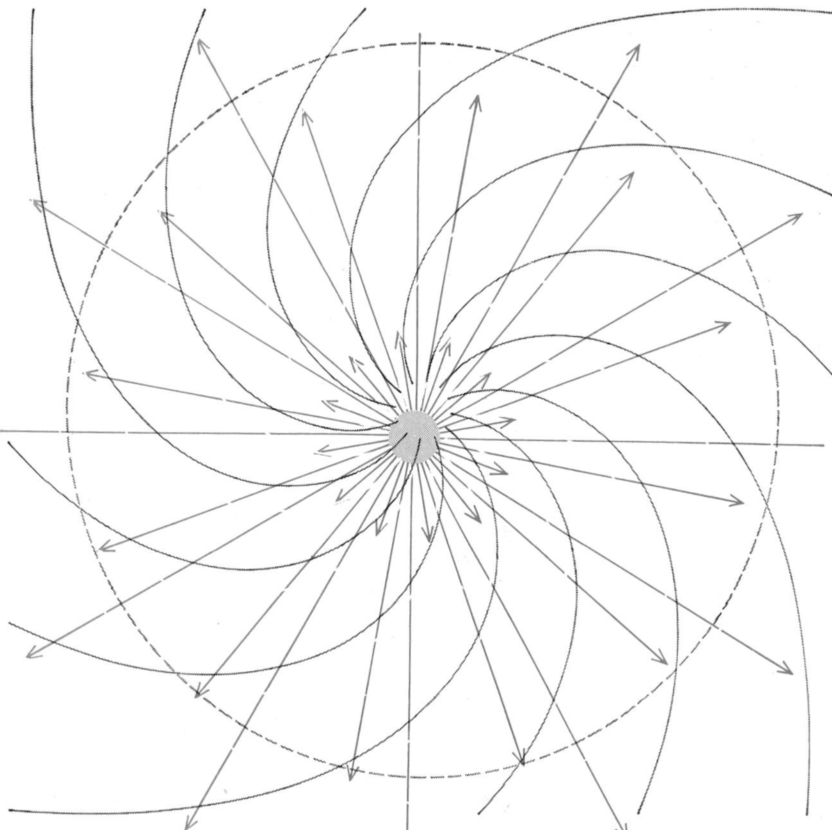

ACTUAL LINES OF FORCE are spirals due to solar rotation. They show how a compass needle would line up at any particular spot. The arrows are paths of solar-wind particles. Here the wind is assumed to be traveling through space at a steady 300 kilometers per second.

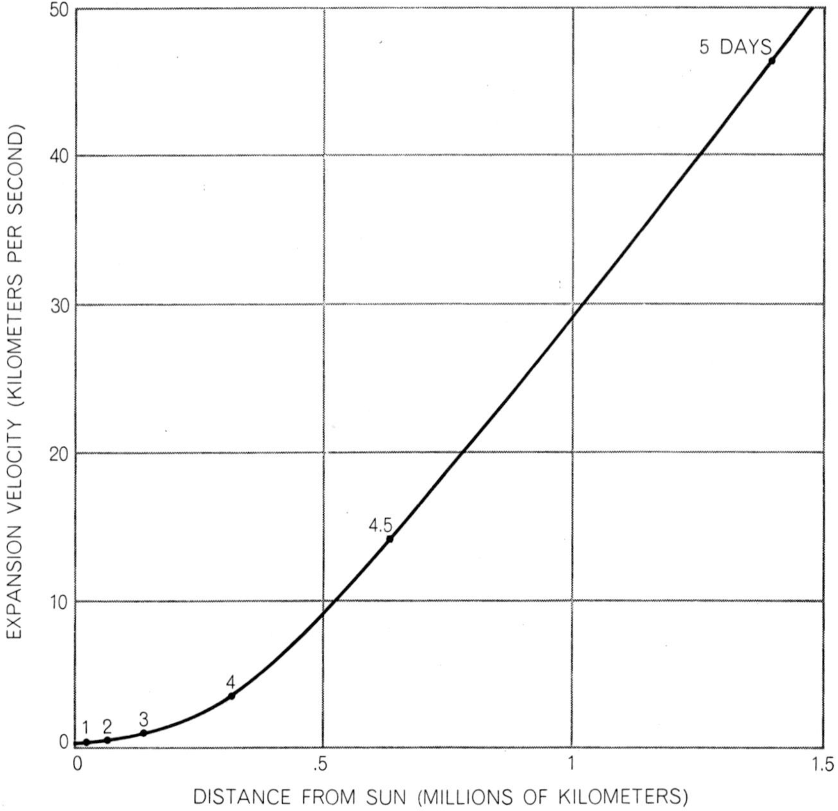

EXPANSION VELOCITY of solar corona near the sun increases rapidly after a relatively slow start. This is because the particles from the sun meet no resistance in space.

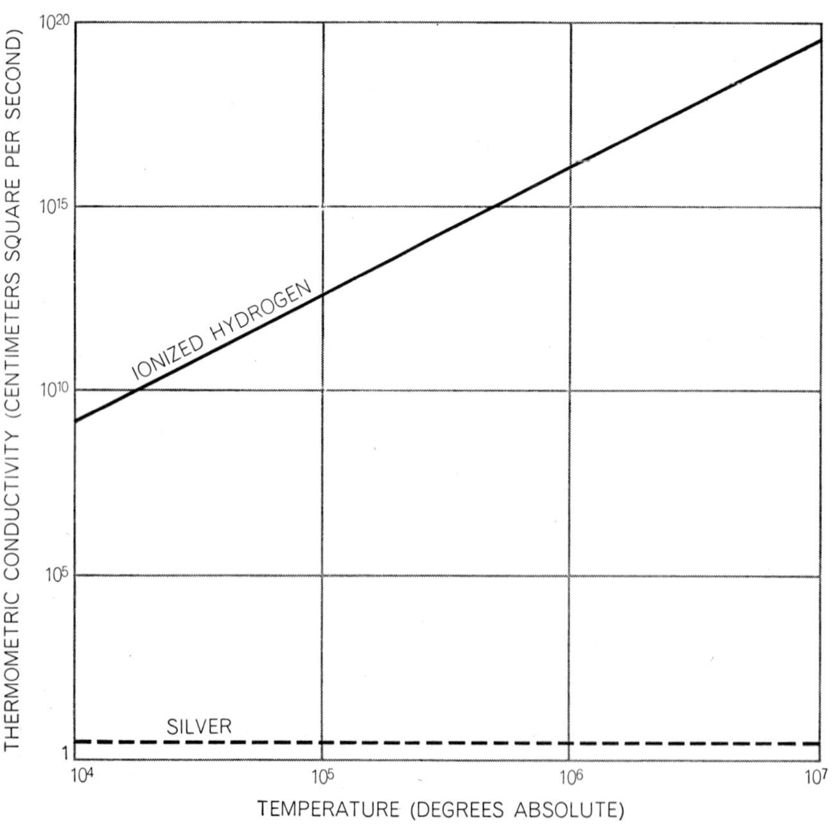

ABILITY TO CARRY HEAT of the solar wind is indicated by the thermometric conductivity of ionized hydrogen at a pressure typical of lower corona. Broken horizontal line is approximate value for solid silver, which could not exist at such temperatures.

the magnetic field carried by the solar wind should be down to about three or four hundred-thousandths of a gauss.

Evidence from Spacecraft

What have space probes shown about the solar wind? Many of the vehicles have carried equipment for recording charged particles encountered in space. In the first place, it can be said that they have unmistakably confirmed the existence of the wind. It was detected and measured by the Soviet vehicles *Lunik I* and *Lunik II* and by several U.S. vehicles, including the Venus craft *Mariner II* and the satellite *Explorer X*. They have shown that the wind blows continuously throughout the space they have traversed and that near the earth it is traveling at the expected velocity of about 400 kilometers per second. It blows straight out from the sun, sometimes steadily and sometimes in gusts. It tends to be turbulent and to move faster when the sun is active. The density of the wind has been hard to pin down. *Lunik I* and *Lunik II* indicated a flow rate of perhaps 100 million protons per square centimeter per second. *Explorer X* and *Mariner II* found that the wind's mean density near the earth lies in the range of one to 10 protons per cubic centimeter most of the time. This is in accord with the model of the corona that assumes that its temperature is close to a million degrees throughout the gas for a considerable distance from the sun.

In addition, the space vehicles' measurements of the magnetic field in interplanetary space bear out the theoretical picture of the solar wind. *Mariner II* and *Pioneer V* measured the field as being a few hundred-thousandths of a gauss, and *Mariner II* indicated that on the average the field had the expected spiral pattern. There were kinks and wiggles in the observed pattern, but this would merely confirm that the solar wind is sometimes gusty.

Fortified with all these confirmations of the nature of the solar wind and with some definite measurements, we can proceed to explore several interesting questions. For example, there is the matter of how much energy and mass the solar wind carries off into space. It can be calculated that it removes hydrogen from the sun at the rate of about a million tons per second. This is not a significant drain on the sun; in the estimated 15-billion-year lifetime of the sun it would amount to only a little more than a hundredth of 1 per cent of the solar mass. Similarly, the energy consumed in expanding the corona to the solar-wind velocity is only about a millionth of the total energy output of the sun. The wind's energy per unit of volume is so slight that an object in space is not warmed significantly by it.

How Far Does the Wind Blow?

There is also the question of how far the solar wind goes into space. This is considerably more interesting than the drain on the sun because it offers the possibility of using the solar wind as a probe into interstellar space.

The density of the wind must drop off in proportion to the square of the increase in distance from the sun. Eventually the wind must become so tenuous that it is stopped by the other thinly dispersed gases and weak magnetic fields in interstellar space. The general magnetic field of space in our galaxy is estimated to be no more than two hundred-thousandths of a gauss. If we take this maximum figure as the strength of resistance to the solar wind and use as our index of the wind's density the smallest value that has been measured near the earth (one atom per cubic centimeter), we can calculate that the solar wind ends at about 12 astronomical units (12 times the sun-earth distance) from the sun—that is, a little beyond Saturn. At the other extreme, if we take the smallest estimate of the resisting magnetic field (one two-hundred-thousandth of a gauss) and the highest measurement of the wind's density near the earth (10 atoms per cubic centimeter), then the wind goes out to 160 astronomical units —four times the distance of the farthest planet, Pluto. These, then, seem to be the lower and upper limits: the solar wind apparently extends to at least 12 but not more than 160 astronomical units from the sun.

Two possibilities are at hand for exploring the outer limits of the wind. One is based on the fact that the hydrogen in interstellar space has been observed to emit faint ultraviolet radiation when it is excited. A recent analysis of such emission by Thomas N. L. Patterson, Francis S. Johnson and William B. Hanson of the Graduate Research Center of the Southwest in Dallas, Texas, suggests that the solar wind ends at perhaps 20 astronomical units from the sun.

The second possibility stems from the fact that the solar wind's magnetic field tends to sweep cosmic rays out of the solar system. During the years of high solar activity the intensity of cosmic rays coming to the earth is cut at least in half. We have calculated that a reduction of this size means that the solar wind extends well beyond Jupiter (five astronomical units from the sun). Simpson recently presented direct evidence that it probably goes out to at least 40 or 50 astronomical units. Analyzing the decline and recovery of cosmic ray intensity during the 11-year sunspot cycle, he found that the increase of the intensity of the higher-energy cosmic ray particles lags at least six months behind the drop in solar activity. The time lag apparently is a measure of the distance to the farthest extent of the solar wind. Just as it takes a certain time for a given ripple started in the middle of a pond to reach the edge of the pond, so will it take a certain time for an increase or decrease in the strength of the solar wind to be communicated to the outer boundaries of the wind. Therefore there is a delay between a drop in the sun's activity, with the consequent weakening of the solar wind, and the arrival of the weakened wind at the limit of the space in which it acts as a barrier to the entry of cosmic rays into the solar system. Since Simpson finds the delay to be at least six months, and the wind travels at the rate of one astronomical unit in four days, simple computation shows that the distance to the borders of the solar wind is at least 40 to 50 astronomical units.

There is a great deal more to the observations than this one distance number. All in all, the fluctuations in cosmic ray intensity provide us with a natural probe for exploring the fields and other conditions in space out to the borders of the solar system and beyond, because the fluctuations bear the mark of the distant fields.

Do other stars have winds like the sun's? Very likely. The main requirement is that the star have a hot corona. Our sun's corona is generated by churning and convection of the gas beneath its photosphere. According to the theoretical picture of the interior of stars, subsurface convection is likely to occur in any ordinary hydrogen star with a surface temperature of less than 6,400 degrees. Most of the stars in our galaxy fall into this class, therefore stellar winds must be rather common.

The light from distant stars cannot tell us whether they have a corona or not. One really has to live in the midst of the wind to detect it. So most of our knowledge of stellar winds in general must come from studying the wind of the nearby star, the sun.

10

Neutrinos from the Sun

JOHN N. BAHCALL

July 1969

Most physicists and astronomers believe that the sun's heat is produced by thermonuclear reactions that fuse light elements into heavier ones, thereby converting mass into energy. To demonstrate the truth of this hypothesis, however, is still not easy, nearly 50 years after it was suggested by Sir Arthur Eddington. The difficulty is that the sun's thermonuclear furnace is deep in the interior, where it is hidden by an enormous mass of cooler material. Hence conventional astronomical instruments, even when placed in orbit above the earth, can do no more than record the particles, chiefly photons, emitted by the outermost layers of the sun.

Of the particles released by the hypothetical thermonuclear reactions in the solar interior, only one species has the ability to penetrate from the center of the sun to the surface (a distance of some 400,000 miles) and escape into space: the neutrino. This massless particle, which travels with the speed of light, is so unreactive that only one in every 100 billion created in the solar furnace is stopped or deflected on its flight to the sun's surface. Thus neutrinos offer us the possibility of "seeing" into the solar interior because they alone escape directly into space. About 3 percent of the total energy radiated by the sun is in the form of neutrinos. The flux of solar neutrinos at the earth's surface is on the order of 10^{11} per square centimeter per second. Unfortunately the fact that neutrinos escape so easily from the sun implies that they are difficult to capture.

Nevertheless, within the past year a giant neutrino trap has begun operating in a rock cavity deep below the surface in the Homestake Mine in Lead, S.D. The neutrino trap is a tank filled with 100,000 gallons of tetrachloroethylene (C_2Cl_4), an ordinary cleaning fluid. The experiment is being conducted by Raymond Davis, Jr., of the Brookhaven National Laboratory, with the assistance of Kenneth C. Hoffman and Don S. Harmer. In 1964 Davis and I showed that such an experimental test of the hypothesis of nuclear burning in stars was feasible. The idea was strongly supported by, among others, William A. Fowler of the California Institute of Technology, Richard W. Dodson, chairman of the Brookhaven chemistry department, and Maurice Goldhaber, the director of Brookhaven. Subsequently the Homestake Mining Company contributed valuable technical help.

The initial results published by Davis and his co-workers have left astronomers and astrophysicists somewhat puzzled because the neutrino flux rate seems low. It is less than half the theoretical value one obtains by assuming certain "standard" values for quantities used in constructing theoretical models of the solar interior. I shall discuss the range of theoretical predictions later. The important initial fact is that one can now use the results of the experiment to improve our knowledge of the sun's thermonuclear furnace.

Neutrinos were first suggested as hypothetical entities in 1931 after it was noted that small amounts of mass seemingly vanish in the radioactive decay of certain nuclei. Wolfgang Pauli suggested that the mass was spirited away in the form of energy by massless particles, for which Enrico Fermi proposed the name neutrino ("little neutral one"). Fermi also provided a quantitative theory of processes involving neutrinos. In 1956 Frederick Reines and Clyde L. Cowan, Jr., succeeded in detecting neutrinos by installing an elaborate apparatus near a large nuclear reactor. Such a reactor emits a prodigious flux of antineutrinos produced by the radioactive decay of fission products. For purposes of demonstrating a particle's existence, of course, an antiparticle is as good as a particle.

In the late 1930's Hans A. Bethe of Cornell University followed up Eddington's 1920 suggestion of the nuclear origin of the sun's energy and outlined how the fusion of atomic nuclei might enable the sun and other stars to shine for the billions of years required by the age of meteorites and terrestrial rocks. Since the 1930's the birth, evolution and death of stars have been widely studied. It is generally assumed that the original main constituent of the universe was hydrogen. Under certain conditions hydrogen atoms presumably assemble into clouds, or protostars, dense enough to contract by their own gravitational force. The contraction continues until the pressure and temperature at the center of the protostar ignite thermonuclear reactions in which hydrogen nuclei combine to form helium nuclei. After most of the hydrogen has been consumed, the star contracts again gravitationally until its center becomes hot enough to fuse helium nuclei into still heavier elements. The process of fuel exhaustion and contraction continues through a number of cycles.

The sun is thought to be in the first stage of nuclear burning. In this stage four hydrogen nuclei (protons) are fused to create a helium nucleus, consisting of two protons and two neutrons. In the process two positive charges (originally carried by two of the four protons) emerge as two positive electrons (antiparticles of the familiar electron). The

NEUTRINO TRAP is a tank filled with 100,000 gallons of a common cleaning fluid, tetrachloroethylene. It is located in a rock cavity 4,850 feet below the surface in the Homestake Mine in the town of Lead, S.D. The experiment is being run by Raymond Davis, Jr., Kenneth C. Hoffman and Don S. Harmer of Brookhaven National Laboratory. Suggested in 1964 by Davis and the author, the experiment was begun last year. The first results showed that the sun's output of neutrinos from the isotope boron 8 was less than expected.

DEEP-MINE LOCATION shields the solar-neutrino detector from the intense flux of energetic particles produced when cosmic ray protons collide with atomic nuclei in the atmosphere or in the solid earth. Here a positive pion (π^+) generated in an atmospheric collision decays into a positive muon (μ^+) and a "muon neutrino." High-energy muons are very penetrating and can knock protons out of atomic nuclei well below the earth's surface. If such a proton entered the neutrino detector, it could mimic the entry of a solar neutrino by converting an atom of chlorine 37 (^{37}Cl) into an atom of radioactive argon 37 (^{37}Ar).

fusion also releases two neutrinos and some excess energy, about 25 million electron volts (MeV). This energy corresponds to the amount of mass lost in the overall reaction, which is to say that a helium nucleus and two electrons weigh slightly less than four protons. The 25 MeV of energy so released appears as energy of motion in the gas of the solar furnace and as photons (particles of radiant energy). This energy ultimately diffuses to the surface of the sun and escapes in the form of sunlight and other radiation.

Bethe and C. F. von Weizsäcker of Germany independently proposed one mechanism for assembling four protons into a helium nucleus. Because it involved nuclei of carbon, nitrogen and oxygen it became known as the CNO cycle [*see illustration on page 113*]. The cycle starts with a nucleus of ordinary carbon, $^{12}_{6}C$. (The symbol specifies a nucleus containing a total of 12 nucleons, of which six are protons and the rest neutrons.) Three protons are added one at a time, culminating in a nitrogen nucleus containing eight neutrons and seven protons ($^{15}_{7}N$). With the addition of another proton a reaction occurs in which two nuclei are produced: $^{12}_{6}C$ (the original nucleus) and $^{4}_{2}He$, which is ordinary helium.

In each cycle two neutrinos are emitted whose maximum energies are greater than 1 MeV. One comes from the radioactive decay of ^{13}N and the other from the decay of ^{15}O. (For simplicity I shall start omitting the subscripts indicating the number of protons in the nucleus.) The rates at which nuclear reactions in the CNO cycle occur in stars have been carefully studied over the past 20 years at the California Institute of Technology in the W. K. Kellogg Radiation Laboratory, first under the leadership of Charles C. Lauritsen and now of Fowler.

An altogether different series of nuclear reactions known as the proton-proton chain, also investigated 30 years ago by Bethe, can accomplish the same fusion of helium from four protons [*see illustration on page 114*]. In the first step of the chain two protons combine to form a deuteron, ^{2}H, the nucleus of heavy hydrogen. The deuteron then combines with a proton to form a light helium nucleus, ^{3}He. The next reaction can go in one of three directions. We estimate that in the sun's interior two nuclei of ^{3}He combine to form an ordinary helium nucleus, ^{4}He, with the release of two protons in about 91 percent of the cases. The other two possible

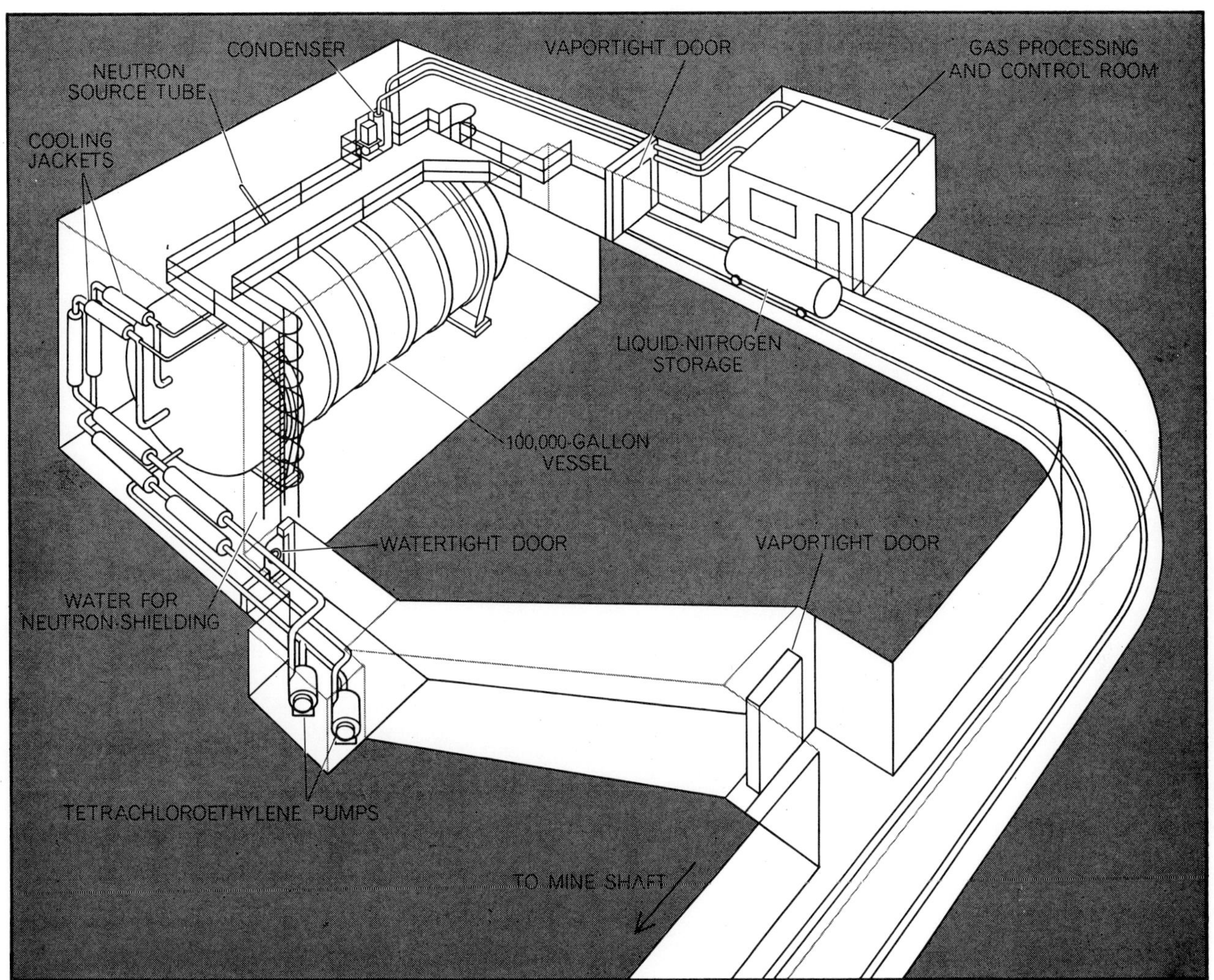

SOLAR-NEUTRINO DETECTOR is a tank 20 feet in diameter and 48 feet long, holding 100,000 gallons of tetrachloroethylene (C_2Cl_4). On the average each molecule of C_2Cl_4 contains one atom of the desired isotope, ${}^{37}_{17}Cl$, an atom with 17 protons and 20 neutrons. The other three chlorine atoms contain two less neutrons and are designated ${}^{35}_{17}Cl$. When a neutrino of the right energy reacts with an atom of ${}^{37}_{17}Cl$, it produces an atom of ${}^{37}_{18}Ar$ and an electron. The radioactive argon 37 is allowed to build up for several months, then is removed by purging the tank with helium gas. The argon is adsorbed in a cold trap and assayed for radioactivity.

branches, or routes, involve the formation of nuclei of lithium, beryllium and boron (${}^7Li$, ${}^7Be$, ${}^8Be$ and ${}^8B$), which give rise eventually to two helium nuclei.

At the time Bethe first investigated the proton-proton chain there was little experimental information on the rates of the relevant nuclear reactions. Since then laboratories all over the world have provided the data for a detailed understanding of the chain and its several branches. At the low energies believed to exist in the solar furnace (a few thousand electron volts) the probability of the occurrence of any given reaction in the proton-proton chain is low and hence difficult to measure. Nevertheless, the experimental group at the Kellogg Laboratory, aided by a succession of able graduate students, has refined the difficult experiments to the point where most of the information necessary for predicting the rates of reactions in the proton-proton chain is now available.

Three of the reactions in the proton-proton chain are of special importance for solar-neutrino experiments. Referring again to the illustration on page 114, these are the basic proton-proton reaction [*Reaction 1*], the proton-electron-proton, or "pep," reaction [2] and the decay of a radioactive isotope of boron, ${}^8B$[*10*]. All produce neutrinos but only the second and third produce neutrinos energetic enough to start a reaction in the tetrachloroethylene detector.

The proton-proton reaction is the slowest in the proton-proton chain, and hence it determines the overall rate at which energy is produced. Unfortunately the rate of the reaction is so slow that it cannot be measured in the laboratory; the "weak" force that governs this reaction is the same force that determines the interaction of neutrinos with matter. Over the years the rate of the reaction under stellar conditions has been estimated by a number of theorists. Last year I collaborated with Robert M. May of the University of Sydney in making a new estimate that we believe is accurate to within 5 percent.

The pep reaction differs from the basic proton-proton reaction only in that it has a negative electron present initially rather than a positive electron present after the reaction. Its rate at solar densities and temperatures is even slower than the proton-proton reaction. May and I estimate that the pep reaction occurs 1/400th as often as the proton-proton reaction under solar conditions. The pep neutrinos, which are 3½ times more en-

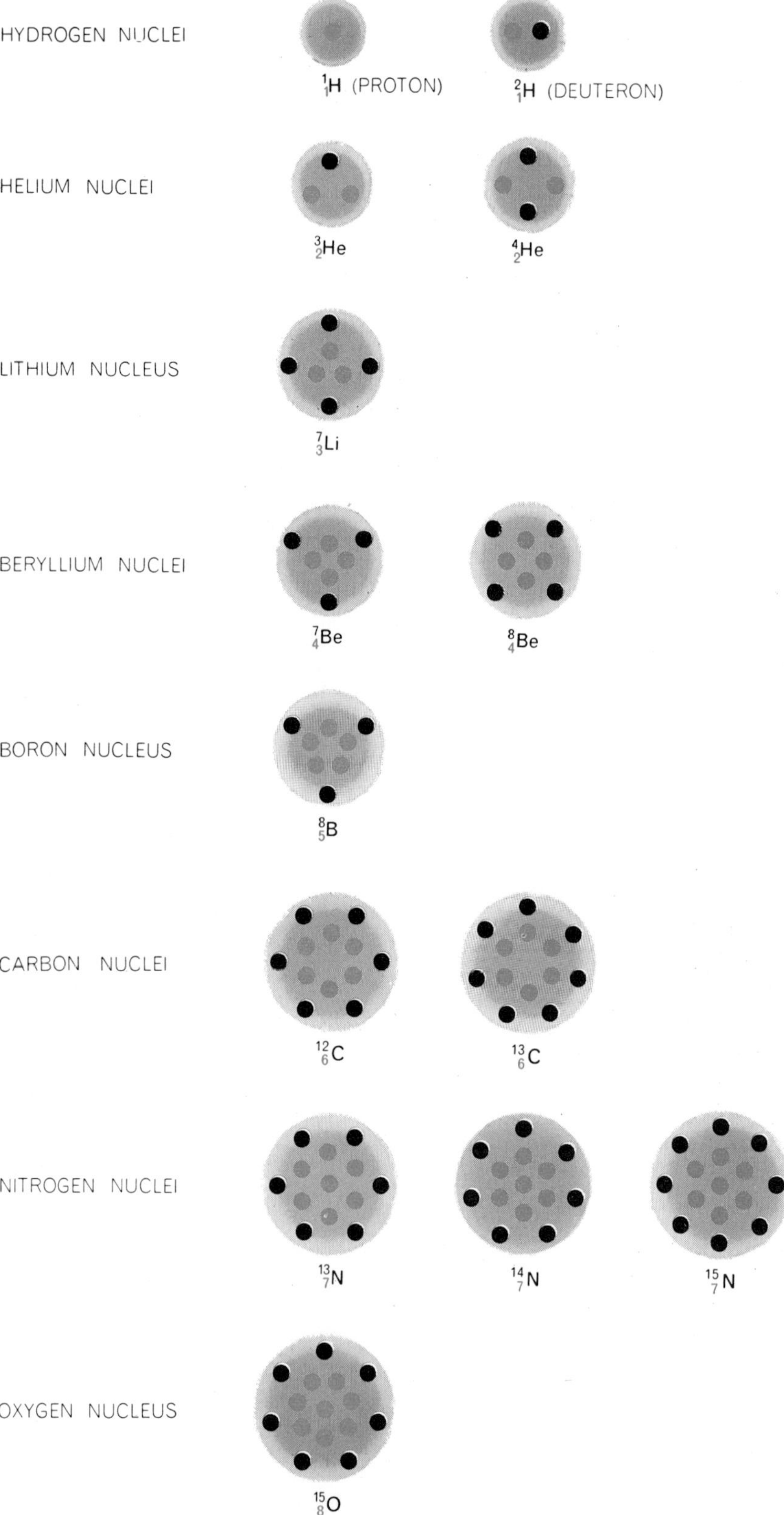

SOURCES OF SUN'S ENERGY are believed to be the atomic nuclei symbolized here, which may be present either as reactants or as products in the sun's thermonuclear furnace. The basic fuel is ordinary hydrogen, $^1_1$H, whose nucleus consists of a single proton. Four protons can be fused into a helium nucleus, $^4_2$He, by two principal mechanisms, one called the CNO cycle because it involves carbon, nitrogen and oxygen nuclei (*see illustration on opposite page*) and the other known as the proton-proton chain (*see illustration on page 114*). Protons are represented by colored dots, neutrons by black dots, arranged in arbitrary patterns. Electrons (positive and negative), photons and neutrinos also take part in the reactions.

ergetic than the most energetic proton-proton neutrinos, should be barely detectable. As we shall see, their capture rate establishes the minimum rate compatible with the hypothesis that the sun has a thermonuclear furnace.

The third reaction of special importance, the decay of radioactive boron, ^{8}B, produces the most energetic neutrinos of all: they have a maximum of 14.06 MeV, or nearly 10 times the maximum energy of the pep neutrinos. The ^{8}B is formed when beryllium, ^{7}Be, adds a proton, a reaction [9] that occurs in a rare branch of the proton-proton chain. This branch begins with the fusion of light and heavy helium nuclei, ^{3}He and ^{4}He, which form ^{7}Be. In 1958 Harry D. Holmgren and R. L. Johnson of the Naval Research Laboratory discovered that this reaction is significantly faster than had been thought. It proceeds at a rate of about once for every 1,000 occurrences in the sun of the more probable ^{3}He-plus-^{3}He reaction. Immediately following this discovery Fowler and A. G. W. Cameron suggested that the decay of ^{8}B might produce a detectable flux of solar neutrinos. I subsequently made some calculations that showed that the capture probability for the energetic neutrinos emitted by ^{8}B was 18 times larger than had been previously estimated. On the basis of this calculation Davis suggested in 1964 the experiment eventually located at the Homestake mine.

We are now ready to ask: How can tetrachloroethylene serve as a detector of solar neutrinos? Some 20 years ago Bruno M. Pontecorvo, then at the Chalk River Nuclear Laboratories in Canada, pointed out that an isotope of chlorine, ^{37}Cl, could capture a neutrino and be transformed into an isotope of the rare gas argon, ^{37}Ar, with the release of an electron. Subsequently the suggestion was discussed in detail by Luis W. Alvarez of the University of California at Berkeley. On the basis of Alvarez' discussion Davis and Harmer attempted to observe the argon produced by antineutrinos from the decay of fission products. (They placed a 3,000-gallon detector near a nuclear reactor.)

The argon isotope produced by neutrino capture is unstable and reverts to ^{37}Cl by capturing one of its own orbital electrons. Fifty percent of a sample of ^{37}Ar atoms will undergo such a transformation in about 35 days. The decay process shakes loose a low-energy electron from the argon atom, and this elec-

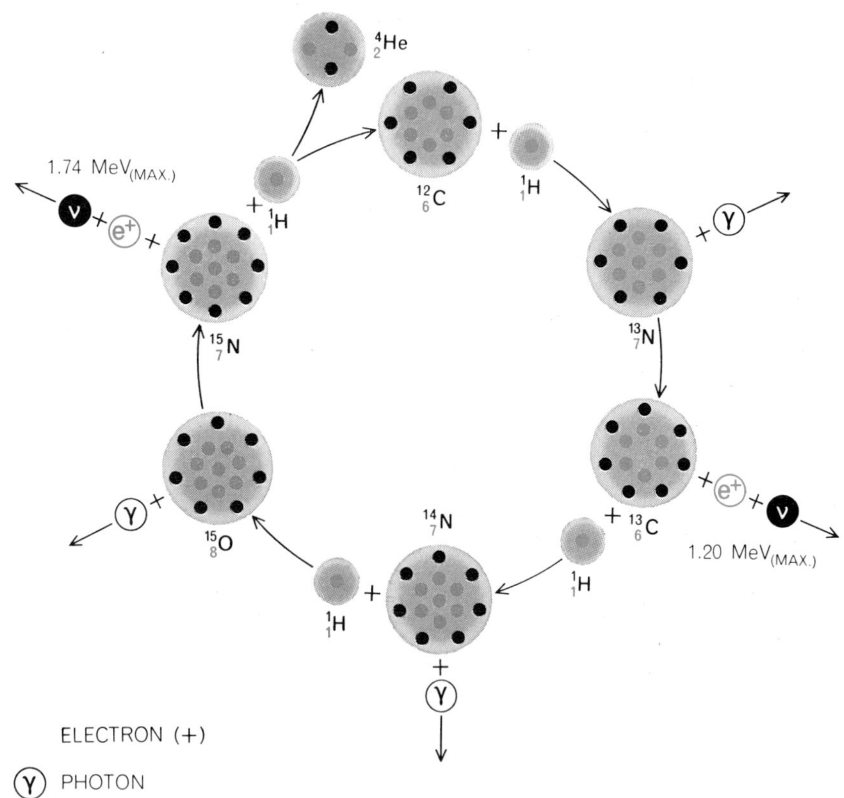

CNO CYCLE for fusing four protons into a helium nucleus employs ordinary carbon, $^{12}_{6}$C, as a catalyst, which is regenerated. Neutrinos are released in the second and fifth steps of the cycle. Because they share energy with the positive electrons that are emitted simultaneously, the neutrinos emerge with a spectrum of energies, whose maximum values are indicated. Unfortunately many of the neutrinos from the cycle lack the energy to trigger chlorine-37 detection system, which has a threshold of .81 million electron volt (MeV).

tron can be detected by counters placed around the sample. The detection of such electrons would be a sign that a few ^{37}Cl atoms had been transformed into ^{37}Ar atoms by neutrinos. The minimum neutrino energy for this reaction is .81 MeV [*illustration on page 115*].

For the detection scheme to work one must end up with at least a dozen or so ^{37}Ar atoms. Calculations suggested that a practical experiment would require a detector consisting of 100,000 gallons of a chlorine-containing fluid, such as tetrachloroethylene. (In natural chlorine a fourth of the atoms are the isotope ^{37}Cl.) In the experiment designed by Davis and his colleagues this volume of tetrachloroethylene is contained in a tank 20 feet in diameter and 48 feet long, located 4,850 feet underground [*see illustration on page 111*].

Why underground? The answer is that the detector must be shielded from the shower of subnuclear particles of all kinds produced when cosmic rays (chiefly high-energy protons) crash into the earth's atmosphere [*see illustration on page 110*]. Several reactions triggered by such particles could simulate the reaction Davis was looking for, but it is particularly important to exclude free protons from the tank because if ^{37}Cl absorbs a proton it can be converted into ^{37}Ar by the release of a neutron. Although one does not expect free protons to penetrate many feet of rock, muons produced by cosmic rays are very penetrating and can cause reactions that will release protons many feet below the surface. As a shield against neutrons, which are another hazard, the entire tank can be surrounded by water.

The tankful of tetrachloroethylene is exposed to the flux of neutrinos from the sun for several months to allow the atoms of ^{37}Ar to accumulate. (I might add that the flux of neutrinos from the rest of the universe presumably bears roughly the same relation to the solar-neutrino flux as starlight does to sunlight; hence it can be ignored.) The ^{37}Ar formed by neutrino capture is then removed from the bulk of the liquid by bubbling large quantities of helium gas through the system. About 10 cubic feet of helium is circulated through the tank per minute. The argon is separated from the helium by adsorbing it in a charcoal trap maintained at the temperature of liquid nitrogen (77 degrees Kelvin). The efficiency of the extraction procedure is determined in each experiment by adding to the 100,000 gallons of tetrachloroethylene a known amount (less than a cubic centimeter) of ^{36}Ar, a rare nonradioactive isotope of argon. Davis finds that purging the tank for 22 hours with helium will usually recover 95 percent of the ^{36}Ar.

The argon that is finally removed from the tank consists primarily of ^{36}Ar, deliberately inserted, together with two other isotopes: a few atoms of ^{37}Ar produced by solar neutrinos and a tiny amount of the ordinary nonradioactive isotope of argon, ^{40}Ar, that might have leaked into the tank from the air. After the sample of argon is purified chemically it is placed in a small counter holding about .5 c.c. of gas. The counter is made small to minimize its exposure to cosmic rays or other unwanted particles. It is protected from outside radiations by a series of shields and by large counters that signal when something has penetrated the outer defenses. The shape of each pulse occurring in a counter is photographed, and the pertinent data (such as the time of occurrence and the energy of the pulse) are stored on computer tape.

Ray Davis tells me that the experiment is simple ("Only plumbing") and that the chemistry is "standard." I suppose I must believe him, but as a nonchemist I am awed by the magnitude of his task and the accuracy with which he can accomplish it. The total number of atoms in the big tank is about 10^{30}. He is able to find and extract from the tank the few dozen atoms of ^{37}Ar that may be produced inside by the capture of solar neutrinos. This makes looking for a needle in a haystack seem easy.

Let me explain now how I calculated the probability that a solar neutrino that enters the tank of tetrachloroethylene will be captured by one of the ^{37}Cl atoms. The fraction of neutrinos with energies in any given range can be determined for a particular neutrino source from laboratory experiments. One can also calculate with the aid of Fermi's theory of neutrino processes the likelihood that an atom of ^{37}Cl will capture a neutrino of a particular energy. From a threshold probability of zero for a neu-

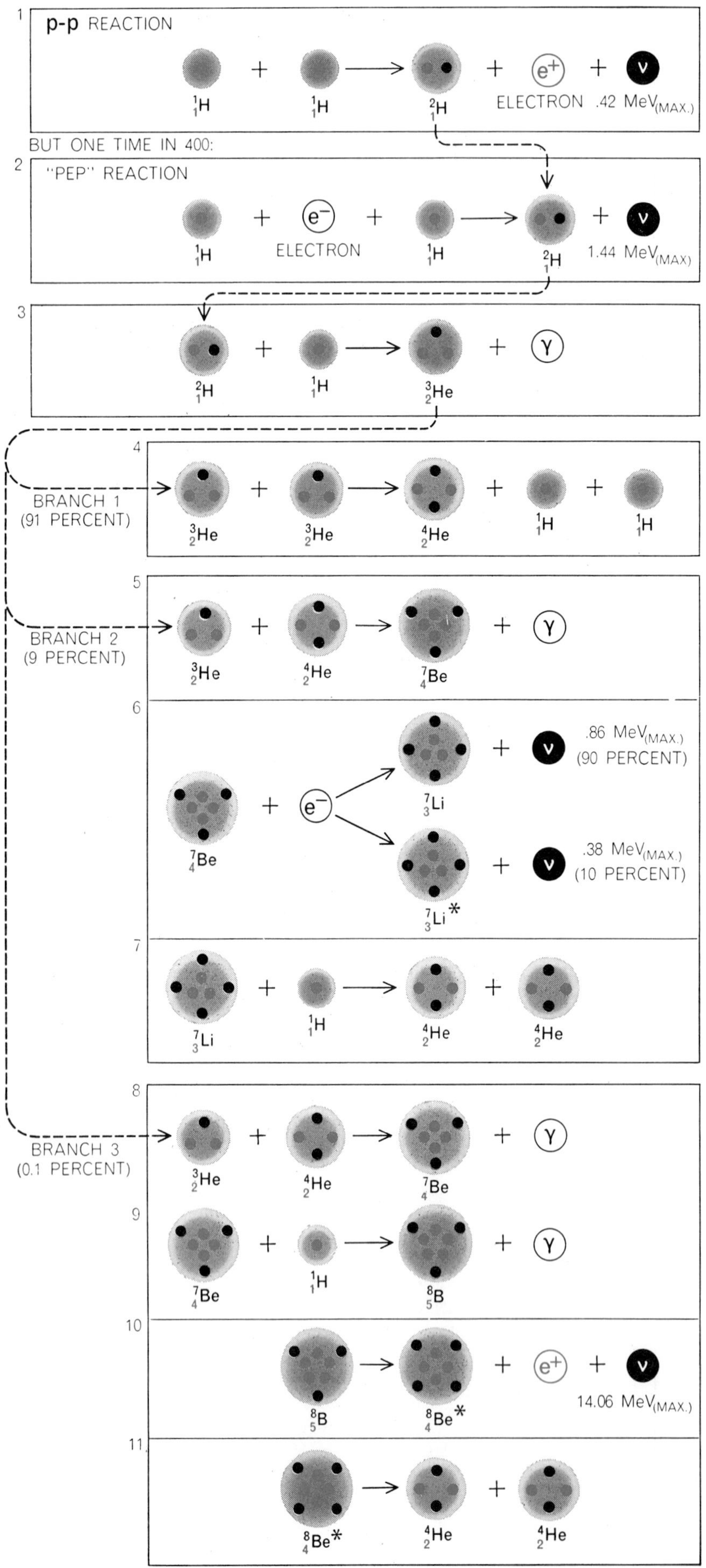

trino of .81 MeV the relative capture probability rises to 100 for a neutrino of 4 MeV, to 1,000 for a neutrino of 7 MeV and to 30,000 for a neutrino of 14 MeV. Thus the probability of capturing a 14-MeV neutrino from ^{8}B decay is some 3,000 times higher than the probability of capturing a 1.4-MeV neutrino from the pep reaction.

Moreover, the neutrinos from the decay of ^{8}B can do something that none of the other solar neutrinos can. They are so energetic they produce a ^{37}Ar nucleus that is in an excited state. This means that the nucleus has more internal energy than it would have in the ground, or normal, state. The significance of this is that favorable nuclear transitions can be caused by ^{8}B neutrinos that are not possible with the lower-energy neutrinos. The most important excited state of ^{37}Ar is quite similar to the ground state of ^{37}Cl; it is the nuclear analogue of the ground state of ^{37}Cl. The consideration of excited states led to an accurate determination of the probability that ^{37}Cl will capture an ^{8}B neutrino.

The chain of argument is based on the symmetry properties of nuclei containing the same number of nucleons and proceeds as follows. The nucleus $^{37}_{17}$Cl (I shall again indicate the protons by subscripts) should behave very much like the calcium nucleus $^{37}_{20}$Ca, which was unknown when I made my nuclear model. The model predicted that $^{37}_{20}$Ca should decay within 130 milliseconds, on the average, into an excited potassium nucleus $^{37}_{19}$K plus a positive electron and a neutrino. In a nuclear sense this is exactly analogous to the capture of a neutrino by $^{37}_{17}$Cl, producing $^{37}_{18}$Ar plus an electron.

About a year later the isotope ^{37}Ca was observed, and its decay rate was found to be within 25 percent of the value predicted. More important, subsequent measurements by Arthur M. Poskanzer and his associates at Brookhaven

PROTON-PROTON CHAIN is thought to be the dominant source of energy generation in the sun. The initial proton-proton reaction (*1*), which produces neutrinos undetectable with ^{37}Cl, establishes the basic rate for all subsequent reactions. Detectable neutrinos are released by the "pep" reaction (*2*), so named because the reactants are proton, electron and proton. Deuterons, $^2_1$H, produced in these two reactions fuse with protons to form a light isotope of helium, $^3_2$He (*3*). At this point the proton-proton chain breaks into three branches. A few barely detectable neutrinos are produced in the second branch by Reaction *6*. The most energetic neutrinos are released (*10*) in the rare branch involving boron 8.

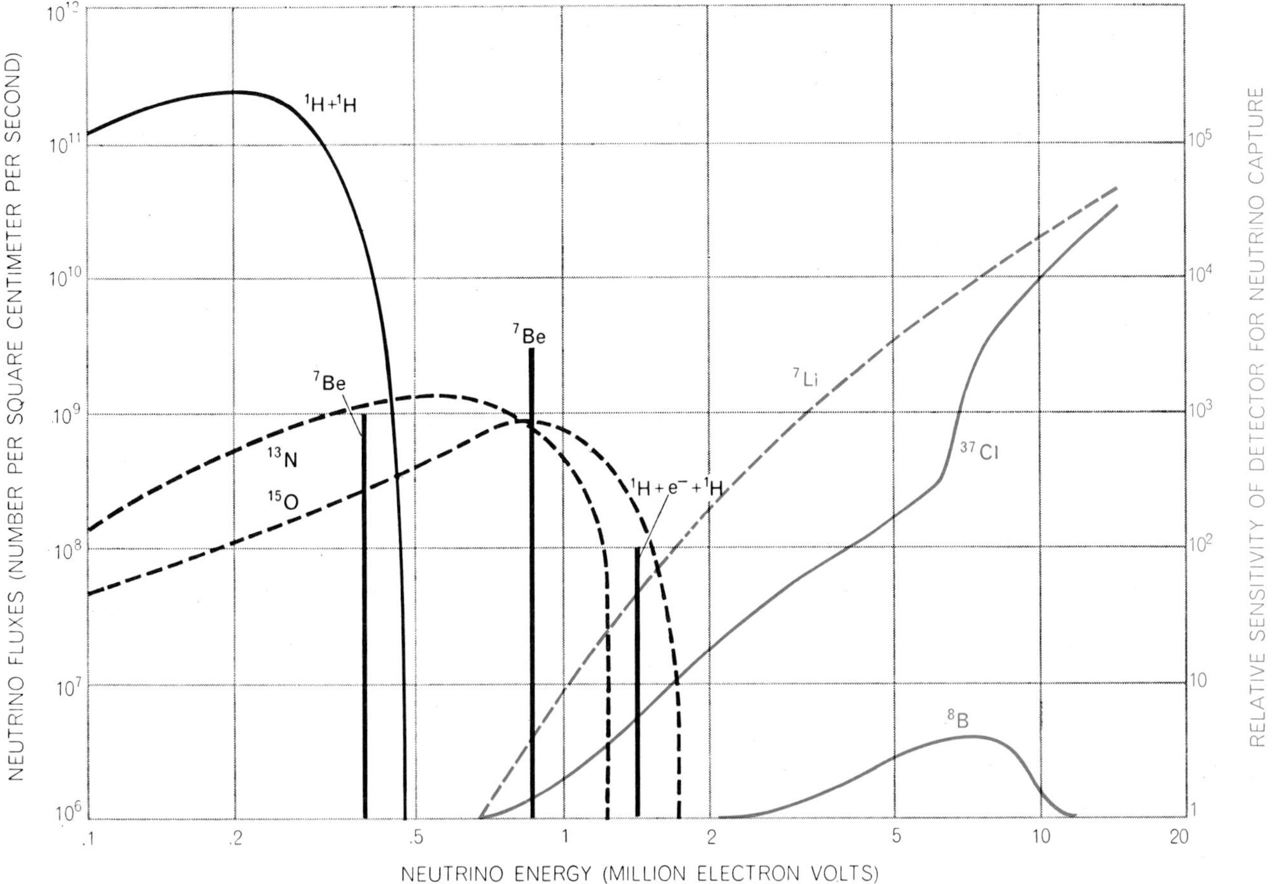

SPECTRUM OF SOLAR-NEUTRINO ENERGIES is plotted with curves showing the sensitivity of the ^{37}Cl detection system now in use (*solid line in color*) and the sensitivity of a proposed detection system employing lithium, ^{7}Li (*broken line in color*). Neither system is sensitive in the region below about .8 MeV, where the energies of most of the solar neutrinos would be found. The lithium system, however, would be more sensitive than the ^{37}Cl system to neutrinos produced by the pep reaction, $^{1}H + e^{-} + ^{1}H$. Most of the neutrinos expected to be captured by ^{37}Cl are those released by the decay of ^{8}B. Neutrinos from the proton-proton chain are indicated by solid black lines, neutrinos from the CNO cycle by broken lines. The neutrino fluxes are plotted as the number per square centimeter per second per MeV for continuum sources and as the number per square centimeter per second for line sources.

determined the fraction of decays of ^{37}Ca that lead to various excited states of ^{37}K. These were precisely the results needed to calculate the neutrino-capture rate of ^{37}Cl with an accuracy of better than 10 percent.

It is convenient to introduce a special unit to express the neutrino-capture rate in solar-neutrino experiments. The unit is the "solar-neutrino unit," or SNU (which we pronounce "snew"). One SNU equals 10^{-36} capture per second per target atom. This implies that an atom of ^{37}Cl would have to wait 10^{36} seconds, or roughly 10 billion billion times the age of the observable universe, before capturing a neutrino. Of course, in the 100,000-gallon tank, which contains about 2×10^{30} atoms of ^{37}Cl, the average waiting time for a single capture when the rate of capture equals 1 SNU is only 5×10^{5} seconds, or about six days per capture.

Let us now see how the capture rate in SNU's varies, depending on which reaction, or combination of reactions, one thinks is responsible for the sun's thermonuclear energy [*see illustration on next page*]. If the CNO cycle is the dominant source of the sun's energy, I estimate that the capture rate is 35 SNU. On the other hand, if the sun derives its energy from the proton-proton chain, as most theorists now believe, the problem of predicting the capture rate becomes difficult. One has to calculate precise models for the interior of the sun and estimate the average temperature of the solar furnace to an accuracy of .1 percent in order to predict the capture rate to an accuracy of a few percent, which is our usual aim.

The equations needed for such models have been known for some time. The first equation states that the gravitational attraction of the solar gas is balanced at each point in the sun by the thermal pressure of moving gas particles and by the pressure of radiation (photons). The second equation states that the total energy emitted by the sun represents the sum of all the energy released by the individual thermonuclear reactions. Finally, there is an equation that describes how energy is transported from the interior to the surface of the sun. This equation requires one to assume a particular chemical composition for the solar material so that one can estimate its opacity (that is, how strongly it impedes photons trying to reach the surface). We make the conventional assumption that the abundances observed spectroscopically at the surface of the sun are the same today as when the sun was formed. This assumption has been questioned, however, by Icko Iben, Jr., of the Massachusetts Institute of Technology, who has pointed out that our calculated primordial helium abundance for the sun is different from the abundance observed in some other stars. The situation is still somewhat unclear, although we are encouraged by the fact

BASIS OF EXPECTATION	ESTIMATED CAPTURE RATE (SOLAR NEUTRINO UNITS)
CARBON-NITROGEN-OXYGEN CYCLE	35
p-p CHAIN (STANDARD S_{17})	6
p-p CHAIN (INDIRECT S_{17})	3
GENERAL IDEAS OF SOLAR INTERIOR	1-3
ABSOLUTE MINIMUM ("PEP" NEUTRINOS)	.3

PREDICTED NEUTRINO CAPTURE RATES in the ^{37}Cl detection system are listed for various assumptions about thermonuclear processes in the sun. One solar neutrino unit (SNU) is defined as 10^{-36} capture per second per target atom or, alternatively, one capture per atom every 10^{36} seconds. If all the sun's energy came from the CNO cycle, the expected rate would be 35 SNU. The text discusses the basis of the various estimates. The first experimental value obtained by Davis and his associates indicated an upper limit of 3 SNU.

that our models enable us to calculate correctly the abundance of helium atoms observed in those cosmic rays that come from the sun.

All the quantities mentioned—pressure, reaction rates and opacity—must be computed at temperatures some 50,000 times higher and densities 100 times greater than those normally encountered on the earth. The central temperature of the sun is believed to be about 15 million degrees Kelvin and the central density about 150 grams per cubic centimeter. The calculation of reasonably accurate values for the opacity of stellar material alone has taken years of effort by Arthur N. Cox and his associates at the Los Alamos Scientific Laboratory.

The calculation of a detailed solar model requires about 10 minutes on a modern high-speed computer. The first calculations of solar-neutrino flux based on detailed models of the sun were published in 1963 by Fowler, Iben, Richard L. Sears and myself. Our 1963 model indicated that the capture rate would be about 50 SNU. Subsequent pioneering work was done by Sears. Since then I have been trying to estimate and reduce the uncertainties in our calculations that arise from imperfectly known parameters. This work has been carried out with a number of able collaborators, including most recently my wife Neta Bahcall, Giora Shaviv and Roger Ulrich. Our best solar model, using what we consider the most likely set of parameters for the proton-proton chain, predicts a capture rate of only 6 SNU—smaller by nearly a factor of 10 than predicted by our 1963 model. The main difference between our 1963 estimate and the present one results from improved measurements of nuclear reaction probabilities and of the sun's composition. (The composition work was done by D. Lambert and A. Warner of the University of Oxford.) We estimate that the uncertainty in the new value of the capture rate is roughly a factor of two or three.

About 80 percent of the expected rate of 6 SNU represents neutrinos from the decay of radioactive boron, ^{8}B, produced when ^{7}Be captures a proton. The parameter describing the rate of this reaction at solar temperatures is usually designated S_{17} (1 stands for proton and 7 for ^{7}Be). The "standard" value for S_{17} was determined in 1968 by Peter D. Parker of Yale University, and is the value that gives 6 SNU. An indirect determination of S_{17}, yielding a lower value, had been made a few years earlier at Cal Tech by Thomas A. Tombrello. He made use of reactions involving ^{7}Li and neutrons instead of ^{7}Be and protons. If this lower value of S_{17} is used, the estimated capture rate falls to 3 SNU. Independent of the uncertainty surrounding S_{17} and other parameters, my wife, Ulrich and I have established that the most general ideas concerning the solar interior predict a probable capture rate of between 1 and 3 SNU.

A minimum capture rate can be ob-obtained by calculating the contribution from pep neutrinos alone. Regardless of what model one selects for the interior of the sun, the ratio of pep reactions to standard proton-proton reactions is in the ratio of about one to 400. Moreover, the observed luminosity of the sun determines the rate of the basic proton-proton reaction. Therefore the capture rate attributable to pep neutrinos is an absolute lower limit (with one imaginative exception that will be mentioned below) consistent with the hypothesis that fusion reactions make the sun shine. Calculations I have made with the help of my wife, Shaviv and Ulrich show that this minimum capture rate is $.29 \pm .02$ SNU.

The results published last year by Davis and his co-workers show that the capture rate with ^{37}Cl is probably less than 3 SNU. The experimental value clearly implies that less than 10 percent of the sun's energy is generated by the CNO cycle. It also implies that the value of 6 SNU, based on "standard" parameters for the proton-proton chain, is at least twice too high. This discrepancy of a factor of two has caused considerable excitement in the scientific community. As we have seen, however, the uncertainties in some parameters are large enough so that Davis' result does not imply a conflict with general ideas about the solar interior.

Nevertheless, several theorists have suggested ways to explain the apparent discrepancy. D. Ezer and Cameron of the National Aeronautics and Space Administration suggest that the inner parts of the sun have somehow been mixed with the outer parts, thus reducing composition differences attributable to nuclear reactions in the solar interior. Their idea has been investigated quantitatively by a number of workers, who have found that mixing can indeed significantly reduce the expected solar-neutrino flux. Their suggestion, however, has not been widely accepted because the required amount and duration of mixing are quite large.

The most imaginative idea has come from the U.S.S.R., where V. N. Gribov and Pontecorvo (who originally suggested the use of ^{37}Cl to trap neutrinos) have proposed that neutrinos have a kind of double identity: approximately half the time they are the ordinary "electron neutrinos" we suppose them to be but half the time they are "muon neutrinos." Muon neutrinos were discovered at Brookhaven in 1962; they are created by reactions involving the production of muons. They seem to be identical in every way with ordinary electron neutrinos except that when they react with a proton or neutron they produce another muon rather than an electron [see "The Two-Neutrino Experiment," by Leon M. Lederman; SCIENTIFIC AMERICAN Offprint 324]. One can also prove that low-energy muon neutrinos will not react with ^{37}Cl. According to Gribov and Pontecorvo, the transformation from one kind of neutrino into another requires times (or distances) that are too large to be obtainable in the laboratory but that are available to neutrinos traveling from the sun to the earth. The neutrinos that arrive in their muon disguise cannot be detected, and thus all capture rates must be divided by a factor of

ARGON-EXTRACTION SYSTEM is deep underground next to the 100,000-gallon neutrino trap. Helium is circulated through the tank to sweep up any atoms of ^{37}Ar that have been formed from ^{37}Cl. The efficiency of the extraction is determined by previously inserting in the tank a small amount of ^{36}Ar, a rare, nonradioactive isotope of argon. The helium and argon pass through the apparatus at left, where the argon condenses in a charcoal trap cooled by liquid nitrogen. This argon fraction is purified in the apparatus at the right. The purified sample is then shipped to Brookhaven, where the content of ^{37}Ar is determined in shielded counters.

about two. The suggestion made by Gribov and Pontecorvo implies that the minimum capture rate due to pep neutrinos should be reduced. It will not be easy to test this unusual hypothesis, but it cannot be lightly dismissed.

Meanwhile Davis is attempting to improve the sensitivity of his experiment. The improvement, if achieved, should indicate clearly whether the present discrepancy between observation and expectation is due to errors in some experimentally determined parameters used in our solar models or is caused by a basic defect in our theories. Beyond that I believe someone should undertake an experiment using a nuclear species that is more sensitive to low-energy neutrinos than ^{37}Cl. Ordinary lithium, ^{7}Li, is such a nucleus; on capturing a neutrino it yields ^{7}Be and an electron. (This is the reverse of Reaction 6 illustrated on page 114.) Lithium would respond much better than ^{37}Cl to the pep neutrinos of 1.4 MeV [*see illustration on opposite page*] and to the low-energy neutrinos produced by the decay of ^{13}N and ^{15}O in the CNO cycle. The combined results of ^{37}Cl and ^{7}Li experiments would constitute a stringent test of present theories of stellar interiors and neutrino reactions. I would not be too surprised to find myself writing another article a few years from now explaining why the results of a successful experiment with ^{7}Li do not agree with our astrophysical expectations.

IV

Stellar Evolution

IV

Stellar Evolution

INTRODUCTION

Why do stars shine? What is the source of their great energy?

In 1800, such questions bothered astronomers very little. Energy conservation was an unformulated concept, and a time scale of a few thousand years appeared to encompass the entire history of the universe.

By 1900, these questions had emerged as a troubling problem. Geologists had found evidence for a time scale stretching back hundreds of millions of years, yet the only even partially adequate energy mechanism for stars then known—gravitational collapse—could offer a lifetime of only a few hundred thousand years. The various temperatures of stars suggested that some pattern of evolution existed. But since the energy sources remained an enigma, studies of stellar evolution (which were of vast potential importance for astronomy) remained for years on a strictly empirical basis.

A physically grounded theory of stellar evolution became possible only when scientists recognized the role of nuclear energy in stellar processes. The temporal calibration of the changes in stars brought about by the progressive burning of nuclear fuels in stellar interiors has been a triumph of the past two decades. It was finally made possible by the introduction of large electronic computers as a principal tool of theoretical astrophysics.

Recent computations have produced a picture of a slow but inexorable evolution. Yet this picture is only approximate. Chemical compositions, nuclear reaction rates, and the atomic opacity parameters that determine the flow of radiation from stellar interiors are incompletely and imperfectly known. The theoretical calculations must be guided by observations of the kinds of stars that actually exist.

Nevertheless, the emerging theory of stellar evolution provides at last a sensible scheme for interpreting the major patterns of the Hertzsprung-

Russell diagram, with its dwarf and giant stars; furthermore, the theory encompasses such bizarre and previously enigmatic stars as FU Orionis and WZ Sagittae.

FU Orionis appeared abruptly in 1936; its interpretation as a star newly collapsed from the interstellar medium is given by George Herbig in "The Youngest Stars." At the other extreme of the stellar life cycle is WZ Sagittae, a recurrent nova on its way to becoming a degenerate white dwarf star, described in "Dying Stars" by Jesse Greenstein.

Two earlier articles in this section, though still timely and exciting, now tell necessarily incomplete stories. Su-Shu Huang's ingeniously reasoned "Life outside the Solar System" provided the impetus for the now-famous "Project Ozma" at the National Radio Astronomy Observatory in Green Bank, West Virginia. After the publication of this article, Green Bank observers spent about 200 hours "listening" to Tau Ceti and Epsilon Eridani with the 85-foot radio telescope, but no signals were received that could be attributed to life elsewhere. Of course such a small-scale experiment barely scratches the surface of the possibilities, and Huang's arguments, closely tied to the current theory of stellar evolution, have stood with comparatively little criticism. Incidentally, one of the important recent advances in theory of stellar evolution, made around 1963 by Chiusuru Hayashi, changes the track of pre-main-sequence evolution as shown on Huang's diagram on page 126; results of Hayashi's calculations are shown in Herbig's article on page 143. However, this adjustment does not alter Huang's basic thesis.

Jan Oort's "The Crab Nebula" still conveys the excitement of an unexpected celestial phenomenon: the discovery of synchrotron radiation. His text is brief, just enough to support the splendid illustrations that carry the thrust of the message. Oort does draw attention to the southerly member of the pair of stars within the nebula: "But if this fragment is still a star, it must be totally different from any star that we have ever observed." He could scarcely have been more prescient! That totally different object has turned out to be a neutron star and a pulsar, but these discoveries are still so new that no article in our collection chronicles this. You can, however, find a wonderful picture of the object and a further description in the Introduction to Section VII on pages 245 to 251.

11

Life outside the Solar System

SU-SHU HUANG
April 1960

The evolution of stars and the evolution of living organisms appear to be completely dissimilar processes. The differences between the two can be explained in terms of how the particles of matter are bound together and how energy is exchanged among them. Indeed, the evolution of living organisms represents one outcome of stellar evolution. It was the steady flow of energy from the sun for four or five billion years that brought about the biological developments on earth, culminating in the emergence of intelligent organisms able to contemplate the whole remarkable story. If the sun had had a different history, life would not have appeared in its immediate vicinity.

Astronomical evidence acquired in recent years indicates that what has happened here is probably not unique. Two decades ago it was thought that the solar system might have originated in a near-collision of the sun and another star, which event supposedly pulled away enough matter from the sun to form the planets. Because such encounters must be rare events, they would give rise to few stars with a company of planets. Today most astronomers believe that a star is formed by the condensation of a cloud of dust and gas [see "The Dust Cloud Hypothesis," by Fred L. Whipple; SCIENTIFIC AMERICAN, May, 1948]. This hypothesis much more readily explains the origin of the solar system and is supported by the observation that more than half of the stars in our galaxy are double or multiple systems. In fact, it now appears that most stars are accompanied by other stars or by planets, though the latter must be so small as to escape sure detection by the present instruments of astronomy. Thus the appearance of life—even the appearance of mind—may be far from unusual events in the universe.

On the other hand, certain critical conditions must be satisfied if life processes are to be initiated and maintained. In the first place the star must shine long enough and steadily enough to permit life to evolve. The star must also be hot enough to warm up a habitable zone deep enough to offer a reasonable chance that a planetary orbit will fall within it. And the planet must ply a stable orbit within this zone. The number of stars that have given rise to life must therefore be considerably smaller than that immediately suggested by the dust-cloud hypothesis.

Taking account of all these factors, what is the probability that man will ever be able to visit the life-bearing planet of another star, or that the earth will receive a visitor from such a planet? Does the possibility justify taking measures now—in advance of a visit—to put existing technology to the task of scanning the sky for signals from intelligent organisms outside the solar system, or for transmitting signals in the hope they may be heard? If so, what sort of signals should be listened for or sent?

A reliable answer to these questions calls for a somewhat closer consideration of the conditions critical for life and an estimate of how often they are likely to be satisfied in the evolution of stars. The first condition is a steady and prolonged flow of energy. How long it takes life to evolve may be judged from the single instance available: Here on earth rational animals evolved from inanimate matter in about three billion years. Biological evolution proceeds by the purely random process of mutation, that is, the unpredictable occurrence of novel chemical processes among the fantastically numerous reactions that constitute life. Since the process is a random one, the laws of probability suggest that the time-scale of evolution on earth should resemble the average time-scale for the development of higher forms of life anywhere. The rate at which mutations occur is of course a variable in the calculation. It is affected by the electromagnetic and corpuscular radiation from the parent star, and the amount of such radiation that reaches the surface of any planet is governed in turn by the magnetic field of the planet, the depth and composition of its atmosphere and other factors. But since mutation itself is a random process, the introduction of a few more variables does not affect the calculation greatly. Furthermore, a higher mutation rate does not necessarily accelerate evolution, because most mutations are harmful. The more frequently they occur, the greater the chance that an individual will suffer an injurious mutation. In order to favor natural selection, mutations should be rare, perhaps

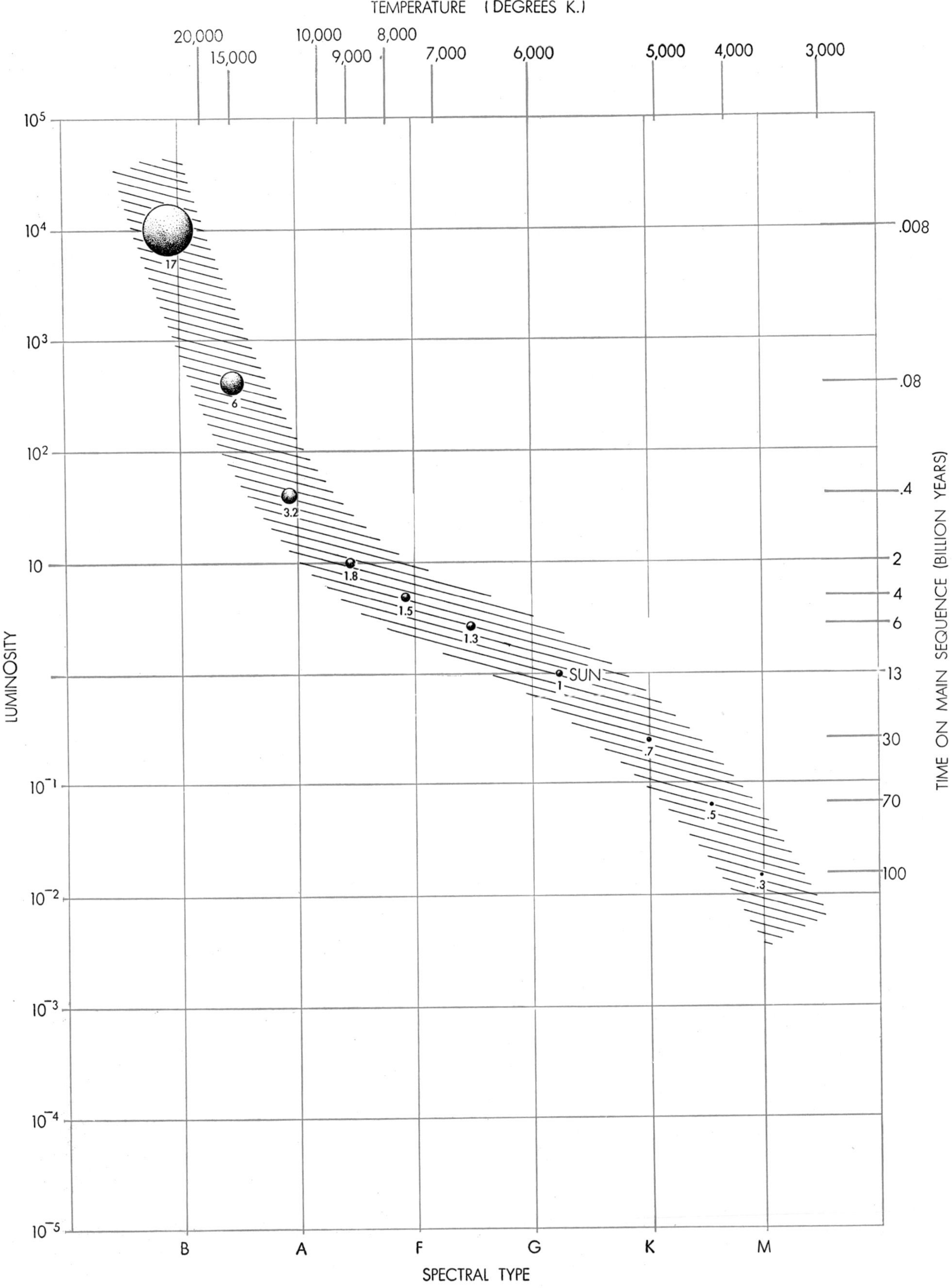

TYPICAL STARS ON THE MAIN SEQUENCE are represented on this diagram. Vertical scale at left indicates luminosity (the sun is unity). Scale at right shows time on the main sequence. The numbers at the top denote temperature in degrees Kelvin; the letters at the bottom, spectral type. Small figures beneath each star give its mass with respect to the mass of the sun.

as rare as in the evolution of life on earth.

It is somewhat easier to estimate the time-scale of stellar evolution. In contrast to the random nature of biological evolution, stellar evolution is governed by the universal law of gravitation and by a relatively small number of thermonuclear reactions. When a star begins to form in a cloud of dust and gas, gravitational attraction among the gas and dust particles causes the cloud to condense until the pressure raises the temperature within it to the point at which the thermonuclear reactions that convert hydrogen to helium begin. The tremendous quantities of energy liberated by these reactions now set up a counterpressure from the center of the star that exactly balances the force of gravitational contraction. In this state of equilibrium the star shines for a much longer time than that required for its condensation out of dust and gas.

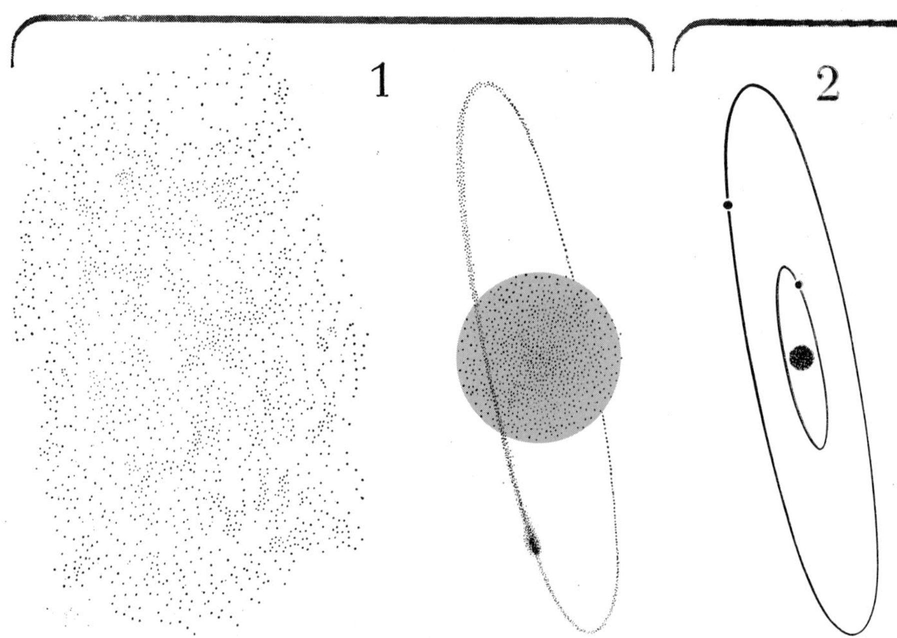

POSSIBLE EVOLUTION OF A STAR 1.2 times the mass of the sun begins with a dust cloud condensing into a star and proto-planets (1), a process taking perhaps 10 million years. Star enters "main sequence" (2) and remains there for approximately eight billion

The vast majority of the visible stars are in this phase of their evolution. They are called "main-sequence" stars because their luminosity (energy output per unit of time) plotted on a graph against their surface temperature places them in sequence in a narrow band [*illustration on page 123*]. What the chart shows is that the hottest stars are also the most luminous. Luminosity depends upon mass, and so the point at which a star appears on the main sequence depends primarily upon the mass of material incorporated in it during condensation. The hottest stars are designated by the letter O, followed in descending order by stars classified B, A, F, G, K and M; these classifications are usually called spectral type. The adjectives "early" and "late," which have nothing to do with the age of the star, are often used before the spectral-type designation to denote further relative temperature differences. An early F-type star is hotter than a late one, which in turn has a higher temperature than an early G-type star such as our sun. The classification is further refined by a number from 0 to 9 written after the letter. Thus the class of early B-type stars includes those from B0 to B4; and the late B-type stars, those from B5 to B9.

The length of time a star remains in the equilibrium state on the main sequence can be calculated from the total mass of hydrogen in its core and the

NOVA IN CONSTELLATION CYGNUS, indicated by pair of white lines, blazed up in 1920. This photograph was made at that time with the 100-inch reflecting telescope at Mount Wilson Observatory. Nova may represent a late stage in the life of a star.

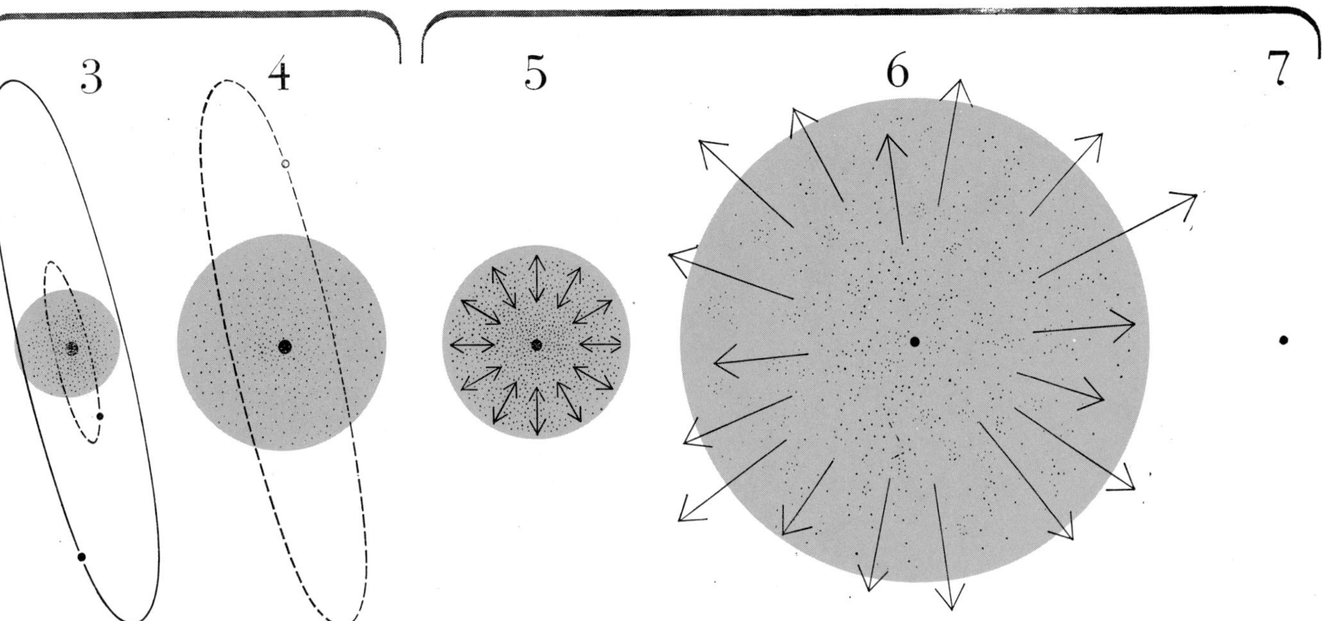

years. Then it expands (3) into red-giant stage (4), first destroying life on its inner planet, then burning up the planets in turn. This period may last about 100 million years. Then star may pulsate in luminosity every few hours (5) for thousands of years, finally exploding into a nova (6) and eventually collapsing into white dwarf (7). Time period for final stages (5, 6 *and* 7) is not known.

rate at which the hydrogen is consumed. Both factors in the calculation are determined by the position of the star on the main sequence. Luminosity, which is an index of the rate of fuel consumption, increases as the fourth power of the star's mass. The most massive stars thus use up their substance most rapidly and so have the shortest lifetimes in the equilibrium state. In general a star will evolve away from the main sequence when the core in which the hydrogen has been consumed has a mass of about 12 per cent of that of the entire star. With the exhaustion of fuel in the central furnace, gravitational contraction takes over again, heating up the interior until the thermonuclear reaction spreads to outer layers. The star now leaves the main sequence and in a comparatively brief period of time evolves into a red giant or supergiant. Its evolution thereafter cannot presently be predicted in detail. But somehow, perhaps through the rapid loss of mass by ejection, it ends up as a hot, faint, dense object known as a white dwarf. A majority of the intrinsic variable stars, of novae and of nova-like objects are apparently in the stage between red giants and white dwarfs.

In leaving the main sequence a star releases so much energy that it would destroy life on any of its planets. Thus

SAME STAR FADED IN 10 YEARS to the faint object visible between the white lines in this photograph made in 1930 with the same telescope. A nova may explode repeatedly before becoming a white dwarf and eventually turning into a cold, dark body.

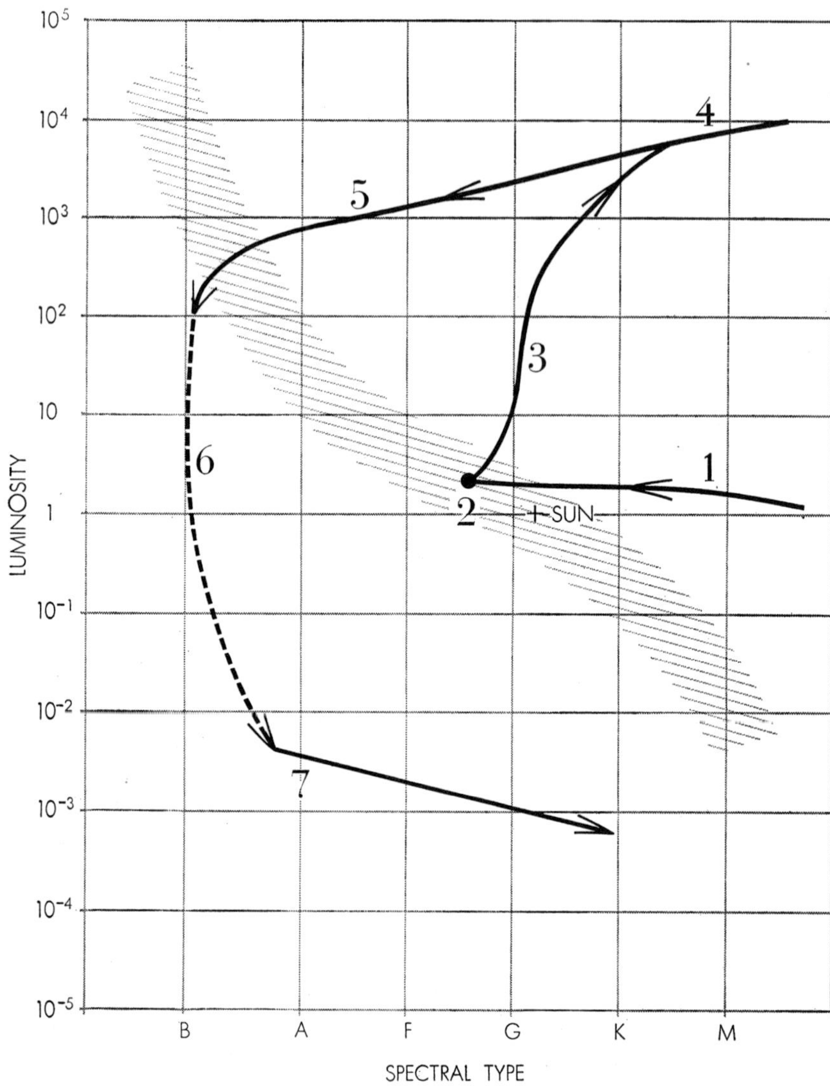

EVOLUTIONARY PATH OF A STAR with a mass 1.2 times that of the sun is traced. The vertical scale is luminosity (the sun is unity); the horizontal scale, spectral type. The main sequence is outlined by the gray hatching. The numbers on the path correspond to the stages of stellar evolution depicted in the illustration at the top of the preceding two pages. At 1 the star is in the stage of gravitational contraction; at 2 it is on the main sequence; at 3 it is expanding; at 4 it is a red giant; at 5 it pulsates. The broken line (6) indicates uncertainty as to the path the star follows in reaching the white-dwarf stage (7).

the main-sequence stage of the star's evolution is the only important one so far as life is concerned. The O and early B stars are the most massive, but since they burn much more rapidly than the smaller, less luminous stars, their life on the main sequence lasts only a million to 10 million years. The small M stars, in contrast, remain on the main sequence for more than 100 billion years. None of the early stars in the O, B and A groups has a stable lifetime longer than three billion years; they cannot therefore sustain biological evolution long enough for intelligent organisms to appear on any of their planets. Closer calculation shows that only the stars farther down the sequence than F4 maintain their equilibrium for a sufficient length of time to bring biological evolution to its culmination.

If time were the only critical condition, then the late-K and M stars would stand the greatest chance of having life-bearing planets. But these stars have low luminosity, and the habitable zones in which planets might travel about them must be quite narrow. A more luminous star can obviously warm up a larger space than a less luminous one [*see illustration on page 130*]. It is like a fire in a field on a cold night: the bigger the fire, the wider the zone around it in which the temperature will be neither too hot nor too cold. Thus the chance of finding a planet and intelligent life within the habitable zone of any particular M or even late-K star is quite small. However, since there are 10 times as many M stars as G stars, the total number of life-bearing planets traveling around the M stars may not be negligibly small.

On the basis of their lifetimes and the depth of their habitable zones, stars of the late-F, G and early-K types seem to offer the most favorable environments for life. Approximately 10 per cent of stars fall in these types. Out of the 200 billion stars in the Milky Way, therefore, some 20 billion might foster intelligent life.

This number is reduced considerably upon consideration of the third critical condition: the maintenance of stable planetary orbits. The dust-cloud mechanism that increases the likelihood of planets has also brought most stars into existence as double or multiple systems. It is obvious that the presence of two or more stars in a system will profoundly perturb the orbits of planets in the system. A few double-star systems have members that are so far apart that if they are sufficiently near to us, the stars can be distinguished by a telescope or even by the naked eye. Such systems are called visual binaries. Much more common are the "spectroscopic binaries." In these systems the stars are so close together that even if they are relatively near to us, they cannot be separated by the largest telescope. We can tell that they are double stars only by regular changes in their spectra [*see illustration on page 127*]. These changes also enable us to calculate their orbits.

The orbits of planets in such systems are so complicated that astronomers have been able to work them out for only a few idealized cases. A life-bearing planet would have to travel on an orbit close to one of the stars in a widely separated system, or on an orbit at a large distance from the stars in a close system. In the case of a hypothetical binary composed of two stars with the same luminosity as our sun and revolving around a common center in a nearly circular orbit, a planet would find the thermally habitable zone dynamically stable only if the stars were more than 10 astronomical units or less than .05 astronomical unit apart (an astronomical unit is the distance between the earth and the sun). With a separation between the stars of .5 to 2 astronomical units there is no overlapping of the habitable zone and the dynamically stable zone. Therefore no inhabitable planets can exist in such systems. Taking everything into consideration, only 1 to 2 per cent of all double and multiple stars may possess inhabitable planets, and perhaps 3 to 5 per cent of all the stars in our galaxy have such planets.

For the present there is no hope of detecting on a photographic plate the existence of a planet of another star. Such planets as may attend even the nearest star are completely lost to view in the brilliance of the star's light. Someday there may be a telescope on a platform in space, free of the interfering effects of the earth's atmosphere. As has been suggested by Nancy G. Roman of the National Aeronautics and Space Administration, the instrument would produce a sharp star image that could be blocked out so that a planet near the

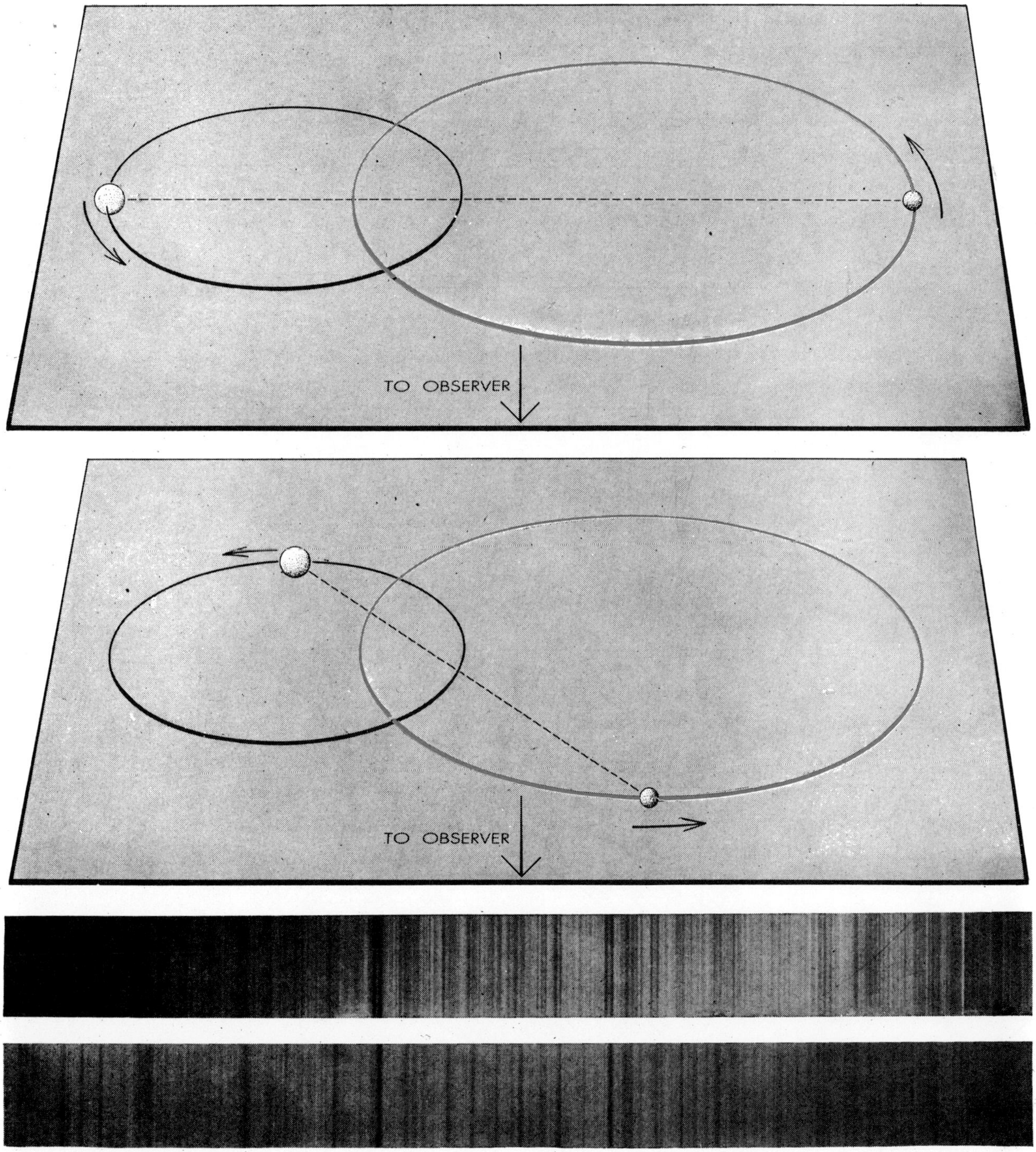

SPECTROSCOPIC DETECTION OF CLOSE BINARIES is depicted in these diagrams and spectra. At top the stars are orbiting around center of mass, the one at left moving toward the earth, the one at right away from the earth. Spectral lines made by star moving toward earth shift toward violet end of the spectrum, while lines from star moving away shift toward red end, splitting spectral lines as seen in upper spectrum below diagrams. In lower diagram the binaries are moving across line of vision as seen from earth. This gives rise to normal single spectral lines seen in the lower spectrum. Spectroscopic binaries are so close together that no telescope can resolve them into separate bodies, and only their spectra enable us to detect them and to calculate their orbital movements.

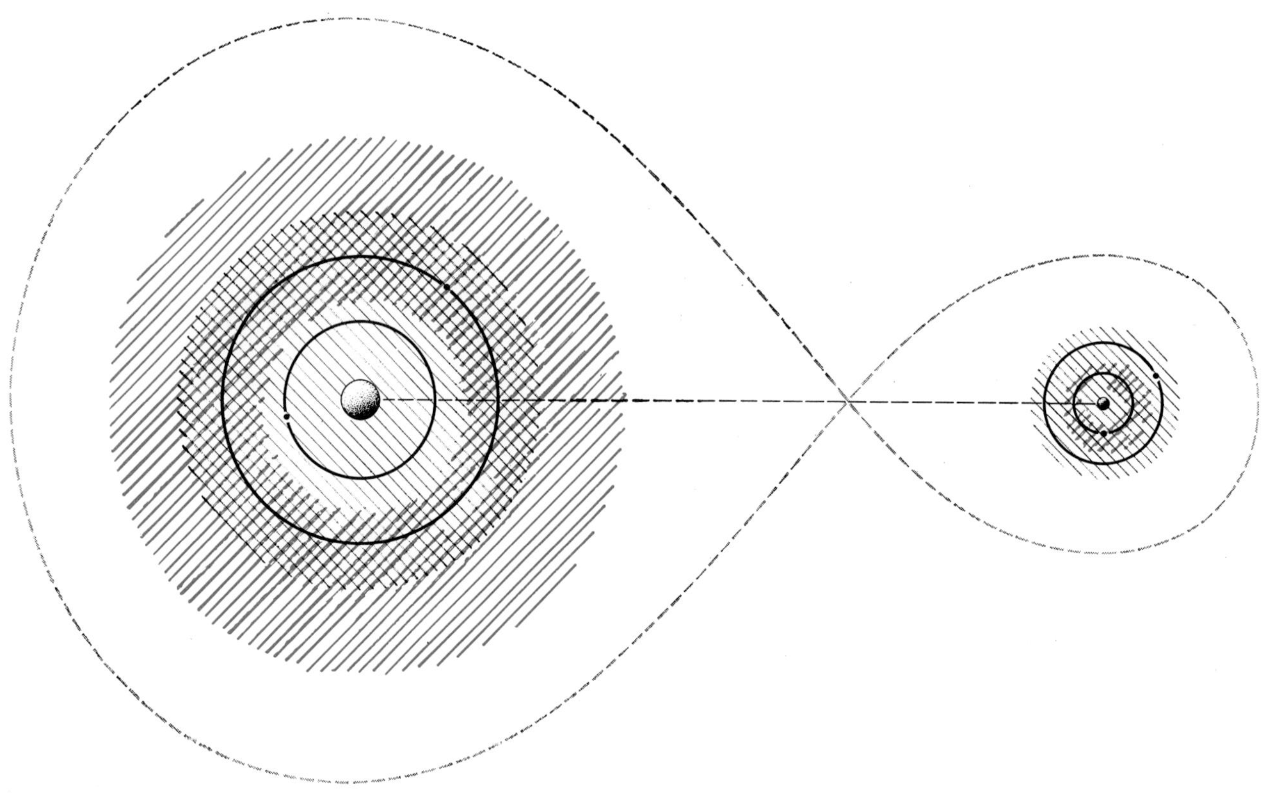

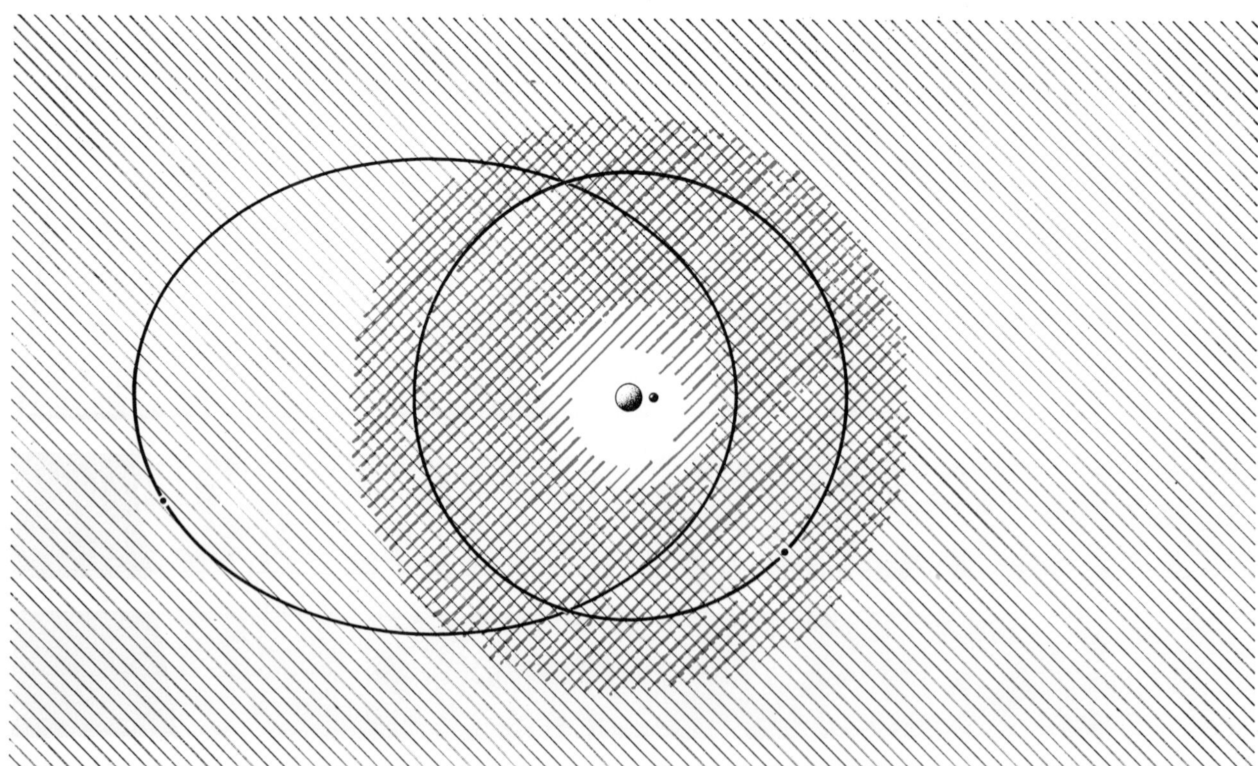

PLANETARY ORBITS AROUND BINARY STARS are depicted here in an idealized manner. The figure eight in the top diagram is simply a mathematical projection from which the dynamically stable planetary orbit is calculated for widely separated binaries. At bottom are close binaries. For a planet to bear life its orbit would have to fall completely within both the dynamically stable zone (*gray hatching*) and the thermally habitable zone (*colored hatching*). At the bottom the highly elliptical planetary orbit is dynamically stable, but it does not fall completely within the thermally habitable zone; the planet therefore could not harbor any life.

star could be detected by means of a long photographic exposure. Even the subtle methods used to detect spectroscopic binaries are not refined enough to find a planet, because the effect of a planet upon the motion of its parent star is so slight.

There is nonetheless other well-established evidence to support the contention that planets are common. In the first place, no sharp distinction can be drawn between binary or multiple stars and stars with planetary systems. According to Gerard P. Kuiper of the Yerkes Observatory, the mean distance of separation between the components of all binaries so far investigated is about 20 astronomical units. This is of the same order of magnitude as the distance between the sun and its major planets (Jupiter, Saturn, Uranus and Neptune). Kaj Aa. Strand at the U. S. Naval Observatory in Washington has studied small perturbations in the orbital motion of a star called 61 Cygni, which has a companion that is too faint to be directly observed. He has found that the mass of this unseen object is about a hundredth of that of the sun. This mass lies between that of stars and of Jupiter. It is therefore reasonable to believe that the masses of small stars in binary systems grade continuously down to the masses of planets. Since binaries are so common in our galaxy, it would seem that many stars that now appear to be alone actually possess planets.

Harold C. Urey of the Scripps Institution of Oceanography has found additional evidence for planets in a certain kind of meteorite. According to Urey, diamonds embedded in these objects show that they must at one time have been under high pressure in a body the size of the moon. Such moonlike objects, known as "prestellar nuclei," would enhance the formation of both stars and planets from a dust cloud. Any irregularity in the motions of a dust cloud should be expected to produce more than one such nucleus, and the formation of planets or multiple stars would follow as a normal consequence.

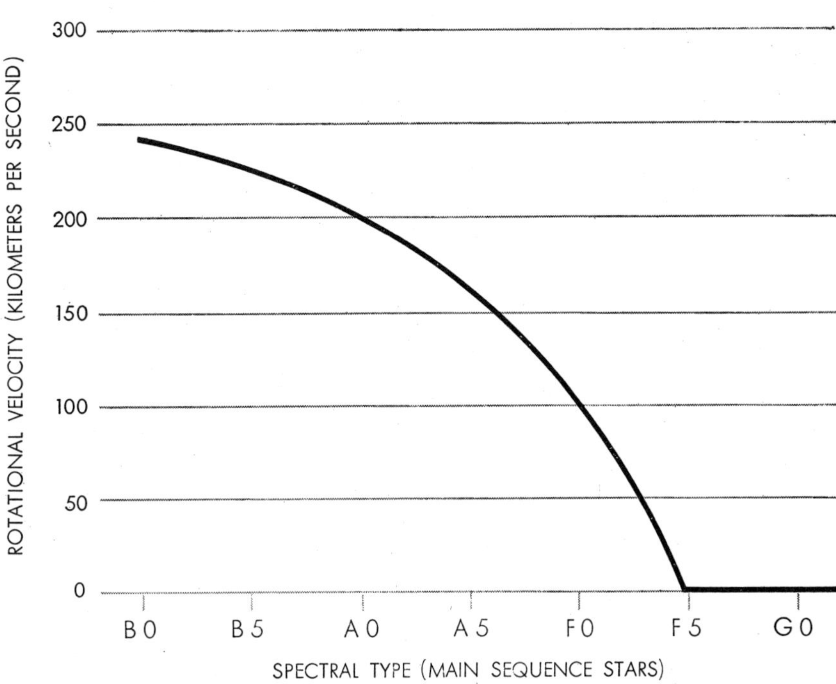

ROTATIONAL VELOCITY OF STARS on the main sequence is diagrammed here according to spectral type. Slow spin after Type F5 may indicate that planets are present.

Finally, measurements of the angular momenta of many stars give every indication that planets exist outside the solar system. Otto Struve of the National Radio Astronomy Observatory has pointed out that main-sequence stars more massive than Type F5 usually rotate rapidly, but starting with this type the rotation of stars slows down abruptly. In other words, the average angular momentum per unit mass of the main-sequence stars exhibits a conspicuous break at Type F5 [*see illustration on this page*]. The most reasonable explanation of this strange phenomenon is that unobservable planets have absorbed the angular momentum, just as Jupiter and other planets of the sun carry 98 per cent of the angular momentum of the solar system, leaving the sun with only 2 per cent and a comparatively long period (27 days) of rotation. If planets do indeed account for the slow spin of these otherwise sunlike stars, then planets appear just where life is most likely to flourish.

Thus it seems that intelligent life may be scattered throughout the Milky Way and the universe as a whole. In our immediate neighborhood, however, we may be its only representatives. The sun's nearest neighbor, Alpha Centauri, is only 4.3 light-years away. It is a triple system with two massive components (a G4 star and a K1) revolving around each other about 20 astronomical units apart; at a considerable distance is a small third star. The two larger bodies have highly eccentric orbits, and if there is a stable zone for a planet in this system, it is extremely hard to compute. Moreover, recent investigations indicate that the system may be much younger than the sun, so that higher forms of life might not have had time to evolve even if a habitable planet does exist in it.

Forty other stars are located within five parsecs (16.7 light-years) of the sun. Only two—Epsilon Eridani (a K2 type) and Tau Ceti (G4)—seem to fulfill the conditions for the existence of advanced forms of life, and Epsilon Eridani may not be exactly on the main sequence. Tau Ceti is 10.8 light-years distant, has an apparent visual magnitude of 3.6, is located on the celestial sphere about 16 degrees south of its equator and appears above the horizon in the northern sky only in the winter.

Since intelligent life is probably not a rare phenomenon, and since at least one star in our vicinity meets the specifications of a life-fostering star, it may seem odd that we have had no visitors from other worlds. The idea would have drawn ridicule 20 years ago, but today it deserves consideration. There are, however, several reasons for believing that we have had no visitors from outer space. For one thing, the 10.8 light-years that separate us from Tau Ceti—astronomically a short distance—is an extremely long distance in terms of human experience. Even traveling at the speed of the artificial satellites that man has launched, space voyagers would need hundreds of thousands of years to traverse it. It is possible that organisms from Tau Ceti might have a far longer life-span than man, but this supposition invokes radical assumptions that cannot be supported by present knowledge. Furthermore, if a meeting is to occur,

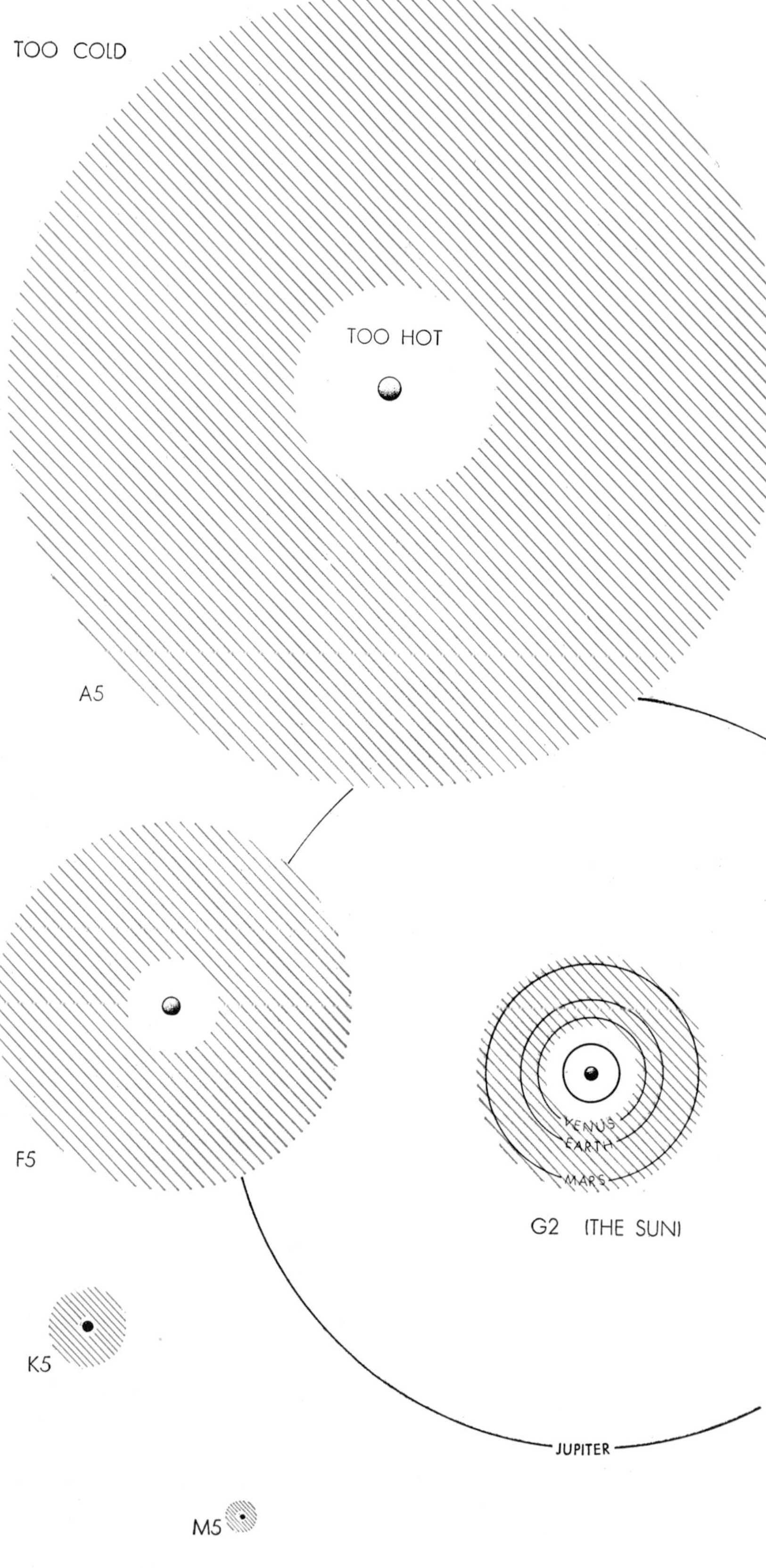

THERMALLY HABITABLE ZONE of various types of star is here represented by hatched area around star. Here the A5 spectral type has largest zone, but star does not remain stable long enough for evolution to take place on a planet near it. Habitable zone of our sun (a G2 type) extends from orbit of Venus to orbit of Mars. Tiny M5 star has the smallest zone.

our high technological civilization would have to be contemporary with that created by intelligent organisms on other planets. The cultural evolution that brought mankind to its present technical competence began only a few centuries ago. And even if human civilization endures for hundreds of thousands of years, it would be a brief episode in the time-scale of biological evolution. Therefore the chance that advanced civilizations might be flourishing at the present time on the planets of one or two nearby stars is excessively small.

Some workers have concluded, however, that the chance is good enough—and certainly intriguing enough—to institute radio surveillance of signals originating outside the solar system. Guiseppe Cocconi and Philip Morrison of Cornell University have pointed out that the most favorable wavelength would be one close to but outside of the 21-centimeter line emitted by hydrogen in space, because in this region of the spectrum the galactic noise and the noise produced in the earth's atmosphere are at a minimum. Moreover, this important wavelength would be of as much interest to astronomers on another planet as to those on earth. Cocconi and Morrison have urged that radiation picked up by the 600-foot radio telescope, now being built by the Navy in West Virginia, be analyzed for the presence of signals. According to them, that telescope will be capable of detecting signals generated 10 light-years away by a technology no more advanced than our own. Meanwhile Frank D. Drake of the National Radio Astronomy Observatory, who makes a more generous estimate of the distance at which signals sent by intelligent organisms could be detected, is in charge of an actual project that is employing a smaller telescope to detect any radio signals transmitted by living beings in other "solar systems."

What kind of signals may we expect to receive or should we send out? Probably the most abstract and the most universal conception that any intelligent organisms anywhere would have devised is the sequence of cardinal numbers: 1, 2, 3, 4 and so on. The most likely signal would be a series of pulses indicating this sequence repeated at regular intervals. Such a signal may upon first consideration appear to be too simple for the sophisticated task of communicating with other beings far away among the stars. It would sound like baby talk. But after all interstellar communication is surely still in the baby-talk stage.

12

The Crab Nebula

JAN H. OORT

March 1957

On the fourth of July in the year 1054 there was an explosion in the heavens which must have been one of the most spectacular in the history of man on this planet. A star in our galaxy which for billions of years had been invisible from the earth suddenly became one of the brightest in the sky—so bright that it could be seen in full daylight. Strangely, although the event must have been witnessed by practically everybody in Europe, not a single mention of it has been found in any European chronicle. But in China and Japan, where heavenly phenomena were watched intently as astrological signs and were recorded in the imperial annals, the "guest star," as the Chinese called it, was described in some detail.

The Chinese chroniclers located the exploding star at a position in the sky "several inches southeast of *T'ien-kuan*" (Zeta Tauri in the constellation of the Bull). After the initial burst, which for three weeks was brilliant enough to be visible by day, the "guest star" gradually faded, and two years later it had disappeared from view to the naked eye.

But in the 18th century, when telescopes had come into general use, astronomers picked up a nebula at the location of the vanished star. There can be little doubt that this object, now called the Crab Nebula, is the debris of the 1054 explosion. The Chinese descriptions also establish that the explosion must have been a supernova of "Class I." Supernovae of this order have been observed by modern astronomers only in galaxies outside our own—too far away to give us much information. The Crab Nebula therefore provides a unique opportunity to see what becomes of a star after a great explosion. What is more, it has presented, upon inspection with modern instruments, a strange and totally unexpected physical phenomenon which is now exciting keen interest as a possible key to some major current mysteries of the universe.

Exactly what sets off the explosion of a star is not known. As George Gamow has pointed out [see "Supernovae," by George Gamow; Scientific American, December, 1949], the star must develop some kind of internal instability which results in the sudden release of an enormous amount of energy, throwing masses of material into space at velocities so high that the debris soon escapes from the star's gravitational pull and goes on expanding indefinitely. The Crab Nebula is expanding at about 1,100 kilometers (680 miles) per second. It now occupies a space about six light-years in diameter. The mass of the material thrown out by the explosion of the star is estimated to be between one tenth and one hundredth of the mass of our sun. At the center of the Nebula we can see two small stars, one of which is thought to be the surviving core of the original star.

The phenomenon that I shall discuss in this article is illustrated by two contrasting pictures of the Crab Nebula, made with different color filters [*see photographs on following page*]. The first was taken through a filter which absorbs much of the light from the Nebula and lets through mainly certain wavelengths emitted by hydrogen and nitrogen atoms. In this picture, based on the selected strong emissions of light, the Nebula seems to consist of a multitude of filaments; actually the filaments form a kind of shell, made up mainly of hydrogen and helium atoms, surrounding the central part of the Nebula.

The second picture bears hardly any resemblance to the first. It was made with filters which absorb all of the stronger lines (wavelengths) of light ordinarily emitted by radiating atoms. What comes through is a continuous spectrum of light spread evenly over a wide band of wavelengths. There is something very puzzling about this "continuous" radiation, as it is called. Rarefied gases in space, when in the glowing state, always emit light with particular intensity at certain discrete wavelengths ("emission lines"). The curious thing is that the light of the structures we see in this picture shows no special emission lines. Yet there can be no doubt that it comes from extremely rarefied gaseous material.

In 1953 a Soviet astronomer, I. S. Shklovsky, suggested that this unusual continuous radiation came not from atoms but from free electrons moving at high speed in a magnetic field. His hypothesis was based on a discovery by the U. S. physicists F. R. Elder, A. M. Guzewitsch, Robert V. Langmuir and H. C. Pollock. They had observed that electrons accelerated to very high velocity in a synchrotron, where their motion is bent into a circular path by a magnetic field, radiated an intense light. This "synchrotron light," according to the classical laws of electromagnetics, is due to the acceleration, or bending, of the electrons from a uniform straight path. Unlike common light, which is generated by the vibration of electrons in their small orbits within an atom, the radiation of the electrons accelerated in the synchrotron has a continuous spectrum instead of discrete emission lines. Shklovsky proposed that the continuous light observed in the Crab Nebula might be generated by the action of a magnetic field in the Nebula upon very high speed electrons.

He suggested also that the same syn-

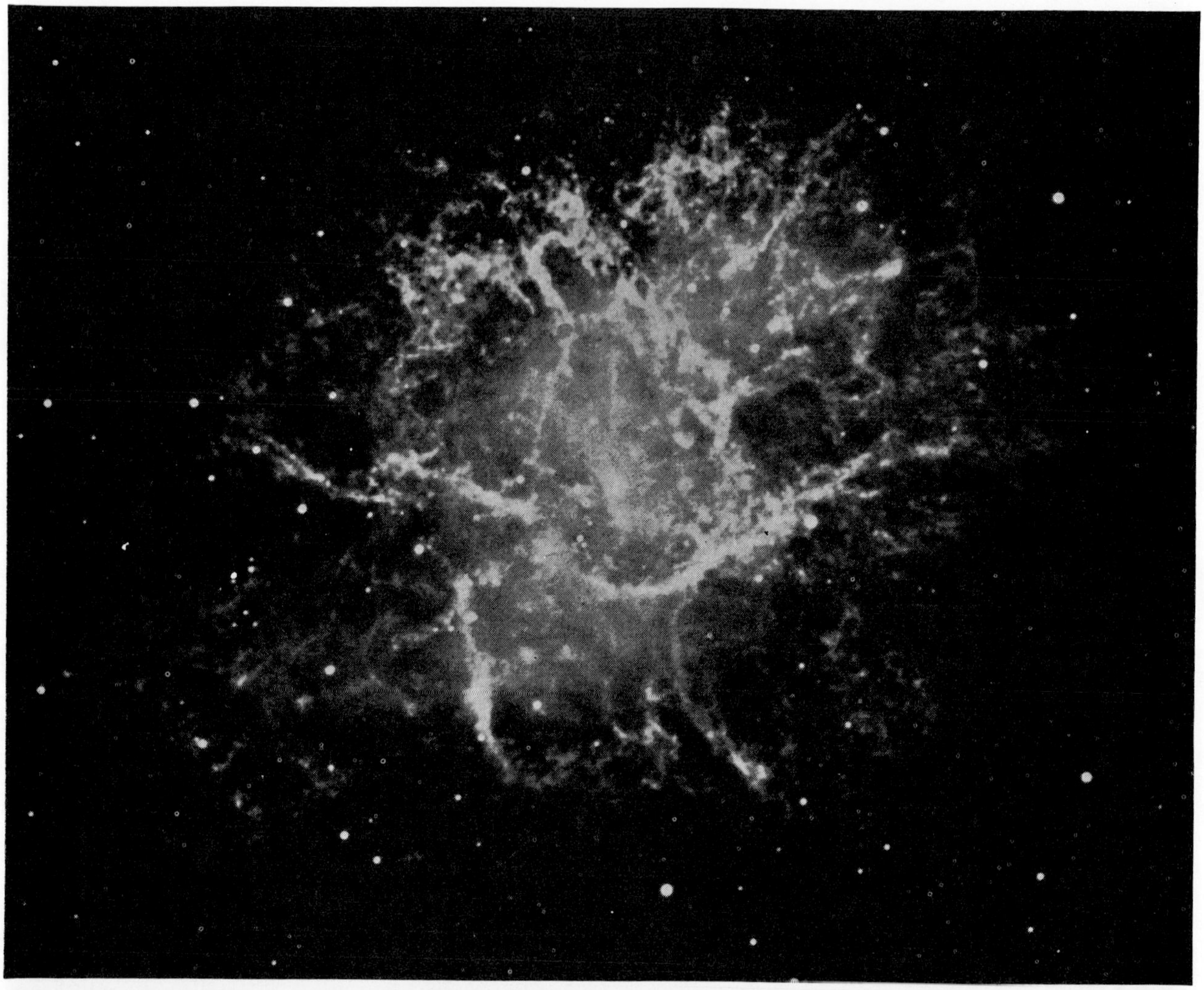

chrotron action could account for the strong radio emissions from the Crab Nebula. The discovery of radio broadcasts from space, and the pinpointing of such emissions in the Crab Nebula and other radio "stars," has been one of the great surprises in astronomy in this century. There has been much speculation about how these radio emissions are generated [see "Radio Galaxies," by Martin Ryle; SCIENTIFIC AMERICAN, September, 1956]. According to Shklovsky's hypothesis, electrons of extremely high velocities in a magnetic field radiate continuous light, and those of slightly lower energies radiate at radio wavelengths. Shklovsky's idea therefore would answer two questions which have puzzled astronomers: the mystery of the strange light from the Crab Nebula and the origin of the radio broadcasts from the Crab and other radio "stars."

It is an attractive theory but at first thought a dubious one, because it seems implausible that a synchrotron mechanism like the very special one created in a laboratory operates in nature. However, there is a way to test whether the light from the Crab Nebula is truly of the synchrotron type. The light from electrons accelerated in a synchrotron is polarized: that is to say, the light waves vibrate only in the direction perpendicular to the magnetic field. Is the continuous light of the Crab Nebula polarized? This is easily determined by photographing it through polaroid screens.

The Soviet astronomer V. A. Dombrovsky was the first to establish that the light is in fact polarized. His pictures showed a considerable polarization of the light from the Nebula as a whole. Later Theodore Walraven of the Netherlands explored the Nebula in more detail and found that the polarization varied greatly in different regions. Pictures made by Walter Baade with the 200-inch telescope on Palomar Mountain portray the polarization in still more detail [*see photographs on the next two pages*]. They indicate that all the light

CRAB NEBULA was photographed through two different color filters with the 200-inch telescope on Palomar Mountain. The filter for the upper picture passed the individual wavelengths that make up the line spectra of incandescent hydrogen and nitrogen. The filter for the lower picture screened out these wavelengths. The light that came through proved to give a continuous spectrum rather than a series of bright lines.

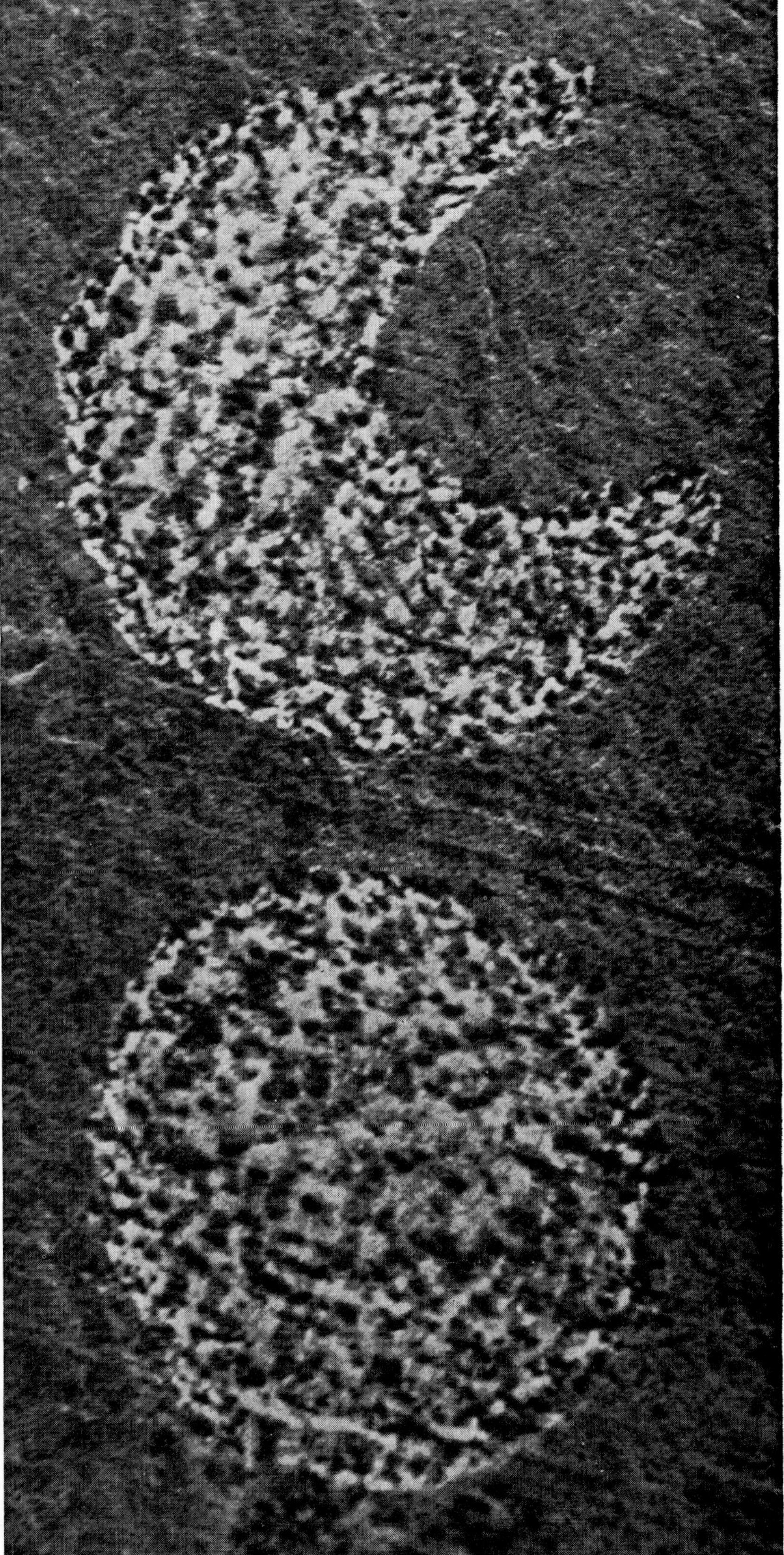

INDIAN ROCK CARVING may very well represent the supernova which gave rise to the Crab Nebula, according to a study made by William C. Miller of the Mount Wilson and Palomar Observatories. The carving was found in Navaho Canyon in northern Arizona. Chinese records indicate that the nova flared up on the morning of July 4, 1054. Before dawn on July 5, 1054, the crescent moon stood two degrees north of the position of the Crab Nebula. Thus the carving seems to represent the supernova below the moon.

in the "continuous light" picture of the Crab Nebula is polarized, and they show further that over the whole of the brighter part of the Nebula the polarization is predominantly in one direction, indicating that the magnetic field in this region likewise must be mainly in one direction. From the detailed pictures we can obtain a rough picture of the structure of the magnetic field in the Crab Nebula [*see photograph at top of page 138*].

We have every reason, then, to believe that the continuous light of the Nebula is actually synchrotron light; it would be extremely difficult, if not impossible, to explain the observed polarization on any other basis. Having reached this conclusion, we can also plausibly assume that the radio emission of the Nebula is generated by the same mechanism. The radio emission should be more intense than the continuous light emission, because radio-emitting electrons (having lower energies) are

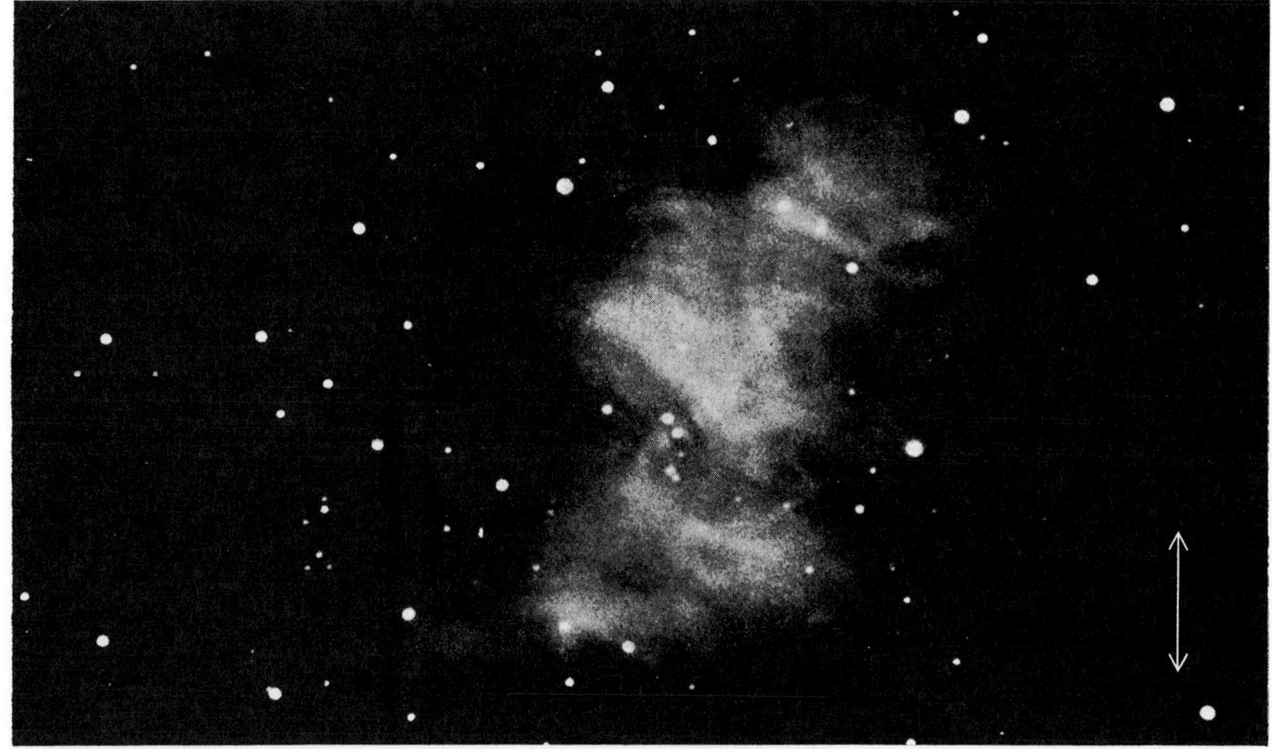

POLARIZED LIGHT from the Crab Nebula is seen in these four photographs made with the 200-inch telescope. The white arrow in the lower right corner of each picture shows the direction of electric vibration of the light waves admitted by a polaroid filter on

likely to be more abundant than the extremely high-energy luminous electrons. Observations show that the radio emission is in fact much stronger.

From the light emission we can, by theoretical arguments, form an estimate of the strength of the magnetic field and of the energy of the luminous electrons. Their energy must be extremely high—considerably higher than the highest attainable in our most powerful laboratory accelerators. But there is one place, directly accessible to us on the earth, where we can find particles with energies comparable to those of the luminous electrons in the Crab Nebula. This place is in the cosmic rays that rain high-speed particles on us from space.

The discovery that the Crab Nebula is a veritable nest of high-energy particles suggests a new explanation of the cosmic rays, whose origin has long been a mystery. Most of the cosmic-ray particles may come from supernovae or

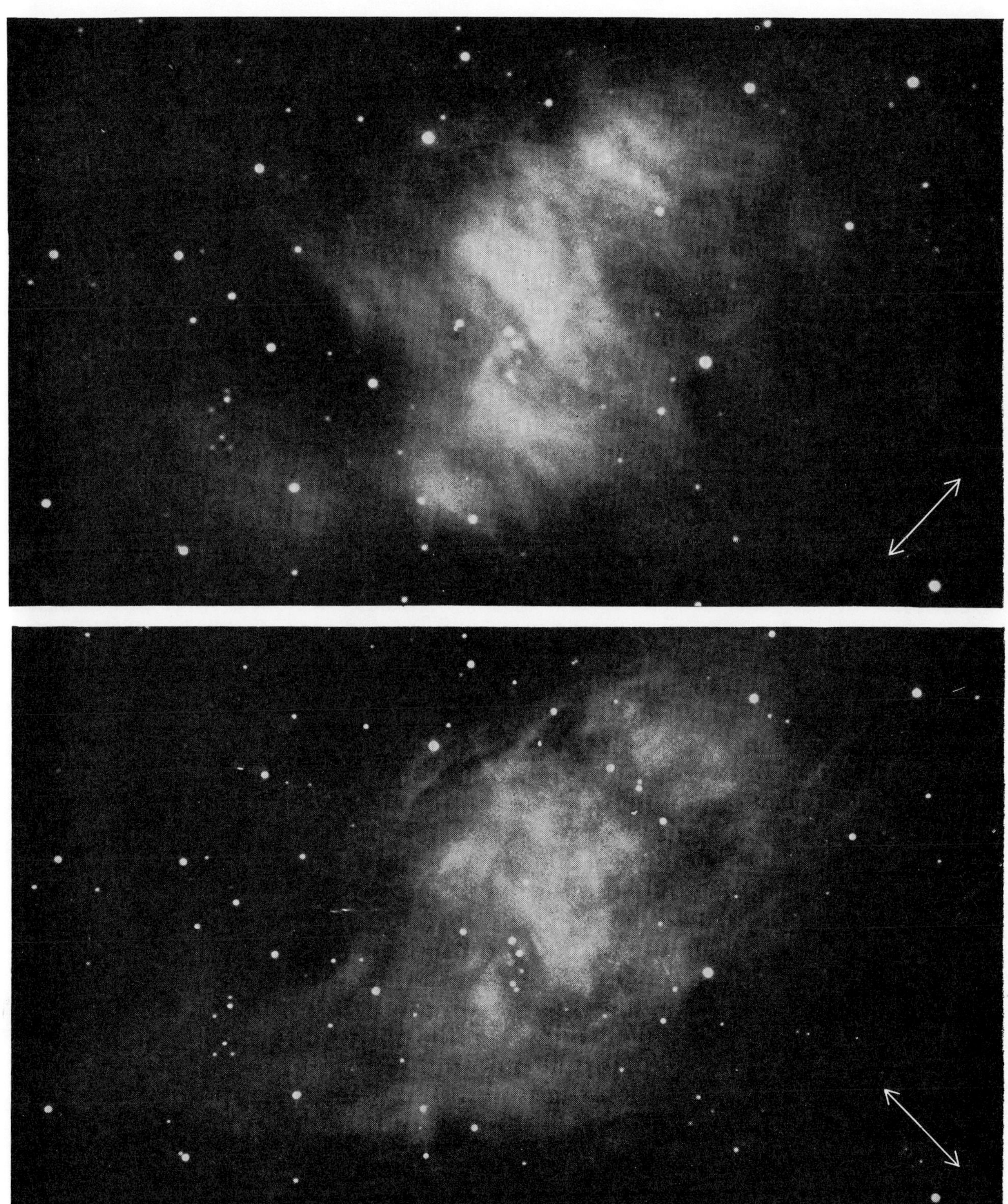

each exposure. The "synchrotron" magnetic field must be perpendicular to this direction of polarization. The linear structures appearing in each photograph are also perpendicular to this direction. They are thus seen to trace out the nebular magnetic field.

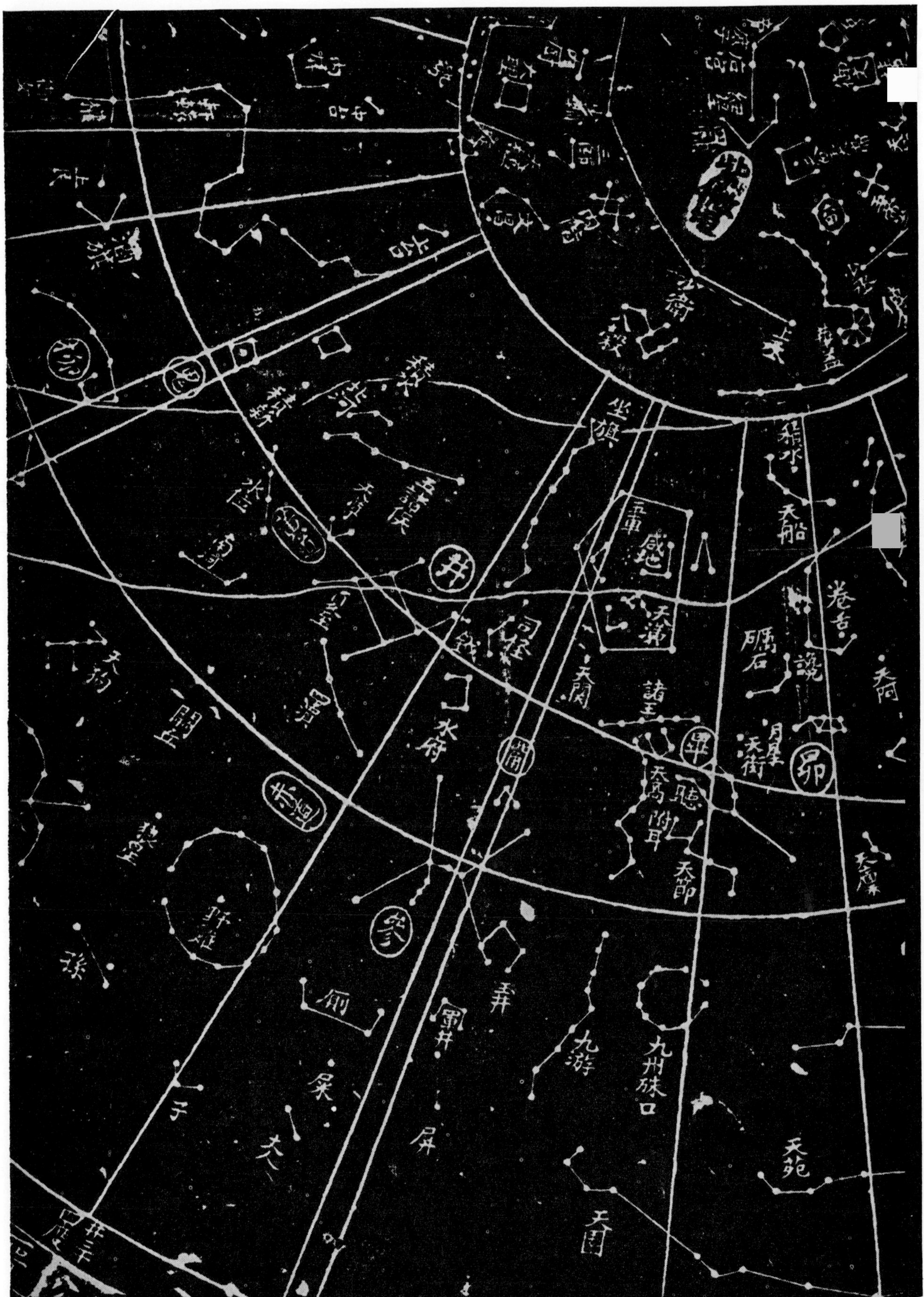

CHINESE STAR CHART was made about 1200 A.D. At right center is a constellation of six stars in a straight line. Above its left end are the Chinese words *T'ien-kuan*, and above them is a dot signifying a star. *T'ien-kuan* is the modern Zeta Tauri, which is immediately adjacent to the Crab Nebula. The Chinese identified the "guest star," or supernova, of 1054 as being close to *T'ien-kuan*.

other unstable stars. It is estimated that explosions of the Crab Nebula type alone could account for about 10 per cent of the cosmic rays bombarding the earth.

Even if we suppose that cosmic-ray particles come from supernovae, it does not follow that we understand how they acquired their tremendous energies. It is highly improbable that the energy of the luminous particles we now observe in the Crab Nebula stems from the force of the explosion in 1054. The particles originally ejected must have slowed down long ago. More likely the supply of luminous particles is continually being replenished by new accelerations.

Our own sun produces cosmic-ray particles, though on an incomparably smaller scale than the Crab Nebula. These particles are associated with the eruptions of the sun's atmosphere called solar flares. Apparently charged particles are accelerated to high speeds in some manner by the strong and rapidly varying magnetic fields which invariably accompany a solar flare. It is interesting that emission of radio waves by the sun also increases greatly during a solar eruption.

We can conceive that the central part of the Crab Nebula is still emitting high-speed particles. This brings us to the question concerning what has become of the exploded star. What exactly is left of the original star?

I have mentioned that at the center of the Nebula we can see two small objects which look like a double star. The southerly member of the pair is believed to be the remnant of the old star. But if this fragment is still a star, it must be totally different from any star we have ever observed. From its vicinity, about every three months, a tiny ripple of light emerges and moves outward through the Nebula, becoming lost to sight a few months later in a strongly luminous region. The ripples move with about one tenth of the velocity of light. They are polarized in a way that identifies them as synchrotron light, and this means that they must contain high-energy electrons. It seems likely that the ripples originate in the atmosphere of the remnant of the old star. It is tempting to suppose that the remnant erupts every three months and emits a new stream of particles which replenishes the supply of luminous electrons giving the Crab Nebula its strange aura of synchrotron light.

What of the Crab Nebula's radio emissions? Astronomers are fully as interested in these as in its synchrotron light. If the acceleration of electrons is

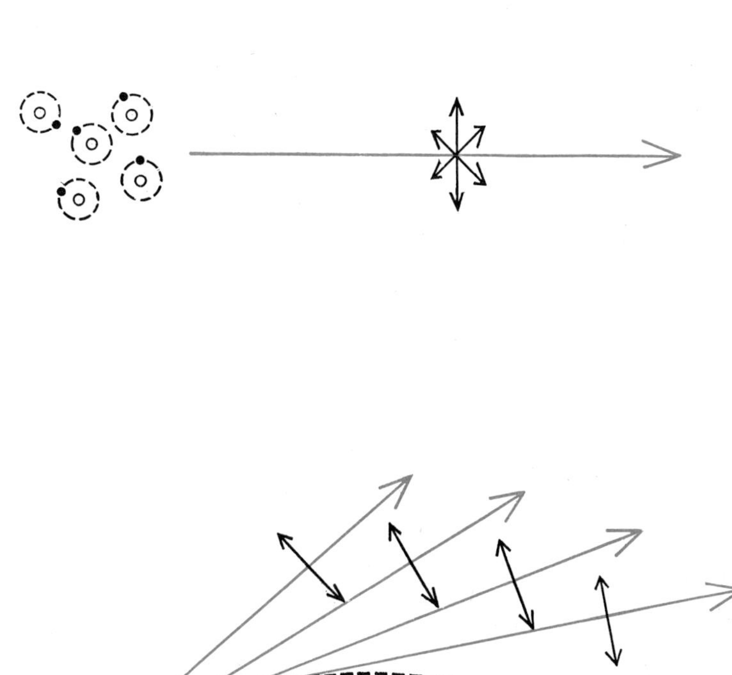

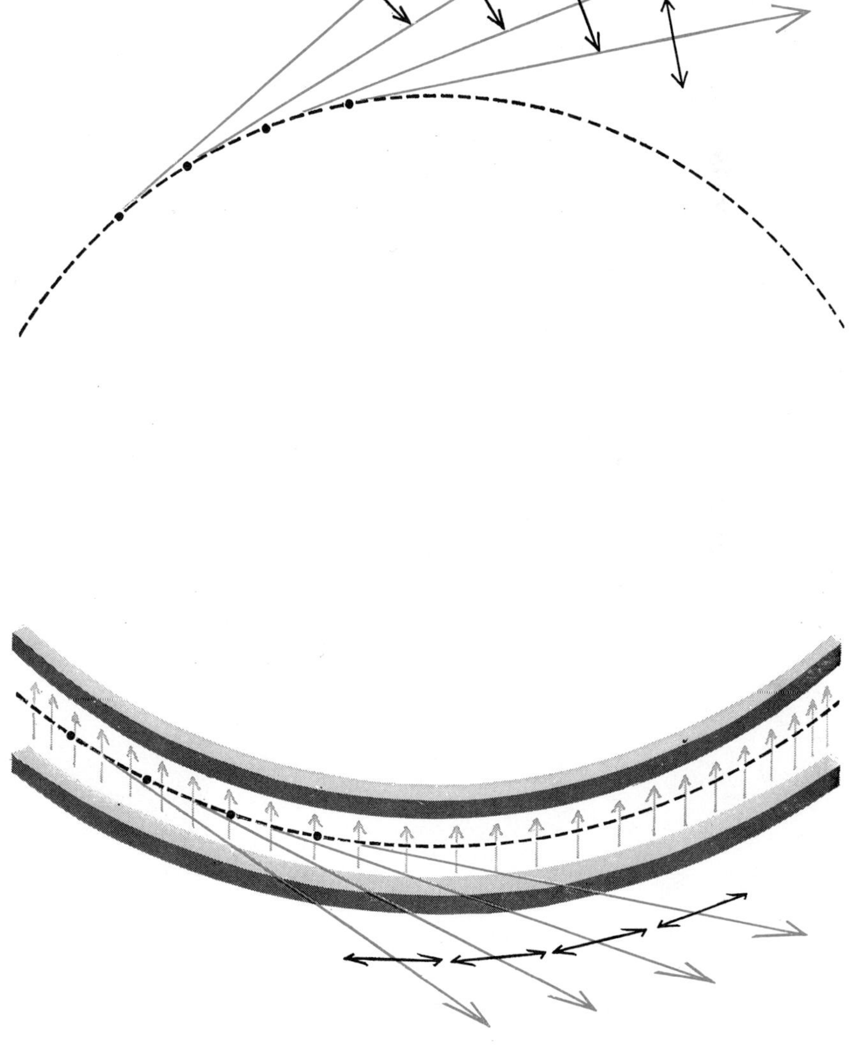

SYNCHROTRON RADIATION arises from free electrons moving in curved paths. When electrons (*black dots*) belong to individual atoms in a gas (*top*), the light they radiate (*colored arrow*) vibrates in all directions perpendicular to the path of the ray (*black arrows*). Electrons traveling a circular path (*center*) radiate light in their direction of motion, and the vibrations are restricted to the plane of the orbit. Diagram at bottom is a schematic view of an actual synchrotron, showing a section of the magnets which hold the electrons on a circular path. Vertical colored arrows give the direction of the magnetic field, long colored arrows direction of radiated light, black arrows direction of polarization.

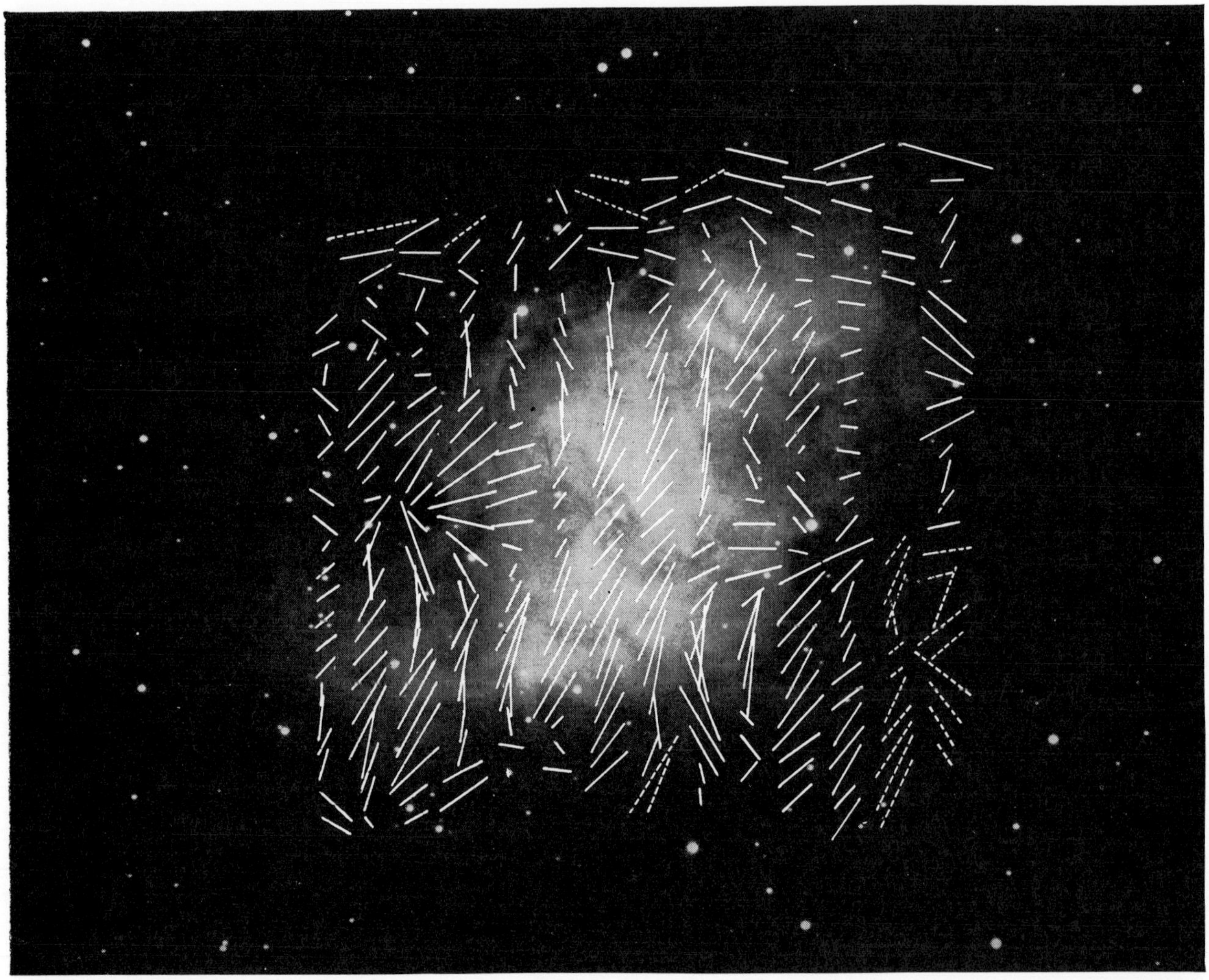

POLARIZATION PATTERN over the Nebula is traced out by lines drawn on this photograph. Each segment gives the direction of polarization in its region, and the length indicates the fraction of light polarized. Broken lines represent uncertain measurements.

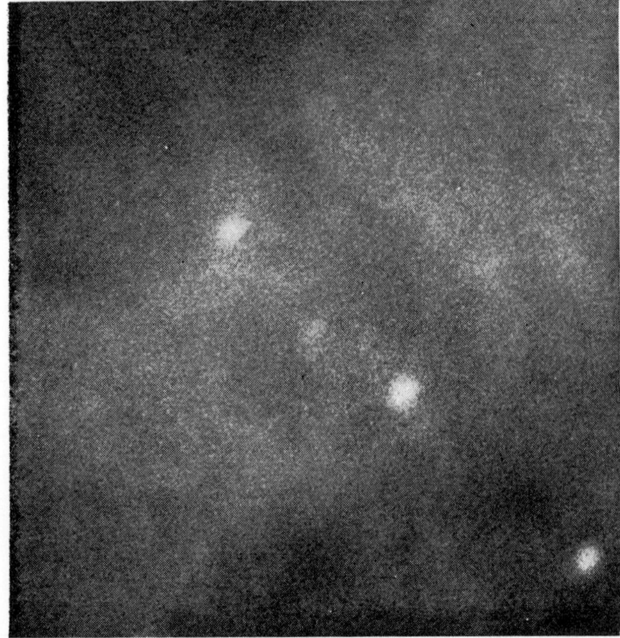

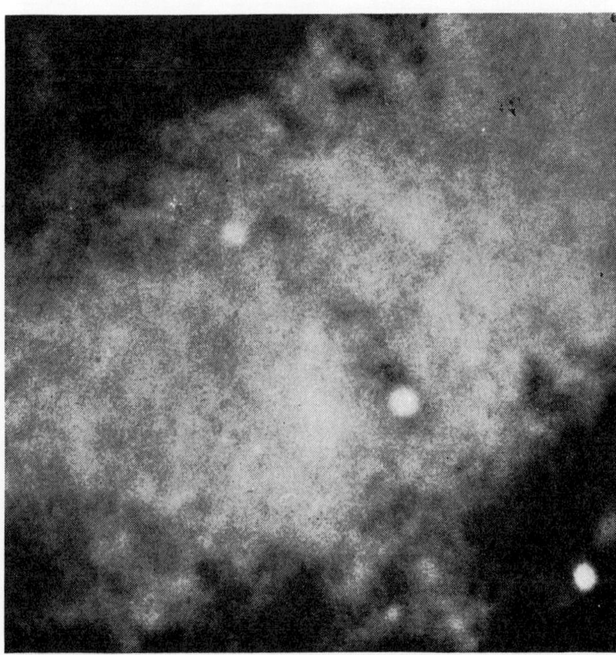

BRIGHTNESS CHANGES in the Nebula can be seen in these photographs of a region in its northwest portion. Both plates were made with the 100-inch telescope on Mount Wilson, the one on the left in November, 1924, and the one on the right in October, 1938.

responsible for the Nebula's radio broadcasts, other radio "stars" in the sky may be generating their broadcasts by the same process; in fact, this is the only plausible explanation of strong radio emissions that has yet been offered. Unfortunately it has not been possible to confirm the hypothesis directly, because the other intense radio sources in the sky, with one exception, show no synchrotron *light* of the sort emitted by the Crab Nebula—possibly because their electrons are not accelerated to the velocity necessary for light emission.

The one exception, however, is remarkably interesting. A giant galaxy in the constellation Virgo that emits strong radio signals has a wisp of continuous light near its center. Baade has established that this light is polarized—that is, it is synchrotron light. The "wisp" in the Virgo galaxy is about 100 times bigger than the Crab Nebula. If it stemmed from the explosion of a star, it must have been a truly gigantic explosion—the disintegration of a superstar at least 100,000 times more massive than our sun!

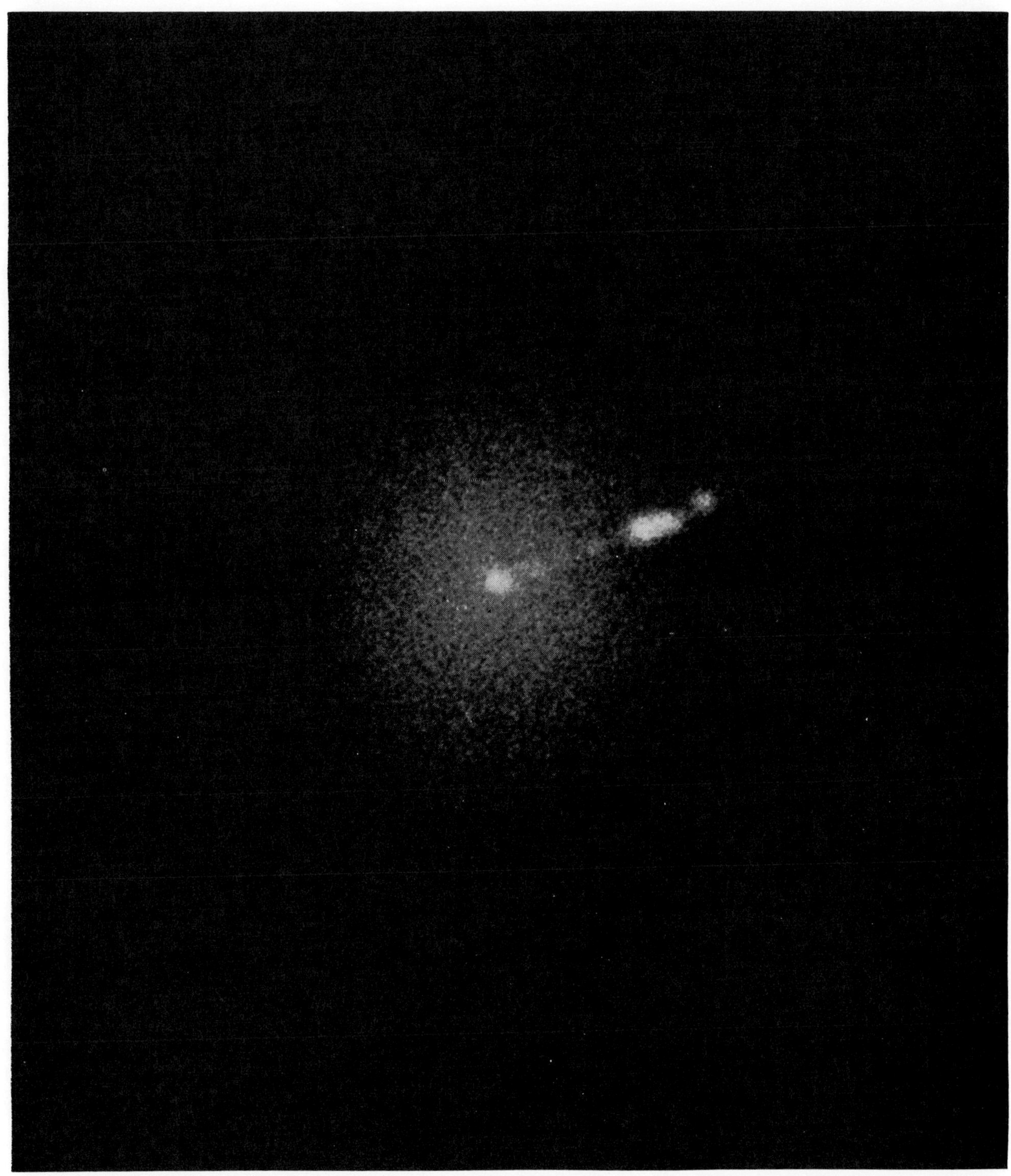

SYNCHROTRON LIGHT is detectable in this giant galaxy in the constellation Virgo, photographed with the 200-inch telescope. The wisp of light near the center has a continuous spectrum and is polarized. This source is some 100 times larger than the Crab Nebula.

13

The Youngest Stars

GEORGE H. HERBIG

August 1967

Until less than a quarter-century ago the stars appeared to astronomers as a bewildering collection of species among which it was difficult to find any systematic interrelations. The sky was filled with an almost zoological diversity of objects whose characteristics could be measured and described but hardly understood. Today, although some of the phenomena remain mysterious and new species keep turning up, the apparent chaos of the sky is being resolved into order. We are at last supplied with a clearly formulated philosophy of stellar evolution that now runs as a powerful theme through every aspect of astronomy and is beginning to fit the structural elements of the universe together into a coherent picture.

The basic concept for understanding the nature and history of stars grew from the discovery (barely 30 years ago) of how they generate their energy. They do so by thermonuclear processes that "burn" hydrogen and convert it into helium. The energy yield is 6×10^{18} ergs per gram of hydrogen converted. On the basis of the known energy output of our sun, this means that the sun must be burning hydrogen at the rate of 700 million tons per second. At that rate the sun's available supply of hydrogen fuel should give it a lifetime of about 10 billion years. When we examine more massive stars, we come to a rather startling conclusion about their life-span. The most massive stars (50 to 100 times greater in mass than the sun) have luminosities that indicate they are burning their hydrogen a million times faster than the sun. Furthermore, in stars of such size only about half of the hydrogen content is accessible for conversion in the nuclear furnace. The inexorable result is that these stars must effectively burn themselves out within a few million years and in a relatively short time thereafter vanish from the scene.

T TAURI, namesake of the class of very young, unstable stars discussed in this article, is shown in this photograph, made with the 120-inch reflecting telescope at the Lick Observatory. Two wisps of the small luminous nebula (called a Herbig-Haro object) in which the star is embedded can be seen protruding above and below the overexposed image of the star. The four sharp spikes are produced by the diffraction of light around vanes inside the telescope tube.

Accustomed as we are to thinking of the stars as permanent fixtures, it is surprising to realize that the brightest present stars—such as Rigel in Orion—cannot have been shining as such when the first men walked the earth. Even more striking are the implications of such a short stellar lifetime. If the population of high-luminosity stars in our galaxy remains approximately constant, as appears to be the case at present, old stars must continuously be replaced by new ones. Considering that there are about 6,000 high-luminosity stars in our galaxy, such replacements must occur at what in astronomical terms is an almost feverish pace—about one bright new star, on the average, somewhere in the galaxy every 500 to 1,000 years. In short, the formation of new stars must be going on *now*—and not so far from us. In the brief lifetime of the very young stars they cannot have traveled far from their places of origin—probably no more than about 100 light-years in the case of a star such as Rigel.

The argument we have just pursued does not, of course, apply only to the massive stars of high luminosity; the numbers are simply more compelling. All the smaller stars, however long-lived, have finite lifetimes, and in the long run they too must die. Star formation must be going on today at all stellar masses, great and small, because of the observed fact that bright stars are usually found together with a large number of less massive but physically related objects formed at about the same time, and always in a region filled with interstellar dust.

Whence comes the raw material that is continuously forming such massive stars or, for that matter, that produces a system such as the Pleiades, a cluster of some 300 stars with a total mass amounting to about 500 times that of the sun. The only known sources that could furnish such huge quantities of matter are the diffuse clouds of dust and gas that lie along the spiral arms of the galaxy. (Another possibility has been suggested: that there may be present in the galaxy very dense and invisible bodies that fission spontaneously and whose fragments become stars. But there is no direct evidence for the existence of such bodies, and there are other serious objections to the proposal.)

The question of the formation of new stars brings us to the particulars of our story. In the early 1940's Alfred H. Joy, the distinguished astronomical spectroscopist of the Mount Wilson and Palomar Observatories, began a systematic study of certain variable stars that exhibited rather peculiar properties. Joy named them the T Tauri stars, after an

T TAURI STARS are among the members of this young star cluster, designated NGC 2264, which was photographed in red light with the 120-inch telescope at the Lick Observatory. The cluster is still involved in the dense cloud of interstellar matter from which it was formed. The cool, dark dusty matter is visible only in silhouette against the background star field; this accounts for the apparent vacancies over part of the area. The hot, bright star just above the center of the photograph ionizes some of the gas in its vicinity, creating the "silver lining" effect on the upper edge of the long finger of dark matter pointing upward from below.

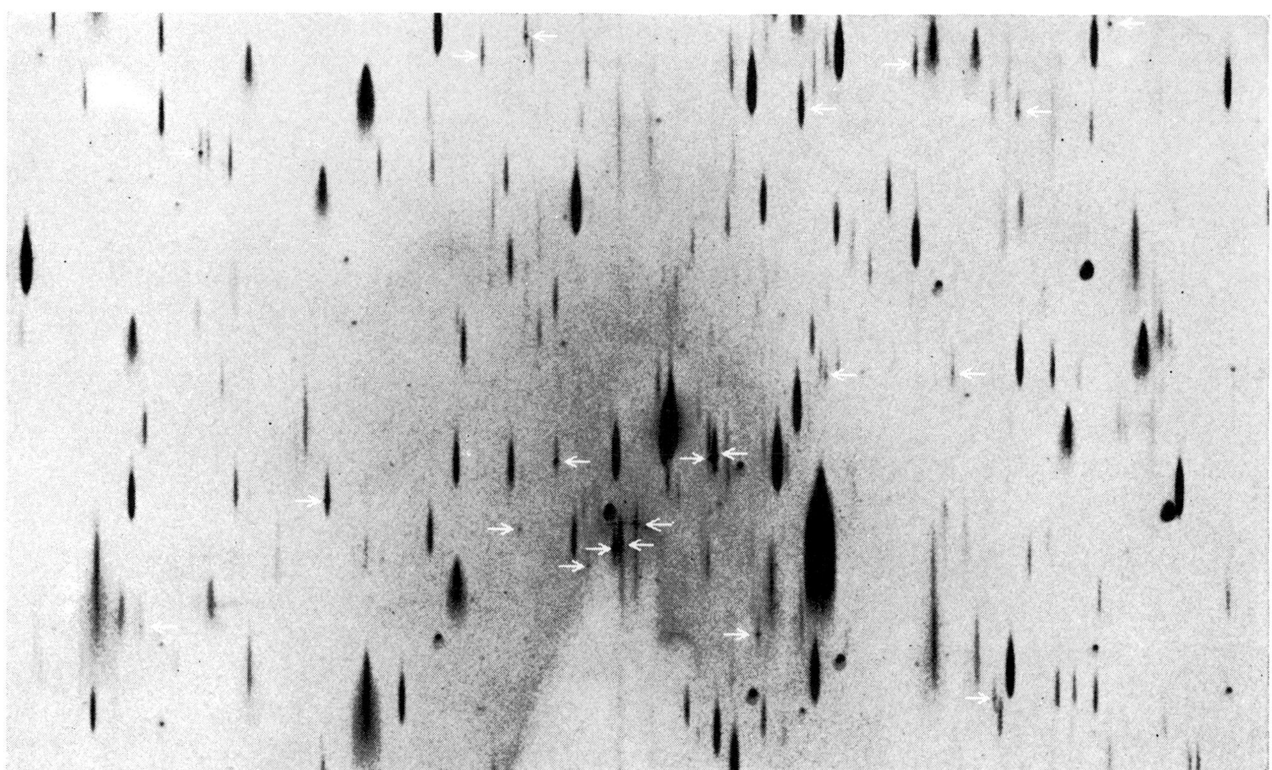

T TAURI STARS ARE DISCOVERED by means of a slitless spectrogram such as the one shown here, which covers the same star field as the photograph of NGC 2264 at the top of the page. In this negative print the thin vertical black streaks are first-order diffraction-grating images corresponding to the red regions of the spectra of the stars in the field. For the T Tauri stars (*arrows*) these streaks represent the red alpha emission line of hydrogen. The extraneous circular spots and hazy streaks are other-order grating images of stars and are to be disregarded. The plate was made with the 36-inch reflecting telescope at the Lick Observatory.

example of this type that had been known for nearly a century. One of their peculiarities was that in spectroscopic examination they showed a spectrum dominated by very intense emission lines. Usually in the spectra of ordinary, well-behaved stars only the dark absorption lines show up, corresponding to wavelengths of light from the star that have been absorbed by the cooler gases of the star's outer atmosphere. We see emission lines in the spectrum of the sun, for example, when a total eclipse by the moon covers the bright solar surface and allows us to detect the light from a thin rim of the solar atmosphere that is not hidden by the moon. This outer fringe of the solar atmosphere was named the "chromosphere" by the early eclipse observers. Now, the distinctive oddity of the T Tauri stars is that they show a fantastic exaggeration of this situation. Their chromospheres are thick, highly active regions whose output of energy is often more than that of the underlying main body of the star; in fact, the emission is often so strong that it masks the ordinary absorption-line spectrum.

The T Tauri stars have other strange properties that we shall consider shortly, but let us note here one circumstance in particular that can be taken as a sign that these stars must be very young. All of them lie in the midst of patches of interstellar dust; they show a striking preference for dust-filled regions in the galaxy, which, as we have seen, are the only places where the raw material for star formation is available in quantity.

When Joy first reported his observations on these stars in the *Astrophysical Journal* in 1945, noting their peculiarities, no thrill of recognition ran through the astronomical world. Astronomy was simply not ready to appreciate the significance of the discovery. It is fair to say, however, that a decade later it would have been necessary to invent the T Tauri stars if Joy had not already discovered them. In the interval it had become clear not only that the formation of new stars should be a fairly common phenomenon in our own vicinity but also that the T Tauris fulfilled the specifications for very young stars.

To begin with, the T Tauri stars are found in precisely those regions, and only in those regions, where one would expect to find young, newly formed stars: in dense clouds of interstellar material. Their motions show that they are not just normal stars that happen to be wandering through those regions, and there are too many for them to have been gravitationally trapped there. Furthermore, their behavior and strange physical characteristics mark them as intrinsically unusual. Another strong piece of evidence is the fact that they are generally clustered with high-luminosity, short-lived stars whose extreme youth is beyond question on the basis of energy-generation arguments. Moreover, the number of T Tauri stars in the neighborhood of our sun is found to be roughly in agreement with the expected birthrate of new stars, on the assumption that the stellar population is to be maintained at approximately a constant level.

The observed size and surface brightness of the T Tauri stars give physical evidence of their youth, according to the present theoretical concepts of the process of star formation and development. To see why, let us briefly discuss the later stages of the formation process as current theory views it.

Consider a cloud of gas and dust, having about the mass of our sun, that is condensing into a star. When the object has over a rather long time shrunk to about the diameter of our solar system, it is still a comparatively cool, dark cloud. But at that point, as gravitational attraction shrinks it further, a new phenomenon appears: some of the energy released by the contraction begins to go not into heating the gas but into internal work such as breaking up hydrogen molecules and ionizing atoms. This diversion of energy results in a reduction of internal gas pressure below the point required to support the outer layers of the cloud. Consequently the cloud quickly collapses. It shrinks from the size of the solar system to a ball with a radius about equal to the distance from the sun to Mercury's orbit—that is, a size about 100 times the size of our present sun. This collapse, occurring with the velocity of free fall, takes less than half a year. It is finally halted by a buildup of heat and pressure within the cloud that restores the system to structural equilibrium. At that point the embryo star has become visible, with a surface temperature of perhaps 4,000 degrees Kelvin and a luminosity about 100 times that of our present sun. In short, the effect is that a newborn star suddenly appears in the sky. It was A. G. W. Cameron, then at the Goddard Institute for Space Studies in New York, who first recognized that this rapid collapse would take place and, so to speak, culminate in the "ignition" of the star. His suggestion has since been supported by the more detailed studies of Chushiro Hayashi and T. Nakano of Kyoto University.

An event that meets the description of a collapse of this kind has actually been observed. In 1936 a new star, since named FU Orionis, made a sudden appearance in a concentration of gas and dust in Orion. A single example cannot, of course, be taken as complete verification of a theory. On statistical grounds we can consider ourselves fortunate to

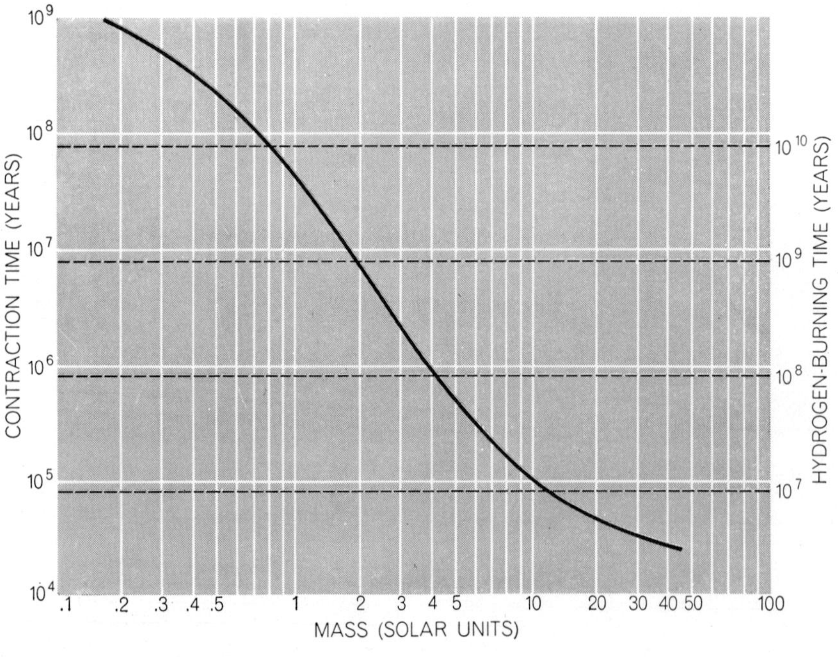

THEORETICAL DEPENDENCE of the time it takes for a star to contract (*scale at left*) on the mass of the star is given by this curve. The same curve is approximately correct also for a reasonable range of hydrogen-burning times (*scale at right*), but these are harder to define precisely, because of the gradual onset of the effects of hydrogen-exhaustion.

be able to witness even one such event with modern equipment, because we expect them to take place in our vicinity on the average of only about once in every 500 to 1,000 years.

After the star's "birth" its initial development proceeds fairly rapidly, although not nearly so rapidly as the transformation from a dark cloud to a visible star. The young star gradually contracts and for a time diminishes in luminosity. Hayashi has pointed out that in very young stars the thermal energy of the interior must be transported to the surface mainly by the convective movement of rising hot gas. The surface temperature remains nearly constant, but the interior grows hotter as the star contracts. Eventually, when the energy supplied by contraction has raised the temperature at the center of the star to about 10 million degrees K., thermonuclear reactions converting hydrogen to helium become an important contribu-

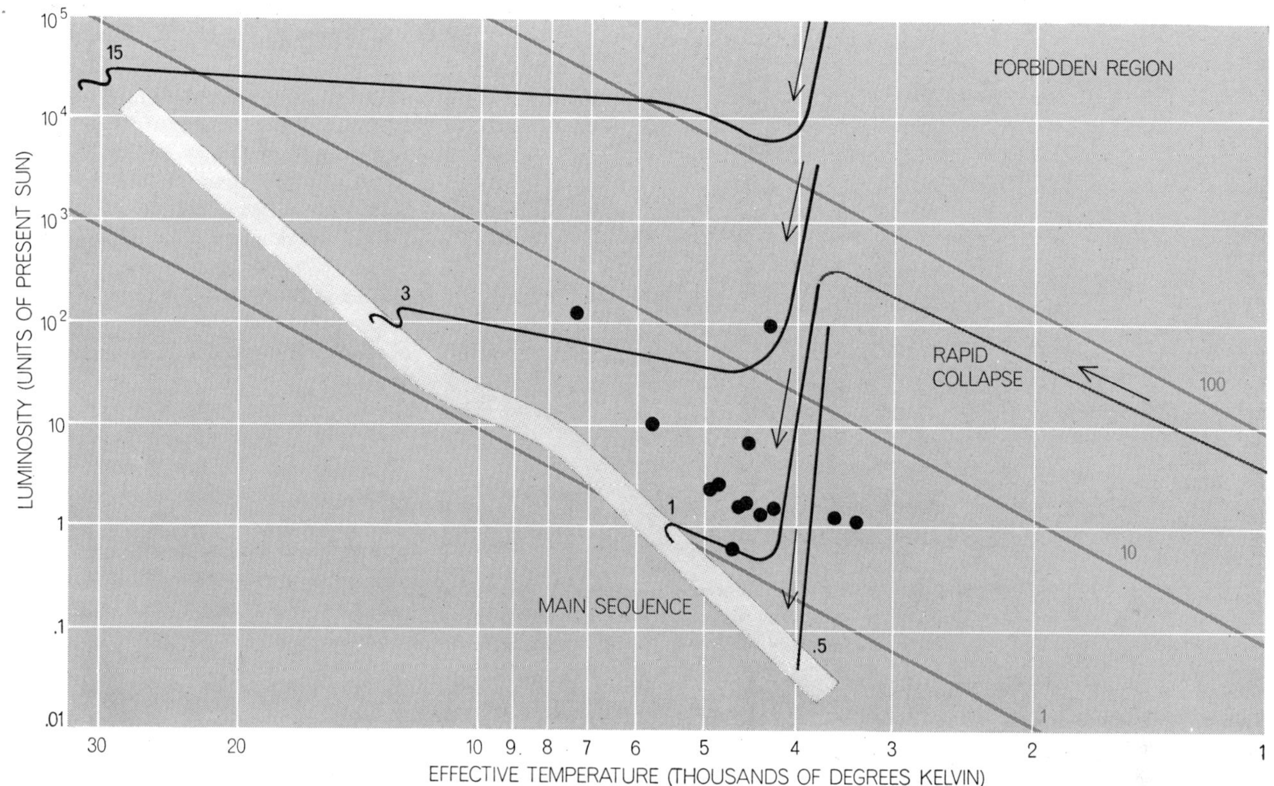

HERTZSPRUNG-RUSSELL DIAGRAM relates the surface temperatures and luminosities of the stars. The black dots give the approximate locations of a number of representative T Tauri stars, which congregate in a region of the diagram populated by contracting stars. The main sequence (*gray band*) is defined by chemically homogeneous stars that have completed their contraction and are now burning hydrogen in their interiors. Along this pathway stars are arranged in order of mass, from small masses at the lower right to large masses at the upper left. The black curves are the tracks followed by stars of various masses as they contract toward the main sequence; the black numbers on these curves indicate the masses in multiples of the sun's mass. The gray line shows the rapid collapse phase of a star of one solar mass during the final 100 days of its roughly 20-year formation period. The slanting colored lines are lines of equal stellar radius; the colored numbers on these lines indicate the radii in multiples of the sun's radius.

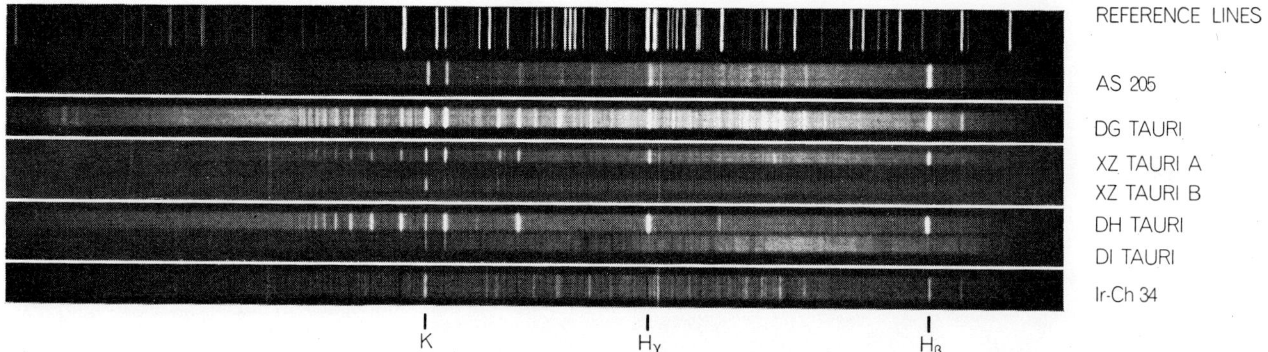

SPECTRA of a number of representative T Tauri stars show the bright emission lines that are produced in the active regions of the stars' chromospheres. In one case (DI Tauri) the emission lines are weak and the dark absorption lines of the underlying star can be seen. The "Hβ" and "Hγ" lines are produced by hydrogen; the "K" line is produced by ionized calcium. In several spectra a few extraneous bright lines run across the spectrogram from top to bottom (for example just to the right of the Hγ line). These lines are caused by illumination of the night sky above the Lick Observatory by the mercury-vapor street lighting of the city of San Jose. The bright reference lines across the top are produced by a helium-argon lamp inside the spectrograph.

VERY YOUNG STAR CLUSTER, designated IC 348, contains a number of T Tauri stars (*arrows*). This infrared photograph suppresses the blue light scattered from dust around the bright stars of the cluster and also brings out many cool, faint members of the cluster.

SAME CLUSTER, IC 348, appears in this photograph made in blue light. The very bright blue star at the top, called Omicron Persei, was quite inconspicuous in the infrared photograph. IC 348 is part of the Zeta Persei association, which is only a few million years old.

tor to the star's energy. Thereafter the star still more slowly approaches its final equilibrium state, in which contraction stops and the star's only source of energy is the burning of hydrogen. In a star such as our sun the time required for this maturation—that is, the "contraction time"—is about 50 million years. For the very massive stars the time is much shorter, only about 30,000 years for a star of 30 solar masses, for example.

The course of this development, tracing the slow structural changes and successive equilibrium states of stars of various masses, is now being calculated with increasing sophistication and attention to detail with the aid of large computers. The results are usually plotted on the well-known Hertzsprung-Russell diagram relating the temperatures and luminosities of the stars [*see upper illustration on preceding page*]; some of the most recent work has been done by Icko Iben, Jr., of the Massachusetts Institute of Technology. Such calculations have produced a chronology of stellar evolution that makes it possible to determine the ages of the stars from their observed location in the diagram. It turns out that the T Tauri variables fall in the domain of the diagram that again identifies them as very young stars: in the age range where they must still be contracting and have not yet begun to produce any significant amount of energy by burning hydrogen. They are indeed stars seen shortly after birth, in this case "birth" meaning the act of condensation from interstellar matter. They are stars in transit, so to speak, between interstellar matter and the settled state of a mature, hydrogen-burning star such as our sun.

Joy had observed in 1945 that the T Tauri stars lay in a region of the temperature-luminosity diagram that was nearly empty of normal stars, but at that time the point was given little attention, being regarded as just one of the oddities of these peculiar objects. In the 1950's E. E. Salpeter of Cornell University recognized, in a rough theoretical exploration of the problem, that just such an effect should exist. Meanwhile a number of astronomers—among them Guillermo Haro in Mexico, M. Dolidze in the U.S.S.R. and Pik-Sin The in Indonesia—undertook a diligent spectroscopic search for T Tauri stars, and today more than 1,000 of these objects have been found in the galaxy. Investigations are now focused on attempts to understand their extraordinary properties and behavior.

We have already mentioned one of these peculiarities: the T Tauri stars' unusually thick and highly active outer atmospheres, in contrast to the thin, faint chromospheres of ordinary stars. A second oddity is the fact that, in the case of those T Tauri stars that show absorption spectra, the spectral lines are abnormally wide, because of Doppler-effect broadening. There is no certain way of determining whether the motion producing the line-broadening is vertical mass movement of portions of the star's atmosphere or rapid rotation of the star as a whole.

A third peculiarity of the T Tauri stars is that they are all rapidly ejecting material into space. Apparently gases rising from the surface of these stars are somehow accelerated and forced beyond the domain of the star's gravitational control. (The force responsible for the expulsion is evidently not centrifugal, because the stars cannot be rotating rapidly enough for that.) In the spectra of many T Tauri stars one can actually see absorption lines of this material that are displaced by Doppler shifts, indicating that the expelled material is flowing out of the star at velocities of as much as 200 or 300 kilometers per second. There is no evidence that this material ever returns to the star. If it is completely ejected, these stars must lose a substantial proportion of their mass. L. V. Kuhi of the University of California at Berkeley has calculated that a typical T Tauri star, when in its most active phase of evolution, sheds material at the rate of one solar mass in about 30 million years, and that by the time it matures into a stable star it may have lost as much as a third of its original mass.

A fourth oddity of the T Tauri variables is that their brightness usually changes in an erratic and unpredictable fashion, showing no regular cycle. Some fluctuate within a few hours; others show no change in brightness for many years. The German astronomer Cuno Hoffmeister has, however, found a tendency to regularity in the fluctuations of several T Tauri stars: they appear to have a cycle length of a few days or a week that is broken occasionally by erratic outbursts but eventually reestablishes itself. Hoffmeister believes, on the basis of close study of these stars, that the cyclical variation in brightness arises from the star's rotation, bringing active areas into view and then hiding them again. If so, the brightness cycle gives an indication of the star's period of rotation, and it is interesting that the rotation rates deduced in this way agree with the rates implied by the hypothesis that rapid rotation is responsible for the Doppler broadening of spectral lines in the T Tauri stars.

A fifth unusual feature of the T Tauri variables is the fact, first discovered spectroscopically by Kurt Hunger of Germany, that they show an extraordinarily high abundance of lithium. The atmospheres of these stars have a lithium content from 80 to 400 times higher than that in the sun's atmosphere, according to estimates by W. K. Bonsack and Jesse L. Greenstein of the Mount Wilson and Palomar Observatories.

What physical processes or attributes could account for the distinctive features of the T Tauri stars: their extremely active and luminous chromospheres, their massive ejections of surface material, their variability in brightness, their high lithium abundance? None of these phenomena are predicted by the modern theory of the contraction of young stars. Each is still a complete mystery.

The sun is a middle-aged star that presumably passed through the T Tauri phase about five billion years ago (some millions of years after the sun came into luminous existence). Does it still show a few feeble memories of its more active youth? Apparently it does: the activity of the solar surface suggests a faint echo of the enormous surface activity that characterizes the T Tauri stars, and the solar wind may be a trace remnant (reduced more than a millionfold) of the immense ejections of matter in its T Tauri youth.

Seeking to reconstruct the decline of chromospheric activity in such stars, O. C. Wilson of the Mount Wilson and Palomar Observatories arranged normal stars of the sun's type in order of age and found that the younger stars indeed show much stronger chromospheric emission lines than the older ones do. Can the strong emission from the younger stars, and the still stronger emission in the very early T Tauri phase, be con-

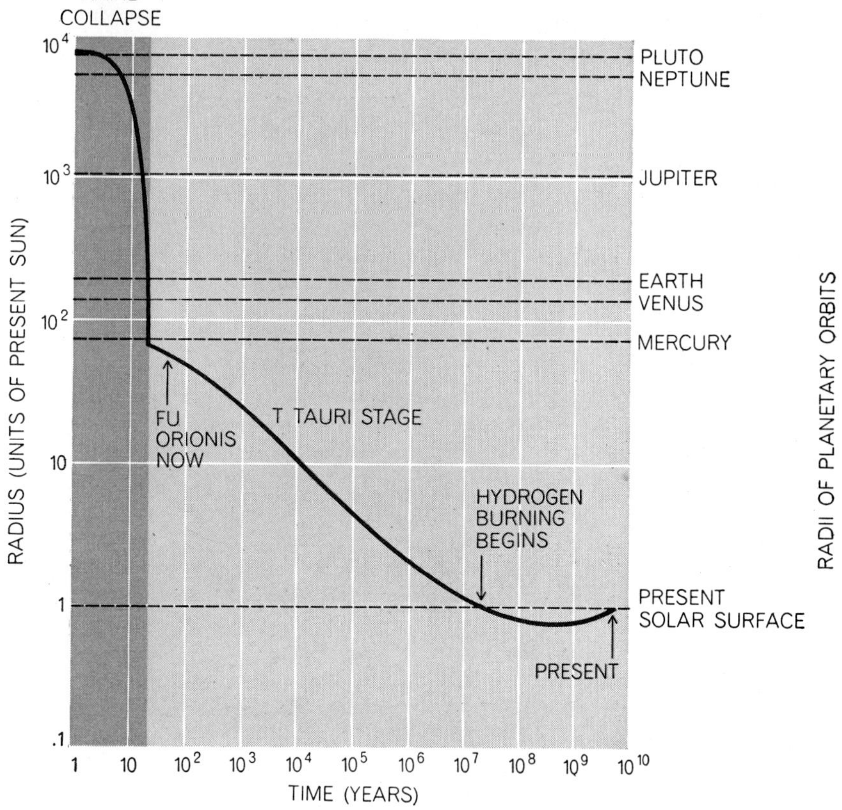

RADIUS OF THE SUN is shown at various stages in its history. The initial stage of rapid collapse was completed in about 20 years. The subsequent slow contraction to a stable stage of hydrogen-burning took about 50 million years, the uncertainty arising from the fact that the conclusion of this stage is not well defined. Thereafter the radius has increased slightly as hydrogen has been consumed in the interior. The position shown for the star FU Orionis (30 years after the end of the collapse stage) is only schematic, since its present radius is only about 20 to 25 times the present radius of the sun, not 60 times as the curve indicates. Since the mass of FU Orionis is unknown, it is not possible to say how significant this difference is. The general features of this diagram are probably about the same for other stars not too different from the sun in mass, except that the time scale is compressed for higher masses, as shown in the contraction-time curve on page 142.

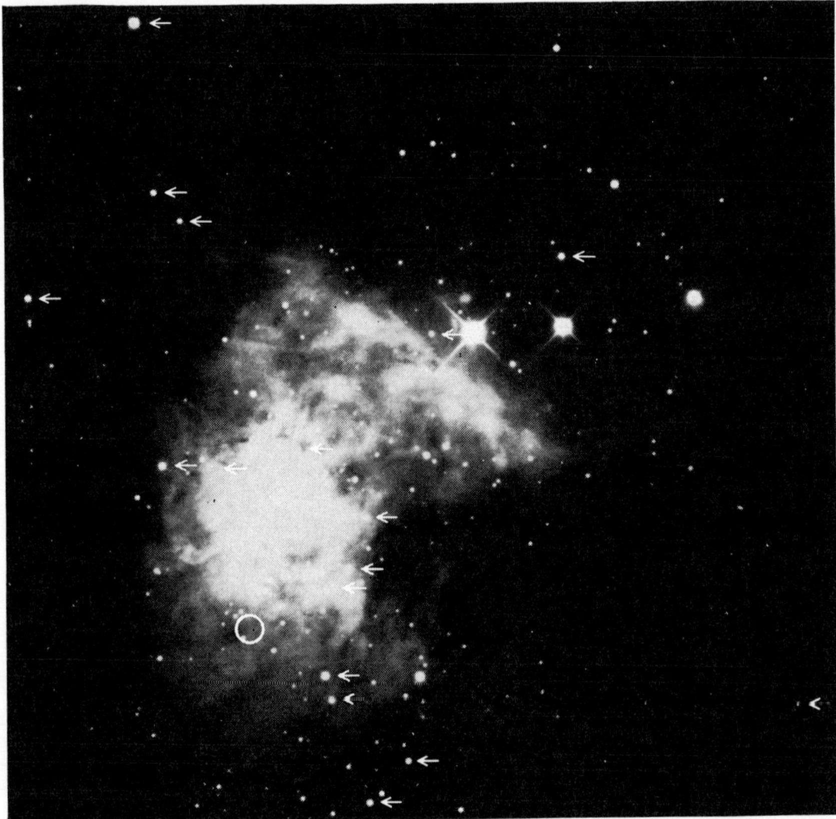

T TAURI STARS IN THE ORION NEBULA (*arrows*) were photographed in near-infrared light with the Lick 120-inch telescope. In this region the nebula is so bright in hydrogen-alpha radiation that the spectroscopic technique usually used for the discovery of T Tauri stars fails. The stars marked here were distinguished by their irregular light fluctuations and are presumed to be T Tauri variables on that basis. The circle marks the location of a large infrared source (called the Becklin-Neugebauer object) that may represent a protostar in the precollapse stage of formation. Theoretically such an object should complete the transition from a dark body to a visible T Tauri star within about 20 years.

nected to a specific cause? On the surface of the sun, Albrecht Unsöld of Germany has pointed out, the areas marked by particularly intense emission lie where local magnetic fields have more than normal strength, and the decline with age of the chromospheric emission in stars may be due to the decay of such surface fields. One is led immediately to look for strong magnetic fields in the T Tauri stars. Unfortunately an attempt by Horace W. Babcock of the Mount Wilson and Palomar Observatories to detect a field in T Tauri was unsuccessful. This does not necessarily mean that high local fields do not exist; since one must work with the average field over the star's surface, intense local fields could well be missed.

Exploration of the lithium problem has proved to be somewhat more productive. What can have happened to all the lithium the sun must have had when it was a T Tauri star? We have indirect evidence that the sun in the beginning probably did contain substantially more of the element than it does now. This evidence is found in certain kinds of stony meteorites, which can be regarded as samples of the outer material of the primeval sun. In these meteorites the lithium content is at least 35 times higher than it is in the present solar atmosphere. If such amounts of lithium were originally present in the sun, how did so much of it vanish? Current opinion is, briefly, that most of the original solar lithium was ultimately converted into helium by the impact of protons at great depths below the sun's surface, as a consequence of the circulation of surface material through the surface convective region.

Evry Schatzman and Christian Magnan of the Institute of Astrophysics in Paris have studied the question of the earlier development of the prestellar cloud when the lithium itself may have been synthesized by nuclear processes. They recently made the interesting suggestion that a strong flux of energetic protons from the center of the condensing cloud would have ionized a considerable volume of gas around the nucleus. The ionized gas, they pointed out, might be visible as a small, luminous nebula. As it happens, in 1946 a number of small, rather peculiar nebulas that fit the Schatzman-Magnan prescription very satisfactorily had been found in certain interstellar dust and gas clouds. About 40 such peculiar nebulas (called the Herbig-Haro objects) have now been detected, and all of them are located in dust clouds among T Tauri stars. T Tauri itself is embedded in one of the brightest of these nebulas. It seems likely that the Herbig-Haro objects do mark the site of some kind of preliminary activity that may culminate in the appearance of a star.

In 1966 the Mexican astrophysicist Eugenio E. Mendoza V, then working at the University of Arizona, discovered that a large amount of infrared radiation is emitted by the T Tauri stars; this energy is greater than the amount that would be expected to be radiated at infrared wavelengths from stars of that surface temperature. It could be explained by assuming that the extra radiation comes from large clouds of dust, with surface temperatures of about 700 degrees K., lying very near the stars. The observation raises the possibility that the outlying dust, if that is the origin of the infrared excess, may represent material left over from the condensation of the star, perhaps even of the kind from which our planetary system formed.

E. E. Becklin and G. Neugebauer of the California Institute of Technology recently discovered in an infrared survey of the Orion nebula an object that could represent a cool infrared-radiating cloud in the precollapse stage of formation. It is emitting infrared radiation similar to that from the T Tauri stars but much greater in amount. The source of this radiation is invisible in ordinary light and could be explained by a cool, dark body about 1,500 times the size of our sun. That corresponds to the predicted size of a prestellar condensation of the sun's mass just prior to collapse. Hayashi and Nakano have calculated that a cloud of this mass should complete its transition from a large, cool body to a visible star within about 20 years.

Further developments are awaited with the greatest interest. If, within the next decade or two, a new star like FU Orionis suddenly makes its appearance at the site of the Becklin-Neugebauer object, it will be a celebrated triumph for the blend of theory and observation that began with Joy's discovery of the T Tauri stars in 1942.

14

Dying Stars

JESSE L. GREENSTEIN

January 1959

Here and there among the tens of thousands of stars in the nearby regions of our galaxy are a few hundred whose fires have gone out. Once they burned as brilliantly as any we now see in the sky. Some had the "normal" size and brightness of the sun; some were giants, with many times the sun's diameter and brightness. Now these stars are approaching the end of the road. They have exhausted their fuel. The inward pull of gravity, no longer opposed by the outward push of pressure generated by heat within, has shrunk their diameters to a tiny fraction of stellar size, to that of the earth and even smaller, compressing their huge masses to unimaginable densities of many tons per cubic inch. In their fading light, detectable only by the instruments and techniques of modern astronomy, they are radiating the heat still left from the past out into the cold reaches of space

We call these stars "white dwarfs." They hold clues to many interesting questions of astrophysics. Until recently, however, much of what we "knew" about them was the fruit of theoretical speculation. They comprise some 3 per cent of all the stars in our galaxy and so must be rated a common type. Yet their luminosities are so low that only a few hundred have been tentatively identified and only 80 observed in any detail. Study of their color and the lines detectable in their spectra is yielding new insight into the synthesis of elements in younger stars. Their densities represent states of matter which we can hardly think of duplicating in terrestrial laboratories. But the white dwarfs have a more general significance. They are a portent. They show us that the laws of thermodynamics, which circumscribe events on the minuscule scale of our planet, hold also as the inexorable plan of the life history of the stars.

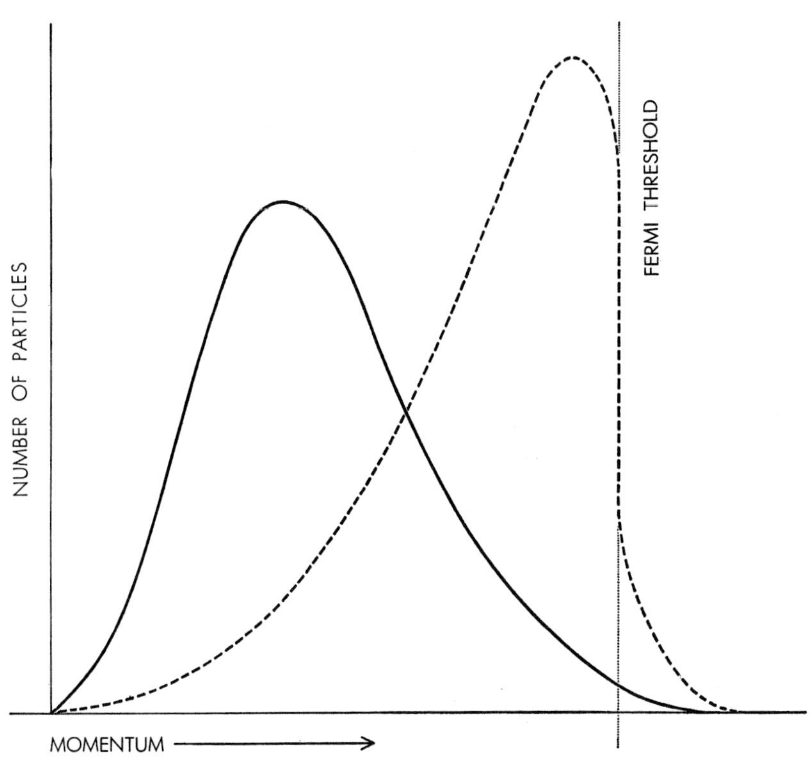

MOMENTUM OF PARTICLES in a "perfect" gas (*solid line*) follows the bell curve of random distribution. In a "degenerate" gas (*broken line*), the curve shows fewer low-momentum states available. Only the few particles above the Fermi threshold move at random.

An irreverent physicist once rephrased the laws of thermodynamics to read: (1) you can't win, (2) you can't even break even, (3) things are going to get worse before they get better and (4) who says things are going to get better?

When it is applied to stellar processes, the first law reminds us that stars do not create energy, but only convert energy from one form to an equivalent quantity of another form; that is, they convert to radiant energy the energy contained in their gravitational potential and in that fraction of their mass which is consumed in thermonuclear reactions. They can never produce more energy than they start out with. In a steady-state star, with a stable balance between its gravitational contraction and the pressure generated by the heat within, the expenditure of thermonuclear energy can go on for a long time—10 billion years in the case of the sun.

But the second law reminds us that this cannot go on forever. A star can never recapture the energy it wastes into the sink of space; its life history is irreversible. As it uses up the hydrogen that comprises the bulk of its substance,

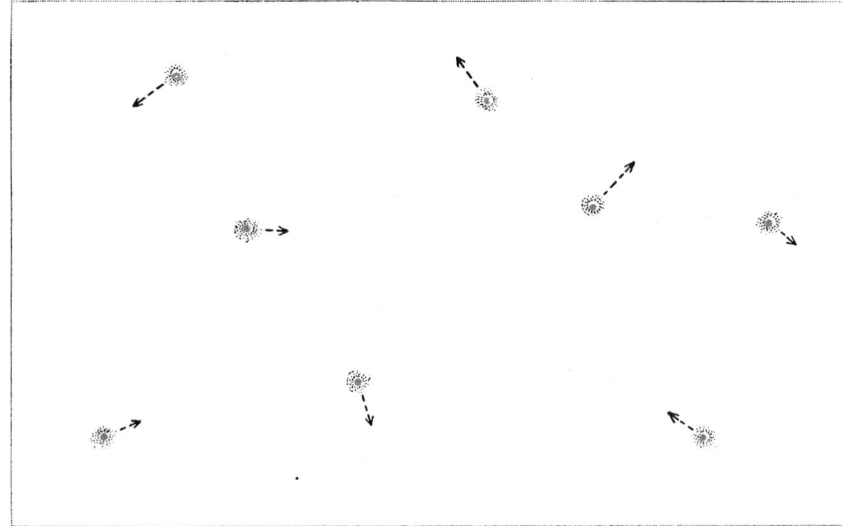

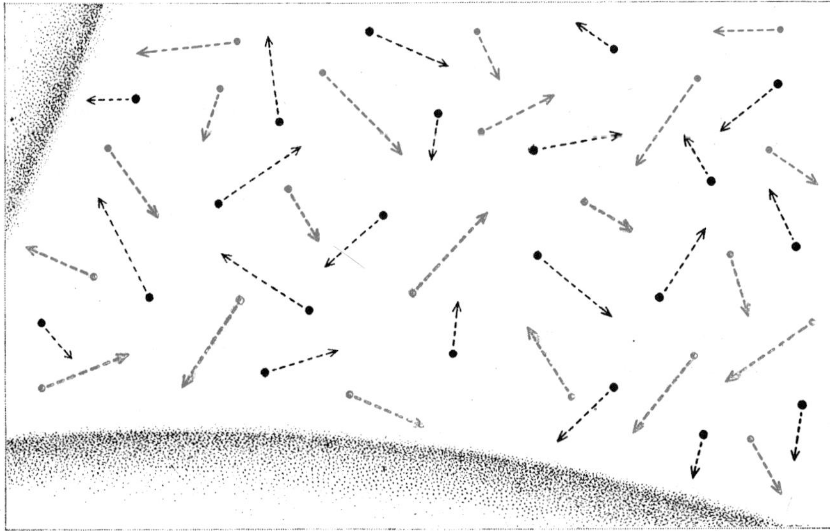

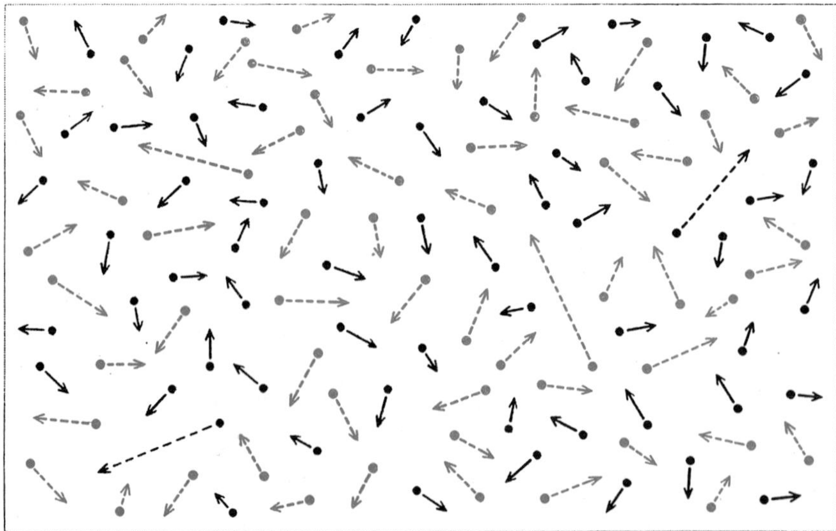

"DEGENERATE" GAS (*bottom*) is contrasted with "perfect" gases, made up of atoms (*top*) and ionized particles (*center*). Space available in gas of normal pressure permits random motion (*broken arrows*) to atoms. In an ionized gas, even at the density of a solid, the obliteration of the structures (*shadowed arcs*) of all but a few atoms opens up space to permit random motion of electrons (*black*) and nuclear particles (*color*). At the extreme density of a degenerate gas the energy states of most electrons are prescribed at low momenta (*solid arrows*). Only the nuclear particles and a few electrons move at random.

the thermonuclear furnace begins to falter. Gravitational contraction restores the equilibrium, converting potential energy into thermal energy. But contraction raises the density of the star, and the new balance between gas pressure, heat transfer, energy production and radiation loss changes the internal structure. The star brightens, its outer envelope grows larger, and stellar "evolution" begins—earlier in the life of brighter stars, later in that of the fainter ones.

As the star enters the last phase of its existence it shrinks to the final, stable configuration of a white dwarf. The third and fourth laws of thermodynamics now assume increasing relevance to its condition. The third law says that the star will ultimately cool down to the temperature of space, and the fourth law declares that it will then no longer give forth light or heat. At this terminal point the white dwarf becomes a black dwarf. Since we could not observe black dwarfs, if there are any, we shall not now give further consideration to them. In any case a star persists as a white dwarf for billions of years. Its structure and condition in this phase is what interests us here.

Matter at white-dwarf density is strange to contemplate by celestial as well as terrestrial standards. A star like the sun has an average density of almost one gram per cubic centimeter, about the same as that of water. Astrophysicists nonetheless find it feasible to deal with the behavior of solar matter as if it were a gas, with its particles free to move about at random. At the high temperatures of the solar interior, hydrogen is 97 per cent ionized; the electrons of nearly all the hydrogen atoms are stripped from their nuclei (protons). This means that the bulky structure of the hydrogen atom, 10,000 times the diameter of its constituent particles, is obliterated. As a result a cubic centimeter of ordinary stellar material is largely empty space. The tiny protons and electrons are free to move in all directions and at all velocities, just as they would in a highly rarefied gas.

In a white dwarf, on the other hand, a mass on the order of the sun, equal to 332,000 earth masses, may be packed into a volume no larger than that of the earth, which has but one millionth the sun's volume. The density ascends to 1,000 kilograms per c.c.—more than 15 tons per cubic inch. Even after a white dwarf has cooled below the temperature needed for ionization, the atoms remain dissociated under the crushing pressure of gravity. The particles are not yet so tightly packed, however, that their volumes overlap; there is still empty space between them. But because each particle has only a small volume of space in which to move, its momentum as well as its position is prescribed. The exclusion principle of physics, which rules that no two particles can occupy the same energy state, rigidly specifies the coordinates and motion for all low-momentum states Since the electrons are the lighter particles, they have the lowest momenta and are frozen in space and velocity. Collisions cannot result in arbitrary changes of momentum, but can only kick the electrons into unoccupied states. A few electrons which attain velocities approaching that of light, above the so-called Fermi threshold, are still free to move, as are the nuclear particles [*see illustration on preceding page*]. The gas has entered the "degenerate" state.

We owe to Subrahmanyan Chandrasekhar of the Yerkes Observatory a beautifully complete theory of a self-gravitating degenerate sphere of gas. Strangely, according to the theory, the greater the mass of a white dwarf, the smaller its radius. This follows, however, from the degenerate-gas law, which predicts a gas pressure, for a given density, sufficient to counteract gravitational pressure only when the star is greatly collapsed. The inverse relationship of mass to radius is not affected, as it is in other stars, by temperature, luminosity or energy production. The mass and hence the radius of a white dwarf is fixed, in the theory, by the elemental composition of the star. For stars of each composition there is an upper limit of mass. Calculation from the theory shows, for example, that a white dwarf composed of hydrogen would have a maximum possible mass 5.5 times that of the sun. On the other hand, a white dwarf made up of heavier elements should have no more than one fourth this mass, or 1.4 solar masses. A more massive star must lose mass or suffer a catastrophe before it becomes a white dwarf. We have few reliable determinations of white-dwarf masses, but all such determinations lie well below the theoretical maximum of 1.4 solar masses. This is important confirmation for the deduction that these stars have exhausted their hydrogen, the principal thermonuclear fuel.

The theoretical picture of the white-dwarf star, extended by other investigators, makes it clear that it will always be difficult to test theory by observation. The dense degenerate mass of the star is surrounded by a sharply differentiated envelope about 65 miles deep; the material here is nondegenerate because of the

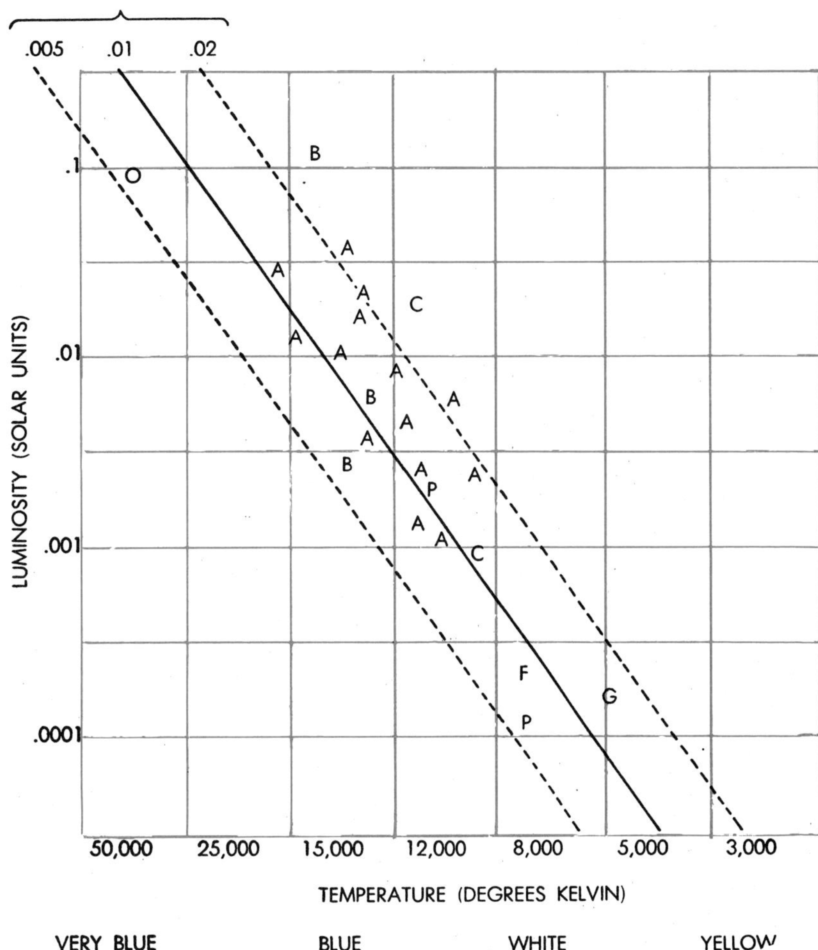

RADIUS AND TEMPERATURE **of white dwarfs show no correlation. Stars of various radii occur at all temperatures as indicated by the positions of the letters standing for various types. This is evidence that dwarf stars cool down without further gravitational contraction.**

lower pressure. Superposed on the envelope is the atmosphere of the star, which is only a few hundred feet deep. This is the only part of the star we can study spectrographically. What we observe in the spectra of normal stellar atmospheres, which are thousands of miles deep, tells us much about their surface temperature and composition, and also a good deal about their interior. The shrunken atmosphere of a white dwarf bears small relevance to the interior and can tell us little about it.

Evry Schatzman of the Institut d'Astrophysique in Paris has shown that white dwarfs cannot have the same composition at their surface as in their interior. In the absence of convection the gas stratifies under the intense gravitational field. The residual hydrogen is squeezed to the surface, while the helium and heavier elements gravitate to the center. Were it not for electrical forces, the electrons would tend to float on top. The electrical fields and nuclear forces set up by the stratification contract the star still further and so reduce the maximum possible mass to 1.25 solar masses.

The fading light that carries off the heat remaining in their interiors has given us the location of several hundred possible white dwarfs. The brightest of them has a luminosity only .01 that of the sun; the faintest known dwarf has only .0001 solar luminosity, so faint that such stars cannot be observed at distances greater than 30 light-years. Their low luminosity, combined with our theoretical knowledge of their internal structure, provides convincing evidence that they have ceased transforming matter into energy. At their high densities thermonuclear reactions would go on at enormously high rates, even if temperatures were as low as 10 to 30 million degrees Kelvin. The reaction rate would be even further increased by the dense packing of the electrons, whose negative charges would partially nullify the mutual repulsion of the nuclei. The only possible explanation of their low lumi-

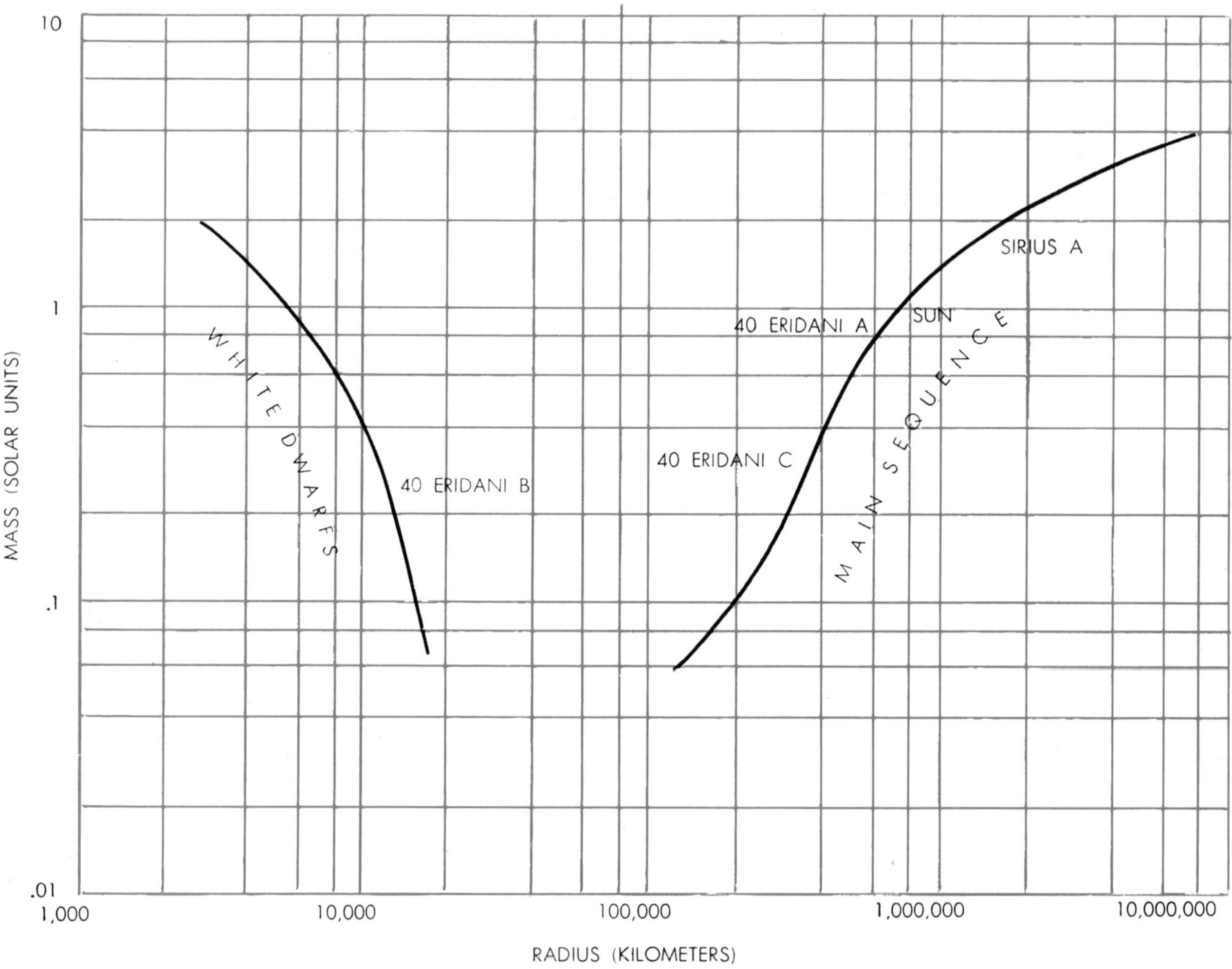

MASS AND RADIUS of white dwarfs show a correlation exactly opposite to that of normal "main sequence" stars (*curve at right*). The latter show increase of radius with increase of mass. White dwarfs, in contrast, have smaller radii at higher mass. The smallest dwarfs have masses which are larger than that of the sun, but these masses are compressed into volumes smaller than that of the earth.

nosity is that hydrogen must now comprise less than .00001 of the mass of a dwarf star. Reactions involving heavier elements—such as carbon, oxygen, nitrogen and neon—require higher temperatures than are likely to occur, though helium might react with these in large concentration at very high densities. However, another set of theoretical considerations argues against the possibility of any energy production at all. In a normal star the thermonuclear reaction-rate is regulated by feedback; with increase in temperature the star expands, and the reaction rate is damped. In a degenerate gas, on the other hand, pressure is unaffected by temperature. Local heating would bring higher temperature and an increase in the reaction rate. The star, in consequence, would explode. We must therefore conclude that the white dwarfs have substantially exhausted their nuclear-energy sources.

Because their luminosity is so low, it is difficult to obtain detailed information about other aspects of the dwarf stars from spectrographic analysis of their light. Only about 80 such stars have been studied in detail. With the light-gathering power of the 200-inch Hale telescope on Palomar Mountain I have observed 50 white-dwarf spectra at a larger scale than any obtained before.

Spectrographic analysis establishes with certainty that the white dwarfs are dwarfs indeed. The derivation of radius from the spectra is somewhat indirect, but it is reliable. Both from photoelectric analysis of the color of the light and study of the behavior of the absorption lines we can determine temperature. From apparent brightness and from independent measurement of distance, we establish the true luminosity. By combining temperature and luminosity, we determine radius. The results are impressively monotonous: the well-determined radii all lie between 3,000 and 10,000 miles. The constancy of dimension is in contrast to the range of size in normal stars, from .1 to 10 times the radius of the sun (430,000 miles) for "main sequence" stars [*see illustration on this page*], and on up to 10,000 times for red giants. The smallest white dwarf known has an estimated radius of only 2,800 miles, much smaller than the radius of the earth. This is close to the theoretical minimum for a star that has exhausted its hydrogen; the radius indicates a mass of 1.2 solar masses and a central density of 150 tons per cubic inch.

One of the most important theoretical predictions is fulfilled with the finding that there is no dependence of radius on surface temperature. The dwarfs we have observed range in temperature from 50,000 to 4,000 degrees K. The hottest is a blue-white star in the earliest phase of white-dwarf evolution; the coolest, a faint, reddish-white dwarf. As plotted in the illustration on the opposite page, stars of the same radius appear down the full range of temperature. Since their initial masses may vary, it is clear that they start with a small spread of radii at the upper left corner of the chart and cool off without further gravitational contraction downward and to the right in straight lines.

Unfortunately it is impossible to match these measurements of radius to equally reliable observational determinations of mass. Newton's laws can give the masses from observed orbital motion only in the case of those stars that are members of multiple systems. Three such dwarfs are known. For two of them, Sirius B and Procyon B, the masses are reliably established at 1 and .65 solar mass respectively. But their major companions, Sirius A and Procyon A, are so bright and so close that the spectrographic plate cannot register an uncontaminated picture of either of these two dwarfs. As a result it is still impossible to measure their radii.

The best-known white-dwarf member of a multiple system belongs to a three-star group: 40 Eridani. Here, fortunately, the distances between stars are wide enough so that good spectra can be obtained, and yet close enough for orbital motion to give reliable measurements of mass. From analysis of the spectrum, I have derived a radius of 6,500 miles, .016 of the solar radius; gravitational measurements establish the mass at .45 solar mass. Calculation from the theoretical mass-radius relationship yields a mass of .39 solar mass, satisfactorily close to observation. Thus, at least in the case of the single star that permits complete test by observation, the well-articulated theory of white dwarfs finds solid support.

The spectra of the white dwarfs also confirm in a general way the theoretical prediction of their elemental composition. One type either shows no hydrogen lines at all, or has hydrogen lines which indicate the presence of relatively tiny residual quantities of hydrogen. Compared to the spectra of normal stars, in which hydrogen lines are universally strong, this anomaly would be enough to identify the dwarfs as a genus apart. The spectra of the commonest type of white dwarf (Type A), however, show only the residual hydrogen and no heavy elements. Here, apparently, gravitational forces have pulled all of the heavier elements, even helium, out of the atmosphere and squeezed the hydrogen to the surface. In dwarfs with surface temperatures below 8,000 degrees, the hydrogen lines vanish completely, and we see only a few lines due to metallic elements. Ross 640 is such a star [*see illustration on page 154*]; it is still hot enough to show hydrogen lines if any hydrogen were present. In general the spectra of white

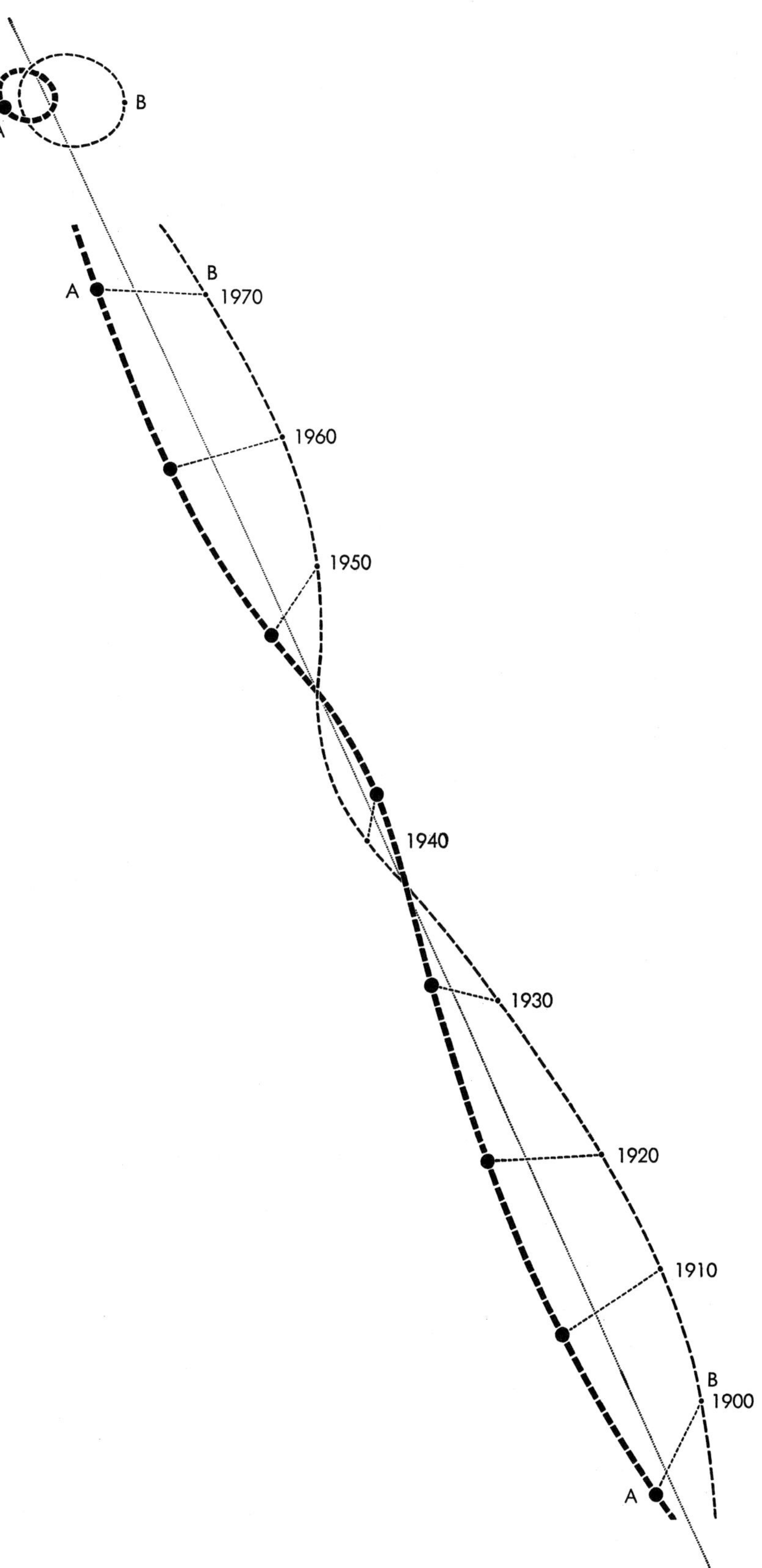

DOUBLE-STAR SYSTEM of Sirius is composed of one of the brightest stars in the sky (A) and a white-dwarf companion (B). Their orbits around the center of gravity of their system is shown at top. The motion of the two stars and of the center of gravity of their system with respect to the earth is indicated by the broken lines running diagonally up this diagram. (*See also the illustration at the top of page 154.*)

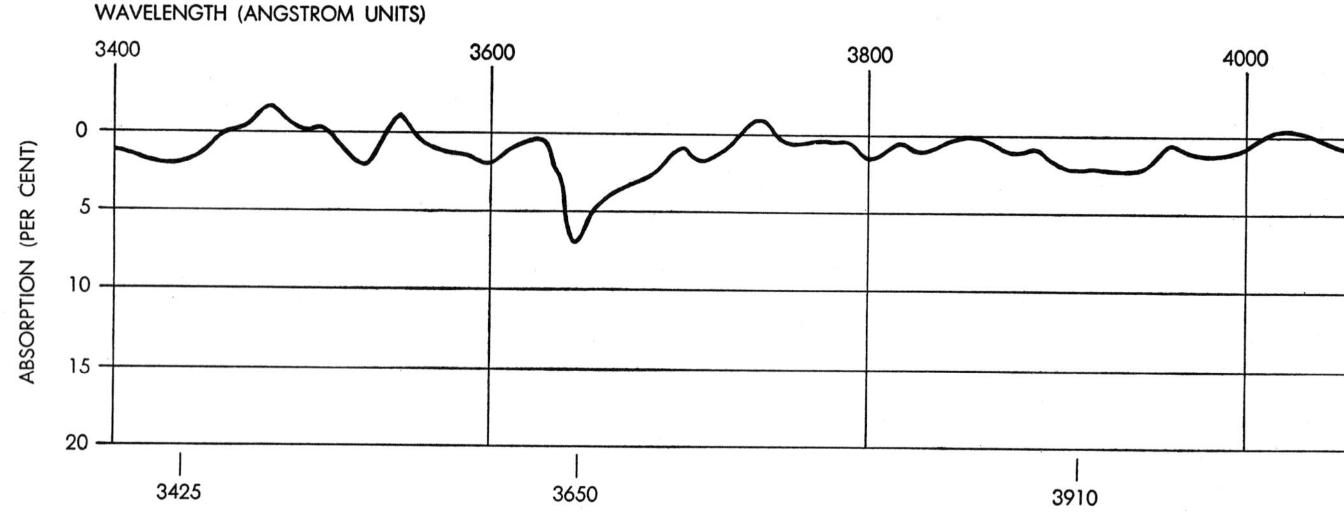

ANOMALOUS SPECTRUM of a white dwarf shows no absorption lines, but does show diffuse bands of absorption at points not associated with any familiar elements or compounds. The spectrum has here been analyzed by a sensitive photoelectric device which meas-

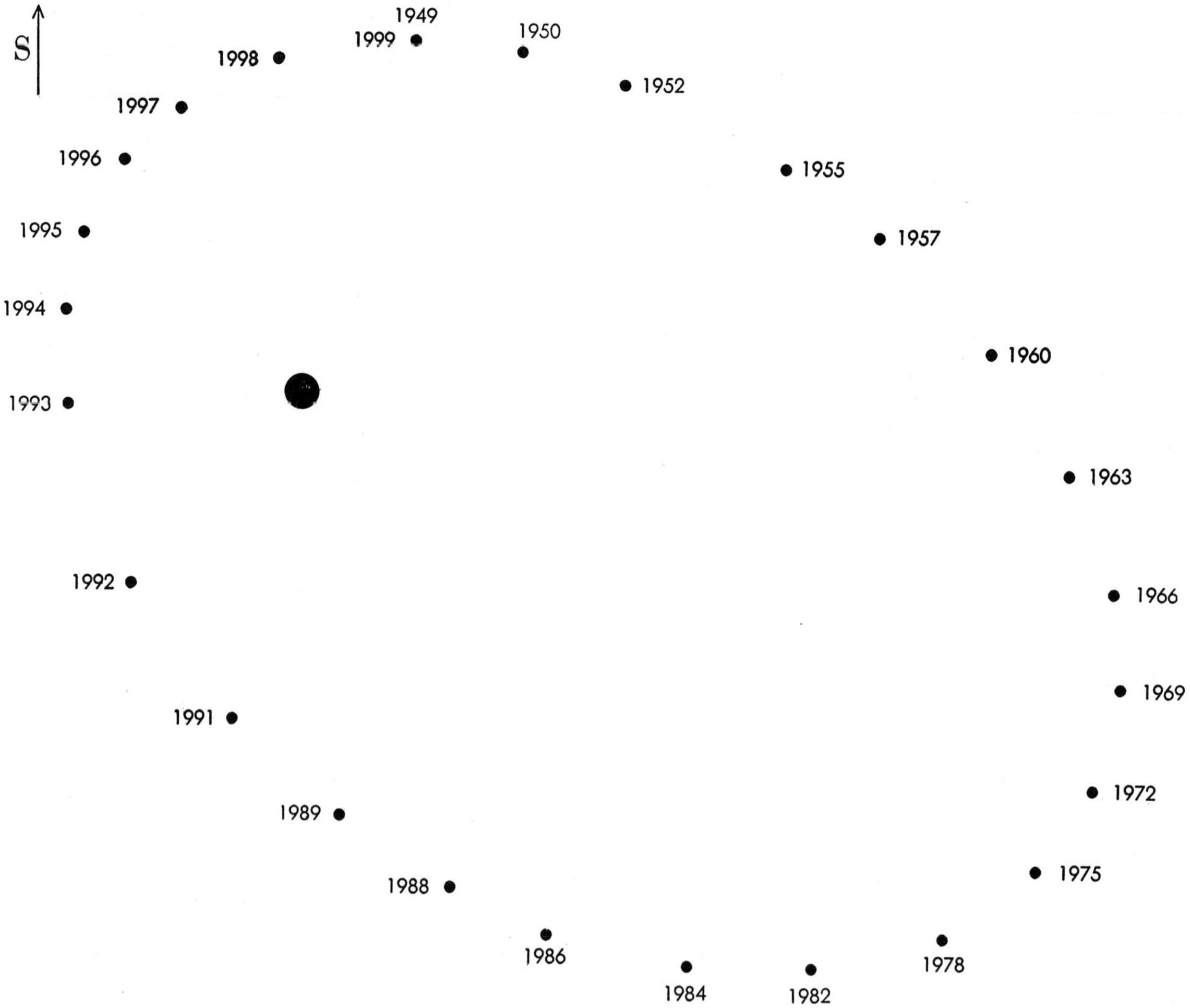

DWARF STAR IN SIRIUS traces the orbit shown here with respect to the large primary star of this double-star system. The dates give the location of the dwarf in its orbit through the second half of this century. Its close approach to the primary star in recent years has made it possible to secure spectrographic images uncontaminated by the light flooding from the 100-times brighter primary star. As the dwarf star approaches the apastron of its orbit during the next 20 years, it may be possible for astronomers to secure better spectra.

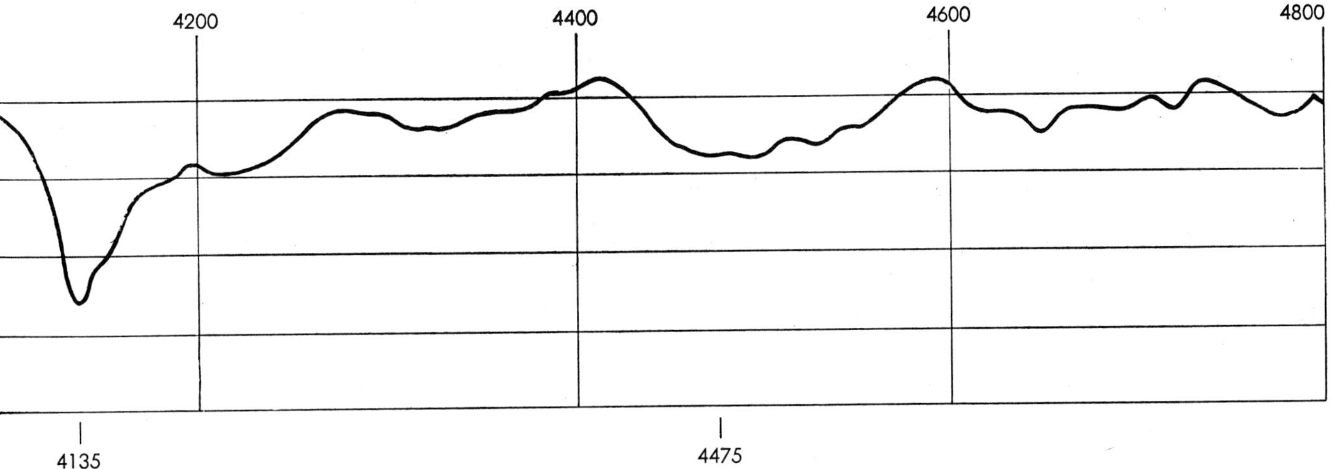

ures the density of the photographic plate from point to point. Only the deep absorption band at 4135 angstroms wavelength is visible to the eye on the spectrographic plate. The absorption bands may be due to presence of elements or free radicals under high pressure.

dwarfs reflect little of the regular correlation between line characteristics and temperature found in normal stars. The varied compositions of their atmospheres therefore may be taken as evidence of their evolutionary history. From the spectrum of Ross 640 we can deduce that this star and other stars like it turned to synthesizing heavy elements from helium after exhausting their hydrogen. The redder and still fainter star called van Maanen 2 (VMa2) is the coolest so far subjected to detailed spectrographic analysis. Its peculiar spectrum [*see bottom illustration on page 154*] indicates that this star began as a metal-poor member of the long-lived, stable Population II family. Since its present low luminosity gives this star an age of four billion years in the white-dwarf phase alone, van Maanen 2 must have lived out its entire life as a brilliant star before the sun and the earth were formed. In a still fainter, cooler and more ancient star, no lines have yet been detected with certainty.

A spectrum without absorption lines might seem to be of academic interest to astrophysicists, who employ these lines as the tools of their trade. But we have spent many nights observing and many months of analysis to establish the real absence of lines in six white-dwarf spectra. Subjected to the most sensitive photoelectric inspection yet possible, the plates show no line, band or absorption depression as deep as 5 per cent. There are a number of possible explanations. Perhaps the most satisfactory will be found upon closer inspection of lines that do appear in other white-dwarf spectra. The extreme broadening and attenuation of the hydrogen lines in some spectra helps to make the complete disappearance of lines at very high pressure more understandable. Such broadening of lines is caused by random electric fields and by collisions between charged particles. In the van Maanen 2 spectrum Volker Weidemann of the Bundesanstalt in Braunschweig, who has been working with us on a grant from the Air Force Office of Scientific Research, has found lines of iron, magnesium and calcium broadened in a way that indicates a rate of particle collision 10,000 times that observed in the sun. He estimates a pressure of 2,000 atmospheres in this peculiar atmosphere—dense enough for some molecules to form. But though metal lines may be thus broadened, it is surprising that they should disappear entirely, as they do in the six spectra that show no lines at all.

To compound the mystery we have come upon several spectra with diffuse, shallow bands that cannot be related to any established laboratory spectral line; the photoelectric tracing of a plate made for one of these is shown at the top of these two pages. These bands may originate from molecules or unstable free radicals under unusual conditions of temperature and pressure. How atoms behave in the strange environment of the white-dwarf atmosphere is not yet known.

Our generation has seen at least one star arrive at the end of the evolutionary road and become a white dwarf. The recurrent nova, WZ Sagittae, which exploded in 1913, exploded again in 1946, brightening about 1,000 times. Its brightness is now about .01 that of the sun, and its spectrum resembles that of white dwarfs in everything but the presence of superposed emission lines. These lines are presumably due to the continued ejection of hot material. WZ Sagittae demonstrates one, though not the only, process by which stars may lose their mass and make the transition to the final stage in their history.

As living things live and die in countless ways, so stars have many possible evolutionary histories and deaths. When we have learned to read the spectra of white dwarfs better, we may see what paths they have traveled. Their faint light may give us evidence which will show what processes went on during ages past in their thermonuclear furnaces.

A white dwarf takes a long time dying. Its light bespeaks the slow leakage of heat from its interior down the temperature gradient set up by the conductive opacity of the degenerate gas. The thermal energy is contained only in the nondegenerate nuclei and the few electrons above the Fermi threshold. Though the initial temperature may be high, this thermal energy is all that is available throughout the entire dying stage. But as the star cools and its luminosity fades, the temperature gradient also declines. The dissipation of energy therewith slows down, and the time scale of evolution toward lower luminosity is greatly extended. According to Martin Schwarzschild of the Princeton Observatory, a white dwarf composed mainly of helium takes three billion years to cool from its initial blue-white stage down to a surface temperature of 7,000 degrees in the yellow-white stage. From yellow down to the 4,000 degrees of the faintest known red-white dwarf, it takes another five billion years. But 4,000 degrees is still red-hot. From red to infrared, the star will fade over fantastic spans of time, large compared to any present estimate of the age of our galaxy.

TRIPLE-STAR SYSTEM in constellation Eridanus is composed of a bright primary normal star (A), a faint late-type star (C) and a white dwarf (B), which appear in the relative positions, but not to the scale, indicated by the small spheres at the top of this diagram. The relative diameters of the three stars are shown across the bottom of diagram, star A (*at left*) having a radius .9 that of our sun; star C (*second from left*) having a radius .4 that of the sun; and the dwarf having a radius .017 that of the sun, or 7,000 miles.

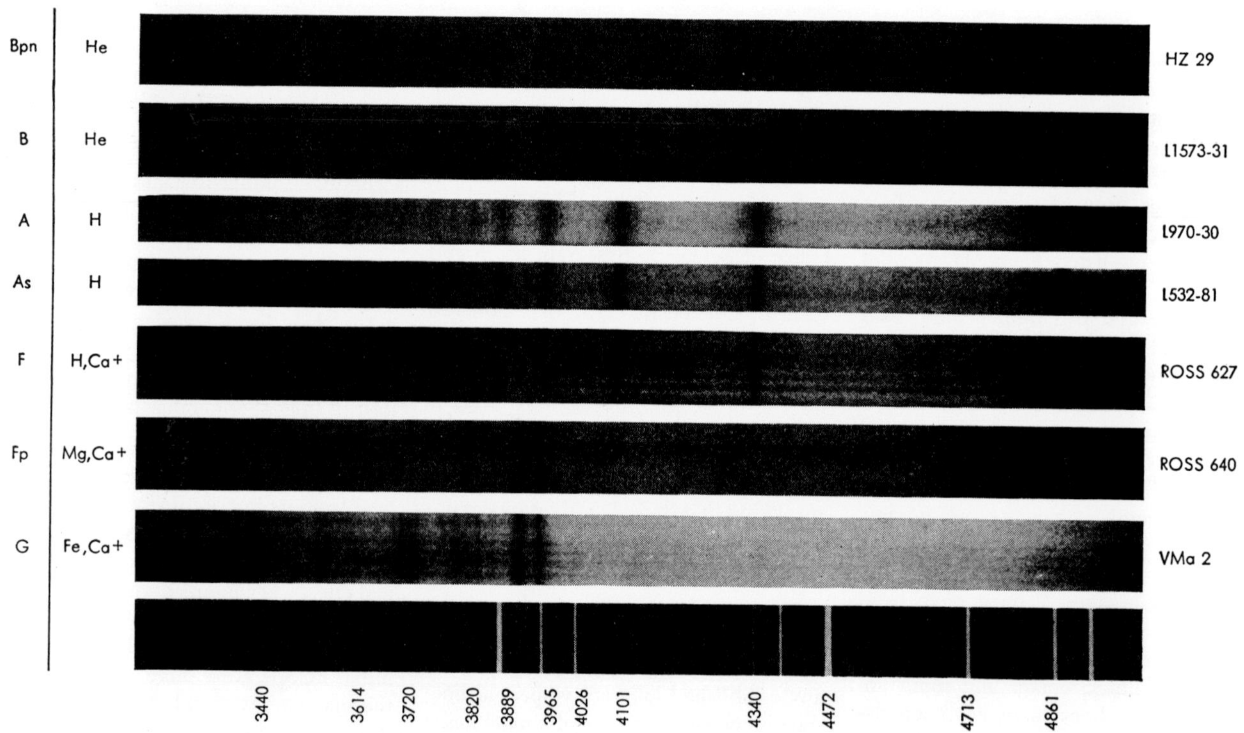

DWARF STAR SPECTRA of various types (identified by initials at far left) show absorption lines for only a few elements (identified by initials in second column at left). The individual stars are identified by their code numbers at right. The spectrum at bottom is the reference spectrum of helium and hydrogen. Absorption lines of Type A dwarfs are characteristically diffuse and broadened.

The fall in temperature brings the degenerate gas phase ever nearer to the surface. The nondegenerate electrons become scarcer and, at a very low temperature, even the nuclei become degenerate. When all the nuclear particles and electrons have occupied the lowest possible energy states, radiation ceases and the star becomes a giant "molecule." This is the end of the irreversible process of evolution—proof of the fourth law of thermodynamics. There are, however, no black dwarfs in our galaxy; it is as yet too young.

On the one-way track described here, all stars eventually fade to extinction. How will the sky look after our sun's evolution is complete, and our dead planets circulate about a dying star? In about seven billion years the sun will be a hot and very blue-white dwarf, too small to show a disk to the unaided eye on earth. The earth's temperature will be about 300 degrees below zero Fahrenheit. The sky at night will no longer be filled with stars, since star formation will have ended, and the high-luminosity stars that comprise our constellations will long ago have disappeared. Probably no star will be visible, except for an occasional faint, red normal main-sequence star that passes by chance near our dying system; such stars are so faint that their nuclear energy suffices for thousands of billions of years. Although the formerly bright stars will have become white dwarfs, they will all be too faint to be seen, and black night will reign supreme. Yet close to one of the faint red stars life might exist on other planets, in forms and for ages unimaginable to us.

V

The Milky Way

V

The Milky Way

INTRODUCTION

The picture of the Milky Way galaxy as a whirling pinwheel of stars interlaced with avenues of nebulous gas and dust has been drawn piece by piece throughout the past five decades. Just over fifty years ago, Harlow Shapley, working at Mt. Wilson Observatory, first demonstrated both the great extent of our galaxy and the sun's off-center position within it. In the 1920's, Lindblad and Oort assembled observational proof for our galaxy's rotation; in the 1930's, Trumpler demonstrated the role of interstellar absorption; in the 1940's, Baade proposed the concept of stellar populations; and in the early 1950's, optical and radio astronomers provided independent evidence for the spiral structure of the Milky Way.

The optical detection of the spiral arms came after years of frustrating and, in retrospect, misguided searches. Ever since the Milky Way was understood to be a comparable object to the beautiful but distant spiral galaxies, astronomers had speculated that our Milky Way ought to exhibit a spiral pattern. They assumed quite simply that the great spiral streams seen in distant galaxies represented regions of high star density, and they supposed that careful star counts might disclose such zones in the Milky Way itself.

The studies made at Mt. Wilson and Palomar Observatories by Walter Baade during the 1940's showed that the spiral structure of the Andromeda Galaxy was not, after all, outlined by large concentrations of stars; instead, it was delineated by comparatively few but immensely luminous younger stars of the sort that he designated as Population I. Baade likened the brilliant but comparatively insubstantial arms to the frosting on a cake—contributing to the appearance but not to the mass.

By 1958, when Margaret and Geoffrey Burbidge wrote "Stellar Populations," this seminal concept had been well honed and refined to encompass five categories. Yet even in the more elementary form of two basic populations, the concept enabled astronomers to recognize that spiral arms would be marked out by Population I "spiral indicators." Armed with this understanding, optical astronomers first detected the Milky Way's spiral structure in 1951—just in the nick of time from their point of view, for within a few months radio astronomers began their investigations of the hydrogen clouds that also outline the spirals.

It is now nearly twenty years since that first mapping of the spiral arms by means of the 21-centimeter radio line of neutral hydrogen took place, and over ten years since Bart J. Bok in "The Arms of the Galaxy" attempted to sort out the various patterns disclosed by both the radio data and the spiral markers viewed optically. In a recent issue of *Sky and Telescope,* Bok has reported on an international symposium on our galaxy's spiral structure, and, curiously enough, a dispute is still raging over opposing interpretations similar to those he described in 1959.

Progress has nonetheless been made in understanding the Milky Way since Bok's lucid account of earlier studies in the *Scientific American;* for example, he no longer believes that magnetic fields play the key role establishing the spiral arms. Instead, a strictly gravitational process advocated by C. C. Lin, F. H. Shu, and C. Yuan has been gaining currency; in their view, the spiral arms are not permanent entities, but rather the loci of concentrations of comparatively slower moving gas atoms. Once a spiral arm is established (by an as yet unspecified method), it will apparently be gravitationally self-perpetuating.

Discoveries of extraordinarily active and even explosive nuclei in remote galaxies has heightened interest and curiosity in the nucleus of our own Milky Way galaxy. "The events taking place at the center of our galaxy are less catastrophic but not less interesting," Brian J. Robinson writes in "Hydroxyl Radicals in Space." His survey of the hydroxyl absorptions includes a discussion of the remarkable hydrogen gas motions in the galaxy's nucleus.

A more recent milestone in galactic studies is the detection of the nucleus in infrared wavelengths. Though the center of our galaxy is totally obscured in visible light, the intervening dust was finally penetrated by infrared techniques a few years ago. For an illuminating discussion of the nature of these obscuring interstellar grains, see J. Mayo Greenberg's article in this section, and for information on the detection of the nucleus, see "The Infrared Sky" in Section VII.

15

Stellar Populations

GEOFFREY AND MARGARET BURBIDGE

November 1958

During the fall of 1943 there were a few nights of almost ideal observing conditions at the Mount Wilson Observatory. The air was exceptionally steady, the temperature nearly constant and the surrounding valley lay in the darkness of the wartime blackout. Walter Baade of the Mount Wilson staff seized the opportunity for another try at a long-standing problem—to photograph separate stars in the central region of the great spiral nebula in Andromeda. All previous photographs with the 100-inch telescope on Mount Wilson, the most powerful instrument then available, had shown this region only as a hazy blur of light.

The combination of favorable circumstances and of Baade's great technical skill and ingenuity was successful. His now-famous plates revealed that the blur was actually a dense mass of faint, reddish stars. He also succeeded in resolving stars in the two small companion nebulae of the Andromeda nebula: M 32 and NGC 205. As Baade himself has explained in Scientific American [see "The Content of Galaxies"; September, 1956], a study of these newly resolved objects suggested that all stars are sharply divided into two classes: one, which Baade called Population I, whose brightest members are hot, blue stars; and one, which he named Population II, whose brightest stars are cool and red, but very large. He concluded that the stars of Population I were relatively young, while the stars of Population II were quite old. Most of the stars near the sun, and those which had long been visible in the arms of the Andromeda spiral, fell into the first group; those in the central region of the Andromeda nebula, and in the so-called globular clusters of our own galaxy, into the second.

After 15 years of investigation our picture of stellar populations has grown less simple and perhaps less surprising. Instead of supposing that all stars fall into just two groups, widely separated in age and location in the galaxies, we now believe that there is a more nearly continuous spectrum of ages, from very ancient stars to those still in the process of birth. We still divide them into classes, but these are more numerous, and one tends to merge into the next. To appreciate how the current view has developed, let us retrace the steps since Baade's original findings.

The brightest stars in his Population I are bluish white, with surface temperature of some 30,000 degrees absolute, and the brightest of them shine with the brilliance of 100,000 suns. As we have said, the great majority of stars near the sun seemed to belong to this group, as do the stars which had been resolved on photographs of the arms of the Andromeda nebula and other nearby spiral galaxies.

On the other hand, no members of Population I were found between the arms or in the central regions of spiral galaxies, or in the so-called elliptical galaxies, which have no spiral arms. In these regions the brightest stars (in Baade's Population II) are 50 to 100 times fainter than those in Population I, and their surface temperatures are relatively low—only 3,000 or 4,000 degrees. Their color is distinctly red, and Baade was struck by their similarity to the brightest stars in the globular clusters of our own galaxy. These dense clusters, each containing some 100,000 stars, are distributed around the galaxy in a roughly spherical volume [*see top illustration on page 163*].

Thus far we have spoken only of the brightest stars in the various regions. Each population also contains a whole array of fainter members, which seemed to fall into the same population grouping. All these stars can also be classified in another way: by means of the well-known temperature-luminosity diagram, in which intrinsic brightness is plotted against temperature. When this is done, the stars fall into a well-defined pattern, with the two populations occupying different parts of the diagram [*see illustration on page 161*]. The band running from upper left to lower right is known as the "main sequence." There are Population I stars along its entire length, but members of Population II are found only below a certain point near the middle.

Now it is known that the brightness of stars on the main sequence depends on their mass. In fact the brightness increases as the square of the mass at the lower end of the main sequence, and as the third or fourth power at the upper end. If one star has twice the mass of another, it will be four to 16 times as bright. But the mass of a star is a measure of the amount of fuel it has available to burn in the thermonuclear reactions which produce its radiant energy, and the brightness is a measure of its rate of burning. Therefore the lifetime of the star (the time required to consume all of its nuclear fuel) is proportional to the

LUMINOUS CLOUD of interstellar matter in the constellation of Scutum Sobieski was photographed with the 200-inch telescope on Palomar Mountain. The matter is made luminous by the hot young stars of Population I embedded in it. In such regions stars are probably being formed at the present time. Small dark patches of dense matter can also be seen; some of them may represent an early stage in the formation of stars.

GREAT NEBULA IN ANDROMEDA was photographed by the 48-inch Schmidt telescope on Palomar Mountain. The arms of this spiral galaxy, the disk of which is seen at an angle, are outlined by the dark lanes that lie between them. Population-I stars are found in the arms; Population-II stars, in the bright central region. The small blobs above and below the disk are small satellite galaxies.

mass divided by the brightness. Because brightness increases so much more rapidly than mass, the bright, hot stars at the upper left in the main sequence must burn themselves out much faster than the fainter stars do. In fact, the brightest and hottest appear to be less than a million years old.

Thus Baade's Population I contains relatively young stars. They are so young as a group that some must even now be in the process of formation. Indeed, we can probably see this happening in our own galaxy. There are some faint, irregularly flickering stars in the Great Nebula in Orion whose unsteadiness is almost certainly due to their youth. They have not settled down to an orderly existence on the main sequence. Some of them may even be growing yet, drawing to themselves more of the surrounding gas and dust.

We can now understand why Population I stars exist only where Baade found them, in the spiral arms of galaxies. Formed recently, they have not had time to move away from the region containing the raw materials out of which they were made. And for some reason, probably having to do with magnetic fields, interstellar gas and dust are concentrated in spiral arms. Between the arms and in the central parts of spiral galaxies, as well as in the whole of elliptical galaxies, there is no dust and very little gas.

These regions are the domain of Population II. There is good reason to suppose that they have been dust-free for a very long time, and so all the Population II stars must be quite old. It is easy to see, then, why their main-sequence members extend only to stars slightly brighter, and about 20 per cent heavier, than the sun. All the brighter ones that must have been on the upper part of the main sequence came to the end of that phase of their lives long ago, and there has been no material to make replacements.

How does a star leave the main sequence? Theories of the nuclear reactions in stars show that most of their hydrogen is consumed and converted to helium deep in their interiors. As time goes on, the core in which the hydrogen has been totally consumed grows larger and larger. When the core comes to contain about a 10th of the whole mass of a star, the star's internal structure becomes unstable. To restore equilibrium its material must be rearranged. In the process the star expands fairly quickly, and its surface layers cool; it becomes a "red giant." This is what has happened to the red stars that are the brightest members of the globular clusters, and to those which appear on Baade's photographs of the Andromeda nebula and of its companions.

Eventually a red giant uses up all the nuclear fuel in its center and comes to the end of its life as a normal star. Then it may become a supernova, disintegrating in a giant thermonuclear explosion, or it may suffer a series of lesser explosions known as nova outbursts. Such must have been the fate of the bright stars of Population II. In Population I there are also some red giants, but they are about 15 times less bright than those of Population II. Examination of their spectra has shown that the chemical compositions of the two types of red giants are not the same. In the Population II red giants the heavier elements such as calcium and iron are only about one one-hundredth as abundant with respect to hydrogen as they are in the red giants of Population I. Thus the members of Population II must have been made out of material that was relatively poor in the heavier elements.

This is just what we should expect from the current theory of the origin of the elements. It tells us that in the beginning there was only hydrogen. All the other chemical elements have been created out of hydrogen by nuclear reactions in stars. Each time a star goes through its explosive death throes it spews out the heavy elements it has manufactured during its life. Hence the dust and gas out of which new stars are made must have been gradually enriched in the heavier elements. Thus it is not surprising that the oldest stars we observe today should contain the lowest proportion of these substances.

All this seems reasonable enough, but it does not explain why there should be only two stellar populations, an old one and a young. Why not some middle-aged stars? As we have indicated, the sequel to Baade's work has resolved the puzzle. Middle-aged stars do indeed exist.

The most revealing indication came from studies of the movements of stars in our own galaxy. The galaxy as a whole is revolving, and the individual stars share in this motion. In addition they have movements of their own. Not only do they travel about in the central plane of the galaxy, but most of them have a component of motion perpendicular to the plane [*see illustration at the bottom of page 163*]. This component

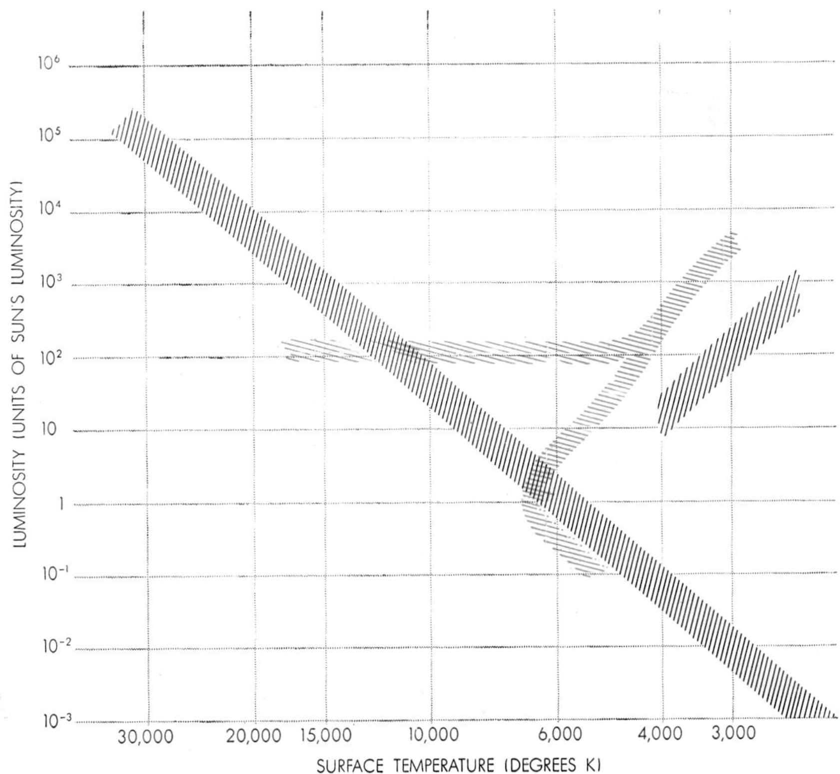

TEMPERATURE-LUMINOSITY DIAGRAM elucidates the relationship between stars of Population I and those of Population II. The black hatched areas are occupied by stars of Population I; the colored hatched areas, by stars of Population II. The red-giant stars are at the upper right. The sun, a star of Population I, is indicated by the colored dot.

GLOBULAR CLUSTER, M 13 in the constellation of Hercules, was photographed with the 200-inch telescope. A member of our own galaxy, it consists only of Population II stars.

ELLIPTICAL GALAXY, NGC 205, photographed with the 200-inch, is composed of Disk-Population or Population-II stars. It appears near the top of the photograph on page 160.

(which, in the reference system conventionally adopted in astronomy, is along the z axis) carries them above or below the central plane. They can travel only so far in the z direction before the gravitational pull of the central mass of stars brings them back. Then they reverse their motion and move back toward the plane. Like pendulums, they do not stop at the equilibrium point, but overshoot and travel an equal distance to the other side of the plane, and so on.

At any given moment the z-speed of a star depends on its position in the oscillatory cycle. But, in general, stars with the highest z-speed should swing farthest from the central plane. And so it turns out. The globular clusters, arranged nearly spherically around the plane, have z-speeds of about 100 kilometers per second. The brightest, hottest stars, which lie in a very flat disk in the central plane have z-speeds of only about five kilometers per second. But the catalog of speeds is not restricted to these extreme values. There is a continuous gradation; and in the case of main-sequence stars the brighter they are, the smaller their speeds and the nearer they lie to the central plane. This fact emerged gradually, in the course of many years of observation by several astronomers. In 1950 the Soviet astronomer P. P. Parenago drew attention to it and suggested that, rather than two clearly defined populations of stars, there must be a full range of populations.

More recently we have found that the chemical compositions of stars show a similar spread. The percentage of heavy elements varies from the very low value characteristic of the globular clusters to the much higher value found in the brightest stars. Thus our galaxy appears to contain stars of various ages, with the older ones lying, on the average, farther from the central plane.

The picture is quite satisfactory because it fits well with the current view of galactic evolution. We suppose that our galaxy began its life as a cloud of hydrogen gas, either pure or slightly contaminated with heavier elements from exploded stars in earlier galaxies. The cloud then began to shrink, pulled together by its own gravitational attraction. As it did so, the first stars or clusters of stars began to form. In the beginning the gigantic cloud probably revolved slowly, speeding up as it shrank. As a consequence of its rotation the sphere gradually flattened, under the same kind of force that makes the rotating earth slightly flattened at its poles.

All the time the cloud was shrinking,

flattening and revolving faster and faster, stars were forming in those regions where, by chance, the density was higher than average. Big stars went quickly through their life histories, cooking up heavy elements and then exploding and scattering them back into the gas out of which new stars were continually condensing. The shape of the cloud at any epoch should be preserved by the stars formed at that time. Once they had coalesced into dense masses, they would move more or less independently of the surrounding gas and dust, and would no longer partake in the general flattening and shrinking. As we have seen, the oldest stars we see today are indeed distributed most nearly spherically.

At a conference in Rome in the summer of 1957 astronomers generally agreed on a convenient classification of stars into five populations. They are as follows:

1. Extreme Population II. This is the oldest group—at least seven or eight billion years old. In our galaxy it is represented by the globular clusters, together with a sparse spherical distribution of stars lying between them. These isolated stars may have escaped from the clusters.

2. Intermediate Population II. Not quite as old as the first group, it occupies a volume not completely spherical, though not flattened very much. Stars which explode as novae, and also the so-called planetary nebulae, apparently belong either to this or to the next group.

3. Disk Population. These stars probably range from three to five billion years old. The Disk Population makes up the great bulk of the stars in our galaxy and in the Andromeda nebula; most of the stars between the spiral arms and in the dense central regions probably belong to it. The sun is probably a member. We know from its chemical composition the sun must be at least a "third generation" star, which indicates that many stars completed their life cycles before the Disk Population was formed.

4. Intermediate Population I. This Population ranges from about a hundred million to a few billion years old. Its members, which include stars like Sirius, lie in or quite near the central plane of the galaxy but are not restricted to spiral arms.

5. Extreme Population I. This is tens of millions of years old or less, and it includes the gas and dust still not condensed into stars. Both the gas and the stars lie in spiral arms. The hot, bright stars in Orion, particularly Rigel and the stars in the Orion nebula, belong to this class.

This division into five populations is a matter of convenience; actually the groups merge into one another and could be further subdivided.

According to the new grouping, stars in the central regions of the Andromeda nebula and of the large elliptical nebulae are Disk Population, while the globular-cluster stars belong to Extreme Population II. The chemical compositions seem to support the classification: globular-cluster stars are considerably

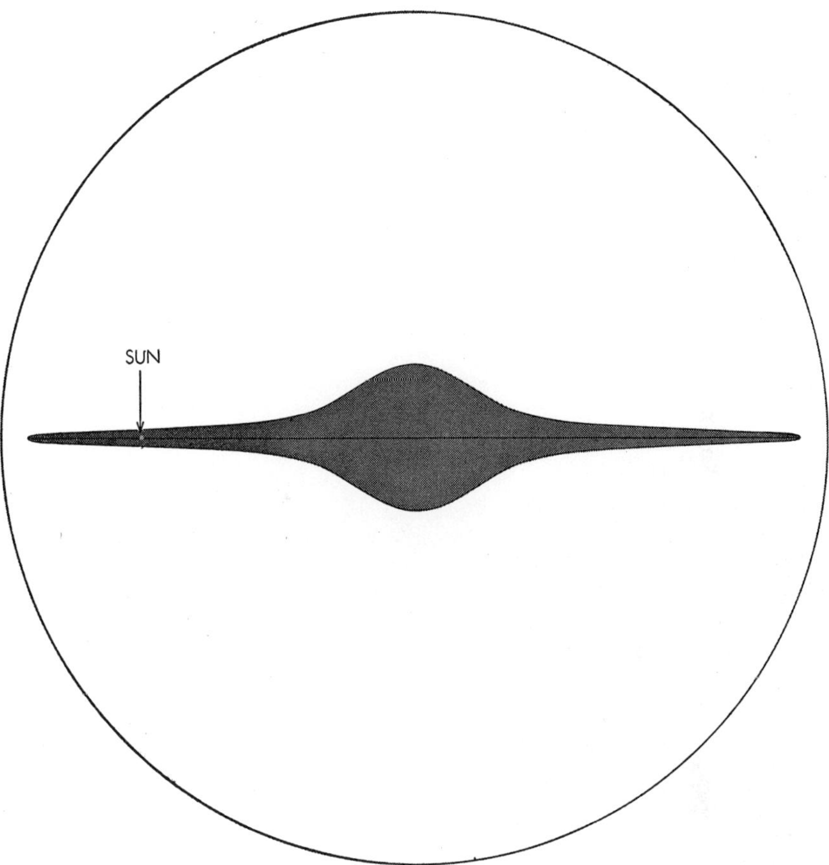

SCHEMATIC EDGE-ON VIEW of our own galaxy shows its central bulge and relatively thin arms. The circle is a cross section of the spherical volume occupied by the globular clusters.

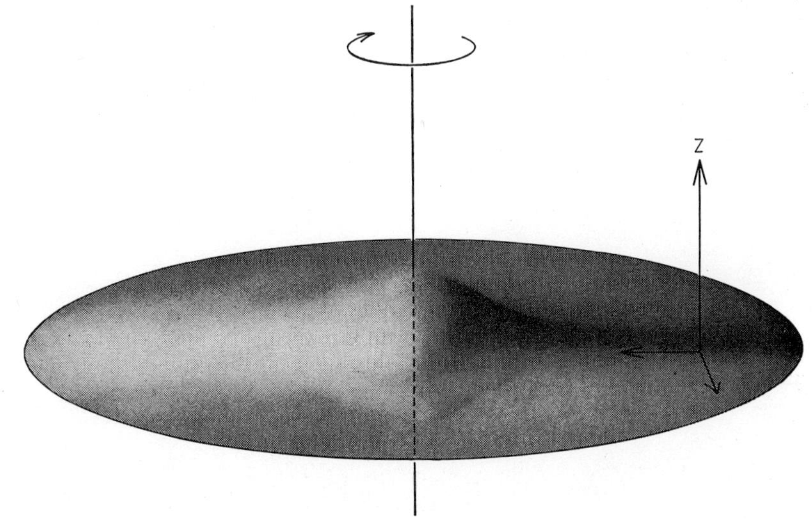

SCHEMATIC THREE-QUARTERS VIEW of the galaxy shows the three components in the motion of a star in its central plane. The star swings like a pendulum in the *z* direction, while it also moves in the *xy* plane, taking part in the rotation of the galaxy as a whole.

CENTRAL REGIONS of two spiral galaxies are compared. At top is the galaxy NGC 5457; its central region is small in comparison to the extent of its arms. At bottom is NGC 4736; its central region is large in comparison to the extent of its arms. Since the central regions of galaxies contain old Disk-Population or Population-II stars, the second galaxy may be older than the first. Both galaxies were photographed with the 200-inch telescope.

poorer in the heavy elements than the others. But there are still some open questions. We have mentioned that, because of their lower concentration of heavy elements, the bright red giants in globular clusters are about 15 times brighter than those near the sun. Yet the brightest red giants in the center of the Andromeda nebula apparently have the same brightness as those in globular clusters. Also, in the central regions of our galaxy there is a large number of variable stars of the same type as those found in globular clusters. Perhaps these regions actually contain a mixture of classes, including Extreme Population II, as well as Intermediate Population II and Disk Population.

Clearly we have only begun to understand the whole problem. When we learn how to classify all the stars we can observe, we shall know a great deal more about the history of the universe. Many intriguing questions suggest themselves. For example, what is the detailed life history of a galaxy? Are the galaxies around us in different stages of development? It seems now that they are. Spiral galaxies with very small central bulges may be much younger than galaxies like ours, in which the central bulges are large. Probably the central region grows larger as a galaxy ages, because its material gradually loses its random movement and falls inward. As the center becomes denser, star formation speeds up. Therefore it is not surprising to find stars covering a considerable range of ages in the central regions.

It is possible that all galaxies do not age at the same rate. Factors such as the mass of a cloud of gas, its initial speed of rotation and the size of its magnetic field, if it had one, may affect its development and the rate of star formation in it. Thus galaxies that were born at the same time may now have reached very different stages in their life histories. This might explain why old- and young-looking galaxies are sometimes found very close together. For example the two Clouds of Magellan, our nearest extragalactic neighbors, seem quite young as compared with our galaxy. Half of their masses are still in the form of gas, whereas the gas in our galaxy comprises only a few per cent of its mass. Whether they are also poorer in heavy elements is not yet certain.

Many different branches of astronomical research are at present converging on the problem of the life histories of galaxies. But the concept of stellar populations, originated by Walter Baade, remains the key to all the approaches.

16

The Arms of the Galaxy

BART J. BOK

December 1959

Once every 200 million years the sun completes one revolution about the center of our galaxy. To an observer in another galaxy ours would appear as a flattened disk of stars, gas and cosmic dust 80,000 light-years in diameter. In the neighborhood of the sun, about halfway between the center of the galaxy and its edge, the disk is only about 5,000 light-years thick; in the center the disk bulges out to about three times this thickness.

Only a fraction of the galaxy is visible from the earth. The view of even the largest telescopes is obscured by the clouds of dust floating in the central plane of the galaxy; in this plane it is impossible to see farther than 20,000 light-years. In other directions we can perceive galaxies millions of light-years away, many of which have handsome spiral patterns. For years most astronomers agreed that our galaxy also has a spiral structure, but there seemed to be no way of proving it.

We tried to tackle the problem by brute force. By counting stars and sorting them according to their brightness, their color and their spectral lines, we attempted to detect statistical "bumps" or patterns of star distribution that might suggest spiral arms. These studies failed so conclusively that in my lectures during the 1930's I often stated that I had no hope of seeing the structure problem solved during my lifetime.

Then came the break-through. In the middle and late 1940's Walter Baade of the Mount Wilson and Palomar Observatories demonstrated that the best way to determine the structure of our galaxy was to find out what kinds of stars most clearly marked the arms of other galaxies. With the 100-inch telescope on Mount Wilson and later with the 200-inch on Palomar Mountain, Baade examined the spiral of the nearby Great Nebula in Andromeda. By skillfully manipulating color filters and fast, red-sensitive photographic emulsions he was able to show that luminous blobs of gas—the so-called emission nebulae—traced out the arms of the Andromeda spiral like a string of street lamps.

Soon additional spiral tracers were found, including cepheid variable stars with long periods of pulsation, clumps of young stars known as galactic clusters, and even the troublesome cosmic dust, which was observed to accumulate at the inner edges of spiral arms [*see illustration on page 172*]. The most interesting tracers proved to be the hot, blue-white supergiant stars, classified (according to their color and spectral characteristics) as O or B stars. These stars, which follow the spiral arms quite closely, are truly mighty radiators, many of them being 10,000 to 100,000 times brighter than the sun. Because they are visible for great distances through the cosmic haze, Baade advised his fellow astronomers that to map the arms of our galaxy they had only to locate its bright O and B stars, measure their distances from the earth and plot their positions accurately on a chart.

The Optical Model

A vigorous search for the supergiants was quickly pressed by W. W. Morgan and his colleagues at the Yerkes Observatory of the University of Chicago. The task of measuring distances accurately was complicated by the fact that such stars are often found in loose groups or "associations" which are embedded in clouds of gas or are partially hidden by drifts of cosmic dust. These drifts, lying in the line of sight between the star and the earth, act like red filters, dimming and reddening the light passing through them. Because we measure the distance to a remote star by applying the law that the intensity of light diminishes inversely with the square of the distance, the "space reddening" caused by dust makes the star appear farther away than it really is.

We can nonetheless determine the distances to partially obscured stars with fair accuracy. We first measure the star's spectral lines, then estimate its true color on the basis of its spectral classification. The difference between the true color and the measured apparent color indi-

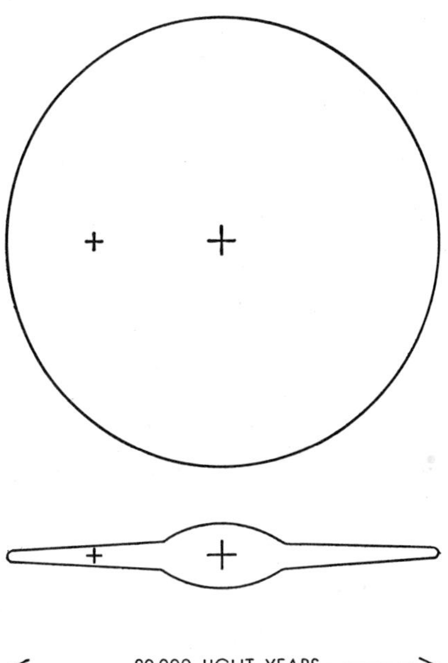

OUR GALAXY, seen from above (*top*) and from the edge (*bottom*), appears as a bulging disk in this schematic diagram. The larger cross indicates the center of the galaxy; the smaller one, the position of the sun.

cates the amount of cosmic dust that lies between the star and the observer. Once this is known we can determine the distance to the star by adding an appropriate correction factor to the inverse-square law.

Using this method, Morgan and his associates Stewart Sharpless and Donald Osterbrock accumulated enough information by the end of 1951 to draw a tentative diagram of the spiral arms in the neighborhood of the sun. When they presented their model at the Cleveland meeting of the American Astronomical Society, its impact on the assembled astronomers was impressive. Here, for the first time, the spiral structure of the galaxy began to unfold.

The model showed that a giant spiral arm extends from the direction of the constellation Orion, enfolds the sun and its planets and swerves off in the general direction of Cygnus. Morgan later called this the Orion Arm. A second arm, well separated from the first and 5,000 to 7,000 light-years farther from the center of the galaxy, was named the Perseus Arm because the famous double star-cluster in Perseus is the most prominent object within it. In the direction of the constellation Sagittarius, Morgan found a few nebulae and associations outlining a possible third arm, located about 5,000 light-years closer to the galactic center than the Orion Arm. Because he had no data for the Southern Hemisphere, Morgan was unable to trace the full extent of the Sagittarius Arm, but he suspected that it was connected to a section of an arm that lay in the direction of the constellations of Carina and Centaurus, near the Southern Cross.

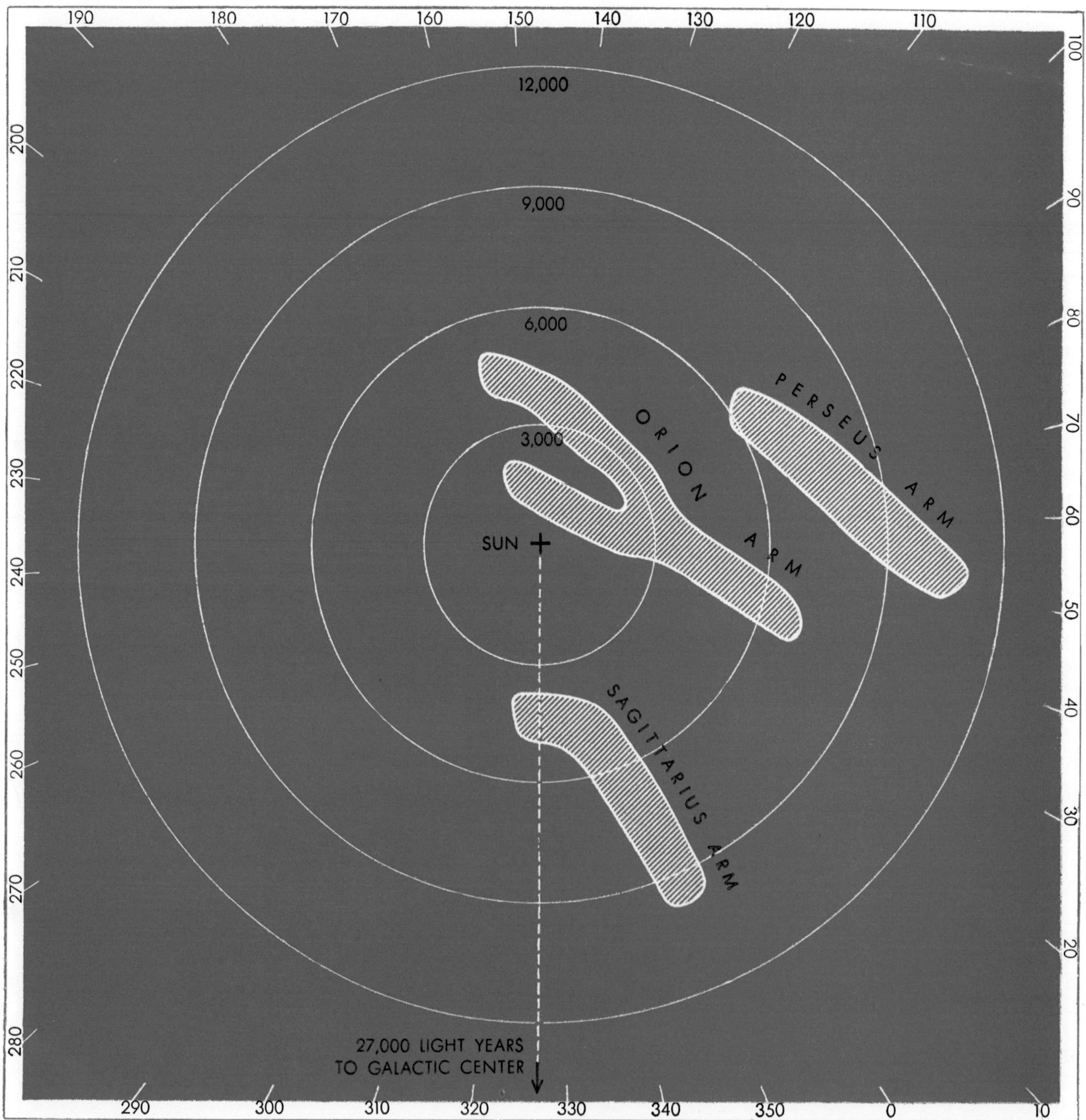

EARLY OPTICAL MODEL traced the structure of the spiral arms near the sun. The numbers around the edge of the map indicate degrees of longitude in the plane of the galaxy. Numbered circles show distance from the sun in light-years. Dotted line shows direction of the galactic center. This drawing is based on the model proposed in 1952 and 1953 by W. W. Morgan of Yerkes Observatory.

After Morgan had presented his dramatic paper there was every hope that the rest of the galaxy's spiral structure might soon be charted, but it quickly became apparent that great difficulties lay ahead. Even with the new method it seemed impossible to map more than our immediate galactic neighborhood; there still appeared to be no way to penetrate more than 20,000 light-years into the cosmic haze. It was a happy coincidence that at just about that time the new techniques of radio astronomy provided a research tool that extended the search for spiral arms to areas apparently closed to optical observation.

The 21-centimeter radio signals emitted by interstellar clouds of neutral (un-ionized) hydrogen in our galaxy were first detected in 1951 by Harold I. Ewen and Edward M. Purcell of Harvard University. The relatively long wavelength of the signals enables them to penetrate clouds of cosmic dust, opening up vistas in all parts of the galaxy—and beyond. By 1952 it had also become clear that the signals were an excellent spiral tracer, since neutral hydrogen is at least 10 times more plentiful in spiral arms than in the "empty" regions between them. By measuring signal strength and slight shifts in wavelength we can locate the hydrogen clouds and hence the spiral arms.

The story of the 21-centimeter studies of spiral structure has been told by earlier articles in *Scientific American*, most recently by Gart Westerhout of the Leiden Observatory [see "The Radio Galaxy"; SCIENTIFIC AMERICAN Offprint 250; August 1959]. The Leiden group, which includes J. H. Oort, H. C. van de

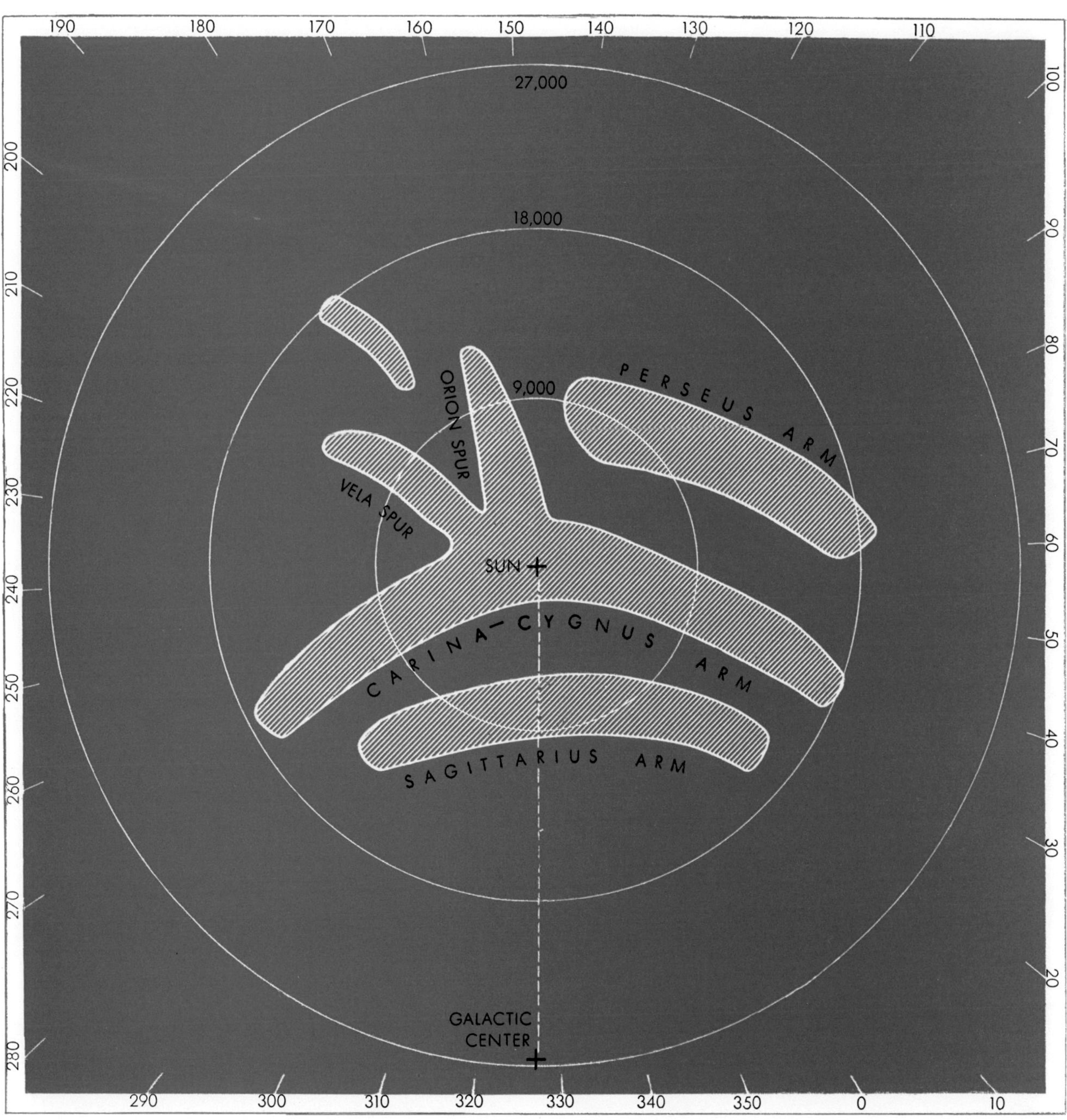

NEW OPTICAL MODEL shows the same region as the earlier model, but here the Orion Arm appears only as a short spur branching from a larger feature: the Carina-Cygnus Arm. In addition, the spiral arms in this model are almost circular, forming an angle of about 80 degrees with a line to the galactic center (*broken line*); the corresponding angle in the earlier model is about 55 degrees.

Hulst, C. A. Muller and M. Schmidt as well as Westerhout, has logged most of the basic observations in the Northern Hemisphere. In the Southern Hemisphere the principal investigators have been F. J. Kerr, J. V. Hindman and C. S. Gum of the Commonwealth Scientific and Industrial Research Organization in Australia. The combined results have made it possible to trace the spiral arms with fair precision to a radius of 25,000 light-years from the sun, and to obtain a picture of the over-all pattern to even greater distances. As the illustration below indicates, radio data have even made it possible to map the arms on the side of the galaxy opposite the sun.

The Model Revised

Up to this point I have told a success story, with Baade, Morgan and the radio astronomers in the Netherlands and Australia as the heroes. In 1953 Morgan presented a revised map of the nearby spiral arms that he believed was fairly complete, except for the Carina region [*see illustration on page 166*]. It looked as though we were making rapid progress. But as more data became available, the mapping of spiral features became increasingly difficult. Spiral tracers were found in low concentrations for thousands of light-years in every direction, making it necessary to extricate the major arms from a confused picture.

To complicate matters further, the radio data, which were expected to support the optical picture, instead contradicted it. In fact, the two pictures conflicted so sharply that Hugh M. Johnson, now working with us at the Mount

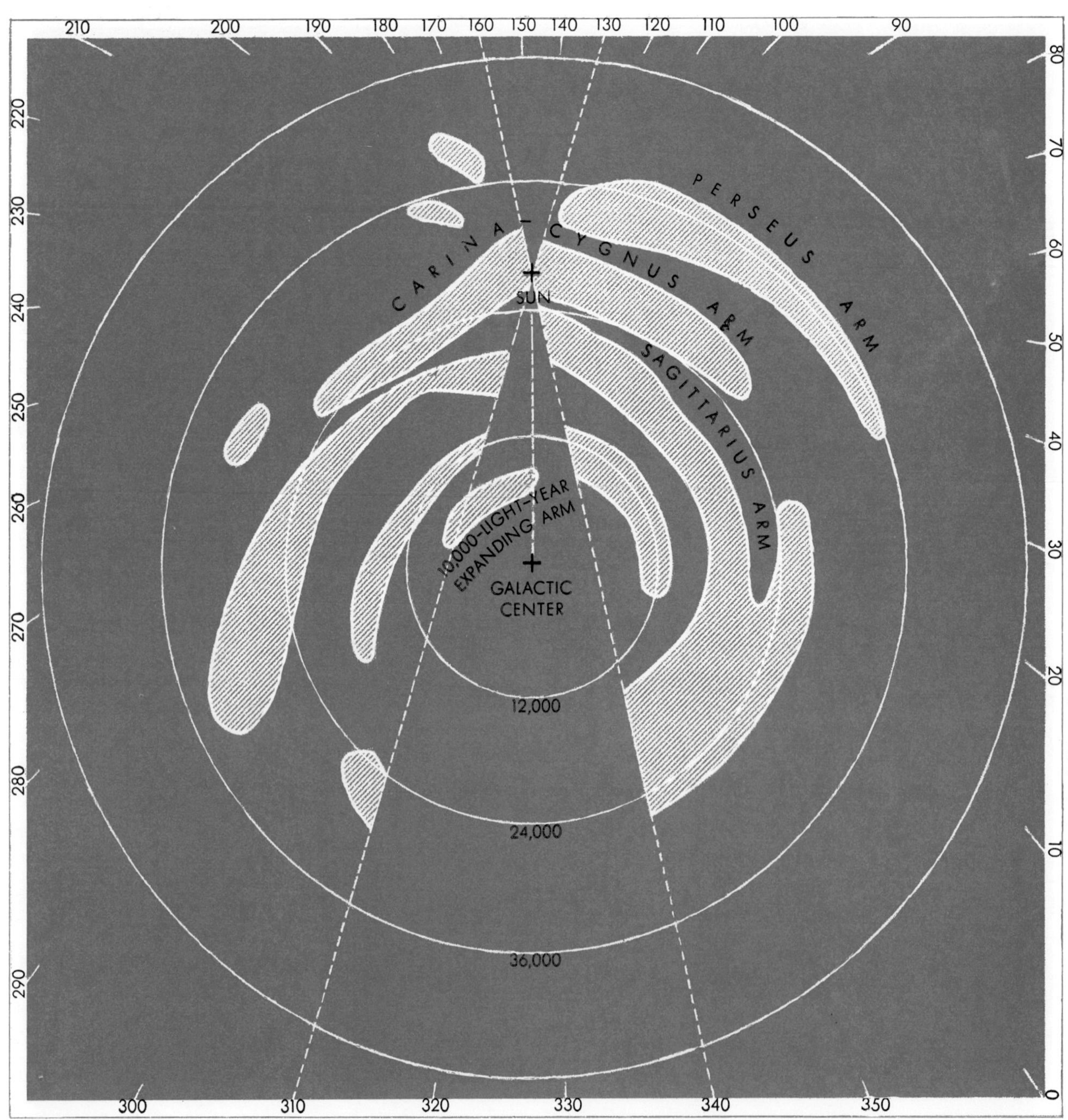

RADIO MODEL of the galaxy was made by charting the 21-centimeter signals emitted by interstellar hydrogen. The coordinates of this chart and the one opposite indicate degrees of galactic longitude in the plane of the galaxy. The numbered circles show distance from the sun in light-years. Data for the fan-shaped areas from longitudes 130 to 160 degrees and from 310 to 340 degrees are inconclusive because most of the gas in these areas moves perpendicular to our line of sight, making its red-shift unobservable.

Stromlo Observatory, commented that they hardly seemed to represent the same phenomena. I believe that this conflict can be resolved by modifying the optical model presented by Morgan as shown in the illustration on the preceding page.

I have two reasons for suggesting the change. The first concerns the over-all shape of the spiral arms. In Morgan's diagram they are almost elliptical, forming an angle of 55 degrees with a line drawn to the center of the galaxy. But photographs of other galaxies show that their spiral arms are more nearly circular, forming an average angle of 80 degrees with the center line. The same picture emerges in 21-centimeter maps of our galaxy, though we must bear in mind that the 21-centimeter picture is a preliminary one and subject to revision.

Further radio evidence that the spiral arms are nearly circular has been presented by B. Y. Mills of the Commonwealth Scientific and Industrial Research Organization. Mills has made an extensive survey of the radio brightness of the southern sky at a wavelength of 3.5 meters, and has found strong signals emanating from the central plane of the galaxy. The signals are strongest in the direction of the galactic center and diminish in intensity on either side of it. They remain constant along the plane of the galaxy for 10 to 20 degrees, then drop sharply to a lower level and remain constant for another 10 to 20 degrees. According to Mills, the sharp decreases in signal strength coincide with the edges of spiral arms.

From his results Mills has constructed a tentative diagram of the over-all spiral pattern of the galaxy [*shown below*].

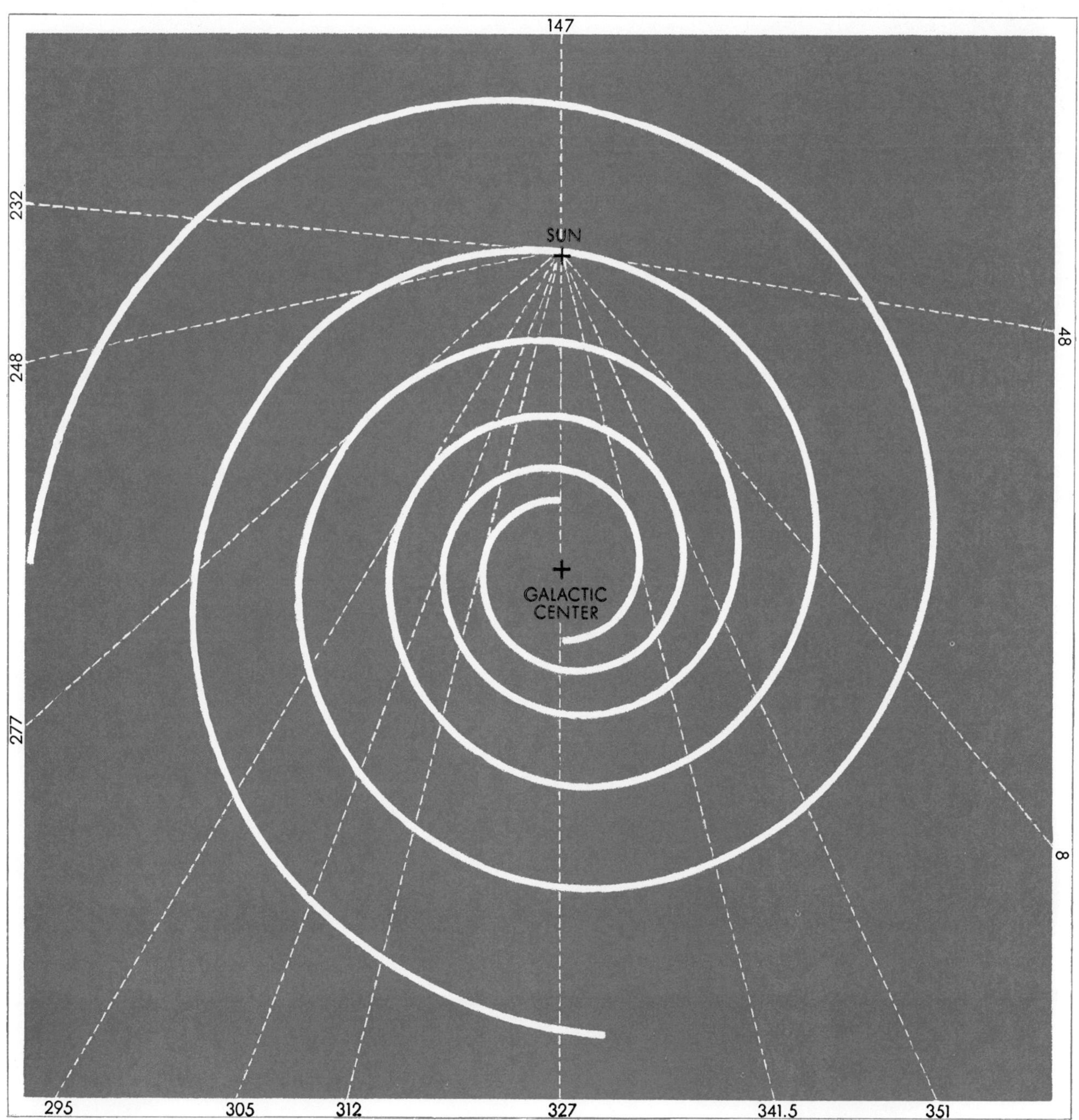

SCHEMATIC RADIO MODEL was proposed by B. Y. Mills of the Commonwealth Scientific and Industrial Research Organization in Australia. Mills drew his tentative diagram of the over-all spiral structure of the galaxy after analyzing radio signals having a wavelength of 3.5 meters. The dotted lines show the line of sight from the sun to the edges of the major spiral arms; no details are shown. The arms in this model and the one opposite are roughly circular, in excellent agreement with the optical model shown on page 167.

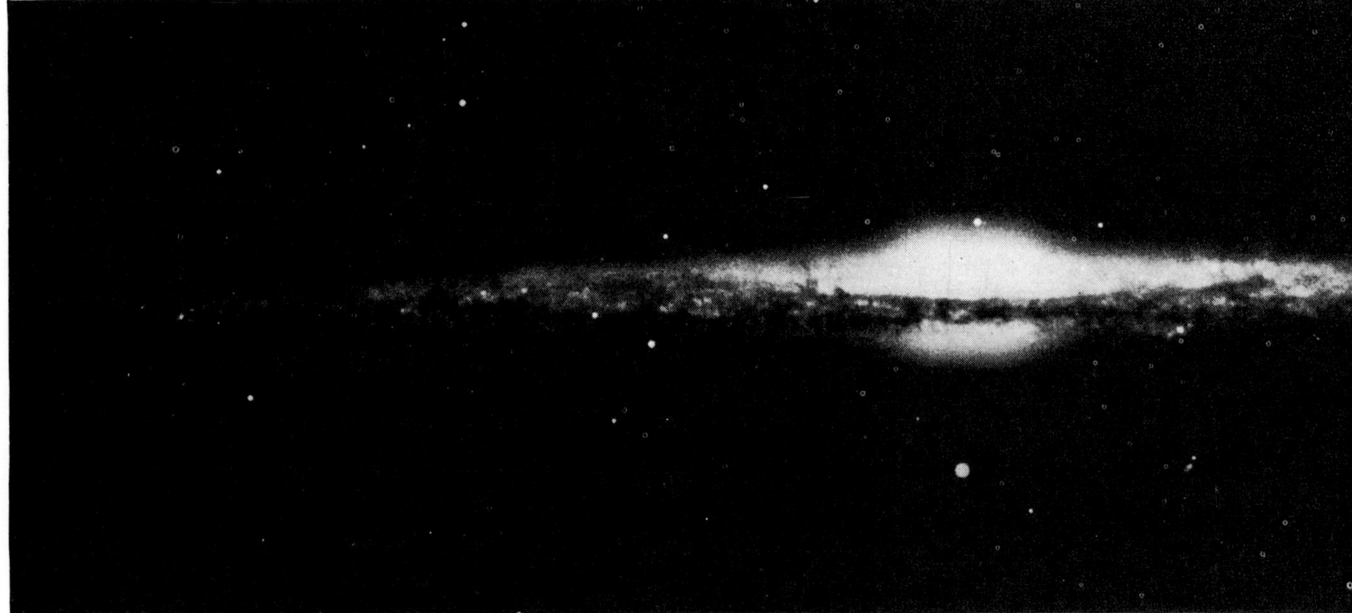

EDGE-ON VIEW of the spiral galaxy NGC 4565 shows the characteristic thin disk of gas and dust lying in the central plane of the galaxy. The dust is visible as a dark band against glowing mass of stars. Our own galaxy would probably appear slightly less flat-

Although this diagram agrees rather well with the new optical model, Mill's approach is at best a generalized one, and does not reveal fine structural details. The true spiral pattern of the galaxy is doubtless complicated by many minor spurs protruding from the major spiral arms.

My second reason for proposing a revision of Morgan's model is that in mapping his Orion Arm, Morgan ignored the long stretch of spiral arm extending from the sun toward Carina and Centaurus. Yet there is no other section of the Milky Way in which we are so obviously looking along a major spiral feature. The arm can be followed for a distance of 15,000 light-years from the sun and is studded with great numbers of spiral tracers: O and B supergiants and associations, young galactic clusters and emission nebulae. And no other region of the Milky Way is so rich in long-period cepheid variables.

Observers in the Southern Hemisphere have long recognized the importance of the Carina Arm, but Morgan and others in the Northern Hemisphere have tended to ignore it. One exception is van de Hulst, who has proposed a galactic model very similar to mine, in which the Carina Arm gets the attention it so richly deserves.

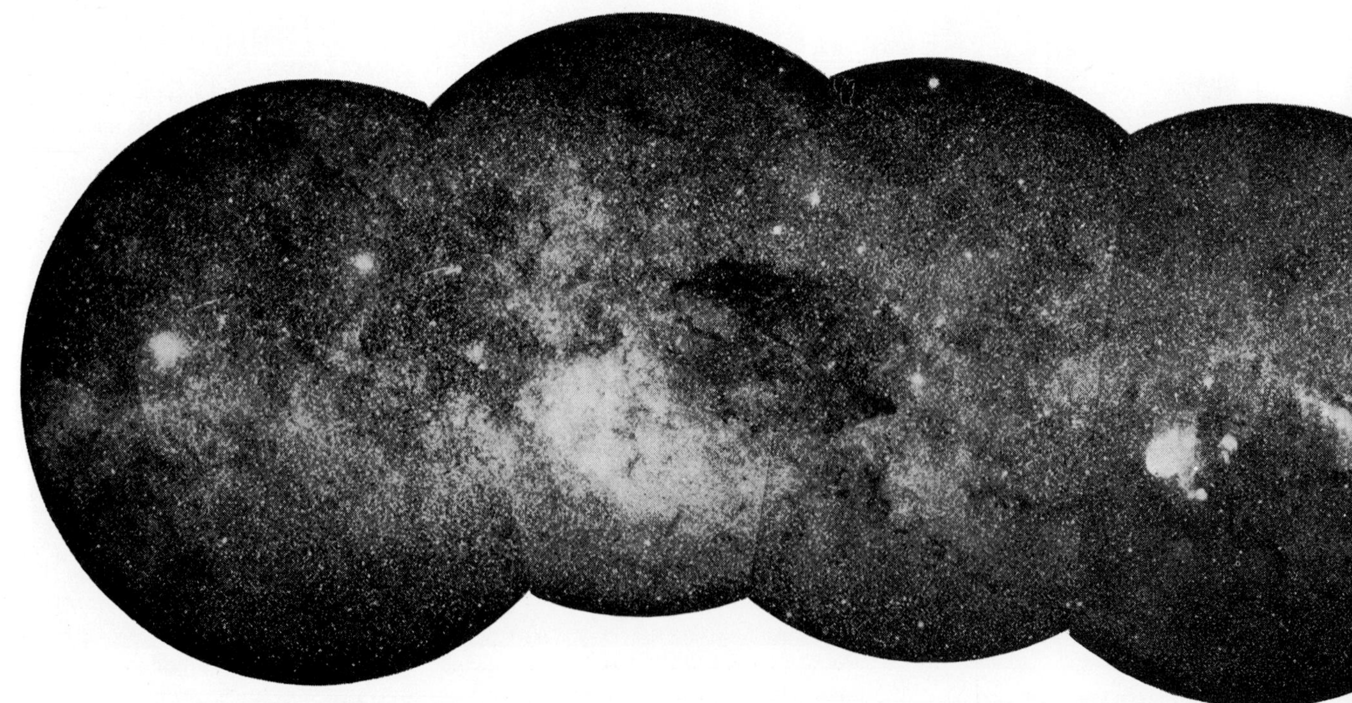

PART OF CARINA-CYGNUS ARM appears in this composite photograph of the Southern Milky Way. The circular section at far left contains the bright stars Alpha and Beta Centauri. The dark clouds in the second and third sections are known as the Coal Sack. The Southern Cross appears at the top of the third section, and the bright clouds in the section at far right form the Eta Carinae Nebu-

tened. The photograph was taken with the 200-inch telescope on Palomar Mountain.

In the new model the Carina and Cygnus arms unite to form the major spiral feature: the Carina-Cygnus Arm. This arm, which contains the sun, replaces the Orion Arm pictured in Morgan's diagram. It now appears that the Orion Arm is not an arm at all but merely a short spur lying partly outside the plane of the galaxy and extending away from the sun to the south. Another spur extends toward the southern constellations of Vela, Puppis and Canis Major, and in the new model is called the Vela Spur.

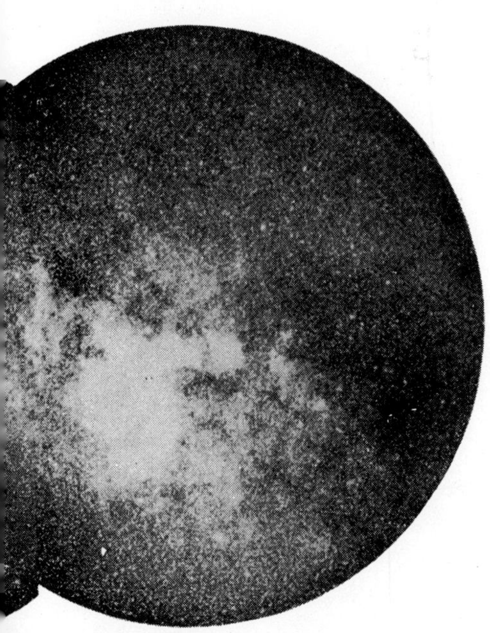

la. The photographs were taken with the eight-inch Schmidt telescope located at the Mount Stromlo Observatory in Australia.

Absorption Lines

In years to come the disparity between the optical and radio models of the galaxy should be further reduced by a number of other techniques that have proved reliable in tracing spiral arms. One such technique is the study of interstellar absorption lines. Clouds of interstellar gas are known to block light of certain wavelengths, producing dark lines in the spectra of remote stars. Because the lines are sharp and narrow they are easily distinguished from the rather fuzzy absorption bands produced by the star's own atmosphere. Observed from the earth, many distant stars have multiple or split absorption lines. The spectrum of one star, for example, has seven such lines, caused presumably by seven separate gas clouds between it and the earth.

Guido Münch of the California Institute of Technology has shown that the split lines enable us to locate the gas clouds responsible for them. We begin by measuring how much each component of the absorption line has been shifted from its normal position. If the cloud is moving toward us, the line will be shifted to a slightly shorter wavelength; if the cloud is moving away, the wavelength will be longer. From this shift we can calculate each cloud's motion with respect to the earth, its rate of rotation around the center of the galaxy and hence its distance from the sun.

Using this method, Münch derived the distances of a number of clouds and found that they fell into two belts: one corresponding to the Carina-Cygnus Arm; the other, to the Perseus Arm. Further work along these lines looks quite promising, especially in the Southern Hemisphere, where great opportunities await astrophysicists with powerful spectrographs.

Still another method of observing spiral arms was discovered some 10 years ago by John S. Hall and W. A. Hiltner at the Yerkes Observatory. They noticed that the light from stars lying in the plane of the galaxy is polarized. The polarization is probably produced by clouds of slightly magnetized particles of cosmic dust aligned by giant magnetic fields. Extensive observations in the Northern Hemisphere by Hall and Hiltner and in the Southern Hemisphere by Elske van P. Smith at the Boyden Station at Bloemfontein, South Africa, indicate that the magnetic fields are associated with spiral arms. Magnetic lines of force appear to stretch along these arms like long rubber bands.

Evolution of the Galaxy

The spiral arms that we observe today were probably formed rather recently and, cosmically speaking, may be rather short-lived. Most of the stars that serve as spiral tracers probably condensed from the interstellar gas in their vicinity less than 100 million years ago, a very short interval in cosmic time. If we define the cosmic year as the time required for the sun to complete one revolution around the galactic center, then the earth and the sun have existed for about 25 cosmic years. The oldest stars in our galaxy are about twice that age, while the very youngest may have ages of less than a 100th of a cosmic year.

The spiral arms, gaseous envelopes filled with stars and dust, are apparently doomed to a brief existence by the law of gravity. Because of the great attraction exerted by the massive core of the galaxy, sections of spiral arm closer to the core should complete their revolutions faster than those at greater distances. In a few cosmic years the difference in rotation rates would render the galaxy's former spiral pattern unrecognizable.

But we should not assign too much importance to purely gravitational arguments. If these forces were the only ones controlling the gaseous arms, the arms would probably never have been formed in the first place; and if they came into existence by some freakish accident, they would have been dissipated in short order. It is highly probable that although gravity is the major force controlling the orbits of stars in the galaxy, it acts in conjunction with magnetic forces in controlling the shape and movement of the gaseous spiral arms.

Apparently these magnetic forces can preserve the spiral pattern more or less intact until the gas in a given arm is exhausted by the condensation and evolution of new stars. The new stars at first lie neatly embedded in the arm, but after one or two cosmic years they begin to migrate from their original positions, smearing out the original spiral pattern. The vague patterns still traceable among the older stars in our galaxy may well be the last remnants of fossil spiral arms that disappeared a few cosmic years ago.

SPIRAL GALAXY resembling our own is NGC 5194, shown in this photograph made with the 200-inch telescope on Palomar Mountain. Cosmic dust appears as dark wisps at the insides of the arms. The bright blob at bottom is NGC 5195, a companion galaxy.

The Origin of Arms

We have almost no information as to how the arms were formed. One clue may lie, however, in the motion of the interstellar gas clouds. According to the estimates of Oort, van de Hulst and Westerhout, the neutral atomic hydrogen in these clouds represents about 2 per cent of the total mass of the galaxy. Ionized atomic hydrogen (dissociated protons and electrons) adds about a twentieth of this amount; other gases, including helium and molecular hydrogen, probably also make a small contribution. The gas forms a wafer-thin disk, 1,200 light-years thick and 100,000 light-years in diameter, in the central plane of the galaxy.

The hydrogen in the disk expands outward from the center of the galaxy at velocities ranging from 30 to 100 miles per second. This expansion may be influenced by gravitational forces set up by the galactic nucleus, but it seems more likely that it is directed by other forces—probably magnetic—that prevent the gas from being flung outward in all directions. Some of the ex-

THE CENTER OF OUR GALAXY lies hidden behind the interstellar dust clouds extending from the upper left to the lower right of this infrared photograph. The photograph was taken with the eight-inch Schmidt telescope at the Mount Stromlo Observatory.

panding gas appears to be concentrated in an arc 10,000 light-years long, called the "10,000 light-year expanding arm." This arm is moving outward from the center of the galaxy at a rate of 30 miles per second [*illustration on page 168*].

From here on everything becomes speculation. Oort and his colleagues have estimated that expanding arms transport roughly one solar mass of matter per year from the nucleus to other parts of the galaxy. This redistribution adds up to 200 million solar masses per cosmic year a quantity of matter sc enormous that we may well ask: Where does it all go?

The radio observations of the 10,000 light-year arm provide a tentative answer. If the arm is accompanied by a magnetic field, it will tend to move outward as a unit, slowing down as it proceeds farther from the galactic center. The outer arms of the galaxy apparently move outward at velocities greater than three miles per second, or 3,000 light-years per cosmic year. As the galaxy evolves further, its arms will continue to expand, and the 10,000 light-year arm may successively come to resemble the Sagittarius, the Carina-Cygnus and the Perseus arms.

The vast amount of material flung outward from the galactic center raises an-

YOUNG GALAXY, the Large Magellanic Cloud, shows at best only an incipient spiral structure. This photograph of one of the two nearest galaxies (the other is the Small Magellanic Cloud) was made with the eight-inch Schmidt telescope at Mount Stromlo.

other question: How does the nucleus replenish its enormous reservoir of gas? There seem to be two possibilities. One is that the gas is supplied by eruptions from some of the many stars in the galactic nucleus. This hypothesis holds so far as it goes, but we have no idea of what kinds of stars might provide the necessary eruptions. A more attractive possibility is that new gas enters the nucleus from the galactic halo, a huge sphere of very thin gas that surrounds the galaxy. The halo is shaped by weak magnetic fields; perhaps these forces tend to pull the gas toward the galactic center, where it is caught by other magnetic fields and pushed outward in puffs resembling smoke rings. The rings, of course, are spiral arms.

So little is known about the magnetic fields that prevail in the halo and the nucleus that fanciful hypotheses like these may survive for a while. I present them mainly to show how far we still are from an understanding of the forces that control the structure of galaxies. The final solution to the problem of spiral structure will probably come from coordinated optical and radio studies of our galaxy, where detailed features obscure the over-all pattern, and of nearby spiral galaxies, where the pattern is clearly visible in all its majesty.

17

Interstellar Grains

J. MAYO GREENBERG

October 1967

For the past 40 years astronomy has been trying to learn more about an important constituent of the universe that cannot be seen with the most powerful telescopes or detected by the most sensitive radio techniques. This constituent consists of fine particles, now commonly referred to as grains, that are known to be widely distributed throughout interstellar space. If they are invisible, how was their presence established and how can they be studied? They are like the purloined letter of Edgar Allan Poe's story: their presence seems obvious when one has properly interpreted their effects on the celestial radiation that *can* be seen and recorded.

Sometimes interstellar grains reveal themselves by reflecting the light of particularly luminous objects or glowing masses of gas. More often they dim the light of luminous objects or completely hide them. Fortunately for astronomy interstellar grains subtract light in a preferential manner, so that the radiation reaching the earth is often reddened and polarized. By measuring the amount of polarization and reddening one can infer much about the nature and distribution of this interstellar material. Given such information, the astrophysicist can begin to construct hypotheses about the origin of interstellar grains and the role they have played (and presumably are still playing) in the evolution of the universe.

Anyone who has looked at the Milky Way on a clear, dark night knows how patchy and uneven it appears. Even astronomers accepted this irregularity at face value until well into this century; they assumed that the stars were simply distributed in uneven clumps. With the recognition that the Milky Way represented the view along the central plane of the great rotating pinwheel of stars in our galaxy it became increasingly obvious that the uneven distribution of visible stars was actually caused by the uneven distribution of interstellar dust.

Considering how effectively this dust obscures our view of the galaxy, one might assume that it represents a sizable fraction of the galaxy's total mass. It is not so, and for an interesting reason that anyone can demonstrate for himself. Shine the light from a slide projector on a screen in a dark room. Now watch the brightness of the screen when the following are introduced separately into the beam: a piece of chalk, an unlighted cigarette, a cloud of chalk dust from a blackboard eraser and a cloud of cigarette smoke. The smoke will cause the greatest reduction in light, even though its total mass is far less than the mass of the entire cigarette and significantly less than the mass of the chalk dust. The critical factor in the extinction of the light turns out to be the size of the particles in the cloud of smoke or dust; for the highest extinction efficiency the particles should be somewhat smaller than the wavelength of the incident light.

If the galaxy were filled with icelike particles the size of basketballs, and the total mass of the particles equaled the mass of all the stars in the galaxy, the particles could not be detected optically. Indeed, their extinction efficiency would be only 1/300,000th the efficiency of the interstellar grains. By the same token the obscuration of starlight by the interstellar gas (chiefly hydrogen) is negligible even though the mass of the gas represents perhaps 10 percent of the total mass of the galaxy. It is estimated that the interstellar grains represent only about a thousandth of the total galactic mass. Within the spiral arms of the galaxy, however, the concentration of interstellar grains rises to perhaps three parts in 1,000. It can also be estimated that if a tenth of the stars in the galaxy have planetary systems like our own, the mass of the interstellar grains would exceed the total mass of the planets by about 10 to one.

The actual size of the grains is not known with any precision. The reason is that their composition is still uncertain. They might, for example, consist largely of ices, perhaps a mixture of ordinary ice and frozen gases such as methane or ammonia. In that case they would tend to be elongated particles—perhaps tiny needles with a thickness of about half a micron, which corresponds to the average wavelength of light. On the other hand, the grains could conceivably consist largely of carbon or metals. If they are carbon in the form of graphite, they would tend to be tiny plates, and they would be smaller than half a micron by a factor of five or 10. The difference in size estimates follows from the composition and optical properties of the grains. If they are optically dielectric, they will absorb very little of the radia-

NORTH AMERICAN NEBULA, embedded in the Milky Way, illustrates some striking characteristics of interstellar grains, or dust. The "Gulf of Mexico" and "Atlantic Ocean" are regions of high dust obscuration. The general luminosity of the nebula, however, is evidently not stellar radiation reflected from interstellar grains, as might be thought. It may represent excitation of gas by bright stars hidden by dust, but more likely it results directly from the collision of interstellar dust clouds. This is consistent with the idea that clouds in the spiral arm of the galaxy are moving at high velocities. The photograph was made with the 48-inch Schmidt telescope on Palomar Mountain by William C. Miller.

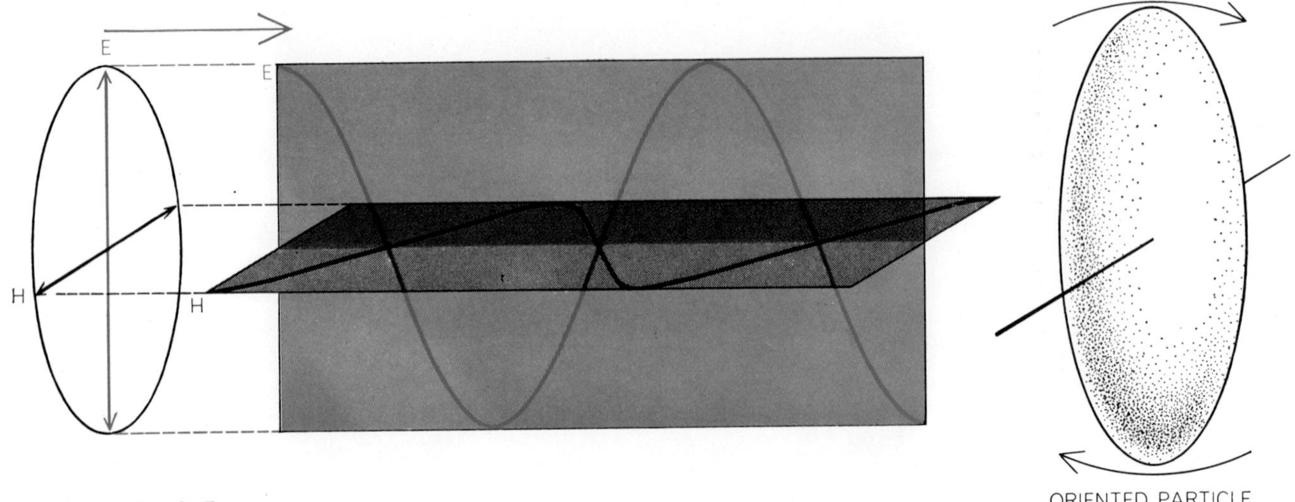

POLARIZATION OF STARLIGHT is believed to occur because interstellar grains are both asymmetric and oriented by weak magnetic fields in space. The ellipsoidal interstellar particle depicted here is viewed as if threaded on a line of magnetic force, around which it twirls. The starlight that reaches the particle from the left is unpolarized, which means it can be resolved into two beams of equal intensity (*E, H*) that are polarized into two planes at right angles to each other. Because of the orientation of the particle the

tion, whereas if they are metallic, they will tend to absorb strongly.

Not quite 20 years ago John S. Hall of the U.S. Naval Observatory and W. A. Hiltner of the Yerkes and McDonald observatories were carrying out a search for a predicted polarization effect when they discovered one that had not been predicted. They independently observed that the light from certain stars was linearly polarized. This means that if one were to look at such a star through an ordinary Polaroid filter of the kind used in sunglasses, their light would dim and brighten as the filter was rotated. It was noted at the same time that the stars exhibiting this polarization were invariably the stars whose light had also been reddened and partly extinguished by interstellar dust. Later studies showed that the degree of polarization is strongly

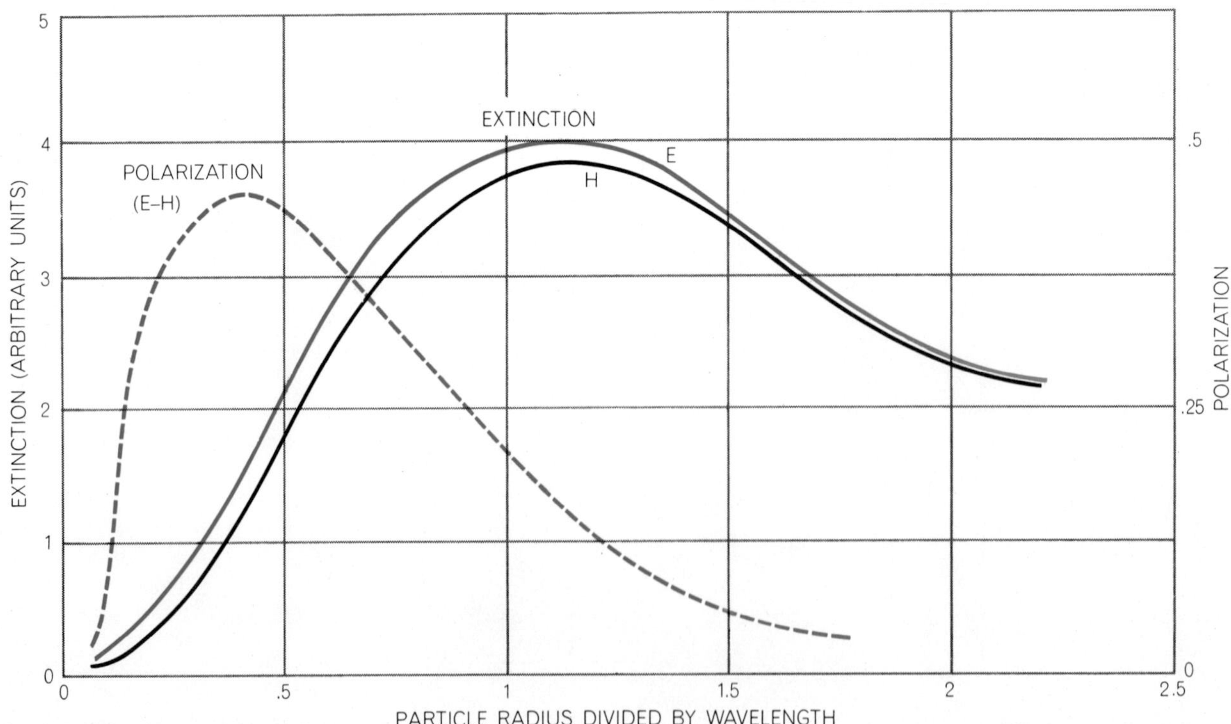

EXTINCTION EFFICIENCIES for long icelike particles, as measured by the author, produce curves of this general type. The particles are oriented as in the illustration at the top of these two pages. Thus the extinction is greater for the *E* beam, polarized vertically, than for the *H* beam, polarized horizontally. Maximum extinction occurs when the wavelength is about equal to the radius of the particle. On the rising portion of the *E* and *H* curves the extinction for a particle of a given small radius is greater the *shorter* the wavelength, which explains why blue light is dimmed more than red. As particles become much larger they reduce the intensity of starlight without altering its color. Maximum polarization (*E–H*) occurs when the wavelength is about twice the particle radius.

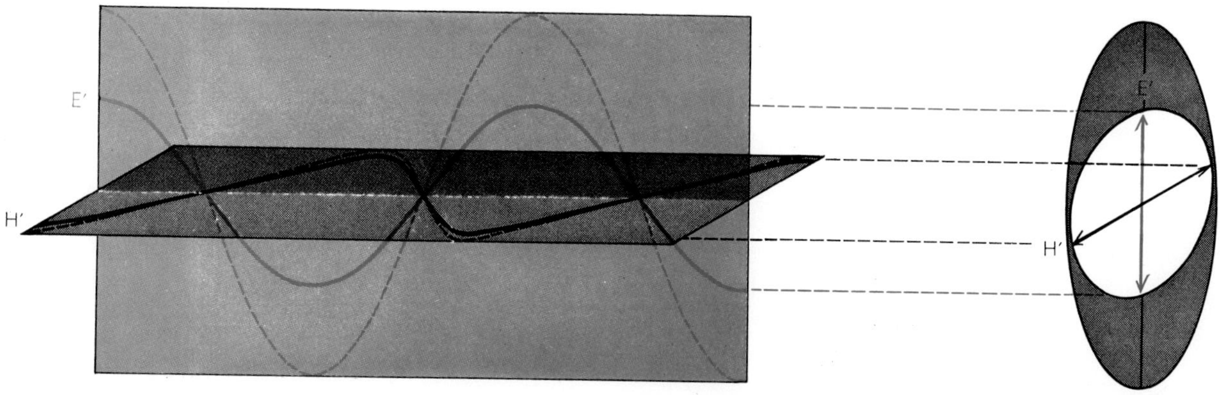

E, or vertical, beam suffers a greater reduction in intensity than the H, or horizontal, beam. Consequently the residual radiation, when viewed from the earth, will be polarized in a plane perpendicular to the direction of orientation of the particle. Using scaled-up models of grains and radio waves to simulate light, the author has found in his laboratory studies at Rensselaer Polytechnic Institute that for icelike grains of this general shape polarization is greatest when the wavelength is about twice the radius of the particle.

correlated with the amount of extinction. It soon became obvious that whatever was responsible for reducing the light must also be polarizing it. This could only be interstellar dust. The problem was to discover the mechanism of polarization.

When light that is unpolarized passes through a polarizing filter, it is reduced in intensity by at least half. It is said to be plane-polarized, and if it is now directed into a second polarizing filter, it can be completely extinguished by rotating the second filter until the filter's plane of polarization is at right angles to the plane of the first filter. Compared with an efficiency of 100 percent for a good polarizing filter, the grains in interstellar space show a polarizing efficiency of at most about 6 percent. Although this may seem a modest amount, it has not made the polarization any easier to explain.

The currently accepted hypothesis is that the polarization results from the orientation of interstellar grains by large-scale magnetic fields. In fact, the polarization of starlight provided early evidence that such fields exist. They appear to play an important role in the motions of gas clouds in the galaxy, and until recently it was conjectured that they might also provide the "ribs" that hold the galaxy's spiral arms outstretched [see "The Magnetic Field of the Galaxy," by Glenn L. Berge and George A. Seielstad; SCIENTIFIC AMERICAN, June, 1965]. More recent studies suggest, however, that the arms develop as a dynamic consequence of the galaxy's motion.

The mechanism by which interstellar grains are oriented by magnetic fields is not the simple one that turns the needle of a compass. The fields in space are much too weak. The strongest interstellar field probably does not exceed a thousandth of the magnetic field at the earth's surface, and the average strength of the galactic magnetic field is only a tenth or a hundredth of the maximum. Such fields would have little effect on the interstellar grains even if they were made of iron, and the fields would almost certainly fail to align particles that were

WAVELENGTH (ANGSTROM UNITS)
10,000 6,500 5,000 4,500
INFRARED VISIBLE ULTRA-VIOLET
EXTINCTION IN MAGNITUDES
0 1 2 3 4
0 1 RED 2 BLUE 3
RECIPROCAL OF WAVELENGTH (MICRONS)

EXTINCTION V. WAVELENGTH can be represented by a family of curves of which two are depicted. The bottom curve shows the characteristic extinction for starlight that has passed through about 2,000 light-years of interstellar space containing a typical amount of dust. The top curve shows the extinction expected after starlight has traveled three times as far through typical interstellar space, or about 6,000 light-years. At a distance of 9,000 light-years a star like the sun could not be seen from the earth with the most powerful telescope.

HORSEHEAD NEBULA is perhaps the most spectacular example of obscuration produced by interstellar grains. Located in the constellation Orion, the Horsehead is about 1,000 light-years distant. In this region there is probably as much matter in the form of interstellar dust as in stars. This photograph was made with the 200-inch telescope on Palomar Mountain.

SPIRAL GALAXY M 51 is probably much like our own, except for its small companion (*left*). Lanes of dust help to delineate the spiral arms, which tend to contain the brightest, bluest and youngest stars. This photograph was also made with the 200-inch Hale telescope.

being steadily bombarded by the atoms of the interstellar gas.

According to a hypothesis proposed by Leverett Davis, Jr., and Jesse L. Greenstein of the California Institute of Technology, the aligning mechanism is paramagnetic relaxation, a force that would make an interstellar grain spin on its short axis, which in turn would be lined up parallel to the lines of force in the magnetic field. Thus the particles would be "threaded" on the field lines much like buttons on a taut string. If the particles are like needles rather than buttons, they would resemble a string of tiny twirling batons.

It is not difficult to visualize how such oriented arrays of button-like or baton-like grains would polarize starlight. If the twirling grains were threaded on magnetic lines of force that ran parallel to our line of sight as we observed a star, they would present a maximum rotationally symmetrical area and therefore would not have any polarizing effect; they would simply reduce the intensity of all the components in the original unpolarized light. But if the lines of force were, say, horizontal and perpendicular to our line of vision, the grains would act as an array of vertical pickets that would reduce the vertical component of the originally unpolarized light and leave an excess of the horizontal component [*see top illustration on preceding two pages*]. If the magnetic lines of force intersected our line of sight at less than 90 degrees, the polarization would be proportionately less. The process is so effective that the direction of optical polarization is extremely uniform over wide ranges of galactic longitude.

Where do interstellar grains come from? It may be that there is more than one source. An early hypothesis was that the grains were similar to the micrometeorites that abound within the solar system. The idea was discarded when it was seen that there would be no way for such particles to be ejected from a solar system—our system or any other—and begin wandering through space.

It was then suggested about 30 years ago by the Swedish astronomer Bertil Lindblad that the grains grow by simple accretion out of the already existing gaseous matter in space. The suggestion was appealing and was consistent with the observation that there seems to be a direct correlation between the presence of interstellar grains and the density of the interstellar gas. Difficulties arose, however, when astrophysicists tried to explain exactly how grains could begin to form in a typical space environment. The proposed mechanisms failed to supply enough grains to satisfy the observations.

Attention was then turned to other hypotheses. It is well known that molecules are formed in the atmospheres of cool stars. Recently it has been shown that it should be possible for small particles of carbon and silica (molecules of silicon and oxygen) to grow under certain stellar conditions and for these particles to be blown out into space by the pressure of stellar radiation. If these small particles should get caught in a reasonably dense cloud of gas, they would serve as nuclei on which atoms could condense. When atoms of hydrogen, nitrogen, carbon and oxygen accidentally collided with such a nucleus, they would tend to stick and ultimately form molecules of various simple compounds: water, methane (CH_4) and ammonia (NH_3). Along the way less abundant atoms present in the gas cloud would also get embedded in the grain; thus one would expect to find, for instance, one atom of iron for every 1,000 atoms of oxygen.

If this process of growth by accretion were allowed to continue undisturbed for several billion years, the interstellar grains would grow to substantial size, perhaps as large as 20 microns, or 40 times the wavelength of light. This is about the size of the ice crystals in the atmosphere that produce the halo sometimes seen around the sun or the moon. Interstellar grains of this size, however, would be much too large to produce the effects we are trying to explain. Accordingly the proposed accretion mechanism is too effective. Might there be some other mechanism at work that limits grain size? There is.

The comparatively dense clouds of gas that are responsible for the growth of the grains could also provide the mechanism for limiting growth. Such gas clouds are in rapid motion in random directions, and about once every million years, on the average, a typical cloud will collide with another one. In the turbulent environment that results the growing grains will be bombarded by high-velocity atoms, will collide with one another and in general will be subjected to attrition. Low-energy cosmic rays and particles traveling with energies of 10 million to 100 million electron volts may also play a part in limiting the size of interstellar grains. Very recently S. B. Pikelner of the Sternberg Astronomical Institute in Moscow has correlated this type of grain attrition with the large numbers of electrons that evidently exist in the cool regions of interstellar space; he believes the electrons are produced when cosmic rays and energetic particles collide with hydrogen gas. The net effect may be to limit the grains to a maximum size of perhaps one micron, which would be about right to satisfy the observations.

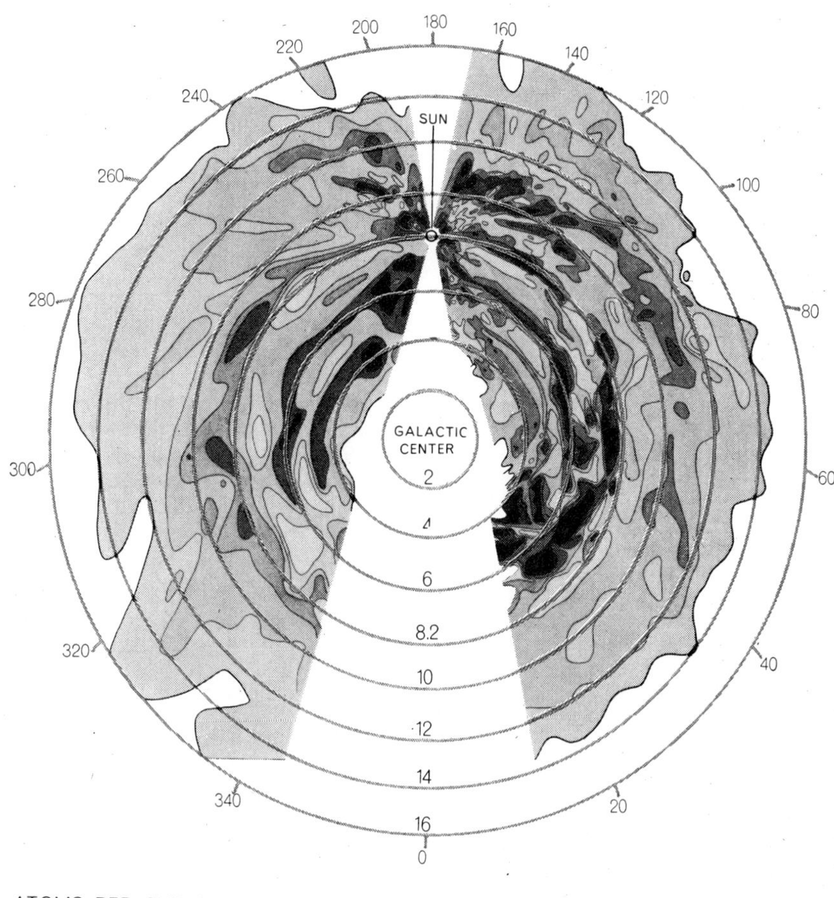

DISTRIBUTION OF HYDROGEN in our galaxy is believed to parallel closely the distribution of interstellar grains as well as young stars. This map of hydrogen distribution is based on measurements of the 21-centimeter radiation emitted by neutral hydrogen, which is not obscured by dust. The map follows one prepared by Jan H. Oort of the University of Leiden. The numbers on the concentric rings indicate kiloparsecs; a kiloparsec is 3,262 light-years. The sun is now thought to be about 10 rather than 8.2 kiloparsecs from the galactic center.

The only growth mechanism that would not satisfy the observations would be one that produced essentially spherical particles. The particles might still twirl around magnetic lines of force and reduce light intensity, but being symmetrical they would have no polarizing effect. The desired asymmetry would result if the grains were platelets of pure carbon in the form of graphite, or if they were produced by accretion on such plates, which would give rise to shapes resembling doorknobs. Alternatively, if molecules such as those of water are heavily involved in the growth of interstellar grains, one would expect the grain to be needle-like or possibly cigar-shaped. One cannot, however, entirely rule out hexagonal plates resembling miniature snow crystals.

In our laboratory at Rensselaer Polytechnic Institute we have developed a technique for measuring quite precisely how grains of various shapes will intercept light. The scattering and absorption of light by particles of wavelength size are not very amenable to calculation, particularly if the particles are not spherical. This is true even for particles of very regular shape, with one exception: infinitely long cylinders of circular cross section. To get at the general problem, including irregular or rough particles, we use an analogue experimental method in which all dimensions are scaled upward by a factor of about 100,000, or 10^5.

Accordingly we replace visible light (wavelength 5×10^{-5} centimeter) with radio waves whose wavelength is about three centimeters. Similarly, we use particle models that are about 10^5 times larger than the assumed interstellar grains and therefore range from about .5

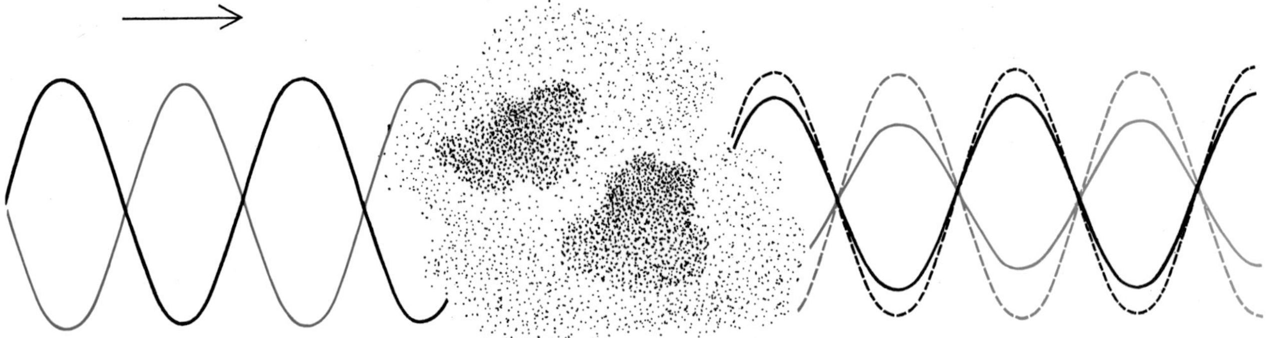

REDDENING OF STARLIGHT depends on both the initial color of the star and the amount of dust its light must traverse to reach the earth. The diagram shows how passage through about 2,000 light-years of interstellar space containing a typical amount of dust will reduce the intensity of red light (*black curve*) from a white star by half a magnitude and the blue light by a whole magnitude.

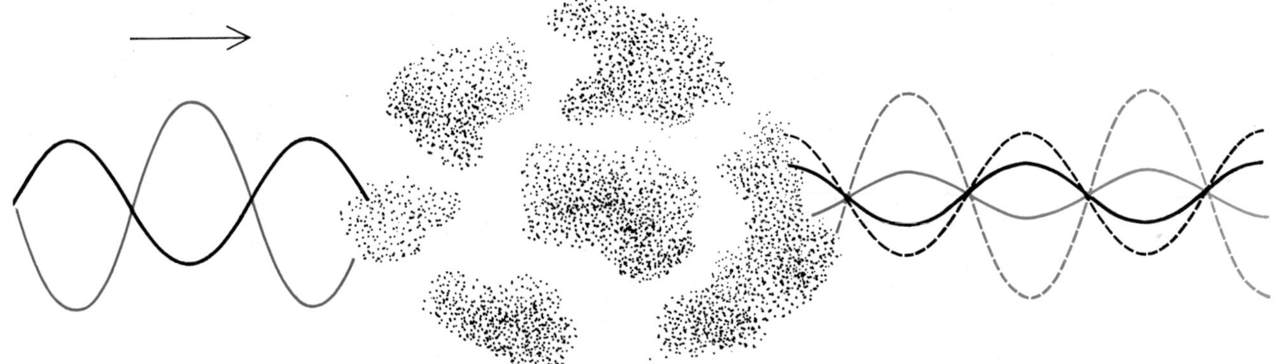

MORE EXTREME REDDENING would be produced if the starlight had to travel about 6,000 light-years through typical interstellar space. This is about the diameter of a spiral arm of the galaxy. It is assumed in this diagram that the light originates from a hot, young star that emits three times as much blue light as red. The interstellar dust reduces the blue component by three magnitudes and the red component by 1.5 magnitudes. As a result the star will appear to contain about 20 percent more red light than blue.

centimeter across to about 10 centimeters. One's commonsense notion is completely unreliable for predicting how radiation will be obstructed by nonspherical particles whose size is less than or about the same as the wavelength [*see illustration on page 183*]. Of course, one can usually find a completely convincing theoretical argument to show how the observed results are really "obvious." In fact, an approximate theoretical calculation had anticipated, in a qualitative way, the experimental results by several years.

For very small particles the cross section, or effective blocking area, is much smaller than the silhouette of the particle, whereas for very large particles the cross section is about twice the silhouette area. Up to a certain grain size the extinction produced by a particle of given radius is greater the *shorter* the wavelength, which explains why blue light is dimmed more than red [*see bottom illustration on page 178*]. The reddened light then passes with relatively small loss through clouds of particles whose size is less than or comparable to the wavelength. It is for this reason that the sun reddens as it sinks toward the horizon: its light passes through increasing amounts of dust-laden atmosphere. The blue portion of the sun's light that does not pass through the dust is scattered and gives rise to the overall blue color of the sky.

Because the degree of scattering varies with wavelength, each wavelength exhibits its own degree of polarization. For very short and very long wavelengths the polarization drops to zero for a particle of any given size.

On the average the attenuation of starlight in the central plane of our galaxy is so great that we can see the stars only within a small region around us. The distance from the sun to the center of the galaxy is about 10 kiloparsecs, or about 32,000 light-years. Assuming an average extinction of 1.5 stellar magnitudes per kiloparsec, it turns out that stars like the sun would be invisible with the largest telescopes at a distance of less than three kiloparsecs. Even the brightest stars at the center of the galaxy are not visible to us.

The reddening of starlight as it passes through clouds of interstellar dust is generally in direct proportion to the amount of extinction. Passage through 2,000 light-years of average interstellar space will reduce the blue component of "white" starlight by a full magnitude and the red component by only half a magnitude. When the light has traveled through 6,000 light-years, the reduction in the two components is respectively 1.5 magnitudes and three magnitudes. These, of course, are only averages; the actual values vary widely. For example, the attenuation per unit distance through the bright nebula around the star Merope in the Pleiades is about 1,000 times greater than the interstellar average.

What role, if any, do the interstellar grains play in the formation of stars? It may be a coincidence that Merope is so near such a dusty place. In other cases, for example in the Orion nebula, the evidence is clear that regions of high gas and dust density are also rich in newborn stars. It is believed new stars condense out of clouds of dust and gas.

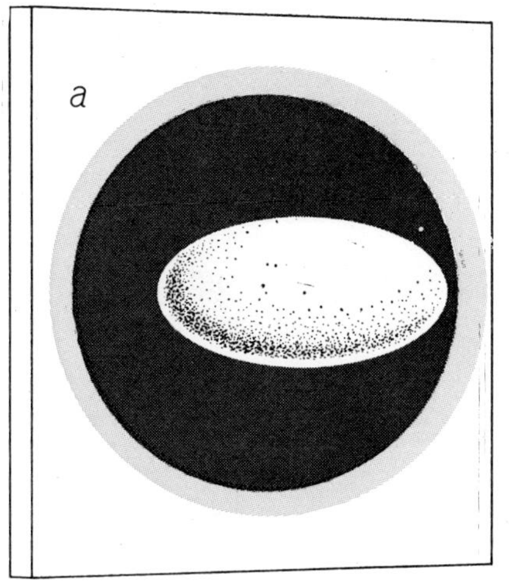

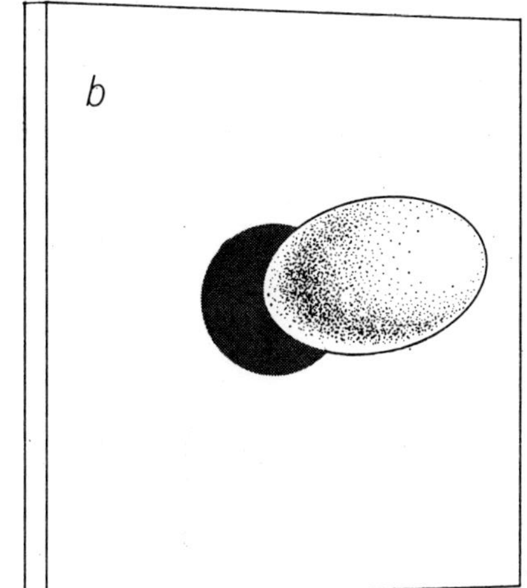

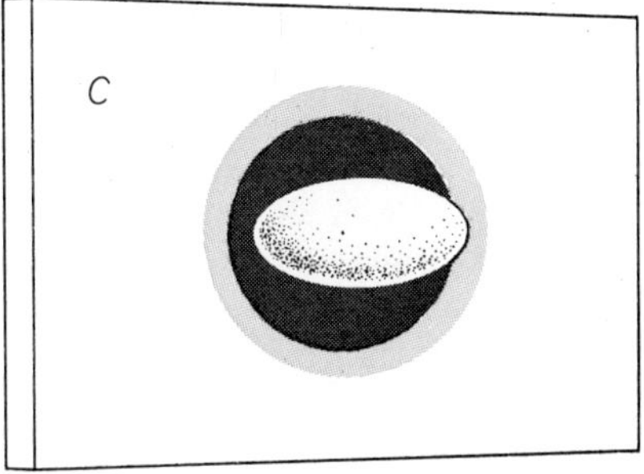

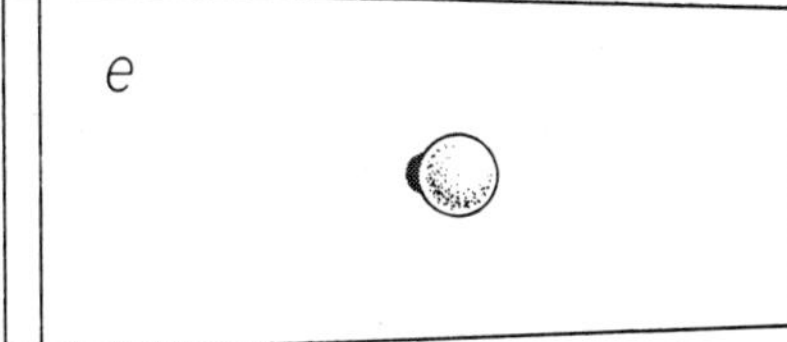

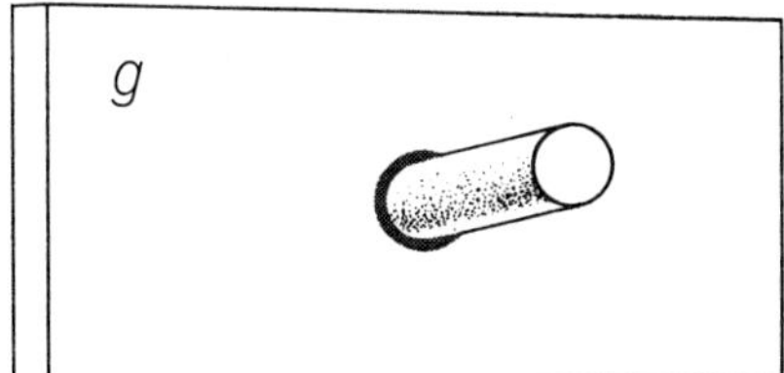

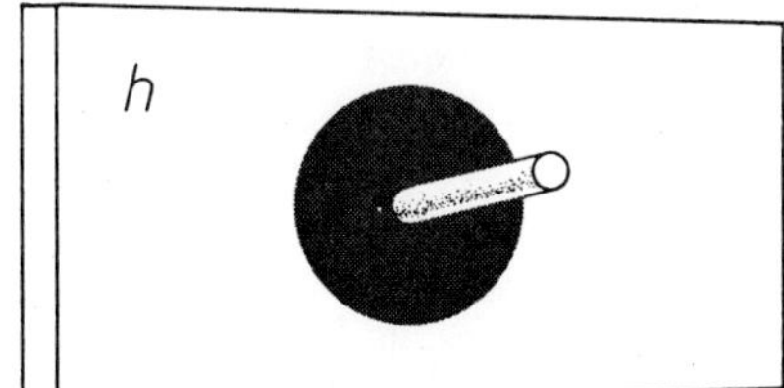

"MODEL" INTERSTELLAR GRAINS, 100,000 times larger in diameter than real ones, have been studied by the author to learn how they interfere with three-centimeter radio waves. As depicted here, the models are a fifth of actual size. The waves strike the models at right angles to the plane on which they are shown mounted. The total cross sections, or blocking areas, of the models are indicated by the various circles behind them. Cross sections depend strongly on the way models are oriented (*a–b, c–d*). Note that model *h* has a much larger cross section than *g*, even though only half the diameter of *g*. The marked difference in cross section of the two small spheres, *e* and *f*, is due to a difference in composition. The effect of polarization shows up only when ellipsoidal grains (*a, c*) are placed with their long axis perpendicular to the radiation. The larger cross section (*gray plus color*) is obtained when the radiation is polarized parallel to the long axis of the interstellar grain. The smaller cross section (gray) is obtained when the radiation is polarized at right angles to the long axis of the particle.

As the condensation proceeds, heat is generated by the release of gravitational energy. Finally the temperature inside the protostar becomes high enough to initiate a thermonuclear reaction. The star bursts forth with radiant energy and what remains of the surrounding dust and gas is vigorously propelled outward. The injection of these high-velocity clouds of gas and dust into the interstellar medium supplies kinetic energy of motion to other gas clouds. Thus are the cycles of interstellar grain growth and attrition continually maintained.

18

Hydroxyl Radicals in Space

BRIAN J. ROBINSON

July 1965

The discovery in 1951 of radio waves emitted by hydrogen in space added an exciting new dimension to astronomy that has proved to be richly informative. It made possible explorations of how hydrogen—the primary building material of all matter in the universe—is distributed in interstellar space. The greatest triumph in the use of the radio signals broadcast by hydrogen has been in mapping the spiral arms of our galaxy. The Doppler shift of these signals, which can be measured far more accurately than the Doppler shift of light, has revealed the motion of hydrogen clouds and has done much to elucidate the dynamics of galaxies in general. Studies of the proportion of hydrogen in distant galaxies have supplied valuable clues as to how they have evolved. The 21-centimeter radio waves emitted by hydrogen have become a potent astronomical tool that has yielded a wealth of new information over the past decade, already reported in several articles in *Scientific American* [see particularly "Radio Waves from Interstellar Hydrogen," by Harold I. Ewen, December, 1953, and "Hydrogen in Galaxies," by Morton S. Roberts, reprinted on page 204].

Now a new signal from interstellar space has joined the signal of hydrogen. Its source is the oxygen-hydrogen molecule OH, known in chemistry as the hydroxyl radical. The hydroxyl signal has been detected only in its muted form—by absorption rather than by emission. In 1963 radio astronomers discovered absorptions by concentrations of the hydroxyl molecule in our galaxy. Unlike the 21-centimeter monotone of hydrogen, the hydroxyl signal consists of four different frequencies. The exploration of hydroxyl clouds in the interstellar medium has become an inquiry of great interest to astrophysicists, because it gives promise of providing information about the formation of molecules in space and about the dynamics of the central region of our galaxy.

Soon after Ewen and E. M. Purcell of Harvard University discovered the hydrogen emission at the 21-centimeter wavelength in 1951, radio astronomers began to search for radio emissions from other atoms in interstellar space. They realized that detection of any such signals would be difficult, because radio emissions from atoms are rare and extremely weak when they do occur. To begin with, there is the rarity of matter in space: the amount of hydrogen averages less than one atom per cubic centimeter, and all other atoms are very much sparser. Secondly, atomic processes that could give rise to detectable radio emissions are considerably less common than those that produce emissions of light. The energy of a quantum of radiation depends on its frequency, and a quantum at radio frequencies is a million times weaker than a quantum at the much higher frequencies of visible light. Consequently radio emissions can arise only from atomic processes that involve comparatively slight changes in the atoms' energy state. Furthermore, because the probability of emission drops sharply with decrease in frequency (or increase in wavelength) of the radiation, any radio signal from a collection of atoms is very weak.

The hydrogen atom's emission of 21-centimeter radio waves arises from an interaction of the spin of the atom's electron and the spin of the nucleus. The two spins may be in opposite directions or parallel. When they are parallel, the electron's spin can flip over so that it is opposite to that of the nucleus, resulting in a state of lower energy. The transition releases a quantum of energy at the frequency of 1,420.4057 megacycles per second (or the wavelength of 21 centimeters).

Few other atoms likely to be reasonably abundant in space have a nuclear spin that would make such interactions possible. The nucleus of nitrogen has a

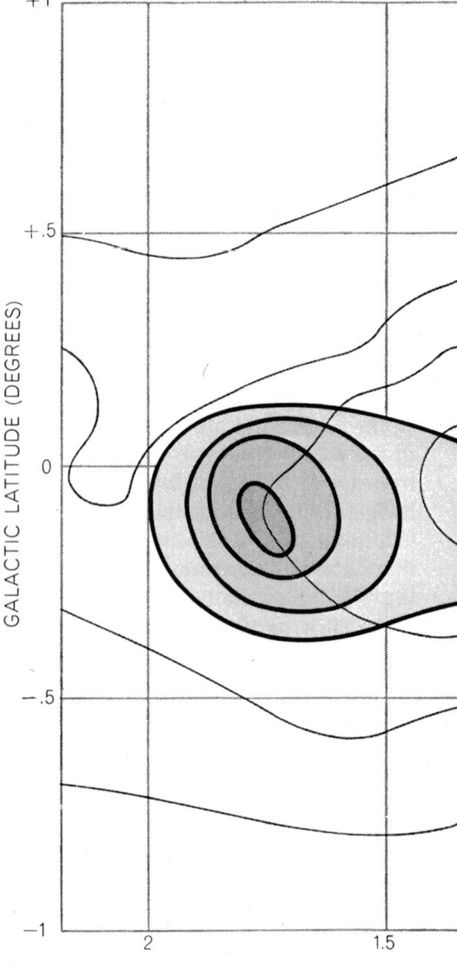

THREE GAS CLOUDS near the center of our galaxy are composed of atoms of oxygen and hydrogen linked together in diatomic molecules of the hydroxyl radical (OH).

spin, but its interaction with the electron spin is too weak to produce radio signals detectable on the earth. On the basis of what was known about atomic radio emissions from laboratory experiments, radio astronomers chose deuterium (heavy hydrogen, or H^2) as their first target of search for a second radio emitter in space. The interaction of the nuclear and electronic spins in deuterium is considerably weaker than in ordinary hydrogen. Moreover, the deuterium isotope must be many thousands of times less abundant in space. But deuterium at least offered a definite objective to aim for, because the frequency of the radio waves it emits (or will absorb) was known with high precision: it is 327.3843 megacycles.

For several years radio astronomers in the U.S.S.R., Australia and Britain made painstaking efforts to discover the 327-megacycle line in the spectrum of radio waves from space. They turned their radio telescopes to strong radio sources, particularly the powerful emitter in the direction of the constellation Cassiopeia called Cassiopeia *A*, and examined the spectrum of each source closely to try to detect absorption of its radiation by deuterium in interstellar space. In this search the investigators developed refined techniques and produced the most accurate measurements that had ever been made in radio astronomy. No sign of the hunted line could be found. The 250-foot radio telescope at Jodrell Bank in Britain, pushing the observations to a level that would have detected absorption of as little as one part in 15,000 of the radiation from Cassiopeia *A*, failed to record any absorption by deuterium. This indicated that the proportion of deuterium atoms to hydrogen atoms was less than one in 4,000. In 1962 Sander Weinreb of the Massachusetts Institute of Technology undertook a new search with a more sensitive technique. In extensive observations of Cassiopeia *A* (totaling more than 1,800 hours) with the 85-foot radio telescope at the National Radio Astronomy Observatory in Green Bank, W.Va., he too failed to detect any deuterium absorption. His technique established that if any deuterium existed in interstellar space, its ratio to ordinary hydrogen there must be less than one atom in 13,000–less than half the ratio of the deuterium isotope in hydrogen on the earth.

In truth, deuterium was not a particularly good candidate for this kind of exploration. It had been chosen for concentrated attention because it was the only plausible candidate whose radio frequency was accurately known. In the meantime, however, accurate measurements of radio emissions from the hydroxyl radical were obtained in the laboratory, and in 1963 Weinreb applied his new techniques to a search of the heavens for that molecule.

In theory radio absorption by the hydroxyl radical and by certain other molecules should be much stronger than that by deuterium. In a two-atom molecule (such as the hydroxyl radical) with

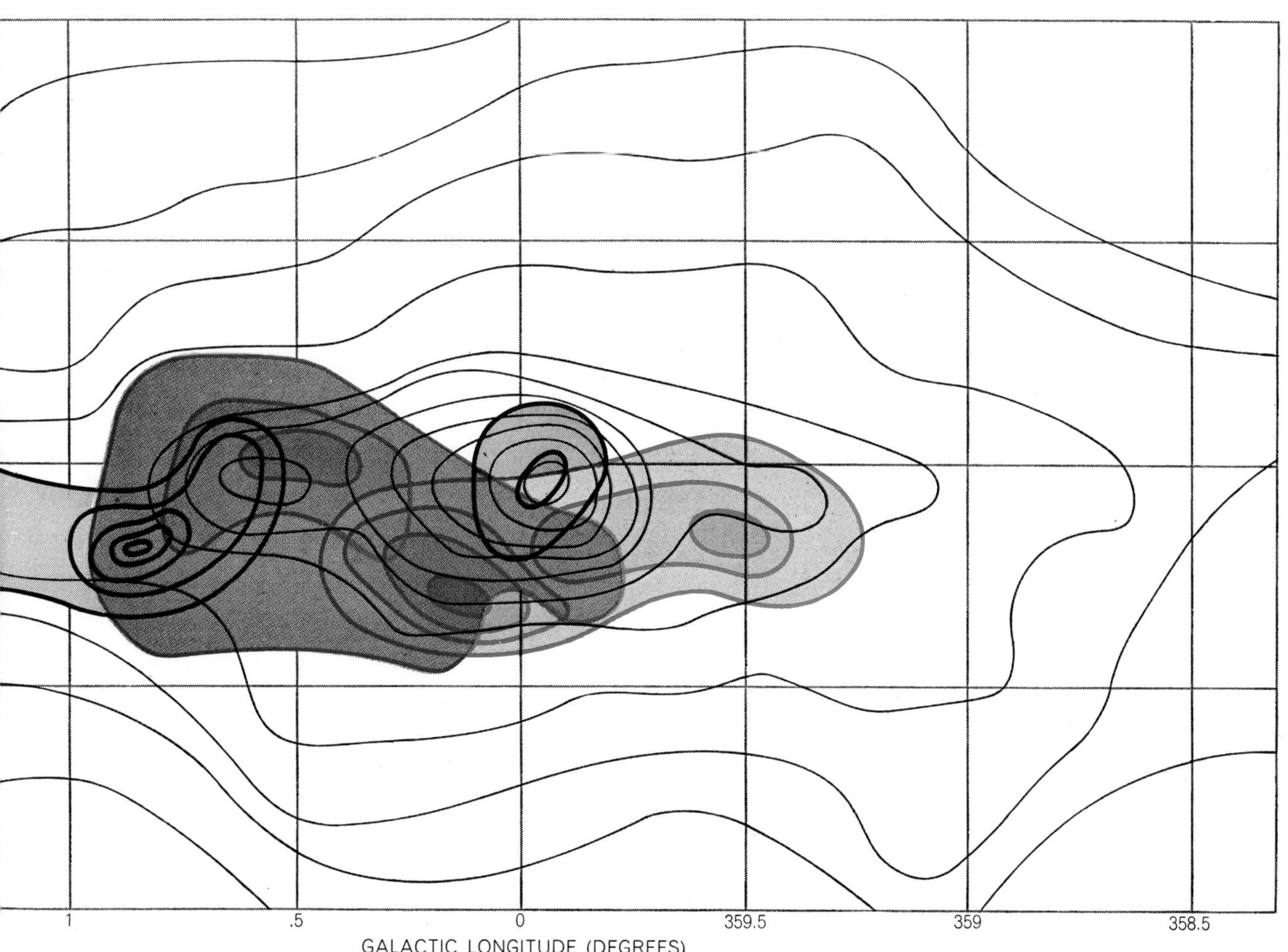

These are a few of the many clouds of such molecules detected here and elsewhere within our galaxy as a result of their absorption of radio frequencies between 1,612 and 1,720 megacycles. The emitter in this region is a complex of radio sources known as Sagittarius *A*; the narrow contour lines show how its radio brightness increases toward the galactic center. Doppler shifts show that two of the clouds (*dark and light color*) are moving outward; the third is receding. OH molecules were first detected in space in October, 1963.

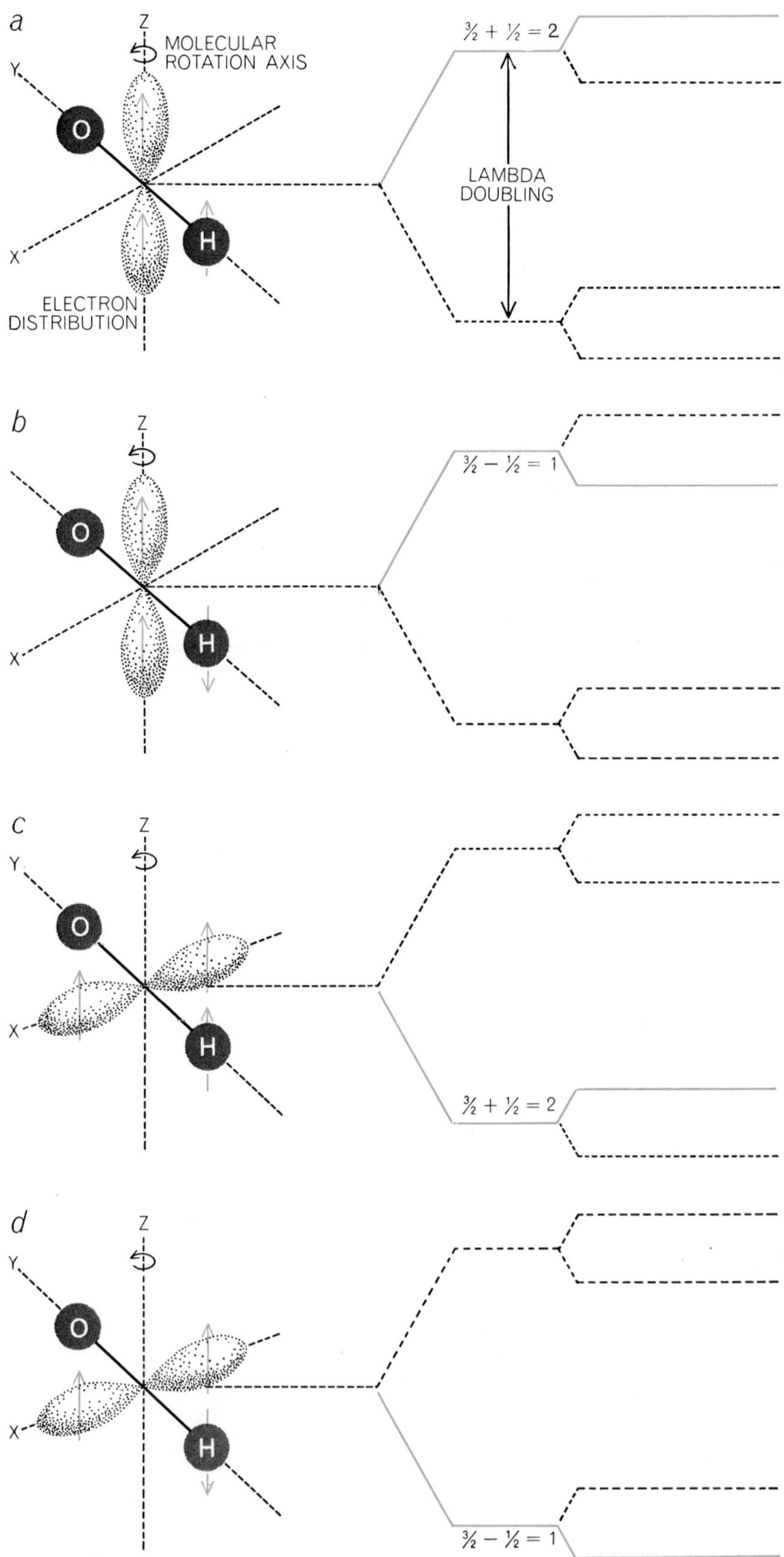

FOUR ENERGY LEVELS of the hydroxyl molecule are shown schematically. The motion of the unpaired electron interacts with the rotation of the diatomic molecule. Two states are possible: *a*, where the electron distribution is along the axis of molecular rotation, and *c*, where it is in the plane of rotation. A transition between the two absorbs or emits a quantum of energy at microwave frequencies. Still another emission or absorption can occur when one nucleus of the molecule (*the hydrogen nucleus in this illustration*) has a magnetic moment opposed to, rather than aligned with, the molecule's internal magnetic moment (*compare "a" with "b" and "c" with "d"*). Numbers show the quantum arithmetic of each level.

an odd number of electrons several interactions giving rise to radio emissions are possible. The rotation of the two nuclei interacts with the orbital motion of the odd electron, and the electron's motion can assume either of two orientations with respect to the rotation axis of the molecule; this produces a "lambda-doubling" of the molecule's rotational states [*see illustration at left*]. The transition from one orientation to the other involves an energy change that corresponds to the frequency of radio waves in the microwave range, well below the frequencies at which transitions in atmospheric molecules absorb waves from space. In the hydroxyl molecule each of these two states in turn has two possible configurations that depend on the direction of the hydrogen nucleus's spin. Thus the hydroxyl molecule has four possible energy transitions and may emit (or absorb) radio waves at four different frequencies. When the motion of the unpaired electron in the molecule is changed by the absorption of radio energy, it is the radio wave's electric field that produces this change, in contrast to the case of the single hydrogen atom, where the reversal of the electron's spin involves only the comparatively weak energy of the magnetic field. Consequently radio absorption by the molecule is a great deal stronger—about 10,000 times stronger—than that by a hydrogen atom. Radio-absorbing molecules in space therefore should be detectable even at concentrations well below one molecule per cubic meter.

It seemed likely that the hydroxyl radical and a few other two-atom molecules (CH, SiH and NO) might be found in detectable concentrations in interstellar space. But the failure to find deuterium, and the lack of precise information about the radio frequencies of the molecules in question, discouraged any thorough search for them. Nonetheless, in 1958 an exploratory attempt was made by Alan H. Barrett and A. Edward Lilley at the Naval Research Laboratory; it was unsuccessful. Then, in 1959, Charles H. Townes and his group at Columbia University did succeed in establishing the frequencies of two radio lines for the hydroxyl molecule; they measured these frequencies in the laboratory and found them to be 1,665.46 and 1,667.34 megacycles, with an estimated possible error of 30 kilocycles. Not until 1963, however, were any serious attempts made to use this information to probe for the hydroxyl radical in space. In the fall of that year Weinreb and his co-workers, applying his new techniques, met success on their first try.

Working with Barrett, M. Littleton Meeks and J. C. Henry, Weinreb attached his ultra-stable receiver to M.I.T.'s 84-foot radio telescope on Millstone Hill in Massachusetts, pointed the dish toward Cassiopeia *A* and sifted its emissions in a band of frequencies around 1,667 megacycles. The first evening's observations, on October 15, 1963, showed a significant absorption at the 1,667-megacycle line of the hydroxyl radical. Soon afterward the receiver also disclosed absorption at the other known OH line, 1,665 megacycles. The relative intensities of absorption by the two lines were in the ratio of nine to five, which agreed well with the values for the hydroxyl radical that had been established by theoretical calculations and laboratory experiments. There could be little doubt, therefore, that the absorption lines signaled the presence of hydroxyl molecules in interstellar space between Cassiopeia *A* and the earth.

The amount of absorption was small and indicated that the average OH density in that direction was about one molecule per 10 cubic meters. From the shapes of the absorption spectra it could be deduced that in the various gas clouds observed the proportion of hydroxyl molecules to hydrogen varied slightly. The observations also yielded precise frequencies for the two hydroxyl lines: 1,665.402 and 1,667.357 megacycles, with a margin of uncertainty of only seven kilocycles plus or minus.

The announcement of the discovery at M.I.T. prompted several other radio observatories to undertake immediate surveys of the same kind. In Australia, John G. Bolton, Frank F. Gardner, Karel J. van Damme and I quickly improvised a suitable receiver and turned the 210-foot radio telescope of the Parkes Observatory to Sagittarius *A*, a radio source at the center of our galaxy lying behind several spiral arms rich in gas clouds. On November 20, a month after the M.I.T. discovery, we detected hydroxyl absorption at the two frequencies in the Sagittarius direction. Three weeks later Ewen and Nannielou H. Dieter of Harvard, using the U.S. Air Force's 84-foot radio telescope on Sagamore Hill in Massachusetts, confirmed the presence of OH in both the Cassiopeia and the Sagittarius directions. In the same week Harold F. Weaver and David Williams of the University of California at Berkeley also observed absorption by hydroxyl molecules in the direction of the center of the galaxy.

What could the hydroxyl radical, as

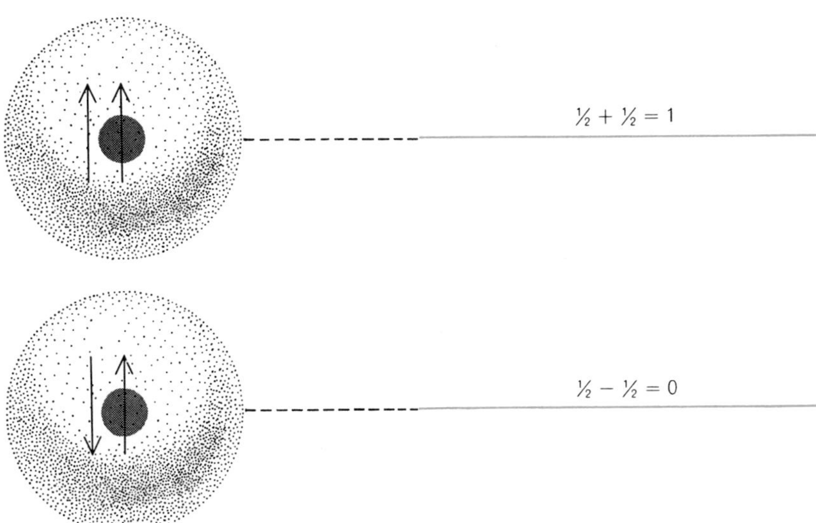

HYDROGEN SPIN-FLIP is the parallel phenomenon on the atomic level responsible for absorption and emission of quanta at the 21-centimeter wavelength. When nuclear and electron spins are parallel (*top*), the atom contains more energy than when the spins are opposed.

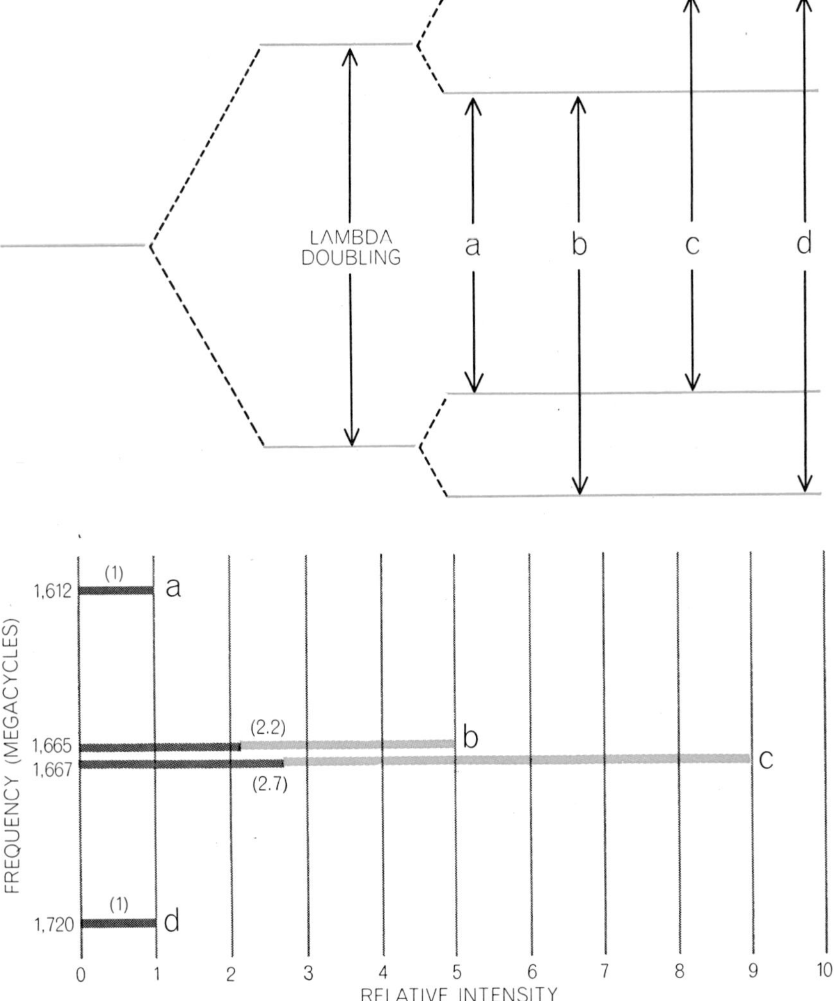

ABSORPTION FREQUENCIES related to the four energy levels of the hydroxyl molecule that stem from "lambda-doubling" are not equally common in occurrence. In theory whole-step transitions ("*b*" *and* "*c*" *at top*) are far more probable than semistep (*a*) or superstep (*d*) ones. Observation shows this is so but not to the extent predicted. Black lines (*bottom*) give the values observed at the galactic center; colored lines, those predicted for whole steps. Observations also show that these values are not alike in all regions of the galaxy.

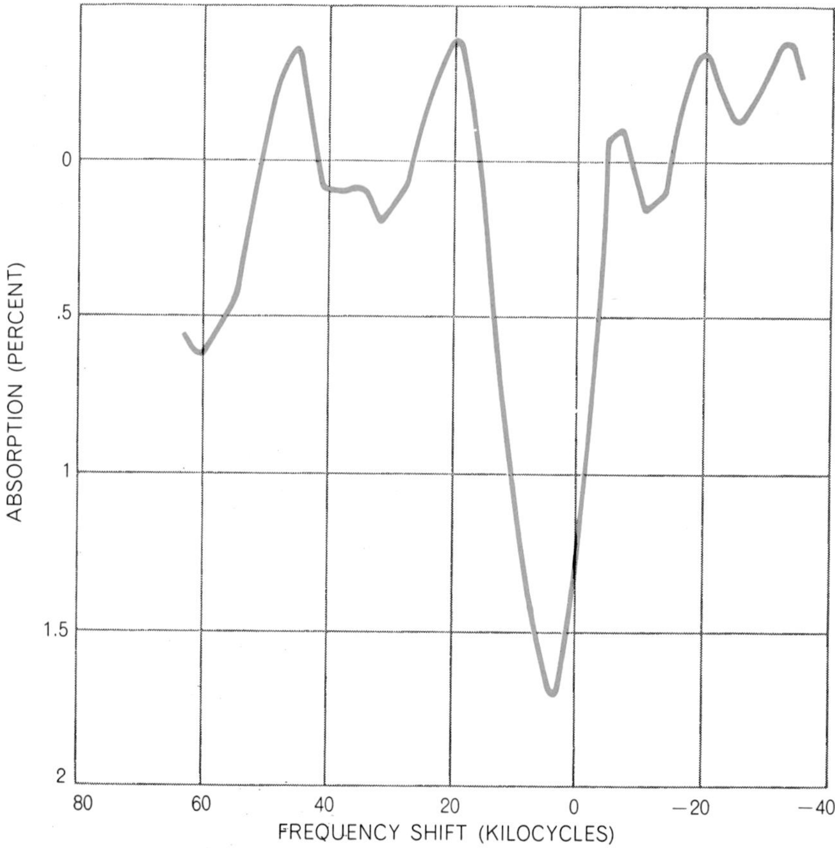

FIRST EVIDENCE that hydroxyl molecules existed in space was this absorption dip recorded by Massachusetts Institute of Technology investigators in the radio spectrum of Cassiopeia *A* near 1,667.357 megacycles (*zero on horizontal scale*). Smaller fluctuations represent "noise." The absorption indicates an OH density of only one molecule per 10 cubic meters.

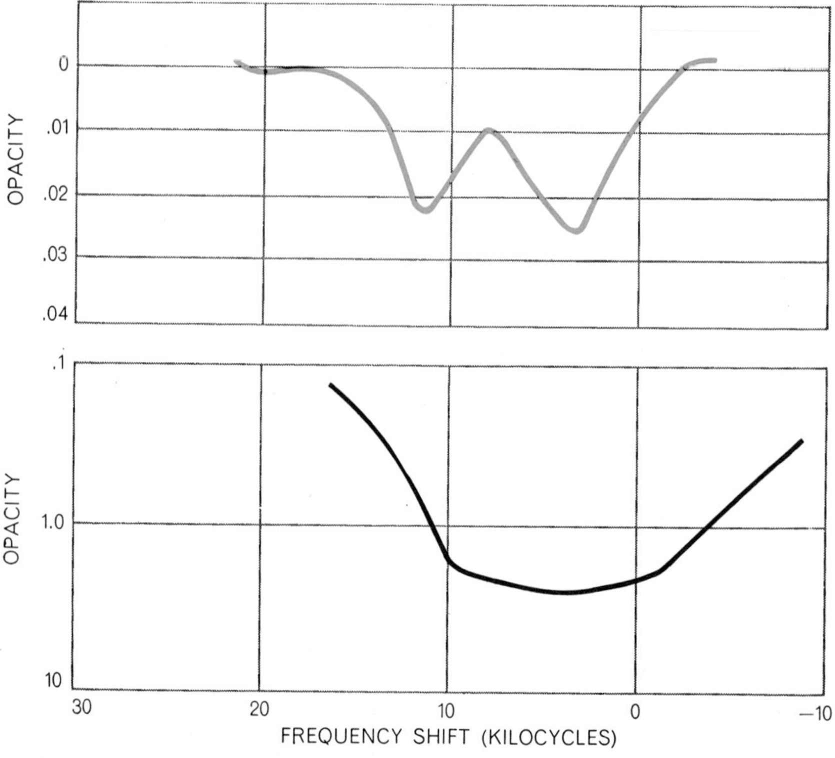

FINER RESOLUTION shows that the OH absorption of Cassiopeia *A* emissions near 1,667 megacycles actually consists of two separate dips (*top*), corresponding to a pair of clouds moving with different velocities. The higher random velocity of hydrogen atoms prevents detection of this motion; 21-centimeter readings produce only a single profile (*bottom*).

a new marker in the sky, tell us about our galaxy and events in interstellar space? Not very much, it was thought at first. No emissions from the molecule could be detected, so observations of its presence were limited to narrow sections of space where it absorbed radiation from strong radio sources behind it. The molecule appeared to be so sparsely distributed in space that it absorbed only 1 to 2 percent of the radiation from these sources. As a tool for astrophysical studies it seemed to hold little promise.

Soon, however, it began to turn up various clues. Barrett and his colleagues at M.I.T. found that OH absorption could serve as a means of resolving different gas clouds in space and estimating their relative speeds and temperatures. Hydrogen atoms, being light, have comparatively rapid thermal motions in a gas cloud. This has the effect of broadening their radio-absorption line—that is, widening the span of frequencies they will absorb. The hydroxyl molecule, on the other hand, has a thermal velocity only one-fourth that of the hydrogen atom, because it is 17 times heavier. An OH absorption line therefore should be sharper than the hydrogen line. Barrett and his colleagues examined one hydroxyl line with fine tuning and found that they could resolve it into two separate absorptions [*see bottom illustration at left*]. They interpreted this to mean that the two dips represented two separate clouds of gas, each moving at a different velocity in the line of sight from the earth and having a different internal temperature—respectively 90 degrees and 120 degrees Kelvin (degrees centigrade above absolute zero).

At the Radiophysics Laboratory of the Commonwealth Scientific and Industrial Research Organisation in Sydney we were struck by the curious shape of the absorption spectrum in our first observations of OH at the center of the galaxy. The general shape of the spectrum was so unusual that we took our receiver apart to see if an aberration of the instrument was responsible for it. After the equipment was modified and reassembled the hydroxyl-radical absorption spectrum in the direction of Sagittarius *A* still showed the same curious shape.

A spectrum of absorption by gas clouds in space represents, of course, a mixture of absorptions by various clouds in the line of sight. They may be moving at different velocities toward or away from the earth, and these differences will produce slightly different Doppler shifts of a given absorption line. Therefore the profile of the absorption spectrum will show a series of separate dips [*see illustrations at right*]. The absorption in the Sagittarius *A* direction by hydrogen shows well-marked dips presumably representing gas in spiral arms moving toward the solar system (that is, outward from the center of the galaxy) at speeds of 30 and 53 kilometers per second. The same "features" (absorption dips) showed up in the hydroxyl absorption profiles. But the hydroxyl spectrum also contained a broad, dominating feature that had no marked counterpart in the hydrogen-absorption picture. It showed a strongly absorbing cloud or clouds of hydroxyl radicals moving *toward* the center of the galaxy at a velocity of 40 kilometers per second! This was completely unexpected, because the hydrogen observations indicated that gas clouds generally move *away* from the center.

Further observations revealed another surprise. The inward-moving OH did not blanket the entire complex of sources of Sagittarius *A* (which extend for several degrees) but covered only the small, intense source near the very center of the galaxy; in that direction, however, the hydroxyl concentration was so high that it absorbed 60 percent of the radiation at its absorbing lines! In contrast, the hydroxyl clouds in the direction of Cassiopeia *A* absorb only 1.6 percent of the radiation at these frequencies. Surveys of other radio sources in our galaxy have since shown, in fact, that the proportion of hydroxyl molecules to hydrogen in different gas clouds varies enormously: by a factor of more than 1,000. The concentration of hydroxyl molecules increases rapidly toward the center of the galaxy, where an appreciable fraction of the oxygen atoms must have formed OH radicals.

Shortly after our observations in Australia, Lilley, Samuel J. Goldstein, Ellen J. Gundermann and Arno A. Penzias surveyed the hydroxyl absorption of Sagittarius *A* over a wider frequency range with the Harvard 60-foot dish and a maser receiver. They found another strong absorption by OH, but this one was moving *away* from the galactic center at 120 kilometers per second. There was no marked absorption by hydrogen associated with it, which again indicated a high proportion of hydroxyl radicals near the center of the galaxy. A month later we confirmed the Harvard observations with our telescope in Australia.

Needless to say, the finding that OH clouds can serve as a probe for the

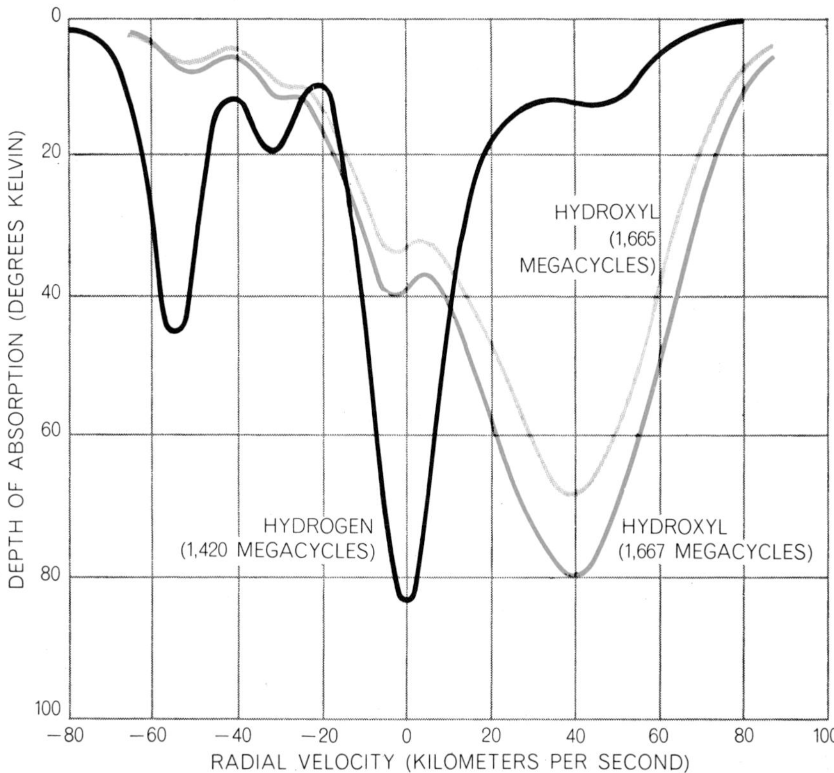

CONTRASTING MOTIONS of hydrogen atoms and hydroxyl molecules in the gas clouds near the galactic center were revealed by Australian studies of Sagittarius *A* in 1964. The two strongest OH lines (*dark and light color*) show positive values on the horizontal scale, or inward movement; the negative value for hydrogen (*left*) means outward movement.

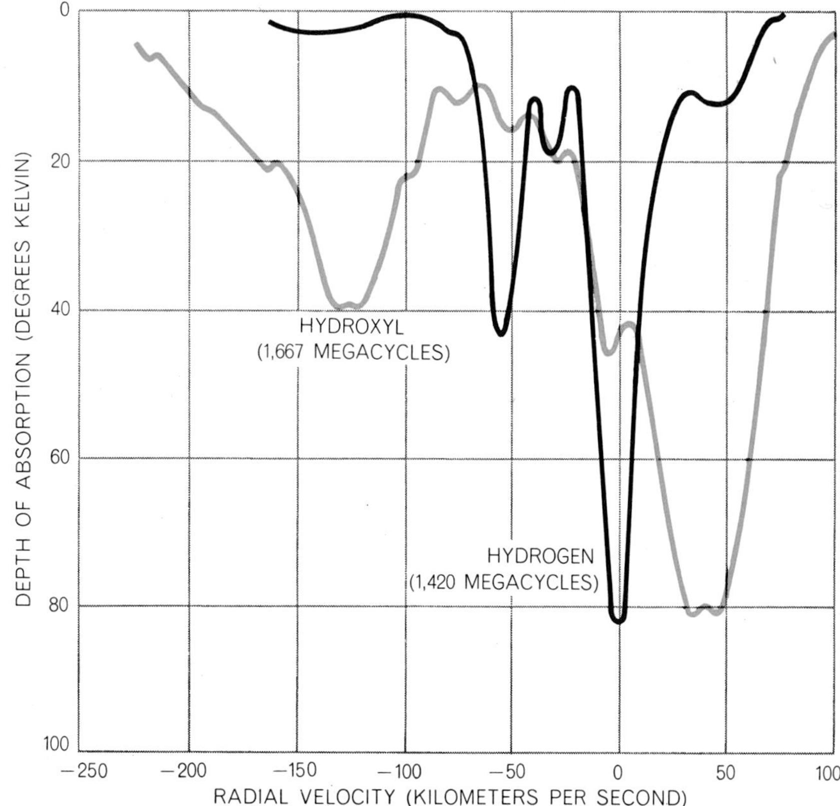

FURTHER EXTENSION of measurements by Harvard University workers revealed an OH accumulation near the galactic center with an outward velocity of 120 kilometers per second. Colored line shows OH absorption at 1,667 megacycles as recorded by Australian radio telescope. Black line shows hydrogen 21-centimeter band at a much reduced scale.

center of our galaxy has aroused keen interest among astronomers. The nuclei of galaxies have recently come in for attention because of remarkable events that have been discovered in some of them [see "Quasi-stellar Radio Sources," by Jesse L. Greenstein, SCIENTIFIC AMERICAN, December, 1963, and "Exploding Galaxies," by Allan R. Sandage; the latter article begins on page 224]. The events taking place at the center of our galaxy are less catastrophic but no less interesting. We cannot see the center because of the heavy banks of dust clouds that lie in the way. The hydrogen line opened the center to radio probes, but it leaves many of the observations unclear because they contain a mixture of emissions by the gas through the full depth of the galaxy and absorptions by the hydrogen on our side of the center. In the hydroxyl radical we have a sharp tool that tells its story only by absorption, without confusing the picture with emissions. In combination with the hydrogen probe it may soon help to unravel some of the complex phenomena taking place at the galaxy's center.

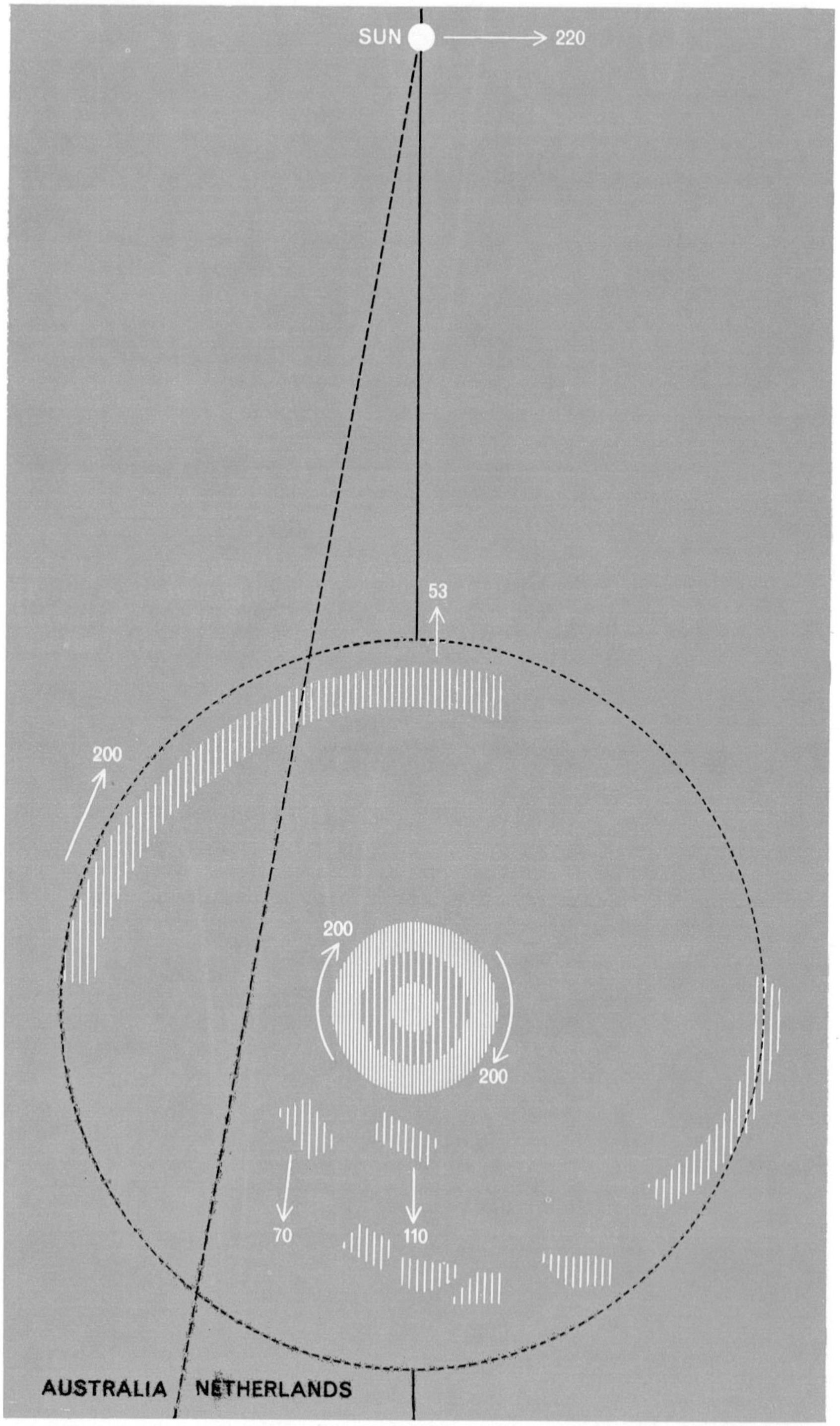

In general the hydrogen clouds in our galaxy revolve in nearly circular orbits around the center under the control of gravitational forces. Near the center, however, the gas is flowing radially outward at a high rate, and at the nucleus itself it seems to be whirling at extremely high speed. What appears to be blowing the hydrogen away from the center? Is it the force of a gigantic explosion that took place sometime in the past? Is it a driving force of the hydromagnetic type? Perhaps close comparison of the distribution and motions of hydrogen and the hydroxyl radical will help to provide an answer.

Our group in Sydney has surveyed the whole central region of the galaxy with the Parkes 210-foot radio telescope. It is an ideal location for such observations, because in Australia the galactic center passes almost directly overhead. The survey shows that, whereas hydrogen covers the entire center, the hydroxyl radical is concentrated in small clouds. Some are traveling away from the center, some toward it, each at its own velocity [*see illustration on pages 184 and 185*]. There are startling differences between the motions of the hydrogen and the hydroxyl clouds, with the movements of the latter corresponding only in the most general way to those of the hydrogen in which they are embedded.

Equally puzzling is the contrast between the high abundance of hydroxyl radicals near the center and the scarcity of these molecules out in the spiral arms of the galaxy. How, indeed, is the OH molecule formed at all? The probability of hydrogen and oxygen atoms uniting by direct collision in space is extremely low. The most likely process is the formation of molecules on grains of dust on which the atoms collect. These "dust" grains are believed to be composed mainly of frozen water; presumably the grains evaporate and

MOVEMENT OF HYDROGEN near the galactic center is of two kinds. Within 2,000 light-years of the center (*radius of inner band*) the gas seems to be rotating rapidly. From that region to a distance 10,000 light-years away from the center (*radius of broken circle*) the hydrogen is streaming radially outward at varying speeds. Arrows show direction of motion; figures are velocities in kilometers per second. The view is from above the galactic north pole; the broken diagonal separates Australian from Netherlands astronomical findings.

release H_2O molecules, which are then dissociated into H and OH. The central part of most spiral galaxies, probably including our own, contains much less dust than the spiral arms. It may be, then, that the hydroxyl molecules in the center are remains of dust clouds that have been evaporated by warming collisions, whereas in the spiral arms the OH is still trapped in dust grains. Near the center of the galaxy there is comparatively little ultraviolet radiation to dissociate the hydroxyl molecule; in the spiral arms, on the other hand, any OH that is released from the dust grains may have only a short lifetime because it is exposed to copious ultraviolet radiation from hot blue stars.

As soon as the OH absorptions at 1,665 and 1,667 megacycles were discovered, efforts were of course made to detect absorption at the other two radio lines expected in the hydroxyl molecule. These minor lines were so weak that no attempt had been made to measure their frequencies in the laboratory. But in April, 1964, on the basis of new theoretical calculations of the frequencies, both lines were detected in the Sagittarius direction with the Parkes telescope: one at 1,612.2 megacycles, the other at 1,720.6 megacycles. These new discoveries brought to light a new anomaly. Theoretical calculations and later laboratory measurements by H. E. Radford showed that the relative intensities of absorption at the four OH lines should be in the ratios 1:5:9:1 (for the lines at 1,612, 1,665, 1,667 and 1,720 megacycles respectively). The actual absorptions recorded in the telescope receiver, however, put the ratios at 1:2.2:2.7:1 for the strong Sagittarius A absorption. These ratios are incompatible with simple self-absorption effects and imply unusual physical conditions at the center. Moreover, it turned out that the ratios vary in different parts of the galaxy. This is an enigma that obviously calls for further investigation; it may lead to better understanding of the conditions in interstellar space and the formation of hydroxyl molecules at the center of the galaxy.

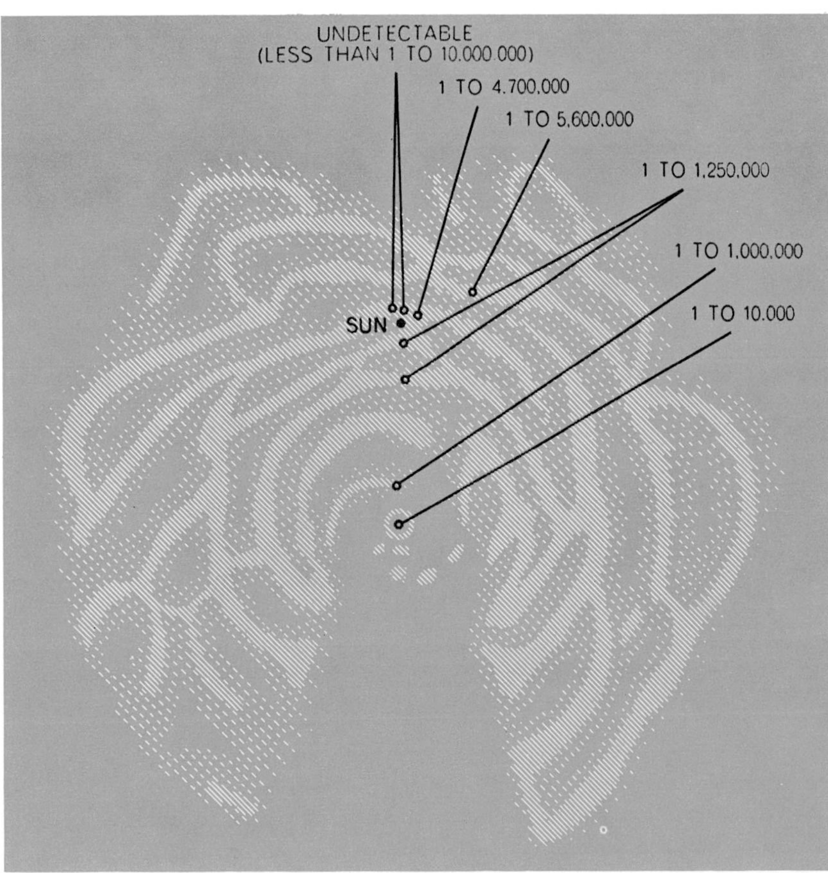

VARYING CONCENTRATIONS of hydroxyl molecules throughout the interstellar medium are plotted on a generalized sketch of the distribution of hydrogen atoms in the spiral arms of the galaxy. The ratio of hydroxyl to hydrogen increases a thousandfold between the region near the solar system and the galactic center. There one hydroxyl molecule is found for every 10,000 hydrogen atoms, but surveys in other directions show less than one per four million hydrogen atoms or are unable to detect any hydroxyl absorption.

What are the prospects for discovering other substances in space by their radio lines? One likely candidate is the diatomic molecule CH, which has a lambda-doubling in its lowest energy state and should absorb radio energy in the region around 3,400 megacycles. Preliminary surveys in that neighborhood have failed to find any absorption, but the molecule may become detectable when a more precise estimate of the frequency has been obtained. Another molecule being hunted is the hydroxyl radical containing not common oxygen 16 but the rare isotope oxygen 18. The four radio lines of $O^{18}H$ should be slightly displaced from those of $O^{16}H$, with the strongest line at 1,639.3 megacycles. The amount of the heavy version of the hydroxyl molecule in space must be very small indeed, but the chances of detecting the strongest line seem good.

What about the hydrogen molecule itself: H_2? We have theoretical grounds for believing that there must be a great deal of hydrogen in the molecular form in interstellar space. One of these grounds is that our galaxy, considering the gravitational motions within it, seems to have about 50 percent more mass than can be accounted for by its known content of stars and atomic hydrogen; the missing mass may be in the form of molecular hydrogen in space. Unfortunately we cannot hope to detect molecular hydrogen in interstellar space, because in its lowest energy state the molecule has no transitions at either optical or radio frequencies. There is, however, a chance that the molecule might be detected close to hot stars, where it may be ionized or excited into states that would cause it to emit detectable radiation.

This general possibility—excitation of atoms or molecules in space by nearby stars—gives rise to a number of interesting speculations. For example, it has been calculated that a hydrogen atom raised to a highly excited state (that is, with its electron in a very high orbit) will emit energy in the microwave range, because the energy difference between adjacent orbits at that height is small. Within the past year Soviet radio astronomers have reported the detection of radio emissions by highly excited hydrogen atoms in the Omega nebula at 5,763.6 and 8,872.5 megacycles. We are waiting with interest for confirmation of these observations.

VI

Galaxies

VI

Galaxies

INTRODUCTION

Ever since the first faint nebulae were found in the 17th and 18th centuries, these enigmatic objects have held a peculiar fascination for astronomers. Yet it is only within the last 50 years—after Edwin Hubble established a preliminary distance of 750,000 light years for the Andromeda Nebula—that astronomers have understood quantitatively the immense scale of galaxies or the vastness of the realm of the nebulae.

Our mental images of galaxies arise for the most part from the beautiful photographs of comparatively nearby spirals—at distances of 10-to-15 million light years. The view is a distorted one: not only are spirals actually in the minority among galaxies, but these splendid "show plates" give little suggestion of the fuzzy, almost structureless clusters of blackened photographic grains that constitute almost our only evidence about the overwhelming volume of the universe. It is virtually impossible to imagine the inky blackness of the immense reaches of intergalactic space.

That an entire section of this volume is devoted to galaxies is not surprising when we consider their intrinsic interest. That five authors can write so informatively about such remote objects is largely a tribute to the giant telescopes—both optical and radio—built within America during the 20th century. These mechanical eyes and ears have extended the senses of ingenious human astronomers, enabling them to glean data from galaxies whose light started out in ages long past, even before the trilobites were swimming in the ancient Cambrian seas. The very unit of measurement, the light year, conjures up a cosmological vision of the vastness of time as well as space.

The task of establishing these distances has been as fundamental as it is treacherous. The entire scheme is built upon a chain of evidence, hypothesis, and conjecture. Yet it would be erroneous to suppose that the distance determinations are as fragile as a house of cards. Though the entire structure has been adjusted occasionally, as Robert Kraft shows in "Pulsating Stars and Cosmic Distances," the procedures are now so interlocked and self-consistent that a further revision as large as a factor of three in the currently accepted distance of the Andromeda Galaxy (that is, 2,000,000 light years) would be unthinkable. At further distances, however, the uncertainty is greater and the cosmological stakes higher, so the quest for more reliable scales goes on.

As astronomers examine increasingly distant nebulae, they are looking further into the past. The question naturally arises, "How do galaxies evolve?" Unfortunately the images of the more distant galaxies are so faint that direct evidence about their evolution is difficult to obtain. Morton Roberts, Halton Arp, and Allan Sandage each approach this question of evolution from a different vantage point. Arp uses the large Mount Wilson and Palomar telescopes to sort out the comparatively nearby galaxies and to study the role of angular momentum in differentiating them. Roberts brings radio astronomy to bear on the relation of hydrogen content to galaxy type (incidentally, the American Astronomical Society has made a movie on the subject of this article, featuring Dr. Roberts). Sandage, in "Exploding Galaxies," explores what may be a transient phase in the evolution of many of them.

One of the important and unexpected findings of the past decade is the discovery that the nuclei of certain galaxies can fluctuate in brightness within a matter of a few weeks. By making the assumption that no coherent phenomena would be observed if the emitter were larger in light time than the period of variation, this variability can be used for estimating the size of the source. Since a galaxy is generally measured in tens of thousands of light years, an active region of only a few light weeks' extent is small indeed, and all the more astonishing when the energy output is considered. Apparently such small but extremely energetic units in the nuclei of certain galaxies are the source of the luminosity fluctuations. Particularly noteworthy in this regard are the Seyfert galaxies described in the article by Ray Weymann. The Seyfert galaxies were originally singled out because of their concentrated but intense nuclei, whose emission line spectra suggested unusual activity. Because their nuclei are similar to the quasars (which also undergo large light variations), the Seyfert galaxies may be an evolutionary link between the quasars and the more quiescent (older?) ordinary galaxies. Even the explosions of galaxies may fit somewhere into this sequence, as a common but short-lived phenomenon. This area of celestial exploration, now in its infancy, will surely produce surprises and clarifications in the 1970's.

19

Pulsating Stars and Cosmic Distances

ROBERT P. KRAFT

July 1959

Our present picture of the universe—its structure, size and age—rests to a large extent upon observations of a few pulsating stars. Each of these stars waxes and wanes as much as one full magnitude (2.5 times) in brightness according to a fixed rhythm ranging in period from less than a day to more than 50 days. In general, the longer the period, the greater the luminosity of the star. Such stars are called cepheid variables after their prototype, star delta in the constellation Cepheus; the most familiar of them is Polaris (the pole star), which brightens and fades in a period of 3.97 days. We do not know what causes the pulsation of cepheid variables, nor what it signifies in the biography of a star. Some 40 years ago, however, by a bold stroke of invention, the variable luminosity of these stars was made to furnish a distance scale that gives astronomy its reach into the cosmos beyond the immediate neighborhood of the solar system.

The new distance-scale at once made it possible to locate the center and to measure the dimensions of our galaxy. A few years later the presence of cepheid variables in celestial objects such as the Great Nebula in Andromeda helped establish that these "nebulae" are themselves galaxies—island universes as large as our own located at immense distances out in space. But in recent years the profound usefulness of the cepheid distance-scale has been almost overshadowed by its defects. Corrections in the scale have made it necessary for the dimensions of the observable universe outside our galaxy to be doubled, and still further revisions may be required. Because the age of an expanding and evolving universe can be deduced from its distance scale, cosmologists have concurrently had to revise the age of the universe upward, from two billion to perhaps 10 billion years. These corrections and further refinements still in progress derive from closer study of the cepheids themselves. It now seems safe to say that the cosmic distance-scale will not again expand so radically, and that it is at last ready for secure calibration.

In all likelihood we shall achieve this objective still without understanding why the cepheids pulsate. Among the 15,000 stars listed in the monumental new Soviet *Variable Star Catalogue*, edited by B. V. Kukarkin, P. P. Parenago, Y. Efremov and P. Kholopov, about 3,000 exhibit the regular pulsation of the cepheids. Spectroscopic observation shows that the surface temperature of these stars varies upward and downward in phase with their light. Apparently they also expand as they brighten and contract as they fade. In the 1920's Sir Arthur Eddington was able to show theoretically that the rate of pulsation must be related to the mean density of the cepheid (its mass divided by its volume), much as the period of a pendulum on earth is governed by its length. But we have no mechanism to explain this behavior, and we cannot say why a star becomes a cepheid.

The most important advance in our knowledge of the cepheids—and the most drastic revision of the distance scale—came a decade ago with Walter Baade's discovery that the stars of the universe may be divided into two major populations. To Population I, made up of young, hot, short-lived stars, he assigned the brighter and longer-period cepheids that appear in the arms of spiral galaxies. The fainter and shorter-period cepheids associated with the globular clusters that swarm around the centers of galaxies Baade placed among the older and longer-lived stars of Population II. While astronomers now believe that Baade's two populations represent an oversimplification and that stars are more continuously graded in age, the cepheids seem mostly to belong to the extreme ends of the population spread.

At present we imagine that the young Population I cepheids represent a phase in the life of any star. If we plot the color (that is, the temperature) of stars against their absolute luminosity (their intrinsic brightness corrected for distance), most of them occupy a rather well-defined "main sequence" [*illustration on page 202*]. To the right of the main sequence is a scattering of other stars, most of them "red giants." Between the main sequence and the red giants is an "instability strip" containing the cepheids. We presently conceive that a star starts out bright and hot, after a very rapid stage of gravitational contraction; then, after the star has consumed a certain amount of its hydrogen fuel, it begins to cool. Thus in terms of the color-luminosity diagram a star spends most of its life on or near the main sequence, but eventually evolves to the right. When it reaches the instability strip, it begins to pulsate. As the star passes through this strip, in the course of a few million years, its pulsation slows and lengthens in period. Upon reaching the end of the strip it ceases to pulsate and becomes a red giant. Ultimately it dims into the graveyard of the white-dwarf stars [see "Dying Stars," by Jesse L. Greenstein, which begins on page 147].

This hypothetical account does not, however, cover the evolution of the old Population II cepheids found in globular clusters. Perhaps these enter the instability strip by evolving "backward" from the red-giant phase instead of from the main sequence. Most of the globular-cluster cepheids have very short periods of less than a day, but even those having longer periods can be clearly distinguished from Population I stars of similar period. Long-period globular-cluster cepheids are on the average 1.5 magnitudes fainter than the younger long-period cepheids, exhibit quite different spectra and have masses only about a fourth as large.

The cepheids are highly luminous stars. Polaris, the nearest of them, is not a particularly bright cepheid, but it is about 600 times brighter than the sun. The brightest Population I cepheids are more than 10,000 times more luminous than the sun! This is a fortunate circumstance so far as the measurement of extragalactic distances is concerned, because it means that such stars make themselves visible at very long range.

In order to understand how pulsating stars can furnish a distance scale, we must go back 50 years to the work of Solon I. Bailey and Henrietta S. Leavitt of the Harvard College Observatory. Bailey carried out an extensive investigation of the cepheids in globular clusters within our own galaxy. He found that almost all had periods of less than a day, except for a few that had periods in the range of 12 to 20 days. Miss Leavitt later studied the cepheid variables that appeared in great numbers in photographs of the Clouds of Magellan, the two small galaxies that are companions of our own; she found that most of these cepheids had periods of more than a day. Even more remarkable was Miss Leavitt's discovery that the average apparent brightness of the Magellanic Cloud cepheids is directly correlated with the length of their respective periods of pulsation. Bailey had found no such dependence of luminosity on period in the globular-cluster cepheids, at least those with a period of less than a day.

Astronomers soon recognized the promise of Miss Leavitt's finding. It was known even then that the Magellanic

LONG-PERIOD CEPHEIDS are found among young Population I stars such as these in the Andromeda Nebula. A small section of one arm of the Nebula was photographed at two different times with the 200-inch telescope on Palomar Mountain. The marked star at upper left is a cepheid variable; the other star (*invisible in bottom picture*) is a nova.

SHORT-PERIOD CEPHEIDS are found in globular clusters of old Population II stars. These photographs of M 3, one of the globular clusters of the Milky Way, were taken 18 hours and 43 minutes apart with the 100-inch telescope on Mount Wilson. Four of the

Clouds are distant congregations of stars. Thus the cepheids in the Clouds are all at virtually the same distance from the solar system, and the light of all is attenuated to the same extent by its journey to the earth. Miss Leavitt's measurements of the varied apparent brightness of these stars could therefore be taken as indications of their relative absolute brightness. Here was a potential yardstick for measuring really long distances in the universe!

It was obvious that, if the distance of the Magellanic Clouds could be ascertained, one could determine the absolute brightness of the cepheids. Miss Leavitt's period-luminosity scale could then be used to find the distance to any stellar system or subsystem containing cepheids by turning the problem around: Measure the period of the cepheid, read off its absolute luminosity from the period-luminosity scale, compare this with the observed apparent luminosity of the cepheid and find the star's distance by applying the law that the intensity of light varies inversely with the square of the distance. Of course the accuracy of such a measuring rod depends on the assumption that cepheids in all parts of the universe obey the same period-luminosity law Miss Leavitt had derived from the cepheids in the Magellanic Clouds. This turned out to be a pivotal assumption.

At the time of Miss Leavitt's discovery there was unhappily no way to ascertain the distance of the Magellanic Clouds. Stellar-distance measurement still depended on direct trigonometric parallax, which is effective only for nearby stars. Against the background of stars distributed in the depth of space at all distances from the sun, a nearby star appears to shift its position as the earth travels from one side of the sun to the other. It is thus possible to measure the distances of such stars by simple trigonometry [*see illustration on page 201*]. Even these distances are so large that it is convenient to describe them with a unit called the parsec. We say that a star is at a distance of one parsec if its parallax, that is, half its shift of position, equals one second of arc. But the nearest star has a parallax of slightly less than .8 second of arc. This corresponds to a distance of slightly more than 1.3 parsecs, or 25,000 billion miles. Sirius, the brightest star in the sky, is 2.7 parsecs away, and the parallax of a star at a distance of 100 parsecs is only .01 second. Such small angles cannot be determined very precisely; a distance of about 30 or 40 parsecs is the practical limit for determination by direct trigonometric means.

The cepheids are so rare in space that the nearest of them—Polaris—is 90 parsecs away. It is clear, therefore, that trigonometry could not be used to determine the distance of a single cepheid, and could yield no information on the absolute brightness of even the nearby cepheids.

How, then, could the distance to any cepheid be obtained? Before Miss Leavitt had made her discovery, astronomers had devised a method for measuring what might be called the "middle distances" of our galaxy. With so many stars on our photographic plates we may assume that many stars in any given group have the same absolute brightness. We may also assume that the motions of these stars, either radially in the line of sight or transversely across the sky, will be at random. Now with the spectrograph we can determine the actual radial velocity of any observable star, independent of its distance from us. The spectrum is shifted toward the violet if the star is approaching and toward the red if it is receding, and the extent of shift gives us the velocity of its motion. On the other hand, the apparent transverse motion across the line of sight (called the proper motion) does depend on distance. If the stars of our given group move, on the average, with the same actual velocities independent of distance, then the proper motions of these stars will appear to get smaller with distance. Of course relatively few stars are near enough to the sun to have exhibited any proper motion during the

more than 200 cepheids in the cluster are marked by pairs of horizontal lines. On comparing the marked stars in the two photographs, one can see a small but perceptible change in their luminosity. The star at top right, for example, becomes brighter.

first century of photographic astronomy. But when we have determined the statistical spread of the radial velocities, it is reasonable to suppose that the proper motions vary in the same range. Since the distribution of proper motions does decrease with distance, the identification of the spread in radial velocities with the spread in proper motions indicates the average distance to the group of stars under consideration.

With the mean distance obtained in this way, one can correct the mean apparent magnitude of the stars for the effect of distance and get the average absolute magnitude. From studies of this sort in 1913 Ejnar Hertzsprung of Denmark found an average absolute magnitude of −2.3 for a cepheid with a period of 6.6 days. (On the magnitude scale the lower number refers to the brighter star; stars brighter than the first magnitude have negative magnitudes.) Hertzsprung's result was based on only 13 nearby cepheids for which the proper motions were known. But astronomers now had the absolute luminosity value needed to convert the apparent luminosity of any cepheid to absolute luminosity by reference to Miss Leavitt's period-luminosity scale.

In 1918 Harlow Shapley of the Mt. Wilson Observatory saw how the scale could be applied to determine the distances of the globular clusters in our galaxy. He fitted the long-period cepheids (periods of 12 to 20 days) of the globular clusters to the period-luminosity scale for the cepheids of the Magellanic Clouds. From this he determined the absolute luminosity and hence the distance of the long-period cluster stars. Using this determination of the distance to the clusters, he deduced that the mean absolute magnitude of the numerous fainter cepheids in the clusters with periods of less than a day was a little brighter than zero (*i.e.*, some 100 times brighter than the sun). Shapley then had a scale to measure the distance to the clusters that contain only faint, short-period cepheids. From the globular-cluster distances thus derived, he deduced that the globular-cluster system was centered on a point about 16,000 parsecs from the sun in the direction of the constellation of Sagittarius. It seemed reasonable to identify this point with the center of our galaxy. Shapley had obtained the first good estimate of the size of any galaxy. Later determination of the luminosities of these shorter-period cluster cepheids, obtained by proper-motion and radial-velocity studies, have verified Shapley's deduction and shown his estimate to be of the right order.

The period-luminosity scale could also be used to estimate the distances to any nearby galaxy that contains cepheids. Edwin P. Hubble and his associates at the Mount Wilson Observatory soon ruled off the distance to the Magellanic Clouds and to the Great Nebula in Andromeda. By the comparison of apparent to absolute magnitude thus effected for these and other more distant galaxies, the cepheid distance-scale made it possible to calibrate the spectrographic shift toward the red for the measurement of distances to the throngs of even more distant galaxies so faint and tiny that the cepheids and other stars in their populations cannot be resolved. Cosmologists working from these data were able to estimate the size of the universe and its age from the time of its initial expansion. All this extrapolated from the observation of the peculiar process of cepheid pulsation that we do not yet fully understand!

In the next 25 years, however, astronomers and cosmologists encountered numerous difficulties that cast increasing suspicion on the period-luminosity relationship upon which the whole edifice was built. All other galaxies, as measured by the cepheid distance-scale, were smaller in size than our own, a peculiarly self-aggrandizing result. As nuclear physicists succeeded in calibrating the rate at which uranium and thorium have been decaying to lead in the rocks of the earth, their "clocks" made the earth ap-

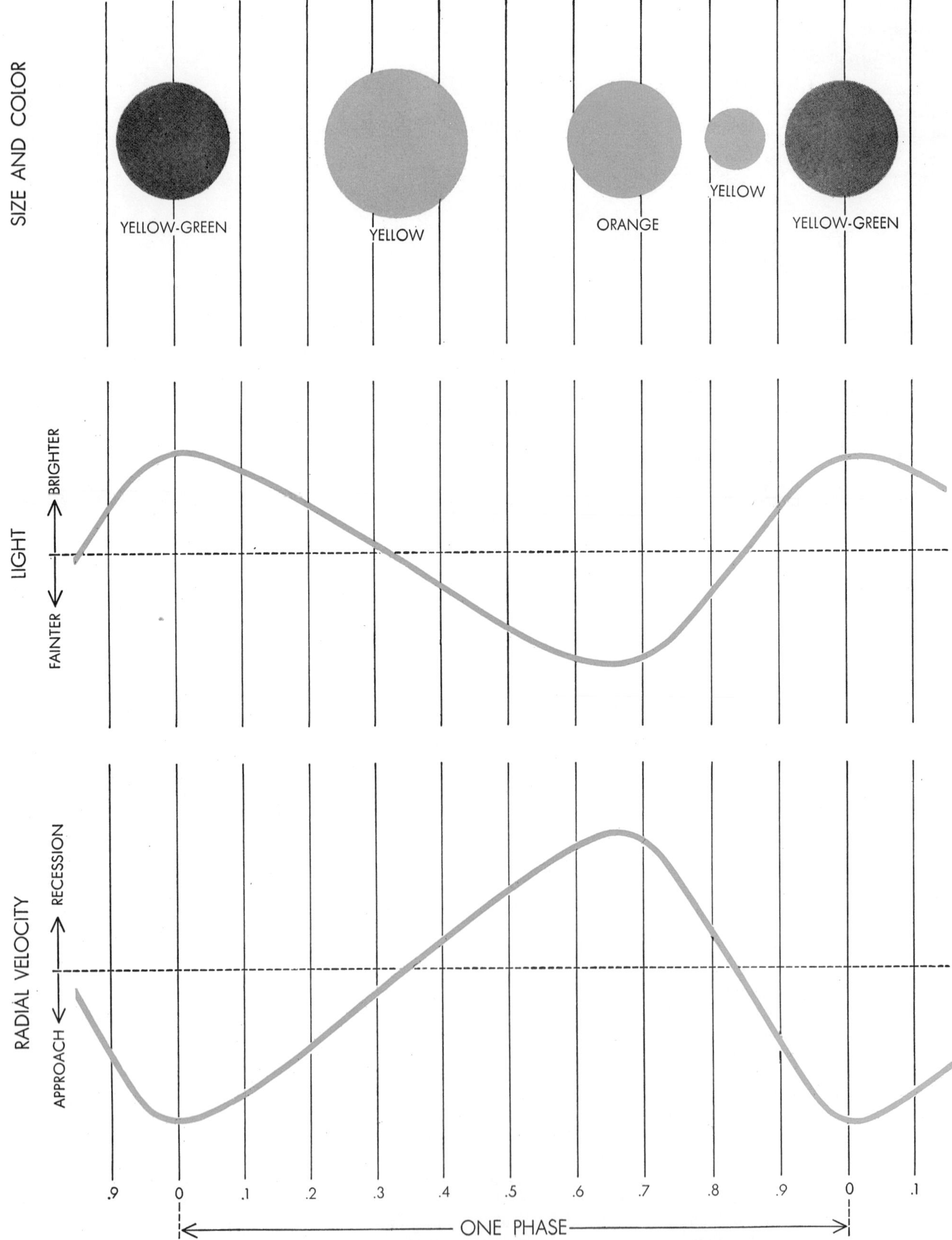

CYCLE OF A TYPICAL CEPHEID includes changes in color (*top*), light (*middle*) and radial velocity (*bottom*). Maximum light coincides with the bluest color (*yellow-green*), that is, with the highest surface temperature. The fluctuations in radial velocity are probably the result of changes in the size of the star such that its radius is largest (*large yellow disk*) midway between maximum and minimum light, and smallest (*small yellow disk*) in the opposite part of the cycle when the light is increasing. The relative sizes have been exaggerated in the drawing for purposes of clarity; the change in radius is never more than 20 per cent.

pear considerably more ancient than the universe. There was difficulty also in reconciling Eddington's calculation of the mean density of the cepheids with density estimates derived from the relationship of the observed luminosity of these stars to their rate of pulsation.

An observation by Hubble and Baade finally opened the way to a test of these suspicions. They pointed out that, if the distance to the Andromeda Nebula had been correctly measured, then the brightest stars of the globular clusters surrounding its central region appeared to be too faint compared to the brightest stars in the globular clusters of our own galaxy. If these bright stars in the Andromeda Nebula were assigned the same absolute brightness as the corresponding stars in our galaxy, then the cepheids visible in the Andromeda Nebula and many of the longer-period cepheids in our own system would also have to be assigned a higher absolute magnitude with respect to the shorter-period cepheids of the globular clusters that had formed the basis of Shapley's scale. Could it be that the globular-cluster cepheids obeyed a period-luminosity law different from that observed for other cepheids?

Such a possibility was foreshadowed in 1940 by an observation made by Alfred H. Joy at the Mount Wilson Observatory. He found a marked difference between the spectrum of a 15-day cepheid in the vicinity of the solar system and a 15-day cepheid in a globular cluster. Then, during the war years, Baade was able to devote the 100-inch telescope on Mount Wilson almost full time to his study of the stellar populations in the Andromeda Nebula. In dividing all stars into two populations he also found a basis for classifying the cepheids into two species.

With the 200-inch telescope in operation on Palomar Mountain shortly after the end of the war, Baade set out to observe the two types of stars "side by side," that is, at the same distance. Unfortunately not even the 200-inch telescope can resolve the faint short-period cepheids in the globular clusters of the Andromeda Nebula. But Baade was able to measure the Population I cepheids of that galaxy with great accuracy against the brightest globular-cluster stars, for which absolute magnitude had been established with the help of the Population II cepheids in our galaxy. Shapley had set the absolute magnitude of these stars at −1.5, based upon his determination that the shorter-period cluster cepheids have an absolute magnitude of zero. The distance to the Andromeda Nebula, calculated from its Population I cepheids in accord with the established period-luminosity scale, predicted that the bright globular-cluster stars should have an apparent magnitude of 20.9. Baade found that these stars were actually magnitude 22.4. In other words, they were 1.5 magnitudes fainter.

This demonstrated that the estimate of the distance to Andromeda was too small by a factor of about two. It also showed that the absolute brightness of the Population I cepheids in the Andromeda Nebula was 1.5 magnitudes brighter than had been indicated by the period-luminosity scale. They have a lower apparent magnitude because they are farther away than had been supposed. With distance to the Andromeda Nebula doubled, its size also doubled, bringing it into line with the size of our own galaxy. These results were dramatically confirmed when A. D. Thackeray and A. J. Wesselink of the Radcliffe Observatory in South Africa discovered short-period cepheids in the Large Magellanic Cloud at exactly the magnitude predicted by Baade.

Hindsight now fully explains the discrepancy in the period-luminosity scale. With the Population I cepheids advanced 1.5 magnitudes in luminosity, there is a discontinuity in the scale that clearly divides the cepheids into two types. We also understand why this distinction was missed in the early part of this century. The young Population I stars in the arms of our spiral galaxy lie close to its central plane; the brighter light of these stars is accordingly dimmed by the clouds of dust and gas in which stars are formed. The older Population II stars, which resemble the stars in globular clusters, have had time to drift above and below the galactic plane, so their dimmer light reaches us without obscuration. By a remarkable coincidence the interstellar absorption of the light from the Population I cepheids almost exactly equals the difference in the actual brightness of Population I and Population II cepheids, that is, 1.5 magnitudes. No such obscuration dims the light of Population I cepheids in the Andromeda Nebula or the Magellanic Clouds; their lower apparent magnitude is now correctly attributed to their greater distance. Thanks to this combination of circumstances Shapley was able to fit the long-period cepheids in globular clusters to Miss Leavitt's period-luminosity curve for the cepheids in the Magellanic Clouds. He could not

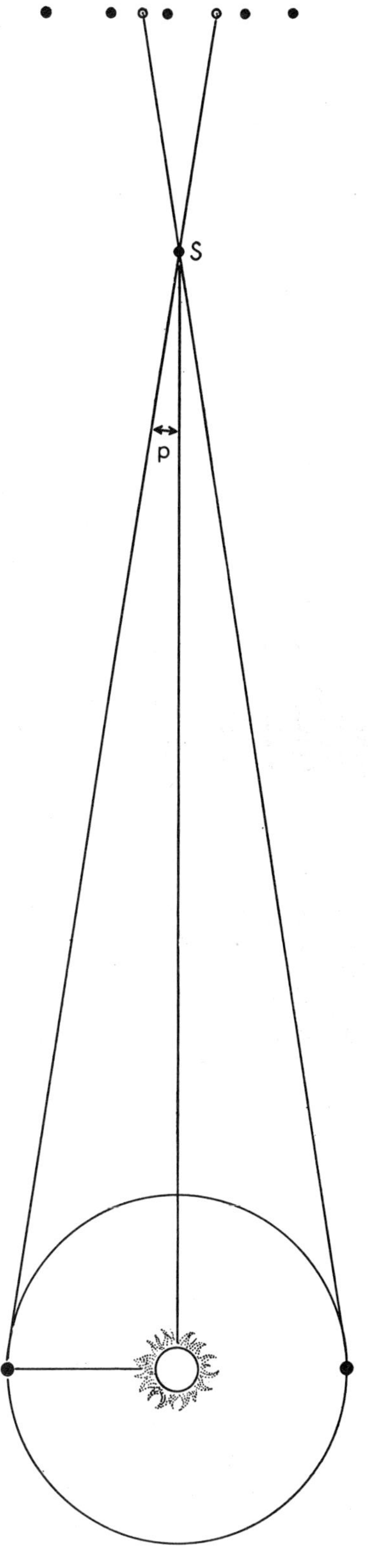

PARALLAX is used to measure the distance of nearby stars. As the earth moves around the sun (*bottom*) a nearby star (S) appears to change its position (*open circles*) in relation to stars much farther away. When the parallactic angle *p* is one second of arc, the star's distance from the sun is 19,000 billion miles, or 3.26 light-years, or one parsec. Here the change in the apparent position is much exaggerated.

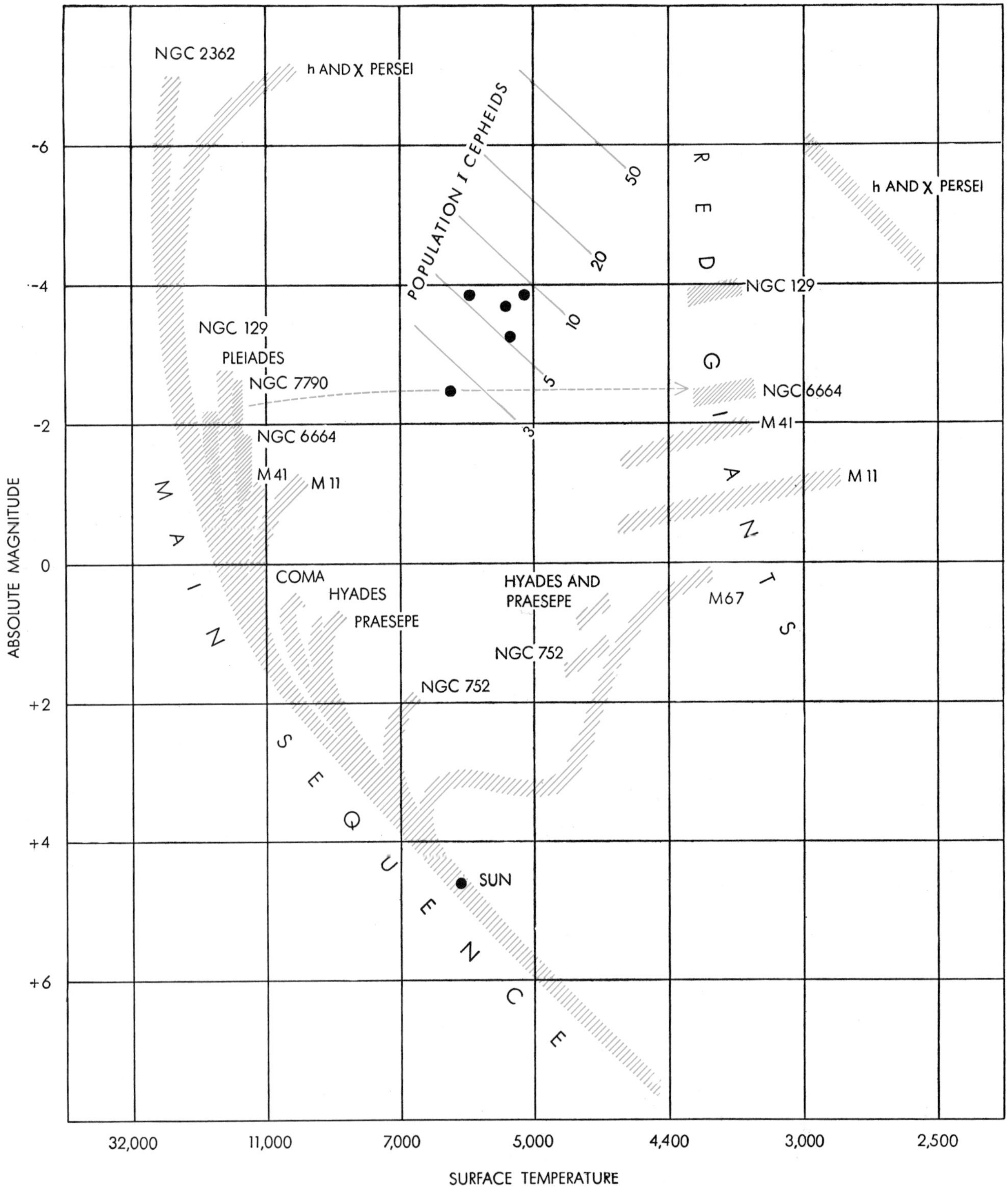

GALACTIC CLUSTERS OF THE MILKY WAY are plotted by determining the absolute magnitude and the intrinsic color of their stars. In this diagram the colors have been converted to approximately equivalent surface temperatures in degrees Kelvin. Galactic clusters are identified by their names or numbers; fine hatching marks those that contain long-period Population I cepheids. Five cepheids of the galactic clusters are shown in the "cepheid domain" which is marked off according to the length of the period in days. The broken line is the evolutionary track of a cepheid of NGC 6664. Originally the star was located among NGC 6664 stars on the main sequence, but about 100 million years ago it increased in luminosity about one magnitude and began to decrease in surface temperature, so that it moved horizontally across the diagram toward the cepheid "instability strip." Later it will become a red giant with a very large radius and a surface temperature of about 4,000 degrees K., like the present NGC 6664 red-giant stars.

have known that the two types of stars are quite different objects.

From the time of Miss Leavitt's first observations the distinction between the two species of cepheids had also been obscured by a scatter of about one magnitude in the positions of the stars along the mean line of the period-luminosity curve. For many years this was attributed to observational error and possibly to internal absorption within the Magellanic Clouds. But the scatter could also result from a bona fide physical departure of a given star from the mean line. This is a point of more than academic interest; such uncertainty in the magnitude of a particular star corresponds to a factor of 50 per cent in the computation of its distance. The range of error is too great if the objective is to measure the distance to a galaxy in which only one or two cepheids are available. Accurate determination of distances to individual cepheids has also assumed new importance in the study of our own galaxy. Population I cepheids might be expected to outline the spiral arms of our galaxy and, being very luminous, to carry our knowledge of the spiral structure to considerable distances from the sun.

We are now certain that the scatter is real. Highly accurate photoelectric measurements of cepheids in the Small Magellanic Cloud by Halton C. Arp of the Mount Wilson and Palomar Observatories have established that the scatter is very much larger than the errors of observation. Allan R. Sandage of the same observatories has offered an explanation. Sandage predicts from the theoretical period-density relation that the period-luminosity law must be amended to take account of a third variable. This variable is the surface temperature of the star.

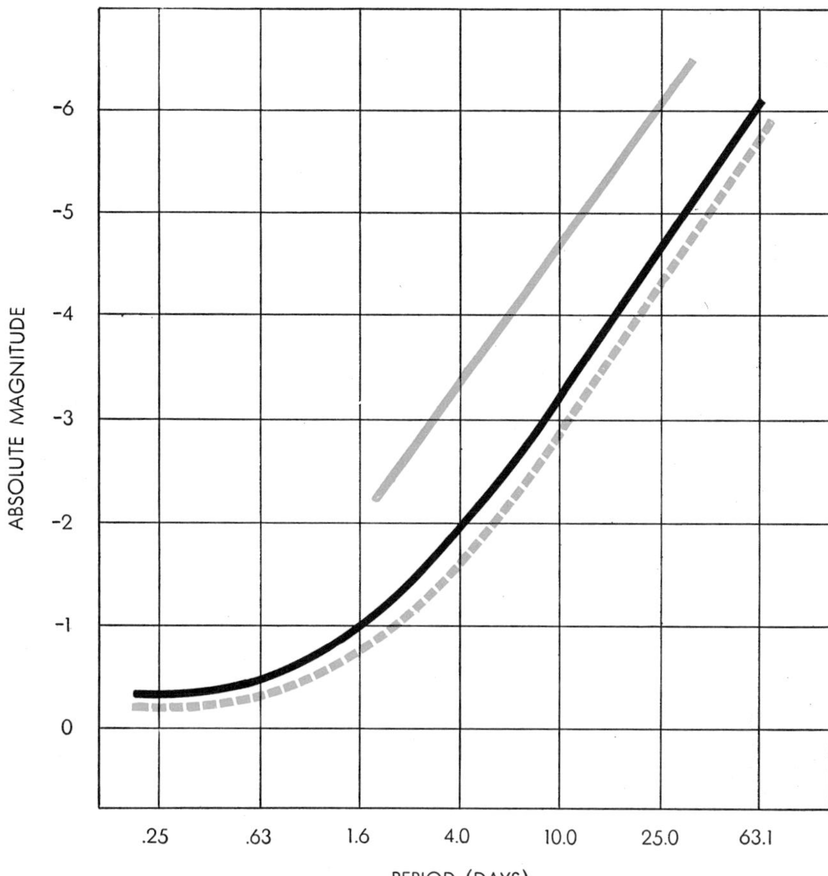

PERIOD-LUMINOSITY RELATION used by Harlow Shapley fitted all cepheids into one curve (*black*), with the short-period cluster variables at the lower end. The period-luminosity relation of Walter Baade divides the cepheids into Population I (*solid color*) and Population II (*broken line*). The latter stars are fainter than Population I stars of the same period. On magnitude scale, brightness increases by a factor of 2.5 from –1 to –2, and so on.

Observations of certain cepheids for which highly accurate surface temperatures and absolute magnitudes can be derived seem to confirm Sandage's theory. These stars are members of loose clusterings of very young Population I stars in our galaxy called open or galactic clusters. The first two were found by John B. Irwin in 1955 at the Radcliffe Observatory. Others were located by Sydney van den Bergh and myself, and the number of such cepheids is now about 10. Their colors (hence surface temperatures) and absolute luminosities are obtained by yet another method for determining distances to stars. We may expect stars that are close together on the color-luminosity diagram, and thus are similar in color and spectral characteristics, to have the same absolute brightness. By matching some of the stars of a cluster to similar stars for which the distance is known, we can derive the distance to the cluster. We can then determine the luminosity of the other stars in the cluster. Unfortunately most of the galactic clusters are obscured by interstellar material. This material not only absorbs light, but also reddens it, making the surface temperature of a star seem lower. By observing these stars in several colors, however, it is possible to derive intrinsic colors and surface temperatures.

With Sandage's period-luminosity-surface temperature relationship apparently well sustained, we can now determine the distance of a single cepheid if we know its surface temperature and period. The procedure may be demonstrated by reference to the color-luminosity diagram on the opposite page. On this diagram the cepheid variables occupy a band that, at a given position, reaches horizontally across a temperature range of about 1,000 degrees absolute and vertically over a factor of about six in absolute magnitude. Sandage has computed the lines of constant period, which slope down diagonally to the right on the diagram. To use the diagram, we locate a given cepheid (not necessarily from a galactic cluster) on its appropriate period line. We then draw a vertical line from the base of the diagram corresponding to the temperature of the star. The intersection of this line with the period line gives us the luminosity with high precision. We can then obtain the distance to the star with what is hoped to be an error of less than 10 per cent.

The final result of these studies of cepheids in galactic clusters should be a useful and accurate period-luminosity-surface temperature chart for the cepheid variables. Astronomers may expect soon to have a much more reliable scale for measurement of long distances inside our own galaxy and beyond.

20

Hydrogen in Galaxies

MORTON S. ROBERTS

June 1963

In 1944 H. C. van de Hulst, a young astronomy student at the University of Leiden, proposed that it might be possible to detect radio emission from hydrogen atoms spread thinly through interstellar space. Because the Netherlands was then occupied by German troops van de Hulst's proposal did not become generally known until the war had ended. Even then few astronomers took it seriously, because it seemed likely that the hydrogen signal would be extremely faint. Finally, in 1951, the hydrogen emission was detected with a crude-looking but sensitive apparatus by Harold I. Ewen, a Harvard University graduate student, and his professor, E. M. Purcell. Almost simultaneously a radio telescope was completed at Leiden and van de Hulst confirmed the Harvard discovery and his own prediction.

Since 1951 radio astronomers at Leiden, Sydney in Australia, Harvard and elsewhere have spent thousands of hours recording the hydrogen emission from within our own galaxy and from some 35 or 40 other galaxies near enough to yield a detectable hydrogen signal. The recordings from within our galaxy have shown that the hydrogen is concentrated in the spiral arms of the galactic structure. What is more significant, the location of the hydrogen and its relative motion in space can be mapped even in distant regions of the galaxy where stars are obscured by dense veils of interstellar dust. The dust blocks visible light but not the radio waves from hydrogen. Much of the mapping of hydrogen concentrations within our galaxy has been done at Leiden and at Sydney. At Harvard my colleagues and I have spent several years measuring the hydrogen content of other galaxies, using a radio telescope with a 60-foot dish completed in 1956. Although it is not large as modern radio telescopes go, the Harvard instrument has been equipped with a very sensitive maser detector.

Our findings, in brief, are that the amount of hydrogen in a galaxy can be correlated with its appearance and structure. In spiral galaxies such as our own about 1 per cent of the total galactic mass is in the form of interstellar hydrogen. In galaxies with more open spirals the hydrogen content runs as high as 14 per cent. Some galaxies with an irregular structure contain as much as 30 per cent. In certain other galaxies, in general those with little flattening or spiral structure, we have been unable to detect any hydrogen at all. These findings are consistent with a recent hypothesis that questions the assumption that galaxies of different appearances represent different stages in galactic evolution. For example, some astronomers have thought that galaxies begin their life as spherical gaseous systems, which become flattened by rotation and later undergo other structural changes. The new view is that galaxies are endowed with different structures from the time of formation and change little with the passage of time. On this view it is reasonable to expect that galaxies may also vary widely in hydrogen content from the outset. The case, however, is far from closed.

Some idea of the difficulty of measuring the hydrogen emission from space can be gained from the following statistics. The amount of radiation from interstellar hydrogen falling on the entire surface of the earth is less than 10 watts. The amount of this energy that can be collected by a radio telescope is an extremely small fraction in which the numerator is the area of the telescope's collecting surface and the denominator is the area of the earth: about 5×10^{15} square feet. Of the total energy of less than 10 watts of hydrogen signal reaching the earth, only about a millionth of a watt is contributed by a galaxy typical of those detectable by the 60-foot Harvard radio telescope. This means that the maser mounted on the telescope is able to detect a signal whose strength is typically 10^{-18}, or a billion-billionth, watt.

What is the source of the hydrogen signal? As van de Hulst recognized, a hydrogen atom floating in the cold void of space can exist in only one of two energy states. In the lower energy state its single electron and proton are spinning in opposite directions; in the slightly higher state the two particles are spinning in the same direction. Every few million years, on the average, the spin axis of the electron will flip over and a hydrogen atom in the higher state will pass to the lower energy state. Simultaneously a photon is emitted and carries off the energy lost in the transition. The emitted photon has a wavelength of 21 centimeters, which is equivalent to a frequency of 1,420 megacycles. This is the 21-centimeter spectral line that Ewen and Purcell were the first to detect. A final point is that hydrogen atoms that have flipped to the lower state will occasionally acquire the energy needed—usually through collision with another atom—to revert to the higher energy state.

The discovery of 21-centimeter radiation has made it possible to carry forward a study of galactic structure that began with the work of Sir William Herschel in the 18th century. A brief account of this quest for knowledge may help to set the stage for a discussion of our own investigations.

It was apparent to Herschel, as to stargazers before him, that the sky is not uniformly populated with stars. Many men must have wondered if the Milky Way, which appears as an irregular band of light across the night sky, is actually made up of stars so densely packed that the eye cannot resolve them. Using a succession of magnificent telescopes, Herschel was able not only to resolve the stars in the Milky Way but also to make a systematic count of the stars in different directions. He assumed that the extent of our stellar system must be greatest in the direction in which the apparent density of stars is highest.

Herschel's method was extended by other observers and finally led, early in this century, to a model known as the Kapteyn universe, named for J. C. Kapteyn, who began his career at the Leiden observatory. On the basis of photographic star counts Kapteyn concluded that the sun is located almost at the center of the Milky Way, which was thought to define the whole universe. The existence of other galaxies was still unrecognized.

Around 1917, using a completely different approach, Harlow Shapley reached the quite different conclusion that the sun must be located some 50,000 light-years from the center of our stellar system. Studying the distribution of globular clusters of stars, Shapley noted that most of them are concentrated in one hemisphere of the sky. He guessed that they are in all probability symmetrically located about the center of our galaxy and that their observed distribution means that the solar system is displaced toward the edge of the galaxy. It turned out that Shapley was basically right but that he had overestimated the distance from the sun to the center of the galaxy. The best current estimate places the distance at about 33,000 light-years, not 50,000.

How could Shapley's observation be reconciled with Kapteyn's star counts,

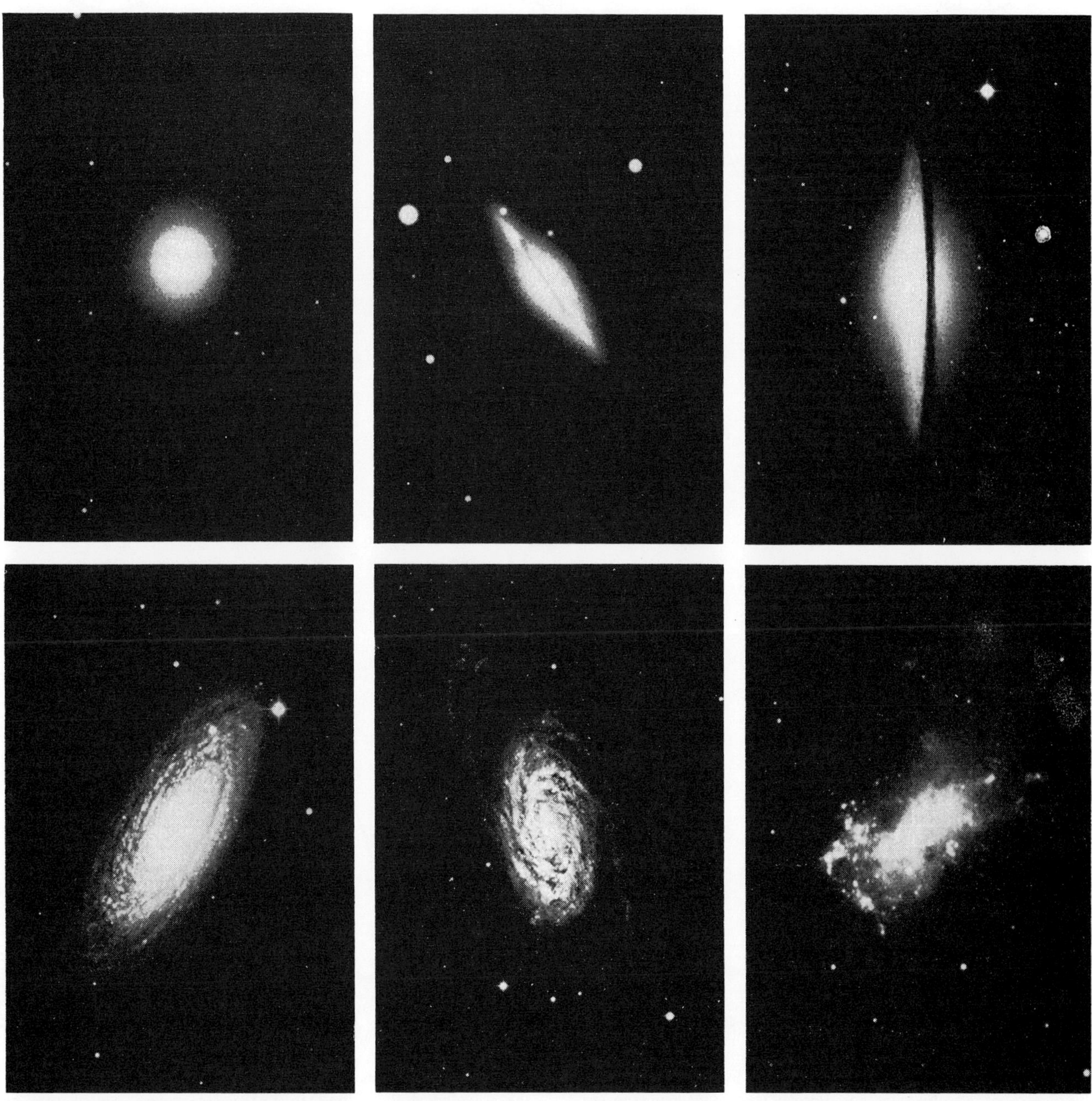

DIFFERENT TYPES OF GALAXIES include ***(left to right, from upper left)*** **NGC 4486, an elliptical galaxy 40 million light-years distant; NGC 5866, an S0 spiral 33 million light-years away; NGC 4594, an Sa spiral at 37 million light-years; NGC 2841, an Sb spiral at 30 million light-years; NGC 2903, an Sc spiral at 23 million light-years; and NGC 4449, an irregular galaxy 11 million light-years away. Amount of hydrogen in galaxies seems to be correlated with structural type. Ellipticals have the least; irregulars the most.**

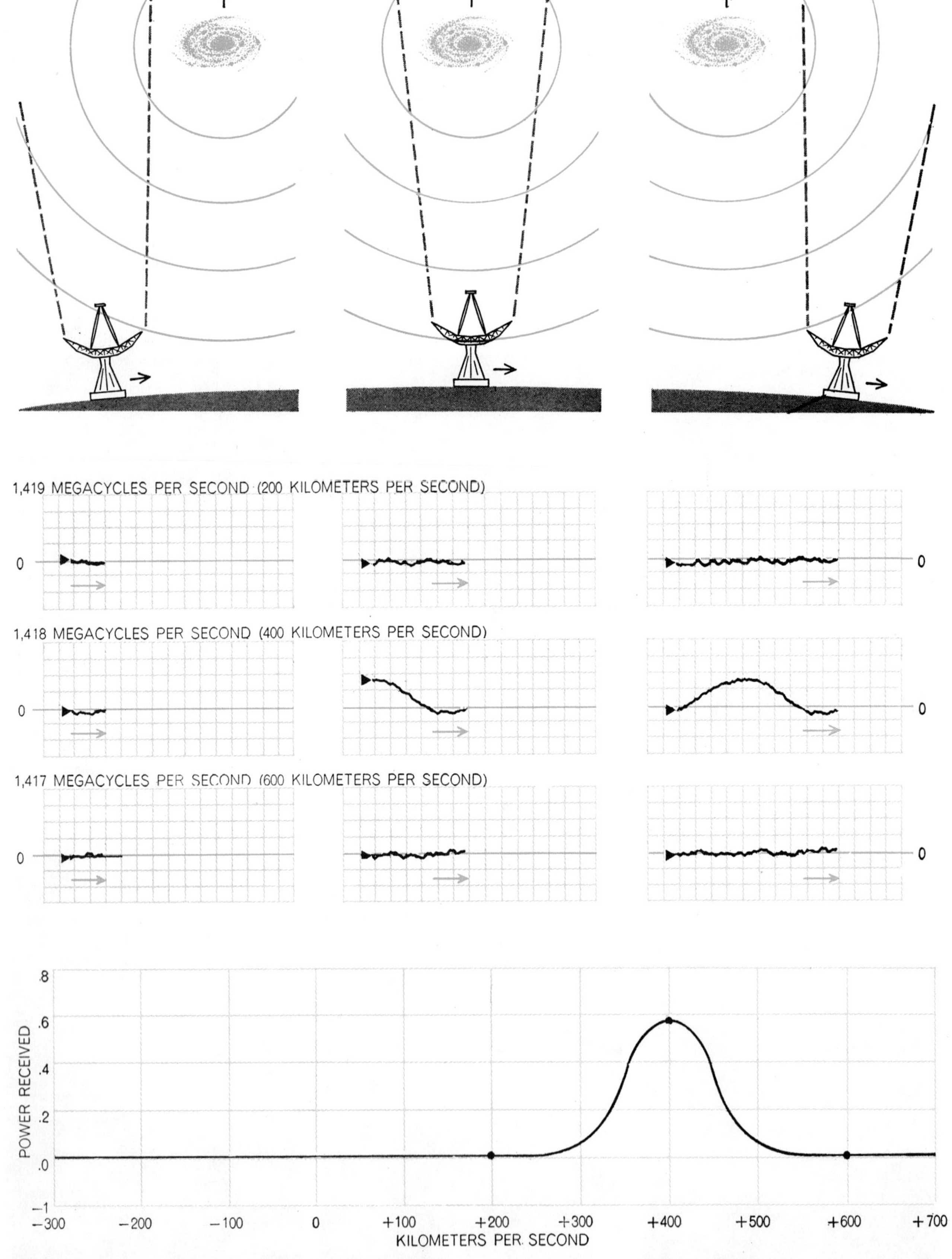

"DRIFT CURVE" METHOD of studying hydrogen in galaxies involves aiming radio telescope directly west of galaxy (*top left*) and letting rotation of earth carry it past the galaxy. Idealized recordings on three channels one megacycle apart are shown. Interstellar hydrogen emits radiation at 1,420 megacycles per second. If the galaxy happens to be receding at 400 kilometers per second, the wavelength is shifted to 1,418 megacycles (*center row of panels*). In this case no emission is recorded at 1,419 or 1,417 megacycles. The curve at the bottom, which plots the power received at various wavelengths, is known as the velocity profile of the galaxy.

which showed roughly equal numbers of stars in all directions? The answer to the riddle was supplied in 1930 by Robert J. Trumpler of the Lick Observatory, and in the process he corrected Shapley's distance scale. Trumpler compared the diameters of several hundred open, or galactic, clusters and reasoned that they should have about the same average size regardless of distance. When he applied the prevailing distance scale, however, he found that the predicted diameters increased with distance. The obvious way to correct this discrepancy was to assume that the distant clusters were actually nearer than the prevailing scale implied. But if they were nearer, why were they not brighter? Trumpler's answer: They were dimmed by interstellar dust.

This also explained why Herschel, Kapteyn and others were misled. In the plane of the Milky Way there is so much dust that they could never see the blaze of stars in the galactic center. By 1930, however, Trumpler's brilliant deduction was overshadowed by the somewhat earlier discovery, due chiefly to Edwin P. Hubble of the Mount Wilson Observatory, that our galaxy is only one in a vast universe of millions of galaxies extending for millions and, as we now know, billions of light-years in all directions.

Although Trumpler's discovery seemed to place an impenetrable veil over large regions of the Milky Way, the veil was simultaneously being lifted—although no one realized it at the time—by a young radio engineer, Karl G. Jansky. An employee of the Bell Telephone Laboratories, Jansky was asked to investigate unexplained radio noises that were plaguing transatlantic radiophone communications. In 1932 Jansky reported that at least some of the noise was "cosmic," originating outside the solar system. This was the accidental beginning of radio astronomy, and for the next 10 years optical astronomers virtually ignored it. Because of poor resolution early radio telescopes had trouble pinpointing the sources of cosmic noise. As a result the first radio maps of the sky resembled topographical contour maps. Closely spaced and roughly parallel lines delineated the plane of the Milky Way. Here and there, like hills on a contour map, a few cosmic hot spots were identified by a series of tightly spaced whorls.

The radio telescopes built before World War II could not possibly have detected the 21-centimeter radiation of hydrogen. They did not operate at such short wavelengths and they were far too

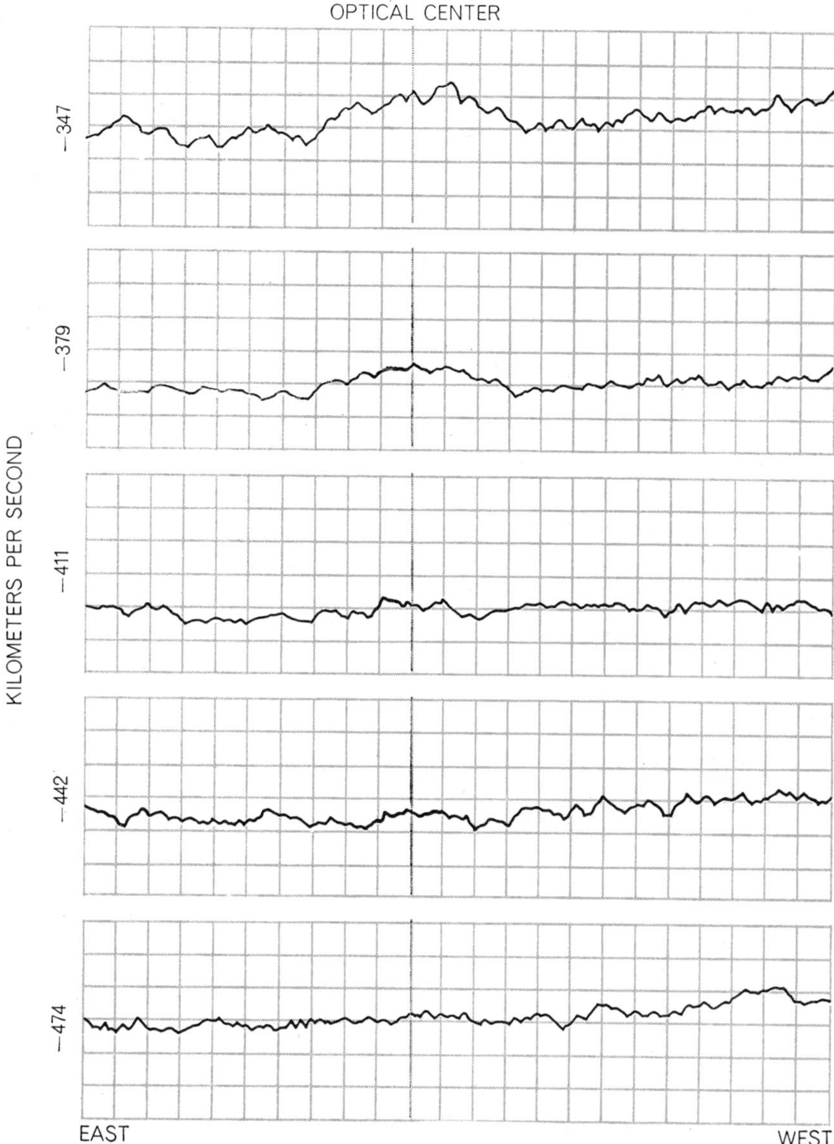

MULTICHANNEL DRIFT CURVES of irregular galaxy IC 10 were obtained in one sweep with maser mounted on the 60-foot telescope of Harvard College Observatory. Different frequencies recorded represent velocity differences of about 30 kilometers per second. Responses here show that this galaxy is receding at approximately 347 kilometers per second.

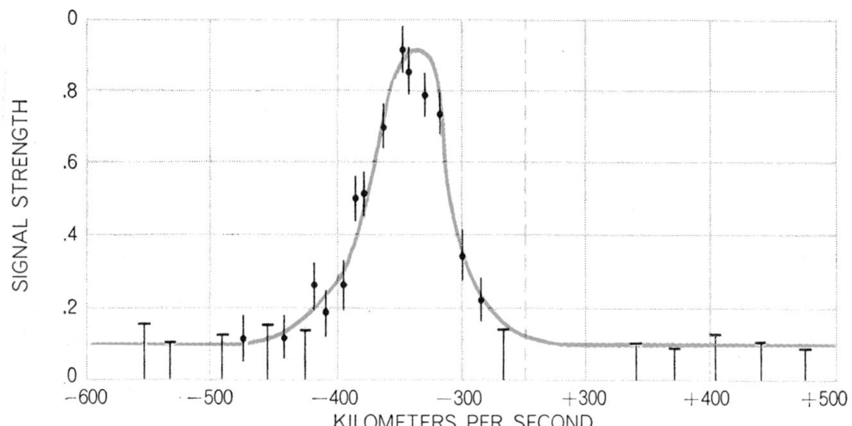

VELOCITY PROFILE of galaxy IC 10 comes from several multichannel sweeps. Dots mark specific observations. Thin vertical bars indicate range of error. Short dashes at top of lowest bars mark maximum strength of signal there. Dashed vertical line indicates break in scale.

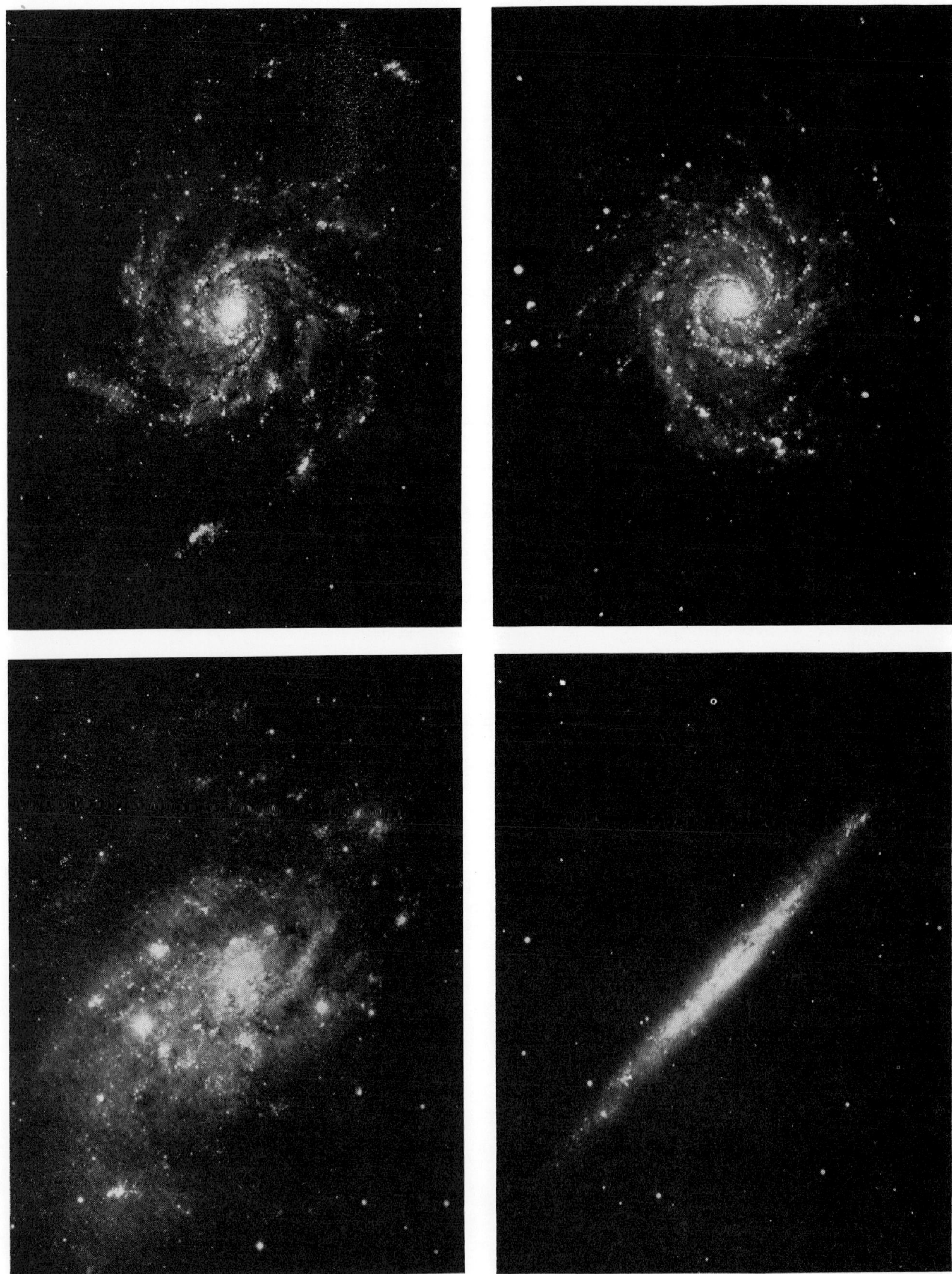

ANGLE OF GALAXY affects velocity profile. All these are Sc spirals. NGC 5457 (*top left*), 12 million light-years away, is seen full face or along axis of rotation. Its angle is thus zero. NGC 628 (*top right*), 25 million light-years away, tilts at an angle of 35 degrees. NGC 2403 (*bottom left*), 11 million light-years away, is inclined at 54 degrees as seen from earth, and NGC 4244, 11 million light-years distant, at 86 degrees. These photographs and those on page 205 were made at Mount Wilson and Palomar.

insensitive. The required sensitivity at short wavelengths was largely an inheritance from wartime radar.

An important virtue of 21-centimeter radiation in astronomy—its ability to penetrate interstellar dust—arises from the fact that its waves are about a million times longer than the dimensions of a dust particle. Light waves, having about the same length as the dimensions of a dust particle, tend to be scattered. Otherwise the two kinds of radiation behave in the same way. Like light waves, the 21-centimeter waves are shifted in frequency if their source has a radial component of velocity—that is, if the source is moving in the observer's line of sight. This Doppler shift is toward a higher frequency if the source is approaching and toward a lower frequency if the source is receding.

A rather complete picture of the central plane of our galaxy, as seen in the "light" of hydrogen radiation, has now been assembled from radio observations made primarily at Leiden and Sydney. The map shows not only the location but also the amount of hydrogen in various parts of the galaxy. Astronomers have been surprised to learn that interstellar hydrogen represents only about 1 per cent of the total mass of the entire galaxy. They had thought the figure would be several times larger.

In 1953 Australian observers recorded the first 21-centimeter radiation from a galaxy outside our own. The faint signal originated from the two small galaxies known as the Clouds of Magellan—our nearest extragalactic neighbors. Before long hydrogen radiation was detected from the Great Nebula in Andromeda, a spiral galaxy only about two million light-years away. It was apparent that to record hydrogen emission from any sizable sample of galaxies one would have to build larger radio telescopes or design more sensitive receivers. Ideally, of course, one should do both.

About five years ago the Harvard radio astronomy group set out to obtain a more sensitive receiver in the form of a maser. The actual design and building was done by two visitors at Harvard, B. Cooper of the Australian Commonwealth Scientific and Industrial Research Organization and J. Jelley of the British Atomic Energy Research Establishment at Harwell. The design followed suggestions made by N. Bloembergen of Harvard and Thomas Gold, now at Cornell University. The maser has been in continual operation since early 1960 and represents one of the major research tools of the Harvard Radio Astronomy Project.

Since the maser had to be located close to the feed horn at the focus of the 60-foot parabolic reflector, a compact, light design was required. The heart of the device is a ruby crystal located between the poles of a powerful electromagnet. Because the maser operates in a vessel of liquid helium, a convenient way to replenish the helium as it evaporated had to be found. The entire system weighs 200 pounds. It is at least five times more sensitive than conventional receivers and has tripled the number of galaxies that can be detected with the 60-foot dish.

In our study of 21-centimeter radiation from galaxies we have looked for correlations with the galactic structure as deduced from photographs. In the classification scheme originally proposed by Hubble galaxies are of three basic types: elliptical, spiral and irregular.

The ellipticals, which are essentially featureless, are classified according to their apparent ellipticity. The spirals are divided into two general categories: "normal" and "barred." In the former the spiral arms appear attached to the nucleus, or central region, of the galaxy; in the latter the arms extend from the ends of a bar that passes through the galactic nucleus. For the sake of simplicity I shall make no distinction between normal and barred spirals.

The spirals are further classified according to the relative size of the nucleus, the degree of openness of the arms and the resolution of the arms. A spiral with a large nucleus and smooth, tightly wound arms is classified Sa. An Sc galaxy has a small nucleus and large, open arms displaying a great deal of structure. Our own galaxy is an Sb type, intermediate between Sa and Sc in plan. A fourth type of spiral, S0, appears at first sight to be an elliptical, but actually it has a quite different profile of luminosity and contains very little dust.

The irregular galaxies exhibit no features by which they can be further classified; their appearance is chaotic and they show little or no evidence of a nucleus. They are not so highly flattened as spirals seen edge on, but they probably do possess a central plane.

The optically determined colors and spectral features of galaxies are known to vary in a systematic fashion with galaxy type; both sets of properties provide information about the stellar content of a galaxy. From 21-centimeter observations we have tried to determine how the hydrogen content varies in galaxies of different types. Evidence for some sort of correlation was suggested by earlier work, primarily that of Dutch

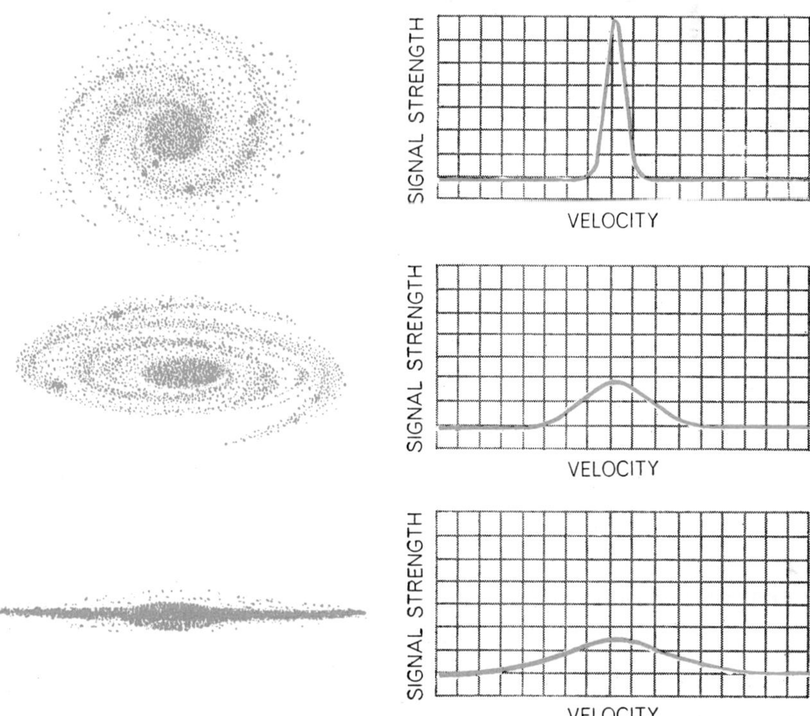

EFFECT OF TILT is illustrated with a single schematic Sc galaxy seen from three different angles. In all three views distance to galaxy, its radial velocity, rate of rotation and hydrogen content do not change. Differences in velocity curves (*right*) are due only to the angle of view. Because other factors are the same the area beneath all three curves is equal.

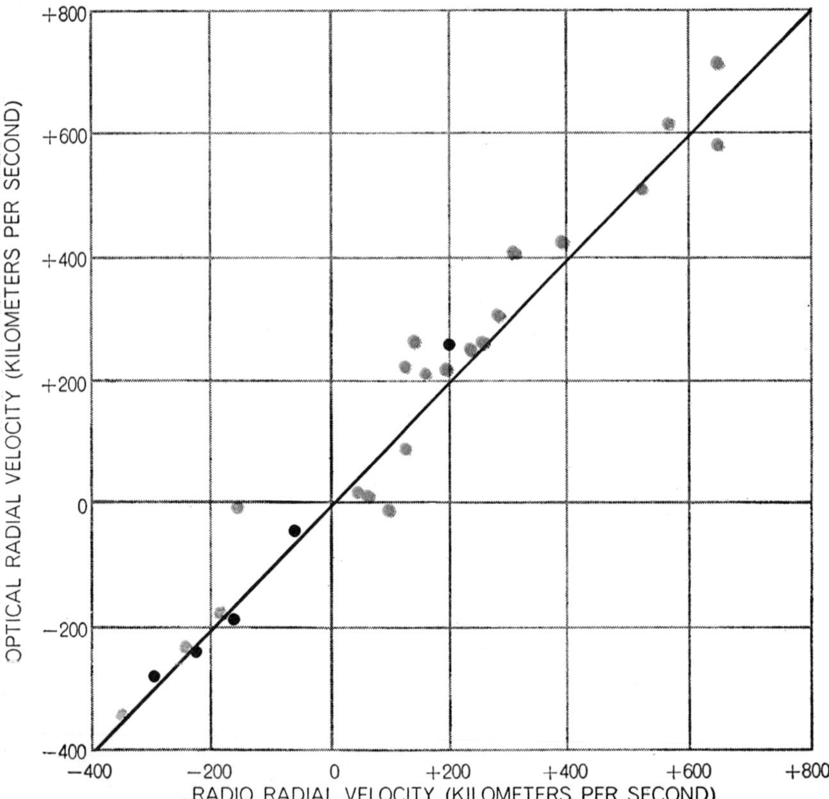

RADIAL VELOCITIES of galaxies as measured with radio telescope are plotted against those calculated from optical study of same galaxies. Colored dots indicate radio observation at Harvard; black dots, at Leiden. Diagonal line shows mathematical fit to the points.

observers. We needed a large sample so that meaningful averages could be obtained.

Our observations are made by the "drift curve" method. The radio telescope is aimed directly west of the galaxy under study and the rotation of the earth carries the telescope past the galaxy. In this manner a base line as well as the signal is obtained with one setting of the telescope. Unlike an optical telescope, which has a precisely defined field, a radio telescope accepts a certain amount of radiation from all directions through its "side lobes." The drift-curve method minimizes any changes that occur in side-lobe reception during the observation. The receiver output, recorded on paper tape, shows how the signal strength varies as the telescope "drifts" across the sky.

In making the observations allowance must be made for the radial velocity of the galaxy. For each increment of one kilometer per second in radial velocity, the "pitch" of the 1,420-megacycle hydrogen signal is shifted by five kilocycles. Because such Doppler shifts can often be determined with an accuracy of plus or minus 25 kilocycles, radial velocities determined by radio methods are inherently more accurate than those determined by optical means. Nevertheless, optically determined velocities are a great aid to the radio astronomer in selecting the frequency range needed to observe a given galaxy.

Because the hydrogen within the galaxy being observed is also in motion, the radiation will be further shifted, this time around the average velocity of the entire system. Observations must therefore be made over a range of velocities. In order to reduce the observing time we usually record simultaneously on five different channels, spaced 150 kilocycles apart. This is equivalent to measuring hydrogen radial velocities in five ranges spaced at intervals of 30 kilometers per second. The total recording range is thus 150 kilometers per second, but even this range is not great enough to encompass the random and systematic motions of hydrogen within some galaxies. In such cases the recording frequencies must be shifted and the galaxy may have to be scanned several times.

The multichannel recordings are converted into a velocity profile, which indicates the amount of radiation received at each velocity and therefore the total radiation falling on the antenna. If one knows the distance of the galaxy from optical measurements, one can relate the measured radiation to the total energy emitted at the galaxy. From this one can compute the total mass of interstellar hydrogen within the galaxy.

The velocity profile also supplies all the information needed to compute the total mass of the galaxy. For this computation one must first take into account how the galaxy is tipped with respect to our line of sight. If a galaxy is viewed edge on, it is evident that galactic rotation will carry the hydrogen in one half of the system toward the observer and that in the other half away from him. This differential motion broadens the velocity profile. If, on the other hand, a galaxy is viewed full face, or along the axis of rotation, all the hydrogen, except for random motion, will have the same relative motion along the observer's line of sight. In this case the velocity profile will be high and narrow. If the same galaxy could be viewed at different inclinations, the area under the profile, which measures the total power received, would remain constant [*see illustration on page 209*].

When we analyze a velocity profile to estimate the total mass of a galaxy, we begin by estimating its inclination as revealed in photographs. For a given inclination two galaxies of different mass will produce velocity profiles of different width. The reason for this is that the total mass of a system determines the strength of its gravitational field, and this in turn determines the motions of particles within the system. In general hydrogen atoms in a low-mass system will not have to travel as fast to maintain a given orbit as hydrogen atoms in a system of high mass.

The largest variation in hydrogen content has been found in Sc galaxies—highly developed spirals. Their hydrogen content varies from 2 to 14 per cent, with an average of about 8 per cent. Although the range seems rather large, we are reasonably confident of our measurements. The range might be reduced, however, if it were found that the galactic distances, which enter into the hydrogen calculations, were in serious error. Irregular galaxies show the highest hydrogen content: the average for 11 systems is about 16 per cent. The results for Sc galaxies and irregular systems are based primarily on observations made by Nannielou Dieter, E. Epstein and the author.

Two spiral galaxies of the Sb classification have been examined for hydrogen and found to contain about 1 per cent. This is the same value established for our own galaxy. The two Sb measurements were made by Leiden observers.

No 21-centimeter radiation has been detected in other types of galaxies studied. These include a number of elliptical systems. In general these systems have been fairly distant ones and their hydrogen radiation, if present, would at best be faint. It seems safe to say that their content of interstellar hydrogen is well below 1 per cent.

Before commenting on these results, I should like to mention that for all galaxies we have studied, the Doppler shifts obtained with 21-centimeter radiation have agreed in every case with shifts found by optical methods. This was not really unexpected, but there was always an outside chance that it might have been otherwise. A comparison of radial velocities obtained by optical and radio methods is shown in the illustration on page 210. The most distant galaxy represented is about 23 million light-years away—a rather short distance by extragalactic standards.

The systematic correlation of hydrogen content with galaxy type offers an important clue in the understanding of the life history of galaxies. The largest amount of hydrogen is found in the galaxies that also contain the highest percentage of very luminous stars. Such stars must have formed recently (within the past few million years), since their energy output is so great that they must exhaust their supply of nuclear fuel in a short period.

The relative number of such luminous stars decreases as one goes along the structural sequence of galaxies: irregular, Sc, Sb, Sa, S0, elliptical. This sequence is just the one in which the hydrogen content decreases, at least in so far as it has been measured. The correlation between hydrogen content and numbers of young stars is not surprising. Stars consist primarily of hydrogen, therefore a plentiful supply of hydrogen is necessary if many stars are to form. In galaxies containing little hydrogen one would expect to find relatively few young stars.

Some astronomers have suggested that this correlation indicates an age sequence, that galaxies with the highest hydrogen content and many young stars are the most recently formed. From exactly the same data, however, one can conclude that these galaxies need not be young; with their high hydrogen content they could support the formation of many fast-burning stars for a long period, with only slight depletion of their hydrogen supply.

Three general theories attempt to account for the various galactic structures. One holds that galaxies are continuously forming and implies that as they age they also change from one type to another. The other two theories assume a common age for all galaxies. The different galactic types are attributed either to evolution along the Hubble sequence or to their initial conditions of formation. This last proposal presumes that no significant structural changes have occurred since the formation of the galaxies. Halton C. Arp of the Mount Wilson and Palomar Observatories is a leading proponent of this viewpoint [see "The Evolution of Galaxies," by Halton C. Arp; the next article in this volume].

Calculations based on the new 21-centimeter measurements indicate that there is ample hydrogen in irregular and spiral galaxies to support the formation of bright stars at the present observed rate for at least 10 billion years, which happens to coincide with the most recent estimate for the age of our own galaxy. During this time the interstellar gas content would have dropped only slightly. These results do not allow us to decide which of the several galactic theories may be correct but they do tell us that the high percentage of bright stars in the irregular and Sc galaxies need not imply that they are young systems. The existing information on the stellar and gaseous content of galaxies is consistent with the view that all galaxies formed at the same epoch and have not evolved from one type to another. Depending on conditions surrounding their formation, different types of galaxies were left with different amounts of interstellar gas. It is this gas that has determined the subsequent rate of star formation, thereby maintaining the structural identity of the galaxy.

RADIO TELESCOPE of Harvard College Observatory has parabolic antenna 60 feet in diameter. The 200-pound maser unit is held at the prime focus where the three legs meet.

21

The Evolution of Galaxies

HALTON C. ARP

January 1963

It is not quite 40 years since Edwin P. Hubble, using the 100-inch reflecting telescope on Mount Wilson, conclusively demonstrated that the Great Nebula in Andromeda was not a mass of glowing gas but a galaxy—a huge system of stars outside our own. Subsequently Hubble examined thousands of galaxies recorded on Mount Wilson plates and classified them according to their appearance. He was careful never to suggest that his classification scheme represented an "evolutionary" series. His terminology of "early" and "late" spiral galaxies, however, suggested to some astronomers that a galaxy originated as a spherical mass of stars, that it gradually flattened by rotation into an ellipsoid and finally into a disk with more and more open and conspicuous spiral arms. Other astronomers felt, particularly in the last decade, that evolution proceeded in the opposite direction, at least in the spirals. Because a galaxy would presumably take billions of years to evolve from a ball to a disk (or vice versa), there seemed no way to establish the direction of galactic evolution.

Astronomers, however, have learned to extract a surprising amount of information from the feeble samples of stellar radiation that reach the earth, and they have not been content to ignore the problem of galactic evolution. Before discussing some recent investigations that bear on this problem let me briefly describe our own galaxy. According to Hubble's classification it is an intermediate form of spiral galaxy, technically designated Sb. Viewed edgewise, it would resemble two thin dishes set face to face. Viewed at right angles to its central plane, it would display a dense, luminous nucleus from which star-speckled arms extend in long sweeping arcs. The galaxy contains nearly 100 billion stars; for light to travel across it takes more than 100,000 years. Our sun is located far out in one of the spiral arms, some 33,000 light-years from the center of the galaxy. Distributed in and around the galactic nucleus are more than 100 globular clusters of stars, which contain from a few thousand to a million stars. These are the largest subunits in the galaxy. Smaller subunits range downward in size from gaseous nebulae and stars to atoms of gas and subatomic particles. Electromagnetic energy is radiated by the stars and is radiated or absorbed by the gases, depending on their temperature and density. The energy is observed mostly as light or as radio waves. Magnetic fields organize the pattern of the observed radiation. Dust and solid particles redden and block the light. All the gas and dust and stars are in swirling motion, with the sun's portion of the galactic disk rotating around the center of the galaxy with a speed of about 260 kilometers per second, or almost 600,000 miles per hour. Even at this velocity it takes more than 200 million years for the sun to make one complete circuit of the galaxy. One can estimate, therefore, that there must have been nearly 50 revolutions, or solar-galactic years, since the supposed birth of the universe 10 billion years ago.

The galaxy in Andromeda, technically known as M 31, looks about the way our own galaxy would look if it were seen from outside, tilted about 15 degrees from edge on [*see illustration on page 219*]. Two million light-years away and the nearest of the larger galaxies, M 31 is barely visible to the naked eye as a small hazy patch of light. If it appeared as bright to the eye as it appears in photographs, its long dimension—as seen in the sky—would be about seven times the diameter of the moon. Another spiral galaxy, M 81, seen more at right angles than M 31, is shown on page 220. Both M 31 and M 81 are classified Sb, a designation in which S stands for spiral and b for an intermediate stage of development, according to the Hubble classification. An Sc, or late-type spiral, is shown on the opposite page. Only about two million light-years away, it is a member of our local group of galaxies. A more distant Sc spiral, M 74, appears on page 221. Sc galaxies have smaller nuclei than Sb galaxies; their spiral arms are less tightly coiled and more conspicuous.

The Orientation of Spiral Arms

A picture of a spiral galaxy suggests that it is in motion and that it is rotating perhaps like a pinwheel or a whirlpool. To test this hypothesis early workers recorded the spectra of galaxies and found that the spectral lines were indeed shifted in such a way as to indicate that one edge of the flattened disk is approaching and the other receding. Unfortunately this determination can be made only when a spiral galaxy can be seen more or less edge on, and in this orientation the spiral pattern is hard to see. Conversely, when the spiral is tipped enough for us to see the arrangement of the spiral arms, the direction of rotation is hard to determine. As a result the obvious and important question of whether the spirals rotate in an unwinding sense or rotate with their arms trail-

GALAXY M 33, two million light-years away, has spiral arms more widely separated than those in our own galaxy. According to the system devised by Edwin P. Hubble, it is a late-type (Sc) galaxy. This color photograph was made by William Miller with the 200-inch telescope on Palomar Mountain.

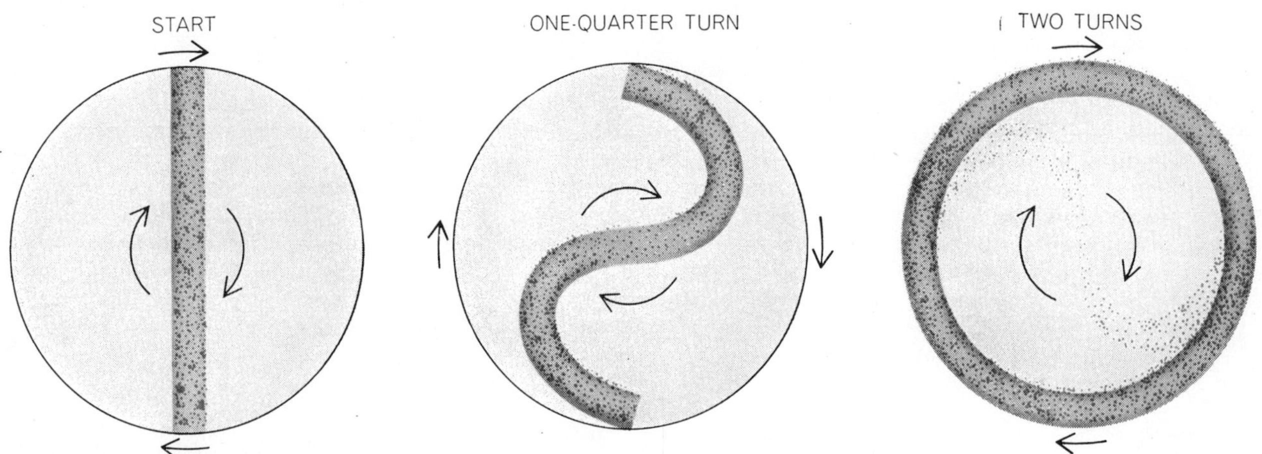

EFFECT OF DIFFERENTIAL ROTATION would be to wind the arms of a spiral galaxy into a ring, unless other forces were at work. The three diagrams show two opposed arms at the start of rotation, after a quarter-turn and after completion of two turns.

ing became a subject of considerable astronomical controversy. Even today a few astronomers maintain that at least some spirals may be rotating in an unwinding sense. Most astronomers feel, however, that the weight of evidence supports the view that spirals rotate with their arms trailing.

A simple physical consideration practically demands that the arms should trail: spirals viewed edgewise are observed to be in differential rotation. That is, the inner parts revolve around the center with a certain speed, but regions farther from the center rotate more slowly. Therefore a radial assemblage of stars forming one of the arms of a galaxy would unavoidably rotate faster toward the center and be drawn out into a spiral form with the outermost stars trailing in the rotation. This concept immediately raises another problem: Any given set of spiral arms will in a few turns be drawn out into what essentially appears as a closed ring around the center [*see illustration above*]. I have already mentioned that our own galaxy has presumably made something like 50 turns since its birth. Inasmuch as there are many spiral galaxies in the universe but few ring galaxies, one must conclude that some unknown agency prevents the spirals from winding themselves up.

A related puzzle is that the spiral arms are marked by very luminous stars that cannot shine for long at their present rate and therefore must be quite young. Evidently some agency replenishes the bright young stars in the arms. In short, the spiral arms are seen to be dynamically fragile and energetically short-lived. A tantalizing puzzle that has occupied astronomers is how spiral arms can be formed and maintained in the face of the forces working to destroy them.

The first step in solving the puzzle has been to try to establish the composition of the spiral arms. As we have seen,

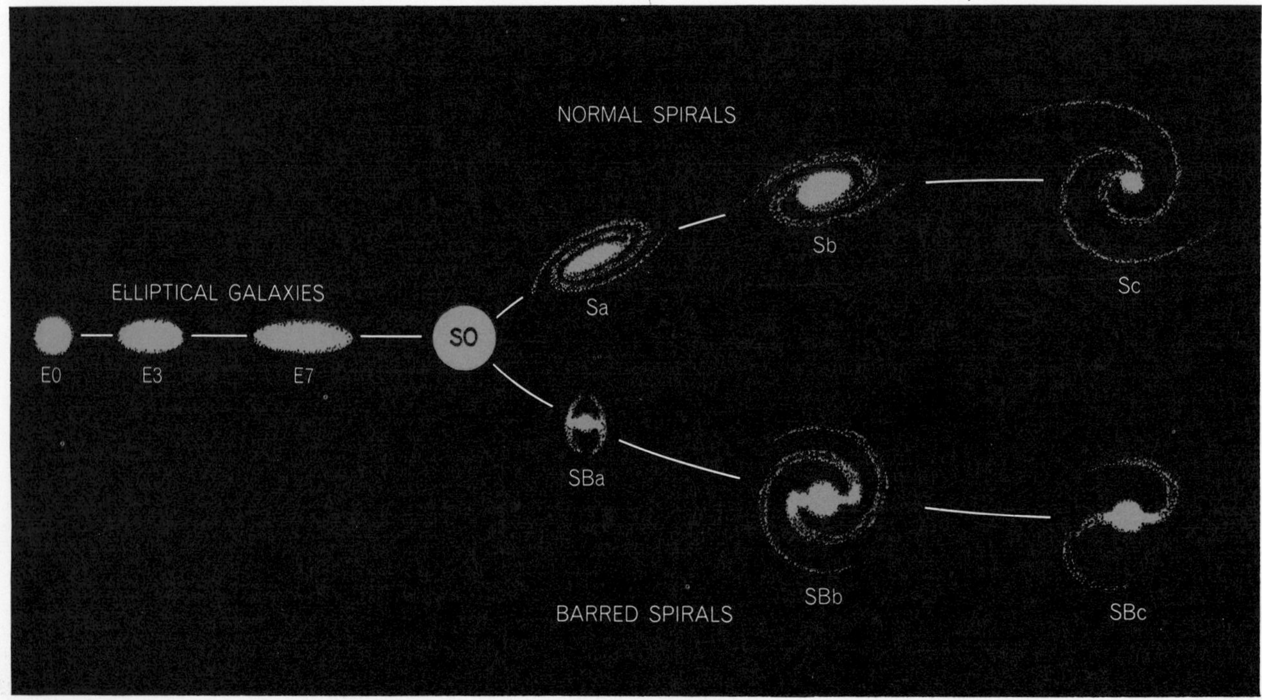

SEQUENCE OF GALACTIC FORMS was proposed by Hubble more than 20 years ago. In his "tuning fork" diagram spherical systems are at the left, followed by elliptical systems, then by disks (SO) and finally by spiral systems of two general types.

the spiral arms sparkle with stars that are hot, blue and from 10,000 to 100,000 times brighter than the sun. Stars can radiate at this rate for only one million to 10 million years, a short time compared with the estimated age of the universe. These supergiant stars are strung out along the spiral arms like beads on a string. Similarly restricted to the arms are great clouds and filaments of dust. Recent radio observations have shown that the spiral arms in our own galaxy are also outlined by hydrogen gas. Clearly this dust and gas furnish material for the formation of the bright new stars that illuminate the spiral arms. In fact, it is possible to observe nearby regions in the spiral arms of our galaxy where stars are in the process of being formed [see "The Pleiades," by D. Nelson Limber; SCIENTIFIC AMERICAN Offprint 285].

But just as the lifetime of very hot stars is limited, so too the supply of dust and gas for creating new stars would seem to be limited. Sidney van den Bergh of the David Dunlap Observatory in Toronto has estimated that at the present rate of star formation the gas in the vicinity of the sun would be exhausted in less than a billion years. He has suggested that the gas was perhaps replenished from the central regions of the galaxy. Subsequently Maarten Schmidt of the California Institute of Technology worked out a galactic model in which sufficient gas was originally present so that a decreasing rate of star formation would still leave about a fifth of the original gas not yet formed into stars. But regardless of whether new material flows in to replenish the old or the spiral arms are simply supplied with a large initial amount of gas, the problem remains of explaining how the material is kept from diffusing out of the spiral arms and away into space.

Evidence for Magnetic Fields

One clue to the solution of the problem may have been provided in 1949, when John Hall and W. A. Hiltner of the Yerkes Observatory observed polarization of the light from nearby arms of the galaxy. Jesse L. Greenstein and Leverett Davis, Jr., of the California Institute of Technology showed that polarization is probably caused by elongated dust particles, all aligned in the same direction, which preferentially absorb the light whose vibration is parallel to the long axis of the particles. The important outcome of this work seems to be that only a magnetic field aligned along the

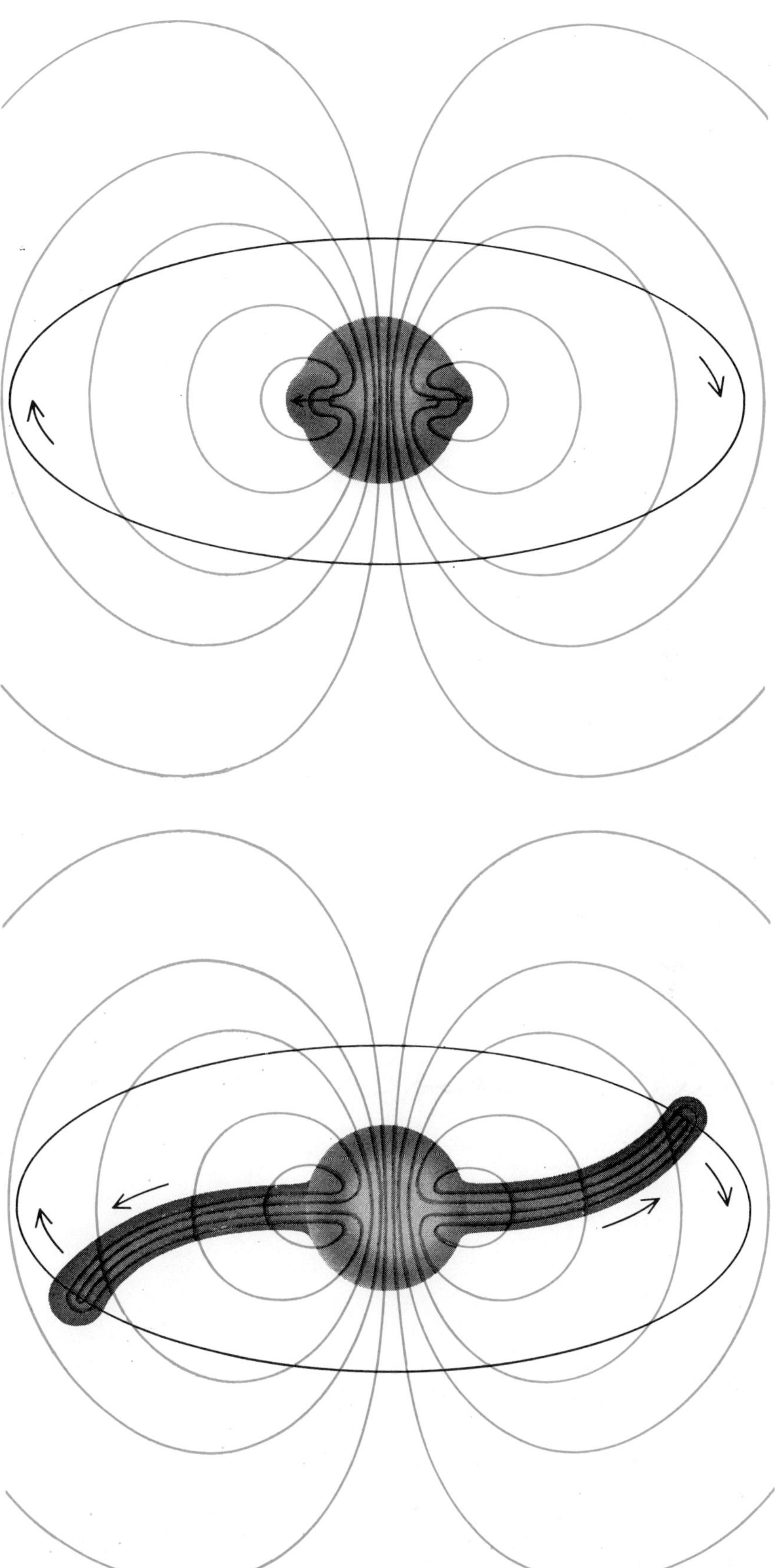

MAGNETIC FIELD OF GALAXY might originally have the general form shown in the top illustration. In the nucleus ionized, or electrically charged, particles of gas would be tightly bound to lines of magnetic force (*color*). If the gas were thrown outward by rotation of the galaxy, the ejected mass would pull the lines of magnetic force with it, as shown in the bottom illustration. Thereafter ejected material would tend to follow these lines and form arms.

spiral arm can explain the preferential alignment of the dust particles. A magnetic field would also serve to trap any atoms of hydrogen that have lost an electron as a result of collisions with photons or other energetic particles. Atoms that lack one or more electrons are said to be ionized. If it is placed in a magnetic field, an ionized particle can move only by spiraling along one of the lines of magnetic force. Even a weak magnetic field may be enough to bottle the gas within the spiral arms of a galaxy. Although only a fraction of the interstellar hydrogen is ionized and thus susceptible to magnetic trapping, the ionized atoms "frozen" to the lines of magnetic force would form a vast web through which un-ionized particles would find it difficult to escape. The actual picture is somewhat complicated because a gas confined in this way is under pressure and therefore tends to expand. Moreover, the magnetic field itself tends to push outward. It can be shown, however, that the gas in the spiral arms probably has enough mass to produce a gravitational force sufficient to offset the dispersive forces.

The proposal that the magnetic field in a spiral arm is held in shape by the gravitational attraction of the mass within the arm was made in 1953 by Subrahmanyan Chandrasekhar and Enrico Fermi of the University of Chicago. They computed that the magnetic-field strength needed to balance the gravitational field was between one millionth and 10 millionths of a gauss. A recent announcement from the Nuffield Radio

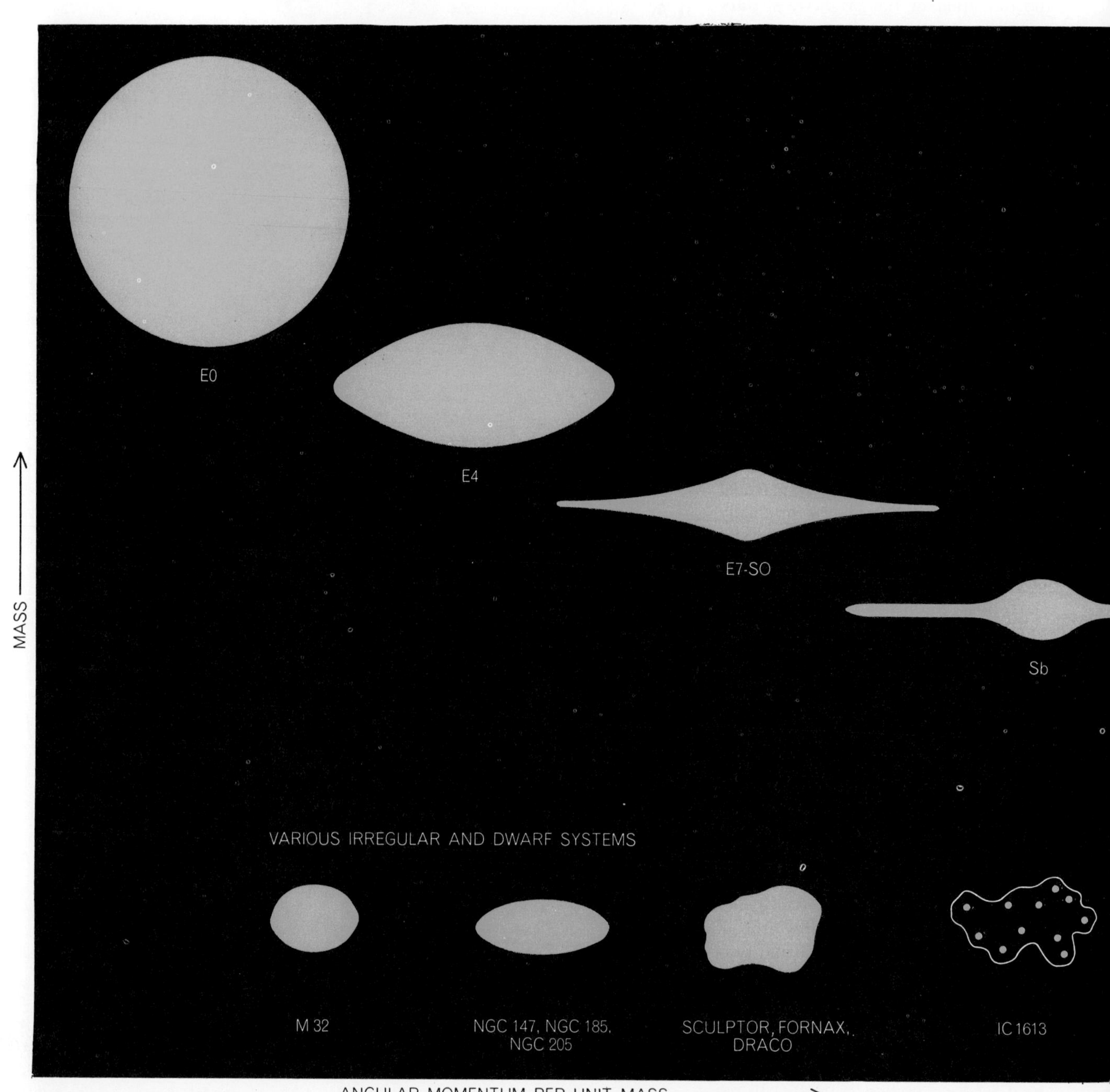

SUGGESTED CLASSIFICATION OF GALAXIES is based on two characteristics: mass and degree of flattening. Presumably the massive spherical and elliptical galaxies possess the least amount of rotational energy, or angular momentum per gram. Spiral galaxies and certain dwarf systems possess high amounts. (Many galaxies fall between the extremes plotted.) It is hard to conceive

Astronomy Laboratories at Jodrell Bank in England reports a directly observed value of 25 millionths of a gauss in certain regions of our galaxy and an over-all galactic field of approximately five millionths of a gauss. (The magnetic field of the earth is about .5 gauss.)

A second clue to the solution of the spiral-arm problem was provided by radio astronomy in 1957. It was observed that there is a flow of gas traveling toward the sun from the center of the galaxy. The gas leaves the galactic center at about 50 kilometers a second and contains enough material to create about one star of solar mass each year. More recently the Australian radio observer Frank Kerr has concluded that the outflow still has a velocity of about seven kilometers a second in the vicinity of the sun. Using the 200-inch telescope on Palomar Mountain, Guido Münch has observed a similar outflow of gas in the center of the Andromeda galaxy.

I should like to propose that at least some of the gas observed to be leaving the center of the galaxy is traveling inside the tube of magnetic force that comprises the spiral arm containing our sun. One solar mass of gas per year is ample to keep the spiral arm glowing with hot new stars. It might also be enough to allow for some leakage of material from the spiral arm.

If the spiral arm has an outward component of motion in addition to its motion around the center of the galaxy, this would tend to keep the spiral from winding up on itself and forming a ring. The mechanism would be similar to that of a Fourth of July pinwheel. The outward component of velocity would have to be an appreciable fraction of the rotational velocity, however, in order to keep the spiral open. It is doubtful that the observed outflow in our galaxy meets this requirement. If the outflow is indeed inadequate, the spiral arm containing our sun seems fated to be wound up in a ring very quickly. Other considerations may be involved, however. For one thing it is difficult to estimate how much rigidity may be imparted to the arms by the magnetic field running through them. There is also the possibility that the arms near the sun are not representative of the arms that give the galaxy its over-all shape. The spiral arms near the sun are only about 5,000 light-years apart, which would make our galaxy a rather tightly wound spiral. In M 31 and other Sb spirals the arms are about twice as far apart. To obtain more information on the fate of spiral arms near the sun, a number of observers have looked for differences in rotation velocity between old stars embedded in the disk of the galaxy and young stars in the arms. Although such differences have been found, they involve about the same uncertainties of interpretation as are encountered in the gas-flow measurements.

Shaping of the Magnetic Field

Although it has not yet been demonstrated conclusively that there is a net outflow of gas in the disk of spiral galaxies, I offer the hypothesis that such a flow is fundamental to their spiral structure. This hypothesis immediately raises the question of the relation, if any, between the flow of matter and the magnetic field. Offhand it is difficult to imagine why the magnetic field should radiate from the galactic center and form a series of spiral arms. This puzzling distribution might be explained as follows. Suppose the over-all magnetic field of a spiral galaxy has its axis through the center of the galaxy and perpendicular to the plane (like the orientation of the earth's magnetic field with respect to the plane of the earth's rotation). It is reasonable to assume also that many of the gas atoms in the center of the spinning galaxy are ionized and spiraling around lines of magnetic force. It is a characteristic of such magnetically trapped particles that if they are forcibly displaced, they drag the lines of magnetic force with them. As a result, if the ionized atoms are being urged out of the galactic center by centrifugal force or by pressure resulting from the inflow of fresh material from outside the nucleus, the initially ejected material will stretch out the lines of magnetic force behind it like taffy. This magnetic tube would then provide a channel for the flow of more ejected material, consisting of both ions and physically entrained un-ionized particles [*see illustration on page 215*].

Of course, what causes the magnetic field in a galaxy is not known, and the over-all configuration of the field can only be guessed at. Almost any general field, however, could be distorted locally by sufficient flow of ionized material to produce a field with a spiral-arm pattern. Other questions remain. Where does the material come from that flows out of the center? What would happen if the outflow were larger? Will it someday stop? Answers to these and other questions must await further knowledge of how galaxies evolve.

The Morphology of Galaxies

Let me return now to the problem that fascinated Hubble: Do the various configurations among galaxies offer a clue to their evolutionary development? There are, first of all, a large number of faint galaxies that bear little or no resemblance to the great spiral galaxies. Many of these galaxies are intrinsically small and faint; they must be quite near in order to be readily observable. Good examples of such galaxies are the two clouds of Magellan, the cloudlike

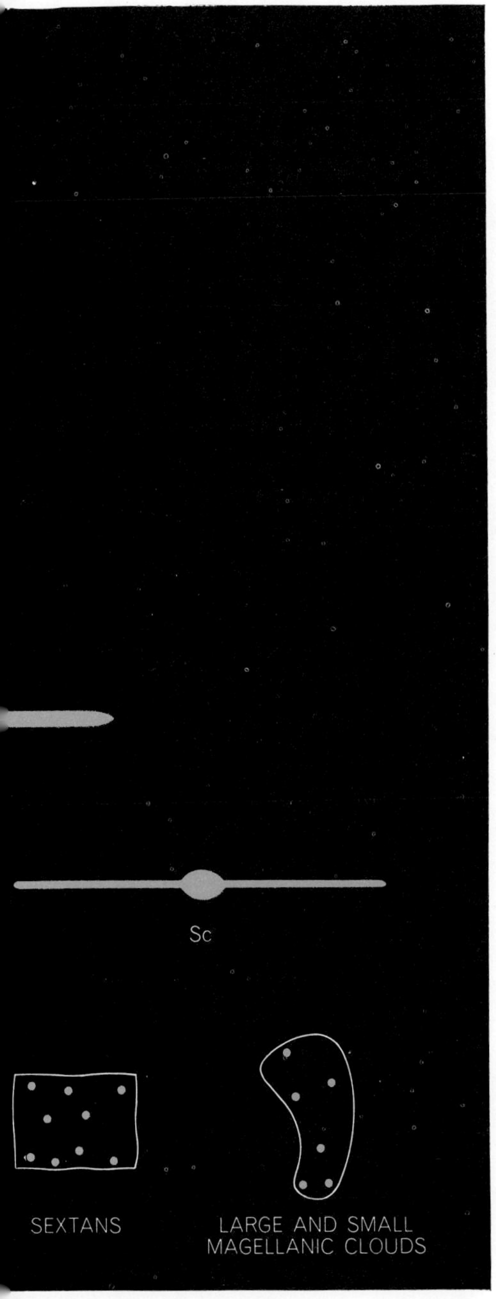

of any mode of galactic evolution that could add (or subtract) mass or angular momentum over the range shown in the chart.

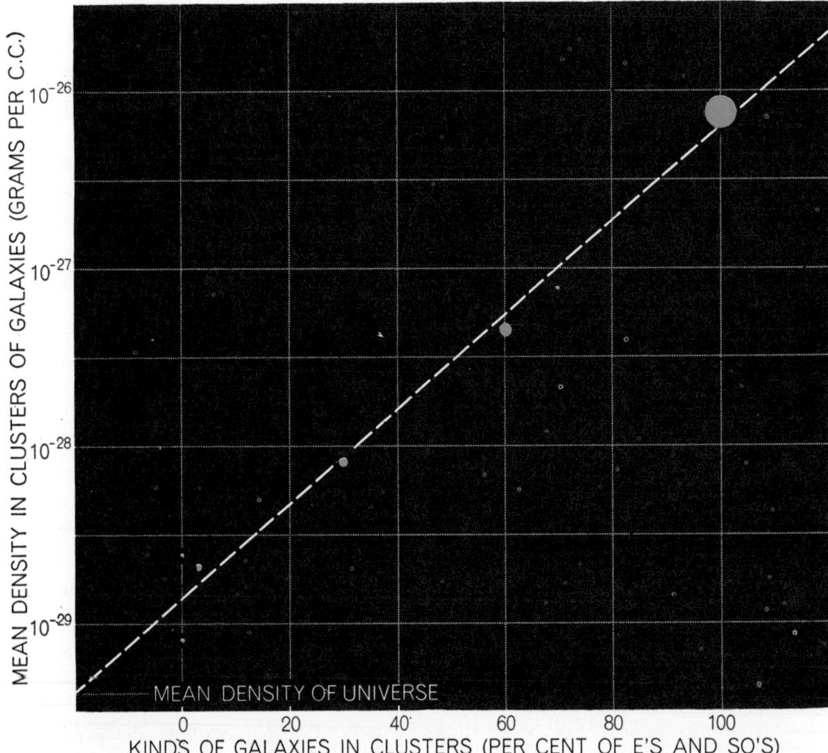

MEAN DENSITY OF CLUSTERS OF GALAXIES is plotted against the kind of galaxy found in the clusters. In the densest clusters the galaxies are all elliptical and disk-shaped (SO). As density decreases, the percentage of spiral galaxies in a cluster increases. At the approximate mean density of the universe there are only isolated galaxies, all spirals.

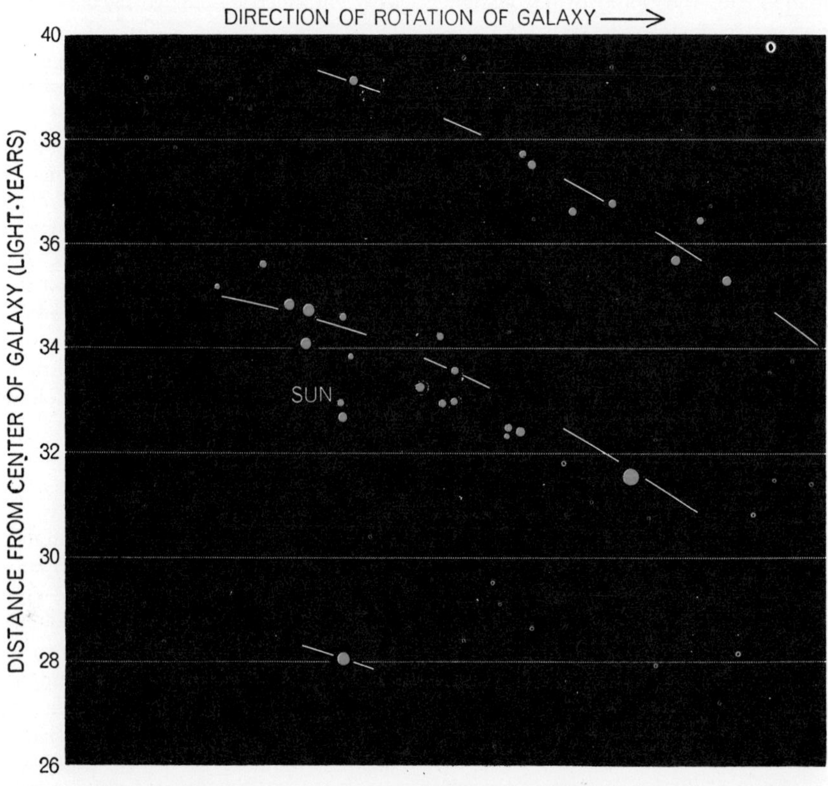

SPIRAL ARMS NEAR SUN can be traced by plotting the location of glowing concentrations of gas known as gaseous-emission nebulae. They indicate the existence of three arms about 5,000 light-years apart, making our galaxy a rather tightly wound spiral. The data are from W. W. Morgan, D. E. Osterbrock and Stewart Sharpless of Yerkes Observatory.

patches observed by Magellan when he first sailed beneath southern skies. These two galaxies, only about 200,000 light-years away, are so close that some observers have been led to search for filamentary bridges between them and our galaxy. Although no such link has been found, it has recently been shown that the two clouds are themselves linked by a tenuous common envelope of hydrogen gas. The large Magellanic cloud is about 40,000 light-years in diameter and the small Magellanic cloud about 30,000 [*see top illustration on page 222*]. The clouds are therefore between a third and half the size of our galaxy and are somewhat irregular in shape. The relative percentage of gas, particularly in the small cloud, is much larger than that in our own galaxy.

There are a number of systems smaller than the Magellanic clouds, such as the one in the constellation of Leo shown at the bottom of page 222. These faint systems lie at distances of 150,000 to 600,000 light-years and have diameters ranging from about 1,000 to several thousand light-years. As far as can be observed they contain no gas or dust—just stars. Still smaller systems are the "intergalactic globular clusters." Roughly 200,000 light-years away, they are only about 100 light-years across and contain even fewer stars and less mass than an ordinary globular cluster lying within our galaxy. It remains to be seen whether these very small star systems exist predominantly in the neighborhood of large galaxies and can be regarded as subunits or satellites or are spread more widely through the universe and must be considered independent galaxies.

If we turn our attention now to the other end of the spectrum of galactic sizes, we find that the elliptical and particularly the spherical galaxies are so big and bright that they can be seen at immense distances. The spherical galaxy shown at the top of page 223 is about 45 million light-years away in a cluster of galaxies in the constellation of Virgo. Being somewhat featureless and nonphotogenic, elliptical and spherical galaxies are not often used to illustrate popular articles on astronomy; they are nonetheless among the most important constituent "particles" of the universe. In the longest exposures with the 200-inch telescope the most distant objects that can be recorded—those whose light required about two billion years to reach the earth—are assumed to be elliptical and spherical galaxies. Spiral galaxies at the same distance would be too faint to register on photographic plates.

The bottom illustration on page 214

shows Hubble's famous "tuning fork" diagram, in which galaxies of different shapes are arranged in morphological sequence. At the left are the spherical EO's followed by the elliptical galaxies of increasing flatness, designated E1 to E7. The diagram branches at SO (a disklike form that lacks spiral arms) into two categories that Hubble called normal spirals and barred spirals. It was not clear 25 years ago where the irregular galaxies such as the Magellanic clouds belonged, but now the sequence of forms has been filled in. As Sc spirals become more and more open, the nucleus disappears, the arms become more and more irregular, the surface brightness becomes lower and finally one reaches systems in which no organized shape or symmetry remains.

Because of the smooth gradation of forms through the tuning-fork diagram, some astronomers felt that it represented an evolutionary sequence. Scientists have a tendency to simplify concepts—to seek "understanding"—by finding a few simple starting blocks and trying to perceive logical steps of transformation that unify a wide range of observations. But the problem of discerning real evolution in the galaxies is a difficult one. It is as though visitors from another planet were given only a few minutes to observe the human race. They might not immediately grasp the fact that many different races exist and that each experiences a separate but analogous aging process—from babies to children to adults to old people. In their attempt to comprehend and unify us, the visitors might guess initially that all people of a certain size were a certain age—dwarfs and children alike. Or they might order us according to color and conclude that the normal course of aging for the human race was from light skin to olive skin to brown skin to black skin. Given only a snapshot of the human population, they would need keen observation and reasoning in order to separate age characteristics from innate characteristics.

Similarly, in an instant of cosmic time, humans try by acute observation and logic to guess the way galaxies are

GREAT NEBULA IN ANDROMEDA, also called M 31, is about two million light-years away and the nearest of the large regular galaxies. It is an intermediate (Sb) spiral that contains more than 100 billion stars and has a diameter of about 125,000 light-years.

formed, how they evolve and the relations among them. It is particularly important to look for critical distinctions among galaxies in trying to decide which ones belong to an evolutionary sequence and which ones perhaps belong to completely different "races" or "species" with innate differences.

The smoothest gradation of forms is from early ellipticals to late-type spirals. If spiral arms were added to ellipticals, they would somewhat resemble spiral galaxies; if arms were removed from spirals, they would superficially resemble ellipticals. For many years, therefore, the question was debated whether the passage of time saw the spirals evolving into ellipticals or the other way around.

The question has been pretty much settled by the patient work of many observers. For example, Thornton Page of Wesleyan University, by measuring the relative velocities of paired galaxies and assuming that they are in gravitationally bound orbits around each other, has been able to calculate the approximate masses of the pairs. He has found that the spheroidals and ellipticals are about 30 times more massive than the average spiral galaxy. It is clear that a spiral or elliptical galaxy aging in an undisturbed state could never generate or get rid of enough mass to pass from one form to the other. Mass is a meaningful physical measure and a stable criterion for arranging galaxies because it is difficult to transform mass into energy, or energy into mass, on a galactic scale. Another physical quantity difficult to increase or reduce on a galactic scale is angular momentum. This quantity is a measure of the rotational energy of a system. In order to change a galaxy's angular momentum, force would have to be applied from the outside.

In the diagram on pages 216 and 217 I have arranged the principal kinds of galaxies according to their approximate mass and angular momentum per unit of mass. (In order to specify the angular momentum of a galaxy accurately it would be necessary to measure the mass and velocity at all distances from the center and integrate them over the whole galaxy. Since such observations are not available, one can take the cross-sectional flattening of the galaxy as a qualitative measure of the relative amount of energy in the rotation at the time when the visi-

GALAXY IN URSA MAJOR, M 81, is a magnificent spiral about nine million light-years away. M 81 and M 31 are near enough so that their distance can be measured quite accurately. They indicate about how our galaxy would look if seen from the outside.

ble stars in the galaxy were formed.) If no outside forces act, one sees that the differences in mass and angular momentum are so great that the galaxies cannot evolve along either co-ordinate of the diagram. Therefore it is extremely unlikely that ellipticals evolve into spirals or vice versa. By extension, if this analysis is correct, most of the distinctively different types of galaxies represent different species and are not one species seen in different epochs of aging.

The Relation of Rotation to Form

Let us now see how one might account for the different galaxy types, accepting the hypothesis that the universe, as we see it, began with a cosmic "explosion" about 10 billion years ago. Immediately after the explosion space was filled with a homogeneous expanding gas. As the gas cooled, local irregularities in density developed and these slowly contracted under the force of gravitation. These clouds of contracting gas were protogalaxies—galaxies in the process of formation. Simply by chance the clouds would have different masses, and most of them could be expected to have at least a small net rotation. As a cloud shrank it would have to spin faster to conserve angular momentum, just as a figure skater spins faster when he draws his arms closer to his body. My belief is that if the shrinking cloud exceeded a certain speed, not all the mass would be able to contract. The galaxy might break in two, or it might throw several fragments out into space. Alternatively it might eject masses of gas along its rapidly rotating equatorial edge. Those protogalaxies that were not rotating, or that were rotating only slowly, could contract without losing mass until the gas became compressed enough to form stars. These became the massive spherical and elliptical galaxies we see today. More rapidly rotating systems presumably would not be able to contract all their mass. Mass in excess of a certain amount would be

GALAXY M 74 is a late-type (Sc) spiral about 20 million light-years away. There is no way to tell in which direction a galaxy is rotating when seen in this orientation. Although the arms probably trail, this cannot be established with certainty in all cases.

SMALL CLOUD OF MAGELLAN, visible in the Southern Hemisphere, is a small irregular galaxy, rich in gas, about 200,000 light-years away. Its diameter is about 30,000 light-years.

DWARF GALAXY is several hundred thousand light-years away in constellation Leo. Approximately 2,000 light-years in diameter, it contains faint stars and little or no gas.

thrown off. The result would be exactly the relation diagramed on pages 216 and 217, which shows that the slowly rotating kind of galaxy can be very massive, but that the faster a galaxy rotates, the smaller its maximum mass can be.

Purely as a bonus, we get a natural explanation of the spiral galaxies. Since they are rotating rapidly, they should be losing considerable matter around their sharp peripheral edges. It is precisely such an outflow of matter along the spiral arms that is needed to explain the riddle of their permanence.

Of course, the scheme just suggested is an oversimplified and tentative explanation for the origin of different kinds of galaxies and their subsequent development. It has by no means gained general acceptance. The hypothesis accomplishes the objective, however, of unifying all the important observational facts into the simplest possible structure. More important, it gives a definite picture that can be tested and discarded if found wanting, or expanded and modified if it continues to be basically satisfactory.

Since there is more going on in the spiral galaxies than in the more massive spherical and elliptical ones, the spirals offer more opportunities for testing the hypothesis. If the spirals are shedding mass at their edges, there are two general possibilities. The ejected matter may leave the galaxy and not be replaced or there may be some replacement. In fact, the matter may be in continuous circulation, flowing out at the edges of the galaxy and back in at the poles. Material (either new or recirculated) could be guided in at the poles by a magnetic field that is also directed inward at the poles.

Regardless of whether the spiral galaxies are ejecting matter or recirculating it, there should be a good deal of observable material immediately outside fast rotating spirals. Indeed, it is in the vicinity of such galaxies that we observe smaller galaxies, irregular galaxies and satellite systems. Around our own galaxy there are the irregular Magellanic clouds and many other small and fragmentary systems. Around other nearby spirals, down to the limit of visibility, one can also observe irregular and satellite companions. These searches are observationally difficult since they involve faint dwarf galaxies around relatively distant systems, but it is fair to say that galactic "remnants" are not so conspicuous about the elliptical and spherical galaxies as they are about the nearby spirals.

Consider, finally, the observation that

galaxies frequently occur not just in pairs but in groups, or clusters. The top diagram on page 218 shows the rather surprising result that the clusters of galaxies that are the densest contain almost exclusively elliptical galaxies and the armless disks known as SO galaxies. Evidently low rotational motion in individual galaxies is accompanied by relatively little random motion of the sort that would tend to disperse a cluster of galaxies. Conversely, it would appear that the higher rotational motions in spiral galaxies are matched by a higher degree of random, or translational, motion. This would explain the observation that the less dense the cluster, the higher its content of spiral galaxies. When the mean estimated density of the universe is reached, "clusters" are found to consist entirely of single galaxies, which are almost 100 per cent spirals. It was Hubble who originally observed that in general the spiral galaxies appear to be field objects rather than cluster objects.

The solution of the problem of the formation and evolution of galaxies may eventually help us to understand why some galaxies are strong emitters of radio waves [see "Radio Galaxies," by D. S. Heeschen; SCIENTIFIC AMERICAN Offprint 278]. For example, the Soviet astronomer I. S. Shklovsky has suggested that certain elliptical galaxies are strong radio sources because matter is still falling into them. This is not unlike the view presented here that matter may be falling into spiral galaxies. If matter is still falling into the galaxies, it is important to discover how it is coming in. For example, one would like to know if it can add angular momentum to a system or if it can add mass without angular momentum.

If the latter were possible, one can conceive that more and more mass could be added to the nucleus of a spiral galaxy until it eventually turned into an elliptical or spheroidal galaxy. Such a phenomenon would require a modification of the picture I have presented. But if it became necessary to consider such massive infall of material, one might argue that one should not even speak of galactic evolution, because the galaxies are in a sense still in the process of formation. In any event, it now seems likely that magnetic fields may strongly control the structure of galaxies, the flow of matter in the vicinity of galaxies and perhaps even the relative motions of the galaxies themselves. In a subject as young as the study of magnetic fields in galaxies, we can be sure only that there will be many surprises.

GIANT SPHEROIDAL GALAXY, M 87, is a member of the Virgo cluster of galaxies, about 45 million light-years away. M 87 contains about 30 times as many stars as our own galaxy.

ELLIPTICAL GALAXY, also in the Virgo cluster, is designated E5, which stands for "elliptical, Type 5" in Hubble's galactic classification scheme. It is flattened by rotation.

22

Exploding Galaxies

ALLAN R. SANDAGE

November 1964

Of the major unsolved problems in astrophysics, none has received more attention over the past 50 years than the origin of cosmic radiation. Since the discovery of cosmic rays in 1911 by Victor F. Hess of Austria a number of theories have been put forward to explain where and how these ultrahigh-energy charged particles (mostly protons, with a small admixture of heavier atomic nuclei) could be accelerated to their enormous velocities. The discovery of supernovae in the decade after 1925 began a train of speculation that attributed cosmic rays to exploding stars in our galaxy. An alternate theory, proposed by the late Enrico Fermi, held that low-energy particles emitted by stars like the sun are gradually accelerated to cosmic ray velocities by repeated encounters with local magnetic fields within the galaxy.

During the past 10 years many astrophysicists have favored still another explanation, which places the source of at least part of the cosmic radiation in events occurring outside our galaxy. There has been strong circumstantial evidence that titanic explosions are taking place in the central regions of certain galaxies—perhaps including our own! The energy released in these explosions could account for the highest observed cosmic ray energies and for at least part of the low-energy flux. This hypothesis has recently received strong support from observations of a neighboring galaxy that indicate that the galaxy was the scene of such an explosion some 1.5 million years ago.

The evidence for the view that cosmic rays originate in galactic explosions has come largely from radio astronomy. In 1946 the first discrete source of radio waves outside our solar system was discovered. It appeared on radio-contour maps as an intense spot in the constellation of Cygnus and was accordingly designated Cygnus A. By 1948 radio astronomers in Australia and Britain had detected additional discrete sources in the constellations of Taurus, Cassiopeia, Centaurus and Hercules. At last count more than 3,000 discrete radio sources had been mapped, and the number will probably pass 100,000 when comprehensive surveys now in progress are completed.

In 1951 Rudolph Minkowski and Walter Baade of the Mount Wilson and Palomar Observatories made the first identification of a discrete radio source with an optically visible object. Their photographs, made with the 200-inch telescope on Palomar Mountain, showed that Cygnus A coincided with the position of a galaxy now estimated to be some 700 million light-years away. Shortly thereafter the radio sources Virgo A and Centaurus A were found to coincide with the giant galaxies designated 4486 and 5128 in the New General Catalogue (NGC). More than 100 of the discrete radio sources have since been identified with visible galaxies, and it is likely that most of the other sources are associated with radio galaxies either too distant to be seen or very near the limit of the 200-inch telescope [see "Radio Galaxies," by D. S. Heeschen; SCIENTIFIC AMERICAN Offprint 278].

The mechanism by which electromagnetic radiation in the radio region of the spectrum is generated by radio galaxies has been the subject of much conjecture. The most plausible explanation was originally suggested in 1950 by Hannes Alfvén and N. Herlofson of Sweden and subsequently developed by the Soviet astrophysicist I. S. Shklovsky. According to this model, radio waves are generated by the interaction of relativistic electrons (that is, electrons moving with velocities close to the speed of light) and a magnetic field. As an electron gyrates around a line of force in a magnetic field, it accelerates and consequently emits energy in the form of electromagnetic radiation. This radiation, which can be generated by any charged particle, is sometimes called synchrotron radiation, since it is identical with the radiation produced in the man-made particle accelerators known as synchrotrons.

The wavelength of any synchrotron radiation depends on the energy of the gyrating particles and on the strength of the magnetic field. Radio waves are produced when the electron energies lie between one billion and 25 billion electron volts (bev) and the magnetic field has a strength of about a millionth of a gauss. These specifications are amply met by radio galaxies. In fact, high as these electron energies are, they may represent only the low-energy "tail" of the actual energy-distribution curve of radio galaxies. For example, M 87 is an intense radio galaxy in which the electron energies appear to be at least 10,000 bev, far higher than the energy of the particles produced in any manmade accelerator.

What does all this have to do with cosmic rays? It seems likely that at least some of the charged particles involved in the production of synchrotron radio waves must eventually escape the magnetic fields of the radio galaxies and fly off into intergalactic space. These particles, together with particles generated by similar explosions in our own galaxy, could account for the flux of primary cosmic radiation that impinges on the earth. (The secondary radiation at the surface of the earth is of course the result of the disruption of atoms in the

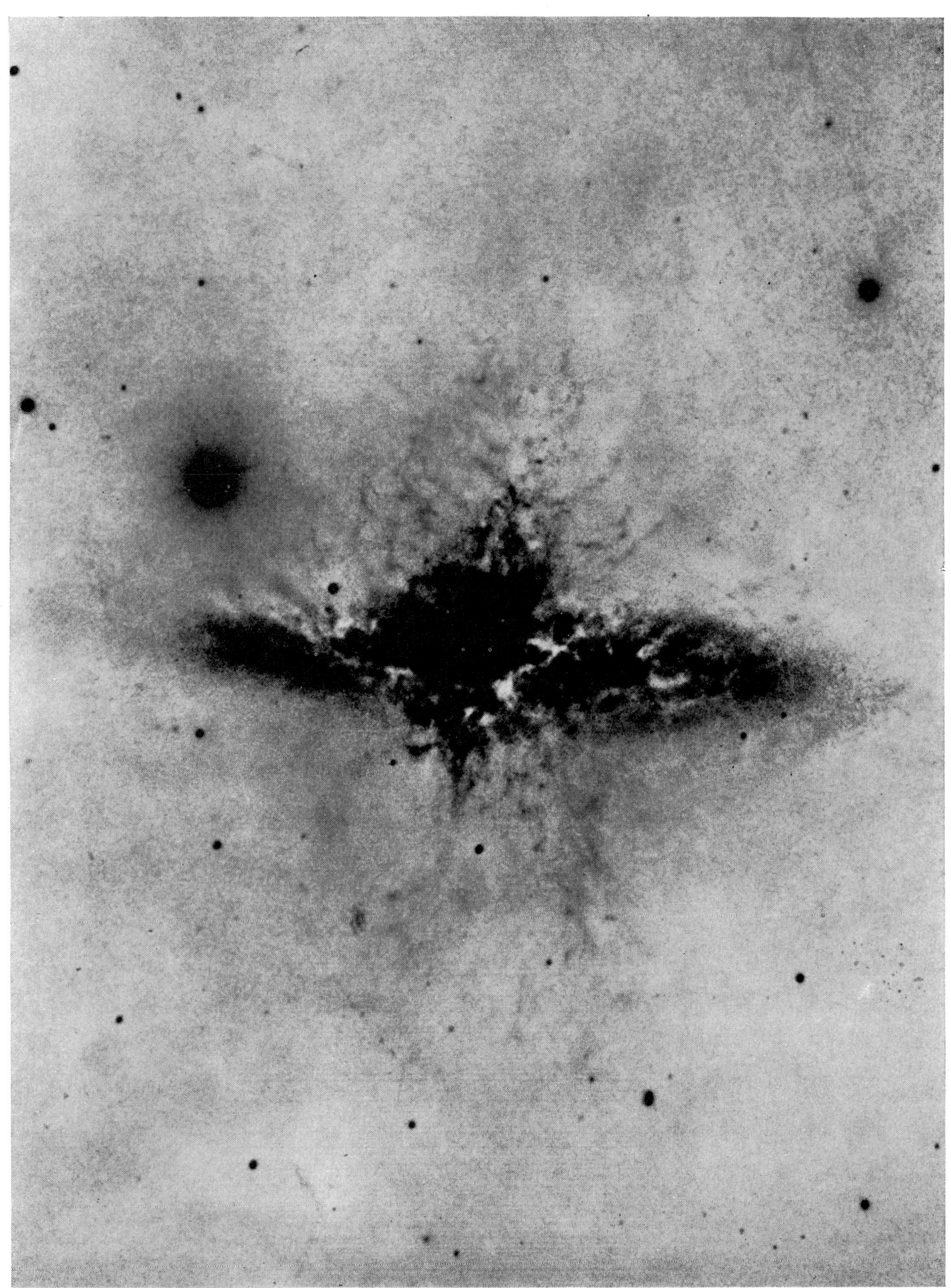

PHOTOGRAPH OF M 82, a nearby radio-emitting galaxy, was made in red light by the author with the 200-inch telescope on Palomar Mountain. This photograph revealed for the first time the spectacular array of hydrogen filaments that extend more than 14,000 light-years above and below the galactic disk of M 82. An explosion at the center of the galaxy presumably ejected the filaments some 1.5 million years ago. Most of the astronomical photographs here are printed as negatives in order to accentuate fine details.

earth's atmosphere by the primary radiation.) Considering the density of radio galaxies in space, their estimated lifetimes and their total output of energy (and also taking into account the production of particles in our own galaxy), this explanation comes close to predicting both the intensity of the observed cosmic radiation and the velocity of its most energetic particles. If we attribute both the discrete radio sources and the cosmic ray flux to synchrotron radiation, we can assume that the immediate source of energy for the synchrotron radiation must be galactic explosions of prodigious magnitude.

As the exploding-galaxy hypothesis gained gradual acceptance during the 1950's it provided a working model for interpreting new data. For example, in 1953 R. C. Jennison and M. K. Das Gupta of the Jodrell Bank radio observatory of the University of Manchester discovered that Cygnus A was *not* a single discrete radio source but rather could be resolved into two separate radio-emitting regions located on opposite sides of the visible galaxy at distances of about 100,000 light-years. Subsequent studies by radio astronomers at Jodrell Bank and at the California Institute of Technology showed that this phenomenon, sometimes called radio doubling, is probably the rule rather than the exception among radio galaxies. Presumably the twin radio regions represent two jets of high-energy particles ejected by the parent galaxy in an explosion millions of years ago. Conserving angular momentum, the jets would move in opposite directions from the center of the galaxy. They would carry part of the galaxy's magnetic field with them, or perhaps encounter another field in intergalactic space. Synchrotron radiation would be produced all along the jets, but it would be most concentrated at the ends, where the lines of force in the magnetic field would be most tightly compressed; in effect, two discrete spots of intense radio emission would be observed.

In spite of the persuasive evidence of the radio observations, the exploding-galaxy hypothesis still lacked optical support. Then, in 1961, a discovery was made that was to lead eventually to the necessary confirmation. Early in that year C. R. Lynds, working at the National Radio Astronomy Observatory in Green Bank, W.Va., was surveying a group of visible galaxies centered on

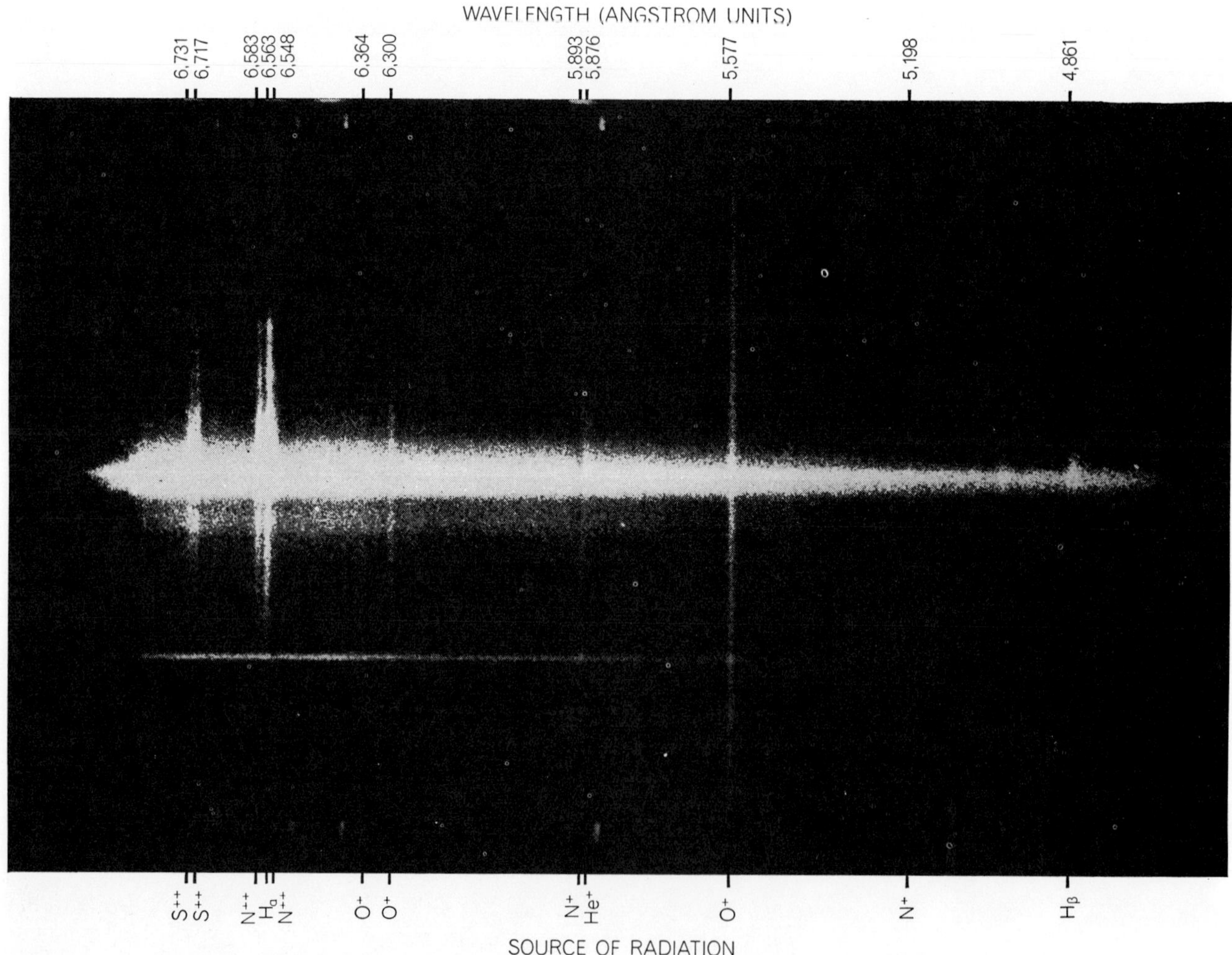

SPECTROGRAM OF M 82 was obtained by C. R. Lynds at the Lick Observatory by placing the slit of a spectroscope attached to the 120-inch telescope along the axis of the filamentary structure, perpendicular to the plane of the galaxy. The broad streak across the center is the continuous spectrum of the galactic nucleus. The vertical lines are characteristic emission lines produced by the recombination of ionized atoms in the filaments; the wavelengths of several of these lines are given at top. The letters at bottom designate the elements responsible for the emission lines; the superscript plus signs after some of the letters signify whether that element is singly or doubly ionized. Two characteristic emission lines of hydrogen are indicated by the subscript Greek letters. All the emission lines are inclined slightly with respect to the laboratory comparison lines (*short lines along top and bottom of plate*). This indicated that the filaments on one side of galactic disk are approaching the earth, whereas those on the other side are receding.

the giant spiral galaxy M 81 in an attempt to locate a weak radio source designated 3*C* 231 in the third Cambridge catalogue of radio sources. This source had previously been identified with M 81 itself, but Lynds's more accurate measurements showed that it actually coincided with the peculiar galaxy M 82, a smaller neighbor of M 81. Older photographs of M 82, made as long ago as 1910 with the 60-inch telescope on Mount Wilson, showed that this galaxy could not be resolved into individual stars, although at its distance normal stars should have been visible within the galaxy. The old plates showed extensive dust lanes across the spindle-shaped image of the galaxy, with a faint filamentary structure extending above and below the galactic disk. In 1949, soon after the 200-inch telescope went into operation on Palomar Mountain, M 82 was rephotographed; the newer photographs showed the filamentary structures more clearly, but optical observations were not pushed further until Lynds's discovery in 1961 that the galaxy was a discrete radio source.

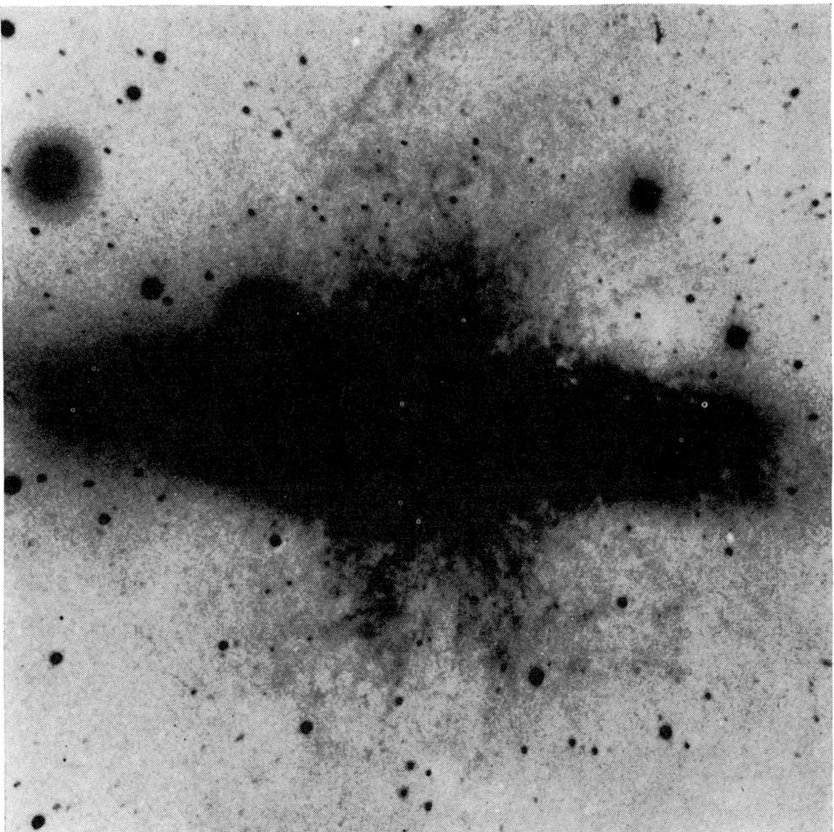

COMPOSITE PRINT OF M 82 was made by William C. Miller of the Mount Wilson and Palomar Observatories by superposing three photographs made in blue light by the author with the 200-inch telescope. The continuous radiation emitted by outer fringe of blue filaments provides strong evidence that the hydrogen gas in the filaments was originally ionized by synchrotron radiation, which in turn was generated by relativistic electrons (that is, electrons moving with speeds close to that of light) produced in explosion of galactic nucleus.

Stimulated by Lynds's findings, I decided to make a new series of photographs of M 82 in March, 1962, using the 200-inch telescope. A special interference filter that admitted only red light with a wavelength of 6,563 angstrom units was employed in order to detect any prominent hydrogen structures that would otherwise go unnoticed. (The spectral line at 6,563 angstroms, known to spectroscopists as the hydrogen-alpha line, is characteristic of radiation produced by the recombination of ionized hydrogen atoms.) To my surprise the new plates showed M 82 in an entirely new aspect. What had appeared on the old plates as inconspicuous filamentary wisps now appeared as vast and intricate hydrogen structures, extending some 14,000 light-years above and below the plane of the galaxy [*see illustration on page 225*].

Meanwhile Lynds had undertaken an independent spectral analysis of M 82, using the 120-inch telescope at the Lick Observatory. He found that when the slit of the spectroscope was placed along the axis of the filamentary structure, perpendicular to the plane of the galaxy, characteristic emission lines of hydrogen, sulfur and nitrogen appeared in the spectra in great strength. Close examination of the plates revealed that the emission lines were inclined slightly with respect to the laboratory comparison lines that had been superposed

POSITIVE PRINT of spiral galaxy NGC 4216 provides an unambiguous example of the rule that for galaxies seen almost edge on, dark dust lanes identify the edge closer to earth.

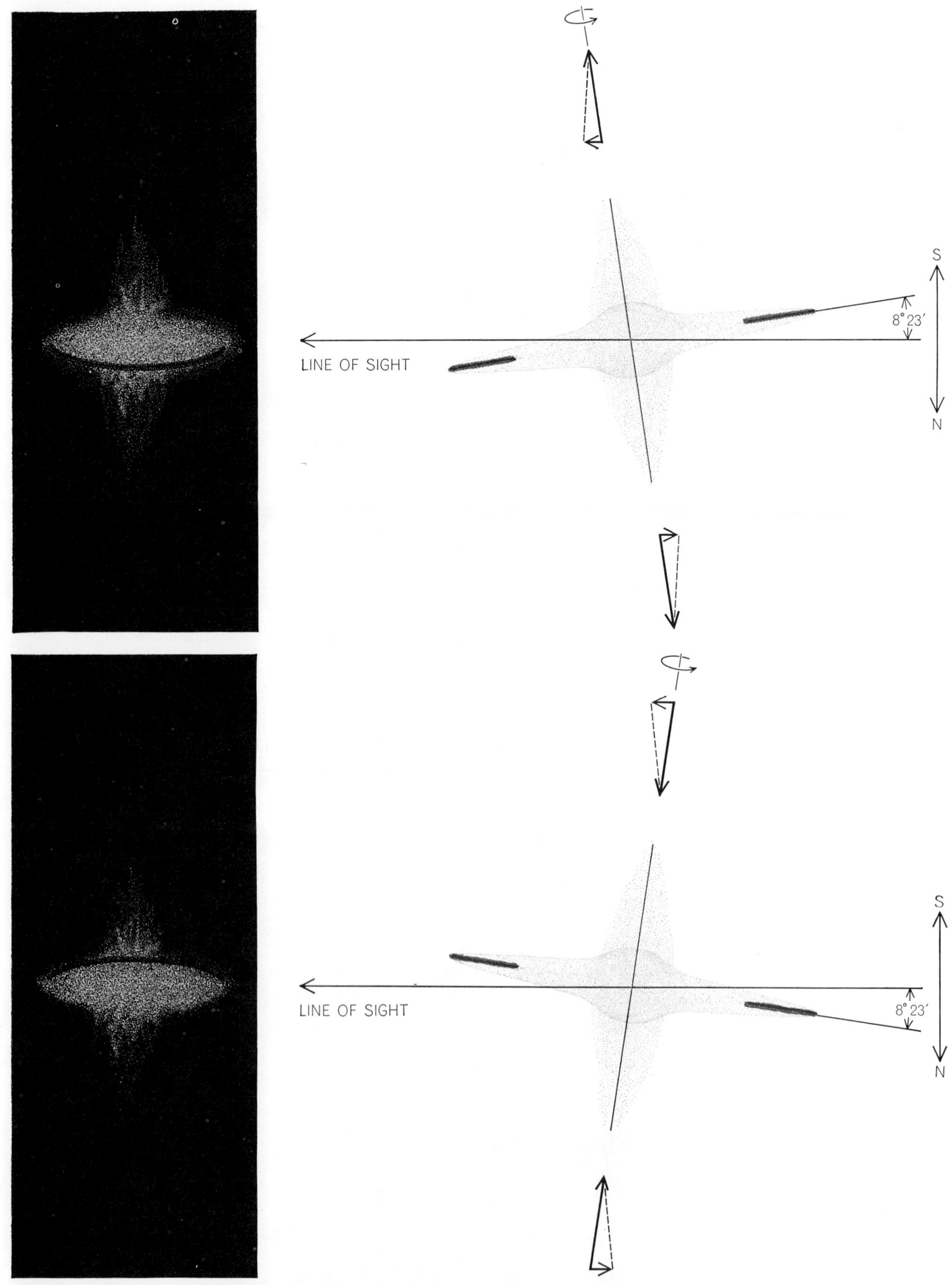

ORIENTATION OF M 82 along the line of sight had to be ascertained in order to decide whether the galaxy was expanding or collapsing. The two possible interpretations of the spectral data are shown here. In each case the picture at left shows the galaxy as it would be seen from the earth; the pictures at right are hypothetical side views of the two possible orientations. If the north edge were closer to the earth (*top*), the spectral data would indicate that the galaxy was expanding. If the south edge were closer (*bottom*), the galaxy would be collapsing. The dark dust lanes that actually appear along the north edge of the galaxy indicate that that side is closer and hence that the galaxy must be expanding. Tilt of galaxy along the line of sight was calculated independently.

on the plates [*see illustration on page 226*]. This tilting of the spectral lines could only mean that the filamentary structure on one side of the galactic disk was approaching the earth, whereas the structure on the other side was receding. Mass motion on so vast a scale had never before been observed perpendicular to a galactic disk.

At first the tilting of the spectral lines could be interpreted as evidence for either an explosion or an implosion. The data merely showed that the filaments on the south side of the galactic disk were approaching the earth and that the filaments on the north side were receding. If the north edge of the galactic disk were nearer the earth, the spectral tilting would indicate that the galaxy was in the process of expanding. If the south edge were nearer, the galaxy would be collapsing.

To decide whether M 82 was exploding or imploding, we adopted a simple criterion established in the 1920's by V. M. Slipher of the Lowell Observatory in Flagstaff, Ariz. Slipher had pointed out that most of the dust in galaxies is confined to a thin sheet coincident with the central plane of the galaxy. Hence for galaxies seen almost edge on, the near edge of the galaxy will be distinguished by dark dust lanes silhouetted against the bright nuclear bulge, whereas the dust lanes on the far edge will be much less conspicuous, since there is no background light for these lanes to obscure. The photograph of the spiral galaxy NGC 4216 at the bottom of page 41 shows the phenomenon to good advantage; there is little doubt as to which edge is closer to the viewer. When we applied this criterion to the hydrogen-alpha photograph of M 82, we found that the north side was almost certainly closer to our galaxy. It follows that the material in the filaments must be moving outward from the center of the galaxy along the axis of rotation.

Lynds's spectral measurements provided even more conclusive evidence for regarding M 82 as an exploding galaxy. He found that the greater the distance from the center of the galaxy along the axis of rotation, the more the spectral lines appeared to be inclined. He concluded that the velocity of expansion of the filaments on each side of the galactic disk must increase linearly with their distance from the center [*see illustration at right*]. Having calculated the tilt of the galaxy along the line of sight to be about eight degrees 23 minutes, Lynds and I found that the velocity of the matter at the ends of the filaments

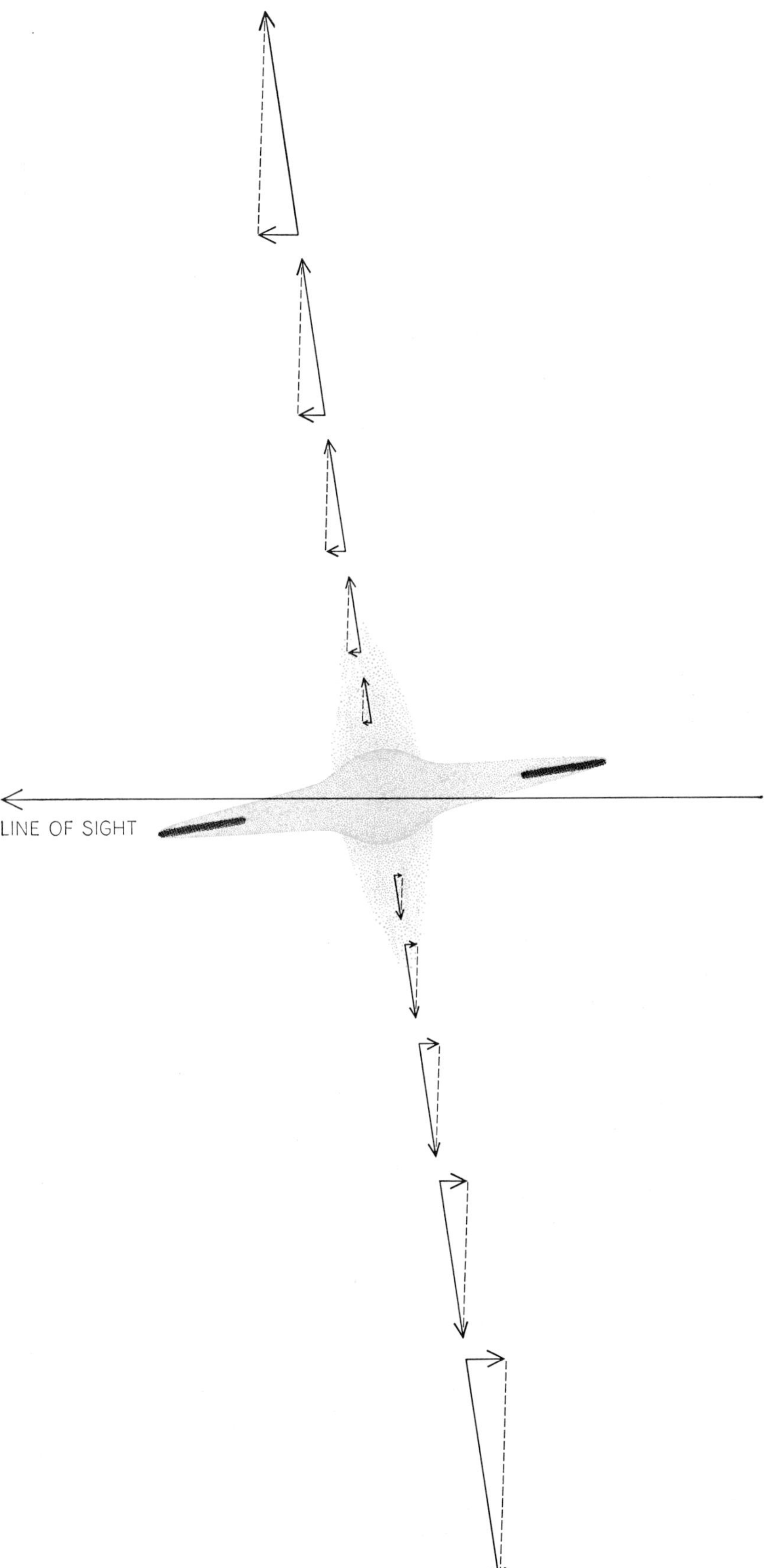

ADDITIONAL EVIDENCE for regarding M 82 as an exploding galaxy was derived from Lynds's spectral measurements. He found that the greater the distance from the center of the galaxy along the axis of rotation, the more the spectral lines appeared to be inclined. This indicated that the velocity of expansion of the hydrogen filaments on each side of the galactic disk must increase linearly with distance from center. Short horizontal arrows represent motion along line of sight; longer arrows represent actual velocity of expansion.

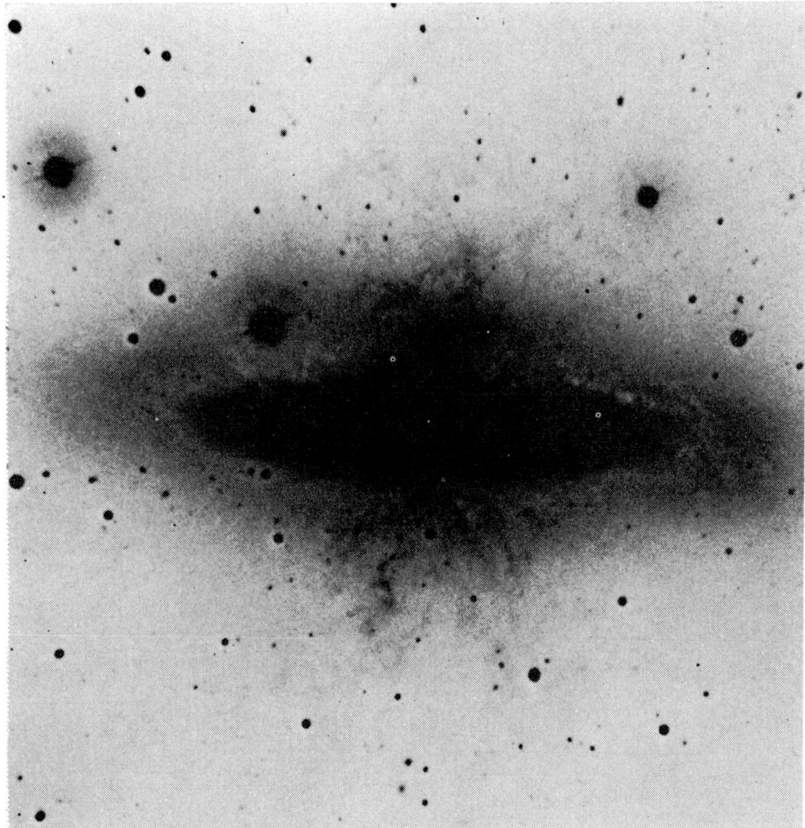

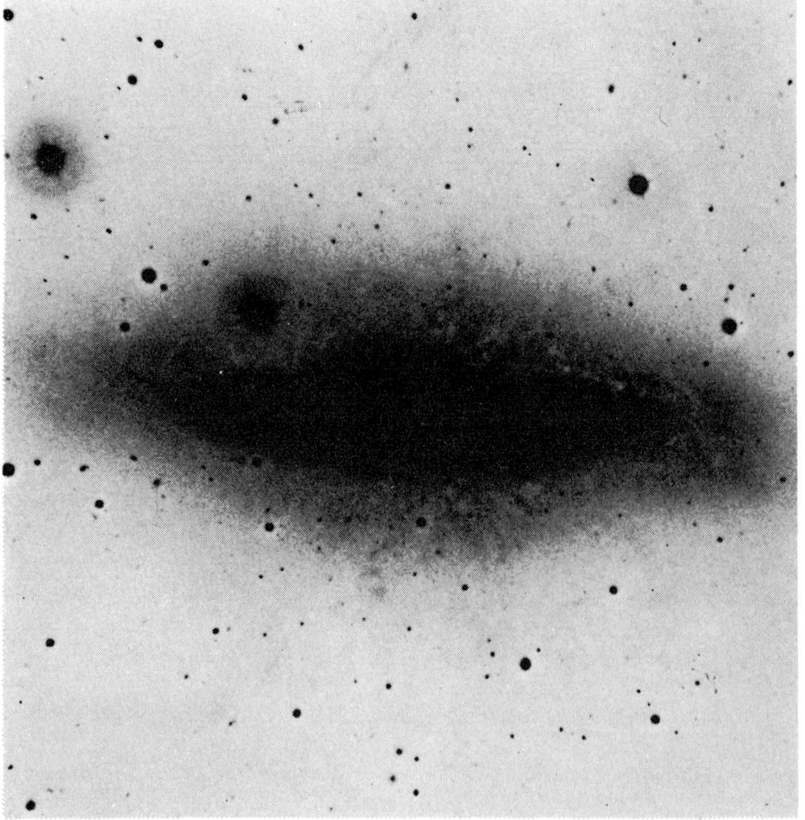

POLARIZED PHOTOGRAPHS OF M 82 support the hypothesis that the gas in the filaments was originally ionized by the synchrotron process. The polarizing filters were set parallel to the plane of the galaxy in the top photograph and perpendicular to the plane of the galaxy in the bottom photograph. The photographs show that synchrotron light from the filaments is highly polarized with the electric vector of the light parallel to the plane of the galaxy. They also show that M 82 possesses a regular, large-scale magnetic field.

must be roughly 600 miles per second.

The direct relation of velocity to distance in the filaments can be taken to mean that the time required for each part of a filament to travel from a common origin to its present position must be a constant; in other words, all the matter in the filaments must have been back in the nucleus of M 82 at a given time in the past. This is strong evidence for regarding the filaments as the residue of a single vast explosion in the nucleus of the galaxy. The date of the explosion can be estimated by measuring the slope of the curve that relates velocity to distance. This method indicates that the explosion occurred some 1.5 million years before the stage we see now. (Since M 82 is roughly 10 million light-years away, the date as measured from the present on earth would be about 11.5 million years ago.) By astronomical standards 1.5 million years is an extremely short interval of time for so vast a change. Even taking into account the deceleration of the filaments due to encounters with the surrounding gas, the date of the explosion can probably be set at not more than two or three million years ago.

The amount of matter moving outward from the center of M 82 can be estimated from the strength of the hydrogen-alpha emission line in the spectrum of the filaments. By this method we were able to calculate that to produce the observed emission roughly 6×10^{63} low-energy protons and electrons must be present in the filaments. This is equivalent to roughly five million times the mass of the sun, or about a two-thousandth of the total mass of M 82. The energy needed to set this huge mass of matter in motion is about 2×10^{55} ergs.

At first it was not at all clear how the hydrogen gas in the filaments became ionized. Ordinarily interstellar clouds of hydrogen are ionized only when they are in the immediate vicinity of extremely hot blue stars. Ultraviolet photons from the stars are sufficiently energetic to overcome the binding energy of the electron-proton pair in the neutral hydrogen atom. Once the gas is ionized in this manner the electrons are free to recombine with the proton nuclei, causing the emission of visible light as the electrons cascade down the various atomic energy levels [*see bottom illustration on opposite page*]. This process could not possibly be taking place in M 82: the galaxy is conspicuously free of hot blue stars, particularly in the filaments. How then could the hydrogen gas in the filaments become ionized?

Again the exploding-galaxy hypothesis provided a tentative answer. Suppose that large numbers of relativistic electrons produced in the original galactic explosion were able to generate by the synchrotron mechanism not only the observed radio flux but also enough ultraviolet radiation to ionize the hydrogen gas in the filaments. Although it seemed incredible in 1962 that enough electrons could be produced at the required energy (10,000 bev), no other energy source appeared remotely possible.

The first evidence of the correctness of this line of reasoning was obtained early last year by Hugh M. Johnson of the Lick Observatory. Working with the 36-inch Crossley reflector, Johnson exposed four blue-sensitive plates to M 82 and printed all four plates on the same piece of photographic paper. This increased the contrast sufficiently to reveal an additional fringe of faint outer filaments on both the north and the south side of M 82. Shortly thereafter I made a new series of photographs with the 200-inch telescope. These were combined in a similar manner by William C. Miller of the Mount Wilson and Palomar Observatories, who used a new technique he had developed for increasing the contrast of the faint details without losing the features of the brighter regions. The resulting composite photographs, one of which is reproduced at the top of page 227, showed for the first time a delicate outer fringe of blue filaments that emit continuous rather than line radiation.

It was now crucial to establish if this outermost fringe of blue light was indeed produced by the synchrotron process. Fortunately it is a well-known fact that synchrotron radiation produces light that is highly polarized, with the electric vector of the light perpendicular to the lines of force in a magnetic field [*see top illustration at right*]. If the magnetic field of M 82 is regular over an appreciable portion of the filaments, photographs made through polarizing filters should not only test the existence of synchrotron radiation but also reveal the orientation of the galaxy's magnetic field. Polarized photographs were made with the 200-inch telescope in February of this year. They show beyond doubt that light from the filaments is highly polarized, with the electric vector of the synchrotron radiation parallel to the plane of the galaxy [*see illustrations on opposite page*]. Evidently M 82 possesses a regular, large-scale magnetic field aligned predomi-

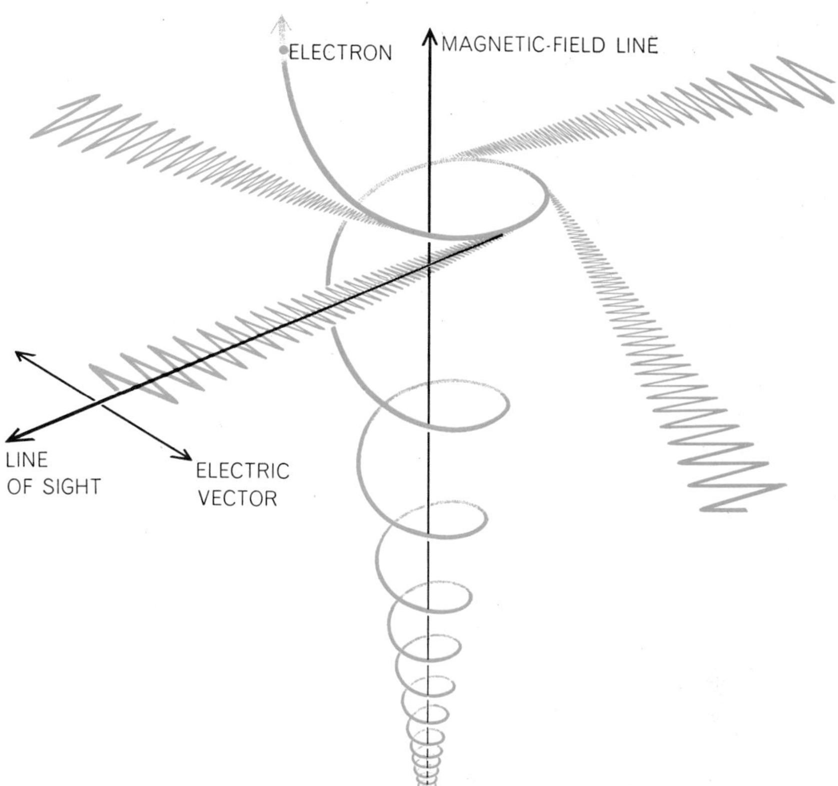

SYNCHROTRON RADIATION is produced by the rapid gyration of high-energy charged particles in a magnetic field. Light generated by this process is highly polarized, with the electric vector perpendicular to both the magnetic-field line and the direction of the particle. Wavelength of the radiation depends on velocity of particle and strength of magnetic field.

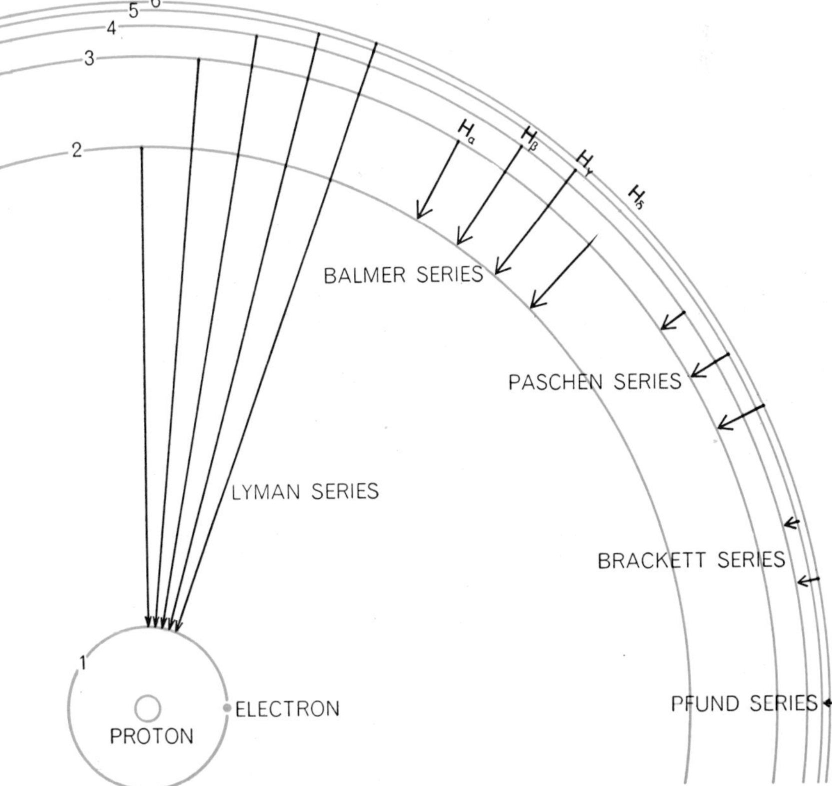

IONIZED HYDROGEN ATOM emits radiation when a free electron recombines with a proton to form a neutral hydrogen atom. The wavelength of the radiation depends on how far the electron travels in each "jump" from a higher to a lower energy level; the hydrogen-alpha (H_α) emission line is produced when electron jumps from the third to second level.

nantly along the axis of rotation on both sides of the galactic disk.

Preliminary measurements indicate that the synchrotron radiation process is sufficiently energetic to account for the ionization of the hydrogen gas in the filaments and hence for the characteristic hydrogen-alpha emission lines. Further confirmation will have to wait until measurements of the intensity of the ultraviolet radiation emitted by the filaments can be made from orbiting astronomical observatories.

Astrophysicists interested in cosmic ray research can only envy the location of their hypothetical colleagues on planets orbiting the stars in M 82. The local cosmic ray flux of electrons alone in that galaxy is about 1,000 times greater than it is around the earth. Cosmic ray protons must also be present in enormous quantities in M 82, although their presence cannot be directly observed because protons are very sluggish emitters of synchrotron radiation.

Although M 82 is undoubtedly a special case, it can in some respects be regarded as a typical radio galaxy. For example, although it does not exhibit two separate regions of radio emission, these may perhaps develop later in its history. What is more to the point, its filaments, which constitute the primary evidence for an explosion in its nucleus, have counterparts in many other radio galaxies. Jets of high-energy gas have been observed in such intense radio galaxies as M 87 and in the "quasi-stellar" radio sources 3*C* 48 and 3*C* 273. A particularly interesting example of this phenomenon can be seen in the spiral galaxy NGC 4651, which has been identified with the radio source 3*C* 275.1 [*see illustration on opposite page*]. In this galaxy two jets extend from opposite sides of a spiral arm to distances of about 50,000 light-years, where they terminate at the edge of a visible halo.

Astronomers have just begun to study exploding galaxies. Yet even at this early stage of inquiry we find that radical new ideas are needed to account for the enormous energies involved in these events. The synchrotron model enables us to calculate the total energy required to produce a galactic explosion; in the case of M 82 this input of energy amounts to about 10^{57} or 10^{58} ergs. The giant radio sources, such as Cygnus A, Hercules A and Hydra A, undoubtedly call for a considerably greater input of energy, perhaps as much as 10^{62} ergs. Herein lies the dilemma. Thermonuclear reactions, which convert hydrogen to helium, are comparatively inefficient, producing only 6×10^{18} ergs for every gram converted. Even if the conversion of mass to radio energy were 100 percent efficient, 10 billion solar masses of hydrogen would be required to produce 10^{62} ergs of radio energy. This figure, however, corresponds to the entire mass of a medium-sized galaxy! (Since con-

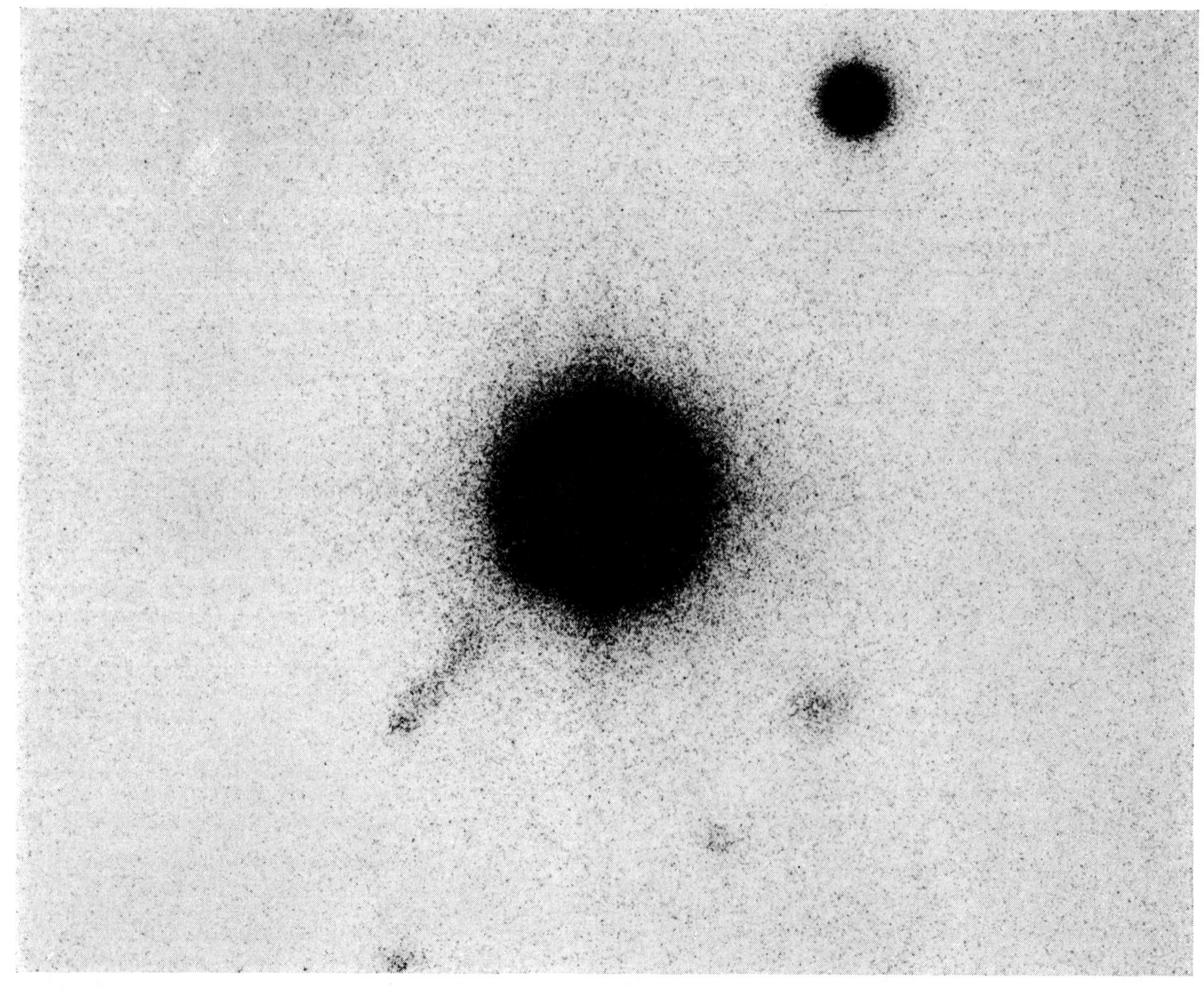

JET protrudes some 150,000 light-years beyond the outer edge of the quasi-stellar radio source designated 3*C* 273 in this photograph made with the 200-inch telescope. The jet resembles the filaments in M 82 and may be the product of a similar galactic explosion.

version efficiencies of mass to radio energy are probably not higher than 1 percent, such a thermonuclear reaction would actually require closer to a trillion solar masses.)

It is obvious that conventional energy sources are not adequate to explain the phenomenon we are observing, and some totally new energy principle may have to be devised. One possible new mechanism has recently been suggested by Fred Hoyle of the University of Cambridge and William A. Fowler of the California Institute of Technology. According to their model, when any scattered mass condenses, gravitational potential energy is released. Under certain conditions this can be a very efficient process, since the potential energy varies directly with the square of the mass and inversely with the final radius. Carried to its limit, this model predicts that if collapse is not countered by rotation, a mass of gas can condense until it ultimately disappears from view!

The disappearance of a massive object into its own gravitational field is a concept inherent in the general theory of relativity, proposed by Albert Einstein in 1916; the details of the concept were worked out a few years later by the German astronomer Karl Schwarzschild. Simply stated, the local curvature of space, which is dependent on the mass of the matter in the immediate vicinity, can ultimately close around itself and isolate its contents from the rest of the universe, provided that the density of the matter is high enough. The ultimate radius at which this envelopment occurs is given by the expression $2GM/c^2$, where G is the gravitational constant, M is the mass and c is the speed of light. This radius is called the Schwarzschild singularity, and when it is reached matter disappears entirely from view. What is important for our purpose is that when matter collapses to this radius, the energy released is equal to $\frac{1}{2}Mc^2$. Without the multiplying factor of ½, this is equal to the annihilation energy of matter, as expressed by the familiar formula $E = Mc^2$. This process is estimated to be about 100 times more efficient than any thermonuclear reaction. An energy of 10^{62} ergs would be produced if only 100 million solar masses were to collapse to the Schwarzschild radius.

It is only fair to say that no one knows if this kind of gravitational collapse is actually possible, or if it is, what mechanism could account for the exchange of energy from the gravitational field to the relativistic particles needed to produce the observed radio emission. In any event the discovery of exploding galaxies has presented both astronomers and physicists with problems of fundamental importance and complexity. Observations over the next few years may illuminate at least part of the mystery presented by these spectacular events, which are by far the most energetic ever perceived by man.

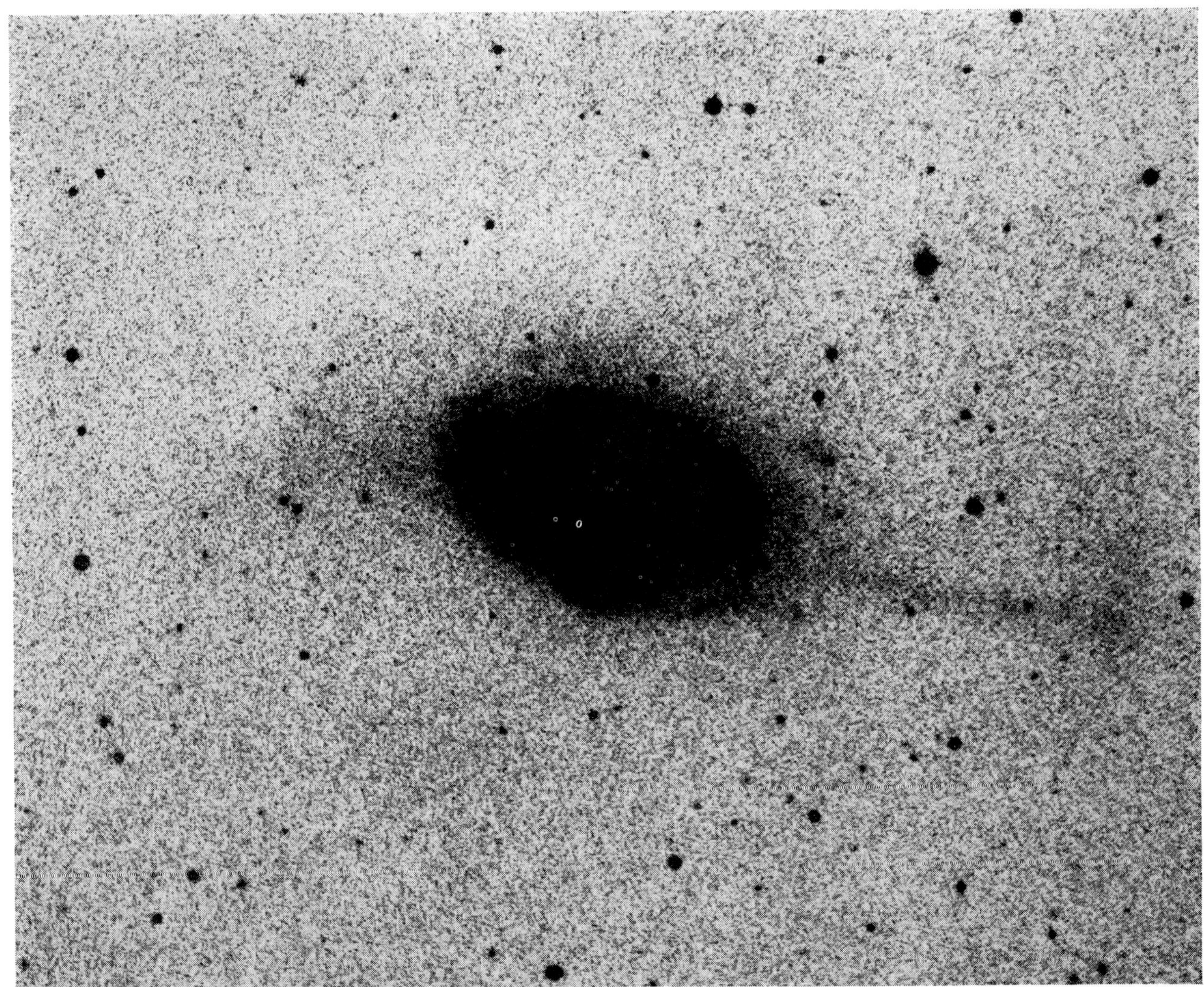

TWIN JETS extending outward about 50,000 light-years from opposite sides of a spiral arm in the intense radio galaxy NGC 4651 also resemble the hydrogen filaments ejected in the explosion of M 82. The jets terminate at the edge of a barely visible halo.

23

Seyfert Galaxies

RAY J. WEYMANN

January 1969

During the second quarter of this century astronomers gradually came to realize that many of the faint nebulas that populate the sky are really great islands of stars far outside our own galaxy. On the basis of their appearance Edwin P. Hubble of the Mount Wilson and Palomar Observatories divided these objects into two broad classes: spiral galaxies, whose general shape resembles a flat pinwheel, and elliptical galaxies, smooth, featureless structures that range in shape from elliptical to spherical. A typical large galaxy, such as our own spiral galaxy, contains about 100 billion stars. In the universe visible to the largest telescopes there are billions of galaxies.

In 1943 Carl K. Seyfert, then working as a postdoctoral student at the Mount Wilson Observatory, described a small class of spiral galaxies that seemed notably different from the hundreds of other spirals they superficially resembled. The most distinctive feature of the Seyfert galaxies (as they are now called) is that they have very small, intensely bright nuclei whose broad emission lines, as recorded in spectrograms, indicate that the atoms present are in a high state of excitation. The spectra of normal galaxies show few, if any, strong emission lines, and those they do show are rather narrow and typical of the emission lines found in the nebulas within our own galaxy.

In addition to their puzzling spectra, the Seyfert galaxies now confront astronomers with a number of other peculiar properties. The light emitted by at least two of the Seyfert galaxies has varied strongly over a period measured in months. Two of them are now known to be powerful emitters of radio energy, and this emission has been changing violently in intensity. Several of the Seyfert galaxies emit an enormous amount of energy in the infrared region of the spectrum by a mechanism as yet unknown. All Seyfert galaxies emit more ultraviolet radiation than can be explained in terms of starlight, and in some instances this radiation is polarized.

Many readers will recognize that these are some of the characteristics that make the quasi-stellar radio sources (quasars) so fascinating. These starlike objects are among the most powerful radio sources in the sky [see "The Problem of the Quasi-stellar Objects, by Geoffrey Burbidge and Fred Hoyle; begins on page 316 in this volume]. There are, however, two striking differences between the nuclei of Seyfert galaxies and quasars. First, the emission lines in the spectra of quasars are enormously shifted from their normal wavelengths to the red end of the spectrum, whereas the emission lines in the Seyfert nuclei have only the very modest red shifts observed in normal, nearby galaxies. Second, the quasars appear as mere points of light with no indication of a surrounding galaxy, whereas the Seyfert nuclei are clearly embedded in the center of apparently normal spiral galaxies.

Thus, in addition to the challenging questions posed by the Seyfert phenome-

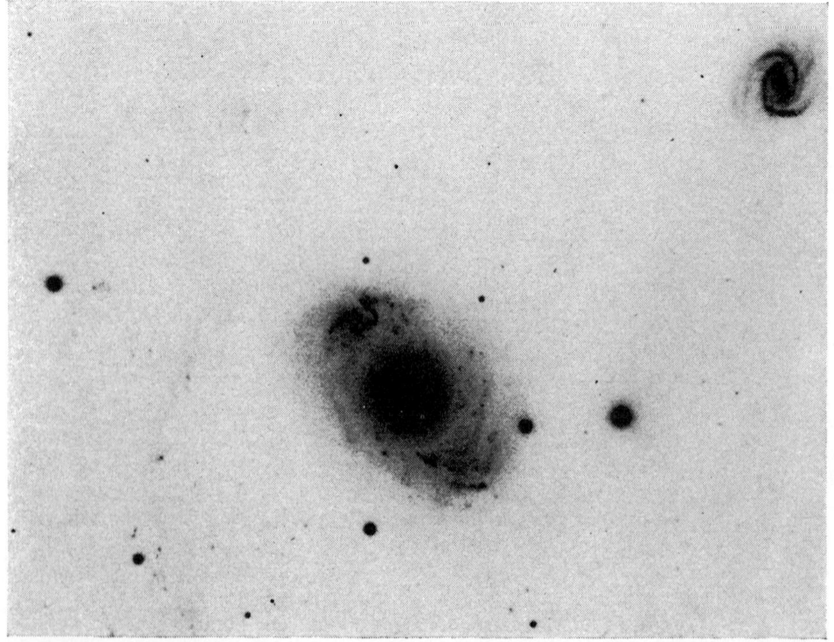

SEYFERT GALAXY NGC 4151, shown here in a negative print, appears to be a fairly typical spiral galaxy. When viewed directly through a telescope, however, it presents a starlike appearance because most of its light emission originates in a dense, central nucleus (*see illustration on opposite page*). Such "Seyfert" nuclei have now been identified in about a dozen galaxies. The picture was taken with the 200-inch telescope on Palomar Mountain.

non itself, there are the larger questions of the relation between the Seyfert nuclei and the quasars and the nature of a whole range of violent activity found in the nuclei of galaxies. At the present time no one knows whether the quasars are "local" (nearby objects having distances comparable to the Seyfert galaxies) or "cosmological" (that is, having the distances implied by assuming the validity of the relation between red shift and distance found for ordinary galaxies), so that their intrinsic luminosities are highly uncertain. The distance to the Seyfert galaxies, however, is known with reasonable precision; thus they offer the astronomer the significant advantage that their distances are not in dispute by a factor of 100 nor their true luminosities by a factor of 10,000.

The number of objects generally recognized as Seyfert galaxies is still only about a dozen; a few that were in Seyfert's original group have now been dropped and a few new ones have been added. Two of the best-studied are NGC (for New General Catalogue) 1068 and NGC 4151. The three photographs of NGC 4151 at the right represent the appearance of the galaxy at three different exposures. The two shortest exposures reveal the presence of the brilliant small nucleus at the center of the galaxy, almost starlike in appearance. Such nuclei are characteristic of Seyfert galaxies and account for most of the remarkable properties of these objects.

It is the spectra, however, that provide the richest source of information about conditions inside the nucleus of Seyfert galaxies [*see illustration on next page*]. In most instances the emission lines are the same as those seen in several "high excitation" planetary nebulas, which are shells of gas surrounding a small, very hot star. The chief difference is that the emission lines are much broader in the Seyfert spectra. The width of the lines provides the first clue to conditions inside the Seyfert nucleus.

In the spectrum of a typical Seyfert galaxy the lines attributable to oxygen ions (oxygen atoms stripped of one or more electrons) and the lines attributable to hydrogen present a puzzle. A curve tracing the intensity of the hydrogen line from one side to the other is a peak flanked by broad "wings"; in curves tracing the intensity of the oxygen lines the wings are usually missing. When astronomers see a broad line in a spectrum, they immediately assume that the emitting gas is in motion, so that some of the gas is moving toward the observer (shift-

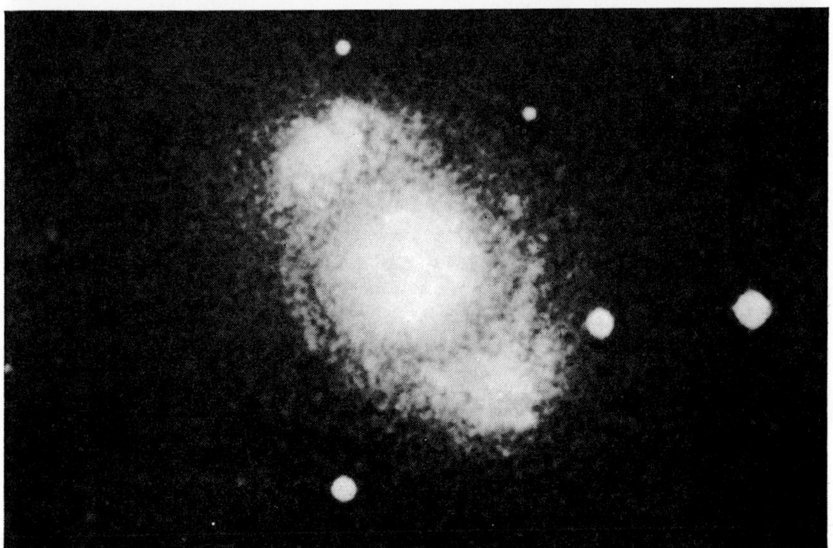

STARLIKE NUCLEUS OF NGC 4151 is revealed by short-exposure photographs (*top and middle*). In the top picture, made at Mount Wilson, the nucleus resembles a slightly fuzzy star. In the middle picture, also from Mount Wilson, the spiral structure has become faintly visible. In the long-exposure photograph at the bottom, made with the 48-inch Schmidt telescope on Palomar Mountain, the spiral structure is fully developed. The negative print on the opposite page, however, shows considerably more detail than any of the positive prints. The series of three photographs was assembled by W. W. Morgan of the Yerkes Observatory.

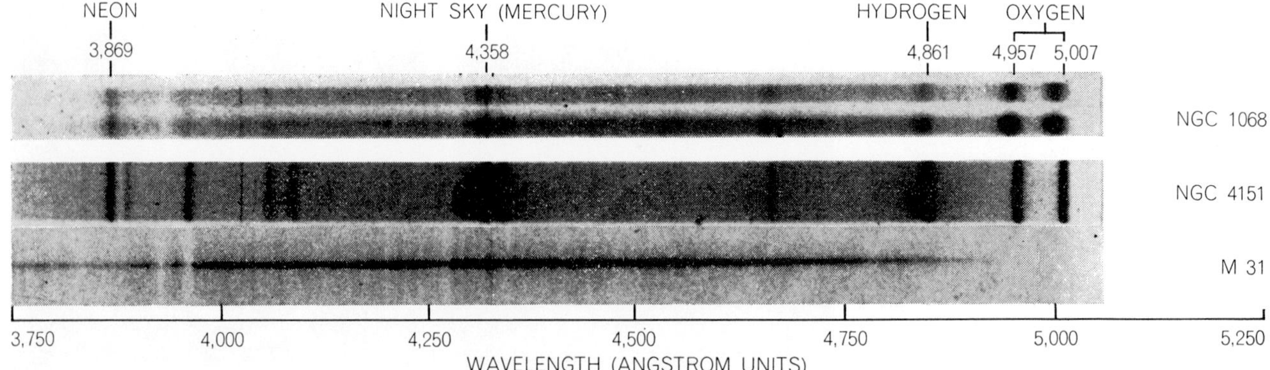

EXCITED NUCLEI OF SEYFERT GALAXIES produce spectra with prominent emission lines, such as those visible in the spectra of NGC 1068 and NGC 4151. In the spectrum of a normal spiral galaxy, for example M 31 in Andromeda (*bottom*), there are no emission lines but there are a number of strong absorption lines, characteristic of the spectra produced by fairly cool, ordinary stars like the sun. Of the two Seyfert spectra shown here, that of NGC 4151 is the more typical in that the beta hydrogen line at 4,861 angstrom units has very broad "wings," whereas wings are lacking on the "forbidden" oxygen lines at 4,957 and 5,007 angstroms. The Seyfert spectra were made by J. Beverly Oke and Wallace L. W. Sargent of the California Institute of Technology; the spectrum of M 31 is by Ivan R. King of the University of California at Berkeley. The line at 4,358 angstroms is produced by the glow of city lights.

ing the line toward the blue) and some of it is receding (shifting it toward the red). These are the well-known Doppler shifts. If the widths of the oxygen and hydrogen lines in the Seyfert spectra are interpreted in this way, one deduces velocities of several hundred kilometers per second for the oxygen and velocities of several thousand kilometers per second for the hydrogen.

Are these true velocity effects? If so, is the motion chaotic and disorganized, or is it orderly and perhaps organized in a manner that might be associated with rotation? If the motion is chaotic, would not the material soon leave the nucleus and eventually the entire galaxy? What mechanism can continuously replace the matter being ejected and accelerate it to such high velocities? Why, in the majority of these objects, should there be such a striking difference between the profiles of the oxygen and hydrogen lines?

Let us consider the last question first. In planetary nebulas the oxygen and hydrogen lines are produced in the same region of space but by quite different processes [*see illustration on page 238*]. Both the hydrogen and the oxygen are kept in a state of nearly complete ionization by the intense flux of high-energy (ultraviolet) photons emitted by the central star of the planetary nebula. Occasionally a bare proton, that is, a hydrogen nucleus, will recombine with a free electron to form a hydrogen atom; the process is called radiative recombination. If the newly formed hydrogen atom is in a high-energy state when it is created, the electron will rapidly jump to the "ground" state, either directly (producing an ultraviolet photon that is invisible to observers on the earth) or cascading downward in smaller jumps, some of which give rise to the observed hydrogen line. Soon after reaching the ground state the hydrogen atom is reionized by the radiation from the central star and the process is repeated.

For oxygen the process is a different one: collisional excitation. The flux of high-energy photons from the star readily strips one or two electrons from each oxygen atom, and it is these ions that give rise to the observed radiation. They usually have electrons in low-energy levels that can be excited to higher levels by free electrons in the immediate vicinity. When the excited ions subsequently return to the ground state, they emit the observed radiation.

In addition to this difference in the way the hydrogen and oxygen emission lines are generated there is a further important distinction. The cascades producing the hydrogen lines take place very rapidly: a hydrogen atom typically spends only 10^{-8} second in one of the excited states. The lifetimes of the excited oxygen ions, on the other hand, are far longer: up to 1,000 seconds. (Because these downward transitions occur so infrequently they are never observed in the laboratory and are thus called forbidden lines.)

This long life in the excited state can be cut short, however, by collisions with other electrons, and if the density of electrons is high enough, the excited oxygen ions will be collisionally de-excited before they have had a chance to radiate.

With this background one can speculate about the conditions that give rise to the contrasting widths of the hydrogen and oxygen lines in the spectra of Seyfert galaxies. If the density of particles is very low, the intensity of forbidden oxygen lines should roughly equal the intensity of hydrogen lines. As the particle density increases one can expect to see a decrease in the intensity of forbidden lines, relative to the intensity of the hydrogen lines, as more and more oxygen ions are collisionally de-excited before they can radiate. What this suggests is that the region responsible for the wings on the hydrogen lines is a rather small, dense core where hydrogen is traveling chaotically at high speed (5,000 kilometers per second) and generating smeared-out emission lines. If oxygen ions are also present and are moving at comparable speeds, they are being de-excited before they can radiate, and hence there are no comparable wings on the oxygen lines.

It has been estimated that the amount of material needed to produce the observed emission lines is only about five solar masses of gas confined to a region perhaps as small as .06 light-year (roughly three light-weeks) in diameter. Hydrogen atoms moving at a speed of 5,000 kilometers per second would traverse such a core and escape in only about four years. To hold this relatively small amount of high-speed gas inside the core indefinitely by gravitational forces would require that some 20 million stars of solar mass also be packed within the core. Because such a density of stellar matter is difficult (although not impossible) to conceive, it has been assumed either that the high-velocity gas must escape or that there are other ways in which the hydrogen lines might be broadened. In any event, it seems likely that the excited oxygen atoms emit their forbidden lines not in the core but in a much less dense region around the core. The gas velocities there are much lower, with the result that the lines produced in that region (by hydrogen as well as oxygen) are much narrower. Indeed, the profile of the hydrogen lines shows a narrow central peak that closely matches the wingless profile of the oxygen lines.

If we accept the fact that the gas inside the tiny core of a Seyfert galaxy is moving at the high apparent velocity indicated by the spectra, and if we assume that the gas is not held within the core by gravitation, we must explain how it is replaced or conclude that the violent activity observed in the core is a rare transient event caused by some explosive outburst. The difficulty with this point of view is that although only a dozen Seyfert galaxies are known, they cannot be considered particularly rare. The Seyfert phenomenon has been observed only in large spiral galaxies, of which between 500 and 1,000 are close enough and sufficiently well classified to be candidates. Thus between 1 and 2 percent of large spiral galaxies are Seyfert galaxies.

Let us assume, for purposes of simple computation, that all spiral galaxies, including our own, have spent 1 percent of their lifetime in the Seyfert condition. One can readily calculate how many times a typical spiral galaxy has experienced a Seyfert outburst during the estimated 10 billion years the galaxies have been in existence. If the duration of the outburst is as brief as 10 years, 10 million outbursts would be needed. This implies that in a 10-billion-year lifetime a spiral galaxy has experienced an outburst on the average of once every 1,000 years.

It would be remarkable indeed if an astronomer in his own lifetime could follow all or a substantial part of a Seyfert outburst, but such may be the case. Recently the French astronomers Y. Andrillat and S. Souffrin remeasured the spectrum of NGC 3516, one of the galaxies Seyfert had studied, and found that marked changes have taken place in the relative strengths of the hydrogen and forbidden oxygen lines. To explain these changes they proposed a model, quite similar to the one described above, that assumes an explosive outburst of gas from the nucleus of NGC 3516.

The problem of confining the gas would be eased somewhat if some mechanism other than high-velocity motions could be found to explain the broad wings on the hydrogen lines. One mechanism that seems attractive is the scattering of photons by electrons. When a gas containing both electrons and protons is in equilibrium, both kinds of particles have about the same amount of thermal kinetic energy, equal to one-half the mass of the particle multiplied by the square of its velocity. Since an electron has only about 1/1,800th the mass of a proton, its velocity must be greater by a factor of about 40 (the square root of 1,800). Thus the protons responsible for the hydrogen lines need not be traveling at 5,000 kilometers per second but could be moving at the much lower speeds associated with a gas whose temperature is about 20,000 degrees. Although the probability of a photon's being scattered by a free electron is low, if there are enough electrons blocking the escape of a photon, each photon will undergo several scatterings, whose net effect would be to smear out a hydrogen line that was intrinsically narrow.

An additional remarkable feature has been observed in the spectrum of the Seyfert galaxy NGC 4151: lines produced by very highly ionized iron atoms. These same lines are found in spectra of the corona of the sun, where the temperature is several million degrees. The energy needed to produce such lines is nearly 30 times greater than the energy needed to produce the hydrogen lines. J. Beverly Oke and Wallace L. W. Sargent of the California Institute of Technology have suggested that the "coronal" iron lines may be produced by a very hot, tenuous gas filling the nucleus of NGC 4151. They further suggest that embedded in this hot gas are cooler, denser clouds moving at high velocity that account for the more familiar lines of hydrogen and oxygen. Occasional collisions between these clouds might produce and replenish the tenuous gas. An alternative hypothesis suggests that a powerful source of X rays is ionizing the iron atoms responsible for the coronal lines. Detection of such X-ray sources by rockets equipped with sensitive detectors now appears feasible and should provide a decision between these hypotheses.

Direct evidence for the existence of the cloudy structure envisaged in Oke and Sargent's model has been found by Merle F. Walker of the Lick Observatory, who observed NGC 1068 and NGC 4151 with an electronic image-amplifier attached to the spectrograph on the 120-inch Lick telescope. Working with the velocity data obtained from many separate observations, Walker plotted a two-dimensional map of the nucleus of NGC 1068 and found what appear to be four clouds moving at different velocities [*see illustration on page 239*].

Let us turn now from the emission lines seen in the spectra of Seyfert galaxies to the way continuous radiation, as distinct from line radiation, is distributed among various wavelengths. A rough index to the character of continuous radiation from an astronomical object is obtained by comparing the amount of energy received through filters passing one of three bands of wavelengths: ultraviolet (U), blue (B) and yellow, or "visual" (V). When ordinary stars and ordinary galaxies are observed through such filters, one finds that the U intensity is less than the B and the B intensity less than the V. In other words,

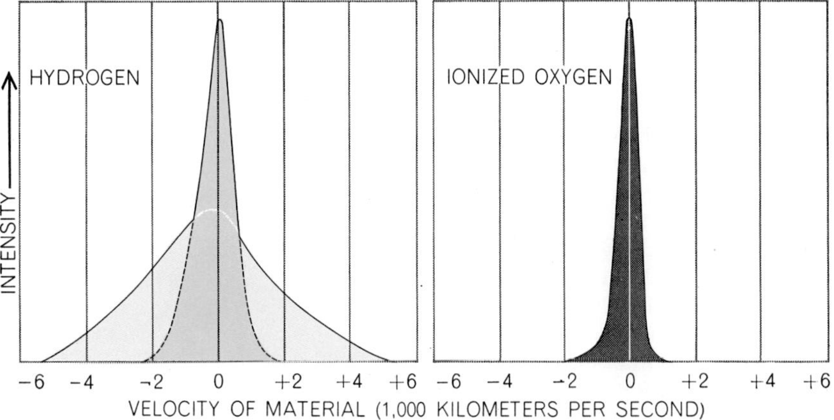

COMPARISON OF INTENSITY PROFILES for hydrogen lines (*left*) and the forbidden lines of oxygen (*right*) in NGC 5548 demonstrates the difference usually seen in Seyfert spectra. If the line-widening is attributed to the motion of gas particles, the hydrogen velocity is four or five times the velocity of oxygen. Negative velocity values indicate motion of gas toward the earth, positive values motion away from the earth. The hydrogen lines could be smeared out, however, by the scattering of photons off rapidly moving electrons. The curves are based on the work of Kurt Anderson of the California Institute of Technology.

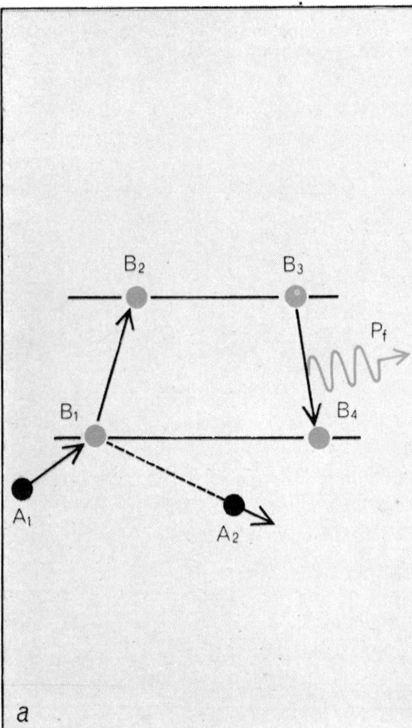

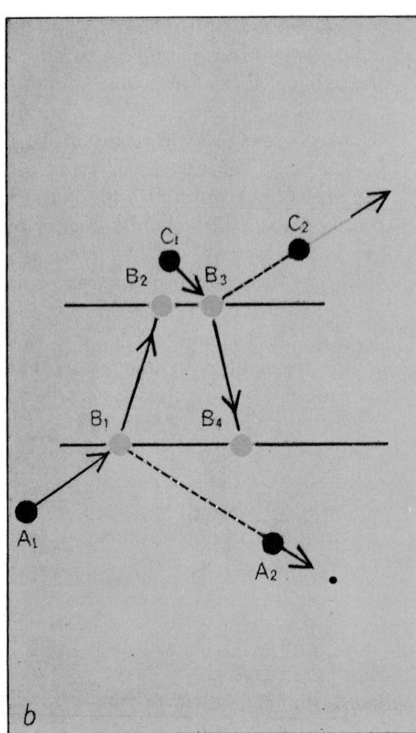

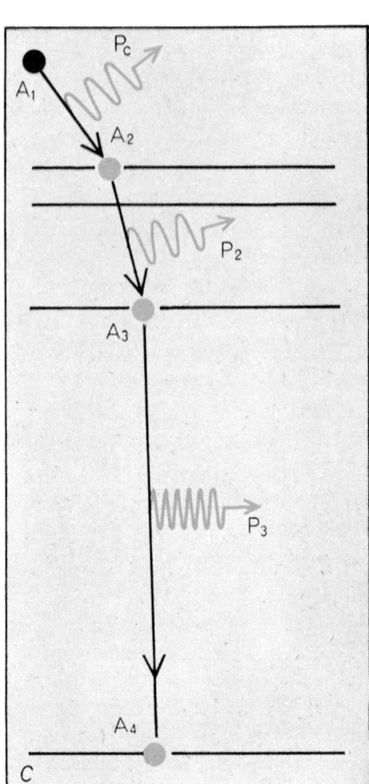

EMISSION LINES OF OXYGEN AND HYDROGEN arise by different processes in Seyfert nuclei. Moreover, the forbidden lines of oxygen appear only when the gas is at low density (*a*). In that case a free electron (A_1) collides with an oxygen ion and excites one of its electrons that is in the lowest level (B_1), raising it to a higher level (B_2). The free electron leaves with reduced energy (A_2). After about 100 seconds the excited electron makes a forbidden transition (B_3 *to* B_4), emitting a photon (P_f). When the gas is at high density (*b*), the electron that has been excited to the higher level (B_2) is de-excited (B_3, B_4) by another free electron (C_1) before it has a chance to emit a photon. The energy the photon would have carried away is acquired by the colliding electron (C_2). Hydrogen-line emission arises differently (*c*). A free electron (A_1) is captured by a proton, forming a hydrogen atom whose electron is in an excited level (A_2). A photon emitted in the process (P_c) is said to be a continuum photon because it can have various wavelengths. Within about 10^{-8} second the electron cascades to a lower level (A_3), emitting an observable Balmer-line photon (P_2), and immediately drops to the ground state (A_4), emitting a Lyman-line photon (P_3), which being in the far ultraviolet cannot be observed.

ordinary stars (such as our sun) emit more yellow light than blue light. This is also the case if one observes a Seyfert galaxy through an aperture that admits most of the light from the galaxy. As the aperture is reduced to accept light only from the central region, however, the ultraviolet and blue part of the spectrum begins to predominate. Conceivably this could be due to a concentration of ordinary hot young blue stars in the nucleus. On the other hand, such stars have spectra with certain characteristic absorption lines, and these lines do not appear in the spectra of Seyfert nuclei. If one subtracts from the spectrum of Seyfert nuclei the yellowish radiation from cool stars, and also the radiation from gaseous hydrogen, one is left with a smooth blue spectrum similar to the one produced by quasars and, within our own galaxy, by the Crab nebula.

The Crab nebula is the remnant of a supernova explosion observed on the earth in A.D. 1054. It is now generally accepted that both the radio and optical radiation emitted by the nebula is produced by synchrotron radiation: radiation from electrons spiraling at virtually the speed of light in a magnetic field. A characteristic of such radiation is that it is polarized. A significant amount of the radiation from the Crab nebula is polarized, and a modest amount has now been detected in some of the Seyfert nuclei and in quasars.

An additional striking similarity between Seyfert nuclei and quasars is their variability in light output. This was first shown for the quasar 3C 273 by Harlan J. Smith of the University of Texas and E. Dorrit Hoffleit of Yale University, who reported variations in the image intensity of 3C 273 in photographs that had been made over a period of several decades.

By this time attention had been called to the similarities between the nuclei of Seyfert galaxies and quasi-stellar sources by a number of astronomers. Consequently a search began for light variations in the nucleus of Seyfert galaxies. Such variations were detected in NGC 4151 by photoelectric measurements at the University of Arizona [*see illustration on page 240*]. Simultaneously Oke reported finding variations in an "N-galaxy," one of a class of objects that have many of the properties of Seyfert nuclei. Subsequently T. D. Kinman of the Lick Observatory found that 3C 120, a radio galaxy that is now classed as a Seyfert galaxy, had varied in light output.

The history of our knowledge of the quasi-stellar sources has been one surprise after another. Indeed, almost without exception, every new line of observational investigation has disclosed something unexpected. Infrared measurements of the quasi-stellar sources were no exception. Observations of 3C 273 at wavelengths from one micron to 20 microns made by Frank J. Low and Harold L. Johnson of the University of Arizona revealed that at the longer wavelengths the amount of energy radiated increases sharply. Its peak cannot be observed because long infrared wavelengths are absorbed by the earth's atmosphere, but it must lie somewhere between a wavelength of 20 and 1,000 microns (one millimeter). In fact, the bulk of the energy is radiated in the infrared. This discovery implies a severalfold increase in the already incredible luminosity of this quasi-stellar source, if its distance is indeed cosmological.

It was logical to search for strong infrared radiation in Seyfert galaxies. In fact, this search was carried out by A. G. Pacholczyk and W. Wisnewski at the University of Arizona before the search for optical variability was made. They found a very steep increase in the infrared radiation of NGC 1068, a phenomenon that has now been found in most of the remaining Seyfert galaxies [*see illustration on page 241*]. Three possible mechanisms have been proposed to account for the infrared output: synchrotron radiation, a process involving high-velocity streams of plasma (free electrons and protons) and thermal radiation from dust. Convincing models of the first two processes are difficult to construct, and the fact that no well-substantiated variations in the infrared light have been detected argues against their correctness. There is, however, some observational support for the dust hypothesis.

If a dust particle is exposed to a distant, intense source of radiation, the temperature of the particle will rise until its own rate of radiation just equals the rate at which it is absorbing energy. For example, a small blackened sphere placed at the earth's distance from the sun will be warmed to a temperature of about 300 degrees Kelvin (about room temperature) by the sun's rays. Such a body will radiate primarily in the infrared with a peak output at a wavelength of about 10 microns. The precise wavelength distribution of the radiated energy will depend on the composition and size of the dust particle.

If there are large concentrations of dust in Seyfert nuclei, and if the dust particles are anything like those with which we are familiar in our own galaxy, they should absorb far more readily in the ultraviolet and blue parts of the spectrum than in the yellow or red. This is merely to say that dust usually "reddens" light passing through it. There is no way to check for this reddening of the continuum radiation in Seyfert nuclei because we have no way of knowing what the intrinsic radiation is like. One can calculate, however, the relative intrinsic intensity in the red and blue of certain sets of emission lines, since we do know the probability of the atomic transitions responsible for the lines. E. Joseph Wampler of the Lick Observatory measured these lines in the Seyfert nuclei and found that the blue lines were very much weakened relative to the red ones. Reddening by dust is the only known explanation. What is disturbing about the dust hypothesis is that it implies that the intrinsic continuous radiation of Seyfert nuclei is far bluer than anything previously encountered.

One additional bit of evidence in favor of the dust hypothesis is the case of the planetary nebula NGC 7027. There is no evidence at all for anything like synchrotron radiation in this object, but it too has intense radiation in the infrared strikingly similar to that in NGC 1068. A quite plausible explanation is that ultraviolet radiation trapped in the nebula is converted into infrared radiation by dust grains.

The final property that some Seyfert galaxies share with quasars is intense radio emission. Here, on first consideration, the differences seem more striking than the similarities. The quasars are powerful radio emitters by definition, whereas the radio emission of several of the Seyfert galaxies either is not detectable or is quite weak. On the other hand, the Seyfert galaxies NGC 1068, NGC 1275 and 3C 120 are strong radio sources; moreover, the last two objects are variable, and at wavelengths around one centimeter they very much resemble quasars. The recent quasar-like variations in 3C 120 are particularly spectacular [*see illustration on page 242*].

It now seems likely that quasars are actually a subclass of a much larger group of objects now known as quasi-stellar galaxies (as distinct from quasi-stellar objects), which emit no detectable radio signal. Similarly, the three Seyfert galaxies that are strong radio sources may simply be a subclass within a larger group of objects most of which do not now have (or perhaps never had) the extended halo of high-energy electrons in a weak magnetic field that seems to be necessary for generating strong radio emission at long wavelength.

Let me summarize briefly the observed properties of the Seyfert galaxies and indicate their points of similarity with some of the quasi-stellar sources. The Seyfert galaxies have compact nuclei, whose emission spectra indicate that the gases in them are in a high state of excitation and are traveling at high speed in clouds or filaments. Outbursts probably occur from time to time, producing new high-velocity material. The continuous spectrum of a Seyfert nucleus, perhaps produced by the synchrotron mechanism, is very blue and may also vary in intensity with strong outbursts. The comparatively short duration of these outbursts requires that the energy be released within a remarkably small volume—not more than a third of a light-year across. There are also outbursts that can trigger strong fluctuations in the radio emission. Much of the energy emitted by Seyfert nuclei is in

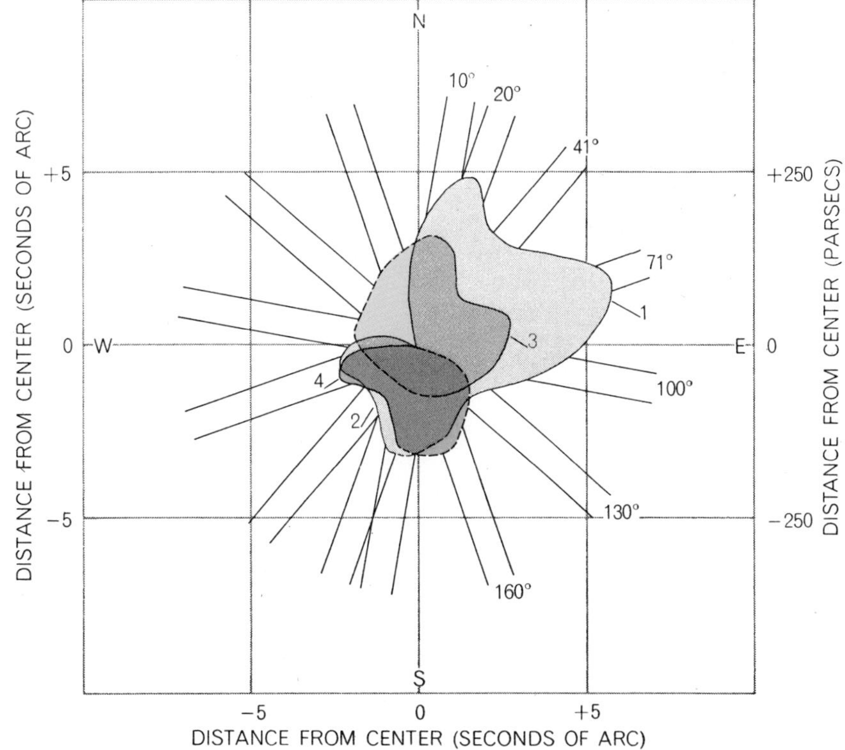

MAP OF CLOUDS IN NGC 1068 was made by Merle F. Walker of the Lick Observatory, who analyzed high-dispersion spectra produced by a spectrograph whose entrance slit was positioned across the nucleus of the galaxy at the seven different positions shown. From his analysis Walker identified four distinct cloud masses, each moving at a different velocity.

the far infrared, possibly because the intrinsic radiation is absorbed by dust and reradiated. Most of the foregoing characteristics are shared by the quasi-stellar radio sources. The most striking difference between the two kinds of object is that the spectra of quasars show a strong red shift, implying that the quasars are receding at high velocity, and hence that, according to the cosmological interpretation of red shifts, they are enormously distant. If they are indeed cosmologically distant, quasars must be extraordinarily luminous to appear as bright as they do.

Therefore, except for an apparent difference in luminosity, Seyfert galaxies and quasars may represent essentially similar phenomena. Astronomers inclined to this view have looked for objects that might bridge the "luminosity gap" between the two. In fact, the gap is nearly bridged by the radio-bright Seyfert galaxy *3C* 120, if one takes into account its great infrared emission. *3C* 120 is then only about 20 times less luminous than the most luminous quasar, *3C* 273, and it approaches the luminosity of some of the dimmer quasars. Nevertheless, the average luminosity of Seyfert galaxies at light wavelengths alone is several hundred times less than the average optical luminosity of the quasars.

Efforts to find objects resembling Seyfert nuclei and quasars—some of which might turn out to be the missing links between the two—have led to the investigation of a bewildering variety of objects with somewhat confusing *ad hoc* designations: N-type galaxies, "blue interlopers," Braccesi objects, Markarian objects, "blue stellar objects," Haro objects, Zwicky compact galaxies and others. Because these objects were discovered by different observers using different techniques no one knows how many distinct species they represent. The most pertinent generalization emerging from a study of these objects is that when they are very blue in color, very compact and essentially stellar in appearance, they often exhibit the kind of optical spectrum we have been discussing. Some are radio sources, some are not. Their starlike nuclei may or may not be embedded in a fuzzy nebulosity, which in turn may or may not represent an ordinary—but very distant—spiral galaxy. Two of the Zwicky compact galaxies recently studied by Sargent have pronounced red shifts and thus seem to be as luminous optically as some quasars.

It should perhaps be emphasized that astronomers have only two ways of estimating the distance to other galaxies. If the galaxies are close enough for individual stars to be visible, one can look for certain stars (for example Cepheid variables) whose intrinsic luminosity is known because they occur in our own galaxy at distances that have been measured directly by triangulation, with the earth's orbit serving as a base line. Knowing the absolute luminosity of individual stars enables one to compute the distance to galaxies that lie within about 30 million light-years. Beyond that individual stars cannot be observed. One must then use the second method, which assumes that the brightest galaxies in a remote cluster of galaxies are as intrinsically bright as the very bright elliptical galaxies whose distance has been determined by the first method. It has also been found that the light from remote galaxies is red-shifted in direct proportion to their distance as estimated by the second method. This relation is named Hubble's law.

One would like to know, therefore, if the various blue, compact objects, ranging from the Seyfert nuclei to the quasars, follow Hubble's law. To answer this question Halton C. Arp of Mount Wilson and Palomar plotted apparent brightness (since intrinsic brightness is unknown) against red shift for a smooth sequence of Seyfert-like objects and quasars. He then compared the slope of the resulting curve with the slope of a similar curve for the bright elliptical galaxies on which the Hubble law is based. Arp found that for his collection of objects the red shift increased more rapidly as magnitude decreased than it did for normal galaxies; Hubble's law was seemingly violated [*see illustration on page 243*].

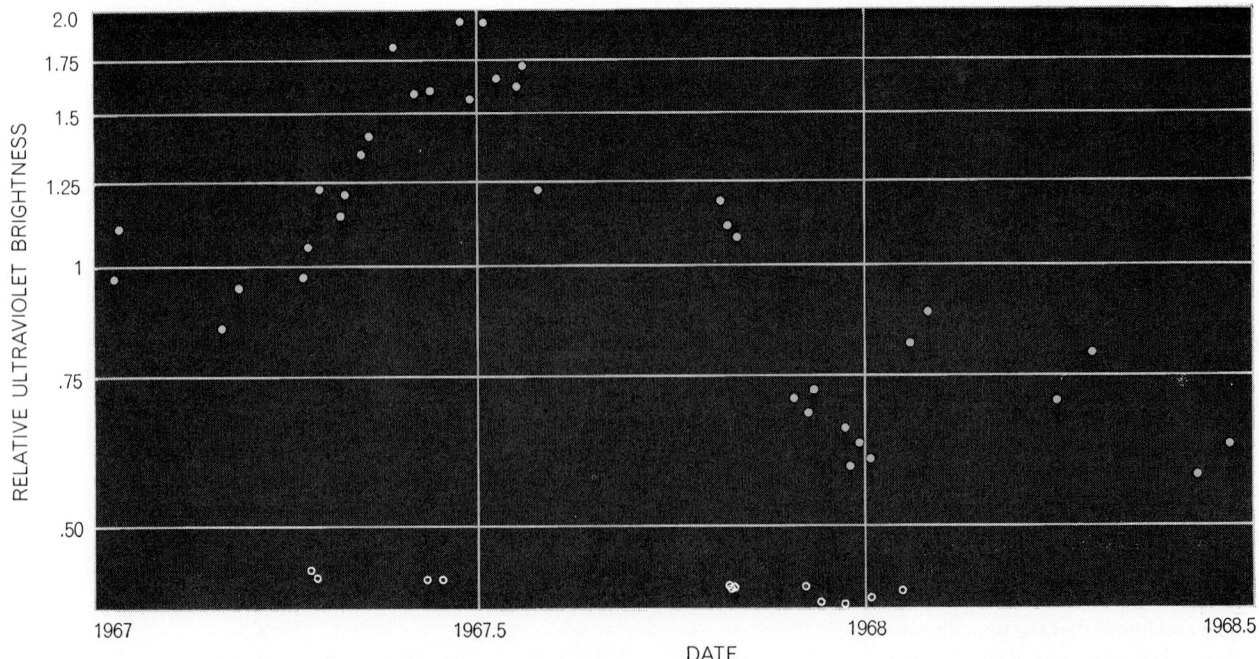

BRIGHTNESS VARIATIONS IN NUCLEUS OF NGC 4151, measured in the near ultraviolet (*color*), covered a range of nearly four to one during a recent 18-month period. The open circles show the brightness of a nearby reference star. Because the aperture used in the study let in starlight from outside the nucleus of NGC 4151, the actual variability must be greater than shown.

Is this violation real or only apparent? One possible explanation is the following. The Arp objects may have a wide range of intrinsic luminosities, with the less luminous objects being the more common; if one selects objects down to a given apparent brightness (knowing nothing, of course, about their distance), then any random selection will be skewed by including too many nearby objects that are below average in luminosity and too many very distant ones that are above average. Even though Hubble's law may actually hold for these objects, it will seem to be violated because the most distant (and thus more red-shifted) objects will tend to be intrinsically much more luminous than the nearby objects.

Another possible explanation for the apparent failure of Hubble's law is that a significant portion of the red shift in high red-shift objects is not cosmological (that is, due to the expansion of the universe) but is due to something else. What this other red-shift-producing mechanism might be has been the subject of intense speculation. The two alternatives most commonly invoked are a "local" Doppler shift, which would be observed if local objects were moving away at extremely high velocity (produced perhaps by gigantic explosions either in our own galaxy or ones nearby) and a gravitational red shift. The latter explanation demands the presence of a very strong gravitational field at the emitting source, produced by a compact, enormously massive object. A few years ago the gravitational explanation was in disfavor, but the total absence of any blue shifts—which one might expect to find if high-velocity objects had been ejected by nearby galaxies and were moving in random directions—has reopened the question of whether satisfactory models involving large gravitational fields might be constructed. If and when blue shifts are found, the whole outlook will of course be drastically changed.

A decisive observation that would rule out the cosmological origin of the red shift would be the discovery of a high red-shift object that is clearly associated physically with other material in space (perhaps a cluster of galaxies) that does not have the same high red shift. Alternatively, the cosmological nature of the red shift would be confirmed if it could be shown that the spectra of quasi-stellar sources exhibit effects of some kind that could be attributed unambiguously to the presence of intervening matter, associated perhaps with an ordinary galaxy at moderate distance.

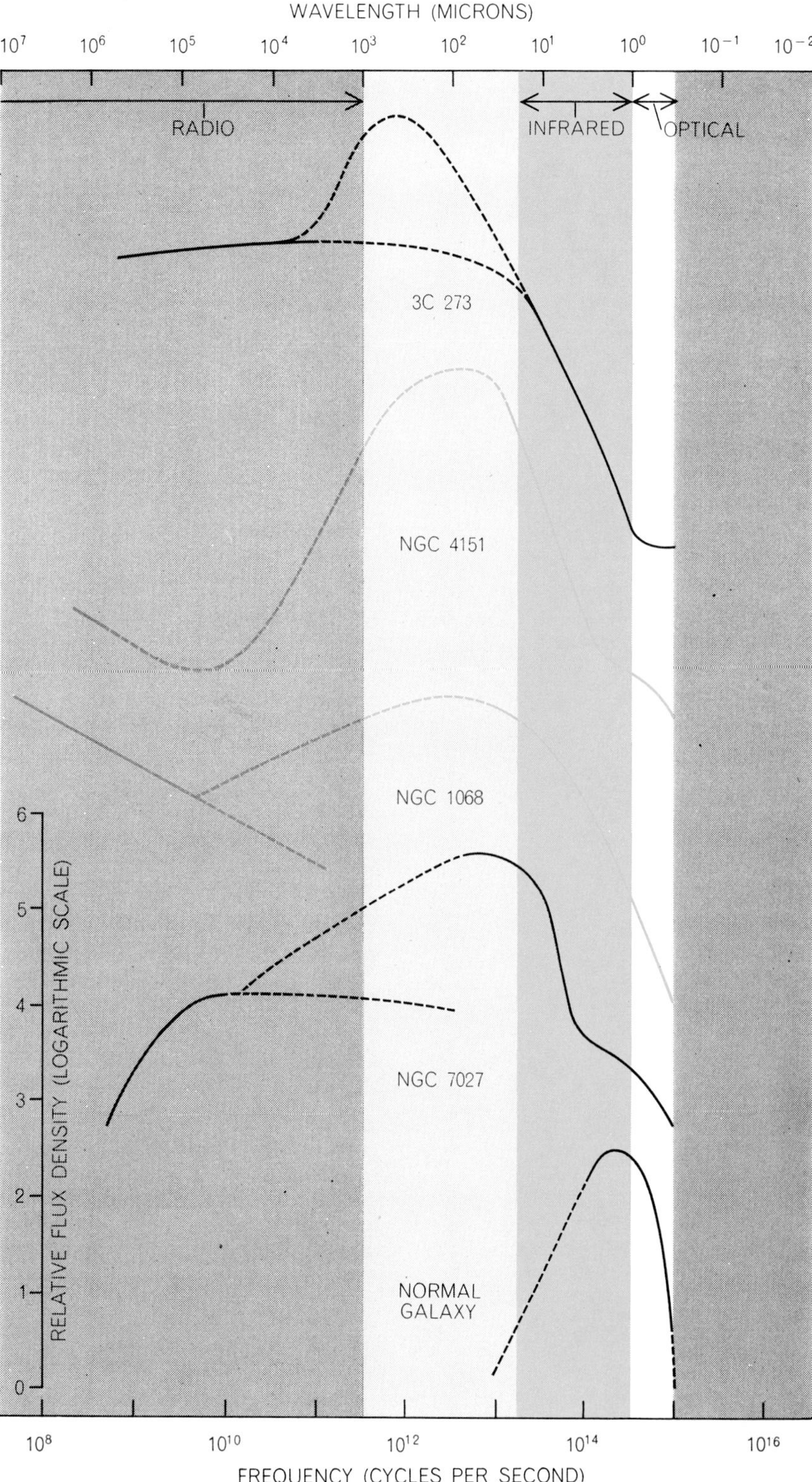

WAVELENGTH DISTRIBUTION OF ENERGY EMITTED by Seyfert galaxies (*colored curves*) closely resembles the distribution characteristic of the powerful quasi-stellar radio sources (quasars), represented here by *3C* 273. Seyfert emission, which is totally unlike that of a normal galaxy, also resembles that of the planetary nebula NGC 7027, a cloud of gas raised to incandescence by a hot central star. Quasar *3C* 273 and the two Seyfert galaxies exhibit a greater energy output at infrared than at optical wavelengths and strong emission at radio wavelengths as far out as the emission has been measured. The broken lines indicate estimates where measurements are lacking or uncertain, or where atmospheric absorption precludes observations. The placement of curves does not signify absolute differences in emission; it simply ranks the quasar, the Seyfert galaxies and the planetary nebula from most energetic (*top*) to least energetic (*bottom*); the radiation curve of a normal galaxy is shown for reference. The scale of relative flux density applies to each curve individually.

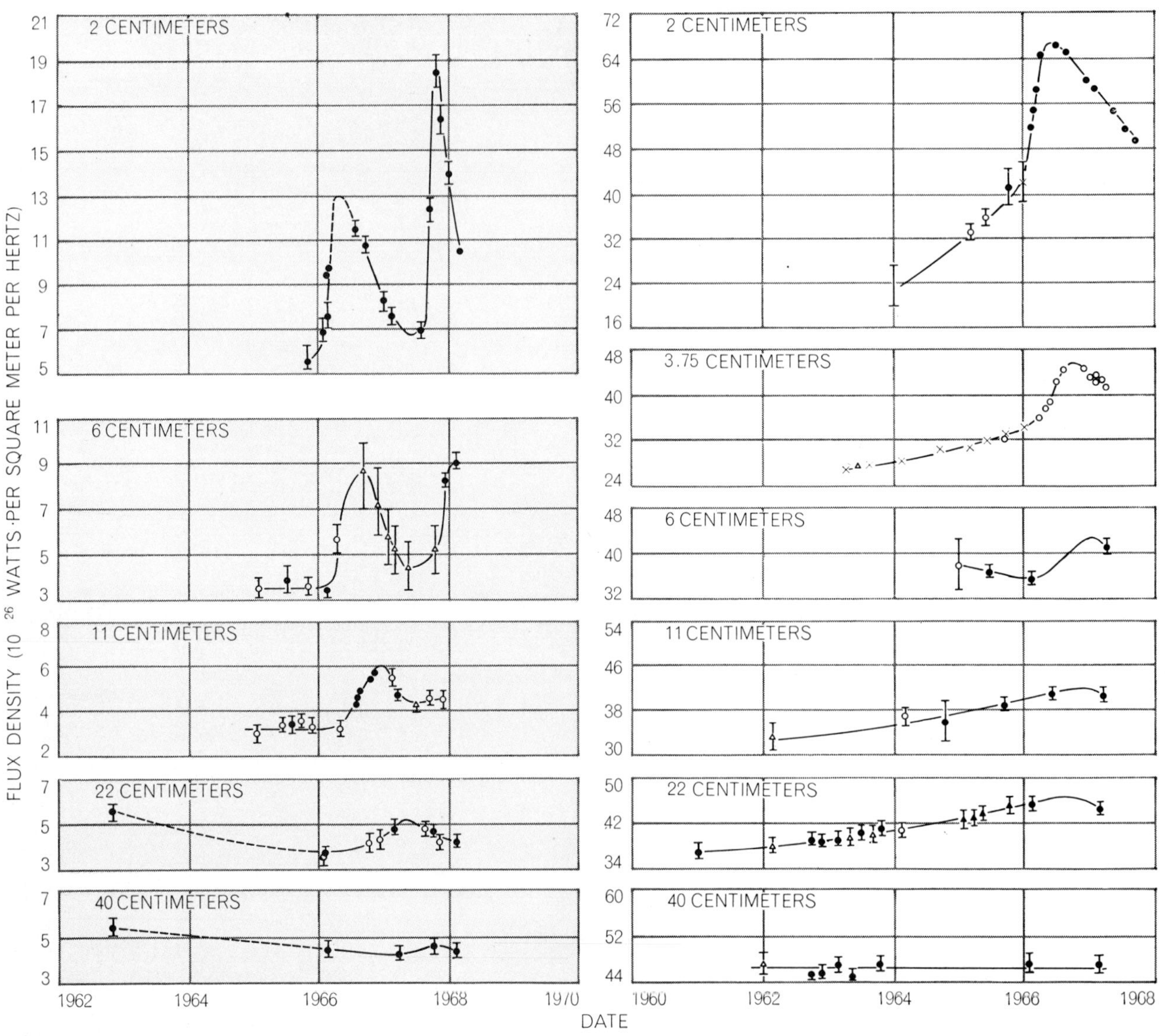

CHANGES IN RADIO EMISSION from Seyfert galaxy 3*C* 120 over the past few years (*curves at left*) are remarkably like those displayed by quasar 3*C* 273 (*curves at right*). In both objects the changes are much more abrupt and steeper at shorter wavelengths, for example at two and six centimeters, than at longer wavelengths. On the basis of the red shift observed in its spectrum, 3*C* 273 may be as remote as two billion light-years and thus among the most distant objects known. Its intrinsic radio luminosity is therefore assumed to be much greater than that of 3*C* 120 or other Seyfert galaxies, which all lie within about 100 million light-years. The emission curves are from a paper published by I. I. K. Pauliny-Toth and K. I. Kellermann of the National Radio Astronomy Observatory.

Because the various strange objects plotted on Arp's diagram have many properties in common, it is tempting to believe we are presented with a whole sequence of blue, compact, intensely bright objects culminating in the quasars and representing various degrees of violent activity in the center of galaxies. Beyond a certain distance the underlying galaxy could not be discerned; only the optically bright nucleus would remain. If this view is correct, then the objects with large red shifts must be quite distant. We still know too little about the so-called compact galaxies to say if it is possible to have an alternative sequence of compact galaxies culminating in the quasars, whose character is wholly different from that of either ordinary spiral or elliptical galaxies.

If there is some kind of sequence of violent activity in galactic nuclei, we should not be surprised to find "baby" Seyfert nuclei. The strong activity we recognize in Seyfert nuclei may well occur in other galaxies on a more modest scale. For example, radio observations indicate that something quite unusual is going on in the center of our own galaxy. Recently strong infrared emission has been found in the direction of the galactic center. Unfortunately the amount of dust in the direction of the galactic center is so great that we are prevented from inspecting the nucleus of our galaxy at optical wavelengths.

Recent observations by Campbell M. Wade of the National Radio Astronomy Observatory have detected small but intense sources of radio emission in the nuclei of a number of ordinary spiral galaxies as well as in several Seyfert galaxies. For example, in M81, one of the nearest spirals, Wade has found a radio source at the center less than 100 light-years across whose intensity is far stronger than the small source at the center of our own galaxy. It appears certain that this source is not just ordinary radio emission from ionized gas.

We turn finally to the question of the cause of this violent nuclear activity. Any model of a Seyfert nucleus must be based on some estimate of how much energy is released; the model must then provide a source for this energy and a mechanism for its release. For 3C 120 the energy release is estimated to approach 2×10^{46} ergs per second; for a more typical Seyfert nucleus the rate is about 100 times less: 2×10^{44} ergs per second. Therefore over the 10-billion-year lifetime of a galaxy the Seyfert process, if it is operating about 1 percent of the time, will have expended about 10^{60} ergs.

This amount of energy could be released by the thermonuclear conversion of 5×10^{8} solar masses of hydrogen to helium. In ordinary stars, however, this is a sedate, well-regulated process, not the violent, explosive kind of event that is needed. The kind of objects able to provide such events are supernovas and extremely massive objects, which yield their energy primarily through gravitational collapse rather than through nuclear reactions.

To produce 10^{60} ergs through conversion of gravitational potential energy would require the collapse of an object with a mass of at least 10^{6} solar masses. The problem here is to find some way to stretch the energy release over a long period of time; single massive objects would tend to collapse in one short, convulsive event. Although such events would not meet the energy needs of Seyfert nuclei, they might provide the energy for quasars, which are far rarer and more luminous and seem to need great bursts of energy at frequent intervals. It may be that the Seyfert nuclei are fueled by the repeated formation and collapse of somewhat less massive objects.

An alternative process visualizes the agglomeration and collapse of considerably smaller objects (on the order of 10 to 100 solar masses) in a restricted volume of space. Such a hypothesis has been presented by Stirling A. Colgate of the New Mexico Institute of Mining and Technology in his stellar-coalescence model. A high-density star cloud in the nucleus of a galaxy evolves to the point where stars begin to collide. At first they collide with such low velocity that they stick together. Eventually some of the agglomerates evolve to the supernova stage, whereupon they collapse with a violent release of energy. Assuming that such supernovas form at random in a reasonable volume of space, say 100 light-years across, it should be possible to ascertain by observation that the center of activity moves about from time to time. Alternatively, one might be able to show that the center of activity remains fixed, which would be a strong argument in favor of a model in which energy is released by single massive objects.

The reader will by now have formed the entirely correct impression that we are still groping for some unifying rational picture of all this activity. Astronomers working today on processes in the nuclei of galaxies are continually startled by the many unexpected observations being secured with both conventional and new techniques. Some very recent observations are even casting doubt on our traditional ideas about the origin of the galaxies. The present situation might be compared to the one in stellar evolution about half a century ago. At that time empirical trends and relations were being revealed through calculations of the mass, luminosity and radius of stars, but a fundamental theory of stellar evolution had to await the key provided by the development of nuclear physics. We are still looking for the key to a real understanding of the violent activity taking place on the fantastic scale of galactic nuclei.

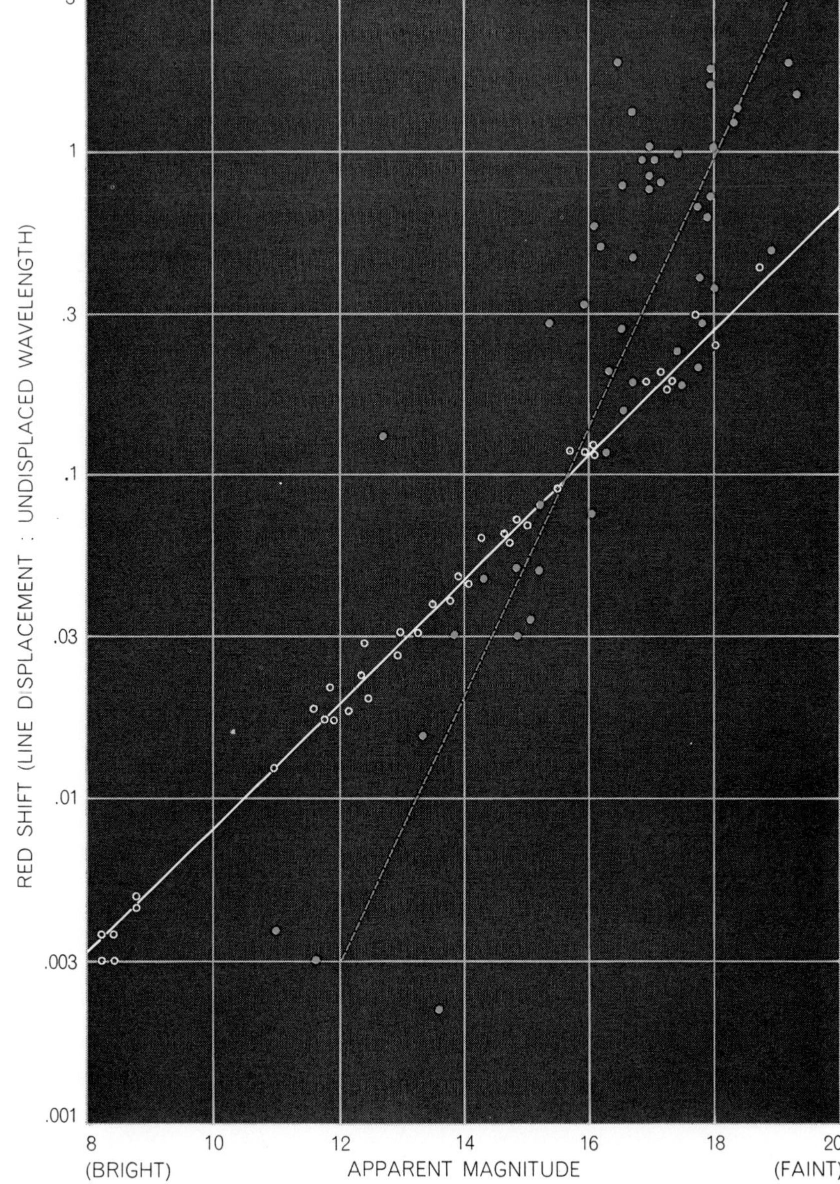

COMPARISON OF "COMPACT" OBJECTS AND NORMAL GALAXIES suggests that the former (*colored dots*) do not follow Hubble's law, which defines the relation between the apparent brightness, or magnitude, of an object and the red shift of its spectrum. The law is based on measurements made on giant elliptical galaxies (*open circles*), which fall along the flatter curve. Data on these giant objects were assembled by Allan R. Sandage of the Mount Wilson and Palomar Observatories. The compact objects represent a smooth sequence of Seyfert-like objects and quasars selected by Halton C. Arp of the same institution. Most of the objects are very blue and have spectra with broad emission lines. They either may have an excess red shift that is not "cosmological" (that is, not related to velocity of recession) or they may represent anomalously luminous objects that do obey Hubble's law.

VII

The New Astronomy

VII

The New Astronomy

INTRODUCTION

A continual stream of unexpected and often serendipidous discoveries, bringing ever vaster horizons, has for several centuries given to astronomy a "new look" for each generation. Kepler's *Astronomia Nova* of 1609 revolutionized celestial physics; Samuel Langley's *The New Astronomy* of 1898 called for a stronger consideration of the emerging discipline astrophysics; and a 1955 *Scientific American* paperback with this title chronicled the revised, larger distance scale of the universe as well as radio astronomy, the new-found spiral structure of the Milky Way, and the first tentative views of element-building in the stars.

Today "the new astronomy" can still be adopted as a title without embarrassment. In fact, it nowadays conveys a particularly distinct meaning, conjuring up an image of the brilliant new astrophysical explorations of the electromagnetic spectrum: the radio sky, the infrared, the ultraviolet, gamma rays, and X-rays. These techniques have touched virtually every section in this volume. Nevertheless under this rubric are grouped several subjects whose very newness seems to defy classification elsewhere.

Yet what is today's new astronomy is tomorrow's history of science. Already progress has marched on, leaving Herbert Freidman's fascinating speculations in "X-Ray Astronomy" only partially accepted. His article includes an exemplary presentation (circa 1964) of the pioneering speculative gropings to understand neutron stars and to capture them observationally. The article ends with the thrilling prospect of an X-ray scan of the Crab Nebula during a lunar occulation, in order to determine the angular size of that famous X-ray source. The rocket was indeed launched at the precise moment the very next month, and this technological *tour de force* of Friedman's group revealed not the point source of a hypothetical neutron star, but an extended X-ray region.

At that time a refined theory of neutron stars was also in the making, which suggested that any phase of intense X-radiation from neutron stars would be very short lived. This interpretation joined with Friedman's findings to portray the extended X-ray source in the Crab as synchrotron radiation from the gaseous nebula itself.

A second key rocket observation for understanding the nature of X-ray sources occurred on March 8, 1966, when Riccardo Giacconi and his colleagues at American Science and Engineering measured the size and position of Sco X-1, the powerful source in the southern Milky Way. Armed with this data, astronomers at the Tokyo Astronomical Observatory and then at the Mount Wilson and Palomar Observatories successfully identified Sco X-1 as a blue 13th-magnitude variable star. "The most provocative fact," wrote Giacconi, "is that this star emits 1000 times more energy in X-rays than in visible light, a situation astronomers had never anticipated from their studies of the many varieties of known stars. There are indications that the X-ray emission of Sco X-1 is equal to the total energy output of the sun at all wavelengths."

How can this enormous output of X-rays be explained? No generally accepted picture of Sco X-1 has been postulated. One thing seems certain:

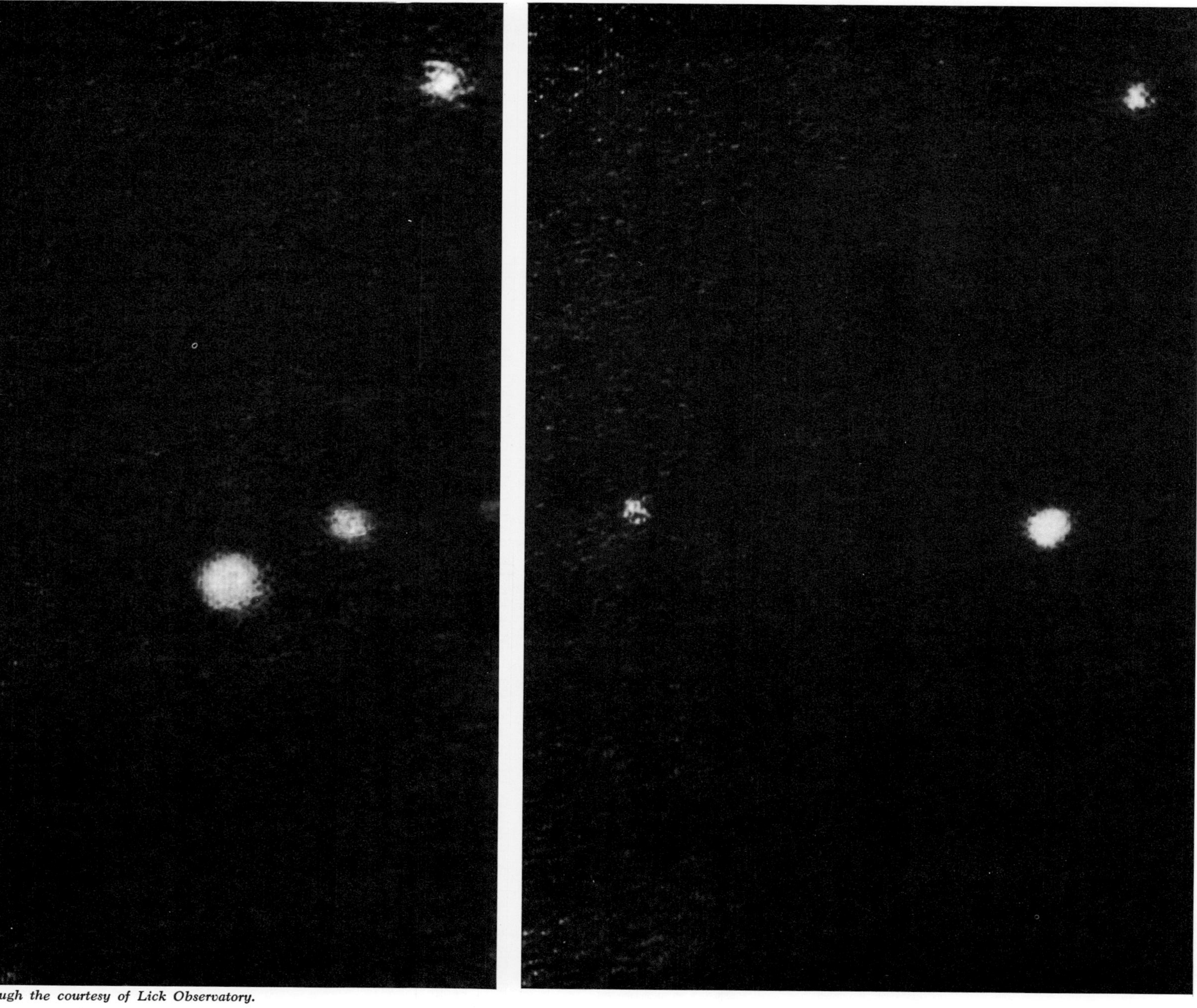

Photographs supplied through the courtesy of Lick Observatory.

FIRST PICTURES OF A PULSAR were made at Lick Observatory with the aid of a sensitive television system operated like a stroboscope in conjunction with the observatory's 120-inch telescope. In the top picture the pulsar is invisible; in the bottom picture it has flashed on. The pulsar is NP 0532, the "south preceding" star of a pair near the center of the Crab nebula. That one of the pair emitted flashes of visible light about 30 times a second was established at the University of Arizona (see "Visible Pulsar," SCIENTIFIC AMERICAN, March, 1969). At Lick the light from NP 0532 was interrupted by a rotating disk perforated with six evenly spaced holes. When the disk was spun at about five revolutions per second, but not quite in synchrony with the pulsar, the light flashes from the object would pass through the holes for several revolutions, then fall out of step and be obscured for several revolutions. When viewed on a television screen, NP 0532 appeared to be pulsing in slow motion. The system was devised by E. Joseph Wampler and Joseph S. Miller.

the speculation summarized in Friedman's "X-ray Astronomy," that Sco X-1 is probably a neutron star, is untenable. Nonetheless, the broad view of the evolution and existence of neutron stars pictured there has been adopted by astronomers as a working hypothesis. Only the identification is wrong. Pulsars, not X-ray sources, may be the sought-for stars.

Of course, pulsars were unknown in 1964. Not until early in 1968, when astronomers despaired of finding neutron stars outside the computation centers, were these extraordinary pulsing radio objects announced to the world by Antony Hewish, whose article is included in this collection.

"No satisfactory explanation for the pulsars has yet been suggested," Hewish declared in a succinct epitomization of the morass of theories and counter theories that prevailed in 1968.

Perhaps the most aggravating and perplexing problem concerned the periods of the pulsars. Ranging from a quarter second to a few seconds, they seemed too short for the expected vibrational periods of white dwarf stars and far too long for the small, incredibly dense neutron stars. An alternative (but at first not completely convincing) explanation was soon suggested: the pulsars could be spinning objects, beaming their radio energy into space like giant rotating beacons. If the sun, for example, conserved its angular momentum as it slowly collapsed to a superdense star 10 kilometers in diameter (the probable size of a neutron star), its rotation period would be about one-tenth of a millisecond. If some angular momentum was lost to a surrounding gas cloud, the period would be longer, but, for the beacon theory to be accepted, at least some pulsars with rotations as fast as a few tenths of milliseconds were required.

A major breakthrough occurred in October, 1968, when workers in Australia found a much faster pulsar, with a period of 89 milliseconds near the midpoint of the extended radio source Vela X. This source is a nebulous shell about five degrees in diameter, believed to be the remnants of a prehistoric supernova explosion. Since theoreticians had previously postulated that neutron stars might be the final core products of supernovae, the discovery of a pulsar within an ancient supernova remnant was exciting news indeed.

As the word spread of the rapidly pulsating Vela source, fresh efforts were made to pinpoint a pulsar within the best documented supernova remnant of all, the Crab Nebula. Within a few weeks D. H. Staelin and E. C. Reifenstein III, working with the 300-foot transit antenna at the

National Radio Astronomy Observatory, found a pair of pulsars in the vicinity of the Crab Nebula. One of them, NP 0532, had an extremely short period, 33 milliseconds. Such a rapid pulse rate would be expected from a small spinning object such as a neutron star, a massive pigmy only about ten kilometers in diameter.

If pulsars are rotating neutron stars, they should gradually slow down as they lose energy. Such observations came forth quickly; by late 1968, radio astronomers had found that several pulsars were increasing their periods by a few billionths of a second per day, that is, by a few nanoseconds per day. The fact that NP 0532, which presumably originated in the comparatively recent supernova explosion that created the Crab Nebula, is not only the fastest, but also one of the strongest pulsars fits in perfectly with this picture. Furthermore, the rotational energy lost by the slowing of the pulsar provides the previously unknown source of energy for the synchrotron radiation of the Crab Nebula itself.

As astronomers became familiar with the radio properties of pulsars, efforts were made to find correlated pulsations in ordinary light. After more than a year of fruitless searching at many observatories, a comparatively amateur team at the University of Arizona captured the long-sought prize by detecting in the Crab Nebula 4-millisecond flashes every 33.095 milliseconds. The Lick Observatory photographs accompanying this introduction (which appeared in the *Scientific American* for March, 1969), show the Crab Nebula pulsar both "off" and "on." Soon after the discovery of the optical pulses, rocket experimenters found X-ray pulses as well; in fact, some of the X-ray photon counts had been taped much earlier, but it had not previously occurred to the investigators to analyze their records for such rapid pulses.

In March of 1969, these strange objects provided yet another surprise, when the Vela pulsar, which had been slowing down steadily like the others, suddenly sped up its period by 200 nanoseconds, only to resume its former slowing. Such an event may have been a "star-quake" when the material of the pulsar rearranged itself into a smaller radius. Alternatively, and perhaps more likely, the magnetic field of the pulsar may have interacted in an unusual fashion with some surrounding nebulosity.

No one dares to predict what surprises the pulsars still hold, although no one can doubt but that further unanticipated discoveries await the observers of these remarkable objects.

In other regions of the spectrum further new phenomena greeted the investigators of the 1960's. Observing with instruments above the earth's atmosphere in balloons, rockets, and satellites, astronomers scrutinized the ultraviolet spectra of hundreds of celestial bodies; a summary of the results, especially for the sun, is found in Leo Goldberg's "Ultraviolet Astronomy." That no new class of ultraviolet objects has turned up is mildly disappointing, but perhaps not unexpected considering the infancy of the discipline and the difficulties of carrying out these observations with remote devices.

Observers in the infrared spectrum have had more luck. In the various infrared wavelengths, cool objects glow with unanticipated brilliancy, outshining all the naked-eye stars except for the sun. Many of these strange objects were discovered by Gary Neugebauer and Robert Leighton, who review their own and other findings in "The Infrared Sky." Mentioned in their article is the Becklin object, a small but intense infrared source embedded in the Orion Nebula; this source also emits radio radiation characteristic of the hydroxyl radical (OH). Just after their article was written, additional infrared sources were correlated with OH emitters. These intense OH objects have several curious properties of their own; for example, they are in many cases extremely small for nebulae, that is, the size of a solar system. Quite possibly infrared and radio astronomers are here observing protostars or solar systems in the making. Undoubtedly an important development of the 1970's will be a better understanding of star formation, complete with a richer set of observations made across the spectrum.

24

X-Ray Astronomy

HERBERT FRIEDMAN

June 1964

The energetic phenomena that take place in stars, galaxies and cosmic space radiate surplus energy in the form of electromagnetic waves ranging in length from a few thousandths of an angstrom unit (a ten-billionth of a meter) to hundreds of kilometers. Until recently, however, astronomers had available for study only those wavelengths able to pass through two principal "windows" in the atmosphere, one in the optical and the other in the radio region of the spectrum. In order to observe the full spectrum of celestial radiation, instruments must be placed above the main mass of the atmosphere. High-altitude balloons are useful for some observations, but the only undimmed view of the heavens is that obtainable from rockets or artificial satellites.

The first celestial body whose radiation in the X-ray region of the spectrum (from .1 to 100 angstroms) was studied intensively by rocket-borne instruments was the sun [see "Rocket Astronomy," by Herbert Friedman; SCIENTIFIC AMERICAN, June, 1959]. Astrophysicists had calculated the amount of X radiation to be expected from various other potential sources and had concluded that X rays from outside the solar system would probably be too faint to be detected by conventional rocket instruments. Exploratory scans of the night sky seemed to bear out this prediction: early flights failed to turn up any new sources of X radiation.

Then in June, 1962, an instrumented rocket launched from the White Sands Missile Range in New Mexico detected an extremely powerful source of X rays in the general direction of the center of the galaxy. Subsequent flights have confirmed the existence of this source and have located other sources elsewhere in the sky. The unexpected intensity and character of the X-ray emissions from these sources have made it necessary to postulate new mechanisms to explain the production of the rays. It now appears that the emissions may originate in the incredibly dense cores of "neutron stars," the tiny, invisible remnants of collapsed supernovae.

The strong X-ray source near the galactic center was discovered by Herbert Gursky, Riccardo Giacconi and Frank R. Paolini of the American Science and Engineering Corporation and Bruno B. Rossi of the Massachusetts Institute of Technology. The discovery was made in the course of an experiment aimed at detecting X rays produced on the moon by impinging solar X rays. For this experiment they equipped an Aerobee rocket with a pair of X-ray-sensitive Geiger counters, each of which had a field of view of about 100 degrees of arc [*see bottom illustration on page 255*]. The rocket was stabilized by rapid spinning, so that its axis pointed toward the zenith. As the flight progressed, the counters repeatedly swept out a great circle of the sky that included the moon.

When the counter records were analyzed, they showed no evidence of X-ray emission from the moon, but they did show a broad peak of X-ray emission somewhere in the direction of the galactic center. The wavelength of these X rays was about three angstroms. When the counters were aimed away from the direction of peak response, the X-ray count fell to a low but steady background level.

The same experiment was repeated twice during the following year by Gursky and his colleagues. A rocket launched in October, 1962, could not confirm the single strong emission source because the galactic center is not visible from New Mexico in the fall, but two possible weaker sources were detected above the steady back-

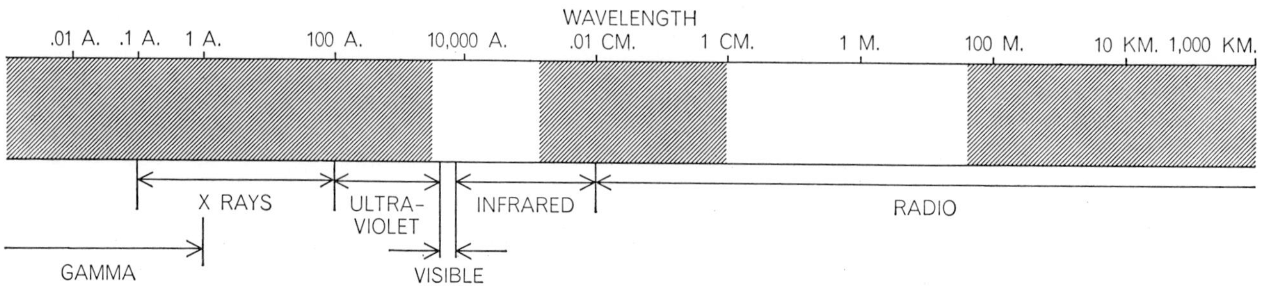

ELECTROMAGNETIC SPECTRUM ranges from short-wavelength, high-frequency gamma rays (*left*) to long-wavelength, low-frequency radio waves (*right*). An angstrom unit (A.) is a ten-billionth of a meter. The X-ray region of the spectrum extends from .1 to 100 angstroms. "Windows" in the earth's atmosphere that admit electromagnetic radiation are indicated by unhatched areas.

ground count. A third flight in June, 1963, succeeded in detecting strong X radiation from the same general region that gave rise to the recording of the previous June.

Meanwhile my colleagues (Stuart Bowyer, Edward T. Byram and Talbot A. Chubb) and I at the Naval Research Laboratory had launched another Aerobee rocket from the White Sands Missile Range in April, 1963. Our rocket was equipped with an X-ray counter about 10 times more sensitive than those flown by Gursky's group, but it covered about the same wavelength range: one to eight angstroms. In front of the counter was a hexagonal honeycomb collimator that limited the field of view to 10 degrees of arc [*see top illustrations on page 255*]. The rocket was deliberately given a slow spin to make it develop a large precession cone, or wobble, around its line of flight. As a result the circle of sky swept out by each spin of the rocket slowly revolved in such a way that almost the entire sky above the horizon was scanned during the flight [*see illustration on page 256*].

At the time the center of the galaxy was below the horizon and therefore invisible to the rocket, but roughly 58 per cent of the celestial sphere was covered. In this large expanse of sky the detector recorded one outstanding X-ray source in the constellation of Scorpius, about 20 degrees away from the direction of the galactic center. Closer examination of the data revealed another source, about an eighth as bright, coincident with the Crab nebula. No other discrete sources were distinguishable above the general background radiation.

As the X-ray counter swept the sky it passed over the Scorpius region eight times. In spite of the relatively rapid rate of scanning, the X-ray signal was strong and clear on each pass. A map of the eight scans was then used to plot equal-intensity contours, which approximated to concentric circles centered at 16 hours 15 minutes right ascension and −15 degrees declination [*see illustration on page 257*]. The detector was not capable of distinguishing between a point source and a diffuse gas cloud as large as two degrees in diameter.

The intensity of the X radiation from the Scorpius source is remarkable. It is comparable to that emitted by the quiet sun in the same wavelength range. Yet the entire neighborhood around the source is devoid of any visibly bright star, nebulosity or radio emission. What kind of celestial object could produce such intense X-ray emission and still re-

AEROBEE ROCKET was launched on April 29, 1963, from the White Sands Missile Range in New Mexico by the author and his colleagues at the Naval Research Laboratory. The rocket was equipped with a sensitive X-ray counter that covered a wavelength range of one to eight angstroms (*see top illustrations on page 255*). In the time that the rocket was above the earth's atmosphere it observed two powerful X-ray sources in the sky: one in the direction of the constellation of Scorpius and another coincident with the Crab nebula.

INSTRUMENT SECTION of an Aerobee rocket has been stripped of its outer skin to reveal its contents. The X-ray counter is behind the circular window at bottom left.

main invisible in the optical and radio wavelengths?

A neutron star seems to meet all the requirements. The concept of a neutron star was first proposed in 1934 by Walter Baade and Fritz Zwicky of the Mount Wilson Observatory and the details of its theoretical structure and evolution were later worked out by J. Robert Oppenheimer and G. M. Volkoff, then at the University of California at Berkeley. Briefly, a neutron star is what remains after the collapse of a large star due to the depletion of its source of energy. A neutron star would contain about as much mass as the sun compressed into a sphere roughly 10 miles in diameter. Its superdense core would consist almost entirely of neutrons and would be up to 100 million times denser than the core of a very compact white-dwarf star. The surface temperature of a neutron star would be about 10 million degrees Kelvin (degrees centigrade above absolute zero) and the temperature in its core as high as six billion degrees. Because of these high temperatures a neutron star would emit about 10 billion times more energy in the form of X rays than in the form of visible light. This is just the reverse of the ratio produced by the undisturbed sun.

Recently Hong-Yee Chiu of the Goddard Institute for Space Studies undertook to develop a model of a neutron star from which some of its radiative characteristics could be determined. He hoped to find out if emissions from such a star could be detected in the ultraviolet or X-ray regions of the electromagnetic spectrum. When we compared his predictions with our observations of the X-ray sources in Scorpius and the Crab nebula, the agreement was surprisingly close. Early this year Chiu and A. G. W. Cameron of the Goddard Institute and Donald C. Morton of Princeton University published more refined calculations of the X radiation to be expected from a neutron-star source; these estimates agreed even better with our observations.

To understand how X rays could be produced in a neutron star, let us consider how such a star may have evolved. A star is supported against the inward force of gravitation by the internal pressure maintained by its high central temperature. If this vital balance is disturbed, the star quickly adjusts to a new equilibrium by shrinking or expanding. Even at the center of the sun, where the density is seven times that of lead, the behavior of the gas never departs from the ideal gas law (pressure is proportional to density times temperature). Inside a white-dwarf star, however, temperatures are not significantly higher than they are inside the sun (about 13 million degrees), yet the density often reaches several tons per cubic inch, and as a result the ideal gas law breaks down. The British astrophysicist Sir Arthur Stanley Eddington suggested in 1924 that the high density of a white-dwarf star could only be explained by supposing that the atoms in the core were completely stripped of their electrons, enabling the bare nuclei to pack tightly together. The sheer squashing of material would produce the condition known as electron degeneracy. Although the nuclei themselves would continue to follow the ideal gas law, the degenerate electrons would now produce a pressure so much higher than the nuclear-gas pressure that the latter could be considered negligible in comparison. If the density in a white-dwarf star were to be increased 100,000-fold, even the nucleons (that is, the neutrons and protons) would be compressed to the point at which they would begin to touch. This condition of nucleon degeneracy must exist in a neutron star.

In the course of evolution of a star such as the sun, the nuclear fuel is consumed slowly. As the source of energy is gradually depleted, the star shrinks and its density increases until electron-degeneracy pressure is approached. Thereafter the rate of cooling becomes progressively slower. Within the age of the galaxy it is not certain that any star has had enough time to cool off completely by this process, but considerable numbers have reached the white-dwarf stage. A typical white dwarf has a density of about 10 tons per cubic inch and a size comparable to that of the planet Jupiter, with only a thin radiating layer at the surface remaining at white heat.

If the mass of a star originally exceeds 1.44 times the mass of the sun (a threshold known as Chandrasekhar's limit) it cannot follow the gradual evolutionary path just described; it is doomed to end its life in a catastrophic supernova explosion. Because the inward force of gravity is too large to be balanced by electron-degeneracy pressure, the temperature of the core must rise to balance gravity. At various temperature levels different nuclear fuels are ignited. Thus the star ages in successive stages, first burning hydrogen to form helium, then helium to form carbon, carbon to form oxygen, neon and magnesium, magnesium to form sulfur and sulfur to form iron. The duration

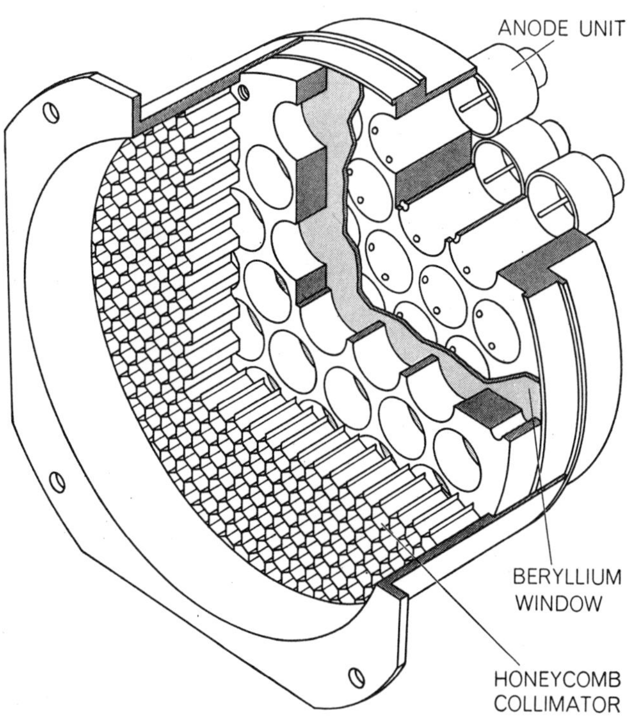

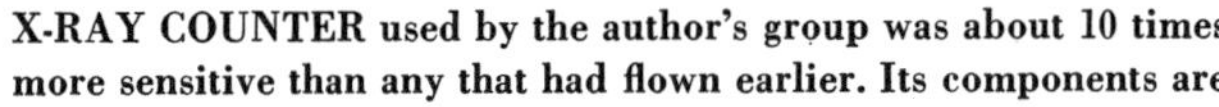

X-RAY COUNTER used by the author's group was about 10 times more sensitive than any that had flown earlier. Its components are shown in the cutaway drawing at left. The honeycomb collimator in front of the counter limits the field of view to 10 degrees of arc.

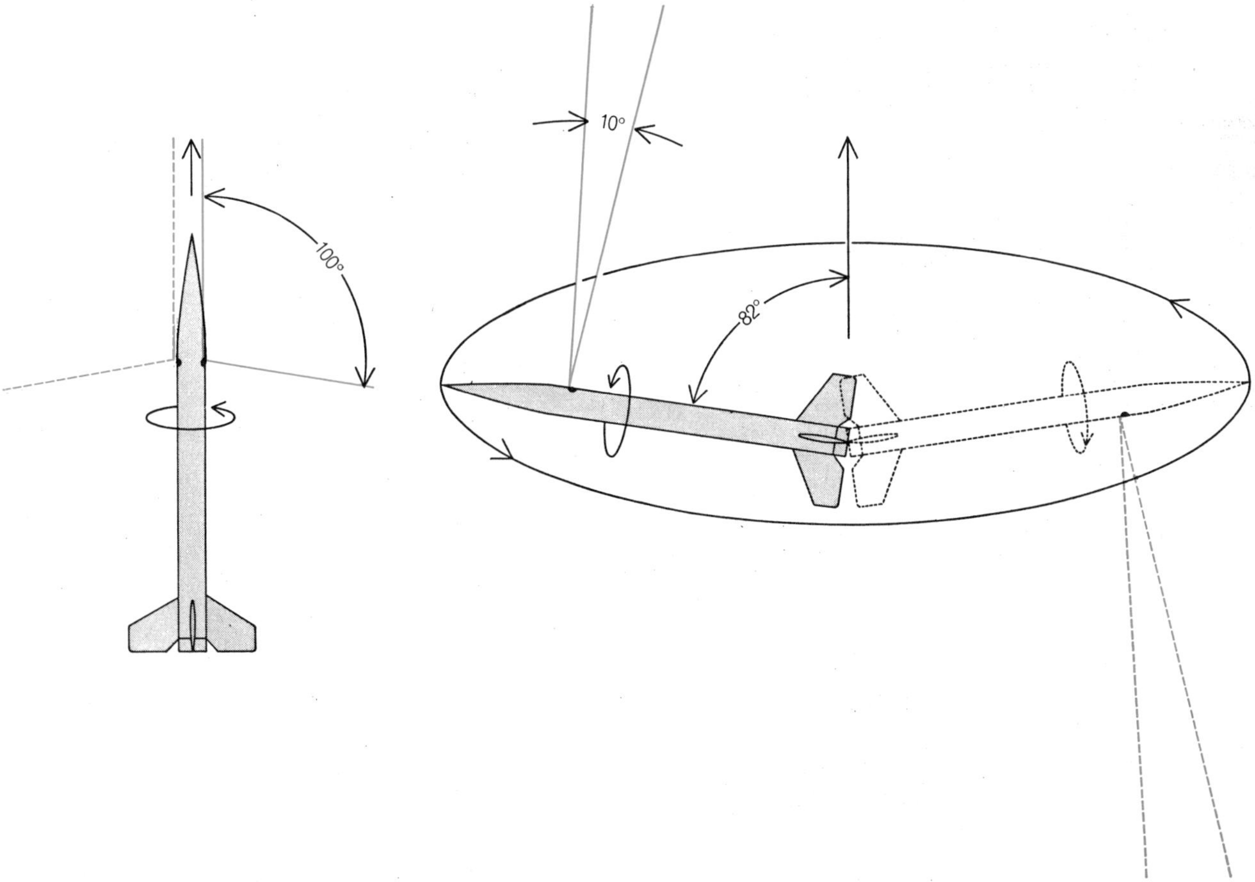

TWO ROCKET-SCANNING TECHNIQUES are illustrated here. The rocket at the left has been stabilized by rapid spinning, so that its axis points toward the zenith. Its X-ray counters sweep out a 100-degree-wide field extending from the horizon to the zenith. The rocket at the right is similar to that flown by the author's group. It has been deliberately given a slow spin to make it develop a large precession cone, or wobble, along its line of flight. As a result the circle of sky swept out by each spin of the rocket slowly revolves in such a way that almost the entire sky above the horizon is scanned during the flight. The 10-degree-wide field of view of the X-ray counter in this rocket enables it to make much more precise measurements than those made by the rocket at the left.

of the various stages of burning may be as long as 10 billion years (hydrogen to helium) or as short as one year (sulfur to iron).

As each stage nears its end, a period of gravitational collapse ensues. These contraction periods may last from 100 to 10,000 years, depending on the mass of the star. The energy derived from each contraction raises the temperature of the core until the next stage of nuclear synthesis is ignited. Temperatures range from a low of 10 million degrees for the fusion of hydrogen up to about five billion degrees for the conversion of sulfur to iron. At the lower temperature levels, surplus energy is dissipated in the form of visible light quanta, or photons. At the higher temperature levels, protons are converted into neutrons with the consequent emission of neutrinos.

If the temperature in the iron core of the star exceeds five billion degrees, electron-positron pairs are produced in such great abundance that their density reaches several tons per cubic foot. At this point a new and still more rapid process may become dominant. Electrons and positrons can collide and annihilate one another to form neutrino-antineutrino pairs. According to Chiu and Philip Morrison of Cornell University, a star can dissipate all its energy by this means in little more than a day. To compensate for the rapid escape

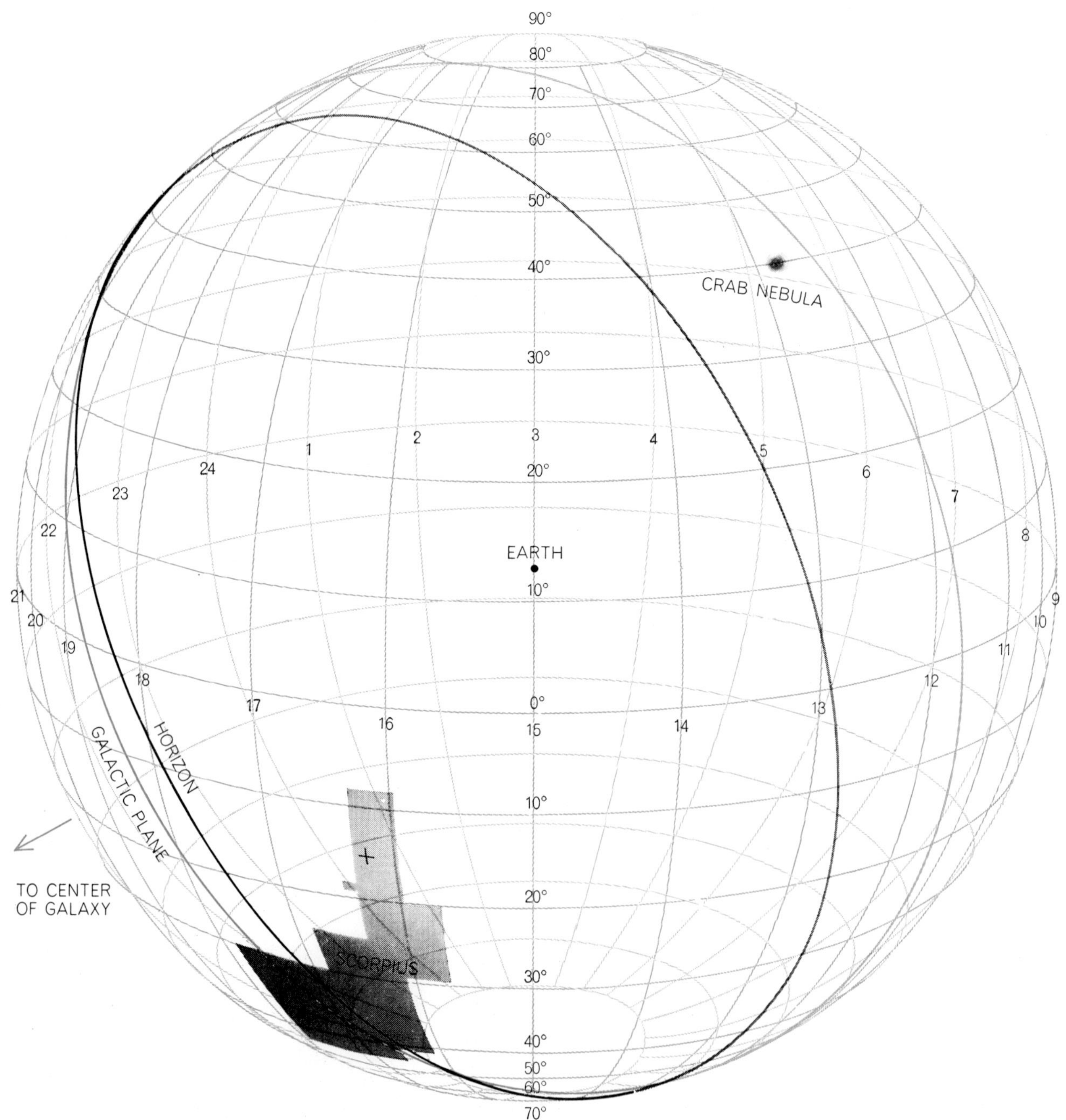

NEARLY 60 PER CENT OF CELESTIAL SPHERE (*area to right of black circle*) **was viewed by the author's X-ray detector during the April 1963 rocket flight. The brightest X-ray source observed is marked by a small cross in the constellation of Scorpius** (*large gray area*)**. Another source, about an eighth as bright, was detected in the vicinity of the Crab nebula** (*gray dot on back of sphere*)**. At the time the galactic center was below the horizon and therefore invisible to the rocket. The numbers down the center of the drawing denote declination (corresponding to geographic latitude) in degrees; the numbers around the middle, right ascension (corresponding to longitude) in hours. An enlarged view of the region around the Scorpius source appears on the opposite page.**

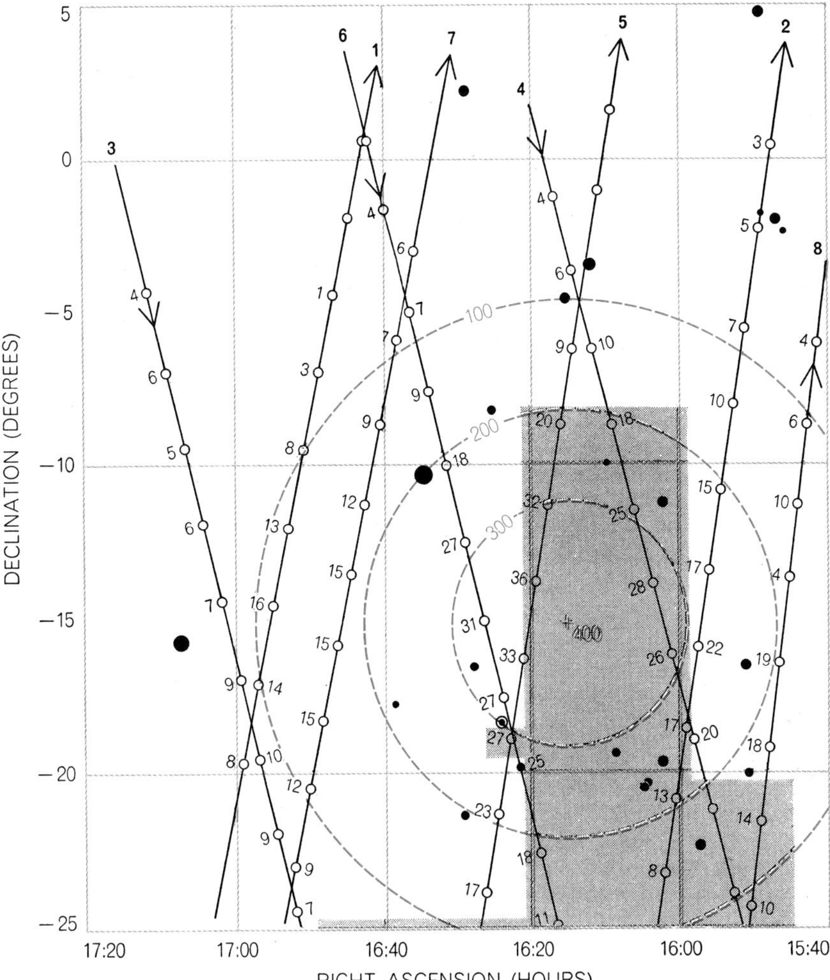

X RAYS FROM SCORPIUS appear to come from a source that is undistinguished by any visibly bright star, nebulosity or radio emission. The source is indicated by a small cross at the center of the three concentric circles. The heavily numbered arrows are tracks made by the X-ray counter as it passed over this region in Scorpius eight times; the lighter numbers next to the small open circles along each track denote the counts recorded per .09 second. The circles approximate equal-intensity contours at 100, 200 and 300 counts per second. The estimate of 400 counts per second at the center is extrapolated from these values. The detector could not distinguish between a point source and a diffuse gas cloud as large as two degrees in diameter. Stars are represented by black dots, which vary in size according to the star's magnitude. A black dot inside an open circle designates a source of radio emission. The gray area represents less than half the total extent of Scorpius.

of energy through neutrino radiation, the star must call on its gravitational energy. It begins to shrink rapidly and the temperature in its core rises accordingly. This rise in temperature in turn increases the rate of neutrino radiation and causes the star to contract still more rapidly; the collapse of the star becomes a runaway process.

As the temperature in the core approaches six billion degrees, the equilibrium shifts suddenly. Iron and related elements are converted into helium, with the release of excess neutrons. The mean binding energy of the 56 nucleons in the common isotope of iron is 8.8 million electron volts (mev) per nucleon and that of the 13 helium nuclei and four neutrons produced by the breakup of iron is only 6.6 mev per nucleon. Thus the breakup process requires an input of energy of 2.2 mev per nucleon. The only available source of energy adequate to support this reaction is the star's gravitation. All the energy the star has used up in going from hydrogen to iron must now be returned to convert iron back to helium. The star resists this rapid process of refrigeration by contracting still further. Gravitational energy is used up without increasing the temperature; the result is a catastrophic implosion. When this total collapse begins, the core has already contracted to a density of between 100 and 1,000 tons per cubic inch. The implosion takes only the time required for free fall: about one second.

With the collapse of the core the pressure that has supported the outer regions of the star is suddenly removed. The outer material also falls inward, almost as rapidly as the core implodes, and the dynamic energy of its fall is converted into heat. The resulting rise in temperature accelerates the nuclear burning of the lighter elements in the outer regions. At three billion degrees the oxygen zone near the surface is burned completely in about one second and a supernova outburst results. If the mass of material in the oxygen zone is equal to one solar mass, the energy released per second is comparable to the normal solar output for a billion years. The exploding supernova shines with a light equal to 200 million suns for two or three weeks.

What happens to the imploded core? When the infalling material exceeds a density of 100 tons per cubic inch and the temperature reaches six billion degrees, protons absorb electrons and are converted into neutrons at such a high rate that the core turns into a mass of neutrons while the collapse is in progress. It is difficult to determine just when the core attains a stable configuration. Present estimates place the limit of stable density at about 100 million tons per cubic inch and the core temperature at about six billion degrees. At this high pressure and temperature, neutrino radiation would dissipate energy so rapidly that the core would not have time to rebound. The result would be a superdense neutron star. (This whole discussion of the formation of a neutron star is, of course, quite general, and many of the details are still in dispute.)

Chiu and Morton have examined the type and intensity of radiation that might be emitted from the surface of a neutron star. Because degenerate matter conducts heat very efficiently, the core must be fairly uniform in temperature. Enveloping the core is a half-mile-thick mantle of electron-degenerate but otherwise normal material similar to that found in the core of a white dwarf. The top few yards of this mantle consist of normal, nondegenerate matter, from which energy is radiated directly into space. The entire thermal gradient from core to surface takes place in this thin topmost layer, which, in Chiu's models, has a temperature of between one million and 10 million degrees.

Morton's models predict a somewhat hotter surface temperature: from about

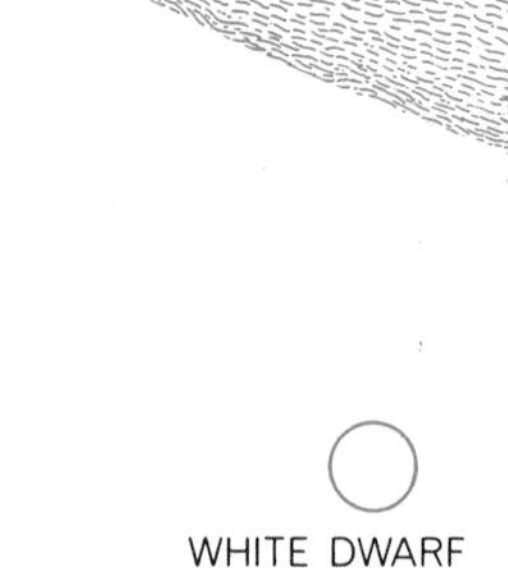

RELATIVE SIZES of the sun, a white-dwarf star and a neutron star are depicted in this drawing. The diameter of the sun (which would be more than 40 inches if the entire sun were shown to scale) is nearly a million miles. The diameter of a white dwarf is about 6,000 miles and that of a neutron star about 10 miles. Yet all three stars have approximately the same mass. The density in the core of

four million to 16 million degrees. X rays produced in the atmosphere of a neutron star at these temperatures would match the measured intensity of the Scorpius source if the star were between 300 and 4,000 light-years away. A neutron star at this distance could not be detected by any existing optical telescope or by any yet contemplated for an orbiting astronomical observatory.

Although the Scorpius X-ray source is detectable only by means of its X-ray emission, the Crab nebula is a conspicuous source of radiation in a wide range of wavelengths. It is the third most powerful source of radio waves in the sky. Its visible nebula is six light-years across and is expanding at a rate of some 700 miles per second. The nebula is the residue of a violent supernova explosion observed by Chinese and Japanese astronomers in A.D. 1054, when it suddenly appeared in the sky with a brightness that temporarily exceeded that of Venus. No visible remnant of the original star has ever been detected in the nebula.

Morton has calculated that a neutron star at the center of the Crab nebula must have a surface temperature of 7.6 million degrees if it is to satisfy the observed X-ray intensity. The apparent visual magnitude of such a star would be 28—far below the limits of detectability.

Although our X-ray observations agree remarkably well with the predictions of neutron-star models, other hypothetical mechanisms for the production of celestial X rays should not be disregarded. It is possible that the observed radiation arises from the rapid gyration of high-energy electrons in a strong magnetic field; such radiation is known as synchrotron radiation [*see illustration "a" on page 260*]. The polarization of light in the tangled filaments of the Crab nebula is characteristic of synchrotron radiation. Whether the X rays from the Crab nebula come from a thermal source, such as a neutron star, or from synchrotron radiation may eventually be settled by more refined spectral measurements.

If we attribute both the Scorpius and the Crab X-ray sources to neutron stars, we are forced to explain why the two sources appear so dissimilar in other wavelengths. In particular, why does the Scorpius source exhibit no radio emission or visible nebulosity? Morton has suggested that since the Crab supernova exploded in the central plane of the galaxy, the galaxy's magnetic field was strong enough to confine the expanding gas to a visible nebula. The Scorpius source, on the other hand, is 20 degrees above the galactic plane and may have exploded in a region where the magnetic field was too weak to confine a nebula. In any case, the statistics on supernova explosions in general are still too few to reveal any clear patterns of appearance and behavior.

Is there any evidence for the occurrence of a supernova explosion in the immediate neighborhood of the Scorpius X-ray source? Oriental records contain several references to brilliant novae or supernovae in Scorpius. Morton has noted four possible supernova explosions in the Chinese and Japanese chronicles that were fairly close to the observed X-ray source. A medieval Arabic reference is particularly intriguing: in 827 Haly and Giafar Ben Mohammad Albumazar in Babylon reported a new star in Scorpius that was as bright as the quarter-moon and was visible for four months. This object, however, may have been a comet.

Assuming that the rate of appearance of supernovae in our galaxy is one a century, there would now be about 100 million neutron stars in the entire galaxy. Of these only a few would have been produced within the past 1,000 years and would therefore have surface temperatures high enough to emit substantial amounts of X radiation. Emission from these sources could be detected at distances of up to 3,000 light-years. Even if there were large dust clouds between the earth and a neutron star, they would absorb very little X radiation with a wavelength shorter than 10 angstroms. Thus, with improved techniques, we may find as many as 50 neutron stars in our part of the galaxy.

The observations of Gursky and his colleagues raised the possibility that the galactic center is itself a source of intense X radiation. According to the Japanese astrophysicists S. Hayakawa and M. Matsuoka, X rays could be produced in gas clouds at the center of the galaxy by the interaction of cosmic ray protons and helium nuclei. X rays produced in this way would be sufficiently intense to be detected by present rocket instruments.

An alternative mechanism for the production of X rays at the galactic center has been proposed by Robert J. Gould and Geoffrey R. Burbidge of the University of California at San Diego. They suggest that various elements, including aluminum and iron, may have

15-ANGSTROM X RAYS

10-ANGSTROM X RAYS

OPACITY OF THE GALAXY to X rays coming toward the earth from extragalactic sources varies with direction. Only X rays

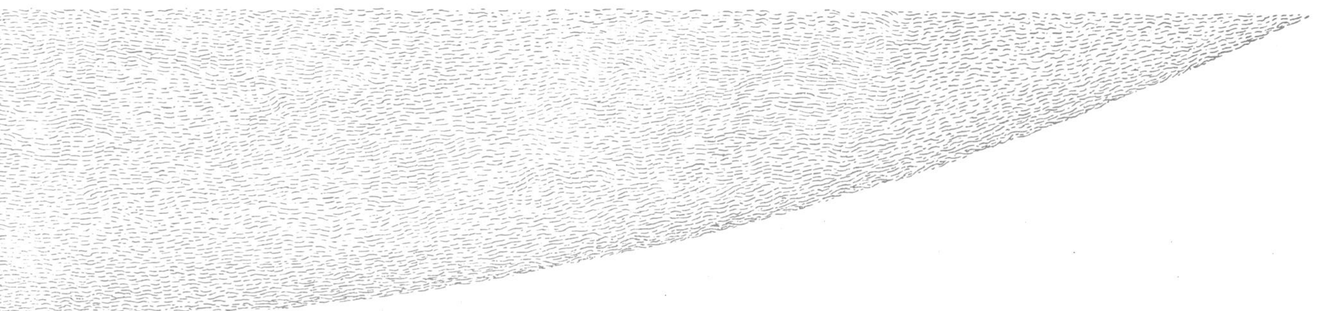

the sun is about two pounds per cubic inch, in the core of a white dwarf about 10 tons per cubic inch and in the core of a neutron star about 100 million tons per cubic inch. In the core of a white dwarf atoms are stripped of their electrons, enabling the bare nuclei to pack tightly together. A neutron star consists almost entirely of neutrons, which are compressed until they touch.

become ionized in this region by losing an electron from one of their inner shells. In this case the ionization would be caused by a shower of protons and electrons emitted by old stars in the neighborhood of the galactic center rather than by cosmic ray particles. The filling of an inner-shell vacancy by an electron from outside the atom or from an outer shell would then lead to a release of energy in the form of an X ray [*see illustrations "b" and "c" on next page*]. The intensity of these X rays would also be within the present limits of detectability.

The origin of the weak but steady background count of X radiation detected on all the flights is difficult to determine. Although this radiation may be produced in the earth's atmosphere, there is substantial evidence that it comes from an external source. In fact, Gould and Burbidge have suggested that it may even originate outside the galaxy. They have calculated the X radiation to be expected from the hypothetical neutron-star populations of all the external galaxies combined; the total intensity is comparable to the observed X-ray background.

It is also possible that the uniform background count may be the result of synchrotron radiation produced by the gyration of cosmic ray electrons in the weak magnetic field of the galactic halo, or corona. The galactic halo is a sphere of low-density gas that surrounds the disk of the galaxy. Computations show, however, that the intensity of X rays produced by this means would be a million times weaker than the observed X-ray background in the direction of the galactic pole and 10,000 times weaker in the direction of the galactic center.

A much more likely source of the weak background X rays is the scattering of cosmic ray electrons by starlight [*see illustration "d" on next page*]. When a cosmic ray electron and a photon of starlight collide, about .01 per cent of the electron's energy is transferred to the photon, which is thereby promoted to the status of an X ray. Morrison and J. E. Felton have computed the intensity of such X rays to be expected from the galactic halo; their figure is between 100 and 1,000 times weaker than the intensity measured in the rocket experiments. If all space were filled with cosmic ray electrons to a density of about 1 per cent of the halo density, however, the computed X-ray intensity would exactly match the observed background intensity.

The steady background of X radiation can be used to test several current cosmological theories. For instance, particularly knotty problem is to expl both the expansion of the universe a the condensation of galaxies if gravity is the only force considered; the energy required to maintain universal expansion against the force of gravity should inhibit local gravitational condensation. One is forced into the unsatisfactory position of postulating vague fluctuation phenomena to initiate condensation. To escape this dilemma Thomas Gold of Cornell and Fred Hoyle of the California Institute of Technology have proposed that a galaxy condenses by local cooling in a hot intergalactic gas, which exerts pressure on the surface of the cooler pocket. As soon as the gas in the compressed region reaches a certain

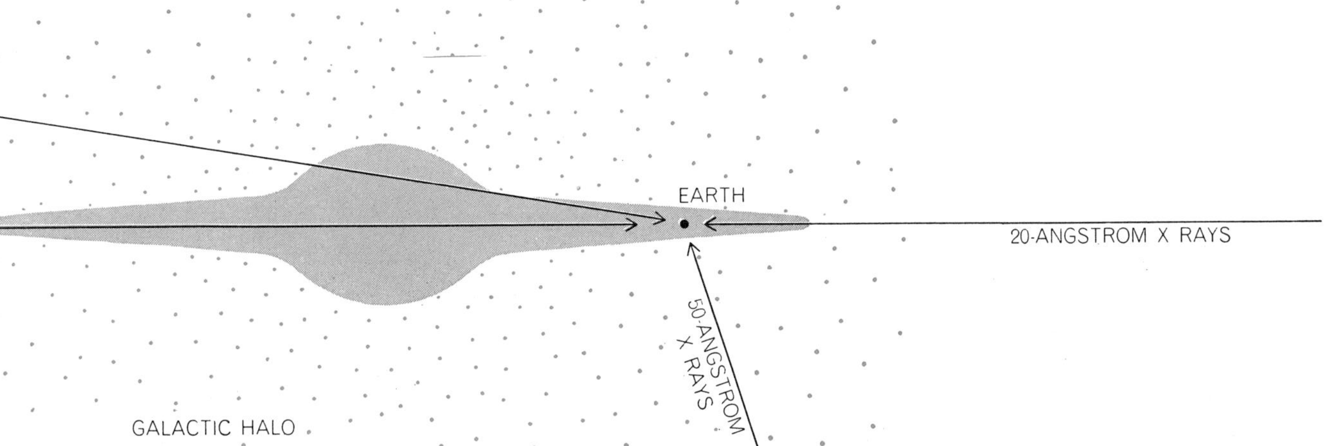

with wavelengths shorter than 10 angstroms can penetrate to the earth from beyond the far edge of the galactic disk. In directions away from the disk the opacity is much less and X rays with wavelengths as long as 50 angstroms may be detectable. Because of the galaxy's spiral structure opacity also varies with direction within the galactic plane. The galactic halo is relatively transparent.

a
MAGNETIC-FIELD LINE
X RAY
ELECTRON

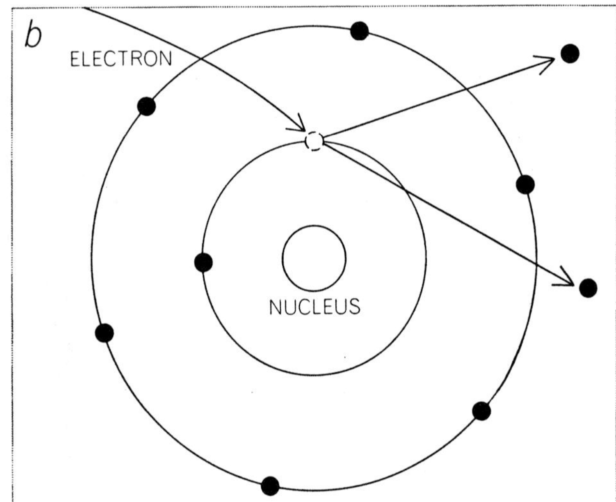

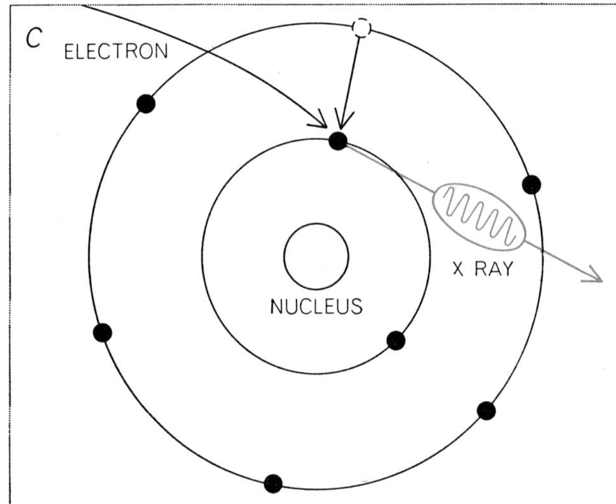

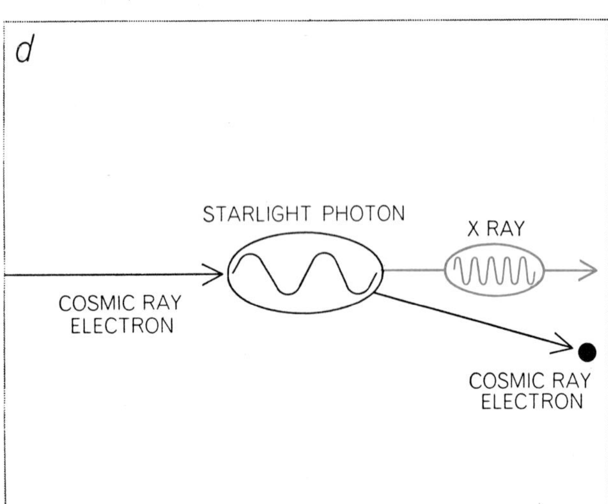

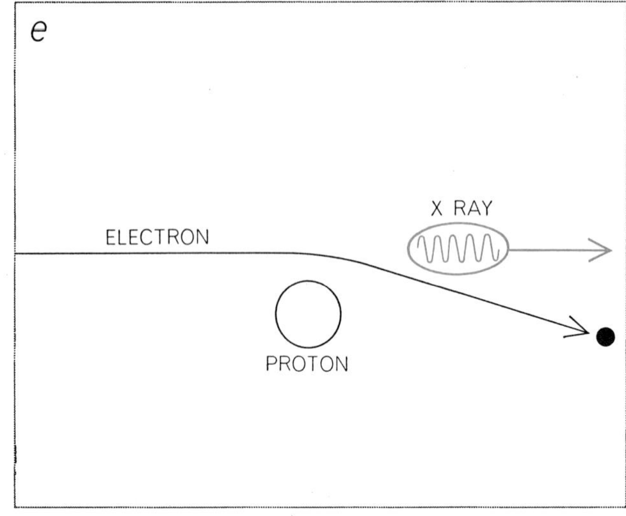

HYPOTHETICAL MECHANISMS for the production of celestial X rays are alternatives to the neutron-star theory. The rapid gyration of high-energy electrons in a strong magnetic field (*a*) throws off a spray of X rays (*colored lines*) in all directions. (Only the X rays emitted in one turn of the electron are shown here.) This process, known as synchrotron radiation, could take place in almost any hot cosmic gas. X rays could also be produced in the gas at the center of the galaxy by a process that begins with the ionization of an inner shell of an atom such as iron by protons or electrons from old stars near the galactic center (*b*). An X ray is emitted when the inner-shell vacancy is filled by an electron from outside the atom or from an outer shell (*c*). (In actuality there would be many more shells than the two shown here.) A likely source for the weak background X rays observed on all the flights is the scattering of cosmic ray electrons by photons of starlight in the low-density gas of the galactic halo (*d*). Cosmic ray electrons emitted by the neutron-decay process would also produce X rays when they interacted with protons in the hot intergalactic gas (*e*). The X rays in *c* through *e* are represented as tiny packets of energy that share some characteristics of both particles and waves.

threshold density, gravitation takes over and completes the process of condensation.

Gold and Hoyle estimate that the temperature of the intergalactic gas required for the first stage of this process must be about a billion degrees. In one model of the steady-state universe, in which neutrons rather than protons are being created continuously, such a temperature can be generated by the decay of neutrons into protons and electrons. This model, called the "hot universe" hypothesis, has the attractive feature of requiring the thermal energy of the intergalactic gas to be comparable to the observed energies of cosmic rays; it thus provides an adequate source of energy for the acceleration of cosmic ray particles to their enormous energies.

The high-energy electrons produced in the hot intergalactic gas by the neutron-decay process would emit X rays when they interacted with intergalactic protons [*see illustration "e" on opposite page*]. The anticipated X-ray intensity from this source, however, is about 20 times higher than the observed background count. The measured intensity would seem to indicate that the temperature of the "hot" universe could be no more than 10 million degrees, rather than the billion degrees suggested by Gold and Hoyle.

The techniques of X-ray astronomy can also be employed to determine the density and distribution of hydrogen gas within our galaxy. The transparency of interstellar gas increases as the wavelength of the transmitted radiation decreases. Hence the wavelength threshold below which radiation is transmitted indicates the density of gas along the line of sight. Just where on the spectrum this threshold will occur depends, of course, on the intensity of the source being observed. For sources of X radiation fairly good wavelength resolution is possible; thus rocket data can be analyzed to determine with fair precision the density of interstellar gas in various directions [*see bottom illustration on pages 258 and 259*].

This technique could also be extended to determine the source of the observed background radiation if it is indeed extragalactic in origin. The intensity of the background radiation at different X-ray wavelengths would indicate whether these emissions originated primarily in galactic halos and intergalactic space or at the centers of external galaxies. A peak at about 50 angstroms would indicate galactic-halo or intergalactic emission; radiation from galactic centers would be filtered down to about 15 angstroms by the gas in these galaxies before being transmitted in our direction.

From the foregoing account it is clear that the first crude results of X-ray astronomy have provided astrophysicists with a wealth of new information. One could hardly imagine a more exciting beginning for the new rocket astronomy than the detection of a neutron star. For the immediate future it is essential to repeat and verify the first observations. Substantial increases in instrument sensitivity are easily possible and should lead to the detection of many new sources. Laboratory versions of large focusing telescopes for X rays have already been built with a resolution better than 20 seconds of arc. To support the neutron-star interpretation it is crucial to prove that the Scorpius and Crab objects are truly point sources. It will take some years, however, before a satellite-borne X-ray telescope can scan a large region of the sky with good enough resolution to pinpoint a neutron star.

In the meantime it may be feasible to take advantage of the lunar-occultation technique that has been applied so successfully for measuring the positions and dimensions of the newly discovered quasi-stellar radio sources [see "Quasi-Stellar Radio Sources," by Jesse L. Greenstein; SCIENTIFIC AMERICAN, December, 1963]. An occultation of the entire Crab nebula by the moon, for example, takes roughly five minutes, or about the same time that an Aerobee rocket remains above the X-ray-absorbing atmosphere. A stabilized Aerobee rocket has recently been developed that can point to a selected object in the sky with an accuracy of better than 2 degrees of arc. It is possible to launch such a rocket, aimed toward the Crab, just before lunar occultation begins. If the X-ray intensity dwindles during the occultation in proportion to the nebular surface blacked out by the moon, we shall have evidence that the source is an extended region emitting synchrotron radiation. But if the source disappears abruptly within one or two seconds of arc we shall have strong evidence that the X-ray source is a neutron star. We shall have an opportunity to perform this experiment next month and we hope to have the rocket and instruments ready. Unfortunately, if this date is missed, we shall have to wait until the next lunar occultation of the Crab nebula, which will not occur until 1972.

CRAB NEBULA in the constellation of Taurus is the second most powerful source of X radiation in the sky. Whether these X rays come from a thermal source, such as a neutron star, or from synchrotron radiation in the expanding gas is not known as yet. The nebula is the residue of a violent supernova explosion observed by Oriental astronomers in A.D. 1054. This photograph was made by red light with the 200-inch telescope on Palomar Mountain.

25

The Infrared Sky

G. NEUGEBAUER AND ROBERT L. LEIGHTON

August 1968

Astronomers have long been aware that if we could see the night sky with eyes sensitive to infrared radiation, it would probably look much different from the sky dominated by the Big Dipper, Orion, Pegasus, Cygnus and other familiar constellations. Until two years ago, however, no one really knew how the night sky would look to an "infrared eye" or what kinds of objects besides stars might appear in it. Although one could be reasonably sure that the few thousand stars visible to the unaided eye would probably not dominate the infrared sky, no one could say how drastically different the infrared sky might look.

To find out we and our associates at the California Institute of Technology have used a specially built 62-inch reflecting telescope on Mount Wilson to conduct a comprehensive survey of infrared sources embracing approximately 75 percent of the celestial sphere. The principal wavelength we used was centered on 2.2 microns, about four times the wavelength of yellow light. The survey disclosed about 20,000 infrared sources in all. As we expected, the great majority are stars. If, in order to proceed with maximum confidence, we count only sources that have 2.5 times the minimum detectable brightness, we are still left with 5,500 sources. For purposes of comparison, a total of about 6,000 stars could be counted in a visual survey made with the unaided eye at the same latitude. In short, our survey has employed an infrared "eye" that detects about as many stars as one can see in the night sky with the unaided eye.

Are they mostly the same stars or mostly different ones? The answer is mostly different. The number of stars visible to both kinds of detector, visual and infrared, is less than 2,000. Thus about 70 percent of the 6,000 stars visible to the unaided eye cannot be detected by our 62-inch infrared eye; similarly, about 70 percent of the 5,500 brightest sources in our survey are not visible to the unaided eye. If man had viewed the night sky with infrared-sensitive eyes, he would have constructed an entirely different assortment of constellations.

Early Infrared Surveys

It is largely an accident of uneven technological development that the sky was extensively surveyed at radio wavelengths more than 20 years before this could be done at two-micron wavelengths in the infrared. Prior to our work infrared studies at two-micron wavelengths were carried out mainly on individual objects such as the sun, the planets and a selection of bright stars [see "Infrared Astronomy," by Bruce C. Murray and James A. Westphal, which begins on page 174 in this book].

As long ago as 1840 Sir William Herschel demonstrated that the sun emits invisible radiation beyond the red end of the visible spectrum. The visible spectrum of course covers only a tiny fraction of the total range of possible wavelengths, which is theoretically infinite. The wavelengths of visible light lie between .4 micron (violet) and .7 micron (red). The peak of the sun's energy falls near the middle of this region, at about .5 micron. Stars cooler than the sun, which has a surface temperature of 5,700 degrees Kelvin, would emit most of their energy in the infrared region beyond .7 micron. Stars as cool as 3,000 degrees K. are well known and a few stars cooler than 2,000 degrees had been found before our survey.

Estimates of the relative numbers of stars of various kinds per unit volume of space have consistently shown that the coolest stars—red dwarfs and the dark companions of hotter stars—account for most of the stellar matter in space. This suggested the possibility that even more matter might be in the form of "dark stars": stars too cool to emit visible light. Astronomers have also believed that protostars (stars in the process of formation) should exist and might be detectable with the appropriate equipment. The temperature of protostars might be as low as a few hundred degrees K. It has been to search for such objects and to extend knowledge of infrared sources in general that astronomers have long sought to push their observations as far into the infrared part of the spectrum as possible.

In the late 1930's Charles W. Hetzler of the Yerkes Observatory searched for cool stars using photographic plates sensitive to the near infrared in conjunction with the 40-inch Yerkes refractor. Perhaps because he never completed the survey, his work did not make much of an impact. He did, however, find a number of stars whose temperatures were between 1,000 and 2,000 degrees K.

Our interest in making an infrared sky survey was more directly stimulated by

ORION NEBULA contains one of the reddest and thus presumably one of the coolest stars yet discovered in infrared sky surveys. It cannot be seen in photographs made with the largest telescopes. The color photograph on the opposite page was made with the 120-inch reflector at the Lick Observatory. The site of the infrared object lies hidden at the northern end of the glowing mass of gas; the exact location is given by the short white lines at the sides of the photograph, which mark the object's vertical and horizontal coordinates. The visible stars in the neighborhood of the infrared object can be seen more clearly in the black-and-white photograph that is reproduced at the bottom of page 271.

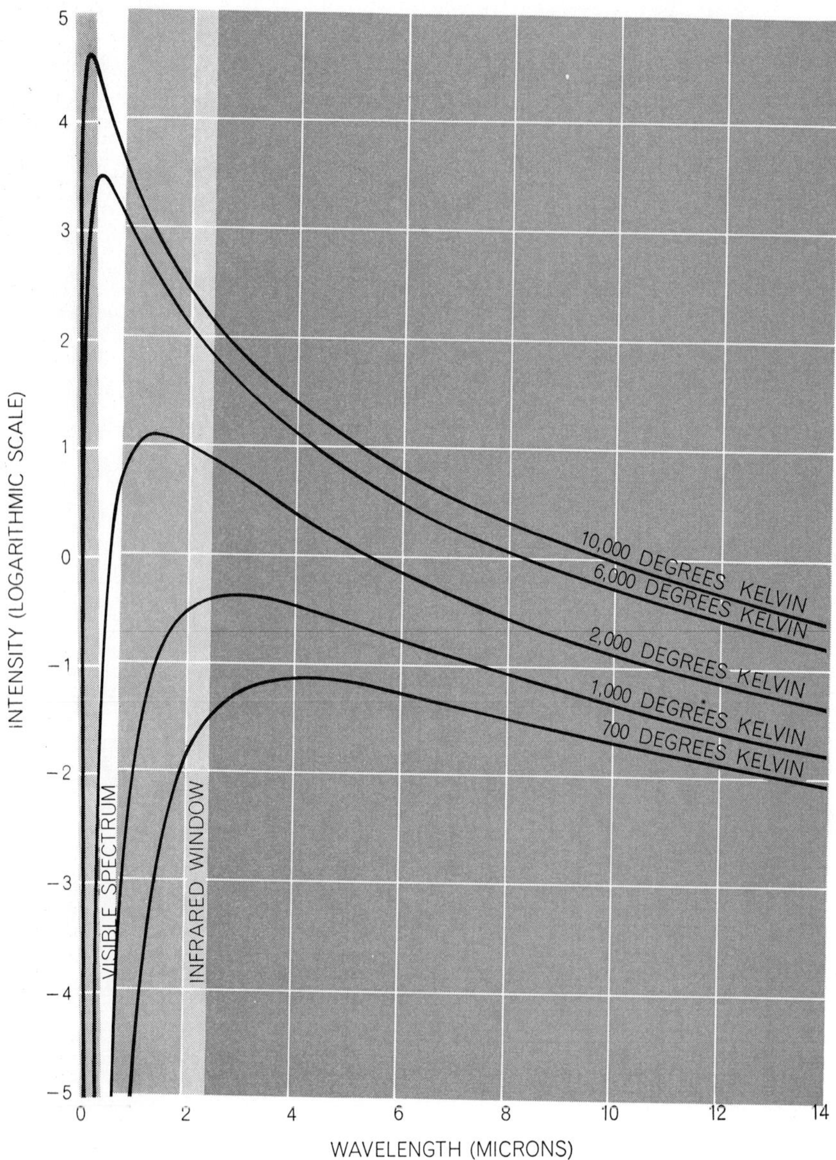

WAVELENGTH OF RADIATION emitted by heated bodies varies with the absolute temperature and follows Planck's law of black-body radiation. Ordinary stars range in temperature from about 2,500 degrees to 25,000 degrees Kelvin. Only about 1 percent of the radiation emitted by a 2,500-degree star is in the visible region, whereas about 10 percent lies in the band between two and 2.4 microns, chosen by the authors for their infrared survey.

the work of Freeman F. Hall, Jr., of the International Telephone and Telegraph Corporation, who mounted an array of infrared detectors cooled by dry ice on a 24-inch reflecting telescope. Using this instrument in the San Fernando Valley, Hall scanned some 20 percent of the Northern Hemisphere sky looking for cool objects. He seemed to find a number of sources not shown on star maps, but he also failed to detect some stars that were expected to be very bright in the infrared.

Building an Infrared Telescope

About five years ago we decided it would be worthwhile to make a comprehensive, and thus unbiased, search of the sky for infrared emission using sensitive detectors coupled to the largest possible telescope. It was evident that the existing large astronomical telescopes were unsuited for such a survey, both because of the large amounts of time that would be needed and because of the telescopes' relatively narrow field of view. We therefore decided to build a telescope specially suited to the problem.

It is well known that the free surface of a liquid, rotating uniformly around a vertical axis, assumes the shape of a paraboloid, which is precisely the surface needed to bring parallel rays to a focus by reflection. Exploiting this fact, we poured a slow-setting epoxy resin on a rotating aluminum "dish" that had first been shaped on a lathe to approximately the right configuration. A constant rate of rotation was maintained for three days while the epoxy hardened. After considerable effort and several attempts the technique provided us with a usable 62-inch mirror having a 64-inch focal length. Although the mirror is somewhat lacking in perfection, its aluminized epoxy surface is satisfactory for the long wavelengths and coarse detectors used in infrared astronomy. With the help of students we built a telescope mounting and a small building to house it on Mount Wilson [*see bottom illustration on opposite page*].

Of equal importance to the design of an efficient telescope is the choice of the spectral region to be studied and the design of the detector system to be used. Astronomers who want to observe the infrared emission from celestial objects face several handicaps. One is the lack of sensitive infrared-detection systems. This limitation has been alleviated in the past 10 years, but infrared detectors are still less sensitive than good visible-light detectors by a factor of at least 1,000. If infrared detectors were as sensitive as visible-light ones, we could have made our survey with a two-inch mirror.

A more fundamental limitation is that gases in the earth's atmosphere, chiefly water vapor, absorb most of the incoming infrared radiation. Beyond the .7-micron limit of the visible spectrum the atmosphere remains fairly transparent out to about 1.3 microns, beyond which it is opaque except for a few transmission "windows." The first few of these are centered at the following wavelengths in microns: 1.65, 2.2, 3.6 and 4.8. There is a relatively broad window between eight and 14 microns and another between 17 and 22 microns. From 22 microns to 1,000 microns (one millimeter) the atmosphere is largely opaque. The last good window available to the infrared astronomer, the region from one millimeter to three millimeters, carries him into the domain of the radio astronomer, who employs antennas and wave detectors rather than the photodetectors that are normally used in conjunction with optical reflecting telescopes.

A final problem confronting the infrared astronomer is that the entire world radiates. Opaque nonmetallic bodies radiate energy whose wavelength is distributed according to Planck's law of black-body radiation [*see illustration on this page*]. The spectral distribution of black-body radiation depends on the absolute temperature (T) of the body in such a way that the maximum energy (per unit wavelength interval) is radiated at a wavelength (λ_{max}) given by Wien's displacement law ($\lambda_{max}T \cong 3{,}000$, where λ_{max} is in microns and T is in degrees Kelvin). Thus the peak of the sun's energy, corresponding to its temperature of 5,700 degrees K., is at about .5 micron, whereas objects at room temperature (300 degrees) radiate with a maximum intensity near 10 microns and emit negligible radiation at visible wavelengths. Therefore the infrared astronomer faces a problem comparable to that of an optical astronomer working in a lighted dome with a luminescent telescope.

We selected the window between two and 2.4 microns for our sky survey. Factors that influenced this choice were the favorable response characteristics of lead-sulfide photoconductive detectors, the good transparency of the atmospheric window at this wavelength and the fact that it provides a significant step beyond the visible. Stars with temperatures around 1,000 to 1,500 degrees K. would emit the peak of their energy in this wavelength region. In order to cover a suitably wide swath in the sky, a linear array of lead-sulfide detectors was used. In addition a silicon photodetector was employed to measure the incoming radiation between .7 and .9 micron, providing a basis for defining a "color" of the detected objects. All the detection is done electronically; no visual observations are made with the telescope.

To eliminate sky emission and the background radiation of objects at room temperature the mirror is rocked gently at 20 cycles per second while the telescope and detectors remain fixed. The background radiation, being constant over a large part of the focal plane, gives rise to a steady, or direct, current in the detector. On the other hand, a small focused source, being shifted alternately on and off the detector by the vibration of the mirror, gives rise to an alternating current. The subsequent amplification of the signal current is arranged to ignore the direct-current component and to enhance the alternating one. This scheme works so well that at two microns it is possible to detect stars almost as effectively during the day as at night. (The Rayleigh scattering of sunlight, which is responsible for the blue color of the sky, is almost 100 times weaker at two microns than at .5 micron.) At .7 to .9 micron, however, there is too much background radiation during the day to allow daytime operation.

The entire telescope is automatically

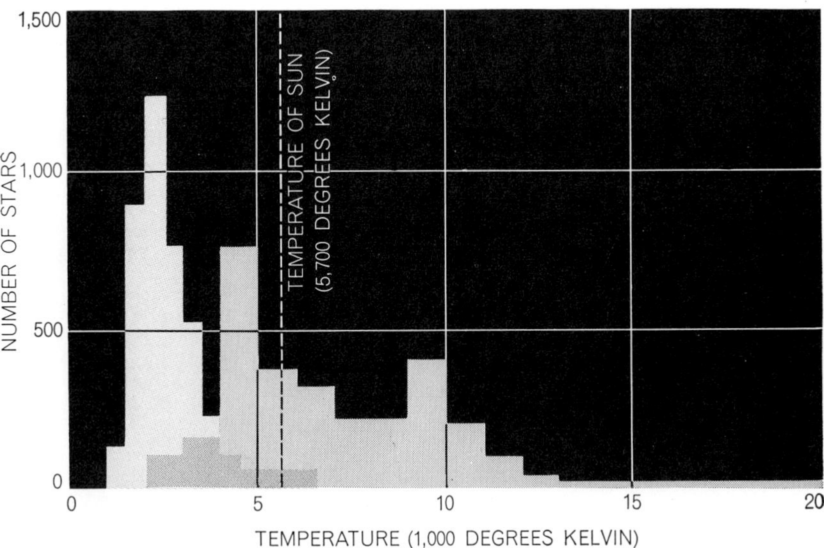

APPARENT TEMPERATURE RANGE of the 5,500 brightest stars in the authors' infrared survey is represented in color. Judged solely by their redness, most fall between 1,500 degrees and 3,500 degrees K. The temperature range of some 6,000 stars visible to the unaided eye is shown by the gray area. The areas overlap slightly between 2,000 and 6,500 degrees K.

INFRARED SURVEY TELESCOPE on Mount Wilson has a 62-inch plastic-on-aluminum mirror with a 64-inch focal length. The mirror's parabolic surface was formed by pouring liquid epoxy resin on an approximately contoured aluminum disk and rotating the disk at constant speed for three days while the epoxy hardened. The epoxy surface was then aluminized. The infrared sensors, chilled by liquid nitrogen, are supported on four legs at the focus of the mirror. When the telescope is not in use, the eight petal-shaped flaps are lowered to protect the mirror. The corrugated roof can be slid forward to provide a weatherproof housing for the instrument. The telescope and the housing were designed and built by the authors with the help of students at the California Institute of Technology.

programmed to scan the sky in a raster pattern. The telescope sweeps east to west through 15 degrees of arc at about 20 times the rotation rate of the sky and then steps north by 15 minutes of arc; it sweeps 15 degrees in the opposite direction, again steps 15 minutes north, and so on. In this way, during each hour of observing, a strip of sky measuring three degrees from north to south and 15 degrees from east to west is covered. During the course of a year the instrument can survey essentially all but the 25 percent of the sky that is too close to, or below, the southern horizon. It has now been used to scan the entire northern sky twice.

Analysis of the Infrared Survey

As we have noted, some 20,000 infrared sources have been detected with our 62-inch infrared telescope. Only the 5,500 brightest sources will be compiled into a catalogue of infrared stars. Those that are less than 2.5 times the minimum detectable brightness will be omitted, because it is important, particularly for statistical work, to have confidence in the completeness of such a survey. Relatively few of the 5,500 (fewer than 30 percent) emit enough visible light to be seen with the unaided eye on a clear night. And approximately a third have visual magnitudes fainter than 10.5, which means they are about 100 times too faint to be seen with the unaided eye.

Whether or not a given star detected in our survey is among the 6,000 that can be perceived visually depends on its temperature—at least its apparent temperature. Typical stars range in temperature from about 2,500 to 25,000 degrees K. For a star with a temperature of 25,000 degrees roughly 2 percent of the total energy output is in the visible region, and less than .05 percent is in the infrared band observed in our survey. In contrast, only 1 percent of the radiation emitted by a 2,500-degree star is in the visible region, whereas 10 percent is in the two-micron infrared band. As a result a star that is on the threshold of detection in our survey would be easily visible as a second- or third-magnitude star if its temperature were about that of the sun, but if its temperature were 1,000 degrees, it would be invisible to a visual observer even with the 200-inch telescope!

These ideas become more concrete if we plot the apparent temperatures of the stars in our infrared survey and the apparent temperatures of the unaided-eye stars [*see top illustration on preceding page*]. It then becomes clear that stars that are bright in the infrared are cooler than stars that are bright visually. These temperatures cannot always be interpreted, however, as kinetic, or actual, temperatures.

A very red and apparently cool star need not be cool. It is well known that interstellar dust will scatter blue light and make a star appear redder, and thus cooler, than it actually is. This effect, however, provided another reason why a search of the sky in the infrared might turn out to be extremely revealing: the extinction of starlight in the infrared is much less than in the visible. In fact, in the direction of the center of our galaxy radiation of two-micron wavelength is attenuated by a factor of only 10, whereas we estimate that visible radiation is attenuated by a factor as large as 10 billion. Not only can stars be reddened by interstellar dust but also their spectral distribution may depart from Planck's law of black-body radiation. The general trends of temperature are nonetheless correct.

The illustrations on these two pages show the distribution in the sky of the brightest stars observed visually and the brightest stars observed at a wavelength of 2.2 microns; about 300 stars in each category are represented. Although the familiar constellations are no longer evident in the infrared, both distributions look more or less random and qualitatively the same. A few of the well-known bright reddish stars, such as Betelgeuse, are present on both charts, but as one would expect the hotter white stars do not stand out conspicuously in the infrared.

The distribution of the 300 faintest stars seen with the unaided eye and of the 300 faintest stars observed in the infrared survey are plotted together in the top illustration on pages 268–269.

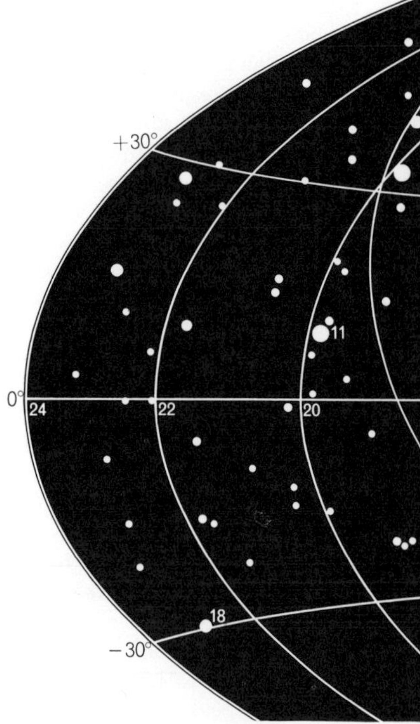

BRIGHTEST VISIBLE STARS, meaning stars visible to the unaided eye, are distributed more or less at random. The numbers

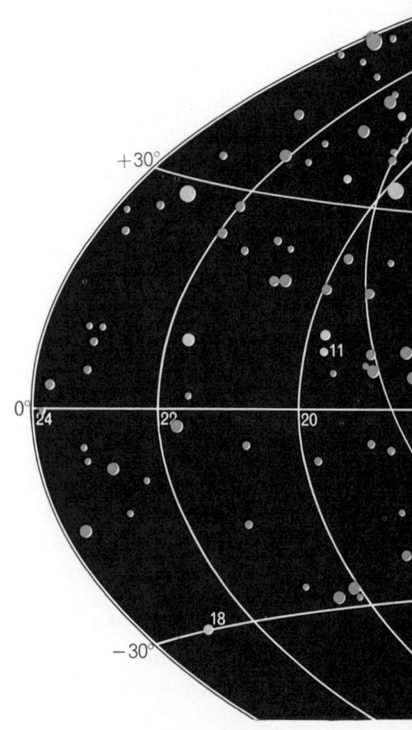

BRIGHTEST INFRARED STARS, meaning stars that are brightest at the survey wavelength of two to 2.4 microns, are also dis-

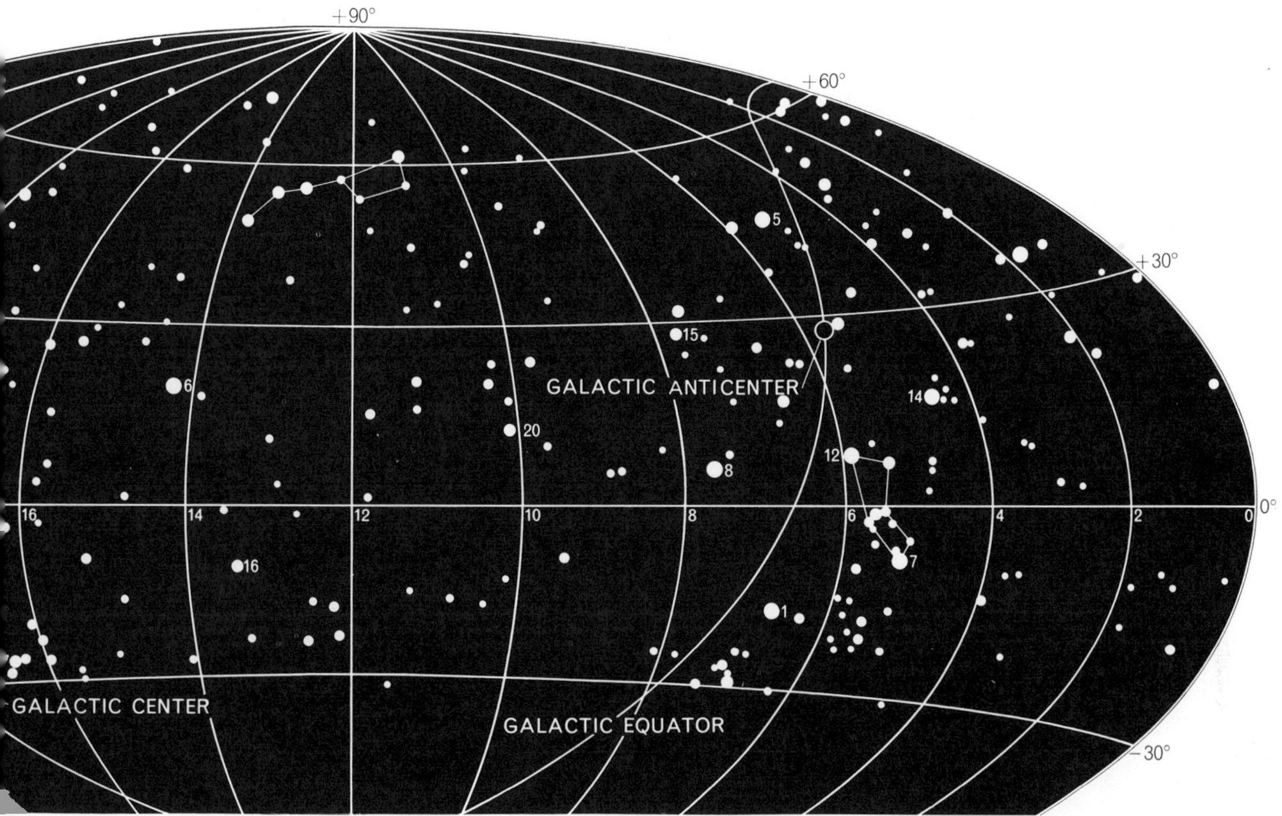

identify 15 of the 20 brightest stars of all; their names are given at the top of the adjacent column. Five, marked by asterisks, are not shown in the sky chart because they are permanently below the southern horizon at the latitude of Mount Wilson. In this and subsequent sky charts the number of stars plotted is about 300. Numerals on the horizontal axis represent right ascension in hours.

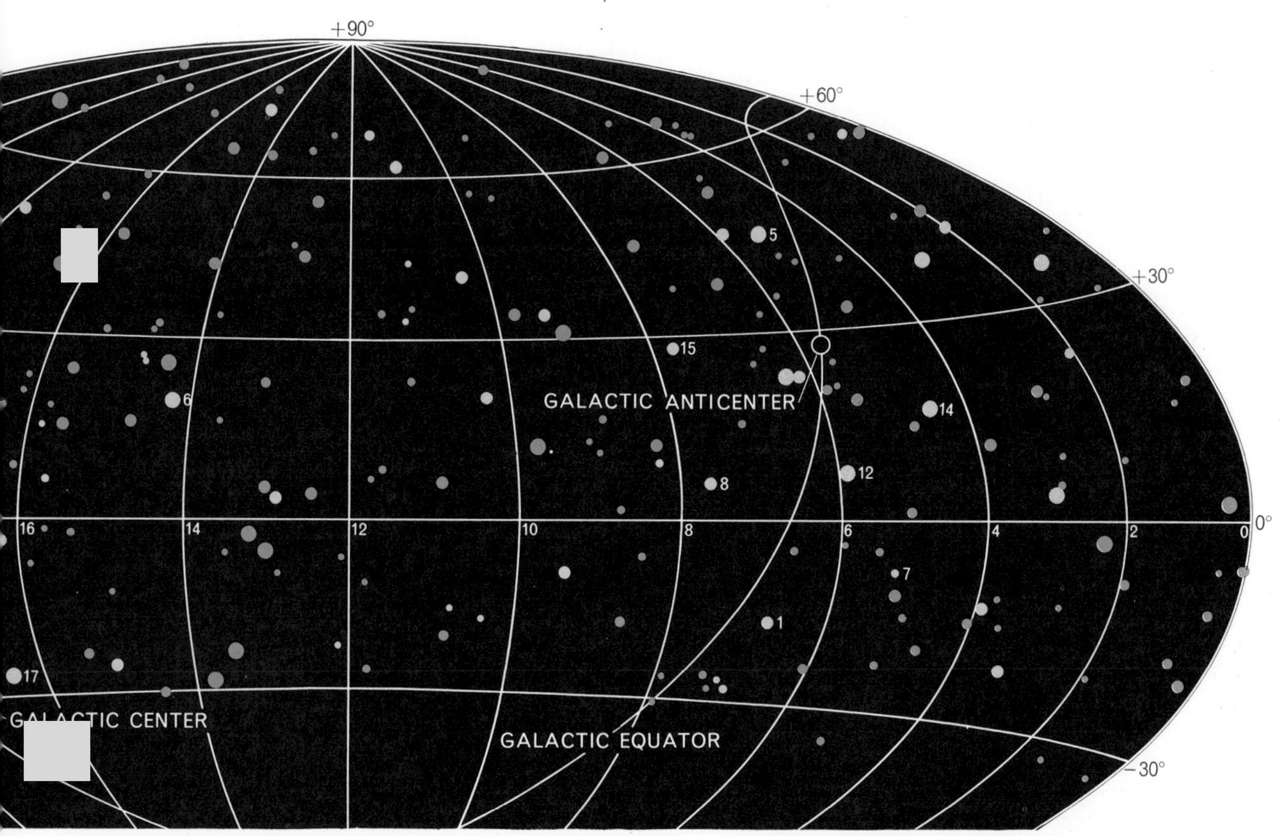

tributed more or less randomly, but only about 20 percent of them (*pale color*) are among the 300 brightest visible stars. Some of the others, however, would be visible as faint stars to the unaided eye. A star may appear red because it is genuinely cool or because its light has passed through interstellar dust, which preferentially scatters and reduces the blue component of the original light.

Only three stars are common to the two categories. Although the distributions are again roughly similar, the randomness is gone and the Milky Way—the central plane of our galaxy—begins to stand out. Perhaps it is even a little more pronounced in the infrared than it is in the visual. The Milky Way becomes very conspicuous if we plot the distribution of stars selected on the basis of redness, as is done in the bottom illustration on these two pages. In this illustration it is apparent that the survey can be used as a powerful tool for probing the structure of the galaxy.

The increasing importance of the Milky Way in the plots of the fainter stars can be explained on a simple basis. On the average the bright stars are relatively close to us and we see roughly the same number in any direction, since within this range of observation the galaxy is nearly uniform. The fainter stars, however, are generally so far away that we can see them well beyond the confines of the galactic disk when we look at right angles to its central plane. In this direction we run out of stars to be seen. On the other hand, when we look along the galactic plane, the density of stars remains high as far as the telescope can see; in fact, the gradual increase in the density of stars in the direction of the galactic center, and the decrease in the direction of the anticenter, stand out clearly.

By means of indirect arguments the facts given above can be used to delineate certain features of the galaxy as it is viewed in the infrared. For example, the cool stars observed in the survey enable us to estimate that the thickness of the galactic disk is approximately 400 parsecs (one parsec is 3.26 light-years), which is about the same as has been estimated on the basis of hotter stars. Another finding is that in the plane of the galaxy our telescope can detect very red giant stars out to an unexpectedly great distance: some 2,000 parsecs. Over the range observed we find that the galaxy thins out with distance from the galactic center: near the sun, which is about 10,000 parsecs from the center, the star density decreases by a factor of 2.5 with each 1,000 parsecs. To be sure, these simple concepts are confused by interstellar reddening, and we have had to make some guess as to how this reddening affects the data. Whereas the effects of the reddening and the question of the intrinsic redness of objects cannot be separated out uniquely, the infrared data should be much less confused by reddening than similar visual data are.

Infrared star counts may also have a bearing on the structure of the galaxy. It is known that at our distance from the galactic center the galaxy makes a complete turn in about 200 million years. On this basis, and from the size and shape of the galaxy, one can compute how much mass the galaxy must have. It turns out that less than half of the mass required by this analysis can be accounted for by detectable stars, gas and dust. One suggestion that has been made is that invisible stars—perhaps infrared stars—may be quite numerous and might account for a significant part of the total mass. The absence of large numbers of faint but randomly distributed stars in our survey indicates that such objects are either absent or lie below the detection threshold of our equipment. The faintest infrared stars we do see, being intrinsically bright stars but at a great distance, account for a very small part of the total mass. Thus the missing mass remains unaccounted for.

The Coolest Objects Observed

Although the statistical results that have come from the survey have been valuable, we have been most interested in the individual very red objects detected. At first we expected that most of these objects would belong to the family of cool variable stars with periods ranging from a few months to several years. (One reason the sky was surveyed in two different years was to estimate how many stars might have been missed on a single survey because they were at a low point of their light curve; another was to obtain a statistical measure of the fraction of survey stars that are variable.) We thought if we were lucky we might also be able to detect protostars. These objects might be in the form of extended blobs of dust and gas at a temperature of only a few hundred degrees K., and might represent a turbulent mass of matter assembled by its own gravitational attraction just before the formation of one or more stars. The survey was designed to detect concentrated objects as cool as 400 degrees K., which meant that we could hope to detect stellar objects that were only a little above the boiling point of water (373 degrees), if any such objects existed in the sky and were near enough to us.

Of the 5,500 stars observed, 450 are so red that at least 97 percent of their total energy output is in the infrared beyond one micron, which corresponds to black-body temperatures below about 1,700 degrees. Almost all these stars are, as we had expected, long-period vari-

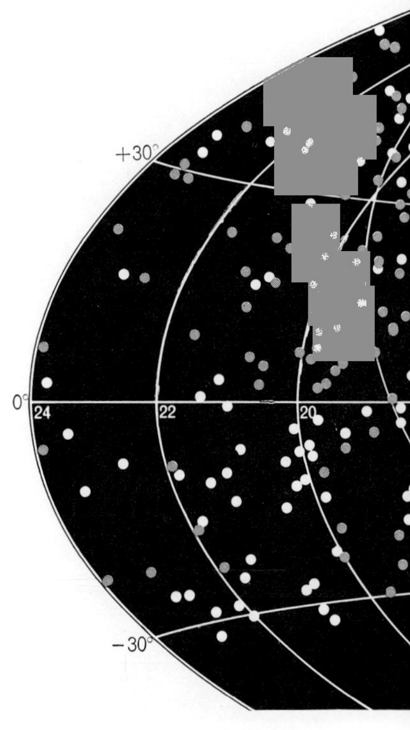

FAINTEST STARS, both visible (*white*) and infrared (*color*), tend to be more numerous near the galactic equator for the

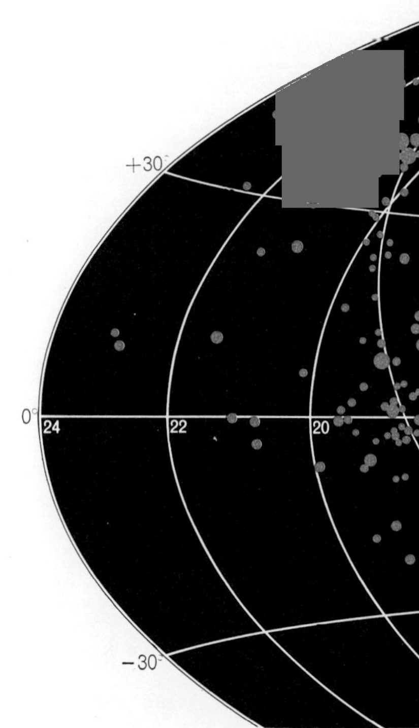

REDDEST INFRARED STARS in the survey show an unmistakable concentration along the galactic equator and particularly

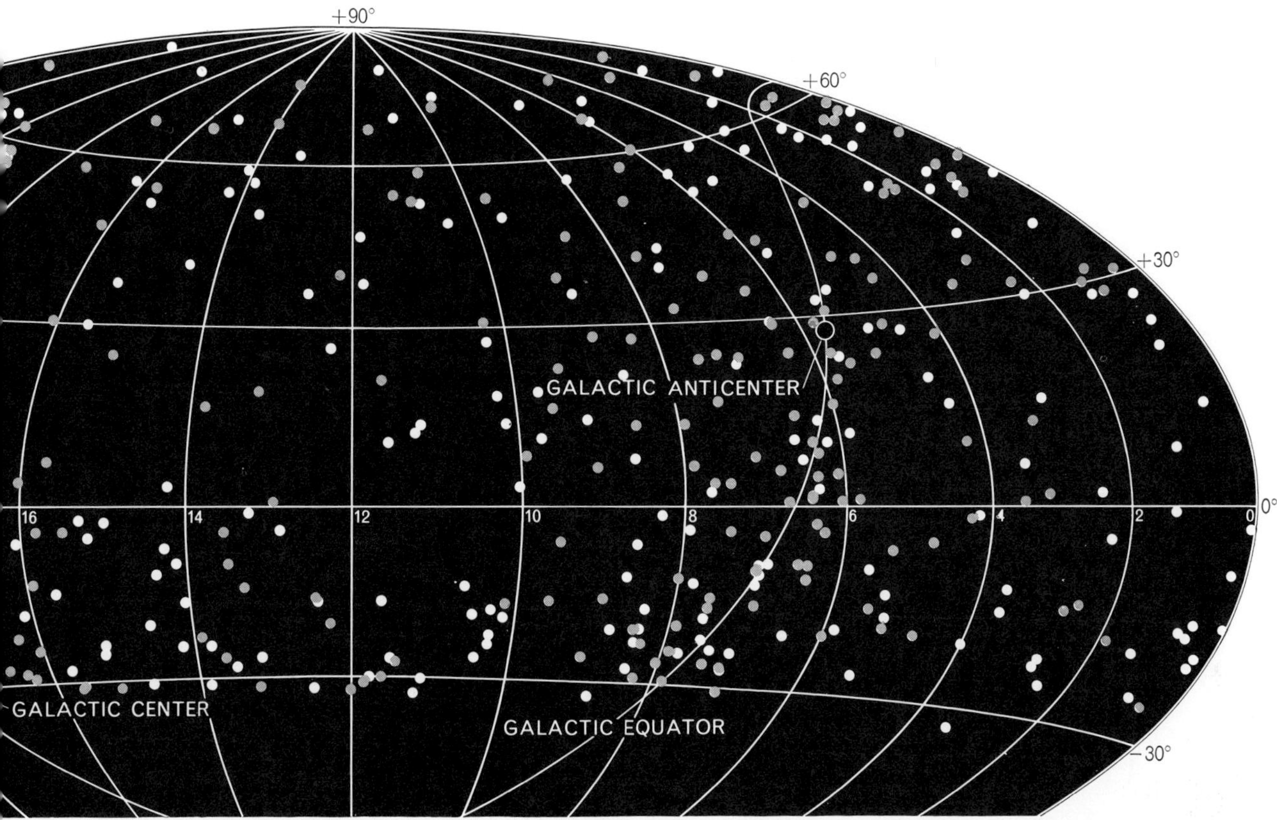

simple reason that there are more stars to be seen when looking into the galactic plane than when looking at right angles to it. Only three of the 300 faintest infrared stars in the survey are included among the 300 faintest visible stars. Among the 5,500 brightest infrared stars in the survey, about 30 percent are among the 6,000-odd stars visible to the unaided eye from the same latitude.

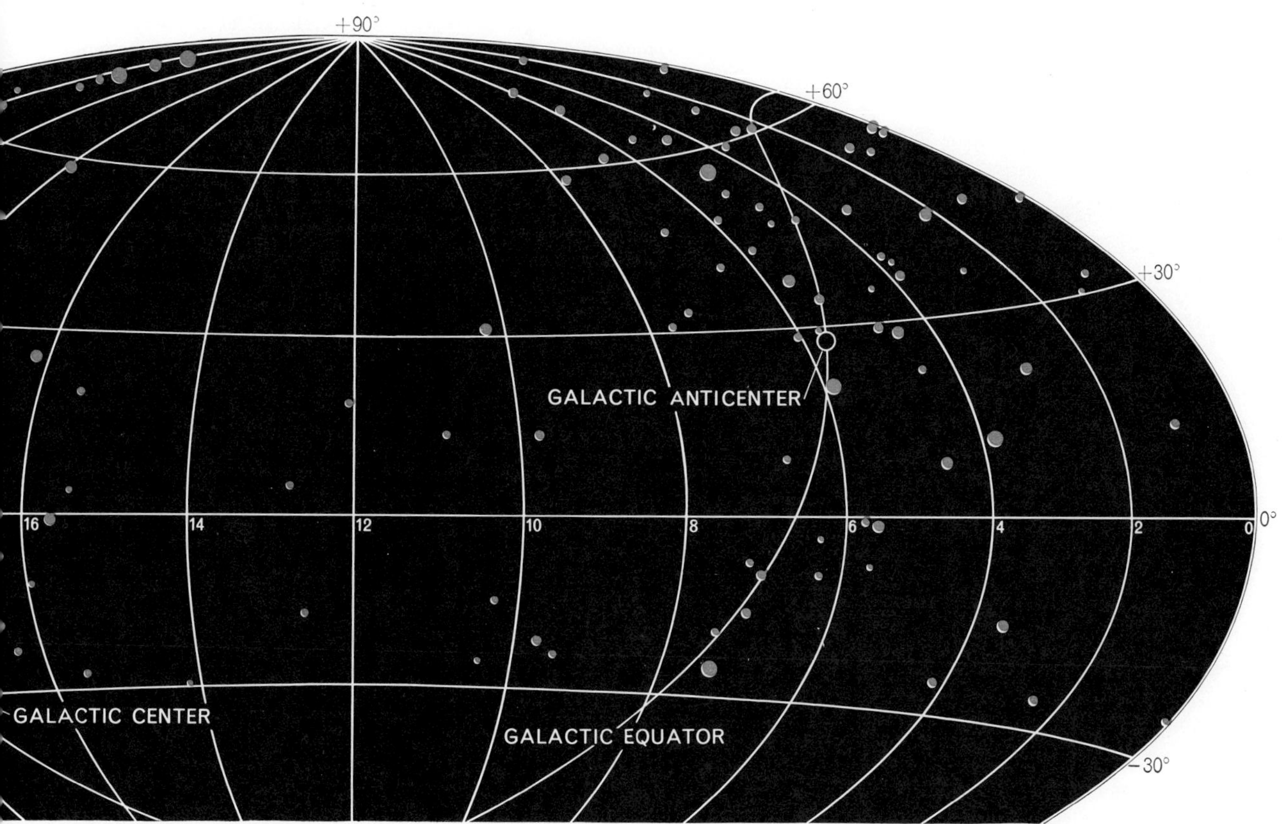

in the general direction of the galactic center. It is estimated that interstellar dust dims visible starlight originating near the center of the galaxy by a factor as large as 10 billion. Infrared radiation of two-micron wavelength, however, is attenuated by a factor of only about 10. The apparent temperature of all the stars plotted here is below 1,700 degrees K. Many are found to be long-period variables.

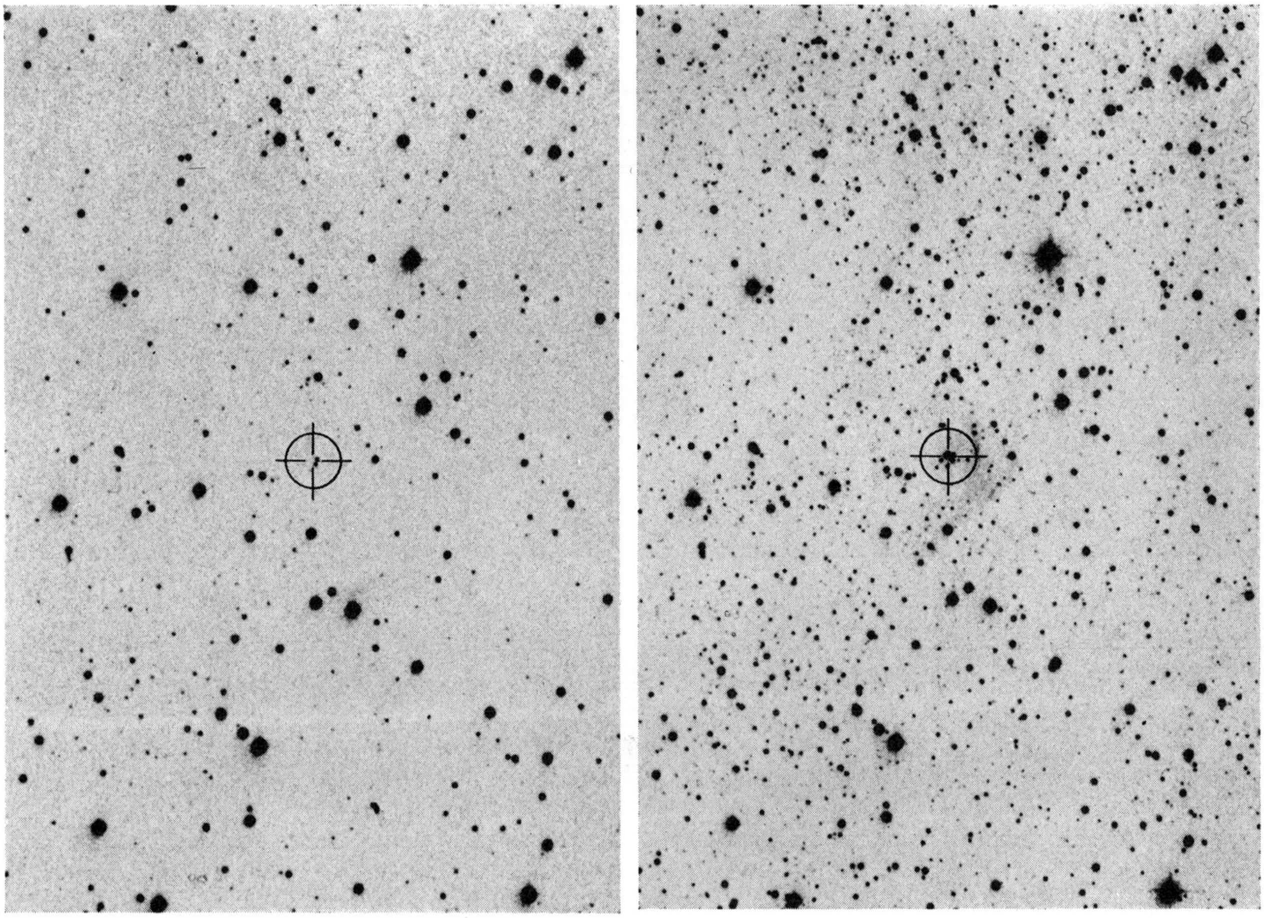

INFRARED SOURCE IN CYGNUS emits as much radiation at two microns as does Vega, the fourth-brightest star. The object is invisible on blue-sensitive plates (*left*) taken with the 48-inch Schmidt telescope on Palomar Mountain. It stands out clearly, however, on red-sensitive plates (*right*). At a wavelength of 20 microns the Cygnus source outshines every known stellar object except the sun.

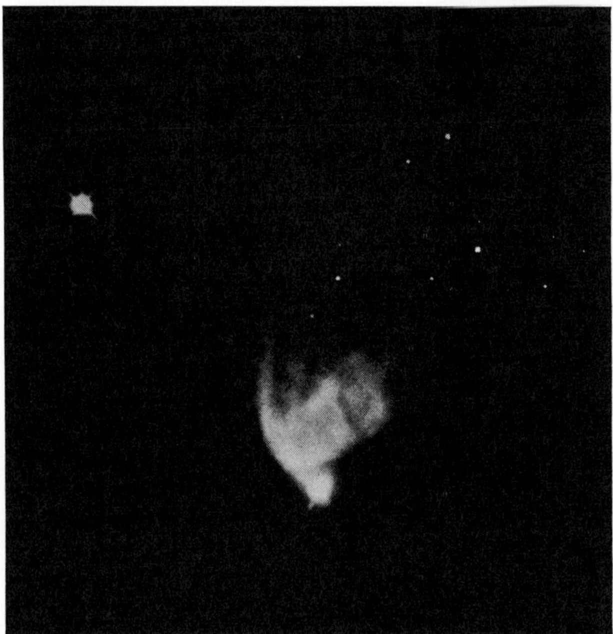

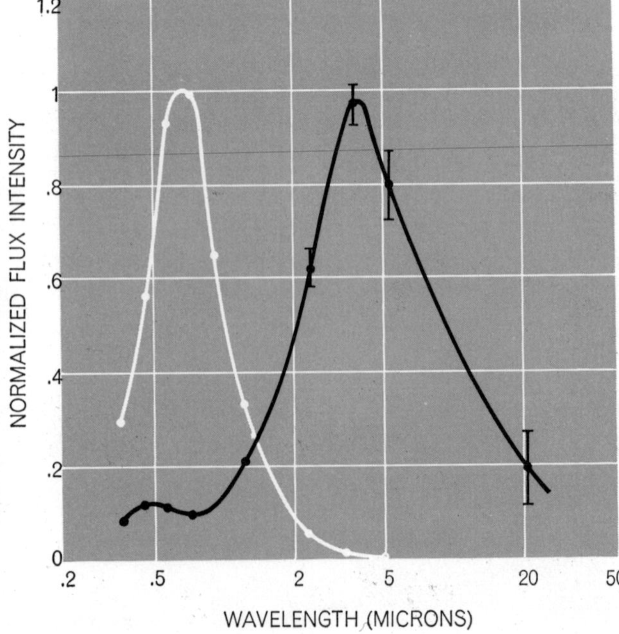

POSSIBLE PREPLANETARY SYSTEM coincides with an object at the head of Hubble's variable nebula (*left*). In this photograph from the Lick Observatory, the infrared object, known as R Monocerotis, looks like the head of a diving bird. The object had been classified as a T Tauri star until Eugenio E. Mendoza of the Tonanzintla y Tacubaya Observatory in Mexico found that the spectrum of R Monocerotis (*black curve at right*) has a large maximum near four microns. White curve shows the spectrum of the sun.

ables. Over three years of observation the brightness of some stars at 2.2 microns has been observed to change by a factor of nearly 10. This in itself is a significant finding because the coolest variable stars previously measured (the Mira variables) change brightness very little in the infrared.

Some of the very red stars are not long-period variables, however. In fact, one of the reddest (and coolest) stars so far detected does not vary. This star, located in the vicinity of Cygnus [*see top illustration on opposite page*], is as bright at two microns in the infrared as Vega, the fourth-brightest star in the sky. Frank J. Low of the University of Arizona has found that at 20 microns the Cygnus source is brighter than any other known stellar object except the sun. Its energy distribution seems to be similar to the one expected from a star whose temperature is about 1,000 degrees. Fred F. Forbes of the University of Arizona has found that its radiation in the near infrared is about 5 percent polarized.

What is this cool object? We do not yet know, except that it seems to be unique in the combination of its great brightness, extreme redness and lack of variability. Harold L. Johnson and V. C. Reddish of the University of Arizona have argued that it may be an extremely bright supergiant star that has been reddened by either interstellar dust or a circumstellar envelope of some kind. If this is the case, the supergiant would be of a kind never before observed. Other workers, notably M. V. Penston of the Royal Greenwich Observatory, have suggested that stars in the process of formation should be surrounded by a shell of cool dust. Hence the Cygnus source may be an example of a protostar, although it is not accompanied by other young stars as one would expect.

Almost all the other bright, very red objects that have been studied more closely after their detection in the survey have been shown either to vary in brightness or to be stars highly reddened by interstellar dust. Thus one important but essentially negative result of the survey has been to establish that extremely red objects that might be associated with star formation, and that are also very bright at two microns, are not common.

A Preplanetary System?

It was our desire to avoid the bias involved in preselecting certain areas of the sky or categories of objects judged to be "interesting" that led us to survey the entire sky. Our goal was to find out what was there. More conventional telescopes equipped with infrared detectors can be much more sensitive than our simple survey telescope. With such instruments several exciting new discoveries have been made on close examination of particular objects.

One of these objects is R Monocerotis, which lies at the head of the comet-shaped nebula called Hubble's variable nebula [*see bottom illustration on opposite page*]. For many years R Monocerotis had been classified as a T Tauri star on the basis of its optical spectrum [see "The Youngest Stars," by George H. Herbig; article begins on page 140]. In 1966 Eugenio E. Mendoza of the Tonanzintlay Tacubaya Observatory in Mexico found that the spectrum of R Monocerotis has a second maximum in the infrared that actually accounts for most of the total energy radiated by the object. The peak of this infrared component is around four microns; the entire energy distribution corresponds roughly to that of a black-body source at 750 degrees K. except for a curious excess at 22 microns. Low, who has made most of the astronomical observations at 10 microns and beyond, and Bruce J. Smith have suggested that this energy distribution may be produced by a dust cloud surrounding the star that absorbs the short-wavelength radiation and reemits the energy at the longer infrared wavelengths. Low and Smith further suggest that this object may be an example of a preplanetary system. It may be significant that T Tauri stars are known to be among the very youngest stars in the galaxy.

A quite similar object has been found in the Great Nebula in Orion, which has

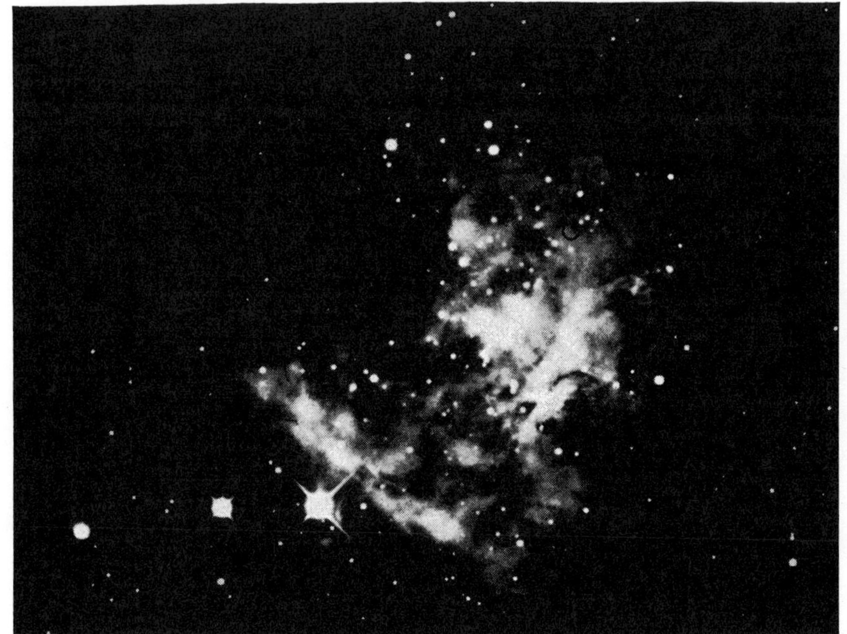

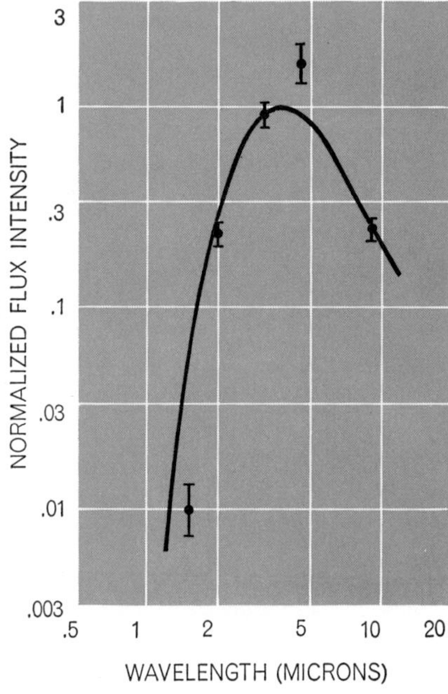

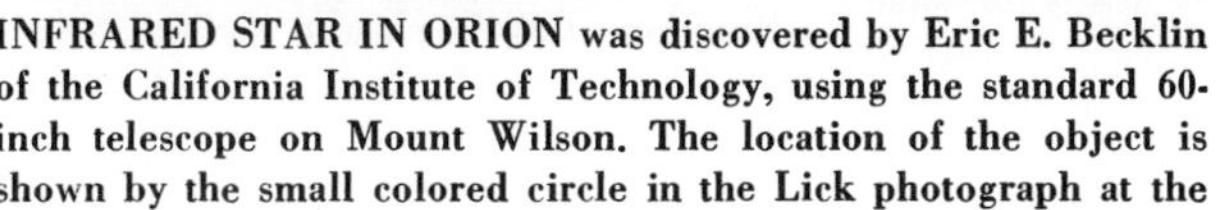

INFRARED STAR IN ORION was discovered by Eric E. Becklin of the California Institute of Technology, using the standard 60-inch telescope on Mount Wilson. The location of the object is shown by the small colored circle in the Lick photograph at the left. Within it no star can be seen even with the 200-inch telescope. The infrared source seems to be a point embedded in a larger region that also emits in the infrared. The radiation curve of the point source (*right*) conforms to that of a black body at 650 degrees K.

been suspected as being a breeding ground for stars [*illustrations on preceding page and on page 263*]. The object ject was found by our colleague, Eric E. Becklin, using the Mount Wilson 60-inch telescope. At photographic wavelengths this object is quite invisible even with the 200-inch telescope, but in the infrared it is almost bright enough to appear on our sky survey. Although the source seems to be a point, it is surrounded by an extended distribution of infrared radiation that is centered on it.

The emission of the point source has been measured at various infrared wavelengths out to 13.5 microns. These measurements yield a radiation curve that conforms closely to that of a black-body whose surface temperature is around 650 degrees K. Again we must consider whether the object really has this temperature or whether it is a normal star whose visible component of radiation has been absorbed and scattered by dust. One can calculate that if it were a red supergiant it would have to be hidden behind enough dust to dim it by a factor of 100 trillion to create the observed appearance. This explanation seems to us most unlikely. We believe the object is a star with an extremely cool surface—quite possibly a protostar. If reasonable guesses are made as to its size and mass, one finds that observable changes should occur in much less than 1,000 years, a short time by astronomical standards.

Last year Low and Douglas E. Kleinmann attempted to study the Orion infrared star at 20 microns. They were unable to detect a measurable signal from the point source but found, adjoining the point source but apparently distinct from it, an extremely bright extended source. Indeed, at 22 microns this infrared "nebula" turns out to be the brightest known object in the sky with the exception of the sun and the moon. We can only guess at its temperature, but from the relative absence of energy at other wavelengths one can estimate that it is less than 150 degrees K., or about 120 degrees below zero Celsius. Again the current belief is that the nebula is an example of a dense cloud in which stars are forming.

A further exciting fact was added recently when Ernst Raimond and Baldur Eliasson of Cal Tech, working with the radio telescope at Owens Valley, Calif., discovered that the Orion infrared point source is also emitting line radiation characteristic of the hydroxyl radical (OH). With this clue, known OH-emission sources are now being examined in the infrared to see if other such instances can be found.

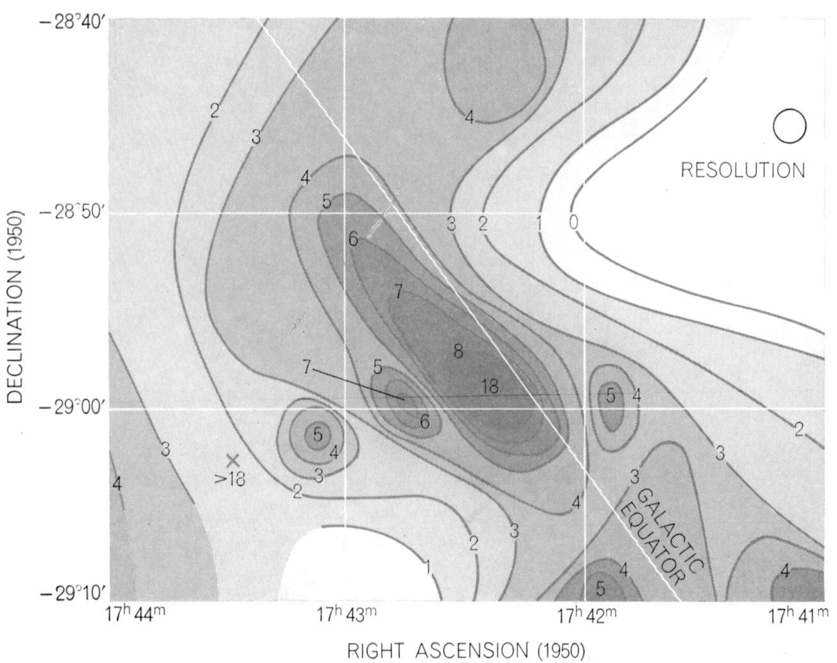

INFRARED MAP OF GALACTIC CENTER shows intense emission at 2.2 microns slightly to the left of the galactic equator. Slightly farther to the left is an intense point source, marked by a colored cross. The zero line represents the sky background radiation. The other lines are labeled in units of 5.2 × 10^{-10} watt per square centimeter per micron per steradian.

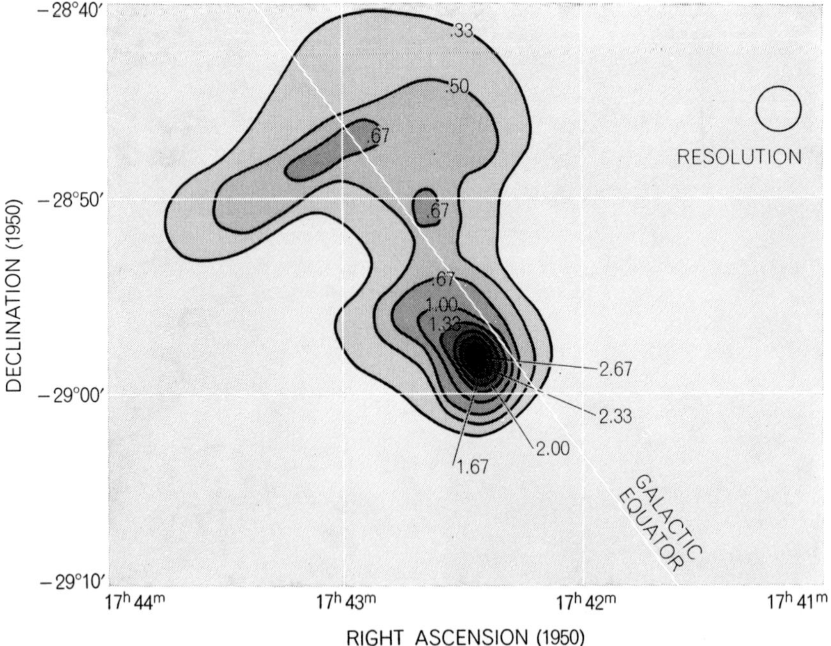

RADIO MAP OF GALACTIC CENTER is plotted to the same scale as the infrared map at the top of the page. Both survey methods locate the galactic center in very nearly the same place. The numbers on the contour lines represent the antenna temperature in degrees Kelvin. A slight correction was made for extinction by dust. The survey was made at a wavelength of 1.9 centimeters by D. Downes, A. Maxwell and M. L. Meeks, using the 120-foot antenna at the Lincoln Laboratory, operated by the Massachusetts Institute of Technology.

The Galactic Center in Infrared

Many other galaxies exhibit a definite nucleus when they are observed at visible wavelengths. In the case of our own galaxy the interstellar extinction is so great that one cannot see into the galactic center, although several studies have shown the presence of a central "bulge." Several years ago, however, radio astronomers found a radio source, Sagittarius A, whose location coincides quite closely with the point identified as the galactic center on the basis of stellar-motion studies. When we started our survey, we hoped to detect the center of the galaxy by its infrared radiation.

An infrared source we now believe to be the center of our galaxy did in fact show up on the survey. It was so inconspicuous, however, that we recognized it only after it had been found by Becklin with a 2.2-micron photometer on a 24-inch telescope on Mount Wilson. After this initial success further measurements were made at .9, 1.65, 2.2 and 3.4 microns using the 60-inch Mount Wilson telescope and the 200-inch telescope on Palomar Mountain. These observations reveal an extended source of infrared radiation centered on the dynamical center of the galaxy or very near it. On a larger scale a weak background is also seen, the pattern of which is quite complex [*see upper illustration on facing page*]. We currently believe most of the secondary features arise from localized changes in interstellar absorption rather than from multiple strong emitters at the galactic center. The gross appearance in the infrared agrees with radio maps of the region [*lower figure on facing page*].

The radiation from the galactic center can be interpreted by comparing it with radiation from the nucleus of the Great Nebula in Andromeda, which is believed to closely resemble our own galaxy. In fact, if infrared observations of our galaxy's nucleus are plotted to the same scale as infrared observations of the Andromeda galaxy's nucleus, the similarity between the two intensity profiles is striking [*see illustration at right*]. If this comparison is valid, and we believe it is, then in all likelihood the infrared radiation recorded from the center of our galaxy is produced by millions of ordinary stars densely packed into a region whose dimensions are only a few parsecs across. Inside the central core, about one parsec in diameter, the number of stars per unit volume may be about 10 million times what it is in the neighborhood of the sun. This means that the stars in the core are 200 times closer together than the stars familiar to us, so that if we lived there the stars in the sky would appear some 40,000 times (more than 11 magnitudes) brighter than those we normally see. Even so, the nearest stars would be, on the average, several thousand times more distant than Pluto, the solar system's outermost planet. As we have mentioned, the galactic center cannot be seen visually because the visible radiation from it is cut down by a factor of 10^{10}.

Close to the center of the main body of radiation there is a particularly strong infrared object that appears to be a point source. If, as seems likely, this source is located at the galactic center 10,000 parsecs away, it must be less than .1 parsec in diameter. Yet it has a total radiative output estimated to equal that of 300,000 suns. The true nature of this source is still open to debate, and no explanation put forward so far is satisfactory. Here we cannot draw an analogy between our galaxy and others because none of them is close enough for us to distinguish such a small point within the nuclear region surrounding it. If the source were a single bright star, it would be among the two or three intrinsically brightest stars ever measured. Alternatively, if it were a cluster of stars rather like the sun, the stars would have to be so close together that two of them would collide every 1,000 years or so, and the cluster would have an expected lifetime of only 1,000 million years—a span much shorter than the age of the galaxy.

The results we have described have come from observations made primarily at the near-infrared wavelengths. What would the sky look like if we used a much longer wavelength such as 20 microns? We cannot tell without an unbiased search, but perhaps a hint is given by observations of unusual objects such as quasars.

Much too faint to be detected by our survey telescope, quasars have been recently examined in the infrared both at the University of Arizona by Low and Johnson and at Cal Tech. Although measurements beyond two microns have been published for only the brightest quasar, 3C 273, it is clear that in this object and quite a few other quasars the bulk of the energy is radiated in the infrared wavelengths. Furthermore, Low and others at the University of Arizona have studied Seyfert galaxies, galaxies that have a starlike nucleus, and have found that here also the bulk of the energy is emitted in the infrared. In fact, Low finds that at least one Seyfert galaxy has its maximum emission intensity at 20 microns. How this radiation is produced is still unknown. Quite possibly it is nonthermal, meaning that the energy originates in a process entirely different from the one that produces the radiation in ordinary stars. Such results make us confident that infrared astronomy holds important clues to understanding the universe.

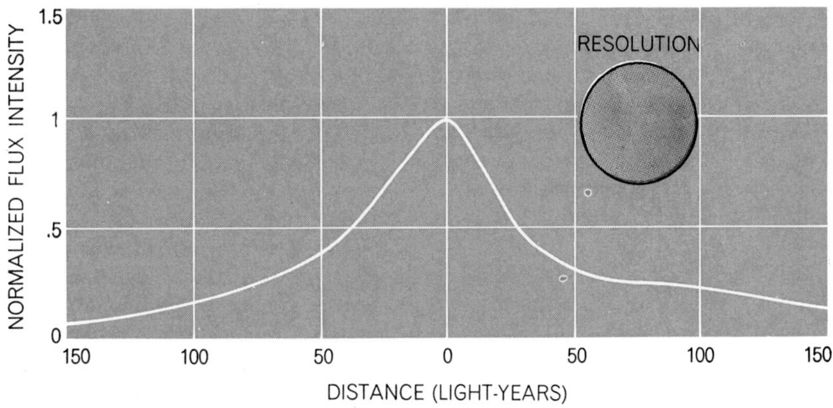

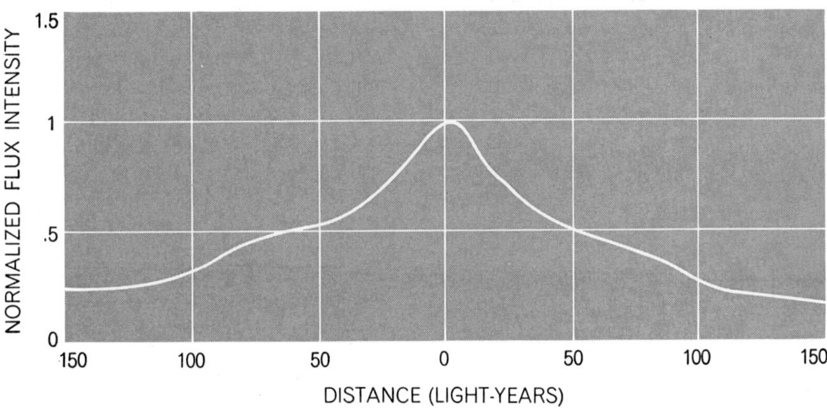

GALACTIC NUCLEI, the Andromeda galaxy at the top and our own galaxy at the bottom, exhibit similar profiles when scanned by infrared detectors at a wavelength of 2.2 microns. The resolution of the lower scan was degraded to match that for the Andromeda galaxy.

26

Ultraviolet Astronomy

LEO GOLDBERG

June 1969

The ability to put telescopes and other instruments in orbit above the earth's atmosphere is giving astronomers an opportunity to observe the sky in a way they could only dream of a dozen years ago. It is well known that the atmosphere blocks out most of the shortwave radiation emitted by the sun and other stars. Thus the radiation reaching the average telescope on the earth has a spectrum only slightly broader in wavelength than the spectrum visible to the eye: 4,000 to 7,000 angstroms. A few infrared "windows" provide glimpses of wavelengths longer than 7,000 angstroms. At the short end of the spectrum the blockage is almost complete below 3,000 angstroms. Consequently astronomers could only guess about the character and intensity of radiation emitted by celestial objects between 3,000 and 300 angstroms in the ultraviolet part of the spectrum. (Wavelengths shorter than 300 angstroms but longer than .1 angstrom are arbitrarily defined as X rays, and wavelengths shorter than .1 angstrom are designated gamma rays.)

The Orbiting Astronomical Observatory (OAO-II), placed in orbit last December, is now providing a systematic sampling of the ultraviolet output of some 50,000 stars. In this article I shall report some of the preliminary findings of OAO-II and explain why astronomers have been looking forward to its results with such keen anticipation. In the broadest sense ultraviolet astronomy, now in its infancy, should help to answer some of the fundamental questions about the universe, including the processes by which stars, and even galaxies, are born, grow old and die. Ultraviolet data should also tell us much about the nature of quasars, the immensely bright starlike objects that seem to be among the most remote beacons in the universe, and about pulsars, the strange stars that flash on and off with precise regularity from once every few seconds to as rapidly as 30 times a second.

In the not too distant future, perhaps by 1980, astronomers hope to establish a great national observatory in space, or preferably an international one. Among the instruments being considered for such an Astronomical Space Observatory are a reflecting telescope at least three meters in diameter for observing planets, stars, nebulas and galaxies; a number of special telescopes for studying the sun, the largest of which might be 1.5 meters in diameter and nearly 40 meters long; large-aperture telescopes and arrays of detectors for monitoring sources of X rays and gamma rays in the universe, and a fantastic long-wave radio telescope shaped like a rhomboid 10 kilometers on a side. The observatory would be designed as a fully automatic permanent facility, but it would probably require intermittent visits by engineer-astronauts to maintain, repair and modernize the instruments. Measured by its potential for fundamental discovery, the proposed space observatory is by far the most rewarding scientific activity that can be envisioned for the U.S. space program in the next two decades.

The utility of telescopes in space goes beyond their capacity to intercept radiation that cannot penetrate the atmosphere. The air around and above us is in constant turbulent motion, with the result that the images produced by earth-based telescopes are severely degraded. In theory the angular resolving power of a telescope is directly proportional to the size of its aperture, but in practice there is little or no gain in resolving power when the aperture exceeds 12 inches or so. The 200-inch telescope on Palomar Mountain resolves features on the moon that are about 3,000 feet across, but a telescope of the same size above the atmosphere could distinguish craters as small as 100 feet in diameter. The brightness of the night sky places another serious limitation on earth-based telescopes. Even in locations where there is very little dust in the air, auroral emission in the upper atmosphere and the airglow ultimately fog photographic plates and drown out the radiation from the faintest and most remote celestial objects. Telescopes in orbiting observatories above the airglow will be able to collect this faint radiation by long exposures. A space telescope 120 inches in diameter should be able to detect stars 100 times fainter than the faintest detectable from the earth. Data

ULTRAVIOLET SOLAR IMAGES, constructed from measurements made by Orbiting Solar Observatory IV, can be compared with a photograph (*top left on opposite page*) taken on the same day at the Sacramento Peak Observatory of the Air Force. The photograph was made by recording a single wavelength: the red line of un-ionized hydrogen. It shows extensive sunspots on the northern hemisphere of the sun's surface and lesser activity in the southern hemisphere. The dark patches are clouds of comparatively cool gas. The five ultraviolet images are numbered in order of increasing temperature, and therefore height, in the solar atmosphere: 10,000 degrees (*1*), 100,000 degrees (*2*), 325,000 degrees (*3*), 1.4 million degrees (*4*) and 2.25 million degrees (*5*). Temperatures are on the absolute, or Kelvin, scale. The last two images respectively correspond to heights of 15,000 kilometers and 200,000 kilometers. The hottest regions are in the palest color; the coolest regions are dark gray. It can be seen that the sunspots exert an influence throughout the solar atmosphere.

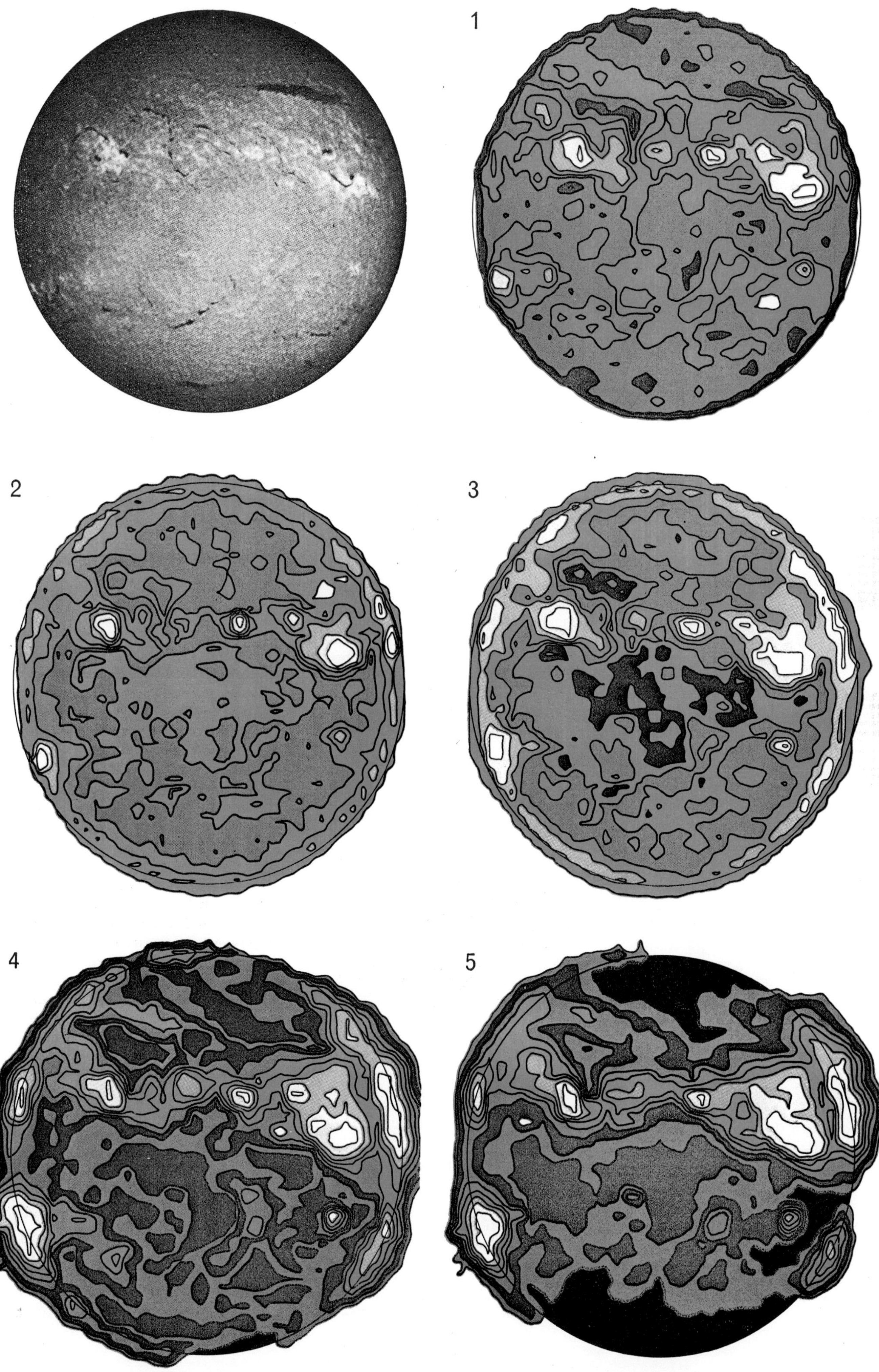
1
2
3
4
5

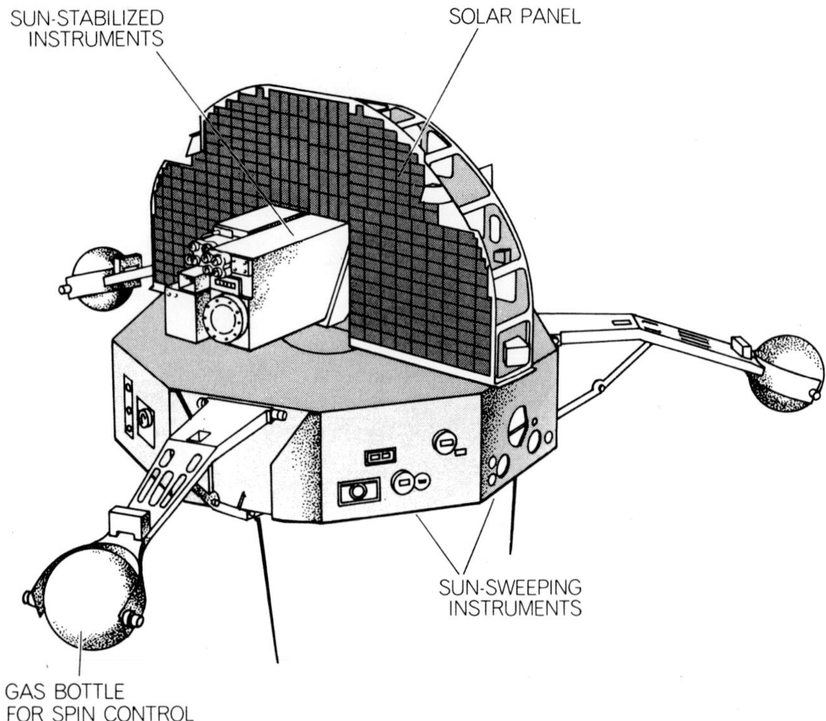

ORBITING SOLAR OBSERVATORY IV, launched October 18, 1967, carries some 250 pounds of instruments in a circular orbit 350 miles high. The base section, which rotates 30 times a minute, has instruments for six experiments. The upper section, which can be aimed at the sun, carries an X-ray telescope, X-ray spectrometer and ultraviolet spectroheliograph designed by the author and his colleagues at the Harvard College Observatory.

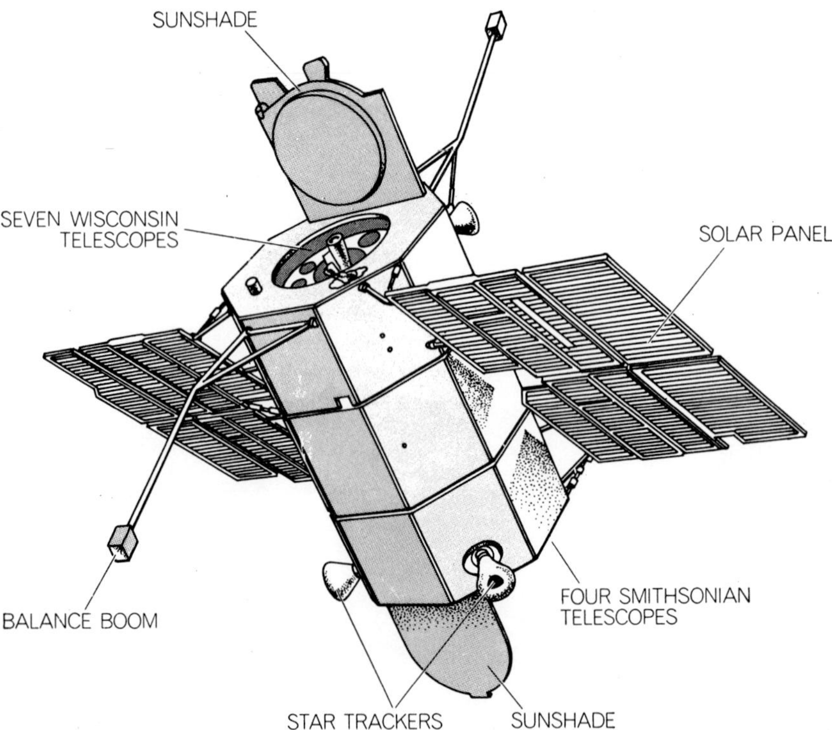

ORBITING ASTRONOMICAL OBSERVATORY II (OAO-II), with a scientific payload of 1,000 pounds, is the most complex unmanned satellite developed by the U.S. It was launched December 7, 1968, into a circular orbit 480 miles high. Designed expressly to measure ultraviolet radiation from stars, OAO-II is equipped with 11 telescopes: three of 16-inch aperture, four of 12-inch aperture and four of eight-inch aperture. Four of the telescopes were provided by the Smithsonian Astrophysical Observatory, seven by the University of Wisconsin.

on faint objects are critical for settling major questions in cosmology, such as whether the universe is infinite or not.

The atoms and molecules responsible for absorbing ultraviolet and infrared radiation in the earth's atmosphere and the heights at which they are found are fully known, so that one can estimate how high it is necessary to go to get above them [*see illustration on opposite page*]. The infrared is absorbed chiefly by water vapor and carbon dioxide, most of which is below 30 kilometers, a height that can be exceeded by stratospheric balloons. On the other hand, ultraviolet radiation is absorbed at much higher altitudes, which can be exceeded only by rocket propulsion. The band of wavelengths between 2,900 and 2,200 angstroms is absorbed by the layer of ozone that is formed by the action of solar ultraviolet radiation on oxygen molecules at an altitude of about 50 kilometers. The spectrum from 2,200 to 900 angstroms is absorbed by molecular oxygen, most of which lies below 100 kilometers. Molecular nitrogen, atomic oxygen and atomic nitrogen absorb ultraviolet radiation of still shorter wavelength, the maximum absorption occurring at about 500 angstroms. At this wavelength very little radiation penetrates below 240 kilometers.

The height of the ozone layer was not well known until the advent of sounding rockets in 1946, and was in fact usually much underestimated. Thus in the late 1920's unsuccessful attempts were made to photograph the solar ultraviolet spectrum from manned balloons at heights up to nine kilometers, and as late as the 1930's it was still thought by some that an unmanned balloon observatory could penetrate the ozone layer and photograph the solar spectrum as far down as 900 angstroms. Space astronomy finally became a reality in October, 1946, when a group at the Naval Research Laboratory headed by Richard Tousey photographed the solar spectrum down to a wavelength of 2,200 angstroms with a spectrograph carried to an altitude of 80 kilometers in the nose cone of a German V-2 rocket. From this modest beginning the frontier of the solar-radiation spectrum has been pushed steadily toward shorter wavelengths by instruments of increasing refinement and complexity mounted in sounding rockets and in artificial satellites.

More recently the development of precision pointing devices has also brought the stellar universe within the range of space telescopes, so that the

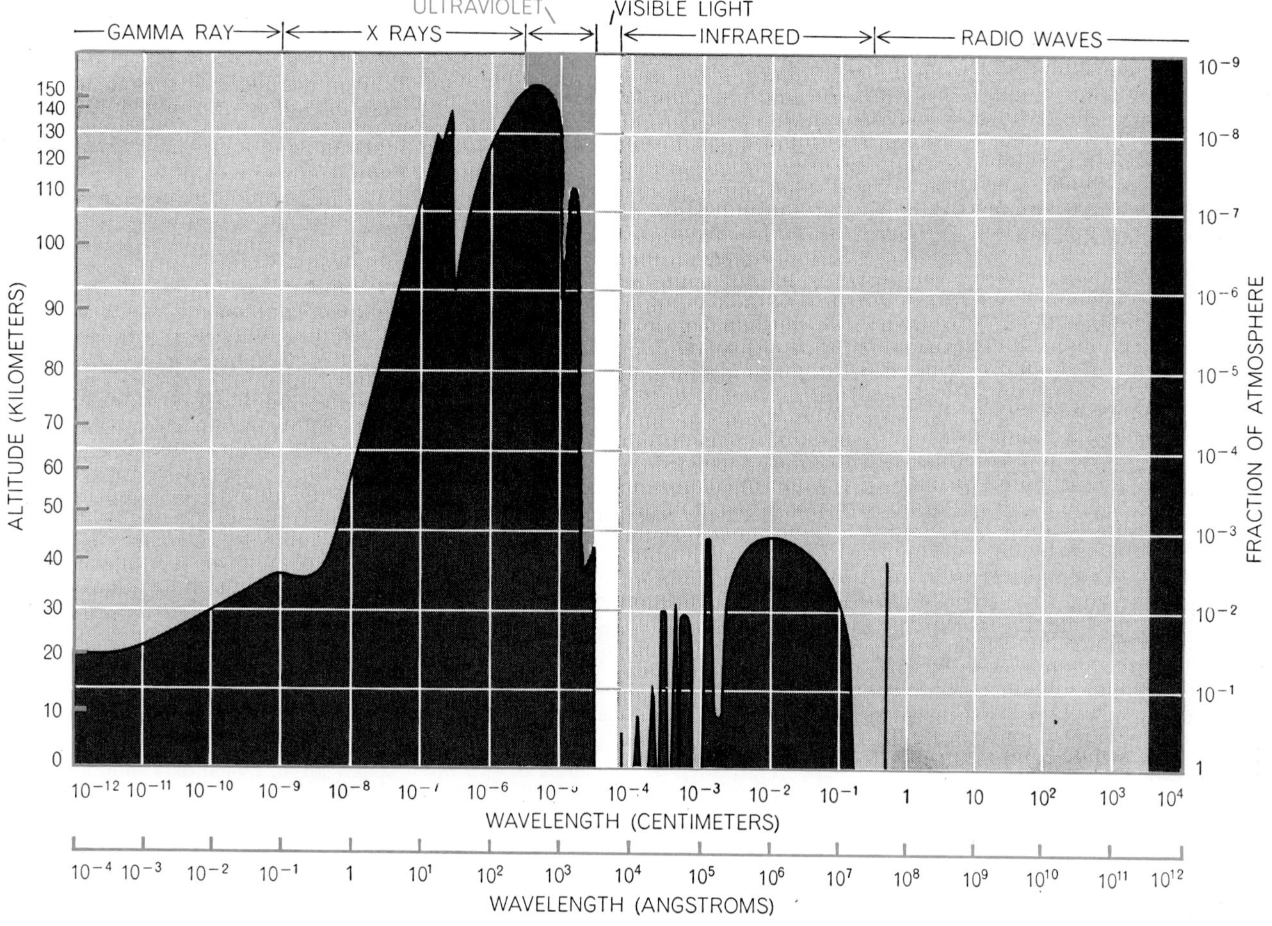

ATMOSPHERIC ABSORPTION of electromagnetic radiation is strong in the ultraviolet region of the spectrum, where the hottest stars emit most of their energy. The upper boundary of the dark gray areas specifies the altitude where the intensity of external radiation at each wavelength is reduced to half its original value. Radio waves beyond 50 meters in length are blocked completely.

possibilities for gaining basic new information from space astronomy are now virtually unlimited. Several new subdivisions of observational astronomy have been created, each concerned with its own region of the spectrum and requiring unique observing techniques and instrumentation. The subdivisions include X-ray and gamma-ray astronomy, ultraviolet astronomy and infrared astronomy (although much important work in the last category can still be accomplished from the ground). A fourth subdivision, particle astronomy, is concerned with the detection of cosmic rays from the sun, the galaxy and the extragalactic universe. I shall deal here only with ultraviolet astronomy, beginning with the sun.

The physical nature of the sun has been under investigation for more than 350 years, ever since the sun was first observed through a telescope by Galileo. There are a number of reasons why the sun continues to occupy a major share of the attention of astronomers. First of all, no other star is close enough to appear as more than a point of light in the most powerful telescopes. Thus the sun is the one star out of billions that we can hope to understand in much detail. In addition, the sun now assumes an important role in geophysics for the reason that the flow of certain types of solar radiation (notably X rays, ultraviolet radiation and charged particles) is not at all steady but often exhibits sudden large transients, which are frequently followed by correspondingly violent disturbances in the earth's upper atmosphere and magnetic field. Finally, the closeness of the sun makes it a unique physical laboratory where the behavior of matter can be studied under an enormous range of physical conditions that have not so far been duplicated on the earth. For example, the temperature at the center of the sun, where thermonuclear reactions convert mass into energy, is on the order of 15 million degrees Kelvin. The relatively new discipline of plasma physics owes much to astronomical studies of the motions of ionized and magnetized solar gases as observed with telescopes and spectroscopes.

Although a great body of knowledge about the sun had been accumulated by 1940, the modern era of solar physics is only a little older than a quarter of a century, dating from the surprising discovery made by Bengt Edlén that the solar corona is an extremely hot ionized plasma with a temperature of about a million degrees K. Since the temperature of the sun's visible surface is only about 6,000 degrees K., the high temperature of the corona presented a puzzle. Studies over the past two decades have revealed that the temperature actually drops to about 4,500 degrees K. at the top of the photosphere, the layer adjacent to the surface that emits most of the visible radiation. The temperature remains roughly constant for a few hundred kilometers and then begins to rise in the region known as the chromosphere, slowly at first and then very rapidly, jumping from 50,000 degrees to 500,000 degrees in a distance of no more than a few hundred kilometers, and then rising more slowly to temperatures exceeding two million degrees [*see illustration on opposite page*].

For heat to flow outward from cool regions to much hotter regions would seem to violate the second law of thermodynamics. In reality, however, there is no violation because work is being done on the atmosphere by turbulent columns of gas moving upward from an unstable region below the photosphere. The evidence for such large-scale convection is provided by the granulated appearance of the sun's visible surface, which is caused by the contrast between the hot, bright material moving upward and the cooler, darker material moving downward [*see illustration on page 284*]. It is generally accepted that the motions of the turbulent gases are transformed into waves of several types that dissipate their energy in the form of heat as they travel up through the atmosphere, but a detailed theory of the heating process has not yet been verified. Such a theory should be able to predict the rates at which the temperature and the density change with height through the chromosphere and corona. Such predictions are difficult, if not impossible, to test by conventional ground-based observational techniques.

Data on the transition region between the chromosphere and corona are particularly elusive. This is the region where the temperature jumps suddenly from 50,000 degrees to a million degrees within a very short distance. In this zone the number of electrons stripped from atoms is determined by the local temperature; the higher the temperature, the higher the degree of ionization. Because the temperature increases rapidly with height, the total volume occupied by any one species of ion is extremely small and therefore only the strongest emission lines of each species can be detected. These lines fall in the far-ultraviolet region of the spectrum. The highly ionized atoms in the transition zone also emit visible light, but it is too feeble to be observed. Only in the corona itself is visible radiation from highly ionized atoms detectable. Even here comparatively few ions make their presence known in this way, and they do so only from the region peripheral to the solar disk. In order to learn more about the structure of the hot transition regions of the sun, where matter radiates chiefly in the ultraviolet part of the spectrum, it is necessary to mount ultraviolet detectors on space platforms.

Many exploratory investigations of the sun's far-ultraviolet spectrum have been carried out by laboratories in this country and abroad, chiefly with the aid of sounding rockets. These studies have paved the way for the much more advanced instruments carried aloft in spacecraft such as the series of five orbiting solar observatories (OSO's) that have been placed in orbit by the National Aeronautics and Space Administration since March, 1962 [*see top illustration on page 276*]. Each OSO provides a stabilized platform on which about 70 pounds of instruments can be mounted and aimed at the sun with an accuracy of within one minute of arc (about a thirtieth of the sun's diameter). In addition to the section that can be trained continuously on the sun the spacecraft includes a rotating section whose instruments survey the sun intermittently.

In 1891 George Ellery Hale of the U.S. and Henri A. Deslandres of France independently conceived the idea of the spectroheliograph, an instrument for constructing monochromatic images of the sun in individual spectral lines radiated by atoms and ions. The device has been one of the principal means for investigating the structure of the lower part of the chromosphere. When the fourth orbiting solar observatory (OSO-IV) was launched in October, 1967, it carried an ultraviolet spectroheliograph designed and constructed by the Harvard College Observatory to record the spectrum of radiation from the center of the sun's disk in the wavelength region from 300 to 1,400 angstroms, and to produce monochromatic solar images at any desired wavelength in this spectral range [*see illustration below*]. Some of the results of the Harvard experiment are presented here as an example of what is now being achieved by the techniques of solar ultraviolet astronomy. In the far-ultraviolet region of the spectrum, radiation from the photosphere is too weak to be detected. Thus instead of the familiar patterns of the visible spectrum, in which dark lines are superposed on a bright continuous spectrum, the spectrum consists entirely of bright emission lines and bands of continuous emission from the chromosphere and the corona [*see illustration on pages 280 and 281*]. Most of the strong bright lines in that part of the spectrum have been identified, but only a few are labeled in the illustration; they suggest the great variety of ionic species that contribute to the sun's ultraviolet emission. The approximate temperatures at which some of these emission lines are formed can be found in the illustration on the opposite page. The great bulge of continuous emission extending to shorter wave-

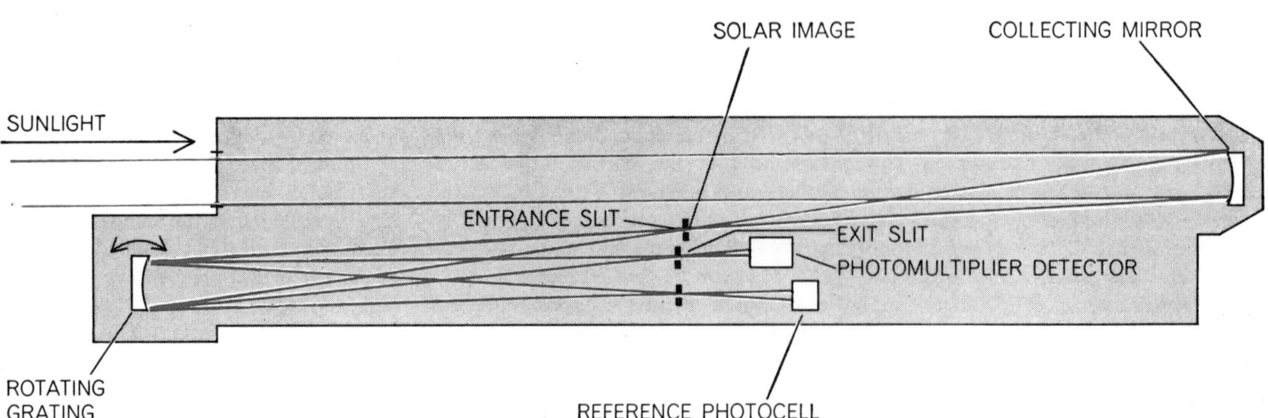

SPECTROHELIOGRAPH aboard OSO-IV supplied the measurements for the illustrations on pages 275, 280 and 281. Sunlight enters at the left, is reflected by the collecting mirror and brought to a focus at the entrance slit. It then strikes an optical grating that separates the radiation into its component wavelengths. By rotating the grating to various positions any narrow band of ultraviolet radiation between 300 and 1,400 angstroms can be centered on the photomultiplier detector. The instrument scans the sun in a

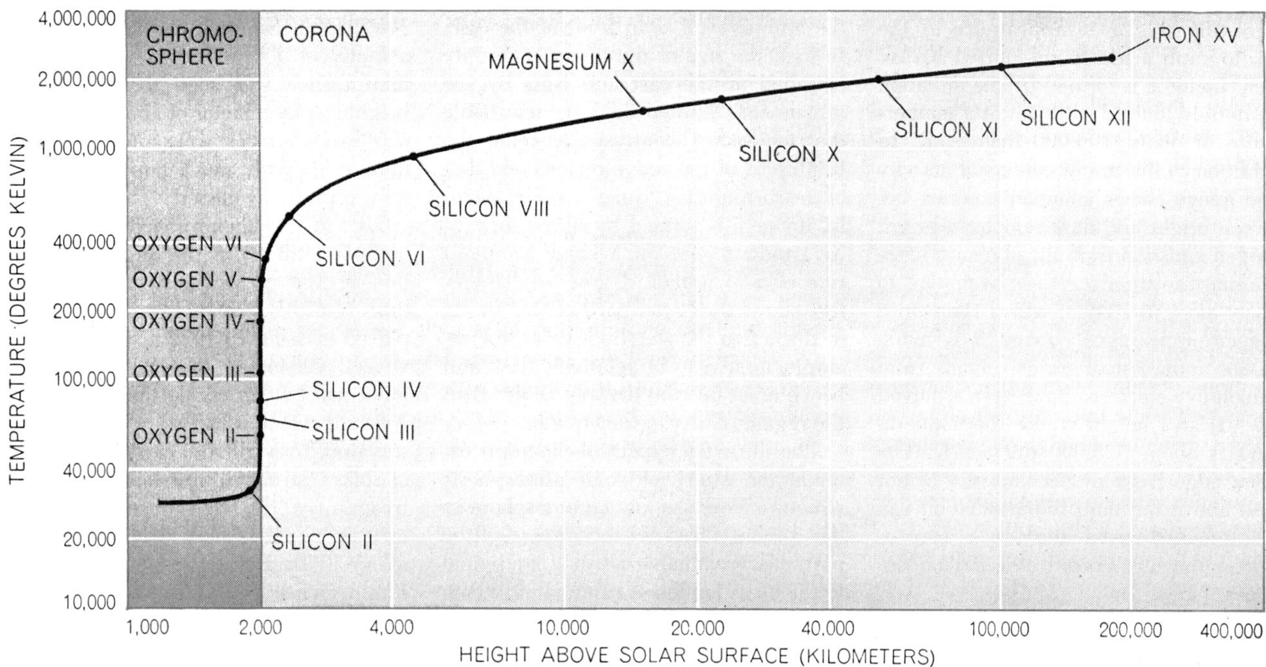

TEMPERATURE OF SUN'S ATMOSPHERE rises steeply some 2,000 kilometers above the photosphere, or visible surface of the sun, which itself is only about 6,000 degrees K. The region where the temperature jumps from 50,000 degrees K. to more than 500,000 divides the chromosphere from the corona. Temperatures are inferred from the spectral lines of various ions, atoms stripped of one or more electrons. The Roman numeral following the name of each element is one greater than the number of missing electrons.

lengths from 912 angstroms is formed when ionized hydrogen recombines with free electrons to form neutral hydrogen at a temperature of about 10,000 degrees K. At the other extreme, bright lines of Fe XV and Fe XVI (iron atoms with respectively 14 and 15 missing electrons) are emitted in highly localized superhot regions of the corona where the temperature may be as high as two million to four million degrees K. Most of the other emission lines are produced by the abundant elements (hydrogen, helium, carbon, nitrogen, oxygen, neon, magnesium, silicon, sulfur and iron) in various stages of ionization; each ion emits radiation at just the height in the atmosphere where the temperature is right for its formation and maintenance.

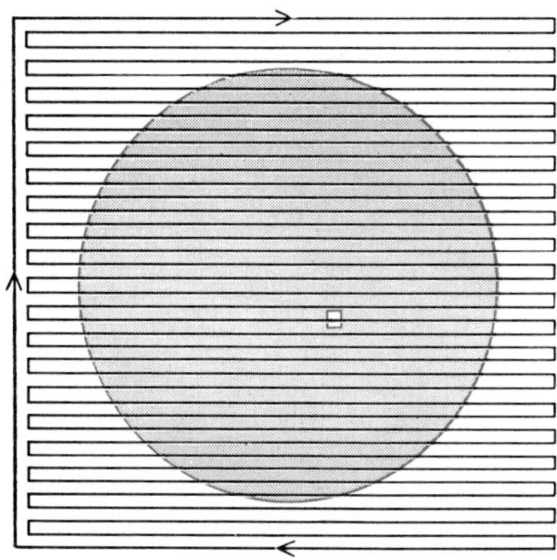

raster pattern (*far right*), providing a total of 1,920 measurements in a period of about five minutes. The area covered by each measurement is indicated by the white square.

When the Harvard instrument is used as a spectroheliograph, it is able on command from the ground to select radiation from any desired narrow band of wavelengths and by scanning the sun in a raster pattern to construct a picture of the solar regions emitting that particular wavelength. Since each emission line is emitted in a narrow height range, pictures of the sun made in this way outline the three-dimensional structure of the solar atmosphere. The pictures are not made by direct photography because the satellite cannot return the film to the ground. Instead images are reconstructed on the ground from a matrix of brightness measurements made at 1,920 points covering the sun's visible disk and the surrounding inner corona. The brightness measurements are made photoelectrically by a photon-counting system incorporated into the instrument. They are then recorded on magnetic tape and transmitted to the ground by radio at the end of each orbit when the satellite passes over one of the several ground stations along a north-south line between Rosman, N.C., and Santiago in Chile.

Approximately a third of the data were received at Fort Myers, Fla., and were then relayed almost immediately to Cambridge, Mass., over a leased telephone line. Hence astronomers in Cambridge were able to examine pictures of the sun 20 to 30 minutes after they had been received on the ground and to make changes in the observing program as desired. The distribution of brightness across the solar image can be represented in a number of different ways. A typical set of five ultraviolet images, recorded on November 22, 1967, are shown on page 275, along with a photograph made on the same day in the red line of un-ionized hydrogen at the Sacramento Peak Observatory in New Mexico. The two zones of activity that coincide with the sunspot zones extend from east to west in both hemispheres and show up clearly as regions of above-average brightness. The dark areas and patches in the un-ionized-hydrogen photograph are comparatively cool clouds of gas seen in projection against the solar disk.

The sequence of ultraviolet images in this illustration is arranged in order of increasing height and therefore of increasing temperature in the solar atmosphere. The first ultraviolet image (Image 1) is formed by the radiation of ionized hydrogen in the chromosphere, where the temperature is about 10,000 degrees K. It shows both the bright active regions and the dark clouds and is generally similar in appearance to the photograph made in un-ionized hydrogen. Image 2 is formed by the radiation of doubly ionized nitrogen at a temperature of about 100,000 degrees K. In addition to the bright centers of activity the image shows a higher contrast between bright and dark regions; a bright ring of emission from the chromosphere-corona transition region is beginning to show along the edge, which takes on a jagged appearance. Image 3 is made in the radiation of oxygen atoms from which five electrons have been removed (O VI) at a height where the temperature is about 325,000 degrees K. The solar edge has now become very bright and above the limb some radiation can be detected from the lower corona, where the temperature is a million degrees.

Image 4 is made in the radiation of magnesium atoms from which nine electrons have been removed (Mg X); this ion is formed at an average temperature of 1.4 million degrees. An image of Mg-X emission is therefore a picture of the corona, which can be seen well beyond the limb as well as in front of the disk. In fact, the size of the image is sharply bounded at the east and west by the instrument's limited field of view. Note the increasing contrast between the brightness of the active regions and that of the surrounding "quiet" corona. Finally, Image 5 is formed by silicon ions (Si XII) radiating at an average temperature of 2.25 million degrees and therefore shows the hotter parts of the corona in the active regions. Both poles, for example, appear to be relatively dark and hence must be considerably cooler than the regions closer to the equator.

The ultraviolet spectroheliograph extends the study of solar atmospheric structure from the low chromosphere to the corona. This new technique will be particularly valuable when it is applied to the investigation of solar flares, which have traditionally been studied with the aid of solar images made in the red line of hydrogen.

Late in 1969 an improved spectroheliograph, with twice the angular resolution of the present instrument and a time resolution of 30 seconds, will be placed in orbit. In 1972 a still more advanced version of the Harvard instrument will be carried aloft on the Apollo Telescope Mount (ATM), a manned observatory that will also carry other types of solar instrument now being prepared by groups at the High Altitude Observatory in Boulder, Colo., the Naval Research Laboratory, the Goddard Space Flight Center of NASA and American Science & Engineering, Inc.

In the case of the sun extraordinary advances in knowledge have been made with very small space telescopes, no more than an inch or two in diameter and with a pointing accuracy limited to about one minute of arc. For stars, however, the requirements are much more stringent: telescope apertures must be

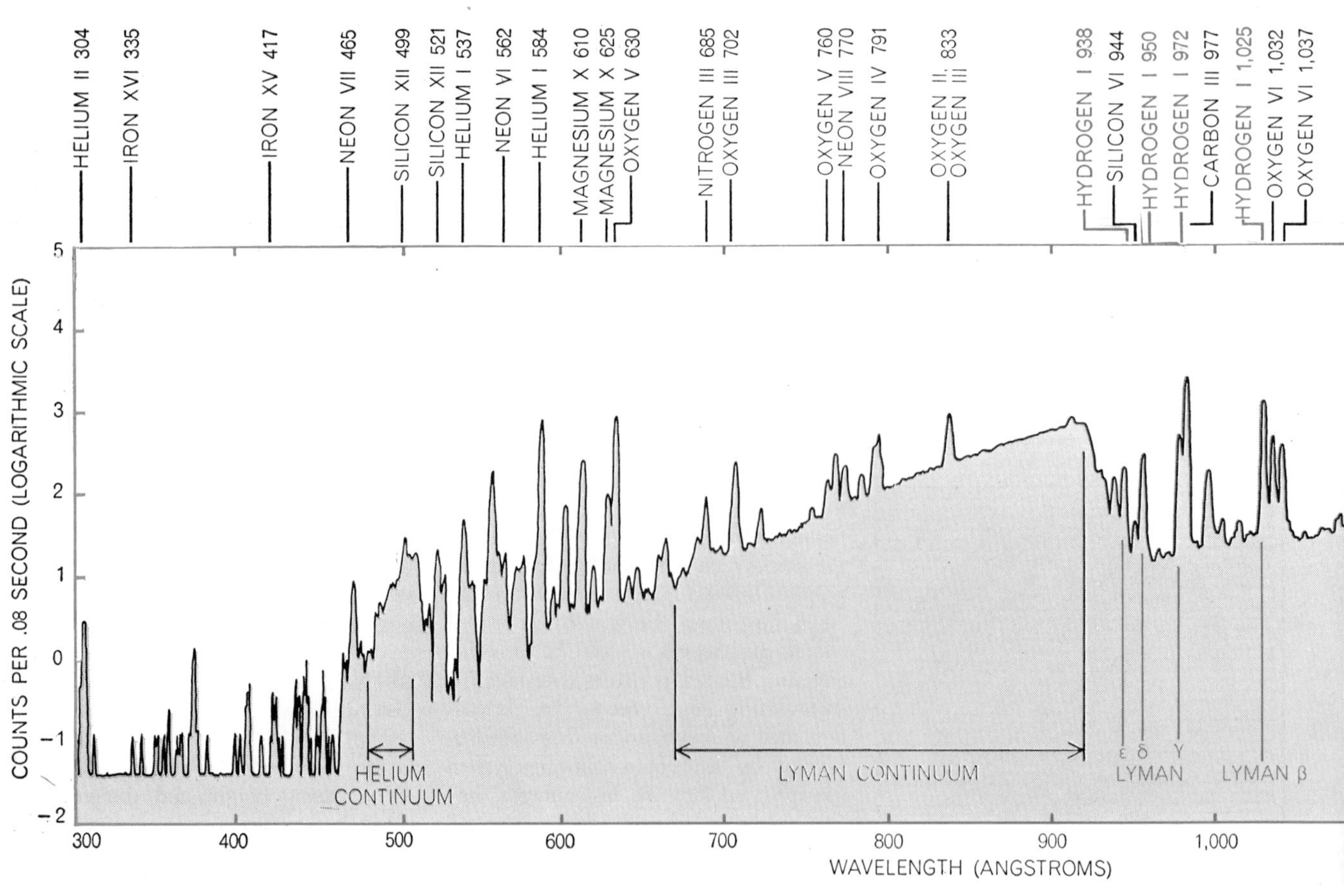

SOLAR SPECTRUM at ultraviolet wavelengths between 300 and 1,400 angstroms was recorded from the center of the solar disk by the Harvard spectroheliograph on OSO-IV. The height of the curve at each wavelength is the number of counts per counting interval of 80 milliseconds. Some of the more prominent emission lines are labeled, including the first five members of the Lyman series of neutral hydrogen (*color*). The great bulge of emission extending from 650 to 910 angstroms is the Lyman series continuum created when

10 inches or more and errors in pointing no greater than a few seconds of arc. This requirement for larger and more complex instrumentation has kept stellar ultraviolet astronomy from developing as rapidly as its solar counterpart, but the recent successful launching (on December 7, 1968) of the Orbiting Astronomical Observatory (OAO-II) has given enormous impetus to stellar space astronomy.

Considering the huge variety of objects in the stellar and galactic universe, it is almost a certainty that the observations of ultraviolet radiation from the new space platforms will transform our ideas in radical and unpredictable ways. Although the element of unexpected discovery is an important one, there are also well-defined scientific reasons for expecting that stellar ultraviolet astronomy will help us to answer basic questions about the universe. Current ideas about the evolution of stars, derived from observations with large optical telescopes on the earth, suggest that a star begins its life as a condensation of interstellar dust and gas. If such a protostar happens to have the right mass (between a hundredth and 100 times the mass of the sun), it will contract further and become a star. During the life of the star thermonuclear processes convert hydrogen into heavy elements with the release of energy; the more massive the star, the more rapid its rate of energy production and hence the shorter its life expectancy. A star with a mass roughly the same as the mass of the sun may last for about 50 billion years. On the other hand, a very hot, massive star—one weighing 10 times as much as the sun—will radiate 10,000 times more energy per unit time, or 1,000 times more energy per unit mass; as a result its life expectancy will be only a thousandth that of the sun, or about 50 million years. Assuming that the age of our galaxy is about 10,000 million years, there has been ample time for several generations of hot, massive stars to have lived and died.

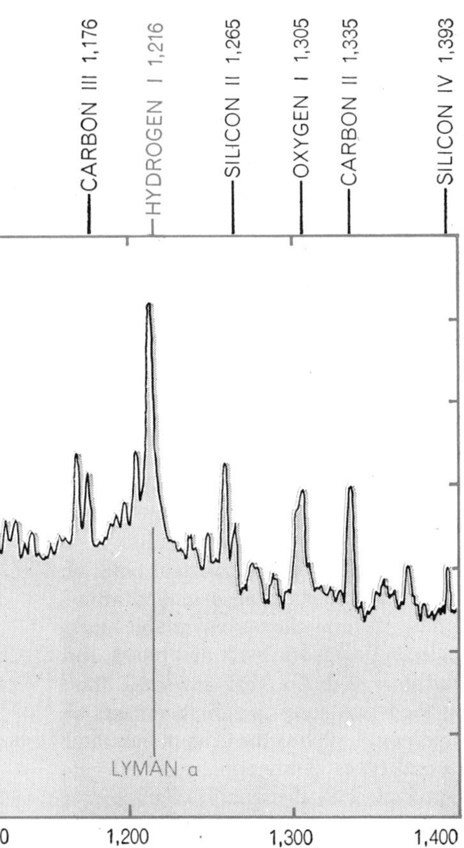

ionized hydrogen atoms (protons) in the lower chromosphere capture free electrons. A similar but weaker continuum is radiated by ionized helium around 500 angstroms.

Thus the hottest stars, having short lives, are more likely to show the effects of aging than the cooler ones. Unfortunately for earth-based astronomy only a small part of the radiation of hot stars is in the form of visible light. For example, two stars that appear equally bright to the unaided eye will emit totally different ratios of ultraviolet and infrared radiation if one star is at 6,000 degrees K., like the sun, and the other is at 25,000 degrees [*see top illustration on next page*]. Theoretically the hotter star should emit more than 80 percent of its energy in the ultraviolet region of the spectrum, whereas the cooler star should emit more than 80 percent of its energy in the visible, infrared and radio regions. One must add that calculating such energy-distribution curves in the absence of observational data is a bit like inferring the shape of an iceberg from the part seen above water. Real stars are likely to radiate in quite a different way, and until their ultraviolet energy fluxes are observed we cannot really determine how fast they are evolving.

When a star is born, it consists almost entirely of hydrogen, the fuel that feeds its thermonuclear furnace. When the fuel gives out, the star may explode into a supernova and eject a large fraction of its mass into interstellar space. The ejected material increases the proportion of heavy elements in interstellar space; therefore the hot, massive stars formed a few tens of millions of years ago should contain a higher percentage of heavy elements in relation to hydrogen than the slowly evolving sun, which we believe was formed a few billion years ago. A crucial test of this theory would be the accurate measurement of the chemical composition both of hot stars and of the interstellar gas. Chemical abundances are derived from measurements of the intensity of emission lines in stellar spectra, or alternatively from the blackness of absorption lines. Some of these lines are formed in the atmosphere of the star, as in the Fraunhofer spectrum of the sun; others are impressed on the spectrum when the starlight passes through the interstellar gas on its long journey to the earth. Although some information on chemical composition can be obtained from measurements of the visible spectrum, the most important lines lie in the ultraviolet. It is there that decisive tests of theories of stellar evolution will be conducted.

In reality the intensity of an interstellar absorption line in the spectrum of a distant hot star tells us only the quantity of a given element that is in a particular state of ionization. It is not always possible to observe spectral lines from all the states of ionization in which a given element may exist in the interstellar medium; therefore the fraction of ions in a given state must be calculated from theory. Since atoms in the interstellar gas are ionized by the action of ultraviolet radiation from nearby hot stars, it becomes essential to measure the flux of this radiation. A valuable by-product of the calculation is the evaluation of the density of free electrons in space. Knowledge of this density not only helps us to determine the properties of the interstellar gas but also is vital for estimating the distance of pulsars.

Another question that ultraviolet astronomy may help to answer is whether or not (and if so, to what extent) other stars have coronas and activity cycles like the sun's. The sunspot cycle and its attendant quasi-periodic variation of solar activity is one of the most spectacular features of the sun's behavior. The number of sunspots visible on the disk of the sun at any one time varies on the average in a cycle of about 11 years. The same regions of the sun where sunspots occur are also the scenes of violent storms and eruptions, the most spectacular of which are the solar flares. The activity is much more widespread and violent near the maximum of the sunspot cycle than near the minimum. The activity zones are not confined to the visible surface but extend up through the

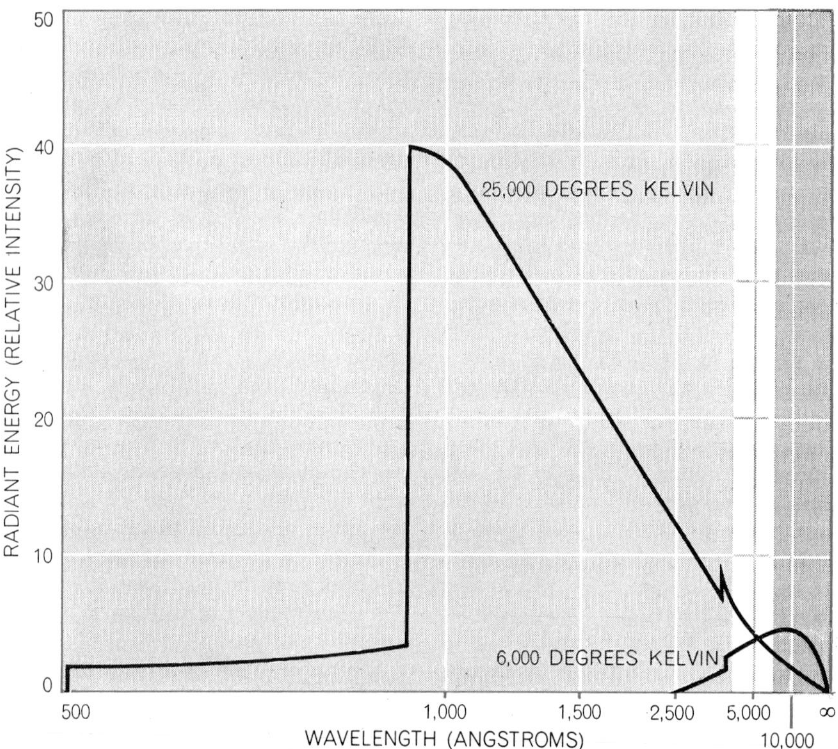

SPECTRAL ENERGY DISTRIBUTION of two stars shows the vastly greater flux of ultraviolet radiation emitted by one whose surface temperature is 25,000 degrees K. compared with one whose temperature is that of the sun. The curves, based on theoretical calculations, are plotted so that the areas under the curves represent the total output of radiant energy. The sharp drop in the hotter star's emission at about 900 angstroms indicates where atomic hydrogen in the stellar atmosphere absorbs ultraviolet radiation emitted from the surface.

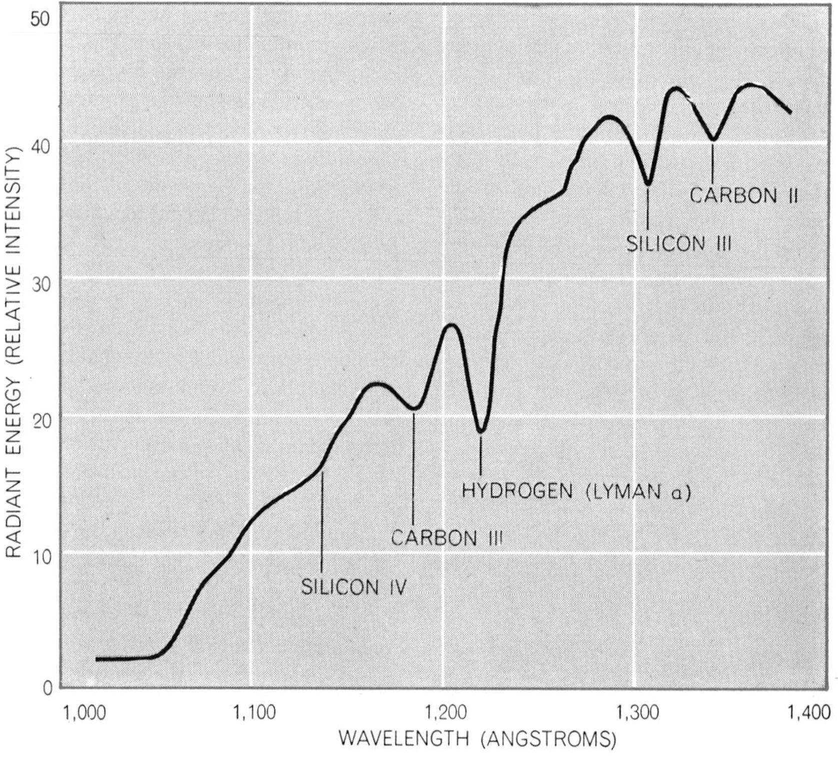

ULTRAVIOLET SPECTRUM of star Alpha Virginis was made by University of Wisconsin telescopes on OAO-II. Four absorption lines are produced by carbon and silicon ions in the star's atmosphere. Hydrogen in interstellar space produced the Lyman-alpha absorption.

corona [*illustration on page 275*]. The material in sunspots is highly magnetized. It is in fact fairly well established that sunspots and their associated activity are caused by the highly localized buildup of the magnetic fields, which have a strength of several thousand gauss in sunspots compared with an average intensity of very few gauss in quiet regions.

Exactly what causes such periodic accumulations of magnetic fields in the sun is not known, but it seems reasonable to suppose that a large fraction of other stars must have similar cycles of activity and that they too are surrounded by extended high-temperature coronas. Since stars are too far away to show a visible disk, it is unlikely that "star spots" can be directly observed in the foreseeable future. Just as the existence of a high-temperature solar corona can be inferred from the observation of emission lines of highly ionized atoms in its ultraviolet spectrum, however, so might similar observations of stellar spectra reveal the existence of stellar coronas. On the average the intensity of these emission lines should go up and down in unison with the star-spot cycle in a period not unlike that of the solar 11-year cycle, although the length of the period might be quite different. It would be extremely valuable if sunlike activity were discovered in other stars, not only because it might explain the variability of our own star but also because it should help to answer basic questions about the physics of a hot magnetized plasma—a gas of ions and electrons.

Much of what has been said here about the importance of observing the ultraviolet radiation of stars applies equally to external galaxies. In addition to the comparatively normal stars and galaxies, a great variety of objects (whose importance in studies of stellar and galactic evolution far outweighs their small number) can be expected to emit spectacular amounts of ultraviolet radiation. They include several types of star known to have an extended atmosphere; several classes of variable stars, including magnetic stars and novas; the planetary nebulas that surround stars on their way to extinction; remnants of supernovas, such as the Crab nebula, and several types of abnormal external galaxy, including the quasi-stellar objects (quasars) and the Seyfert galaxies (rather normal-looking galaxies with bright, starlike nuclei).

Since 1965 sounding rockets have been equipped with pointing systems capable of aiming a telescope at an individual star so precisely that its spectrum can be photographed or recorded by photoelectric cell during an exposure lasting several minutes. Early measurements made in this way by Donald C. Morton, Edward B. Jenkins and their collaborators at Princeton University with ultraviolet spectrographs flown in Aerobee sounding rockets were a harbinger of the scientific discoveries and surprises that were expected from the more extensive and systematic observations to be made with telescopes aboard large orbiting observatories. The Princeton astronomers found, for example, that the average density of hydrogen between the sun and the stars in Orion's Belt is about .1 atom per cubic centimeter, which is about a tenth the density of un-ionized hydrogen derived from measurements of the hydrogen radio line at a wavelength of 21 centimeters. They have also discovered that certain hot supergiant stars in Orion and Puppis are ejecting a shell of gas containing significant amounts of mass at a speed of about 1,500 kilometers per second. Such definite indications of mass-loss are important in considerations of stellar evolution.

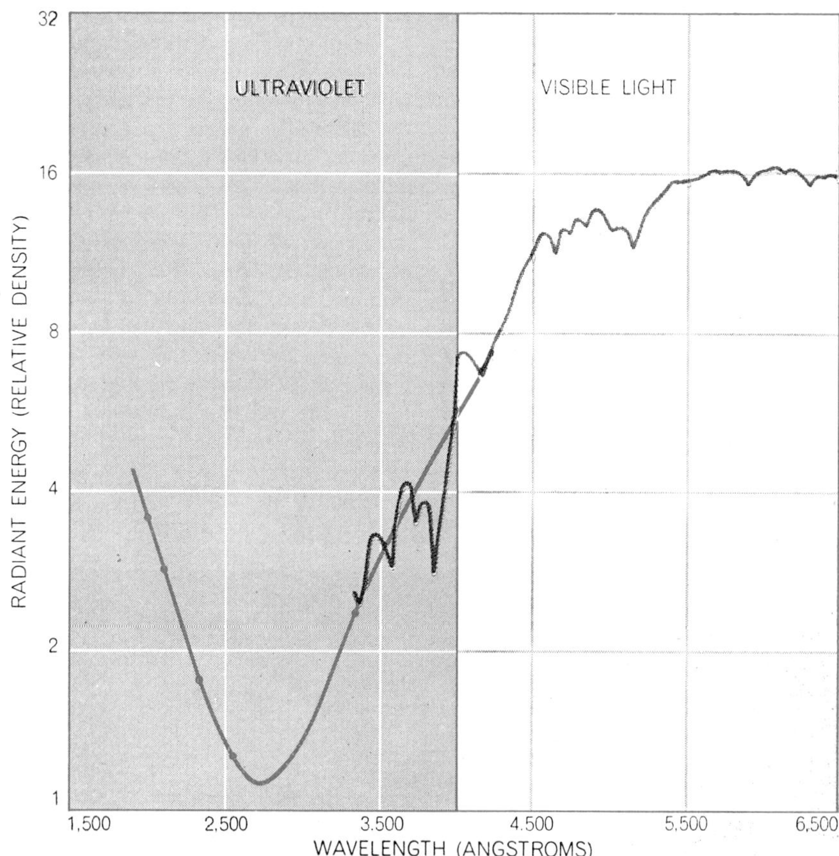

SPECTRUM OF NUCLEUS OF ANDROMEDA GALAXY is a composite of measurements made by ground-based telescopes (*gray curve*) and by a University of Wisconsin telescope on OAO-II (*colored curve*). The rise in the ultraviolet flux was unexpected and implies that the nucleus contains more hot stars than had been suggested by studies from the ground.

The first satellite measurements of stellar ultraviolet radiation were made by the Russians with automatic equipment carried by *Cosmos 51*. Shortly thereafter, in 1966, stellar spectrograms were made with small hand-held instruments by astronauts in *Gemini X* and *Gemini XI*. The first comprehensive results from satellite ultraviolet astronomy began to arrive in December, 1968, with the launching of OAO-II. (The first OAO vehicle was launched in 1966, but a power failure occurred before any measurements could be made.) The second OAO has now been in continuous operation for more than five months, and its performance has been essentially perfect.

OAO-II is the first of three astronomical observatories that were planned in 1958–1959 as one of the earliest programs initiated by NASA during the first year of its existence. It carries 11 small telescopes, four of them supplied by the Smithsonian Astrophysical Observatory and seven by the University of Wisconsin. The objective of the Smithsonian Observatory's Project Celescope is to map the entire sky in four different ultraviolet wavelength bands centered on 1,400, 1,500, 2,300 and 2,700 angstroms. The mapping is being carried out with four telescopes of 12-inch aperture, each equipped with an appropriate combination of optical filter and television camera tube to isolate and record one of the four wavelength bands. It is estimated that some 50,000 stars will be recorded by the Smithsonian telescopes. Three of the Wisconsin telescopes have a diameter of 16 inches, and four smaller telescopes have a diameter of eight inches. These telescopes are being used to make accurate photoelectric measurements of the ultraviolet spectra of stars, of nebulas within our own galaxy, of other galaxies and of planets.

The OAO-II experimenters were rewarded with unexpected and provocative findings within the first few weeks of the satellite's launching. According to Robert J. Davis, principal experimenter of Project Celescope, about 1 percent of the stars observed so far are between six and 40 times brighter in the ultraviolet than had been expected. The stars in the Pleiades, believed to be a collection of very young stars and therefore bright in the ultraviolet, turn out to be brighter than anticipated by a factor of from three to six.

Arthur D. Code, director of the Washburn Observatory of the University of Wisconsin, has generously given me two series of spectral scans made by the Wisconsin instruments aboard OAO-II. The first is a scan in the region between 1,000 and 1,400 angstroms of the spectrum of the bright star Alpha Virginis (Spica) [*see bottom illustration on opposite page*]. Although the resolution of the spectrum is comparatively low, a number of lines can be clearly identified. Note particularly the Lyman-alpha line of hydrogen, which is absorbed out of the spectrum when the starlight passes through a column of interstellar hydrogen gas between Spica and the solar system. The average density of the un-ionized hydrogen in this path is about one atom per cubic centimeter, or about 10 times greater than the density found by Morton and Jenkins in the directions of the constellations Orion and Puppis.

The second ultraviolet scan supplied by Code [*see illustration above*] is a plot of the absolute energy radiated by the nucleus of the great spiral galaxy in Andromeda in the region between 2,000 and 4,000 angstroms. The more detailed curve depicting wavelengths longer than 3,300 angstroms was obtained from ground-based observations, which had suggested that most of the visible radiation from the nucleus, or central region, of the Andromeda galaxy is to be attributed to cool and old giant stars with a surface temperature of between 4,000 and 5,000 degrees K. If the galaxy's nucleus were populated solely by such stars, the energy curve would be expected to continue its downward trend from the visible to the ultraviolet. The reversal of the trend reveals the presence of a certain number of much bluer and therefore hotter stars. These hotter stars are presumably younger than the majority of the stars in the galactic nucleus.

The ultraviolet radiation from the Andromeda galaxy is also of much importance in accounting for the ionization of the interstellar gas in the galaxy's nucleus and in attacking problems of cosmology. The external galaxies comprising the observable universe are receding from our own local group of galaxies, of which the Andromeda galaxy is a member, with speeds up to a substantial fraction of the velocity of light. Furthermore, the velocity of recession, as measured by the shift of the spectrum toward the red, is proportional to the distance, as would be expected if all the galaxies had originated in a gigantic explosion 10 to 20 billion years ago.

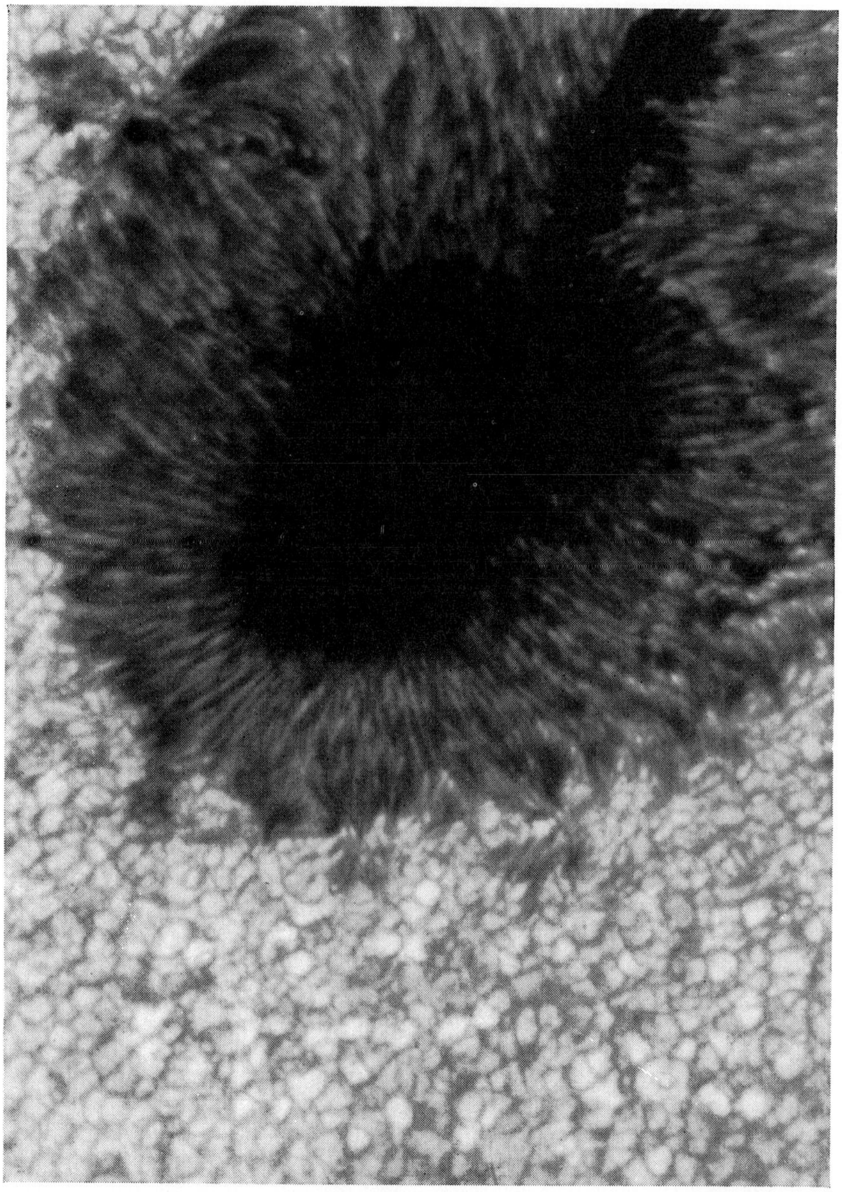

The effect of a red shift, of course, is to displace a galaxy's ultraviolet radiation toward the visible region. Code comments that if all galaxies radiate as strongly in the ultraviolet as the Andromeda galaxy does, the most distant ones, whose red shifts are greater than 60 percent of the rest wavelength, would appear bluer at visible wavelengths than nearby galaxies with very small red shifts. This means that if all galaxies are about equally old, the more distant ones would appear younger than those in our local group by an amount equal to the difference in the travel time of the light from the galaxies, which could amount to a billion years or more. Measurement of the colors of distant galaxies should therefore tell us whether these objects have spectral characteristics different from those of the Andromeda galaxy, and thus whether they were formed at an earlier time or whether they are significantly different in their physical makeup. In either case the implications for cosmology are profound.

Neither the Smithsonian instruments nor the Wisconsin ones have sufficient resolution to record the number of spectral lines required, for example, to study the relative abundances of the elements. Investigations of line spectra will be conducted on the two subsequent missions of the OAO program. Both OAO-III and OAO-IV will carry 36-inch telescopes and their associated spectrographs. OAO-III is being instrumented by the Goddard Space Flight Center and OAO-IV by Princeton. The Princeton project is aimed at measuring the relative abundances of the elements in the interstellar medium.

The emergence of so many new opportunities for exploring the universe has also focused attention on the greatly increased importance of the traditional methods of observation with optical and radio telescopes on the earth. The cost of space telescopes is such that they must not be used to make observations that can be accomplished more efficiently and cheaply on the ground. Although certain kinds of problem can best be attacked by one technique or the other, the most rapid advances will surely be made by coordinated observing programs that exploit both ground and space telescopes to observe the same object over the entire range of its spectrum, from gamma rays to long radio waves. It is this prospect and not merely the establishing of an observatory in space that makes the future of astronomy so appealing not only to astronomers but also to a growing number of physicists.

SUNSPOT, photographed from a balloon at an altitude of 80,000 feet, has a cool center that shows up dark in white light. The granulated appearance of the surrounding disk is caused by convection currents in the sun's atmosphere. This photograph is one of several thousand made by Project Stratoscope, directed by Martin Schwarzschild of Princeton University.

27

Pulsars

ANTONY HEWISH

October 1968

It is ironical that astronomy's latest discovery, the pulsars, should have been stumbled on unexpectedly during an investigation of quasars, those starlike radio sources whose origin is still one of the outstanding problems of astrophysics. Almost exactly a year ago a small group of workers operating a new radio telescope at the University of Cambridge were surprised to find that weak and spasmodic radio signals coming from a point among the stars were, on closer inspection, a succession of pulses as regularly spaced as a broadcast time service. With skepticism bordering on incredulity, the Cambridge group began systematic observations intended to reveal the nature of these strange signals even as they undertook to explain them away in terms of man-made radio interference. After all, seasoned radio astronomers do not make the mistake of supposing that every queer signal on their records is truly celestial; in 99 cases out of 100 peculiar "variable radio sources" turn out to be some kind of electrical interference —from a badly suppressed automobile ignition circuit, for example, or a faulty connection in a nearby refrigerator.

As the days went by excitement rose when we found that the pulses were coming from a body no larger than a planet situated relatively close to us among the nearer stars of our galaxy. Were the pulses some kind of message from another civilization? This possibility was entertained only for lack of an obvious natural explanation for signals that seemed so artificial. It soon declined in attractiveness with the discovery of similar pulses coming from three other directions in space, and with the absence of any planetary motion associated with the sources. (Presumably another civilization would have to occupy a planet.) We finally concluded that the only plausible explanation for these mystifying radio sources was that they were caused in some way by the vibrations of a collapsed star, such as a white dwarf or a neutron star.

The publication of these findings, about eight weeks after the pulses were first detected, stimulated intensive theoretical and observational work at research centers the world over. Whereas the true nature of the pulsars is still far from clear, it may be an opportune moment to take stock of the situation.

One obvious question to ask is why pulsars remained undiscovered for so long although powerful radio telescopes have been in operation for at least 10 years. The short answer is that the signals are not only very weak but also have an energy that falls off at the shorter wavelengths that radio astronomers mostly work with—wavelengths from a few centimeters to a meter. To find the pulsars it is necessary to use a sensitive radio telescope operating at meter wavelengths with the additional requirement that observations of the same areas of sky must be repeated with a recording system that has a sufficiently rapid response to show up the radio flashes. By a lucky coincidence these instrumental facilities were just those planned for a quasar hunt at Cambridge. The search was also designed to exploit the phenomenon of interplanetary scintillation.

The scintillation phenomenon was first observed in 1964. We found that the radio waves from very compact radio galaxies—objects whose angular dimensions are less than one second of arc—showed a characteristic rapid and irregular fluctuation of intensity. This effect, similar to the twinkling of visible stars, is caused by the distortion of the incoming radio waves as they pass through clouds of the electrically charged particles sprayed out by the sun (the "solar wind"). These "plasma clouds" cause scintillation only if the radio waves are sufficiently coherent, that is, if their source has extremely small angular dimensions. Ordinary radio galaxies have angular dimensions considerably larger than quasars and do not scintillate; therefore the effect gives a first-rate indication of whether a radio galaxy is likely to be a quasar or not. Now, just as the earth's ionosphere bends long wavelengths more than short ones, so do the plasma clouds in space. This means that scintillation is more pronounced at meter wavelengths than it is at the centimeter and decimeter wavelengths for which most radio telescopes are designed.

The radio telescope built at Cambridge for seeking possible quasars by the scintillation technique operated at a

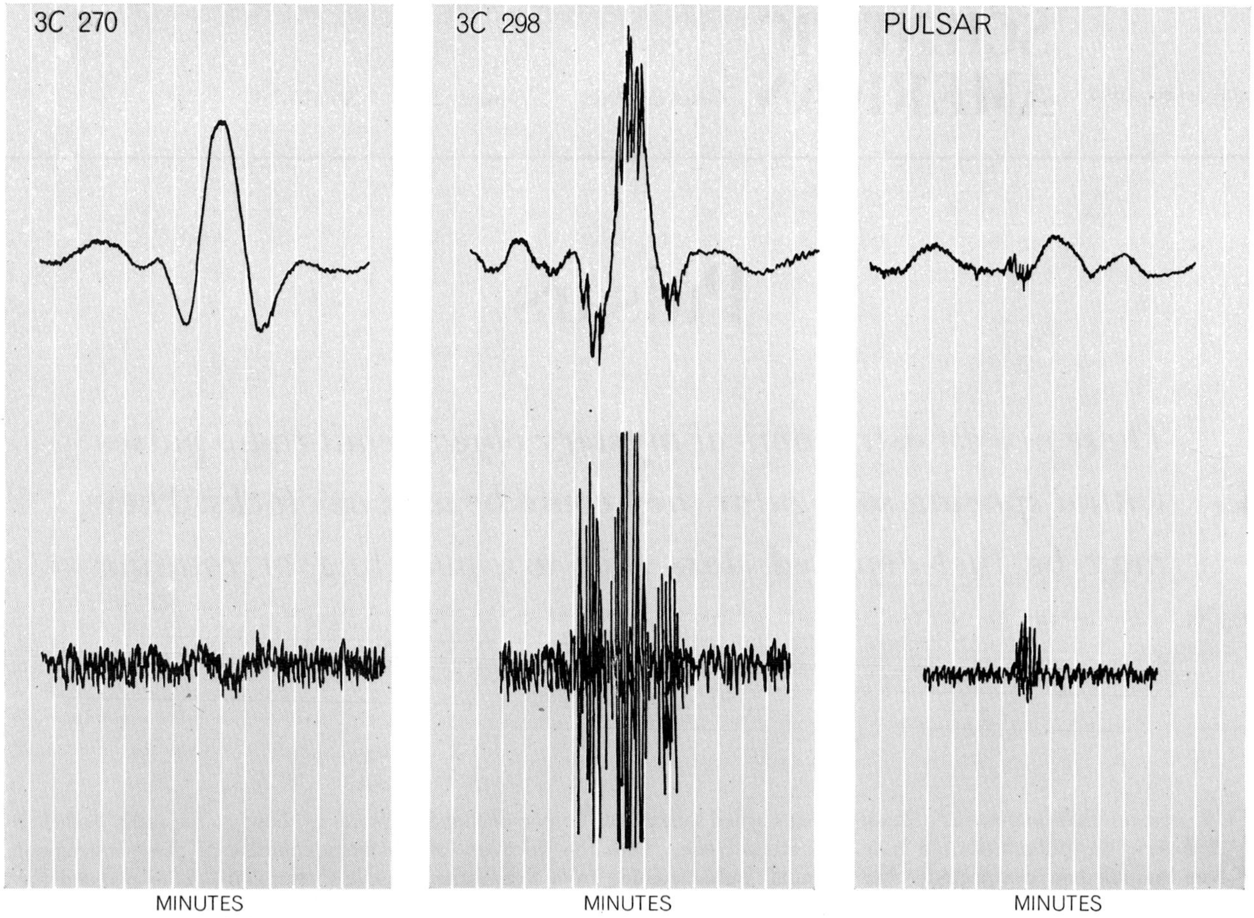

3C 270 3C 298 PULSAR

MINUTES MINUTES MINUTES

CHARACTERISTIC RECORDINGS show how interplanetary scintillation can be used to distinguish between a radio source of very large size, such as a normal radio galaxy (*left*), and sources that are the size of stars or even smaller (*middle and right*). The recordings were made by the author and his co-workers with the large array radio telescope of the Mullard Radio Astronomy Observatory at the University of Cambridge. Such a telescope can be adjusted only for elevation; the rotation of the earth then carries the radio source past the antenna's field of view. The normal radio galaxy 3*C* 270 produced the smooth "envelope" curve at the top left. The lower trace is an amplified version of the fluctuations of the upper trace. The middle pair of traces was produced by 3*C* 298, a quasi-stellar radio source, or quasar. Interplanetary scintillations are clearly visible. They are produced when the coherent waves from a compact source travel through the nonuniform clouds of electrically charged particles emitted by the sun, the "solar wind." The pair of traces at right provided the first indication of radio emission from an object later identified as the first pulsar, CP 1919.

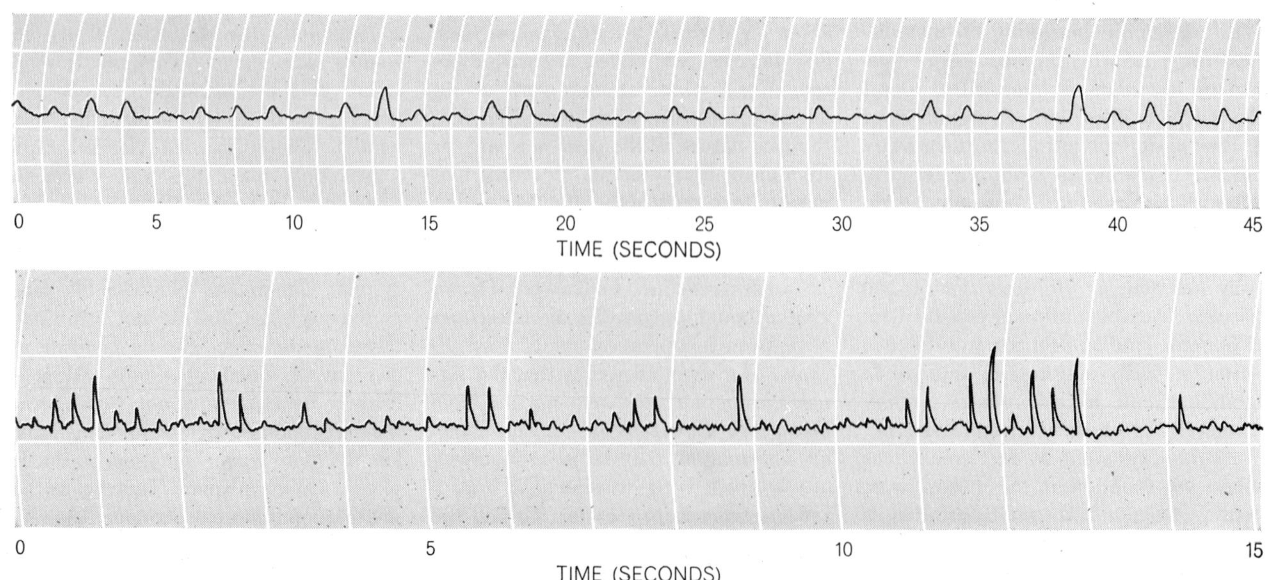

EARLY PULSAR RECORDS show the regular but variable-strength emissions of CP 1919 (*top*) and CP 0950 (*bottom*). In the trace of CP 1919, 15 pulses can quite clearly be distinguished in each 20-second interval, corresponding to one pulse approximately every 1.3 seconds. The exact pulse interval is now established as 1.33730113 seconds. CP 0950 emits a pulse every .2530646 second.

wavelength of 3.7 meters (equivalent to a frequency of 81.5 megahertz). Unlike many other radio telescopes, which are paraboloidal dishes, our instrument has a rectangular array of 2,048 dipole antennas spread over an area of 4.5 acres. By utilizing phase-scanning to steer the reception beam in elevation (declination) and the rotation of the earth to sweep the sky from west to east (right ascension) it was possible to survey a large fraction of the sky in one week.

Observations were begun in late July, 1967. Thereafter the task of analyzing the 400 feet of recorder chart paper that rolled from the equipment each week fell to a young graduate student, Jocelyn Bell. For the first time in the history of radio astronomy a large area of sky was repeatedly surveyed with an extremely sensitive radio telescope tuned to meter wavelengths. The sky was soon found to be liberally scattered with radio galaxies that exhibited plasma-cloud scintillation. These radio sources were carefully studied week by week to ascertain how the phenomenon changed in magnitude as the position of the sources shifted with the motion of the earth around the sun. Since the plasma clouds are embedded in the solar wind, which blows outward past the earth, scintillation becomes more intense as the line of sight to a radio source moves closer to the sun.

The recordings were mostly as we had expected, but one day in August, Miss Bell noticed something odd. What looked like rapid scintillation involving a very weak source was taking place in the middle of the night, a time when scintillation is usually low [*see top illustration on opposite page*]. Another curious feature was that the signals were present for only a fraction of the time required for the reception beam of the antenna to swing past a source in the sky. Had this happened only once we would undoubtedly have disregarded it and supposed that some interfering signal was present. By the end of September, however, the same thing was clearly apparent on six different occasions, although at other times it simply failed to show up. The constancy of position exhibited by the signals showed that they were probably from a true celestial source, which led us to imagine that we had found some kind of flare star. When the sun is active, its radio emission is often enormously enhanced; we calculated that if such a burst of activity occurred in a nearby star, it would be just about detectable with our radio telescope.

Now that we were really interested the source responded by virtually disappearing from the records for the next six weeks. One day late in November, Miss Bell undramatically announced: "It's back." We immediately arranged to make high-speed recordings to see if the signals had the same characteristics as signals from the sun. To our amazement the signals came as a regular succession of pulses at intervals of just over one second [*see bottom illustration on opposite page*]. It was simply unbelievable that such signals could be from a genuine astronomical object, but repeated observations showed that there could be no doubt about it. Naturally this mysterious radio beacon rather dominated our lives for the next few weeks as we carried out more observations in an effort to understand the true nature of the source.

One of our earliest surprises was the discovery that each pulse was associated with radio signals of continuously changing wavelength: the signals swept through the one-megahertz pass band of the telescope's receiver from high frequencies to low. This was revealed when two receivers tuned to slightly different frequencies were operated simultaneously and each pulse arrived a fraction of a second earlier in the receiver tuned to the higher frequency. A measurement of the instantaneous bandwidth of the radiation then showed that the actual duration of the pulse was only from 10 to 20 milliseconds.

This provided the first vital clue to the source itself: the body emitting the pulses had to be extremely small. Its radius could not be larger than a few thousand kilometers. The reason is that a large body cannot emit a pulse of radiation in a time shorter than the time required for light to travel across it. Suppose, for example, the sun could be instantaneously switched off; how would it look from the earth? First we should see a dark spot at the center of the sun, since this is the part nearest to us. The dark area would then enlarge outward until the sun was a bright ring; finally even its outer edge would disappear. The entire sequence would take about two seconds, so that if the sun were to flash on and off like a pulsar the flashes could not have a duration shorter than that.

The second clue was provided by the frequency distribution of the pulses, which gave an important indication of the distance of the emitter. It is known that interstellar space is filled with very-low-density gas consisting predominantly of hydrogen atoms. Light from hot stars ionizes, or dissociates, some of these atoms, giving rise to free electrons. In a perfect vacuum radio waves of all wavelengths travel at the same speed—the speed of light—but this does not hold in an ionized gas. In such a gas the longer the wavelength, the lower the velocity. Thus a sharp pulse of radio waves containing a spread of wavelengths becomes stretched out as it travels through the ionized interstellar gas, with the short wavelengths arriving at the earth slightly before the longer ones [*see top illustration on next page*].

This effect, called dispersion, is a general property of wave motion. It happens, for example, with the ripples in a pond into which a stone has been dropped, and with the seismic waves that travel outward from an earthquake. If the density of electrons in interstellar space were accurately known, the difference in arrival time of a pulse at different wavelengths would give the distance of the source immediately. Unfortunately the density is not easily measured, but it is probably around one electron in a volume of 10 cubic centimeters. If this estimate is adopted, the distance of the first pulsar to be discovered is about 130 parsecs. (One parsec is 3.26 light-years.) When one considers that the diameter of our galaxy is about 30,000 parsecs, it is evident that this pulsar must be regarded as one of our neighbors in space.

The next big surprise came when we began timing the pulses and the truly staggering accuracy of the pulse rate became apparent. We found that from day to day the arrival of the pulses could be predicted to better than a tenth of a second, but we also observed that the interval between pulses was gradually decreasing. The explanation is that during this period the orbital motion of the earth was carrying us in a curving path toward the source. Thus there was a progressive decrease in period resulting from the Doppler effect. When due allowance was made for the Doppler variation, the pulses kept time within the experimental accuracy of about 10 milliseconds for several weeks.

We now know that the regularity of the pulses exceeds one part in 100 million, so that the pulsar can be regarded as a very good clock indeed. The period of the first pulsar is, in fact, 1.33730113 seconds. This period differs slightly from the value we originally gave; we had made a counting error of one pulse per day because we were unable to track the source for long periods. Our mistake was soon revealed when the large steerable radio telescopes at Goldstone in California and at Parkes in Australia were trained on the source. Just what the true

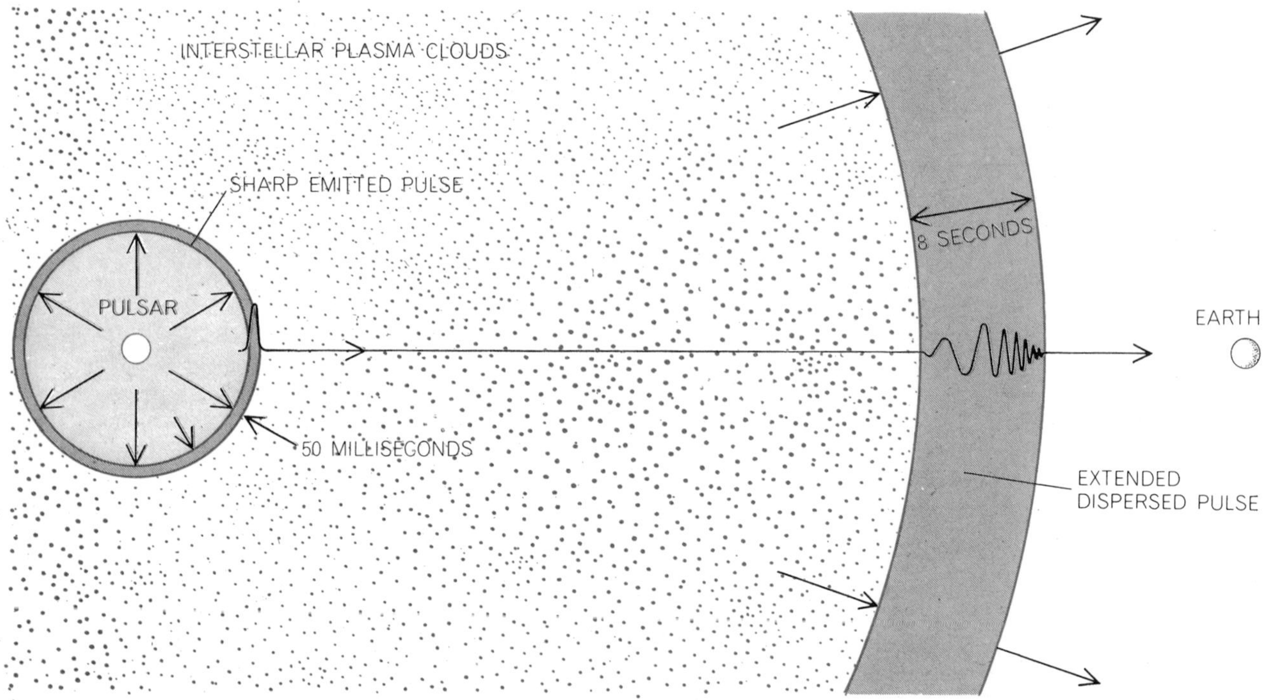

INTERSTELLAR CLOUDS OF ELECTRONS retard the velocity of radio waves in direct proportion to their wavelength: the longer the wave, the lower the velocity. Thus the broad-spectrum pulse that may be emitted within 50 milliseconds by a pulsar is stretched out by electrons in space so that the longest waves may lag as much as eight seconds behind the shortest when received on the earth. By measuring this delay and making reasonable assumptions about electron density one can estimate the distance to the source.

regularity of the pulses is we shall not know until timing experiments have been conducted for several years; the present accuracy is limited by irregularities in the shape of the pulses. The first pulsar is now designated CP 1919, which stands for Cambridge pulsar at right ascension 19 hours 19 minutes.

Faced with the problem of accounting for pulsed radio emission from a body no larger than a planet stationed relatively close to us in the galaxy, the possibility of communication from another civilization cannot be ignored. There is strong evidence, however, that the signals are not radiated from a planetary body in orbit around a star. Just as the orbital motion of the earth gives rise to a Doppler variation of period, so would any planetary motion of the source itself. We searched carefully for such effects, but when the orbital motion of the earth had been allowed for, no additional Doppler variation could be detected. Although it would be possible for an orbit to lie in a plane perpendicular to the line of sight, thus giving rise to no Doppler effect, we felt that this was rather unlikely.

If the source is a natural emitter, possibly a collapsed star we reasoned that there should be others like it in the sky. A careful inspection of all the records made in the quasar survey was undertaken from which we obtained a number of positions where pulses were possibly present. The radio telescope was trained on these places, and after some weeks three more pulsars had been added to the list. By this time we felt reasonably confident that the pulsars were a natural phenomenon. We published our findings on the first pulsar while checking the properties of the other three. One of the three has a pulse interval of only .25306 second, the briefest period yet observed. Nine pulsars are now known. The fifth was found by Harvard University radio astronomers working with the 300-foot

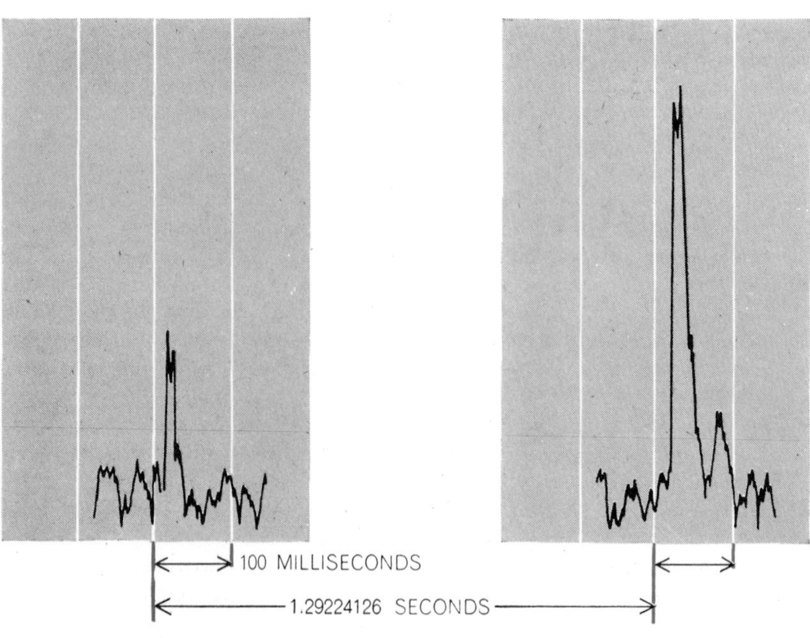

CONSECUTIVE PULSES FROM CP 0808 show marked differences in fine structure at the recording frequency of the large Cambridge antenna array: 81.5 megahertz. This corre-

dish at the National Radio Astronomy Observatory in Green Bank, W.Va., two more have been turned up at Cambridge and two have been located in the southern sky by a team at the Cornell–Sydney University Astronomy Centre at the University of Sydney. One of the southern pulsars, PSR 2045, has the longest period yet observed: 1.9616633 seconds [*see top illustration on page 295*].

During the past six months strenuous efforts have been made at radio and optical observatories throughout the world to penetrate the pulsar mystery. It is now known that pulsars radiate over a very wide band extending from 40 megahertz to 3,000 megahertz. The spectrum of the radiation is difficult to establish because of large pulse-height variations that are quite unrelated at the different wavelengths. It appears that the radiation is strongest at low frequencies and drops off steeply at frequencies above 1,000 megahertz. The rapid and irregular fluctuation of pulse strength from instant to instant is an important property of pulsar radiation and shows up in all the sources [*see illustration below*].

Pulse trains have been carefully studied to see if any kind of strength pattern is present, but without success. Amplitude modulation of the pulses would, of course, be an obvious method of coding a message if the signals were an attempt at communication by another civilization. One feature common to all the pulsars is that the amplitude variations are more rapid at lower frequencies. For example, at 81.5 megahertz CP 1919 radiates relatively strong pulses for only about one minute at a time, whereas at 408 megahertz the bursts of strong signals last for about half an hour. It has been suggested that these fluctuations arise from scintillation caused by plasma clouds near the pulsar, and it does seem quite likely that such clouds would be blown out from the source.

When individual pulses are examined in detail, they show a remarkable complexity of structure. Each pulse is found to consist of a number of subpulses, the exact form of which may change radically between one pulse and the next [*see bottom illustration on next page*]. Subpulses lasting only .2 millisecond have been reported from a group working with the giant 1,000-foot paraboloid at Arecibo in Puerto Rico. Since this figure is at the detection limit of the Arecibo receiving system, it may be that even finer structure exists.

An important property of the subpulses is that they often exhibit considerable polarization. Strong circular polarization with the sense of rotation varying rapidly from one subpulse to the next has been observed at Arecibo, and long-lived linear polarization has been detected at the Jodrell Bank radio observatory in England. At Cambridge we have also found that the sweeping radio-frequency signal that constitutes the pulse has a complicated spectrum. The energy in a single pulse may be quite different at frequencies separated by only 300 kilohertz.

Although the shape of individual pulses varies considerably, it turns out that the "envelope" of the pulse—a smoothed-out form of its detailed trace obtained by superposing pulses for a minute or so—settles down to a characteristic form that is different for each source [*see top illustration on next page*]. CP 1133 has a double-humped envelope, whereas CP 0950 shows a slow rise followed by a rather narrow peak. It is interesting, however, that all the pulse-envelope traces last about 50 milliseconds. Referring back to our earlier consideration of the physical size of the pulsars, this shows that in all cases their diameter cannot exceed about 15,000 kilometers. The subpulses must of course be emitted from bright patches that are even smaller than this upper limit.

Radio astronomers will clearly be busy for some time searching for more pulsars and unraveling the properties of those already found. Another line of investigation is the possibility of observing pulsars with optical telescopes. The trouble here is that the sky is so densely packed with stars that very precise radio positions are necessary in order to be sure of identification. The Cambridge dipole array that originally detected the pulsars was not designed for accurately pinpointing sources, but at the same observatory Sir Martin Ryle operates a large radio telescope system that employs the principle known as aperture synthesis. Using three 60-foot dishes, one of which is mounted on rails, and feeding the radio signals into a computer,

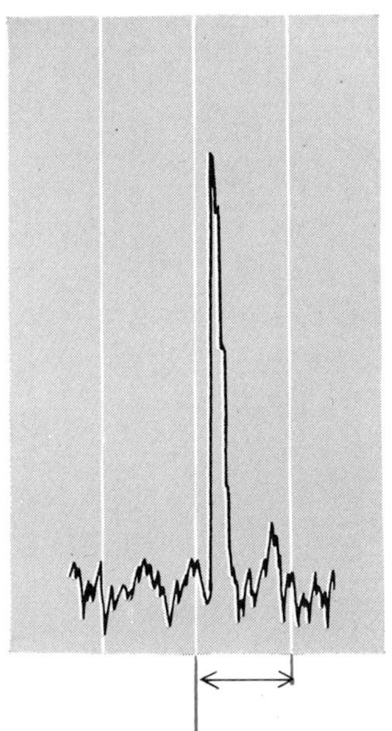

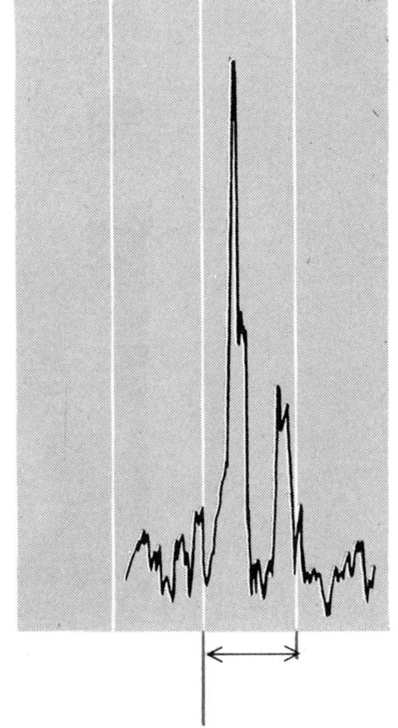

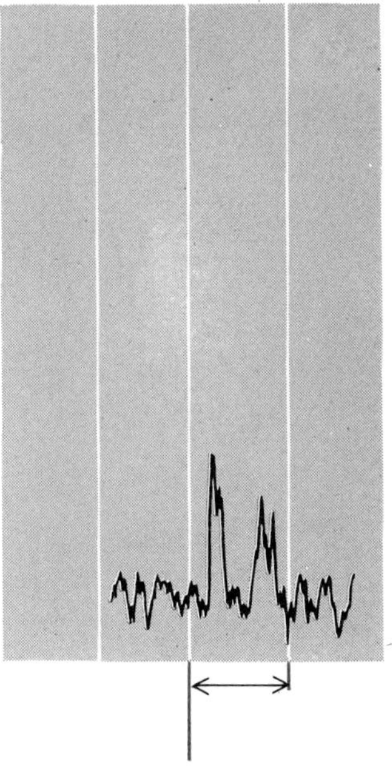

sponds to a wavelength of about 3.7 meters. At centimeter and decimeter wavelengths the emissions from pulsars are much weaker, which explains in part why they were not discovered earlier by the large radio telescopes that usually work at the shorter wavelengths.

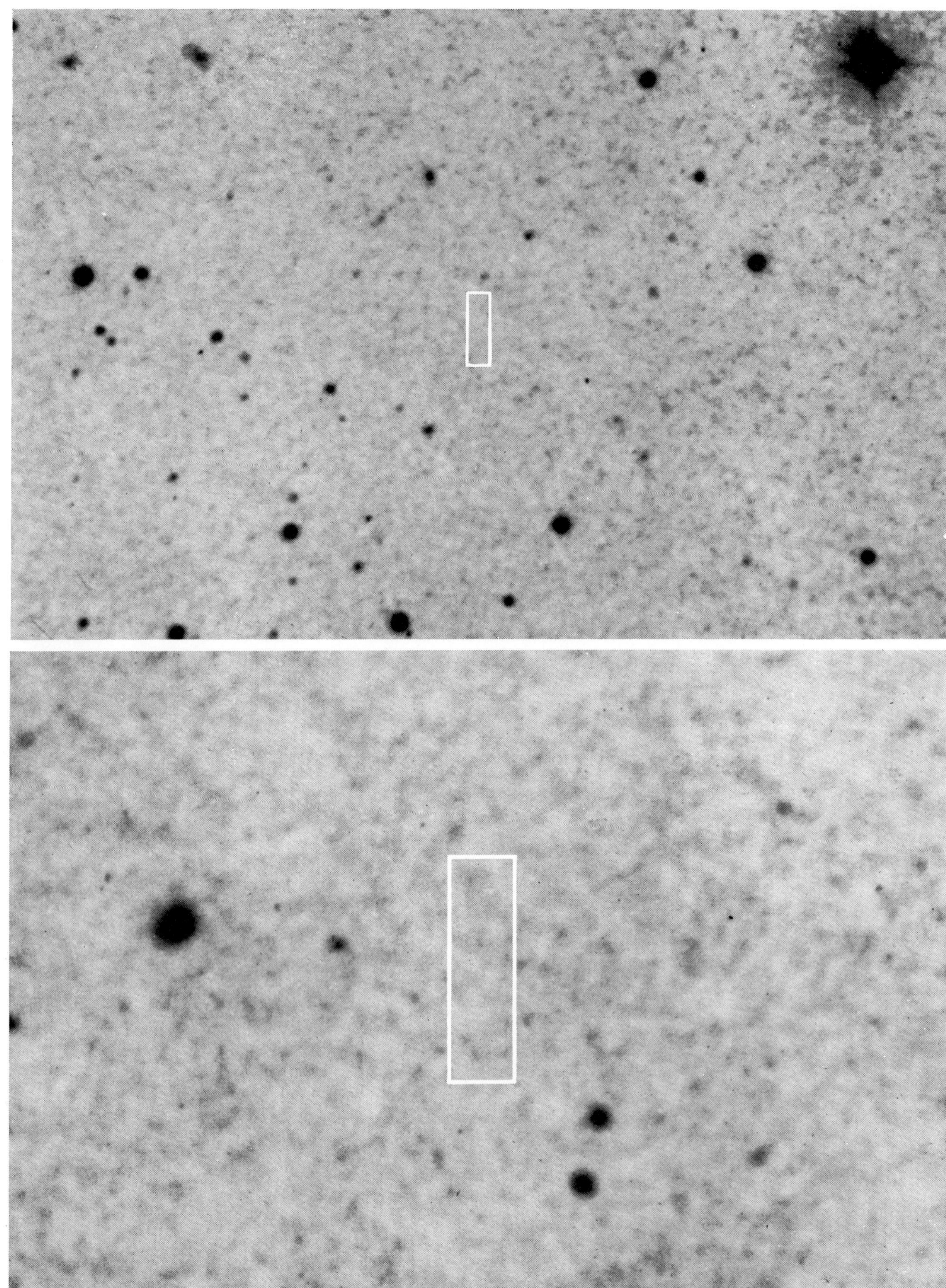

OPTICAL SEARCH for pulsars has so far been unrewarding. The rectangles in these two sky photographs delineate the radio positions of CP 0950 (*top*) and CP 1133 (*bottom*). The long side of the rectangle equals two-thirds of an arc minute in the case of CP 0950 and one arc minute for CP 1133. One would expect that if either pulsar were a normal white dwarf it would be visible in such pictures if it were as nearby as the radio evidence indicates. The prints are from the National Geographic Society–Palomar Observatory Sky Survey.

the system constitutes an instrument equivalent in capability to a steerable dish a mile in diameter [*see bottom illustration on page 295*].

Accurate positions have now been obtained with this instrument for CP 1919, CP 0950 and CP 1133. Inspection of prints from the National Geographic Society–Palomar Observatory Sky Survey reveals no object within the rectangles of probable pulsar position except for CP 1919 [*see illustration on page 292*]. The pulsars CP 9050 and CP 1133 radiate no visible light down to the limit of the photographic plates. In the case of CP 1919 a faint yellow star is seen near the pulsar position, but we cannot be sure that this is not a coincidence.

Much effort has been expended in attempts to detect light flashes from this star. Observers at the Kitt Peak Observatory in New Mexico report that they have found a small optical variation at, curiously enough, *twice* the pulsar period. A similar result was also reported from the Lick Observatory in California, but the report was subsequently withdrawn. Other observations have failed to show any fluctuation of light from the star. In view of the absence of light from at least two of the pulsars, and their close similarity in all other respects, the optical evidence obtained so far must be treated with caution.

No satisfactory explanation for the pulsars has yet been suggested. The energy requirement, combined with the immense stability of the timekeeping mechanism, implies that some object with the mass of a star is involved. We are currently aware of only two types of star, white dwarfs and neutron stars, that are as small as the pulsars must be [*see top illustration on page 293*]. White dwarfs are stars that have collapsed after exhausting their nuclear fuel. Their mass, roughly equal to the mass of the sun, is packed into a volume the size of a planet. Neutron stars have been proposed on theoretical grounds, but none has yet been found. They would have roughly the same mass as white dwarfs, but their diameter would be only on the order of 10 to 100 miles. Calculations indicate that neutron stars might be hot enough to emit detectable X rays.

The nature of these compact stellar bodies and the reason why only the two types are plausible depends on the properties of matter at enormously high density. A typical star such as the sun maintains its size by balancing internal pressure against gravity; the pressure results from the large amounts of radiant energy

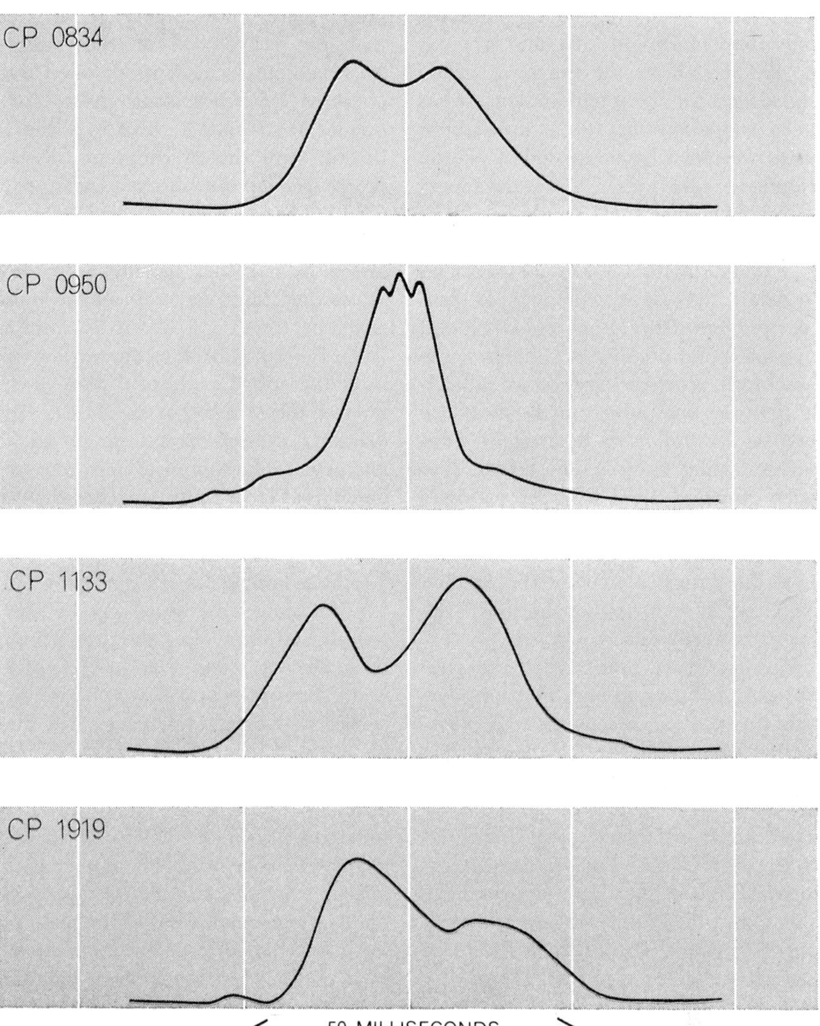

AVERAGE PULSE ENVELOPES of four pulsars were determined by A. J. Lyne and B. J. Rickett with the 250-foot dish telescope at the Jodrell Bank observatory of the University of Manchester. The fact that all emit pulses of roughly the same duration, 50 milliseconds, suggests that the emitting objects are all close to the same size. CP 0950 may be the smallest.

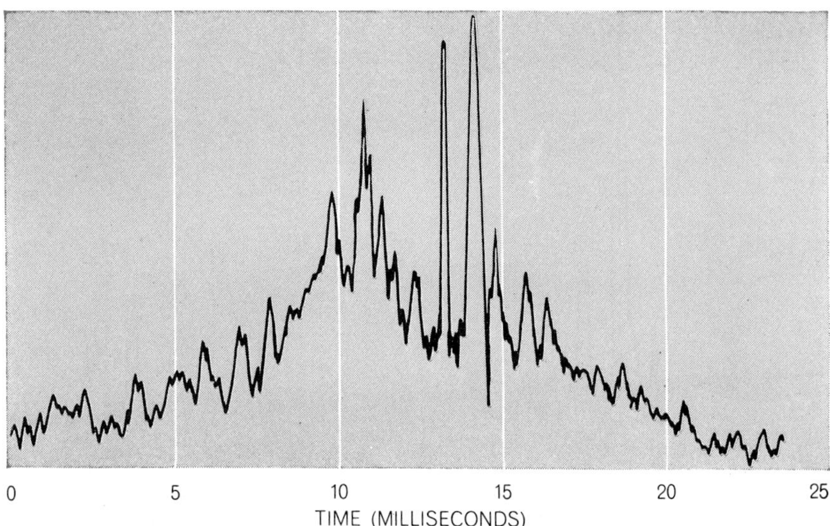

FINE DETAIL IN CP 0950 was recorded at a frequency of 195 megahertz (1.55 meters wavelength) with the 1,000-foot dish telescope at Arecibo, Puerto Rico, operated by Cornell University. J. M. Camella, H. D. Craft, Jr., and Frank D. Drake made the recording.

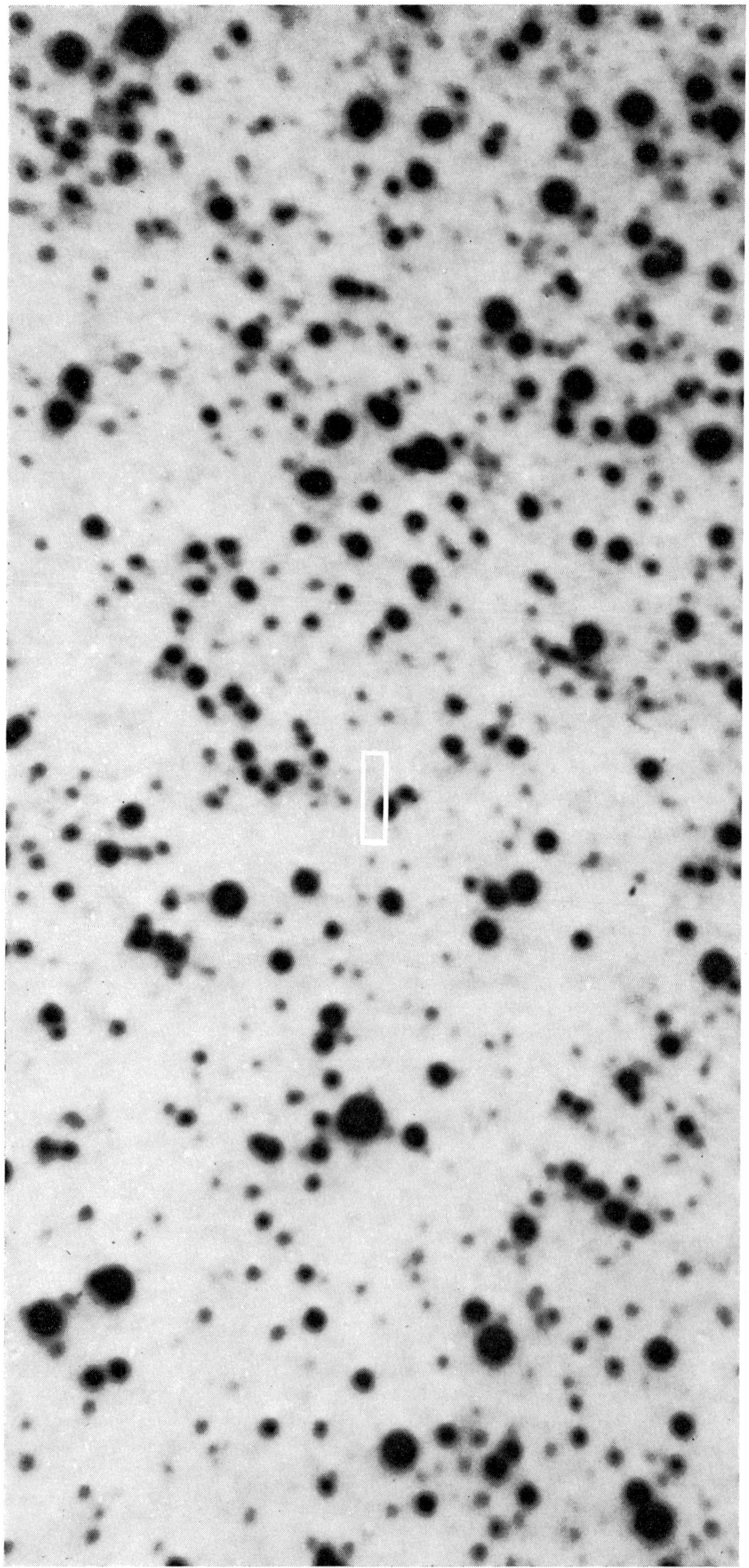

OPTICAL SEARCH FOR CP 1919, first pulsar to be discovered, has been conducted in the region defined by the rectangle. Unconfirmed reports from optical astronomers suggest that some object in this region is emitting light in a fluctuating and possibly regular manner. But the star lying on the boundary of the rectangle exhibits no unusual properties.

released by nuclear reactions deep inside the star. As the nuclear fuel is used up the radiation pressure must eventually decrease and gravity will then contract the star to a smaller volume. A new equilibrium becomes possible, even for a cool star, when the pressure due to the motion of electrons balances gravity. For a star as massive as the sun this happens only when gravity has shrunk the material into a sphere with a radius of some 6,000 miles. The visible white dwarfs are presumed to be old, burned-out stars of about this mass and size.

We can imagine a star much more massive than the sun collapsing in the same way, but an interesting situation arises when gravitational forces compress the spent matter into too small a volume. Quantum effects, and the "degeneracy" of the electrons resulting from the Pauli exclusion principle, dictate that the electrons weaving in and out among the atomic nuclei can be packed into a smaller volume only if their energy is increased. The more massive the star, the greater the inward gravitational force and the greater the electron energy. Eventually the energy is so great that the electrons react with protons to form neutrons—the reverse of the normal process, in which a free neutron decays into a proton and an electron within a few minutes. When the electrons disappear, so does the pressure caused by their motions; gravity then contracts the star still further until it becomes a neutron star. According to current ideas a heavy star is likely to explode before it reaches the white-dwarf stage, leaving a neutron star at its center. Attempts to find neutron stars in the debris of supernovas, however, have not succeeded.

Since the time of Sir Arthur Eddington astrophysicists have contemplated the possibility of pulsating stars that alternately expand and contract over their entire surface. These ideas are of interest in the theory of variable stars such as the cepheid and RR Lyrae types. As we groped for some explanation for the pulsars we were impressed by the fact that the pulsation periods calculated for white dwarfs and neutron stars came fairly close to the observed periods of the pulsars. For the neutron star it had also been calculated that considerable quantities of energy could be stored in the vibrations, implying that the pulsation might persist for a long time. It therefore appeared that the pulsation of an entire star might provide just the required clock mechanism for triggering the emission of the radio pulses.

Precisely how an up-and-down vibra-

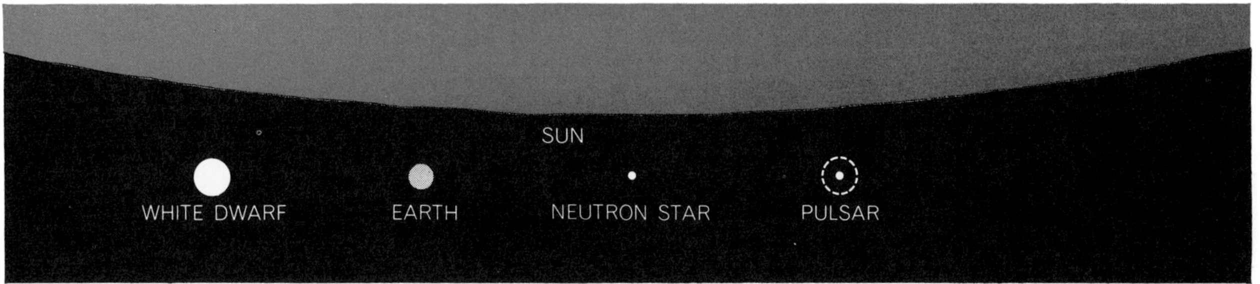

CONJECTURED SIZE OF PULSARS is compared with relative sizes of the sun, the earth, white dwarfs and neutron stars. The sun is 864,000 miles in diameter, the earth not quite 8,000 miles. White-dwarf stars incorporate a mass roughly equal to that of the sun in a volume of planetary dimensions. Hypothetical neutron stars also have a mass roughly equal to that of the sun but a diameter of only 10 to 100 miles. The brief pulse interval suggests that pulsars fall somewhere in size between neutron stars and white dwarfs.

tion of the surface of a star would give rise to radio pulses is far from clear. One can, however, visualize a pressure wave being launched into the atmosphere by the heaving surface; then, as the atmosphere thinned out, the pressure wave would be transformed into a shock wave. The outward-moving shock front might accelerate electrons to high speeds and the electrons in turn might generate radio waves as they rushed out through the ionized gas surrounding the star [*see illustration below*]. Events of this kind certainly occur in the solar atmosphere, giving rise to radio outbursts. We may imagine that similar activity, linked to a pulsating surface, accounts for the regular radio flashes from a pulsar.

When we first considered this possibility, there was one important snag. Calculations indicated that neutron stars would pulsate rather too rapidly, with periods of milliseconds. White dwarfs, on the other hand, would pulsate too slowly; if they were compressed sufficiently to vibrate with periods shorter than a few seconds, they would be collapsed by gravity. Faced with this challenge, several theorists have recently calculated that a spinning white dwarf can be made to vibrate considerably faster, particularly if the interior of the star rotates more rapidly than its equatorial region does. Periods as short as .1 second are reported for this model, which would certainly account for the known pulsars. Other modifications of white-dwarf theory assume that the star pulsates at some overtone of the fundamental period, or that the atmosphere vibrates at some fairly high rate that is coupled to a slower vibration of the interior.

Many other ideas have been suggested to explain pulsars, some of which can be classified as "lighthouse" theories. If radio waves continuously emitted from some active region on a rotating star are suitably formed into a beam, regular flashes at the rotation period of the star will be observed on the earth [*see top illustration on next page*]. This picture presents difficulties if the star is a white dwarf; if the star spun fast enough to account for the most rapid pulsations, it

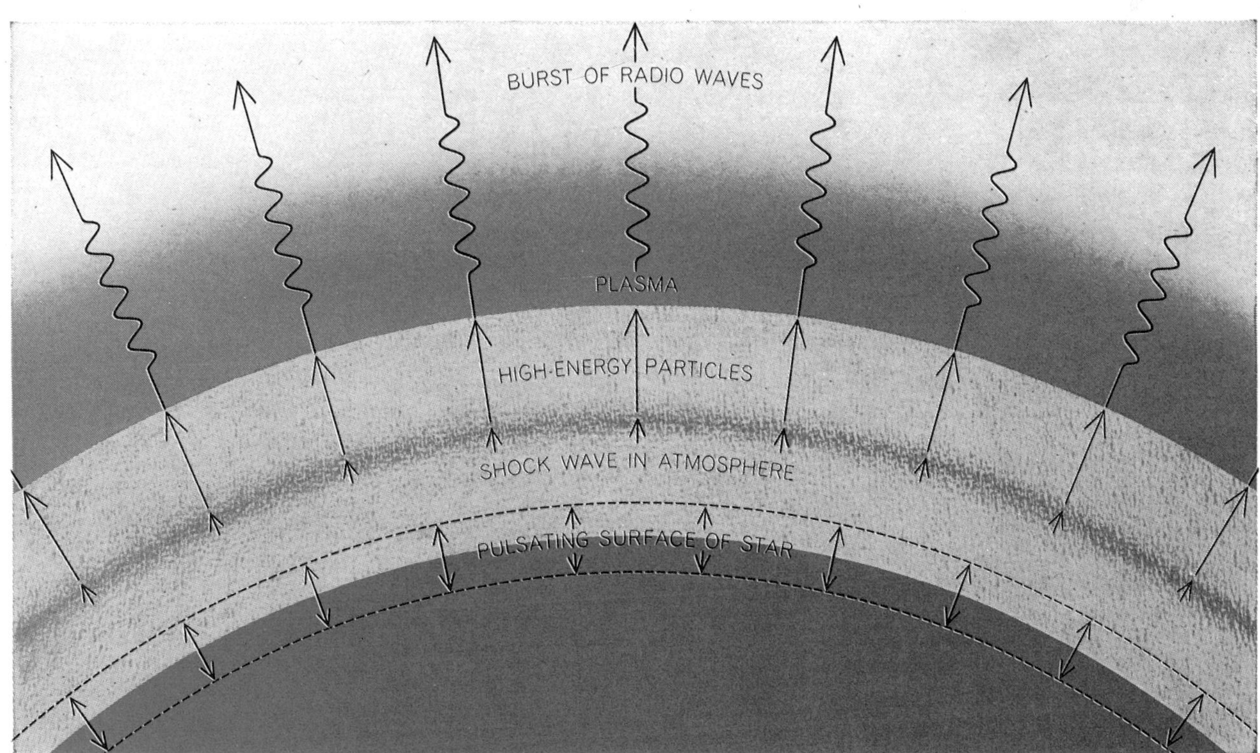

MODEL OF PULSATING STAR assumes that the entire surface of a white dwarf (or possibly a neutron star) is oscillating at the frequency characteristic of the pulsars so far observed. The regular motion of the surface sends out shock waves that generate bursts of radio waves in all directions when they strike the plasma, or electrically charged envelope of gas, in the star's upper atmosphere.

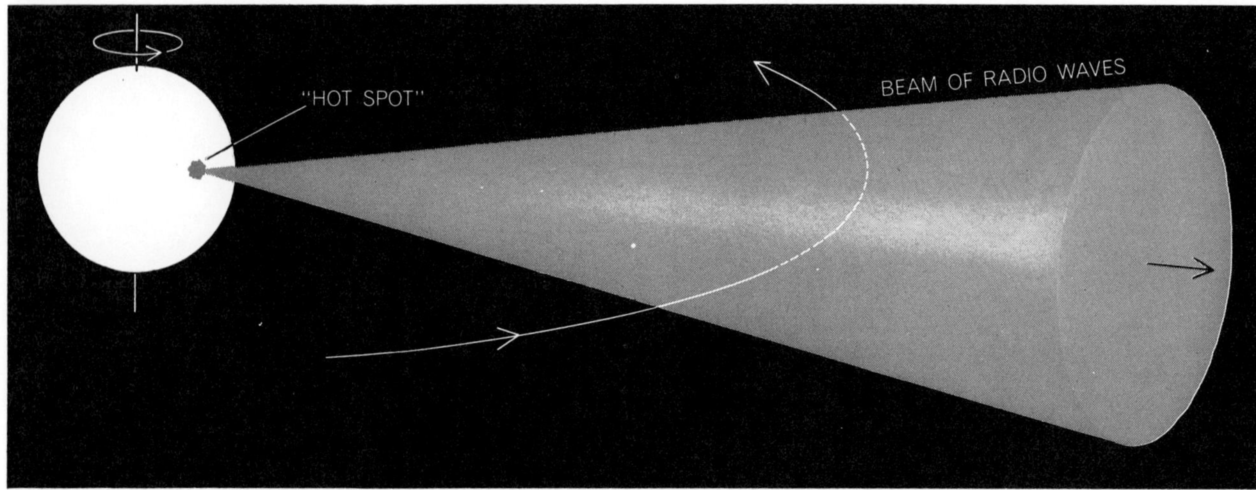

ROTATING WHITE DWARF is one of several "lighthouse" models put forward to explain the frequency and extreme regularity of pulsar signals. In the model illustrated, proposed by J. P. Ostriker of Princeton University, radio waves are continuously emitted from some active region, or hot spot, on the surface of a white dwarf that is rotating at a rate equivalent to the pulse interval.

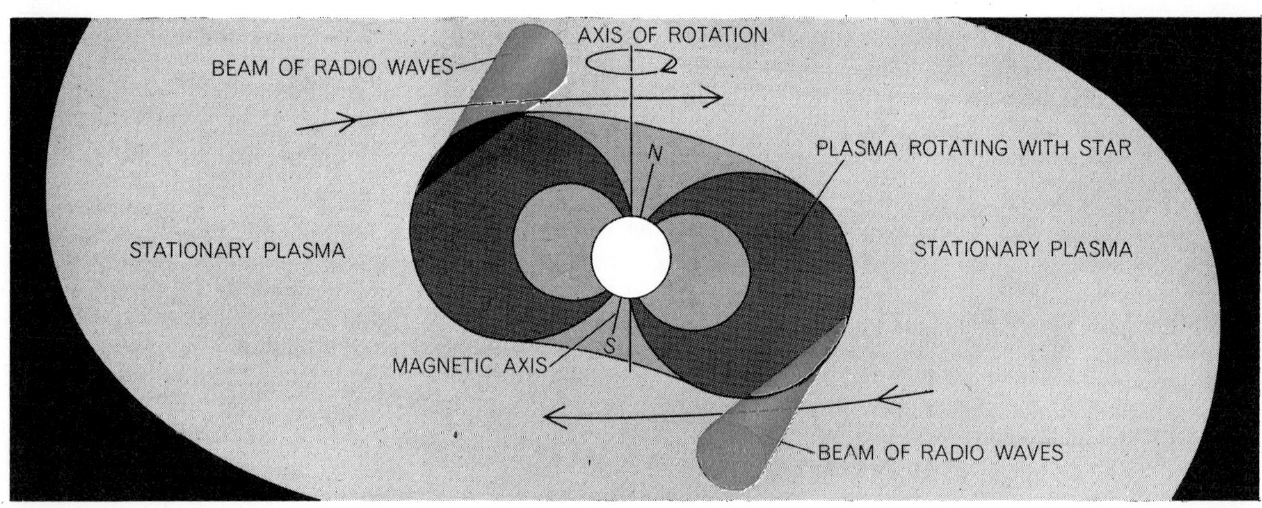

ROTATING NEUTRON STAR is a lighthouse model proposed by Thomas Gold of Cornell University. It presupposes that the star's magnetic field is sufficiently powerful to grip the star's envelope of plasma out to a great distance, forcing it to rotate at the same rate as the star itself. At the periphery where the spinning plasma finally breaks away, highly directional radio beams will be emitted.

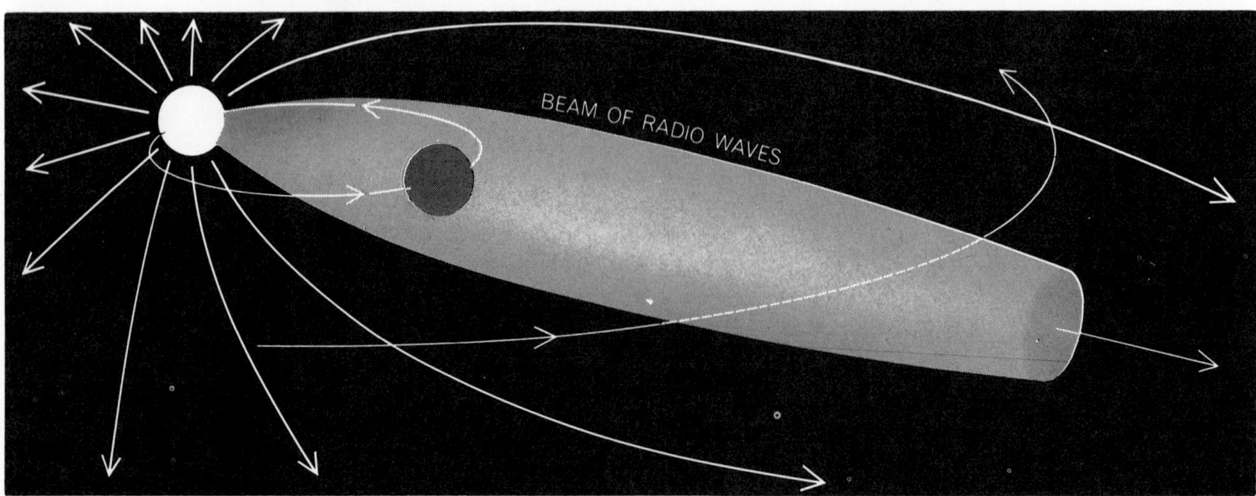

PAIR OF NEUTRON STARS traveling rapidly in orbit around each other provides still another lighthouse theory. The gravitational field of a neutron star is so immense that it would bend and focus radio waves if they were being continuously emitted by the companion star. The various lighthouse theories require that the earth be fortuitously located in line with the rotating radio beams.

might fly apart or rotate so unevenly over its surface that accurate timekeeping would be impossible. Neutron stars, however, could spin at more than 600 revolutions per minute before centrifugal forces disrupted them.

Neutron stars are also likely to have a powerful magnetic field, and such a field would provide the basis for another lighthouse theory. The magnetic field would cause the ionized atmosphere of the star (if there is one) to whirl around with the star, even if the atmosphere were quite extended. Thomas Gold of Cornell University has suggested that the far reaches of such an atmosphere might move at a speed approaching the speed of light before it became disconnected from the magnetic field. At the point where the spinning atmosphere broke away, radio waves would be generated and formed into a beam by relativistic effects [*see middle illustration on opposite page*].

Still another lighthouse model, proposed by W. C. Saslaw, J. Faulkner and P. A. Strittmatter of the University of Cambridge, calls for a pair of neutron stars in orbit around each other. Many typical stars exist in pairs held together by gravitational attraction; if neutron stars exist at all, they too may exist in pairs. Now, the gravitational force near a single neutron star would be immense, which means that light waves (and therefore radio waves as well) would be bent toward the star as they pass by. Consequently if one of the neutron stars is continuously emitting radio waves, some of them would be focused by the companion star: its gravitational field would act like a gigantic lens. The result would be a rotating beam that might extend in the direction of the solar system [*see bottom illustration on opposite page*]. A severe difficulty with this idea is that the orbital velocity of the neutron stars would be slowed down rather rapidly by the loss of energy to gravitational radiation. It may be, however, that bodies in free fall do not radiate gravity waves; that would save the theory.

One could continue to describe many other pulsar models that have been proposed. Difficulties arise with most of them, and clearly theoreticians have much work to do before the truth about pulsars is known. Current opinion seems to be about equally divided between white dwarfs and neutron stars as the source of the periodic emission. Quite apart from what pulsars actually are, however, they are likely to become important in widely different realms of astronomy. For example, there is considerable interest in the question of whether or not the astronomical time derived from the motion of celestial bodies remains in step with the atomic time derived from atomic clocks in terrestrial laboratories. It may be that pulsars are sufficiently accurate astronomical clocks to supply the answer. Already the constancy of pulsar timekeeping is almost good enough to provide a direct test of the theory of general relativity. Pulsars are exciting objects and we can expect a great deal from them.

PULSAR	RIGHT ASCENSION	DECLINATION	PERIOD (SECONDS)	DISTANCE (PARSECS)
CP 0328	03h 28m 52s	54°23′	0.714518563	268
CP 0808	08h 08m 50s	74°42′	1.29224126	58
CP 0834	08h 34m 22s	06°07′	1.2737642	128
CP 0950	09h 50m 29s	08°11′	0.2530646	30
CP 1133	11h 33m 36s	16°08′	1.187911	49
HP 1506	15h 07m 50s	55°41′	0.739677626	196
CP 1919	19h 19m 37s	21°47′	1.33730113	126
PSR 1749	17h 49m 49s	−28°06′	0.5626451	509
PSR 2045	20h 45m 48s	−16°28′	1.9616633	114

POSITION, PERIOD AND DISTANCE of nine known pulsars are listed. The hour and minute values of the position in right ascension provide the four-digit number that follows the designation CP (for Cambridge pulsar), HP (for Harvard pulsar) or PSR (simply for pulsar). The last designation was suggested by the group in Sydney, Australia, that recently discovered the two pulsars in the Southern Hemisphere. The distance estimates are based on an assumed mean electron density in space of one electron per 10 cubic centimeters.

APERTURE-SYNTHESIS TELESCOPE at the Mullard Radio Astronomy Observatory of the University of Cambridge employs three 60-foot dish telescopes, of which only two are shown here. One of the three is mounted on rails. The signals from the three dishes are fed into a computer, where they are synthesized into a signal equivalent to that obtainable from a steerable dish telescope a mile in diameter. The instrument has been used by Sir Martin Ryle and his colleagues to establish the precise positions of pulsars in the northern sky.

VIII

Cosmology

VIII

Cosmology

INTRODUCTION

A comparative dearth of *Scientific American* articles on any particular scientific topic might well indicate that the subject is becoming moribund. In cosmology, it speaks far more about the precarious situation of cosmologists, who currently are inundated with new and unexpected data of cosmological significance, than about the quiescence of the study itself.

In 1956, when the editors of *Scientific American* assembled their landmark September issue devoted to the universe, they persuaded not one, but two cosmologists to defend entirely different viewpoints: George Gamow, with "The Evolutionary Universe" and Fred Hoyle, with "The Steady State Universe." Gamow's article is reprinted here; Hoyle's is not. This is not to say that Gamow was entirely right, or Hoyle entirely wrong. Certainly in terms of the significant astrophysics that resulted, particularly the key ideas about element formation in stellar interiors, the "heretical" steady state was by far the more seminal theory. Yet in the intervening decade new evidence has, at least for the present, sealed the fate of the steady-state theory. Even Fred Hoyle, in a now famous contribution to *Nature* (October 9, 1965, pp. 111–114), recanted his former opinion, although more recently he has advocated a modified steady-state cosmology.

The impact of this new evidence—essentially the quasars and the 3° background microwave radiation—is here discussed by Dennis Sciama in the guise of a book review. "If the red shifts of the quasars are cosmological in origin, and if the universe is filled with black-body radiation," Sciama states sadly, "then the chances of the steady-state theory surviving are very small indeed."

Two articles in this section, not directly on cosmology, address themselves to these newly observed phenomena. Neither Geoffrey Burbidge nor Fred Hoyle, who write on "The Problem of Quasi-Stellar Objects," are fully convinced that the quasars lie at cosmological distances, although they concede that most astronomers prefer to place the quasars near the fringes of the observable universe. This majority preference hangs by fragile threads: first, the more familiar interpretation of large red shifts as high velocities (and great distances); second, the smooth morphological classification that ensues if the quasars are related to Seyfert galaxies and to radio galaxies. Somehow the quasar phenomenon seems to make more sense (though still leaving the enigmatic puzzles of their high energy and rapid variation) if they are fit into the more familiar astronomical patterns. Certainly speculative cosmology is nothing if not an attempt to find sense and coherence in the universe at large.

P. J. E. Peebles and David T. Wilkinson, who write on "The Primeval Fireball," began their researches with the "big bang" evolutionary universe in mind. Their findings not only confirmed their own predictions, but have convinced most former adherents of the steady-state theory that only an ancient super-dense state of the universe can account for this radio radiation which pours in from every side.

Armed with this new knowledge of the primeval fireball, Martin Rees and Joseph Silk have addressed themselves to the earliest stages of galaxy formation; their theory, reported in "The Origins of Galaxies," attempts to describe the scale of the star clusters, galaxies and clusters of galaxies formed as the initial ball of radiation begins its expansion.

And now a note about Allan Sandage's "The Red Shift." Though written in 1956, before the quasars and primeval fireball radiation, it describes the foundation for the single most important observational aspect of cosmology. The most immediate and elementary explanation of the red-shift data has been since the late 1920's an expanding, evolutionary universe. The simplest interpretation of the red shift leads to an age of around 15×10^9 years for the universe, and this accords remarkably well with ages obtained independently both from the distribution of radioactive isotopes and from the evolution of the oldest star clusters. Consequently, a convincing argument and analysis exists for an evolutionary cosmology (with its singular, super-dense universe in times past), even in the absence of the recent discoveries of the quasars and the 3° background radiation.

28

The Red-Shift

ALLAN R. SANDAGE

September 1956

In the nature of things it is a delicate undertaking to try to discern the general structure and features of a universe which stretches out farther than we can see. For more than a quarter of a century both the theoreticians and the observers of the cosmos have been making exciting discoveries, but the points of contact between the discoveries have been few. The predictions of the theorists, deduced from the most general laws of physics, are not easy to test against the real world—or rather, the small portion of the real world that we can observe. There is, however, one solid meeting ground between the theories and the observations, and that is the apparent expansion of the universe. Other aspects of the universe may be interpreted in different ways to fit different theories, but concerning the expansion the rival theories make unambiguous predictions on which they will stand or fall. There is now hope that red-shift measurements of the universe's expansion with the 200-inch telescope on Palomar Mountain will soon make it possible to decide, among other things, whether we live in an evolving or a steady-state universe.

Let us begin by considering just what issue the measurements seek to decide, as between the theories expounded in the two preceding articles by George Gamow and Fred Hoyle. The steady-state theory says that the universe has been expanding at a constant rate throughout an infinity of time. The evolutionary theory, in contrast, implies that the expansion of the universe is steadily slowing down. If the universe began with an explosion from a superdense state, its rate of expansion was greatest at the beginning and has been slowing ever since because of the opposing gravitational attraction of its matter, which acts as a brake on the expansion—much as an anchored elastic string attached to a golf ball would act as a brake on the flight of the ball.

Now in principle we can decide whether the rate of expansion has changed or not simply by measuring the speed of expansion at different times in the universe's history. And the 200-inch telescope permits us to do this. It covers a range of about two billion years in time. We see the nearest galaxies as they were only a few million years ago, while the light from the most distant galaxies takes so long to reach us that we see them at a stage in the universe's history going back to one or two billion years ago. If the explosion theory is correct, the universe should have been expanding at an appreciably faster rate then than it is now. Since the light we are receiving from the distant galaxies is a flashback to that earlier time, its red-shift should show them receding from us faster than if the rate of expansion had remained constant.

The red-shift is so basic a tool for testing our notions about the universe that it is worthwhile to review how it was discovered and how it is used.

An astronomer cannot perform experi-

JIGGLE CAMERA smears the images by moving the photographic plate in a rectangle during exposure. It is mounted in the prime-focus cage at the upper end of the 200-inch.

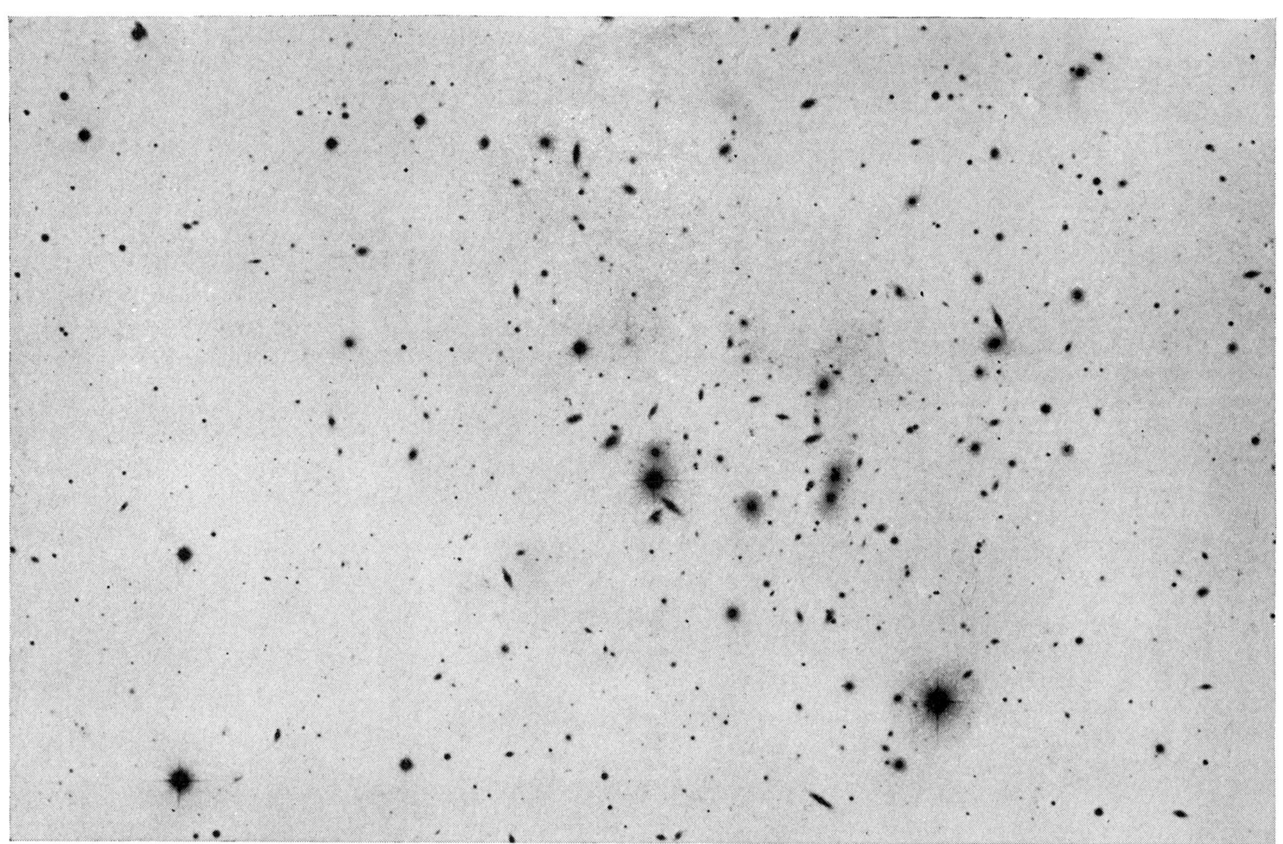

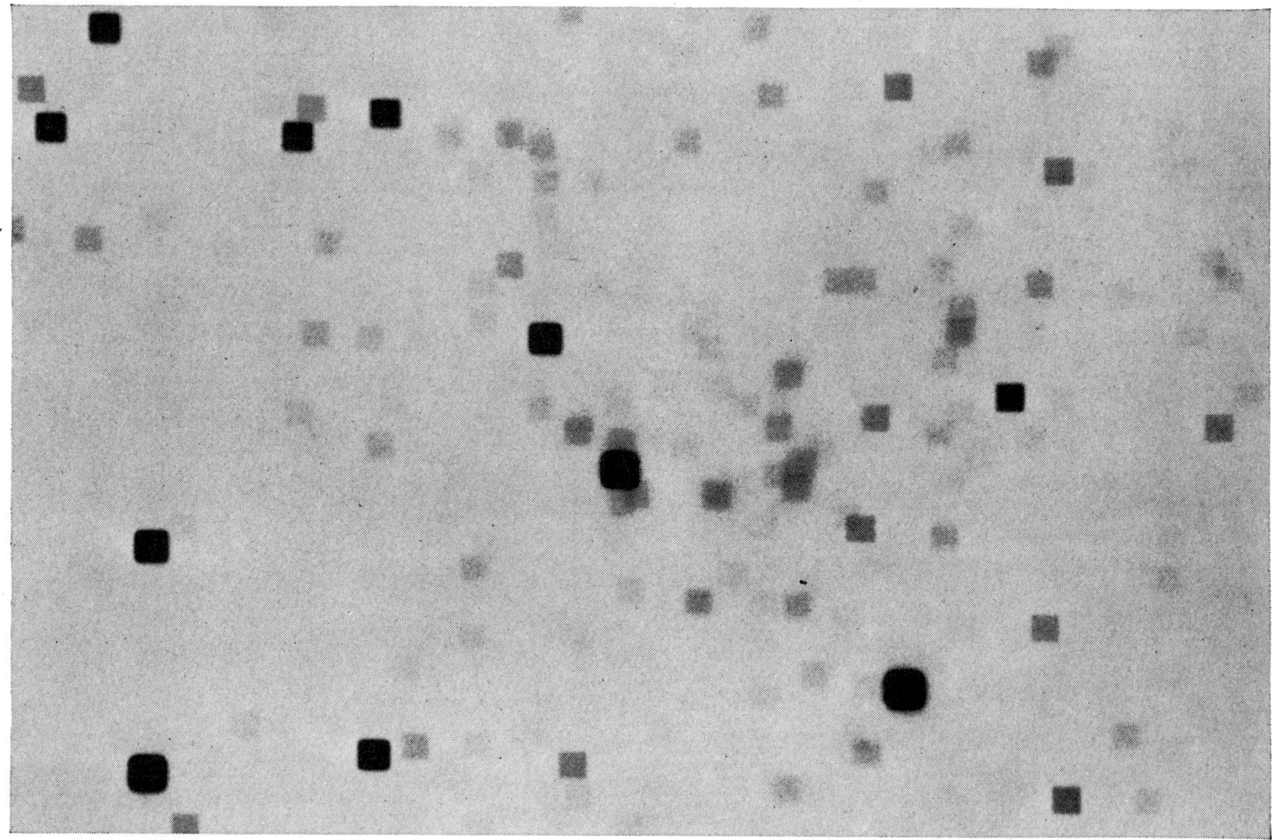

BRIGHTNESS OF GALAXIES may be measured with the help of the jiggle camera (*see photograph on the opposite page*). At the top is a negative print of a 200-inch telescope photograph showing nearby stars and a cluster of galaxies in Corona Borealis. Although the stars have made the brightest images in the photograph, they are essentially point sources of light. The galaxies, on the other hand, are extended sources of light. To measure the brightness of a galaxy by comparing it with the known brightness of a star, the two images must be made the same. This is done by smearing the images as shown at bottom in a jiggle-camera photograph of the same area.

ments on the objects of his study, or even examine them at first hand. All his information rides on beams of light from outer space. By sufficiently ingenious instruments and equally ingenious interpretation (we hope), he may translate this light into information about the temperatures, sizes, structures and motions of the celestial bodies. It was in 1888 that a German astronomer, H. C. Vogel, first demonstrated that the spectra of stars could give information about motions which could not otherwise be detected. He discovered the Doppler effect in starlight.

The Doppler effect, as every physics student knows, is a change in wavelength observable when the source of radiation (sound, light, etc.) is in motion. If it is moving toward the observer, the wavelength is shortened; if away, the waves are lengthened. In the case of a star moving away from us, the whole spectrum of its light is shifted toward the red, or long-wave, end.

This spectrum, made by means of a prism or diffraction grating which spreads the light out into a band of its component colors, is usually not continuous. Certain wavelengths of the light are absorbed by atoms in the star's atmosphere. For example, most stars show strong absorption, by calcium atoms, at the wavelengths of 3933.664 and 3968.-470 Angstrom units. (An Angstrom unit is a hundred-millionth of a centimeter.) The absorption is signaled by dark lines in the spectrum, known in this case as the K and H lines of calcium. Now if a star is moving away from us, these lines will be displaced toward the red end of the spectrum. In the spectrum of the star known as Delta Leporis, for in-

VIRGO

22 MILLION LIGHT-YEARS

1,200 KILOMETERS PER SECOND

CORONA BOREALIS

400 MILLION LIGHT-YEARS

21,500 KILOMETERS PER SECOND

BOOTES

700 MILLION LIGHT-YEARS

39,300 KILOMETERS PER SECOND

HYDRA

1.1 BILLION LIGHT-YEARS

60,900 KILOMETERS PER SECOND

RED-SHIFT of four galaxies on this page is depicted in the spectra on the opposite page. The galaxies are centered in the photographs. The spectra are the bright horizontal streaks tapered to the left and right. Above and below each spectrum are comparison lines from the spectrum of iron. Near the left end of the spectrum at the top of the page are two dark vertical lines: the K and H lines of calcium. If the galaxy did not exhibit the red-shift, these lines would be in the position of the broken line running vertically down the page. The amount of their shift toward the red, or right, end of the spectrum is indicated by the short arrow to the right of the broken line. The larger shift of the K and H lines of the three fainter galaxies is indicated by the longer arrows below their spectra. The constellation, approximate distance and velocity of recession of each galaxy is at left of its photograph.

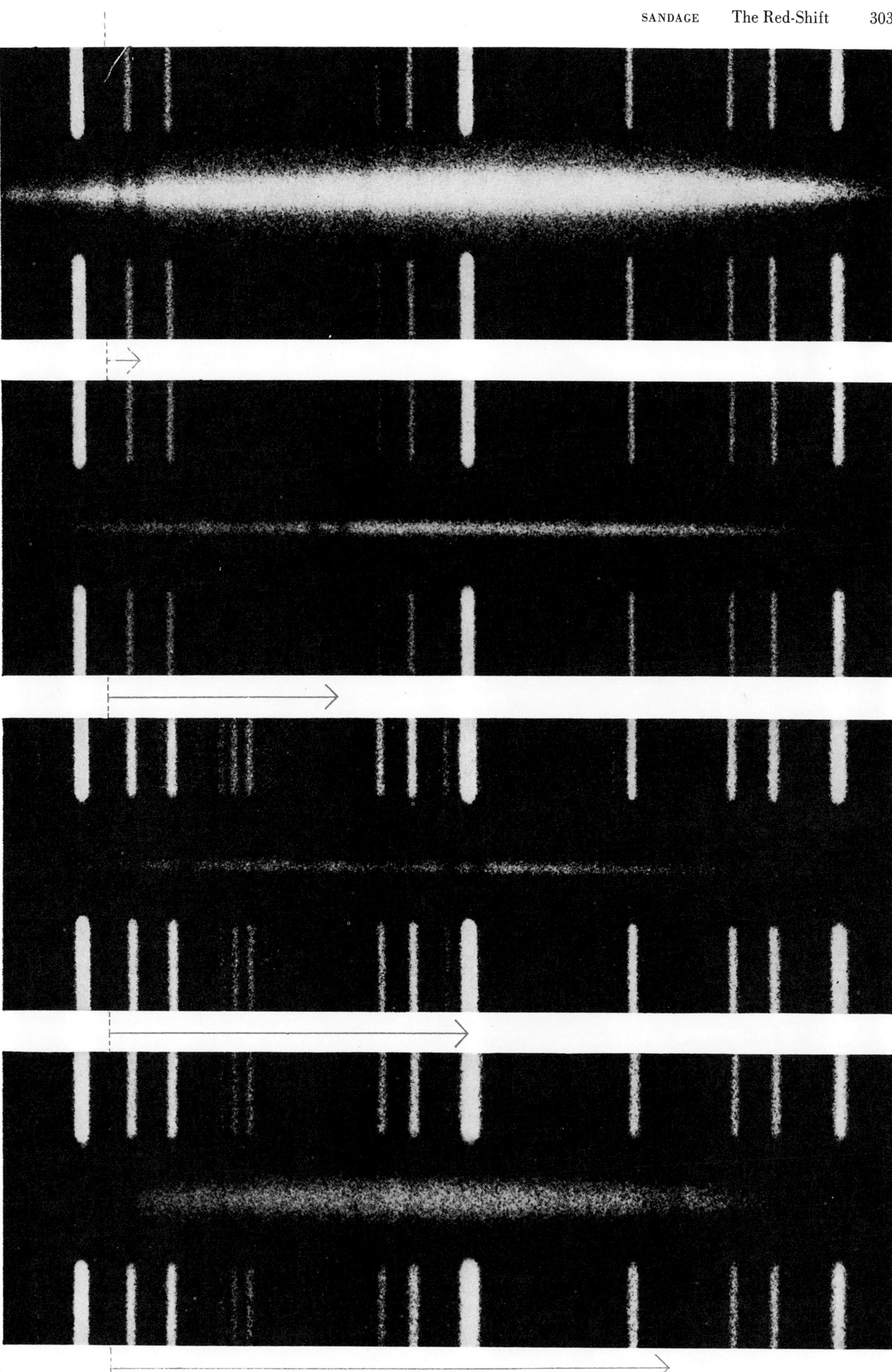

stance, the K line of calcium is displaced 1.298 Angstroms toward the red. Assuming the displacement is due to the Doppler effect, it is a simple matter to calculate the velocity of the star's receding motion. Dividing the amount of the displacement by the normal wavelength at rest, and multiplying by the speed of light (300,000 kilometers per second) we get the speed of the star—in this case 99 kilometers per second. The calculation on the basis of displacement of the H line gives the same figure.

Equipped with this powerful tool, many of the large observatories in the world spent a major part of their time during the early part of this century measuring the velocities of receding and approaching stars in our galaxy. At first it was a work of pure curiosity, no one suspecting that it might have any bearing on cosmological theories. But in the 1920s V. M. Slipher of the Lowell Observatory made a discovery which was to lead to a completely new picture of the universe. His measurements of redshifts of a number of "nebulae" then thought to lie in our galaxy showed that they were all receding from us at phenomenal speeds—up to 1,800 kilometers per second. Edwin P. Hubble at Mount Wilson soon established that the "nebulae" were systems of stars, and he went on to measure their distances. The method he used was the one developed by Harlow Shapley, employing Cepheid variable stars as the yardstick. Shapley had found a way to measure the intrinsic brightness of these stars, and therefore their distance could be estimated from their apparent brightness by means of the rule that the intensity of light falls off as the square of the distance. Hubble observed that the galaxies nearest our own system, including the Great Nebula in Andromeda, contained Cepheid variables, and when he computed their distances he came out with the then astounding figure of about one million light-years! He next tackled the problem of finding the distances of Slipher's nebulae. Since variable stars could not be detected in them, he used their brightest stars as distance indicators instead. He found that these nebulae were at distances ranging up to 20 million light-years from us, and what was more remarkable, their velocities increased in strict proportion to their distances!

Hubble made the daring conjecture that the universe as a whole was expanding. He predicted that more remote galaxies would show larger redshifts, still in proportion to their distance. To test Hubble's speculation, Milton L. Humason began a long-range program of spectral analysis of more distant galaxies with the 100-inch telescope on Mount Wilson. In these faint galaxies it was no longer possible to distinguish even the bright stars, and so the relative brightness of the galaxy as a whole had to be taken as the measure of distance. That is, a galaxy one fourth as bright as another was assumed to be twice as far away. Hubble reasoned that while individual galaxies might deviate from this rule, statistically the population of galaxies as a whole would follow it. The prin-

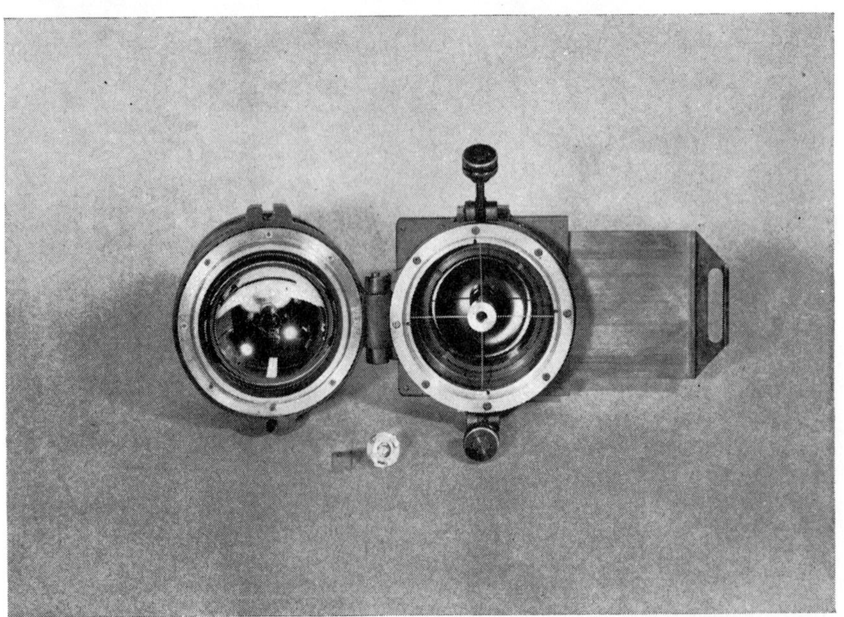

SPECTRA ARE MADE with the spectrograph at the top, which is mounted at the prime focus of the 200-inch telescope. Inside the spectrograph the converging rays of the 200-inch mirror are made parallel by a concave mirror. The light is then dispersed by a diffraction grating. At the bottom is a Schmidt camera used to photograph the spectrum. It has an optical path of solid glass and a speed of $f/.48$. The plateholder and plate are below the camera.

ciple is still the basis of distance determinations today.

Humason laboriously photographed spectra of galaxies, and Hubble measured their apparent brightness, from 1928 to 1936, when they reached the limit of the 100-inch telescope. The history of the red-shift program in those years is a story of extreme skill and patience at the telescope and of steady improvement in instrumentation. It was a long and difficult task to photograph spectra then; the prisms used required long exposures, and it took 10 nights or more to obtain a spectrum which with modern equipment can be recorded in less than an hour today. The improvement in equipment includes not only the 200-inch telescope but also diffraction gratings, faster cameras and a vast improvement in the sensitivity of photographic plates, thanks to the Eastman Kodak Company. Astronomers the world over, and cosmology, owe a large debt to the Eastman research laboratories.

Humason's first really big red-shift came early in 1928, when he got a spectrum of a galaxy called NGC 7619. Hubble had predicted that its velocity should be slightly less than 4,000 kilometers per second: Humason found it to be 3,800. By 1936, at the limit of the 100-inch telescope's reach, they had arrived at a cluster of galaxies, called Ursa Major No. 2, which showed a velocity of 40,000 kilometers per second. All the way out to that range of more than half a billion light-years the velocity of galaxies increased in direct proportion to the distance. In a sense this was disappointing, because the various cosmological theories predicted that some change in this relation should begin to appear when the observations had been pushed far enough. Further exploration into the distances of space had to await the completion of the 200-inch Hale telescope on Palomar Mountain.

In 1951 the red-shift program was resumed, with a new spectrograph of great speed and versatility placed in the big telescope's prime focus cage, where the observer rides with his instruments. The spectrograph has to be of very compact design to fit into the cramped space of the cage. The photographic plate itself, mounted in the middle of a complex

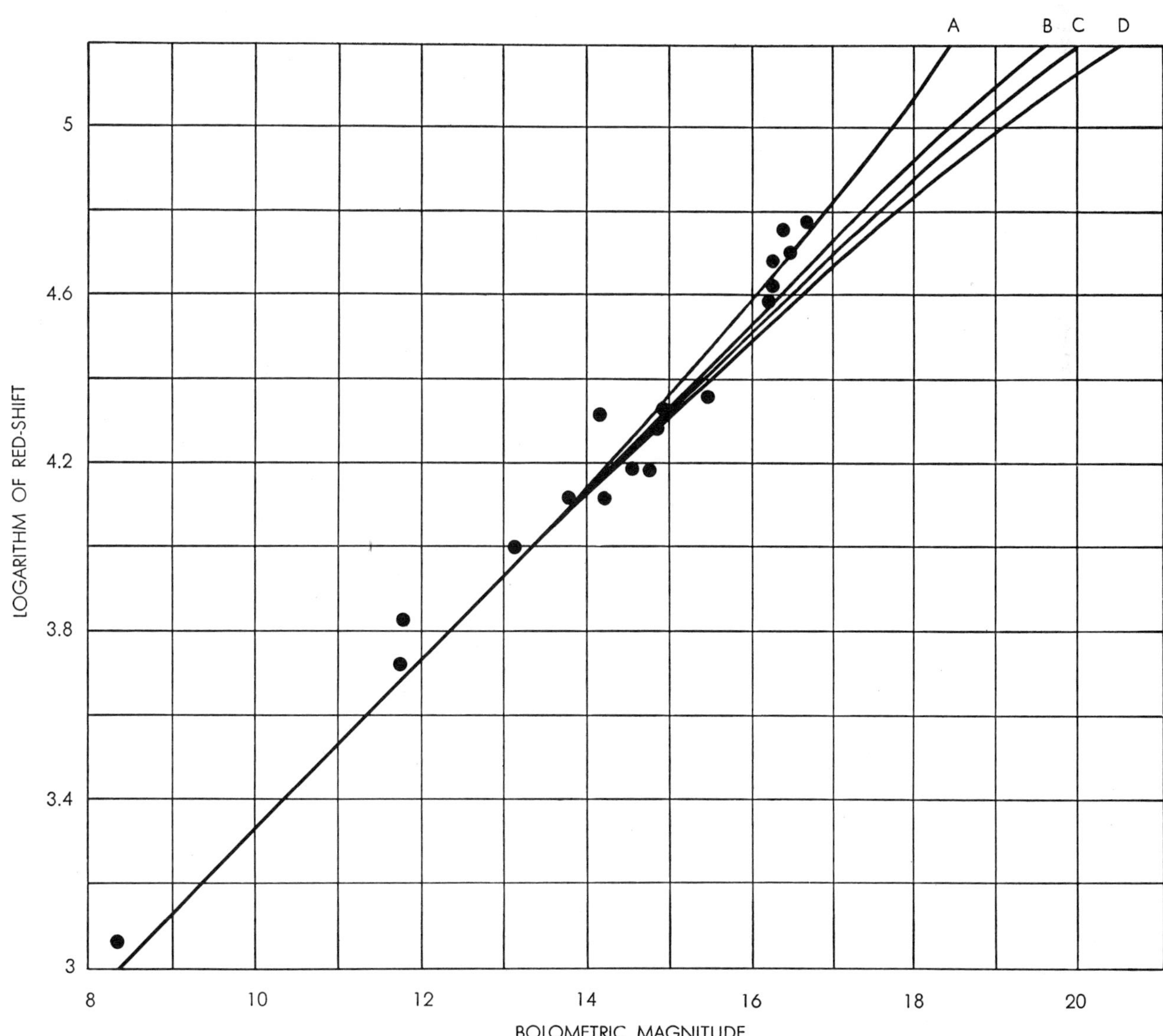

EIGHTEEN FAINTEST CLUSTERS of galaxies yet measured are plotted for their red-shift (or speed of recession) and apparent magnitude (or distance). Line C is a universe expanding forever at the same rate. Line D is a steady-state universe. If the line falls to the left of C, the expansion must slow down. If it falls between C and B, the universe is open and infinite. If it falls to the left of B, the universe is closed and finite. If it falls on B, it is Euclidean and infinite. A is the trend suggested by the six faintest clusters.

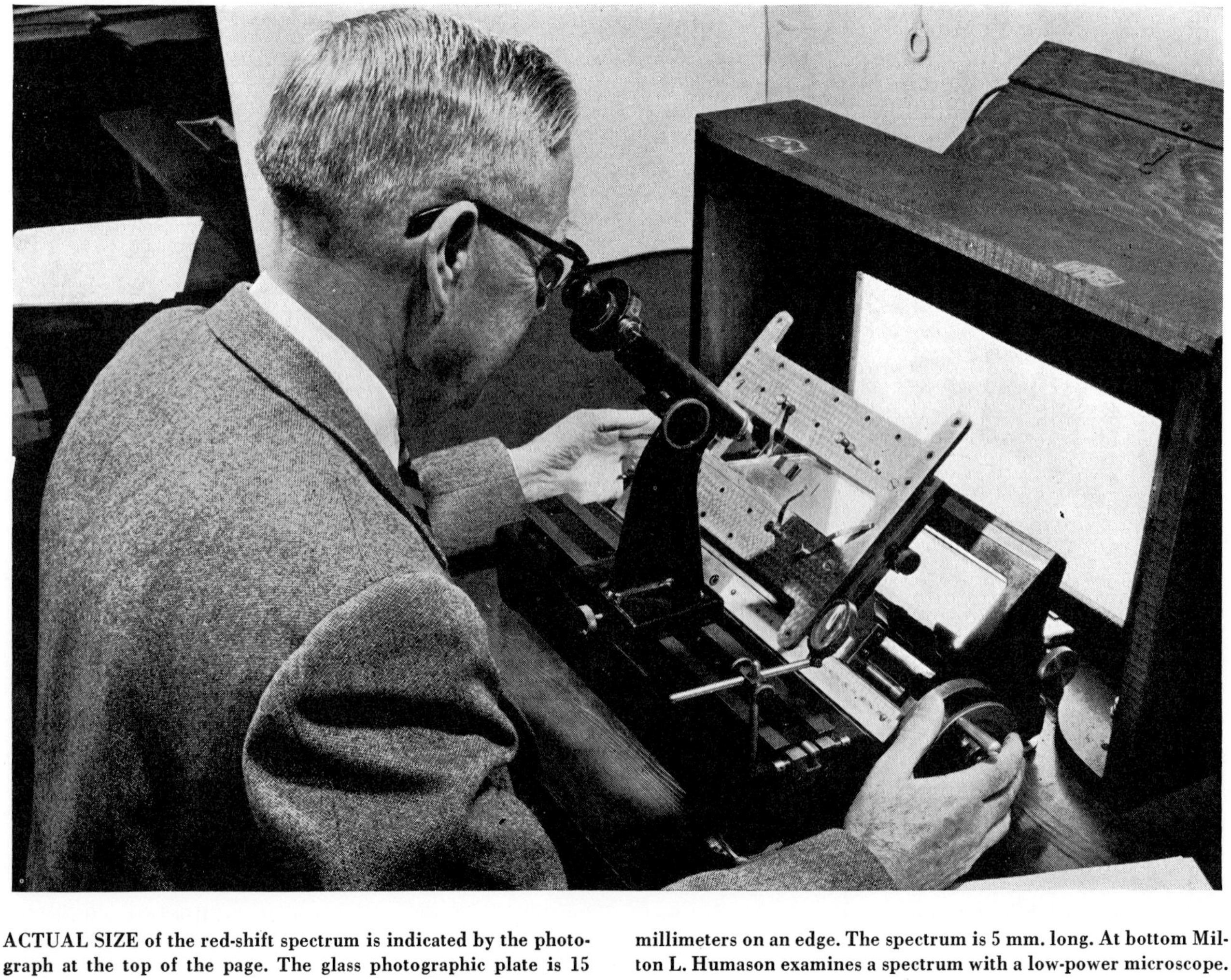

ACTUAL SIZE of the red-shift spectrum is indicated by the photograph at the top of the page. The glass photographic plate is 15 millimeters on an edge. The spectrum is 5 mm. long. At bottom Milton L. Humason examines a spectrum with a low-power microscope.

optical arrangement, is only 15 millimeters (about half an inch) on a side. The cutting and handling of such small pieces of glass in complete darkness (to avoid exposure of the plate) is a tricky business. The spectr m recorded on the plate is a tiny strip only a fifth of an inch long, but it is long enough to measure red-shifts to an accuracy of better than one half of 1 per cent.

The most distant photographable galaxies are so faint that they are not visible to the eye through the telescope: they can be recorded only by extended exposure of the plate. The observer guiding the telescope must position the slit of the spectrograph by reference to guide stars within the same field as the distant object. Another great difficulty in recording the red-shift of extremely distant galaxies arises from the magnitude of the shift. The displacement of the calcium dark lines toward the red is so large that the lines move clean off the sensitive range of blue photographic plates, which astronomers like to use because of their speed. So slow panchromatic plates must be used, and Humason has been forced to return to exposure times as long as 30 hours or more.

The other part of the program—measuring the distances of the galaxies—also has been helped by improvements in technique. For measurement of their brightness the Mount Wilson telescopes employ photomultiplier tubes, which amplify the light energy by electronic means. Such equipment was not available for the 200-inch telescope when the present program began. Instead the intensity of the light from very faint galaxies was measured by a tricky method which compares it with that of stars of known magnitude. No direct comparison can be made, of course, between the picture of a star and that of a galaxy or cluster of galaxies, because the star is a point source of light while a galactic system is a spread-out image. To make the images comparable, a region of the sky is photographed with a "jiggle" camera which moves the plate around so that the images of stars and of galaxies are smeared out in squares [*see photograph at bottom of page 301*]. They can then be compared as to brightness—just as one may use color cards to find a match to the color of a room.

Humason has now measured red-shifts of remote clusters of galaxies with recession velocities up to 60,000 kilometers per second. What do they show? Is the velocity still increasing in strict proportion to the distance?

The information about 18 of the faintest measured clusters is given in the accompanying chart [*see page 305*]. Their velocities are plotted against their apparent brightness, or estimated distances. If velocity increases in direct proportion to the distance, the observed velocity-distance relation should be "linear" (*i.e.*, follow a straight line). But as the chart shows, the very faintest clusters have begun to depart from that line. These clusters, about a billion light-years away, are moving *faster* (by about 10,000 kilometers per second) than in direct proportion to their apparent distance. In other words, the data would be interpreted to mean that a billion years ago the universe was expanding faster than it is now. If the measurements and the interpretation are correct, this suggests that we live in an evolving rather than in a steady-state universe.

The observed change in the curve buys us much more information. To begin with, it tells us something about the mean density of matter in the universe. The rate at which the expansion of the universe is slowing down (if it is) depends on the mean density of its matter: the higher the density, the greater the braking effect. The amount of departure from linearity indicated by the measurements thus far calls for a mean density of about 3×10^{-28} grams of matter per cubic centimeter (about one hydrogen atom per five quarts of space). Now this amounts to about 300 times the total mass of the matter estimated to be contained in galaxies: that figure comes out to a mean density of only 10^{-30} grams per cubic centimeter. If our present tentative value for the slowdown of the expansion should be confirmed, we would have to conclude that either the current estimates of the masses of the galaxies are wrong or that there is a great deal of matter, so far undetected, in intergalactic space. Matter in the form of neutral hydrogen (*i.e.*, normal hydrogen atoms consisting of a proton and an electron) might be present in space and still have escaped detection until now because it is not luminous. The giant radio telescopes now under construction or on the drawing boards perhaps will detect the hydrogen, if it exists in the postulated quantities.

Once we know the rate at which expansion of the universe is slowing down, it becomes possible to determine not only the mean density of matter but also the geometry of space—that is, its curvature. Models of the evolving universe take three forms: the Euclidean case, in which space is flat, open and infinite; a curved universe which is closed and finite, like the surface of a sphere; and a curved universe which is open and infinite, like the surface of a saddle. In the accompanying velocity-distance chart [*page 305*] curves to the left of C represent evolving models, and curve D represents the steady-state model. If the curve of the velocity-distance relation lies between C and B, the universe is open and infinite. Line B is the Euclidean case of flat space. If the curve is left of B, the universe is closed and finite, the radius of its curvature decreasing as we move farther to the left.

According to our present observations, the actual relation follows a curve left of B (curve A on the chart). Although our data are still crude and inconclusive, they do suggest that the steady-state model does not fit the real world, and that we live in a closed, evolving universe.

Humason has gone beyond 60,000 kilometers per second and attempted to measure the red-shifts of two faint clusters whose predicted velocity is more than 100,000 kilometers per second. So far these efforts have not yielded reliable results, but he is continuing them. These two remote clusters may well hold the key to the structure of the universe. We stand a chance of finding the answer to the cosmological problem. The red-shift program will continue toward this goal.

If the expansion of the universe is decelerating at the rate our present data suggest, the expansion will eventually stop and contraction will begin. If it returns to a superdense state and explodes again, then in the next cycle of oscillation, some 15 billion years hence, we may all find ourselves again pursuing our present tasks.

Although no final answers have yet emerged, big steps have been taken since 1928 toward the solution to the cosmological problem, and there is hope that it may now be within our grasp. The situation has nowhere been better expressed than in Hubble's last paper:

"For I can end as I began. From our home on the earth we look out into the distances and strive to imagine the sort of world into which we are born. Today we have reached far out into space. Our immediate neighborhood we know rather intimately. But with increasing distance our knowledge fades . . . until at the last dim horizon we search among ghostly errors of observations for landmarks that are scarcely more substantial. The search will continue. The urge is older than history. It is not satisfied and it will not be suppressed."

The Evolutionary Universe

GEORGE GAMOW

September 1956

Cosmology is the study of the general nature of the universe in space and in time—what it is now, what it was in the past and what it is likely to be in the future. Since the only forces at work between the galaxies that make up the material universe are the forces of gravity, the cosmological problem is closely connected with the theory of gravitation, in particular with its modern version as comprised in Albert Einstein's general theory of relativity. In the frame of this theory the properties of space, time and gravitation are merged into one harmonious and elegant picture.

The basic cosmological notion of general relativity grew out of the work of great mathematicians of the 19th century. In the middle of the last century two inquisitive mathematical minds—a Russian named Nikolai Lobachevski and a Hungarian named János Bolyai—discovered that the classical geometry of Euclid was not the only possible geometry: in fact, they succeeded in constructing a geometry which was fully as logical and self-consistent as the Euclidean. They began by overthrowing Euclid's axiom about parallel lines: namely, that only one parallel to a given straight line can be drawn through a point not on that line. Lobachevski and Bolyai both conceived a system of geometry in which a great number of lines parallel to a given line could be drawn through a point outside the line.

To illustrate the differences between Euclidean geometry and their non-Euclidean system it is simplest to consider just two dimensions—that is, the geometry of surfaces. In our schoolbooks this is known as "plane geometry," because the Euclidean surface is a flat surface. Suppose, now, we examine the properties of a two-dimensional geometry constructed not on a plane surface but on a curved surface. For the system of Lobachevski and Bolyai we must take the curvature of the surface to be "negative," which means that the curvature is not like that of the surface of a sphere but like that of a saddle [*see illustrations on page 310*]. Now if we are to draw parallel lines or any figure (*e.g.*, a triangle) on this surface, we must decide first of all how we shall define a "straight line," equivalent to the straight line of plane geometry. The most reasonable definition of a straight line in Euclidean geometry is that it is the path of the shortest distance between two points. On a curved surface the line, so defined, becomes a curved line known as a "geodesic" [see "The Straight Line," by Morris Kline; SCIENTIFIC AMERICAN, March].

Considering a surface curved like a saddle, we find that, given a "straight" line or geodesic, we can draw through a point outside that line a great many geodesics which will never intersect the given line, no matter how far they are extended. They are therefore parallel to it, by the definition of parallel. The possible parallels to the line fall within certain limits, indicated by the intersecting

Five contributors to modern cosmology are depicted in these drawings by Bernarda Bryson.

lines in the drawing at the left in the middle of the next page.

As a consequence of the overthrow of Euclid's axiom on parallel lines, many of his theorems are demolished in the new geometry. For example, the Euclidean theorem that the sum of the three angles of a triangle is 180 degrees no longer holds on a curved surface. On the saddle-shaped surface the angles of a triangle formed by three geodesics always add up to less than 180 degrees, the actual sum depending on the size of the triangle. Further, a circle on the saddle surface does not have the same properties as a circle in plane geometry. On a flat surface the circumference of a circle increases in proportion to the increase in diameter, and the area of a circle increases in proportion to the square of the increase in diameter. But on a saddle surface both the circumference and the area of a circle increase at *faster* rates than on a flat surface with increasing diameter.

After Lobachevski and Bolyai, the German mathematician Bernhard Riemann constructed another non-Euclidean geometry whose two-dimensional model is a surface of positive, rather than negative, curvature—that is, the surface of a sphere. In this case a geodesic line is simply a great circle around the sphere or a segment of such a circle, and since any two great circles must intersect at two points (the poles), there are no parallel lines at all in this geometry. Again the sum of the three angles of a triangle is not 180 degrees: in this case it is always *more* than 180. The circumference of a circle now increases at a rate *slower* than in proportion to its increase in diameter, and its area increases more slowly than the square of the diameter.

Now all this is not merely an exercise in abstract reasoning but bears directly on the geometry of the universe in which we live. Is the space of our universe "flat," as Euclid assumed, or is it curved negatively (per Lobachevski and Bolyai) or curved positively (Riemann)? If we were two-dimensional creatures living in a two-dimensional universe, we could tell whether we were living on a flat or a curved surface by studying the properties of triangles and circles drawn on that surface. Similarly as three-dimensional beings living in three-dimensional space we should be able, by studying geometrical properties of that space, to decide what the curvature of our space is. Riemann in fact developed mathematical formulas describing the properties of various kinds of curved space in three and more dimensions. In the early years of this century Einstein conceived the idea of the universe as a curved system in four dimensions, embodying time as the fourth dimension, and he proceeded to apply Riemann's formulas to test his idea.

Einstein showed that time can be considered a fourth coordinate supplementing the three coordinates of space. He connected space and time, thus establishing a "space-time continuum," by means of the speed of light as a link between time and space dimensions. However, recognizing that space and time are physically different entities, he employed the imaginary number $\sqrt{-1}$, or i, to express the unit of time mathematically and make the time coordinate formally equivalent to the three coordinates of space.

In his special theory of relativity Einstein made the geometry of the time-space continuum strictly Euclidean, that is, flat. The great idea that he introduced later in his general theory was that gravitation, whose effects had been neglected in the special theory, must make it curved. He saw that the gravitational effect of the masses distributed in space and moving in time was equivalent to curvature of the four-dimensional space-time continuum. In place of the classical Newtonian statement that "the sun produces a field of forces which impels the earth to deviate from straight-line mo-

From left to right they are: Nikolai Lobachevski, Bernhard Riemann, Albert Einstein, Willem de Sitter and Georges Lemaître

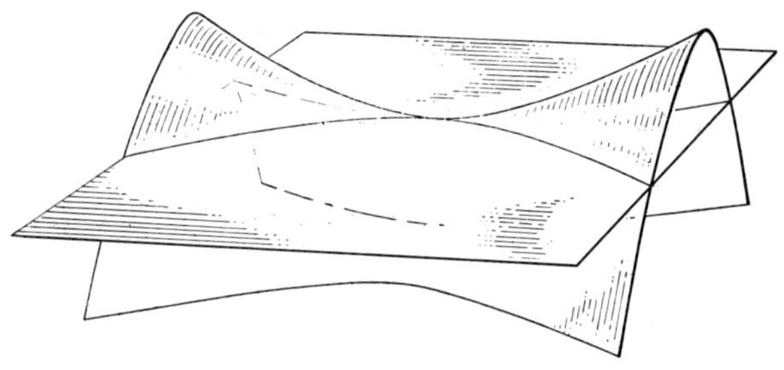

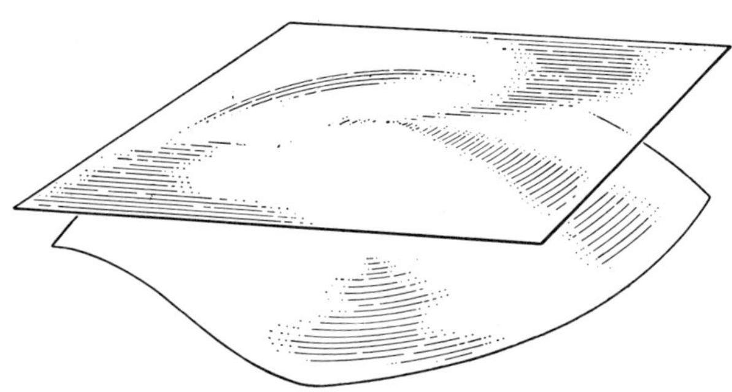

NEGATIVE AND POSITIVE CURVATURE of space is suggested by this two-dimensional analogy. The saddle-shaped surface at left, which lies on both sides of a tangential plane, is negatively curved. The spherical surface at right, which lies on one side of a tangential plane, is positively curved. If space is negatively curved, the universe is infinite; if it is positively curved, the universe is finite.

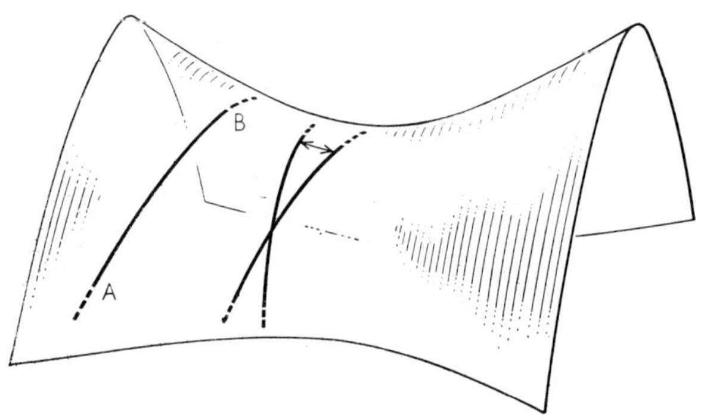

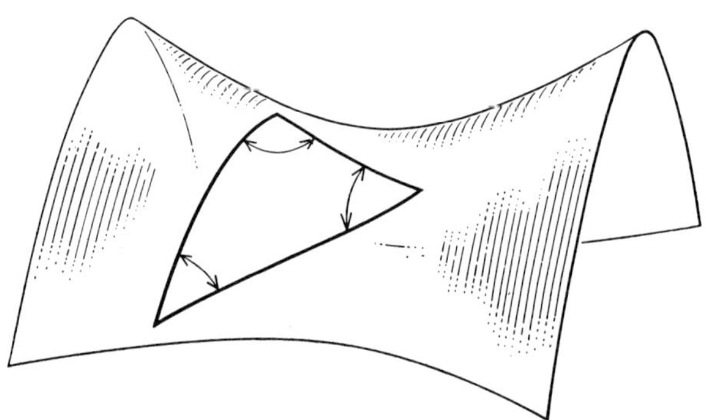

ON A NEGATIVELY CURVED SURFACE the shortest distance between two points is not a straight line but a curved "geodesic," such as the line AB at the left. On a plane surface only one parallel to a given straight line can be drawn through a point not on that line; on a negatively curved surface many geodesics can be drawn through a point not on a given geodesic without ever intersecting it. These "parallel" lines will fall within the limits indicated by the arrow between the intersecting lines at left. On a plane surface the angles of a triangle add up to 180 degrees; on the negatively curved surface at the right, they add up to less than 180 degrees.

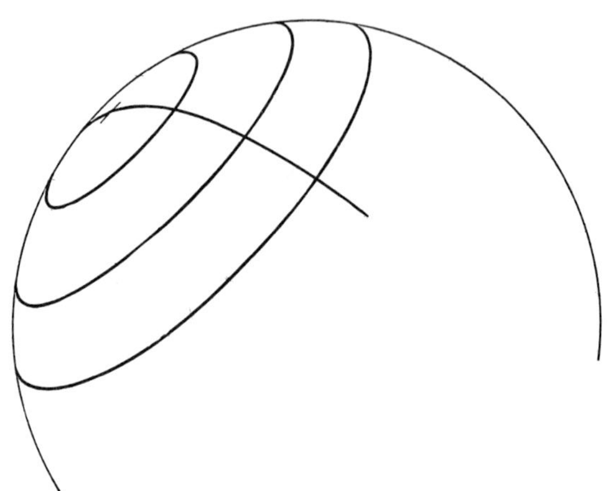

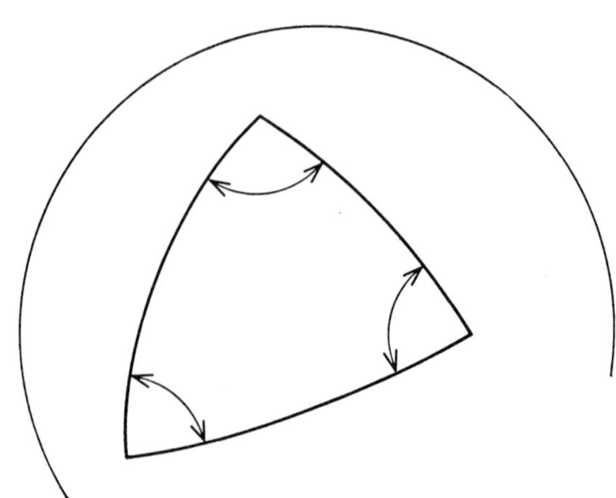

ON A POSITIVELY CURVED SURFACE the shortest distance between two points follows a great circle, a closed line passing through opposite points on the surface (*single curved line at left*). In this geometry there are no "parallel" lines because any two great circles must intersect. The circumference of a circle increases more slowly with diameter than on a flat surface, and the area similarly increases more slowly (*concentric circles at left*). The angles of a triangle on the surface (*right*) add up to more than 180 degrees.

tion and to move in a circle around the sun," Einstein substituted a statement to the effect that "the presence of the sun causes a curvature of the space-time continuum in its neighborhood."

The motion of an object in the space-time continuum can be represented by a curve called the object's "world line." For example, the world line of the earth's travel around the sun in time is pictured in the drawing on this page. (Space must be represented here in only two dimensions; it would be impossible for a three-dimensional artist to draw the fourth dimension in this scheme, but since the orbit of the earth around the sun lies in a single plane, the omission is unimportant.) Einstein declared, in effect: "The world line of the earth is a geodesic in the curved four-dimensional space around the sun." In other words, the line ABCD in the drawing corresponds to the shortest *four-dimensional* distance between the position of the earth in January (at A) and its position in October (at D).

Einstein's idea of the gravitational curvature of space-time was, of course, triumphantly affirmed by the discovery of perturbations in the motion of Mercury at its closest approach to the sun and of the deflection of light rays by the sun's gravitational field. Einstein next attempted to apply the idea to the universe as a whole. Does it have a general curvature, similar to the local curvature in the sun's gravitational field? He now had to consider not a single center of gravitational force but countless centers of attraction in a universe full of matter concentrated in galaxies whose distribution fluctuates considerably from region to region in space. However, in the large-scale view the galaxies are spread fairly uniformly throughout space as far out as our biggest telescopes can see, and we can justifiably "smooth out" its matter to a general average (which comes to about one hydrogen atom per cubic meter). On this assumption the universe as a whole has a smooth general curvature.

But if the space of the universe is curved, what is the sign of this curvature? Is it positive, as in our two-dimensional analogy of the surface of a sphere, or is it negative, as in the case of a saddle surface? And, since we cannot consider space alone, how is this space curvature related to time?

Analyzing the pertinent mathematical equations, Einstein came to the conclusion that the curvature of space must be independent of time, *i.e.*, that the universe as a whole must be unchanging (though it changes internally). However, he found to his surprise that there was no solution of the equations that would permit a static cosmos. To repair the situation, Einstein was forced to introduce an additional hypothesis which amounted to the assumption that a new kind of force was acting among the galaxies. This hypothetical force had to be independent of mass (being the same for an apple, the moon and the sun!) and to gain in strength with increasing distance between the interacting objects (as no other forces ever do in physics!).

Einstein's new force, called "cosmic repulsion," allowed two mathematical models of a static universe. One solution, which was worked out by Einstein him-

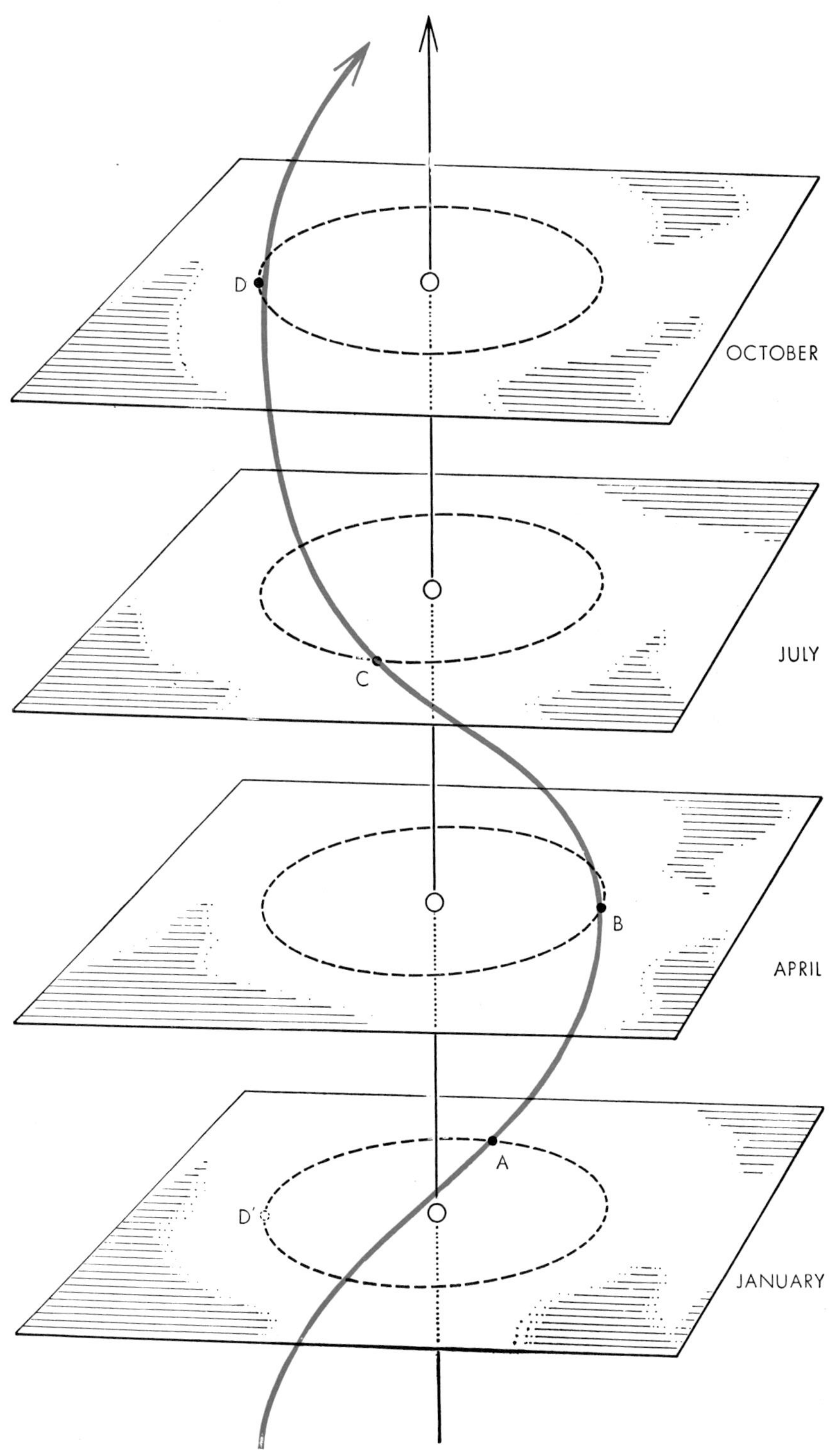

MOTION OF BODY in the curved "space-time continuum" of Albert Einstein is represented by the "world line" of the earth's motion around the sun. Here the sun is the small open circle in each of the four planes. The earth is the black dot on the elliptical orbit. Each plane shows the position of the earth at a month of the year. The world line is in color.

self and became known as "Einstein's spherical universe," gave the space of the cosmos a positive curvature. Like a sphere, this universe was closed and thus had a finite volume. The space coordinates in Einstein's spherical universe were curved in the same way as the latitude or longitude coordinates on the surface of the earth. However, the time axis of the space-time continuum ran quite straight, as in the good old classical physics. This means that no cosmic event would ever recur. The two-dimensional analogy of Einstein's space-time continuum is the surface of a cylinder, with the time axis running parallel to the axis of the cylinder and the space axis perpendicular to it [*see drawing at left on page 145*].

The other static solution based on the mysterious repulsion forces was discovered by the Dutch mathematician Willem de Sitter. In his model of the universe both space and time were curved. Its geometry was similar to that of a globe, with longitude serving as the space coordinate and latitude as time [*drawing at right on page 145*].

Unhappily astronomical observations contradicted both Einstein's and de Sitter's static models of the universe, and they were soon abandoned.

In the year 1922 a major turning point came in the cosmological problem. A Russian mathematician, Alexander A. Friedman (from whom the author of this article learned his relativity), discovered an error in Einstein's proof for a static universe. In carrying out his proof Einstein had divided both sides of an equation by a quantity which, Friedman found, could become zero under certain circumstances. Since division by zero is not permitted in algebraic computations, the possibility of a nonstatic universe could not be excluded under the circumstances in question. Friedman showed that two nonstatic models were possible. One pictured the universe as expanding with time; the other, contracting.

Einstein quickly recognized the importance of this discovery. In the last edition of his book *The Meaning of Relativity* he wrote: "The mathematician Friedman found a way out of this dilemma. He showed that it is possible, according to the field equations, to have a finite density in the whole (three-dimensional) space, without enlarging these field equations ad hoc." Einstein remarked to me many years ago that the cosmic repulsion idea was the biggest blunder he had made in his entire life.

Almost at the very moment that Friedman was discovering the possibility of an expanding universe by mathematical reasoning, Edwin P. Hubble at the Mount Wilson Observatory on the other side of the world found the first evidence of actual physical expansion through his telescope. He made a compilation of the distances of a number of far galaxies, whose light was shifted toward the red end of the spectrum, and it was soon found that the extent of the shift was in direct proportion to a galaxy's distance from us, as estimated by its faintness. Hubble and others interpreted the red-shift as the Doppler effect—the well-known phenomenon of lengthening of wavelengths from any radiating source that is moving rapidly away (a train whistle, a source of light or whatever). To date there has been no other reasonable explanation of the galaxies' red-shift. If the explanation is correct, it means that the galaxies are all moving away from one another with increasing velocity as they move farther apart.

Thus Friedman and Hubble laid the foundation for the theory of the expanding universe. The theory was soon developed further by a Belgian theoretical astronomer, Georges Lemaître. He proposed that our universe started from a highly compressed and extremely hot state which he called the "primeval atom." (Modern physicists would prefer the term "primeval nucleus.") As this matter expanded, it gradually thinned out, cooled down and reaggregated in stars and galaxies, giving rise to the highly complex structure of the universe as we know it today.

Until a few years ago the theory of the expanding universe lay under the cloud of a very serious contradiction. The measurements of the speed of flight of the galaxies and their distances from us indicated that the expansion had started about 1.8 billion years ago. On the other hand, measurements of the age of ancient rocks in the earth by the clock of radioactivity (*i.e.*, the decay of uranium to lead) showed that some of the rocks were at least three billion years old; more recent estimates based on other radioactive elements raise the age of the earth's crust to almost five billion years. Clearly a universe 1.8 billion years old could not contain five-billion-year-old rocks! Happily the contradiction has now been disposed of by Walter Baade's recent discovery that the distance yardstick (based on the periods of variable stars) was faulty and that the distances between galaxies are more than twice as great as they were thought to be. This change in distances raises the age of the universe to five billion years or more.

Friedman's solution of Einstein's cosmological equation, as I mentioned, permits two kinds of universe. We can call one the "pulsating" universe. This model says that when the universe has reached a certain maximum permissible expansion, it will begin to contract; that it will shrink until its matter has been compressed to a certain maximum density, possibly that of atomic nuclear material, which is a hundred million million times denser than water; that it will then begin to expand again—and so on through the cycle *ad infinitum*. The other model is a "hyperbolic" one: it suggests that from an infinitely thin state an eternity ago the universe contracted until it reached the maximum density, from which it rebounded to an unlimited expansion which will go on indefinitely in the future.

The question whether our universe is actually "pulsating" or "hyperbolic" should be decidable from the present rate of its expansion. The situation is analogous to the case of a rocket shot from the surface of the earth. If the velocity of the rocket is less than seven miles per second—the "escape velocity"—the rocket will climb only to a certain

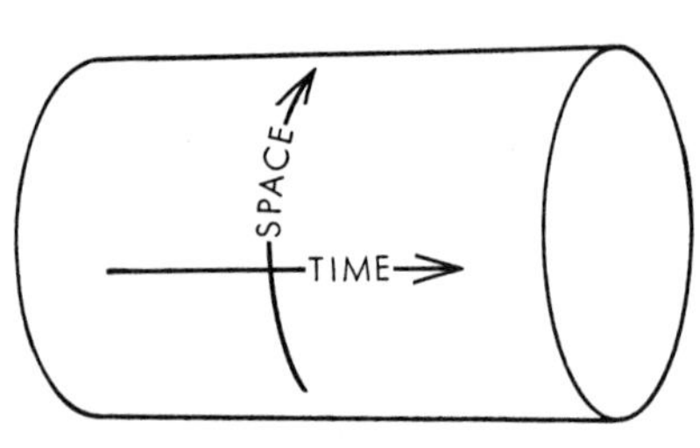

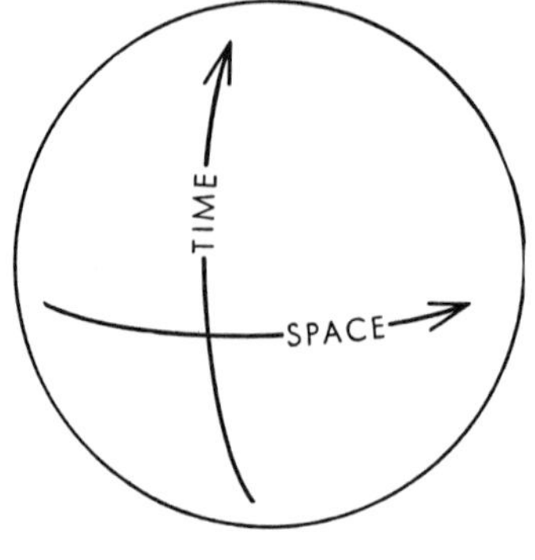

SPHERICAL UNIVERSE of Einstein may be represented in two dimensions by a cylinder (*left*). Its space coordinates were positively curved but its time coordinate was straight. The spherical universe of Willem de Sitter had positively curved coordinates (*right*).

height and then fall back to the earth. (If it were completely elastic, it would bounce up again, etc., etc.) On the other hand, a rocket shot with a velocity of more than seven miles per second will escape from the earth's gravitational field and disappear in space. The case of the receding system of galaxies is very similar to that of an escape rocket, except that instead of just two interacting bodies (the rocket and the earth) we have an unlimited number of them escaping from one another. We find that the galaxies are fleeing from one another at seven times the velocity necessary for mutual escape.

Thus we may conclude that our universe corresponds to the "hyperbolic" model, so that its present expansion will never stop. We must make one reservation. The estimate of the necessary escape velocity is based on the assumption that practically all the mass of the universe is concentrated in galaxies. If intergalactic space contained matter whose total mass was more than seven times that in the galaxies, we would have to reverse our conclusion and decide that the universe is pulsating. There has been no indication so far, however, that any matter exists in intergalactic space, and it could have escaped detection only if it were in the form of pure hydrogen gas, without other gases or dust.

Is the universe finite or infinite? This resolves itself into the question: Is the curvature of space positive or negative—closed like that of a sphere, or open like that of a saddle? We can look for the answer by studying the geometrical properties of its three-dimensional space, just as we examined the properties of figures on two-dimensional surfaces. The most convenient property to investigate astronomically is the relation between the volume of a sphere and its radius.

We saw that, in the two-dimensional case, the area of a circle increases with increasing radius at a faster rate on a negatively curved surface than on a Euclidean or flat surface; and that on a positively curved surface the relative rate of increase is slower. Similarly the increase of volume is faster in negatively curved space, slower in positively curved space. In Euclidean space the volume of a sphere would increase in proportion to the cube, or third power, of the increase in radius. In negatively curved space the volume would increase faster than this; in positively curved space, slower. Thus if we look into space and find that the volume of successively larger spheres, as measured by a count of the galaxies within them, increases faster than the cube of the distance to the limit of the sphere (the radius), we can conclude that the space of our universe has negative curvature, and therefore is open and infinite. By the same token, if the number of galaxies increases at a rate slower than the cube of the distance, we live in a universe of positive curvature—closed and finite.

Following this idea, Hubble undertook to study the increase in number of galaxies with distance. He estimated the distances of the remote galaxies by their relative faintness: galaxies vary considerably in intrinsic brightness, but over a very large number of galaxies these variations are expected to average out. Hubble's calculations produced the conclusion that the universe is a closed system—a small universe only a few billion light-years in radius!

We know now that the scale he was using was wrong: with the new yardstick the universe would be more than twice as large as he calculated. But there is a more fundamental doubt about his result. The whole method is based on the assumption that the intrinsic brightness of a galaxy remains constant. What if it changes with time? We are seeing the light of the distant galaxies as it was emitted at widely different times in the past—500 million, a billion, two billion years ago. If the stars in the galaxies are burning out, the galaxies must dim as they grow older. A galaxy two billion light-years away cannot be put on the same distance scale with a galaxy 500 million light-years away unless we take into account the fact that we are seeing the nearer galaxy at an older, and less bright, age. The remote galaxy is farther away than a mere comparison of the luminosity of the two would suggest.

When a correction is made for the assumed decline in brightness with age, the more distant galaxies are spread out to farther distances than Hubble assumed. In fact, the calculations of volume are changed so drastically that we may have to reverse the conclusion about the curvature of space. We are not sure, because we do not yet know enough about the evolution of galaxies. But if we find that galaxies wane in intrinsic brightness by only a few per cent in a billion years, we shall have to conclude that space is curved negatively and the universe is infinite.

Actually there is another line of reasoning which supports the side of infinity. Our universe seems to be hyperbolic and ever-expanding. Mathematical solutions of fundamental cosmological equations indicate that such a universe is open and infinite.

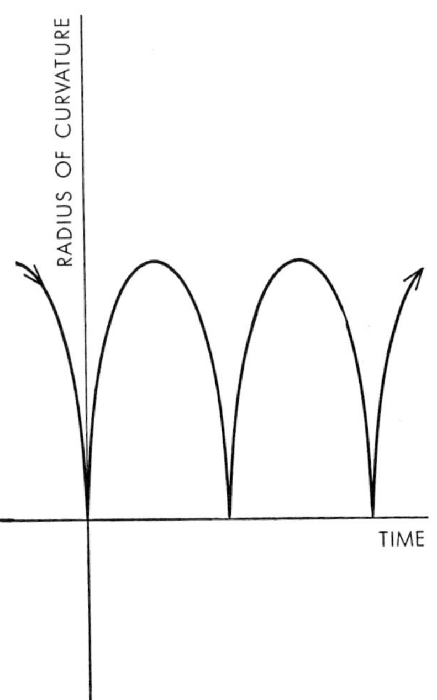

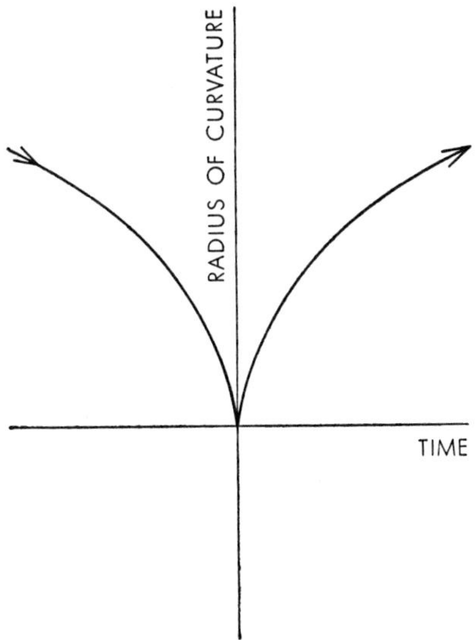

PULSATING AND HYPERBOLIC universes are represented by curves. The pulsating universe at the top repeatedly expands to a maximum permissible density and contracts to a minimum permissible density. The hyperbolic universe at the bottom contracts and then expands indefinitely.

We have reviewed the questions that dominated the thinking of cosmologists during the first half of this century: the conception of a four-dimensional space-time continuum, of curved space, of an expanding universe and of a cosmos which is either finite or infinite. Now we must consider the major present issue in cosmology: Is the universe in truth evolving, or is it in a steady state of equilibrium which has always existed and will go on through eternity? Most cosmologists take the evolutionary view. But in 1951 a group at the University of Cambridge, whose chief spokesman has been Fred Hoyle, advanced the steady-state idea. Essentially their theory is that the universe is infinite in space and time, that it has neither a beginning nor an end, that the density of its matter remains constant, that new matter is steadily being created in space at a rate which exactly compensates for the thinning of matter by expansion, that as a consequence new galaxies are continually being born, and that the galaxies of the universe therefore range in age from mere youngsters to veterans of 5, 10, 20 and more billions of years. In my opinion this theory must be considered very questionable because of the simple fact (apart from other reasons) that the galaxies in our neighborhood all seem to be of the same age as our own Milky Way. But the issue is many-sided and fundamental, and can be settled only by extended study of the universe as far as we can observe it. Hoyle presents the steady-state view in the following article [*page 157*]. Here I shall summarize the evolutionary theory.

We assume that the universe started from a very dense state of matter. In the early stages of its expansion, radiant energy was dominant over the mass of matter. We can measure energy and matter on a common scale by means of the well-known equation $E=mc^2$, which says that the energy equivalent of matter is the mass of the matter multiplied by the square of the velocity of light. Energy can be translated into mass, conversely, by dividing the energy quantity by c^2. Thus we can speak of the "mass density" of energy. Now at the beginning the mass density of the radiant energy was incomparably greater than the density of the matter in the universe. But in an expanding system the density of radiant energy decreases faster than does the density of matter. The former thins out as the fourth power of the distance of expansion: as the radius of the system doubles, the density of radiant energy drops to one sixteenth. The density of matter declines as the third power; a doubling of the radius means an eightfold increase in volume, or eightfold decrease in density.

Assuming that the universe at the beginning was under absolute rule by radiant energy, we can calculate that the temperature of the universe was 250 million degrees when it was one hour old, dropped to 6,000 degrees (the present temperature of our sun's surface) when it was 200,000 years old and had fallen to about 100 degrees below the freezing point of water when the universe reached its 250-millionth birthday.

This particular birthday was a crucial one in the life of the universe. It was the point at which the density of ordinary matter became greater than the mass density of radiant energy, because of the more rapid fall of the latter [*see chart on page 148*]. The switch from the reign of radiation to the reign of matter profoundly changed matter's behavior. During the eons of its subjugation to the will of radiant energy (*i.e.*, light), it must have been spread uniformly through space in the form of thin gas. But as soon as matter became gravitationally more important than the radiant energy, it began to acquire a more interesting character. James Jeans, in his classic studies of the physics of such a situation, proved half a century ago that a gravitating gas filling a very large volume is bound to break up into individual "gas balls," the size of which is determined by the density and the temperature of the gas. Thus in the year 250,000,000 A. B. E. (after the beginning of expansion), when matter was freed from the dictatorship of radiant energy, the gas broke up into giant gas clouds, slowly drifting apart as the universe continued to expand. Applying Jeans's mathematical formula for the process to the gas filling the universe at that time, I have found that these primordial balls of gas would have had just about the mass that the galaxies of stars possess today. They were then only "protogalaxies"—cold, dark and chaotic. But their gas soon condensed into stars and formed the galaxies as we see them now.

A central question in this picture of the evolutionary universe is the problem of accounting for the formation of the varied kinds of matter composing it—*i.e.*, the chemical elements. The question is discussed in detail in another article in this issue [*see page 82*]. My belief is that at the start matter was composed simply of protons, neutrons and electrons. After five minutes the universe must have cooled enough to permit the aggregation of protons and neutrons into larger units, from deuterons (one neutron and one proton) up to the heaviest elements. This process must have ended after about 30 minutes, for by that time the temperature of the expanding universe must have dropped below the threshold of thermonuclear reactions among light elements, and the neutrons must have been used up in element-building or been converted to protons.

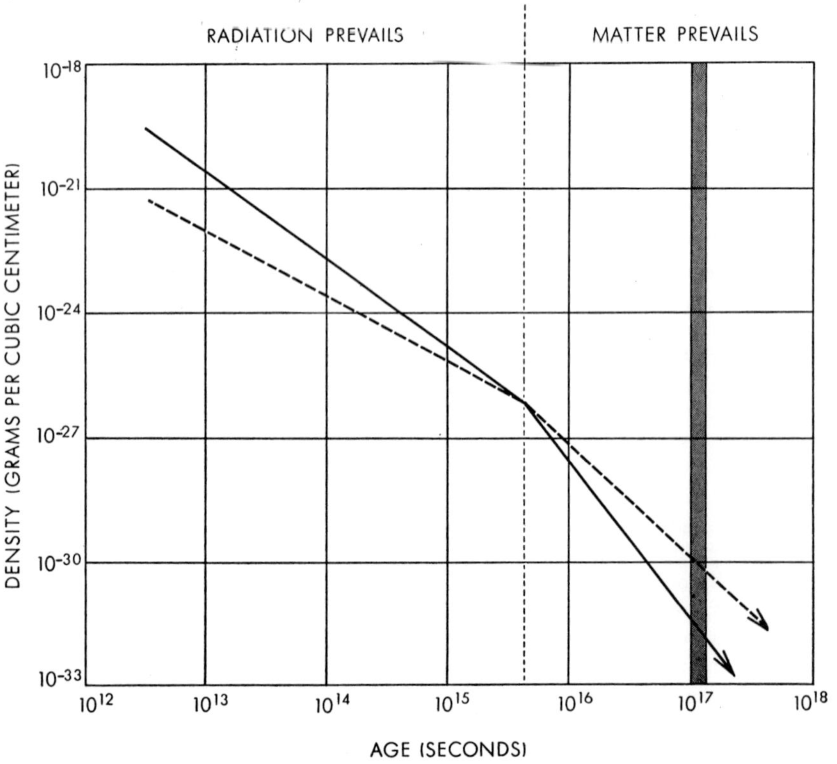

RELATIVE DENSITY OF MATTER AND RADIATION is reversed during the history of an evolutionary universe. Up to 250 million years (*broken vertical line*) the mass density of radiation (*solid curve*) is greater than that of matter (*broken curve*). After that the density of matter is greater, permitting the formation of huge gas clouds. The gray line is the present.

To many a reader the statement that the present chemical constitution of our universe was decided in half an hour five billion years ago will sound nonsensical. But consider a spot of ground on the atomic proving ground in Nevada where an atomic bomb was exploded three years ago. Within one microsecond the nuclear reactions generated by the bomb produced a variety of fission products. Today, 100 million million microseconds later, the site is still "hot" with the surviving fission products. The ratio of one microsecond to three years is the same as the ratio of half an hour to five billion years! If we can accept a time ratio of this order in the one case, why not in the other?

The late Enrico Fermi and Anthony L. Turkevich at the Institute for Nuclear Studies of the University of Chicago undertook a detailed study of thermonuclear reactions such as must have taken place during the first half hour of the universe's expansion. They concluded that the reactions would have produced about equal amounts of hydrogen and helium, making up 99 per cent of the total material, and about 1 per cent of deuterium. We know that hydrogen and helium do in fact make up about 99 per cent of the matter of the universe. This leaves us with the problem of building the heavier elements. I hold to the opinion that some of them were built by capture of neutrons. However, since the absence of any stable nucleus of atomic weight 5 makes it improbable that the heavier elements could have been produced in the first half hour in the abundances now observed, I would agree that the lion's share of the heavy elements may well have been formed later in the hot interiors of stars.

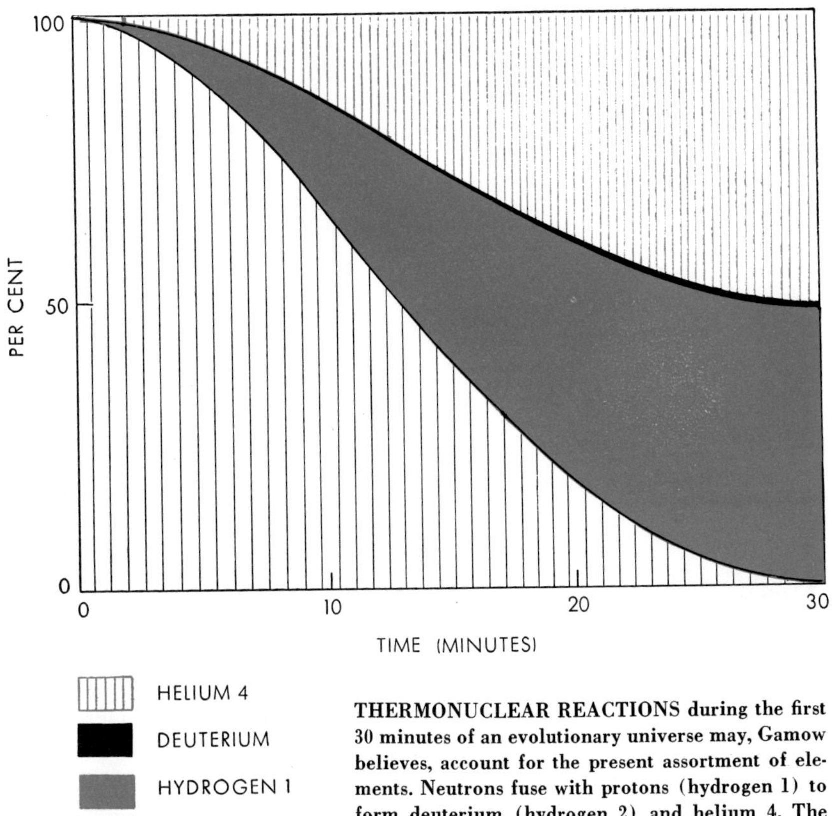

THERMONUCLEAR REACTIONS during the first 30 minutes of an evolutionary universe may, Gamow believes, account for the present assortment of elements. Neutrons fuse with protons (hydrogen 1) to form deuterium (hydrogen 2) and helium 4. The deuterium is then built up into the heavier elements.

All the theories—of the origin, age, extent, composition and nature of the universe—are becoming more and more subject to test by new instruments and new techniques, which are described in later articles in this issue. In the article on the red-shift investigations, Allan Sandage reports a tentative finding that the expansion of the universe may be slowing down. If this is confirmed, it may indicate that we live in a pulsating universe. But we must not forget that the estimate of distances of the galaxies is still founded on the debatable assumption that the brightness of galaxies does not change with time. If galaxies actually diminish in brightness as they age, the calculations cannot be depended upon. Thus the question whether evolution is or is not taking place in the galaxies is of crucial importance at the present stage of our outlook on the universe.

30

The Problem of Quasi-stellar Objects

GEOFFREY BURBIDGE AND FRED HOYLE

December 1966

It is now nearly four years since certain starlike objects, previously identified as strong sources of radio energy, were shown to be receding at velocities comparable to those of the most distant galaxies. The combination of small size, enormous energy output and apparent remoteness quickly made these quasi-stellar radio sources, as they first were called, the most interesting and widely discussed subject in astronomy. As they were observed more closely a further mystery developed: many of them were found to vary in brightness and in radio output over periods as short as months, weeks or even days. If they were as distant as they seemed to be, and therefore so enormously radiant, how could their energy output fluctuate so rapidly? Some astronomers began to question if the quasi-stellar objects were really as far away as the observational evidence seemed to indicate. Perhaps they were not at such "cosmological" distances at all but were comparatively close to us—objects ejected, perhaps, at high velocity from our own or neighboring galaxies.

The evidence that the quasi-stellar objects are receding from us at high velocity is inferred from their spectra, which show that the light they emit is shifted in wavelength toward the red, or long-wavelength, end of the spectrum. This can be determined because each chemical element, when heated to incandescence, emits light at a series of characteristic wavelengths. If the radiating elements are in a star or other object that is moving away from the observer, these characteristic wavelengths will appear to have lengthened. If, on the other hand, the object is approaching the observer, the wavelengths will appear to have shortened and thus to have shifted toward the blue end of the spectrum. So far no quasi-stellar object has exhibited a blue shift. As we shall explain, if a blue shift were to be detected, a large part of the mystery surrounding these objects would be dispelled.

As matters stand, all one can be sure of is that the quasi-stellar objects, remote or not, are powerful emitters of radiant energy. They are so small as to be almost indistinguishable from stars, and if they are indeed very remote, they are smaller by far and much brighter than normal galaxies observed at the same distance.

The problem of quasi-stellar objects arose just six years ago this month when Allan R. Sandage of the Mount Wilson and Palomar Observatories announced that his photographic plates showed what seemed to be a star at the precise position assigned to a strong radio source known as 3C 48 (source No. 48 in the Third Cambridge Catalogue). If Sandage's object was indeed a star, it would be the first radio star ever discovered. Most radio sources are either galaxies or extended regions of very hot gas within our own galaxy. To see what they could learn about this strange object, Sandage and subsequently other astronomers recorded its spectrum and found it to be quite unlike that of any known star.

In early 1963 another radio source, 3C 273, was identified with a starlike object even brighter than the one identified with 3C 48. The accurate radio position for 3C 273 that made possible this second discovery was provided by Cyril Hazard, then at the University of Sydney, and his colleagues, who worked with the 210-foot radio telescope of the Commonwealth Scientific and Industrial Organisation in Australia [see "Locating Radio Sources with the Moon," by R. W. Clarke; Scientific American, June]. Maarten Schmidt of Mount Wilson and Palomar obtained a spectrum of 3C 273 and perceived that several faint lines in the spectrum occupied positions that would coincide with those of a well-known series of hydrogen lines (the Balmer series) if they had been shifted toward the red by 16 percent. With this clue he identified lines of magnesium and oxygen similarly displaced. Immediately Jesse L. Greenstein and Thomas A. Matthews of Mount Wilson and Palomar identified the features in the spectrum of 3C 48 that had been so puzzling and showed that the object had a red shift of 37 percent [see "Quasi-stellar Radio Sources," by Jesse L. Greenstein; Scientific American, December, 1963].

Since 1963 the identification of quasi-stellar objects has proceeded so rapidly that more than 120 such objects are now known. By the latter part of 1965 only about 10 red shifts had been published; today the number of known red shifts exceeds 65.

Nonastronomers should not worry if they have found the terminology confusing. The objects were first called quasi-stellar sources or quasi-stellar radio sources. The term "quasar" was also coined and applied to them, but this rather ugly word has not become popular with astronomers. Along the way Sandage isolated a class of objects that were not listed as radio sources but were in all other respects similar to the quasi-stellar sources. At first he called these objects "interlopers," then "blue stellar objects" and later "quasi-stellar galaxies." In this article we have adopted the term "quasi-stellar objects" to cover all objects that are starlike and have spectra with large red shifts, whether or not they are strong radio sources.

Because quasi-stellar objects radiate strongly in the ultraviolet part of the spectrum, Sandage attempted to make a general search for them using filter photography and photoelectric photometry to pick out objects with an abnormally strong flux of ultraviolet radiation. The next step was to see if the objects so identified had red-shifted spectra. Of six objects Sandage selected as candidates on the basis of accurate color measurements, three were found to have red shifts, and of these two were sufficiently starlike to qualify as quasi-stellar objects; the third must be disqualified because it has a slightly fuzzy image.

On the basis of coarse sky surveys that had previously been made for blue objects, Sandage estimated that quasi-stellar galaxies (as he referred to them at the time) might be 500 times as plentiful as the quasi-stellar objects that are known to be radio sources. These estimates were severely criticized and it is now generally agreed that Sandage's estimate was too high. A reasonable estimate is that quasi-stellar objects that are not known to be radio sources may be about 100 times as plentiful as the objects that are strong radio emitters. This implies that if one were to count all quasi-stellar objects down to a photographic magnitude of 18, one might find such an object for every square degree of sky, or some 40,000 in all. (A change of five magnitudes represents a hundredfold change in brightness; thus an object of zero magnitude, corresponding to the brightest stars, is 16 million times brighter than one of 18th magnitude.) So far very few of these "silent" quasi-stellar objects have been investigated in detail, and red shifts are still known for only three or four of them. These are included among the 65 or so for which red shifts have been determined.

The known red shifts range from the 16 percent found for 3*C* 273 up to values greater than 200 percent. (The greatest red shift yet observed for a non-quasi-stellar object is about 46 percent.) A 200 percent shift means that the Lyman-alpha line of hydrogen, which has a laboratory wavelength of 1,216 angstrom units in the ultraviolet part of the spectrum, is shifted to a wavelength of 3,648 angstroms (1,216 plus 2 × 1,216), which lies in the ultraviolet just beyond the extreme blue end of the visible spectrum. The shortest wavelength normally recordable by a spectrograph attached to an optical telescope is about 3,100 angstroms. Thus a red shift of at least 145 percent is needed to make the Lyman-alpha line observable. During the first year following Schmidt's discovery of the red shift of 3*C* 273 all the spectroscopic work on quasi-stellar objects was carried out with the 200-inch telescope on Palomar Mountain. Starting in 1964, however, E. Margaret Burbidge and T. D. Kinman began recording the spectra with the 120-inch reflector at the Lick Observatory and C. R. Lynds and his associates began using the 84-inch reflector at the Kitt Peak National Observatory near Tucson, Ariz.

What are the characteristics of the spectra of quasi-stellar objects? In general they exhibit broad, weak emission lines [*see illustration on next page*].

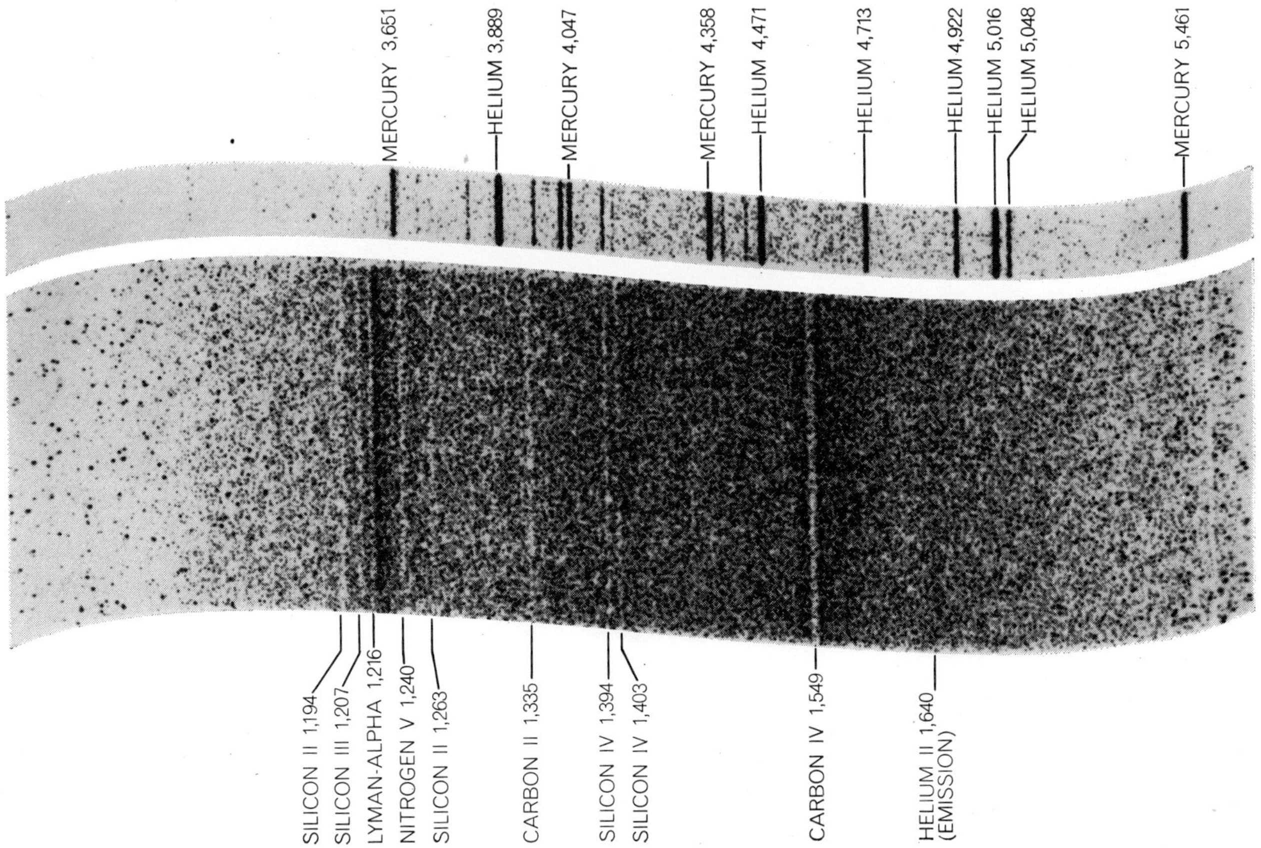

SPECTRUM OF QUASI-STELLAR OBJECT 3*C* 191 was among the first to show strong absorption lines, raising the question of where the absorbing material is located. The spectrum indicates that the absorbing material is a shell of cooler gas expanding outward from the central object. The absorption lines are shifted to the red end of the spectrum only slightly less than the emission lines in the spectrum, which are less easily seen. The numbers beside the names of the various elements give the laboratory, or undisplaced, wavelength in angstrom units for the various absorption lines. Because 3*C* 191 is evidently receding from us at nearly 80 percent of the speed of light, these lines are all red-shifted sharply, as indicated by the wavelengths in the reference spectrum (*top*). The magnitude of the red shift is normally computed by dividing the line displacement by the undisplaced wavelength; for 3*C* 191 this value is 1.946. This can also be regarded as a red shift of 194.6 percent. The spectrum was made by C. R. Lynds and A. N. Stockton of the Kitt Peak National Observatory, who used an image-tube spectrograph attached to the observatory's 84-inch reflecting telescope.

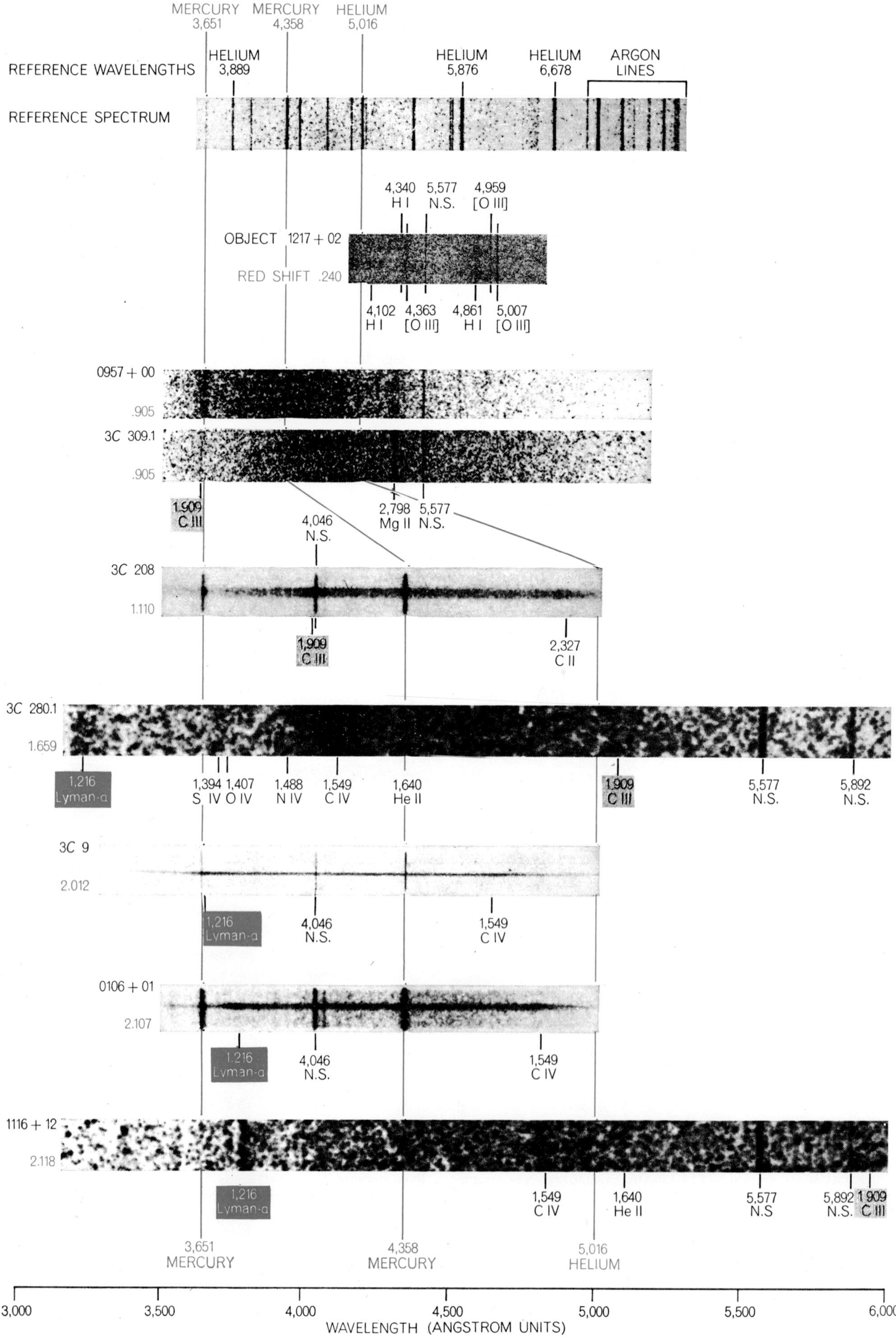

MERCURY 3,651
MERCURY 4,358
HELIUM 5,016
HELIUM 3,889
REFERENCE WAVELENGTHS
HELIUM 5,876
HELIUM 6,678
ARGON LINES
REFERENCE SPECTRUM
4,340 H I
5,577 N.S.
4,959 [O III]
OBJECT 1217 + 02
RED SHIFT .240
4,102 H I
4,363 [O III]
4,861 H I
5,007 [O III]
0957 + 00
.905
3C 309.1
.905
1,909 C III
2,798 Mg II
5,577 N.S.
4,046 N.S.
3C 208
1.110
1,909 C III
2,327 C II
3C 280.1
1.659
1,216 Lyman-α
1,394 S IV
1,407 O IV
1,488 N IV
1,549 C IV
1,640 He II
1,909 C III
5,577 N.S.
5,892 N.S.
3C 9
2.012
1,216 Lyman-α
4,046 N.S.
1,549 C IV
0106 + 01
2.107
1,216 Lyman-α
4,046 N.S.
1,549 C IV
1116 + 12
2.118
1,216 Lyman-α
1,549 C IV
1,640 He II
5,577 N.S
5,892 N.S.
1,909 C III
3,651 MERCURY
4,358 MERCURY
5,016 HELIUM
3,000
3,500
4,000
4,500
5,000
5,500
6,000
WAVELENGTH (ANGSTROM UNITS)

Among the elements identified in these spectra are hydrogen, helium, carbon, oxygen, nitrogen, neon, magnesium, silicon, argon and sulfur. The strength of the lines provides a crude indication of the temperature and density of the gas in which the lines are produced. The temperatures indicated by the electrons in the gas are on the order of a few tens of thousands of degrees Kelvin (degrees centigrade above absolute zero). The density of the gas ranges from about 10^4 to about 10^7 particles per cubic centimeter. For purposes of comparison, the temperature of the surface of the sun is about 6,000 degrees K. and the density of particles in the same region is about 10^{16} particles per cubic centimeter. One unexpected feature of quasi-stellar objects is that they appear to have a normal composition of elements, rather like that found in stars and gaseous nebulas in our own galaxy.

An important new feature of the quasi-stellar objects, discovered only in the past year, is that some of their spectra show absorption lines, indicating that radiation is being absorbed by cooler material lying somewhere between the emitting source and our galaxy. As we shall see, the location of this absorbing material is of fundamental importance. It bears directly on the nature of the quasi-stellar objects and may help to decide whether or not they are at cosmological distances.

Let us therefore describe briefly the possibilities concerning the nature of the quasi-stellar objects. When their large red shifts were first discovered, it was realized that there were only two possible explanations compatible with the known laws of physics. The red shifts meant either that the objects were receding at high speed or that the light was being emitted in gravitational fields far stronger than any previously known. The second possibility follows from the general theory of relativity, which predicts that radiation must "do work" in escaping from a gravitational field and is thereby lengthened in wavelength. The kind of gravitational fields needed to produce the observed reddening would result if the mass of the sun were compressed into a sphere a few kilometers in diameter. In that case, however, the gravitational field would be so enormous that the particles would be drawn quickly into a sphere of smaller and smaller radius, the phenomenon known as gravitational collapse. As an explanation of the red shift observed in quasi-stellar objects, gravitational collapse is unsatisfactory on two main counts. First, the duration of the phenomenon would be too short. Second, the density of the objects would far exceed the 10^4 to 10^7 particles per cubic centimeter indicated by the spectroscopic evidence.

Thus the red shifts almost certainly indicate that the quasi-stellar objects are receding from us. Since the time of the pioneering work on Mount Wilson of Edwin P. Hubble and Milton L. Humason, who showed that the red shifts of galaxies increase with their distance, it has been clear that the galaxies are receding from us and from one another as part of the general expansion of the universe. The constant in the velocity-distance relation is called the Hubble constant. The most distant galaxy we know (the radio galaxy 3C 295) shows a red shift of 46 percent, indicating that it is about five billion light-years away.

It was natural, therefore, that the large red shifts of the quasi-stellar objects were quickly taken to mean that some of the objects were even more distant in time and space than the faintest galaxies for which red shifts had been measured. It follows that they must be extraordinarily bright, since they are similar in appearance to the fainter stars in our own galaxy. If their red shifts are taken as a measure of their distance, they must be emitting on the average about 40 times as much energy in the photographic part of the spectrum as the brightest galaxies. This is most astonishing, since a typical large galaxy contains 100 billion stars.

There is, however, another explanation of the red shifts that does not require the quasi-stellar objects to be so luminous. In 1964 James Terrell of the Los Alamos Scientific Laboratory proposed that they might be comparatively nearby objects that somehow had acquired enormous speed. The "somehow" is not a trivial matter; an object with a red shift of 200 percent must be flying away from us at 80 percent of the speed of light. Terrell proposed that the quasi-stellar objects had been ejected by some unknown force from the center of our own galaxy. In this connection, we had ourselves suggested somewhat earlier that an explosion had occurred in the center of our galaxy about 10 million years ago.

About a year ago we were led to consider this "local" hypothesis from a different standpoint and suggested as an alternative to the cosmological hypothesis that perhaps the objects were ejected from one or more of the nearby radio galaxies in which there is evidence that highly energetic explosions have occurred. We accordingly suggested that the quasi-stellar objects might lie at distances measured only in millions or perhaps tens of millions of light-years rather than in billions of light-years. On this hypothesis they would be less luminous by a factor of at least 10^4, that is, they would have the brightness of only about 100 million suns or less. Obviously the local hypothesis presents its own problems.

In any case it is clear that what we need is an observational test that can show unambiguously whether quasi-stellar objects are local or cosmological. Schmidt's discovery in 1965 that the Lyman-alpha emission line was shifted to 3,663 angstroms in the spectrum of 3C 9 immediately raised the hope that such a test was at hand. Peter Scheuer of the University of Cambridge and two investigators at the California Institute of Technology, James E. Gunn and Bruce Peterson, pointed out that if 3C 9 is indeed very distant, some of its ultraviolet radiation shorter than 1,216 angstroms (or 3,663 angstroms as received) would be absorbed in traveling through the intergalactic medium, provided that the medium contains neutral atoms of hydrogen. These atoms, being in a low-energy state, would readily absorb photons whose wavelength on reaching them corresponded to that of the Lyman-alpha wavelength of 1,216 angstroms. For regions of space close to 3C 9 the radiation absorbed by the intergalactic hydrogen would have left 3C 9 at a wavelength only slightly shorter than 1,216 angstroms. But the hydrogen just outside

SERIES OF SPECTRA of quasi-stellar objects is arranged from top to bottom in order of increasing red shift. The magnitude of the red shift and the name of the object are given at the left of each spectrum. A reference spectrum appears at the top of opposite page. Note that the scale of the first three spectra is different from that of the last five. The first, second, third, fifth and eighth spectra were made by Lynds and Stockton at Kitt Peak. The others were made by E. Margaret Burbidge, using the 120-inch reflector at the Lick Observatory. The letters "N.S." stand for "night sky" and indicate lines emitted by the earth's upper atmosphere and by gases in city lamps. The quasi-stellar object 3C 280.1 is the first in this series that has a red shift large enough so that the Lyman-alpha line of hydrogen (rest wavelength, 1,216 angstroms) is shifted into the visible region. It appears at 3,233 angstroms. In the three succeeding spectra it is shifted still farther: to 3,663, 3,776 and 3,792 angstroms. These three quasi-stellar objects have the largest red shifts yet recorded: 2.012, 2.107 and 2.118.

our galaxy would absorb radiation that had originally been much shorter (in the vicinity of 400 angstroms) and that had been reddened to 1,216 angstroms as a consequence of the expansion of the universe during the several billion years from the time the radiation began its journey. Thus what one should see in the spectrum of 3*C* 9 is a general deficiency of ultraviolet radiation shorter than 3,663 angstroms. After examining spectra of 3*C* 9, Gunn and Peterson thought they could detect the beginning of this long absorption trough in the region between 3,663 angstroms and about 3,100 angstroms, where the ultraviolet absorption of the earth's atmosphere intervenes and cuts off the record. The absorption trough was significantly less than expected, however, so they concluded that only a fantastically small amount of intergalactic gas was in the form of neutral hydrogen: about one atom per 10,000 cubic meters, or 10^{-34} gram per cubic centimeter. Most theoretical models of the universe require a density roughly 100,000 (10^5) times higher for the inter-

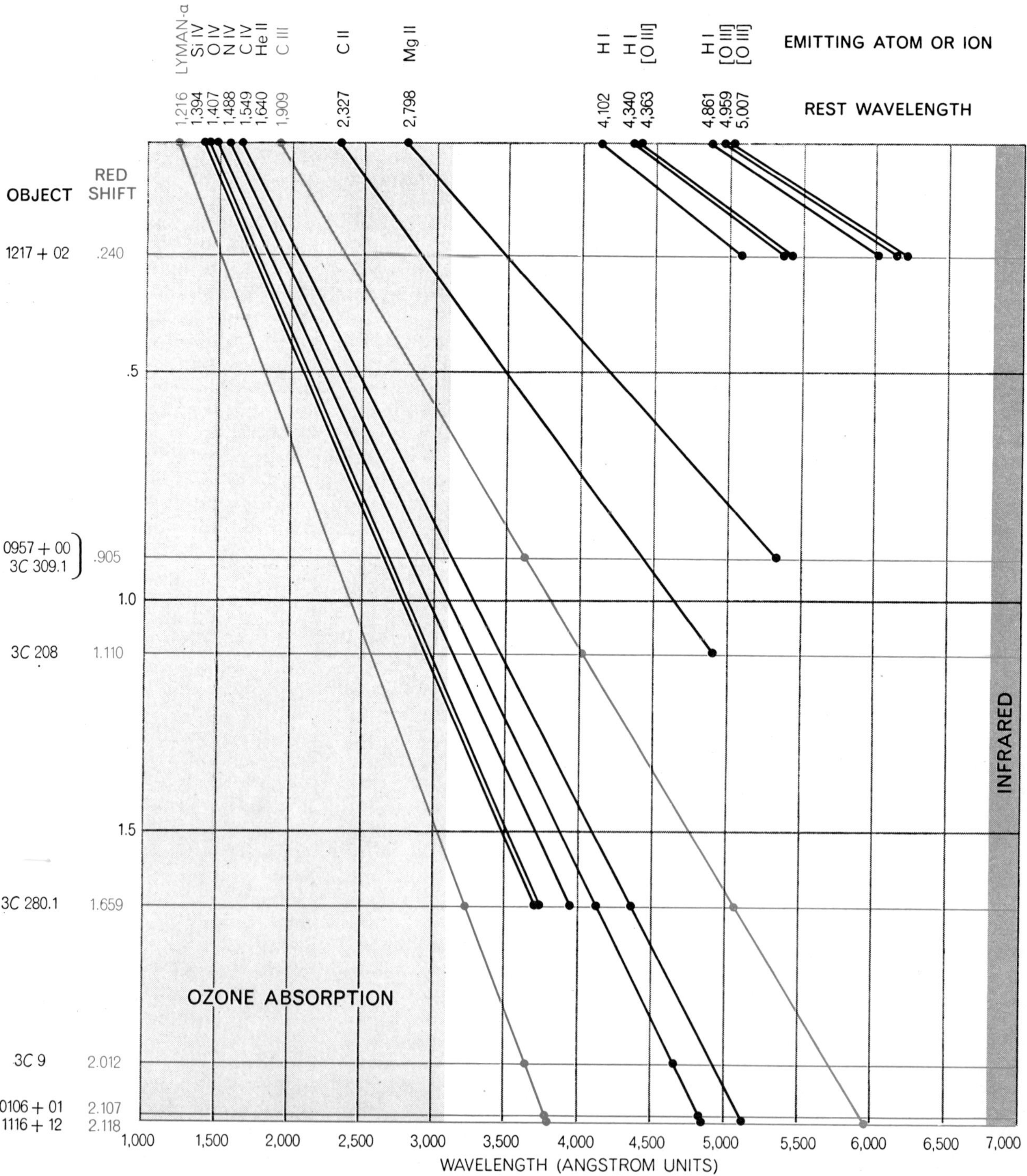

RED-SHIFT DISPLACEMENT for the principal emission lines found in the spectra on page 318 can be more easily visualized with the aid of a diagram in which the undisplaced wavelengths are plotted along the top and the displacements, corresponding to various red shifts, are plotted below. The Roman numerals indicate the state of ionization of the emitting element; the numbers are one greater than the number of electrons missing from the atom. Lyman-alpha and carbon III lines (*color*) show up in four spectra.

galactic medium, unless large amounts of matter are present in a form not easily detected, for example in stars too dim and too dispersed to be seen. The Gunn-Peterson finding might also be explained if the intergalactic gas were present not as neutral hydrogen but as ionized hydrogen, which cannot absorb Lyman-alpha radiation. Indeed, some theoreticians have accepted the Gunn-Peterson results as evidence that most of the hydrogen is in ionized form.

Subsequent observations of 3C 9 and other quasi-stellar objects with comparable red shifts have led to the conclusion that there is really no evidence for Lyman-alpha absorption at all and that the early interpretation by Gunn and Peterson is incorrect. The new studies indicate that the density of intergalactic neutral hydrogen is incredibly low (less than 10^{-35} gram per cubic centimeter); this implies either that virtually all of it is ionized or that the quasi-stellar objects are so near that no significant absorption can be expected. To some the improbability of the first alternative would seem to provide a convincing argument that the objects are local and not cosmological, but enough other possibilities remain so that the debate continues.

One such possibility was pointed out by John N. Bahcall of the California Institute of Technology and E. E. Salpeter of Cornell University. They suggested that the intergalactic gas might not be distributed uniformly but might exist as condensations, for example in association with clusters of galaxies. In that case one might find a number of separate absorption lines in the spectra of quasi-stellar objects, indicating where their radiation had intersected one or more clusters. One might also expect to find absorption lines due to energy transitions in atoms other than hydrogen.

Soon after this proposal was made the quasi-stellar object 3C 191 was observed for the first time by Margaret Burbidge at the Lick Observatory and by Lynds and A. N. Stockton at Kitt Peak. Its spectrum exhibits many absorption lines, and it was thought at first that they might support the Bahcall-Salpeter hypothesis. It was shown, however, that all the lines could be attributed to absorption by ionized atoms associated with the object itself [*see illustration on page 317*]. The evidence for this conclusion was that the absorption lines and the emission lines had almost the same red shift: 195 percent. Evidently the absorption lines are produced in a slightly cooler envelope of material that is part of 3C 191, probably a shell of gas that has been ejected from the quasi-stellar object.

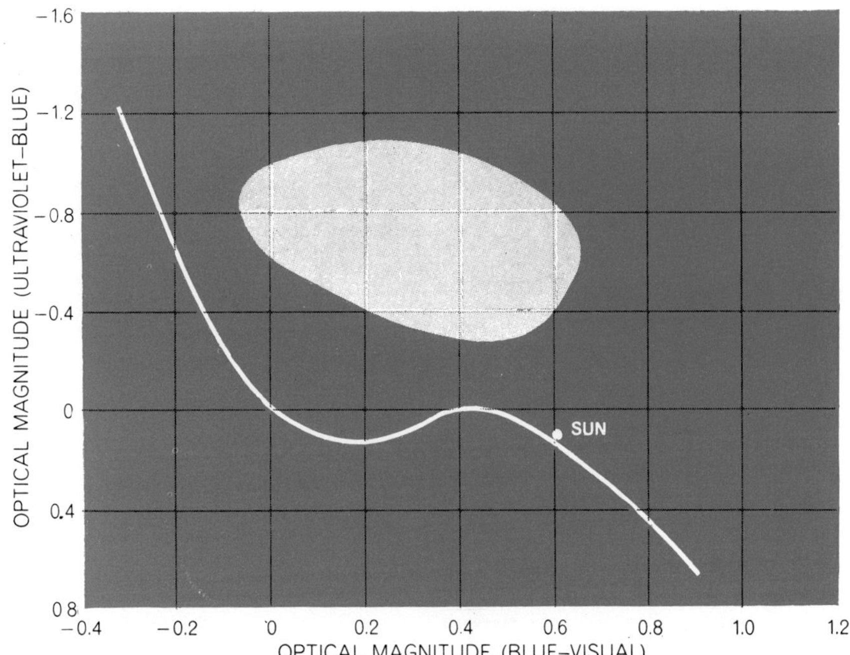

COLOR-MAGNITUDE DIAGRAM shows the general location of quasi-stellar objects (*light area*) compared with the general location of normal stars (*white curve*). Astronomers measure the magnitude, or brightness, of stars through three filters: ultraviolet, blue and yellow (or visual). When differences between the ultraviolet and blue magnitudes are plotted against differences between blue and visual magnitudes, the normal stars over the entire range of temperatures lie on or near a well-defined curve. The hottest stars lie at the upper left corner of the curve, the coolest at the lower right. The quasi-stellar objects lie well above this curve, separated from it by their intense ultraviolet radiation, which moves them upward in the diagram. Some white dwarfs also lie in this region, but since they do not have large red shifts they can be distinguished from quasi-stellar objects.

Absorption lines have now been seen in other quasi-stellar objects. In all cases but one the lines can be interpreted as being produced by a shell of gas whose velocity differs from that of the emissive core by no more than a few thousand kilometers per second—velocities that are no larger than those observed in the explosions of supernovas. The one exception is a spectrum in which the velocities appear to be considerably higher, presenting the possibility that the absorbing medium is not part of the quasi-stellar object but is somewhere in intergalactic space. The arguments for this possibility are in our opinion not conclusive. Thus although absorption features may eventually demonstrate that quasi-stellar objects are at cosmological distances, all the evidence to date is compatible with the conclusion that they are nearby and that when absorption lines are seen, they are due to material associated with the object.

We turn now from the spectroscopy of the quasi-stellar objects to their very important property of variability. Following the initial identification of 3C 273, Harlan J. Smith and E. Dorrit Hoffleit, then at Yale University, examined the Harvard College Observatory collection of astronomical plates, which dates back more than 70 years, to see if the "star" now identified as 3C 273 had varied in brightness. They found not only that it had varied significantly over the years but also that some bright flashes had lasted a month or less [*see top illustration on next two pages*]. Apart from 3C 273 and 3C 48 (also found to vary), none of the quasi-stellar objects is bright enough to be traced back in a similar fashion. In the brief period since they have been discovered, however, almost all the quasi-stellar objects that have been observed on several occasions have exhibited variations in brightness. In one extended study Kinman and D. Goldsmith of the Lick Observatory followed 3C 345 from June to October, 1965, and found that its brightness varied as much as 40 percent in a period as short as a few weeks. A single observation made later in October by Sandage showed its brightness to be three times the brightness measured in June. Subsequently Sandage observed that another quasi-stellar object, 3C 446, has brightened by three magnitudes, or a factor of about 15, in a few months. Recently Kinman and his colleagues at the Lick

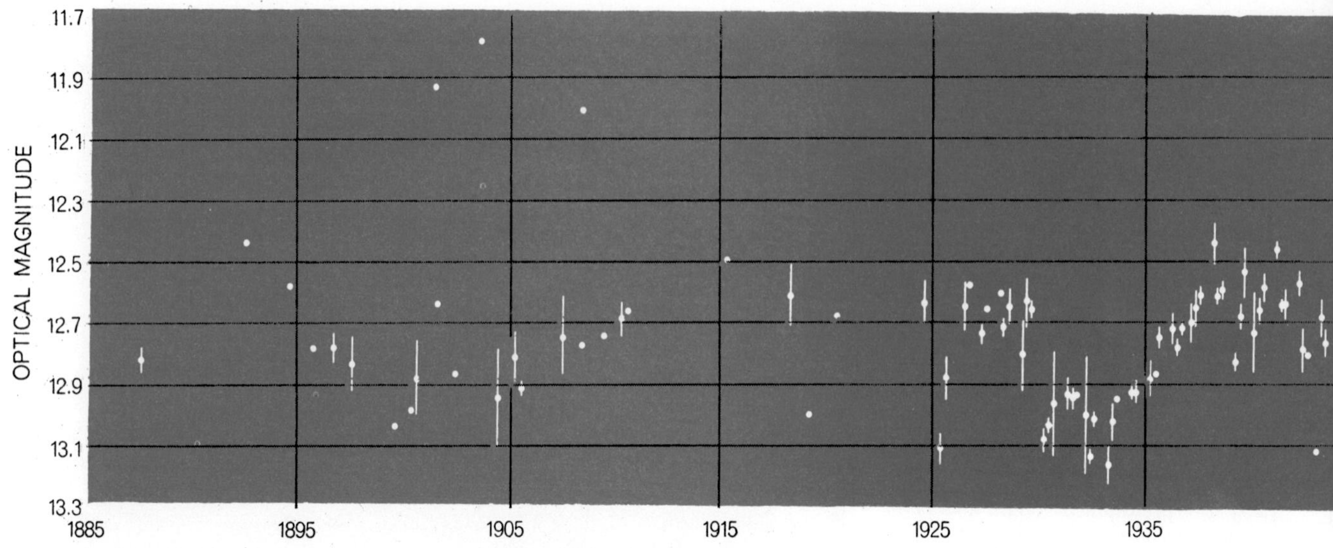

BRIGHTNESS CHANGES IN 3C 273, the second quasi-stellar object to be identified, were plotted by examining plates of variable stars in the collection of the Harvard College Observatory. The record of fluctuations in 3C 273, the brightest quasi-stellar object yet found, was traced back more than 70 years by Harlan J. Smith and E. Dorrit Hoffleit, then at Yale University. The ver-

Observatory have observed a twofold change in brightness in 24 hours.

In the summer of 1965 a variation in the radio emission of 3C 273 was reported by William A. Dent of the University of Michigan. (An earlier Russian report of a variation in another quasi-stellar object, CTA 102, was not confirmed by other observers and is generally discounted.) Dent's measurements, made at a wavelength of 3.75 centimeters, showed that the intensity of 3C 273 had increased by about 40 percent in two and a half years. Since then it has been found that quite a number of quasi-stellar objects show variations at wavelengths of a few centimeters [*see illustrations on pages 324 and 325*].

Only 3C 273 is bright enough so that its radiation can be measured in the infrared region of the spectrum. Harold L. Johnson and Frank J. Low of the University of Arizona, who have studied 3C 273 at wavelengths of 2.2 and 10 microns (a micron is 10,000 angstrom units), have detected no variation in its infrared luminosity. Surprisingly their measurements show that a large part of the object's total energy is being emitted in this region of the spectrum. In the adjacent millimeter-wavelength region of the spectrum, which has been observed by Eugene E. Epstein of the Aerospace Corporation and his colleagues, the emission of 3C 273 is again found to be varying rapidly.

The importance of the observed variability in quasi-stellar objects is that it sets an upper limit to the size of the region from which their energy can be emitted. Unless the objects are in a state of continuous explosion at enormous speeds (and there is no hint of this in their spectra), the period of variation is fixed by the time it takes for light to travel the diameter of the radiating object. The observed variations in quasi-stellar objects mean that the size of the region where much of their radiation originates must be no greater than the distance light can travel in a month—in the case of 3C 446, in a day. A light-month is equal to about 10^{17} centimeters, or roughly 100 times the diameter of the solar system. Moreover, this dimension is independent of distance; it is the same whether quasi-stellar objects are nearby or extremely remote, and thus it has a great bearing on the models one can devise to explain their energy output. If they are remote, they must radiate something like 10,000 times more energy per unit of volume than if they are close to us.

If 3C 273 is a remote object, it is emitting about 1.5×10^{47} ergs of energy per second across the entire electromagnetic spectrum; if, on the other hand, it is a local object, its emission is reduced to about 1.5×10^{43} ergs per second or less, depending on the actual distance. The higher value represents a radiation density comparable to that in a laser, the most intense artificial light source known. The problem of explaining such huge fluxes of energy is very difficult and has not been solved. With rather simple arguments, however, one can limit the models that are compatible with the observations.

From lines in the spectra produced by quasi-stellar objects one can infer the presence of a hot plasma, one whose temperature is perhaps 30,000 degrees K. Although such a plasma will give rise to some continuous radiation as well as to discrete lines, it cannot account for the strength of the background radiation found in the spectra of quasi-stellar objects. And ordinary thermal processes certainly cannot account for the energy emitted by such objects in other parts of the electromagnetic spectrum. The bulk of the energy is almost certainly emitted by the "synchrotron" process, in which highly energetic electrons emit radiation while spiraling around the lines of force in a magnetic field.

Now, the conditions under which the synchrotron process can be effective are narrowly restricted if quasi-stellar objects are at cosmological distances in addition to being small in diameter. In this case the density of radiation in them must be extremely high: about 10^{16} photons per cubic centimeter. If high-energy electrons are injected into a radiation field of such high intensity, they may lose most of their energy through collisions with low-energy photons before they can radiate much energy through the synchrotron process. The synchrotron process can be dominant only if the magnetic field is on the order of 100 gauss, or some 50 times as strong as the magnetic field on the surface of the sun. (This is the result if the object has a diameter of about a light-month. If it is only a light-day across, the field would have to be about 10,000 gauss.) Moreover, the electrons injected into such a strong field will radiate for only a few seconds or minutes before their energy is depleted by the synchrotron process. That is far too brief a time for them to

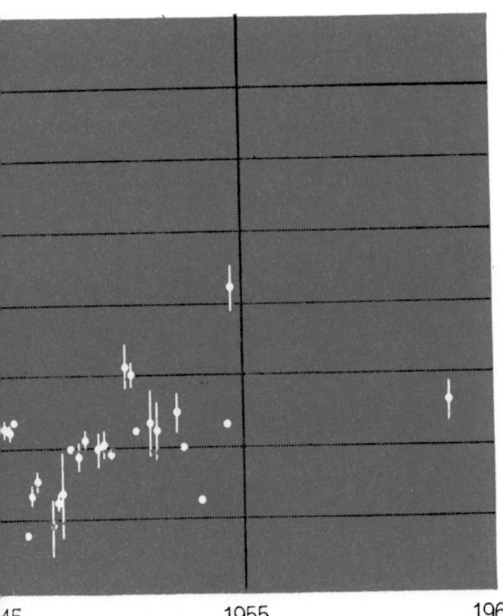

tical bars show the range in brightness for series of plates taken within 100-day periods. Single dots represent isolated observations.

travel across the diameter of a quasi-stellar object, a trip that would take from one to several weeks. This means that quasi-stellar objects must contain not one or even a few hundred but millions of separate points where electrons are injected into a strong magnetic field.

It is possible, of course, that the synchrotron process is not responsible for the bulk of the energy radiated by quasi-stellar objects. One can imagine that some kind of strange "machine" generates a radiation field of low-energy photons even more intense than the one we have described. If enough electrons with the right spectrum of energy were injected into such a field, collision processes alone might give rise to the required flux. This appears to us, however, to be a most unlikely possibility.

Before considering the theories that have been proposed to account for the quasi-stellar objects, we should like to touch briefly on other observational data that bear on the fundamental point of whether they are local or cosmological objects. At one time it was hoped that by plotting the red shift of quasi-stellar objects against their apparent optical brightness one might be able to tell something about the large-scale structure of the universe and the kind of cosmological model that best fits the observations. To be useful such a plot would have to yield a more or less straight-line relation between red shift and brightness, and would itself provide prima facie evidence that the red shifts are indeed a function of distance. It has turned out, however, that the plot has a large scatter in which only a weak trend is apparent [*see top illustration on page 326*]. The scatter evidently arises from the fact that quasi-stellar objects differ greatly in intrinsic brightness. Therefore no simple relation between red shift and brightness can be expected, even if the objects are at cosmological distances.

The scatter does not necessarily lend support to the local hypothesis, because then one must account for the absence of blue shifts. One way to do this is to imagine that most, if not all, of the quasi-stellar objects were ejected from a nearby radio galaxy. If enough time has elapsed since the explosion, even those objects initially headed in our direction will have passed through (or close to) our galaxy and will now be receding on the other side. It should be noted, however, that if blue-shifted objects exist, it may be more difficult to identify their spectral lines than is the case for red-shifted objects. There are fewer lines in the red than there are in the blue part of the spectrum and they are intrinsically weaker; thus they would be more easily hidden by the continuous radiation in the background.

Recently Halton C. Arp of the Mount Wilson and Palomar Observatories has presented evidence indicating that radio sources, including several quasi-stellar objects, tend to lie in the vicinity of certain unusual-looking galaxies called "peculiar" galaxies. He suggests that the quasi-stellar objects were ejected from them. One consequence of Arp's suggestion is that we should see more blue-shifted objects than red-shifted ones, even if equal numbers were headed toward us and away from us. The explanation is that those approaching should appear brighter than those receding and thus seem to be more numerous when counted down to the same limiting magnitude. Since no blue-shifted objects have been seen at all, some astronomers have summarily rejected Arp's idea. The rejection may be too hasty. One must consider the possibility that an object may not radiate spherically in symmetrical fashion if it is moving through the intergalactic medium at relativistic speed: a speed close to the velocity of light. Then there would be no guarantee that approaching objects will be easier to see (that is, brighter) than receding ones. If Arp's hypothesis were to be rigorously established, one might be forced to find still other explanations for the absence of blue shifts. It is possible, for example, that the red shifts do not represent a recession of the objects but are due to the relativistic motion of gas contained within them.

Another kind of test is to see if quasi-stellar objects are to be found within clusters of galaxies. It has been established, for example, that many radio galaxies lie in such clusters and that they are frequently the brightest objects present. If even a few quasi-stellar objects were found in clusters, one could try to see if the red shift of the quasi-stellar object matched that of the cluster. If so, the cosmological nature of these objects would be established. The observational difficulty associated with this test is that the fainter quasi-stellar objects are so much brighter than the clusters of galaxies in which they might be situated that the surrounding galaxies would be difficult to detect and their spectra more

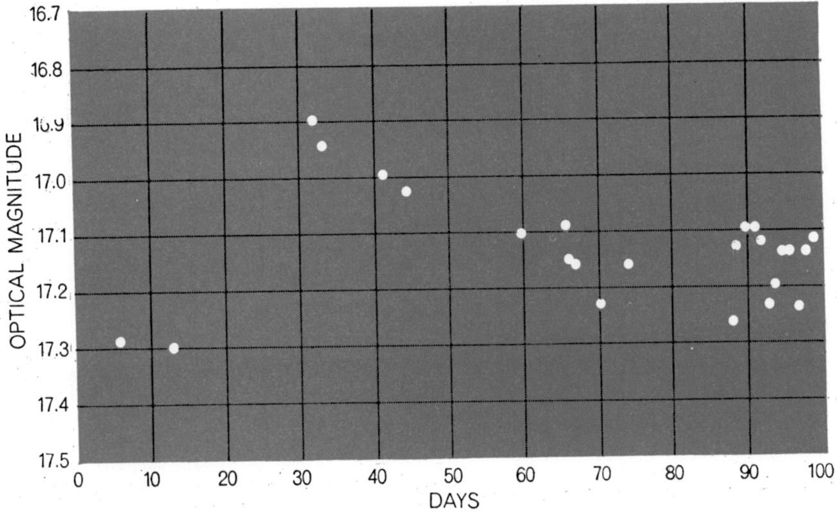

BRIGHTNESS CHANGES IN 3C 345, a quasi-stellar object some 4.5 magnitudes fainter than 3C 273, were recorded over a period of about 100 days in mid-1965 by T. D. Kinman and D. Goldsmith at the Lick Observatory. The maximum variation is about 100 percent.

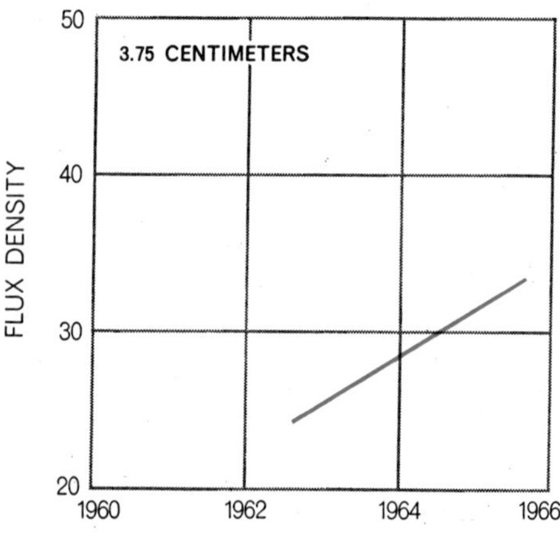

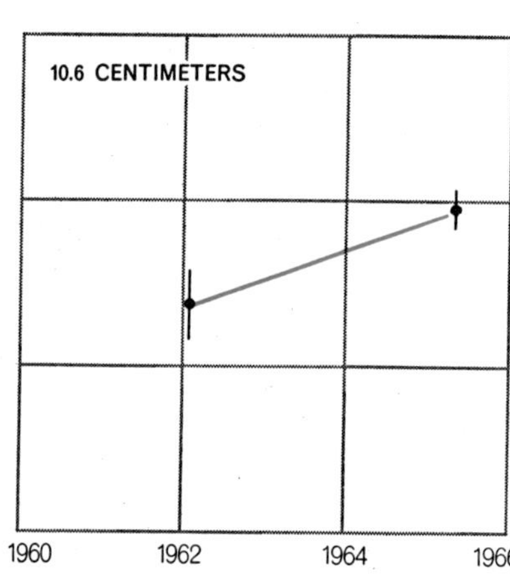

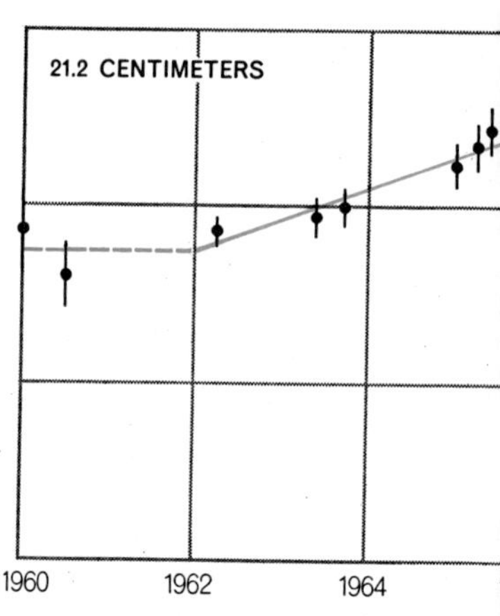

RADIO-FLUX CHANGES IN 3C 273 were compiled from a number of sources by Alan T. Moffet and P. Maltby of the California Institute of Technology. The measurements at 3.75 centimeters were made by William A. Dent of the University of Michigan, who

difficult still to record. In the case of the brighter objects such as 3C 273 and 3C 48, however, the test can be made. It is negative: the objects are not associated with clusters of galaxies. There is optimism that the test can be extended to fainter objects by using sensitive new photographic plates that enable the 200-inch telescope to photograph objects with an apparent magnitude of 24.

We turn now to a discussion of diagrams in which one plots the red shift of quasi-stellar objects against radio magnitude, expressed as the radio flux (*S*) at a particular frequency. We have selected a frequency of 178 megacycles because the most extensive catalogues of radio sources have been compiled at that frequency. One can see from the diagram at the bottom of page 326 that there is little or no correlation between red shift and radio magnitude. This is understandable on the basis of the local hypothesis because it implies no strong correlation between red shift and distance, unless all the quasi-stellar objects were ejected at one particular moment from one galaxy.

How can the diagram be understood if one assumes the cosmological hypothesis? The fact that there is no significant drop in radio flux with increase in red shift can be interpreted to mean that the more distant objects are intrinsically more powerful than the nearer ones. If this is true, it should have important significance for cosmological theories. Let us see if it is a realistic possibility.

For some years now radio astronomers have made counts of sources and presented them in the form of "log *N*–log *S*" curves, where *N* is simply the number of galaxies with a radio flux greater than some value S. As one would expect, there are many more galaxies of weak flux than there are of strong flux. In fact, radio astronomers have found an essentially

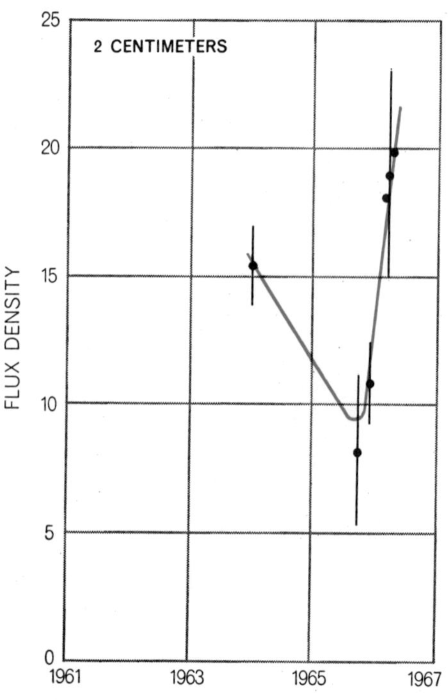

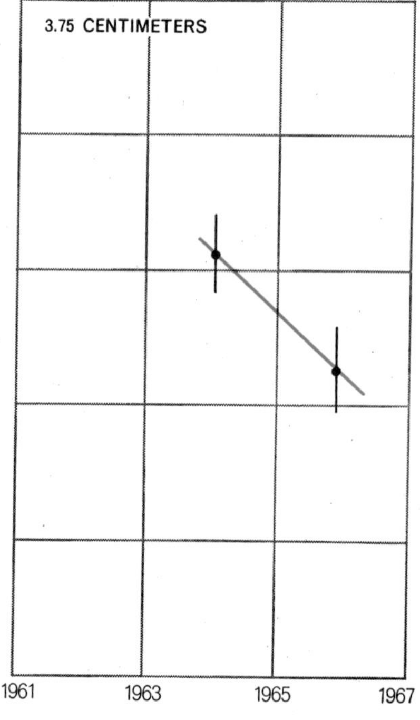

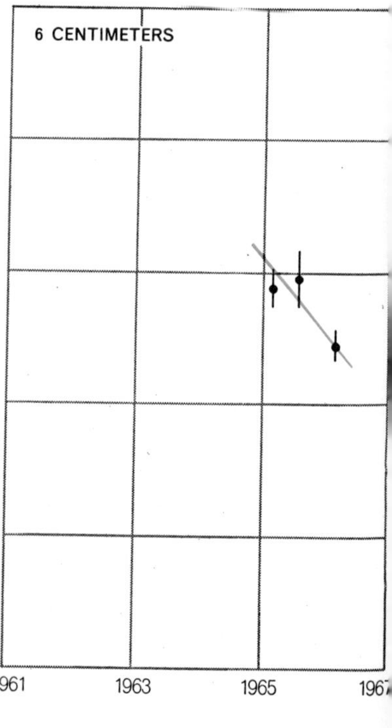

RADIO-FLUX CHANGES IN 3C 279 were recently reported by K. Kellerman and K. Pauliny-Toth of the National Radio Observatory in Green Bank, W.Va., who assembled their own measurements with those made by Dent. The sharpest change took place at a

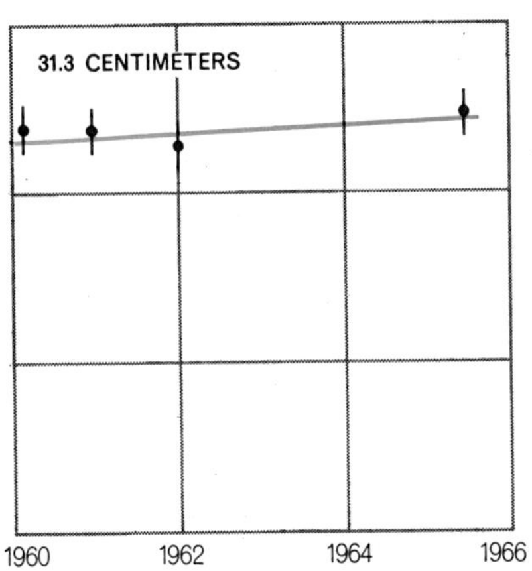

obtained the first conclusive evidence that quasi-stellar objects vary in radio emission.

straight-line dependence between the logarithm of N and the logarithm of S. The precise slope of this line is of great interest because it should help to provide a test between different models of the universe. For example, a slope of -1.5 is the one expected if all sources had the same intrinsic power and were uniformly distributed in a Euclidean universe. The value -1.5 results from the following considerations. The number N of sources brighter than a particular value of S is proportional to the volume of space, that is, to d^3, where d is such that a source at distance d has a flux equal to the assigned S. The flux S, in turn, varies according to the inverse square law; thus S is proportional to d^{-2} (that is, to one over d^2). The ratio of log N to log S, therefore, is simply the ratio of the exponents of d, or the ratio of 3 to -2. This gives the value -1.5.

It is easy to show that the same value, -1.5, is also the one to be expected in a steady-state universe. The slope that is actually observed for all extragalactic radio sources is about -1.8, and although this value is hard to explain it has been used as evidence against the steady-state cosmology.

One of us (Hoyle), who has been closely associated with the steady-state cosmology, has always objected to this argument on the grounds that a counting of a class of objects whose physical nature is unknown cannot be regarded as a satisfactory procedure. Before the discovery of quasi-stellar objects, however, it could be argued that all radio galaxies were more or less alike and thus could be counted as objects of a single class. This position is no longer tenable. One must consider the slope of the log N–log S curve for radio galaxies alone and for quasi-stellar objects alone. When that is done, what is the result?

A critical study of this kind was recently conducted at the Mount Wilson and Palomar Observatories by the French astronomer Philippe Véron. He classified all the sources in the Third Cambridge Catalogue according to the nature of their optically visible counterparts in every case where an optical identification had been made. The revised version contains 328 radio sources, of which 32 probably lie within our own galaxy and another 42 are so obscured by galactic absorption that little of value can be deduced from them. Among the remaining 254, Véron finds 144 radio galaxies (100 certain and 44 probable) and 60 quasi-stellar objects (39 certain and 21 probable), leaving only 50 sources that have not yet been identified. One can safely assume that these 50 objects, since they are only 20 percent of the total, are not likely to upset Véron's conclusions regardless of their ultimate classification. We should like to emphasize, moreover, that Véron's study embraces the only "first class" data that exist at the moment relevant to the source-count problem. By "first class" we mean data that are both reasonably complete and technically good.

Véron's finding is this. The radio galaxies, certain and probable together, essentially fit a slope of -1.5. Since the radio galaxies are not extremely far away (maximum red shift, 46 percent), this is the slope to be expected in any cosmological theory. The steep slope of -1.8 shown by all 3C sources is due in large measure to the quasi-stellar component, which has the very steep slope of -2.2 [*see illustration on page 327*].

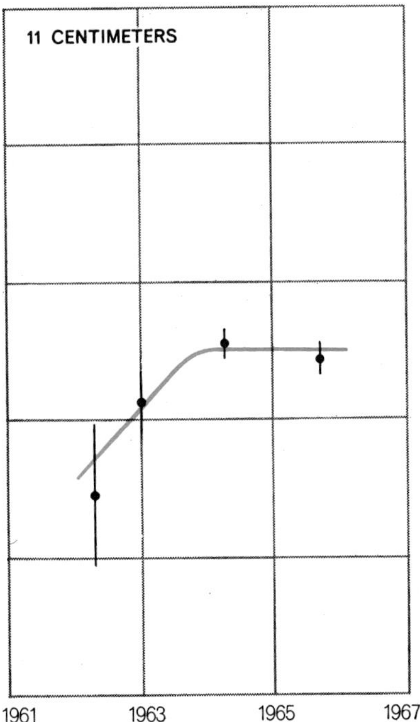

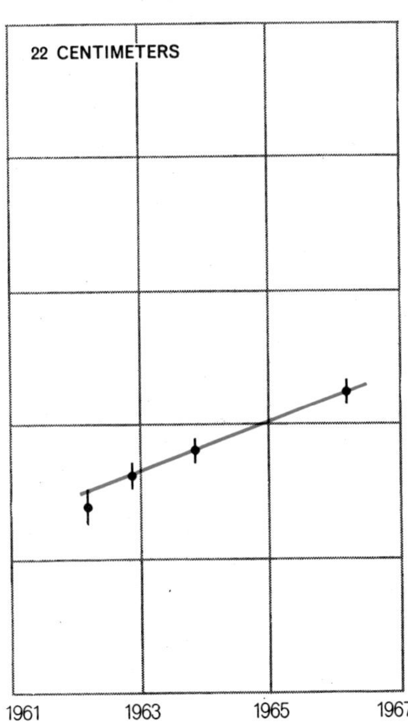

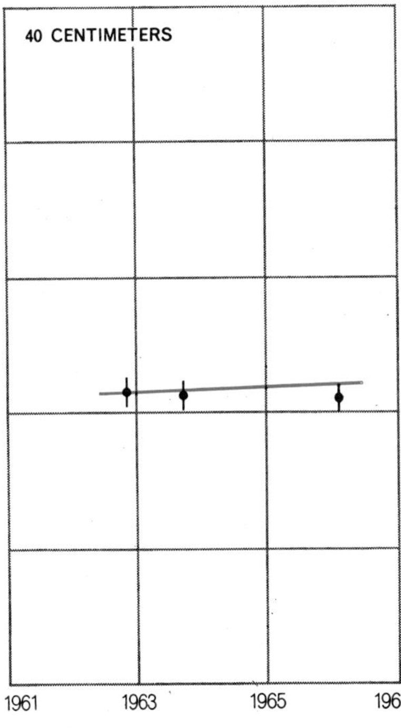

wavelength of two centimeters; the radio flux approximately doubled over a period of a few months between late 1965 and early 1966. In this illustration and in the one at the top the flux density is given in units of 10^{-26} watt per square meter per cycle per second.

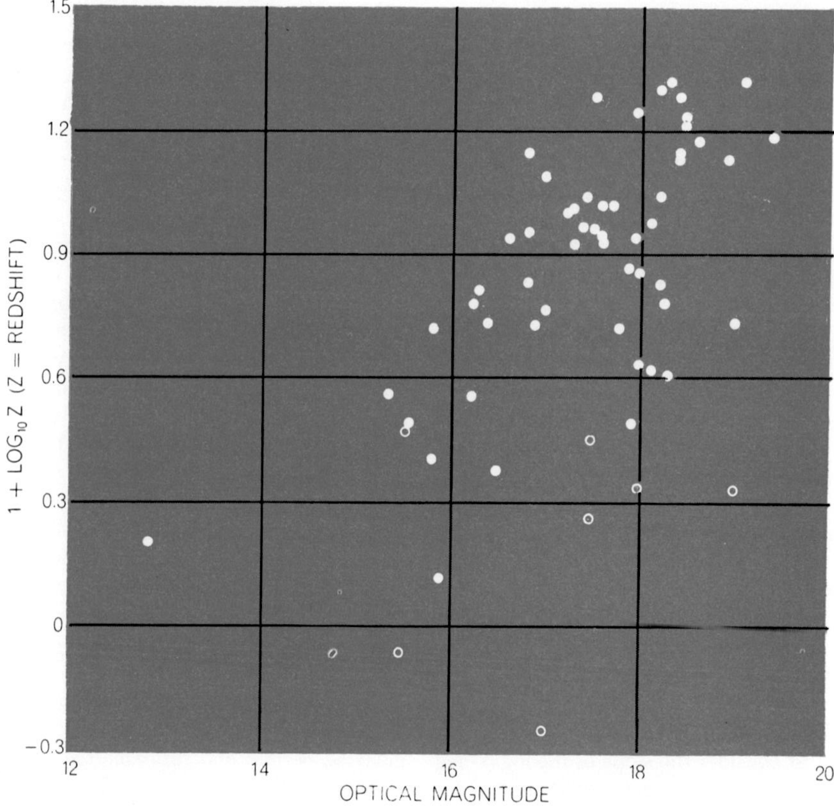

RED SHIFT V. OPTICAL MAGNITUDE is plotted for all quasi-stellar objects (*dots*) and *N*-type galaxies (*open circles*) for which data are available. *N*-type galaxies resemble quasi-stellar objects in spectral features and in having large red shifts, but they are not quite star-like. According to the cosmological hypothesis the greater the red shift of quasi-stellar objects, the more distant they are. Thus if quasi-stellar objects all had about the same intrinsic brightness, one would expect to see a strong correlation between red shift and apparent magnitude. Its absence suggests that the objects differ considerably in intrinsic brightness. An alternative hypothesis suggests that they are not distant objects at all but local ones, in which case the scatter could indicate either variations in brightness or random distribution.

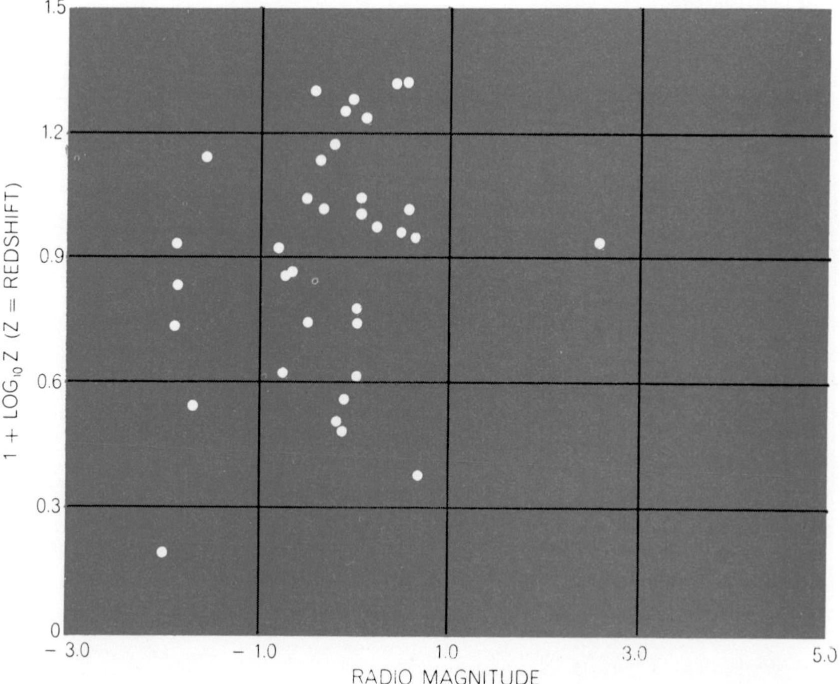

RED SHIFT V. RADIO MAGNITUDE is shown for 35 quasi-stellar objects, most of those for which information exists. The correlation is slim between red shift and magnitude.

It is obvious that if quasi-stellar objects are actually nearby, their slope of −2.2 has no bearing on the cosmological problem.

But—and this is the surprise toward which we have been working—the slope is equally irrelevant to cosmology even if quasi-stellar objects are at cosmological distances. This follows from the scatter diagram at the bottom of this page, which shows that there is at best only a weak correlation between the radio flux of quasi-stellar objects and their distance—assuming now that distance is proportional to the observed red shifts. It follows that the −2.2 slope of the log *N*–log *S* curve for quasi-stellar objects cannot be interpreted as a distance effect. Rather the curve exhibits a "luminosity function" effect, which can be described as follows.

Imagine a group of objects of different intrinsic brightness all placed at exactly the same distance from our galaxy. If there are many more faint objects than there are bright ones, *N* will be significantly larger for a small observed radio flux, *S*, than it is for a large flux, and we shall obtain a curve that runs from lower right to upper left in the log *N*–log *S* diagram. By adjusting the distribution of objects with respect to brightness (that is, the luminosity function) we can indeed obtain any slope we please. The fact that quasi-stellar objects are actually arrayed at various distances (in all likelihood) does not alter the principle involved; it is important only that they have a spread of intrinsic luminosities. In fact, they can have a spread that produces a slope of −2.2 in a log *N*–log *S* curve.

Finally we turn to the theoretical models that have been proposed to explain the quasi-stellar objects. We should first say something, however, about ordinary radio galaxies. By 1956 it was shown that these sources frequently lie in clusters of galaxies whose red shifts are well established; radio galaxies are therefore cosmological objects. The source of their radio emission is assumed to be the synchrotron process. On this basis one can calculate that the minimum total energy released by a large radio galaxy is on the order of 10^{61} ergs. To obtain that much energy by the most efficient conversion process known—the annihilation of matter—would require the disappearance of a mass equivalent to the mass of 10^7 suns for the duration of the outburst. Moreover, there are a number of reasons why it is probable that far larger amounts of energy are actually released.

A variety of processes were suggested to explain energy releases of such magnitude. The earliest suggestion was collisions between galaxies, but this could be ruled out on observational grounds. A later idea was that multiple explosions of supernovas might be responsible. This does not seem to be efficient enough. Then in 1962, before the discovery of quasi-stellar objects, one of us (Hoyle) and W. A. Fowler of the California Institute of Technology considered how the collapse of a large mass might account for the energy released in strong radio sources. It was supposed that the energy released in such a gravitational collapse might be a significant fraction of the rest-mass energy.

With the discovery of 3*C* 273 in 1963 Fowler and Hoyle proposed that such objects, if cosmological, might represent objects with a mass of about 10^8 suns—superstars—whose luminous energy results from standard thermonuclear processes. (The stability of such giant stars and their possible evolution have since been considered by Fowler and others.) It was suggested that the superstar spends about a million years in this highly luminous phase and then collapses, releasing a large amount of gravitational energy that in some fashion is converted into high-energy electrons together with a suitable magnetic field so that the synchrotron mechanism can operate. According to this model a quasi-stellar object is really two objects: a still-burning superstar to provide luminosity and a collapsed superstar to supply the vast flux of particles needed to account for the radio emission. The proposal presents many difficulties, chief among which is understanding how a collapsing object can emit energy amounting to a significant fraction of its rest mass. Since the conventional theory of relativity suggests that this cannot happen, Hoyle and J. V. Narlikar have considered modifications of the field equations that may permit the possibility.

A number of theorists have considered how a superstar of the type needed might be formed. George B. Field, now at the University of California at Berkeley, has proposed that quasi-stellar objects are galaxies in the process of formation and that much of the radiation we observe is gravitational energy released by the contraction of a vast cloud of gas. Along the way stars form, release thermonuclear energy and finally explode as supernovas. The quasi-stellar object, Field believes, is a manifestation of all these processes and at the end the mass will probably condense to form a superstar. It does not appear to us, however, that this model can supply the required energy.

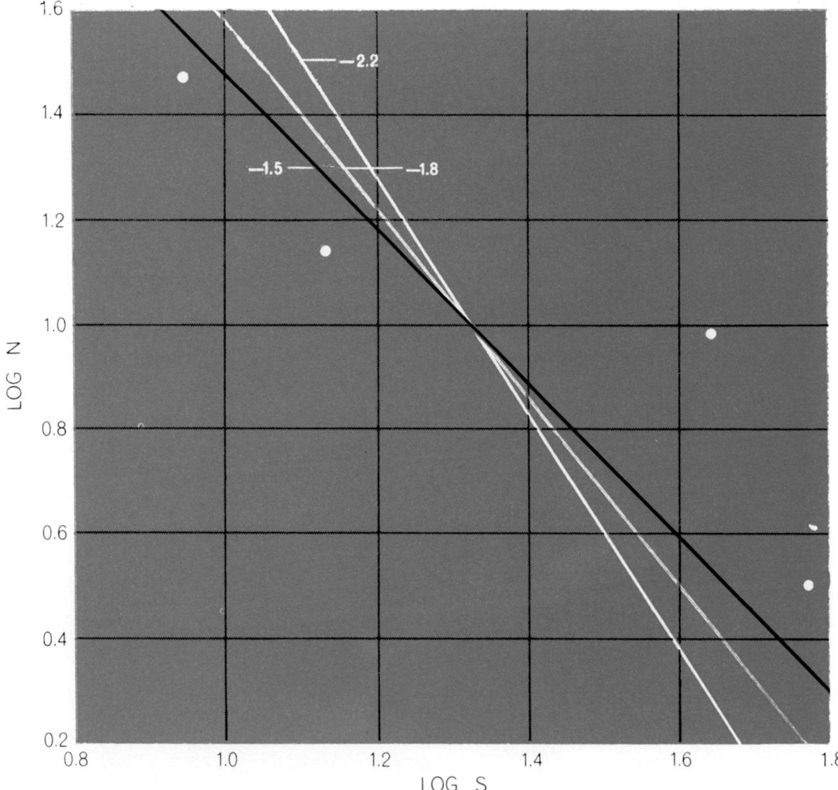

"LOG *N*–LOG *S*" CURVES tend to show a simple straight-line dependence between log *N*, the number of objects above a certain radio magnitude, and log *S*, the selected magnitude in units of radio flux. Thus the lowest point on the quasi-stellar curve (*black*) shows that there are only two objects with a radio magnitude, or flux, above 60 units ($\log_{10} 60 = 1.778$). The point at upper left shows that there are 30 objects with a flux above 9 units ($\log_{10} 9 = .954$). The slope of the log *N*–log *S* curve for these 30 objects is close to −1.5. For all quasi-stellar objects, however, the slope is about −2.2. For all radio sources in the Third Cambridge (3*C*) Catalogue the slope is −1.8. In a simple Euclidean universe the expected slope is −1.5. Evidently the −1.8 value for all radio sources reflects the steeper slope of the quasi-stellar component. The implications of these slopes are discussed in the text.

Another provocative idea is the one proposed by Thomas Gold and W. Ian Axford of Cornell University and Stanislaw M. Ulam of the Los Alamos Scientific Laboratory. They suggest that in some galaxies the density of stars in the nucleus has become so high that the stars begin to collide with increasing frequency. Conceivably the energy released in such collisions could account for cosmological quasi-stellar objects. We do not consider this very likely for a number of reasons; for one, the time during which these events might take place would be quite short. On the other hand, the collisions might lead to an agglomeration process that could produce the kind of superstar we have in mind for our model.

Recently Stirling A. Colgate of the New Mexico Institute of Mining and Technology has proposed a model in which large stars (about 10 solar masses) are formed by the collision of smaller ones. These giant stars evolve and finally explode as supernovas. Colgate believes that such explosions might release much more energy—through gravitational collapse—than is normally observed in a supernova explosion and that the sum would be enough to explain the energy released by quasi-stellar objects placed at cosmological distances. According to Colgate, however, the initial density of small stars that is required in this model is about 500 million per cubic light-year.

In all these theories it is recognized that quasi-stellar objects are both small and very massive. No one really knows if such an object can be formed from low-density matter spread throughout the universe. Thus some people have suggested that quasi-stellar objects are truly cosmological in the sense that their origin is connected with the way in which matter enters the universe. For example, they may be miniature expanding "universes" that have only recently begun to expand. Alternatively, on the steady-state hypothesis, matter may be created in regions of the highest density, such as the nuclei of galaxies; according to this model quasi-stellar objects are blobs of new matter that have been ejected from galactic nuclei. A radio galaxy might be a galaxy that is giving birth to such objects. Models of this kind would serve to explain quasi-stellar objects whether they were at cosmological distances or much closer.

It is still necessary, however, to explain how the energy released is chiefly in the form of particles moving at relativistic speeds. An attractive possibility is to suppose that matter as originally created, whether in a single "big bang" or in a steady-state universe, is largely in the form of the hypothetical massive particles known as quarks. It has been suggested that such particles may form the basic building blocks of the more familiar elementary particles: the baryons that decay into nucleons and the various mesons. If quarks do exist, they are more massive than baryons; three quarks are needed to create a baryon and a quark-antiquark pair is needed to create a meson.

If primordial matter appears in the form of quarks, one can see how immense energy is released when the quarks give rise to lighter particles, including photons and neutrinos. The particles will be endowed at birth with tremendous amounts of kinetic energy, which could account for the relativistic speeds of matter that are observed in quasi-stellar objects.

KITT PEAK 84-INCH REFLECTOR is the one used by Lynds and Stockton in studying the spectra of quasi-stellar objects. Examples of their work appear on pages 317 and 318. Margaret Burbidge records quasi-stellar spectra with the 120-inch reflector at the Lick Observatory. The first quasi-stellar object, 3*C* 38, was discovered in 1960 by Allan R. Sandage with the 200-inch Hale telescope on Palomar Mountain. With the same telescope in 1963, Maarten Schmidt identified the first red shift in a quasi-stellar object, 3*C* 273.

What if the quasi-stellar objects are local rather than cosmological phenomena? In that case the energy they release can be less by a factor of about 10^4, a condition much easier to satisfy. Now, however, a plausible model must offer a mechanism for setting the objects in motion at the velocities indicated by the red shifts—and this requires considerable energy. Let us imagine that galaxies explode from time to time and in the process are transformed into radio galaxies. Conceivably the explosion could throw a cloud of objects out into space, and it is these we see as quasi-stellar objects. Such a process might require on the order of 10^{62} ergs per explosion; moreover, the energy must be emitted in coherent lumps so that each object is provided with a dense enough core to hold itself together when ejected at relativistic speeds. We do not know precisely how this can occur, but it seems that in the process of gravitational collapse very large amounts of energy must be released in relativistic form, and this is required in any case to explain the radio galaxies.

Some have objected to this admittedly sketchy model on the grounds that when they calculate the energy required to explain the cloud of objects moving at relativistic speeds, it is too high by some criterion (of unspecified nature) concerning what is reasonable. We reply that any theoretical arguments marshaled for or against one or another quasi-stellar hypothesis on the basis of energy considerations are entirely fallible. In the absence of any real understanding of the mechanism that gives rise to the colossal energy output of radio galaxies, which are known to be at cosmological distances, we are in a poor position to say what is reasonable or even possible.

There can be no doubt that at present most astronomers would like to conclude that the quasi-stellar objects are cosmologically distant. As we have tried to emphasize, the question is still open because there is no conclusive argument one way or the other. In fact, we may be in for many more surprises on the observational side. Regardless of the outcome, the problem of understanding the quasi-stellar objects and the immense fluxes of energy they and the radio galaxies release is one of the most important and fascinating tasks in all of physics.

31

The Primeval Fireball

P. J. E. PEEBLES AND DAVID T. WILKINSON

June 1967

Modern cosmology undertakes to substitute observational science for myth and speculation in dealing with such issues as: How did the universe originate? What is it like now? What will be its fate? Unfortunately the observational evidence is meager. There is a wealth of data but one becomes lost in detail; there is need for observations of simple and large-scale phenomena, the essential bases of theory. As a matter of fact, most contemporary cosmologies stem from just one such observation: Edwin P. Hubble's discovery that other galaxies are moving away from ours, and are doing so at speeds that are greater the more distant the galaxy. This general recession is the basis for such widely different concepts as the "big bang" cosmology (which holds that the universe originated in a superdense state some seven billion years ago) and the "steady state" one (in which the universe looks exactly the same through all time—past, present and future).

It now appears that radio astronomers have discovered another basic cosmological phenomenon that, like the recession of the galaxies, provides a view of the universe on a truly universal scale. It is low-energy cosmic radio radiation that apparently fills the universe and bathes the earth from all directions. Intense enough to be received by conventional radio telescopes, it has undoubtedly been detected, but not recognized, for years; indeed, it accounts for some of the "snow" seen on a television screen. When it was discovered by Arno A. Penzias and Robert W. Wilson of the Bell Telephone Laboratories about two years ago, they realized that it could not have originated in the earth's atmosphere or in our galaxy. It did fit in well, however, with an earlier suggestion by Robert H. Dicke of Princeton University that one ought to be able to detect a new kind of cosmic radio radiation: a "primeval fireball" of radiation surviving from the earliest days of the universe, when the universe was enormously hot and contracted. The theory and observation of this primeval fireball has been the subject of considerable work and excitement for us and several colleagues at Princeton: Dicke, P. G. Roll and R. B. Partridge.

The discovery and identification of this radiation must be considered a revolutionary development in cosmology. If, as we now believe, it is indeed the primeval fireball, it provides a view of the very early universe, just as optical radiation provides a look at the universe of more recent times. Our colleague John A. Wheeler has suggested an analogy: Compare man's observations of the evolving universe with the view downward from the observation platform of the Empire State Building. Street level corresponds to the beginning of the expansion of the universe. The most distant galaxy discovered so far then corresponds to a view down to the 60th floor, and the most distant quasi-stellar sources are at about the 20th floor. The fireball radiation is equivalent to a glimpse of something just half an inch above the street! With this expanded view of early events in the universe one may hope for a corresponding improvement in cosmological theory.

The concept of the primeval fireball is grounded on Hubble's observation of the general recession and the idea that flows from it: that the universe is in a state of rapid expansion. If this is so, according to big-bang theories at some time in the distant past—about seven billion years ago—all the matter in the universe must have been packed together in an inferno of particles and radiation. As the universe expanded out of this holocaust the matter cooled and condensed to form galaxies and stars. The radiation, which had started out as enormously energetic gamma rays, was also "cooled" by the expansion; its wavelength increased and it now appears mostly in the radio and microwave bands. The idea of a "fireball" dating from the big bang can be somewhat misleading, because what we have in mind is not radiation from some localized explosion off in one corner of the universe. The earth is immersed in this fireball; the radiation comes at us from every direction, and any observer anywhere in the universe should detect it as coming equally from all directions.

This is consistent with the basic theoretical framework developed between 40 and 50 years ago by Albert Einstein, Willem de Sitter, Alexander Friedmann, Georges Lemaître and others. Basic to the work of all of them was the picture of an evolving universe that looks the same to all observers, no matter where they are. In particular such a universe has no boundary, no edge. It is also isotropic, which is to say that it looks much the same in any direction. The presence of matter causes a uniform curvature of space.

A good two-dimensional analogy to this uniformly curved three-dimensional space is the surface of a balloon. The galaxies are inelastic polka dots pasted on the surface. Since the universe is expanding, imagine that the balloon is being inflated. As the balloon expands, a bug standing on any dot would see all the spots around it moving away, and it would see the more distant dots moving away more rapidly. The model thus reproduces the general recession of galaxies and even Hubble's law: that the speed of recession is proportional to the distance of the galaxy. It also points up the fact that the universe has no preferred center. Although the bug sees all

INSTRUMENT with which the primeval fireball is observed at Princeton University is a recent version of the Dicke radiometer, seen here from above. The antenna horns extend to the left and right and are directed upward to collect sky radiation; a switch, microwave receiver and amplifier are at the center. The instrument is operated both in this configuration and as illustrated below.

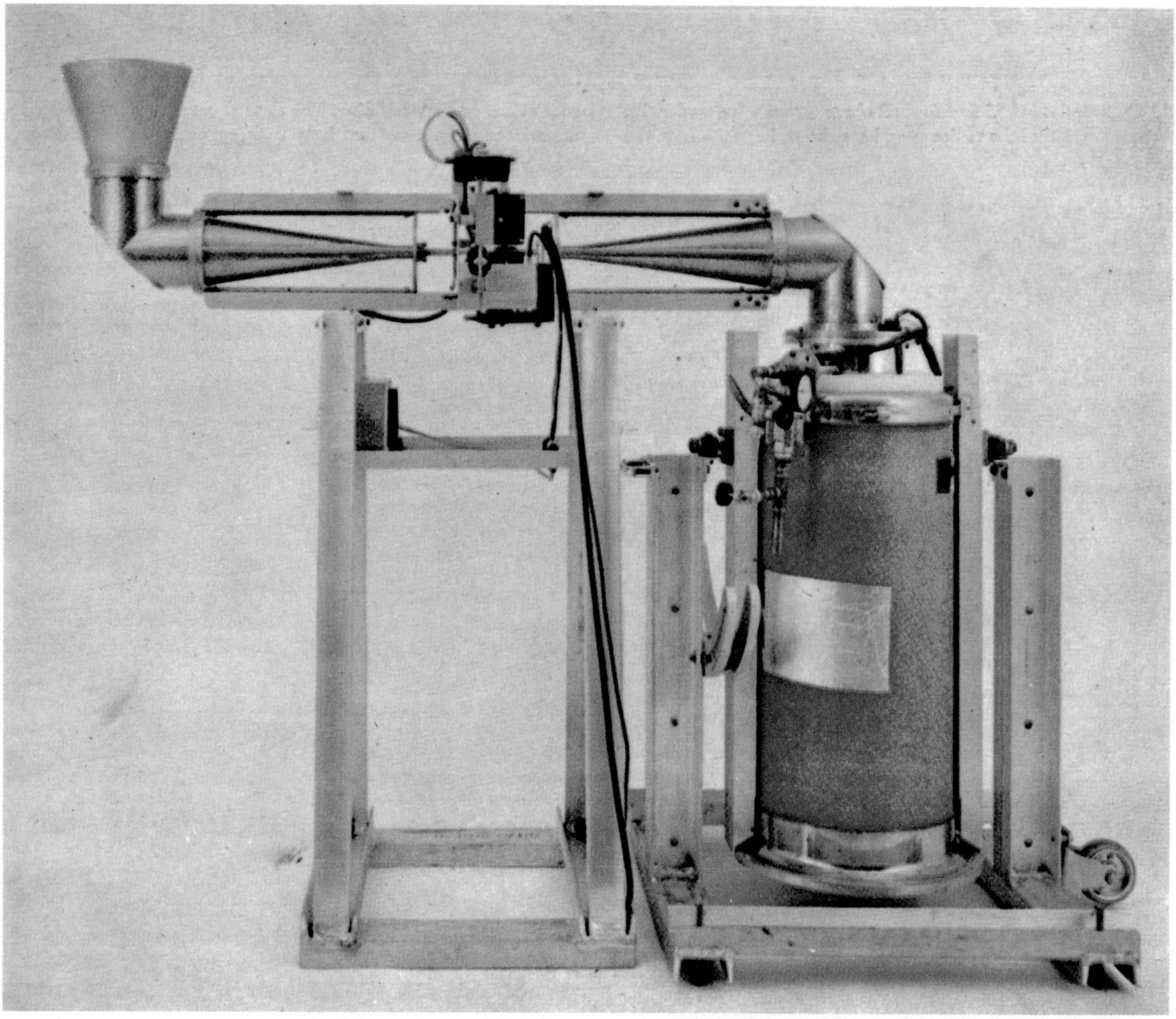

RADIOMETER is seen in a side view with one of the horns in position to receive radiation from the sky. The other horn of the radiometer is coupled to a wave guide leading to a reference source inside the orange Dewar flask. The source is immersed in boiling liquid helium and is therefore known to be radiating at 4.2 degrees Kelvin (degrees centigrade above absolute zero). The receiver input, is switched back and forth between sky antenna and reference source and the intensities of the two are compared.

the spots moving away from it, the bug should not conclude that it is on a preferred spot; another bug on another spot sees the same thing. Similarly, the general recession of the galaxies does not mean that the earth is at the center of the universe; an observer in any other galaxy would see the same general movement away from him.

On this model the primeval fireball radiation might be represented by a number of ants crawling over the surface of the balloon. They are uniformly distributed, and they crawl about in all directions. The number of ants in any given area of the surface decreases as the balloon is blown up. In the same way the density of photons in the primeval fireball decreases as the universe expands. Note also that no matter which way the ants move they will always move toward polka dots that are receding from them, and they must continuously lose energy as a result of this chase. In the real universe the photons of the fireball are always chasing galaxies that are receding from them, so that the photons undergo a continuous energy loss that accounts for the increase in their wavelength.

Based on this picture of the expanding universe it was possible to make two predictions about the nature of primeval fireball radiation. The first was that because it was emitted by a source (the condensed universe) in thermal equilibrium, its intensity should vary with wavelength in the manner characteristic of an ideal thermal radiator, or "black body." A severe test of whether the newly discovered radiation was indeed the primeval fireball would therefore consist in tracing out the observed intensity as a function of wavelength and seeing if the measurements fell on the black-body curve. The second major prediction was simply that the fireball radiation should be isotropic; that is, since the radiation presumably fills the universe and the earth is immersed in it, the observed intensity of the radiation should be the same in every direction.

There is a "window" through which one can observe the fireball radiation: the range of wavelengths from about one to about 20 centimeters. (At longer wavelengths radiation from our own galaxy is so strong that it submerges extragalactic signals; at less than one centimeter the earth's atmosphere radiates too strongly.) Radio astronomers have been observing through this window for many years but they overlooked the fireball because the methods that ordinarily enable one to separate signals of interest from background noise do not work in the case of the fireball radiation. For example, one can detect a weak signal when it is concentrated in a characteristic line in the electromagnetic spectrum. This is the case for the 21-centimeter emission of atomic hydrogen in interstellar clouds. Unfortunately the fireball radiation would have a smooth spectrum, much like that of terrestrial background noise, so that it would be hard to isolate in this way. One can also isolate extremely weak signals by scanning the antenna beam across a suspected localized source. The fireball was expected not to be localized, however, but to be spread across the entire sky as a uniform "glow."

It was clear that the search for a primeval fireball called for a new and different kind of radio telescope, and in the fall of 1964 our group undertook to build such an instrument. The heart of our telescope is a modified microwave receiver known as a Dicke radiometer. Designed by Dicke in 1945, this instrument bypasses the noise generated within the receiver itself, which is about 1,000 times more intense than a weak signal such as the fireball. Dicke overcame the receiver noise problem by putting a switch between the antenna and the receiver that periodically—say 100 times a second—shifts the receiver input from the antenna to a reference source and back again [*see illustration on page 333*]. The receiver output therefore contains a 100-cycle-per-second signal whose strength depends on the difference between the radiation power collected by the antenna and the power emitted by the reference. Since the power of the reference source is known, the strength of the 100-cycle signal becomes a measure of the antenna power. This signal is still buried in receiver noise but is easily separated and measured by an amplifier sharply tuned to 100 cycles per second. In this way one can easily measure antenna power thousands of times weaker than the receiver noise.

Two more sources of terrestrial noise had to be overcome. One was thermal radiation from the ground, which fills half the space around an antenna and tends to leak into the usual parabolic antenna. This problem can be largely avoided by using a horn-shaped antenna,

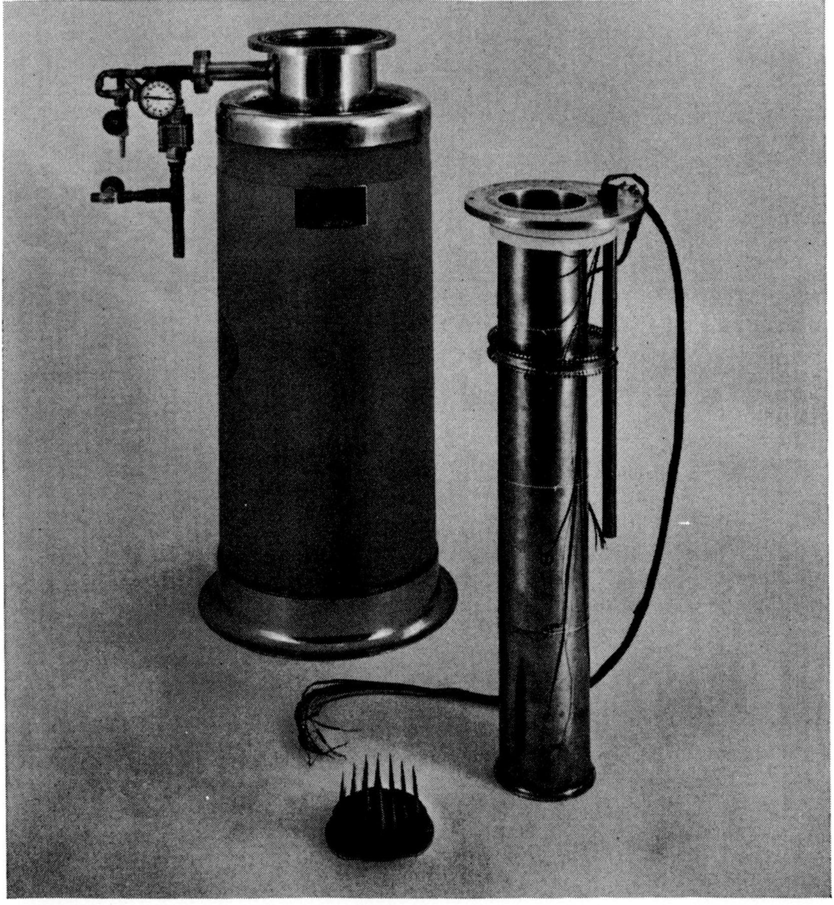

REFERENCE SOURCE (*foreground*) for the Princeton radiometer is made of metal-coated fiber-glass spikes. It has been removed from the Dewar flask and the pipelike wave guide that is normally coupled to the antenna horn. Wires on wave guide lead to thermocouples.

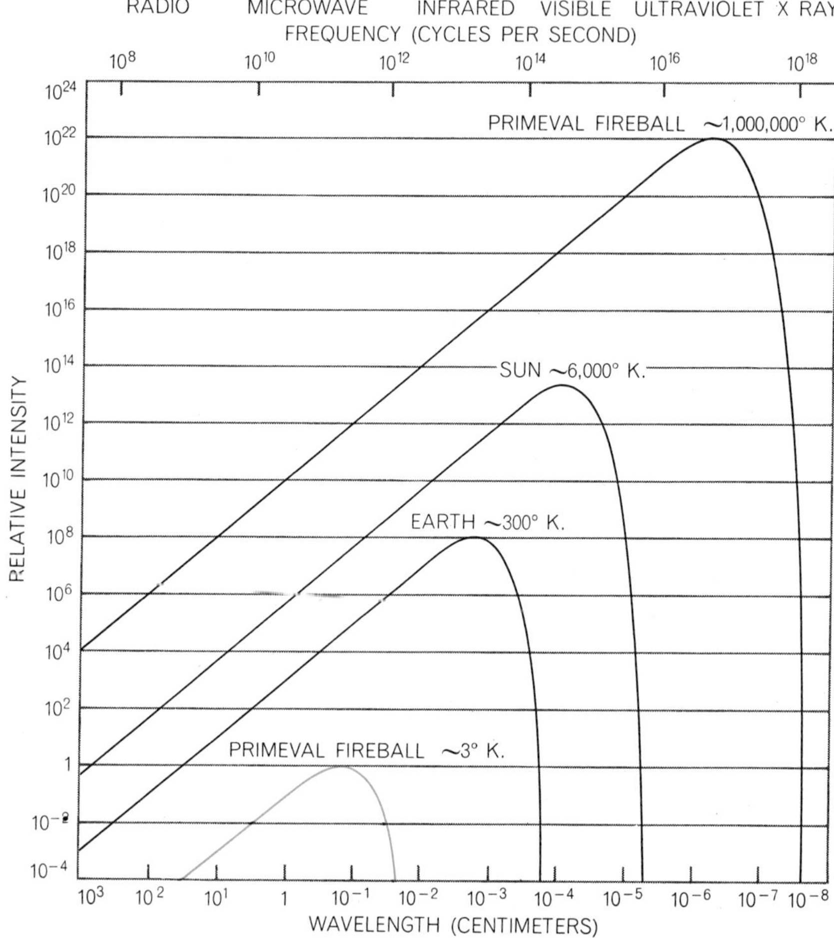

"BLACK BODY" SOURCES of thermal radiation emit across a broad spectrum, the intensity of the radiation varying with wavelength as shown here for several sources. The shape of the curve persists as its position changes according to the temperature of the source. The top curve is for the fireball radiation billions of years ago, the bottom one for the radiation now.

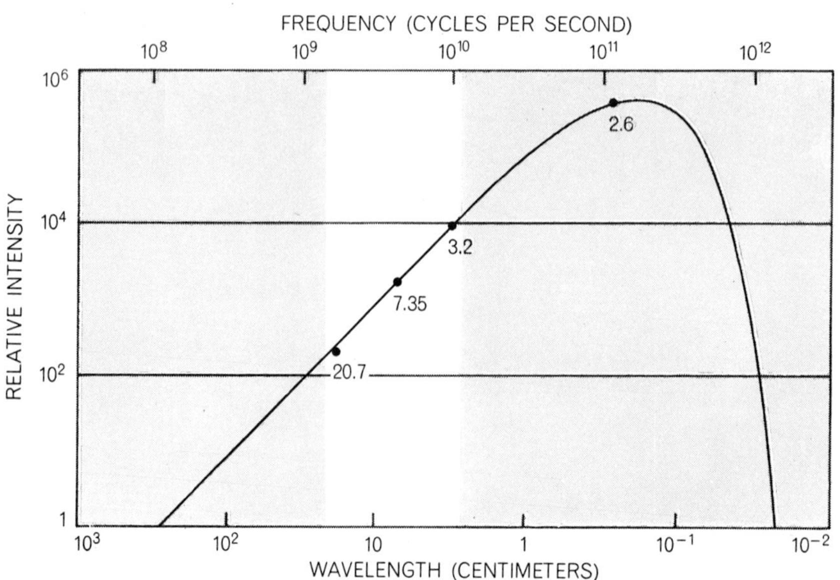

FIREBALL'S INTENSITY has been measured at four wavelengths and conforms to a three-degree black-body curve. Radiometer observations are hampered on either side of a central "window" (*white area*) by galactic (*left*) and atmospheric (*right*) radiation. Measurements at 2.6 millimeters can be obtained by observing light absorption of molecules in space.

which is less sensitive to ground radiation. The other source of terrestrial noise is radiation from oxygen and water molecules in the atmosphere. This emission can be measured and subtracted out if one tips the antenna beam to various angles from the vertical, thus increasing the length of the path through the atmosphere and therefore changing the atmospheric radiation component in a predictable way.

As it happened, the Bell Telephone Laboratories facility at Holmdel, N.J., already had a horn-shaped antenna, originally designed to receive signals reflected by the Echo satellites. Penzias and Wilson had modified the receiver for radio astronomy. Their instrument had all the properties necessary to uncover a primeval fireball, and it now developed that Penzias and Wilson had been attempting for some time to track down excess radio noise thought to be originating within the instrument itself. When Penzias heard what we were doing at Princeton, he invited us to visit Holmdel. What we saw there left us in little doubt that the Holmdel workers' excess noise was in fact extraterrestrial radiation—and was probably the fireball.

As we have already mentioned, the most crucial test of whether or not this new radiation is the primeval fireball is to trace out the spectrum and see if it is that of a black body. The Holmdel result constituted a first measurement of the possible fireball, at a wavelength of 7.35 centimeters. Fortunately the Princeton instrument had been designed to detect radiation of a wavelength different from the one received at Holmdel. We continued our work and about six months later measured a cosmic radiation intensity at 3.2 centimeters that fit in perfectly with the concept of a primeval fireball. Since that time still more measurements have been made at other wavelengths, including a radiometer measurement at 20.7 centimeters by T. F. Howell and J. R. Shakeshaft at the University of Cambridge. All the points so far fall on a typical black-body curve, one appropriate for a source with a temperature of three degrees Kelvin (degrees centigrade above absolute zero), and so the evidence is strong that we are indeed observing the primeval fireball [*see bottom illustration at left*].

The nature of the radio window, however, is such that direct observation at short wavelengths—in the most interesting region where the black-body curve rises to a peak and then falls off steeply—is almost impossible. At such wavelengths one encounters technical problems in building a sensitive radiometer, and atmospheric emission becomes too strong for ground-based observations. These limitations have now been bypassed by an ingenious scheme for measuring the radiation temperature by reading a "molecular thermometer" in interstellar space.

The method depends on the fact that molecules of the carbon-nitrogen compound cyanogen (CN) in interstellar gas clouds are being bathed, along with everything else in the universe, in the black-body radiation of the primeval fireball. It happens that the cyanogen molecule is excited from its ground, or lowest-energy, state into its first excited state by radiation at a wavelength of 2.6 millimeters—a rather long wavelength for such a transition. A certain fraction of the molecules of cyanogen in a cloud exposed to 2.6-millimeter radiation will therefore be in the excited state rather than the ground state, and the size of the fraction is a measure of the intensity of the radiation. The fraction can be measured because the absorption of light by cyanogen molecules accounts for absorption lines in the spectra of certain stars, and light absorbed by molecules in the ground state has a slightly different wavelength from that absorbed by molecules in the excited state. A cloud of partially excited cyanogen molecules therefore causes two or more absorption lines to appear in the spectrum. The relative strength of the absorption lines characteristic of various states therefore gives the proportion of the molecules that are in each state. As long ago as 1941 Andrew McKellar of the Dominion Observatory in Canada used this method to calculate the degree of excitation of cyanogen molecules absorbing light from the star Zeta Ophiuchi. He reported that the molecules were excited as if by radiation with a temperature of 2.3 degrees K. The connection between this finding and a possible primeval fireball was not recognized; the molecules were assumed to be excited by collisions with other particles.

When the fireball hypothesis became generally known, George B. Field of the University of California at Berkeley and Neville J. Woolf of the University of Texas independently pointed out that interstellar cyanogen could be used as a probe to test for fireball radiation in space—and that McKellar's excitation temperature of 2.3 degrees was remarkably close to the three-degree temperature obtained from direct measurements. Field and John L. Hitchcock then reported a new value for the cyanogen excitation temperature. Working with spectra for Zeta Ophiuchi and Zeta Persei made by George H. Herbig of the Lick Observatory, they calculated a temperature range of from 2.7 to 3.4 degrees. The fact that two clouds in very different parts of the sky showed about the same excitation temperature provided an important check on the universality of the excitation mechanism—a necessary feature of fireball excitation.

At the same time Patrick Thaddeus and John F. Clauser of the Institute for Space Studies made some new measurements of Zeta Ophiuchi and obtained a result of 3.75 degrees. Clauser then developed a technique for summing, in digital form, the faint spectra on large numbers of star plates from the Mount Wilson Observatory. The technique has been used to examine the cyanogen absorption lines in the spectra from eight widely distributed stars. In every case the excitation temperature is about three degrees.

The cyanogen measurements are important results. Not only do they pin down the crest of the black-body curve but also they help to eliminate any cause of cyanogen excitation other than fireball radiation. The frequency and energy of particle collisions, for example, would be expected to vary from cloud to cloud depending on local conditions. The results for eight different clouds argue against such local excitation. Note, further, that we have here a very strong test for the fireball hypothesis: If just one cloud is found with a strong ground-state

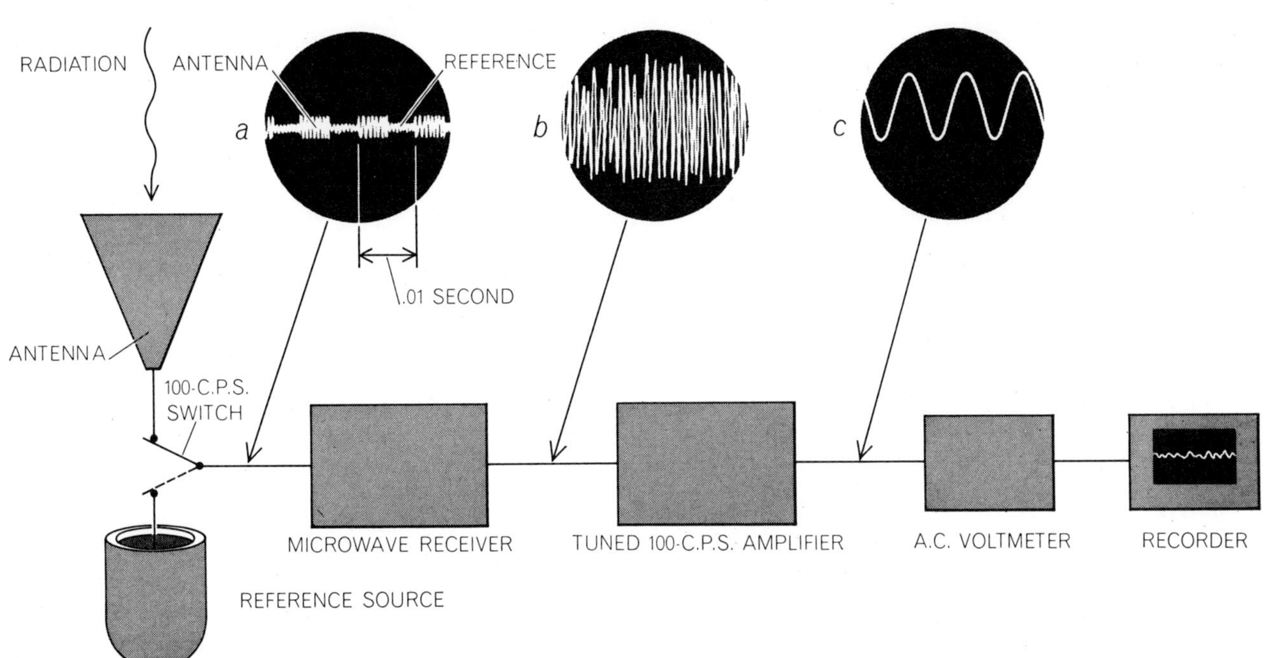

DICKE RADIOMETER can detect signals far below the level of receiver-generated noise. A switch shifts the receiver input from antenna to reference source and back at, say, 100 cycles per second, producing a signal whose amplitude varies at 100 cycles according to the level of the antenna and the reference-source power (*a*). This small signal is obliterated by receiver noise, in which the 100-cycle signal becomes buried (*b*). The desired signal is recovered by filtering out the unwanted frequencies and amplifying the 100-cycle component. The resulting signal (*c*) is fed to a voltmeter that drives a recorder. The displacement of the recorder trace is proportional to the difference between the radiation power being collected by the antenna and the power emitted by the reference source.

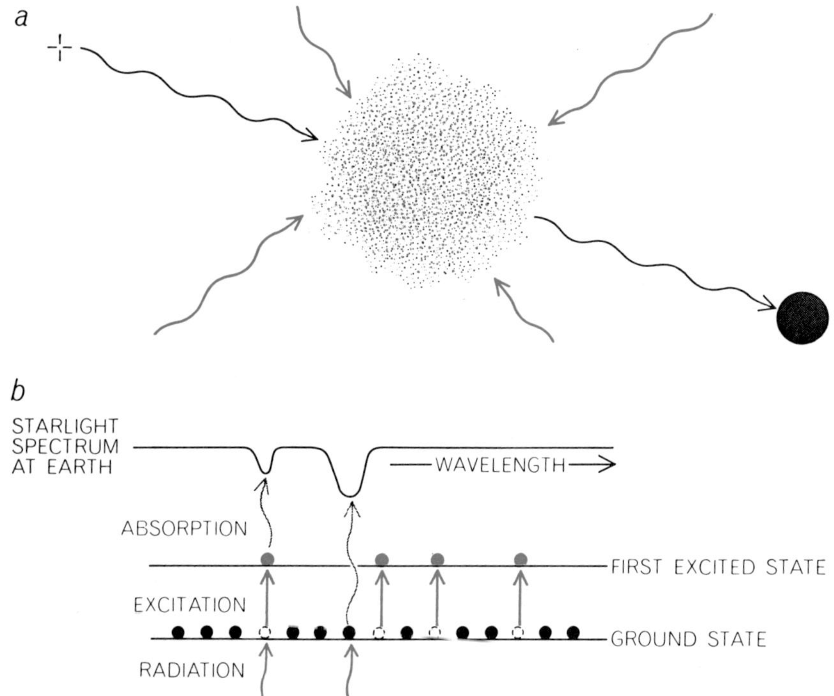

CYANOGEN MOLECULES in space are bathed in fireball radiation (*colored arrows*) and absorb optical radiation (*black arrows*) on its way to the earth from a star (*a*). The wavelength absorbed depends on whether a molecule is in the "ground" state or in an excited state to which it is raised by fireball radiation (*b*). The amount of absorption at each wavelength depends on the number of molecules in each state. The starlight spectrum indicates the fraction of molecules in each state of excitation and thus the intensity of the fireball.

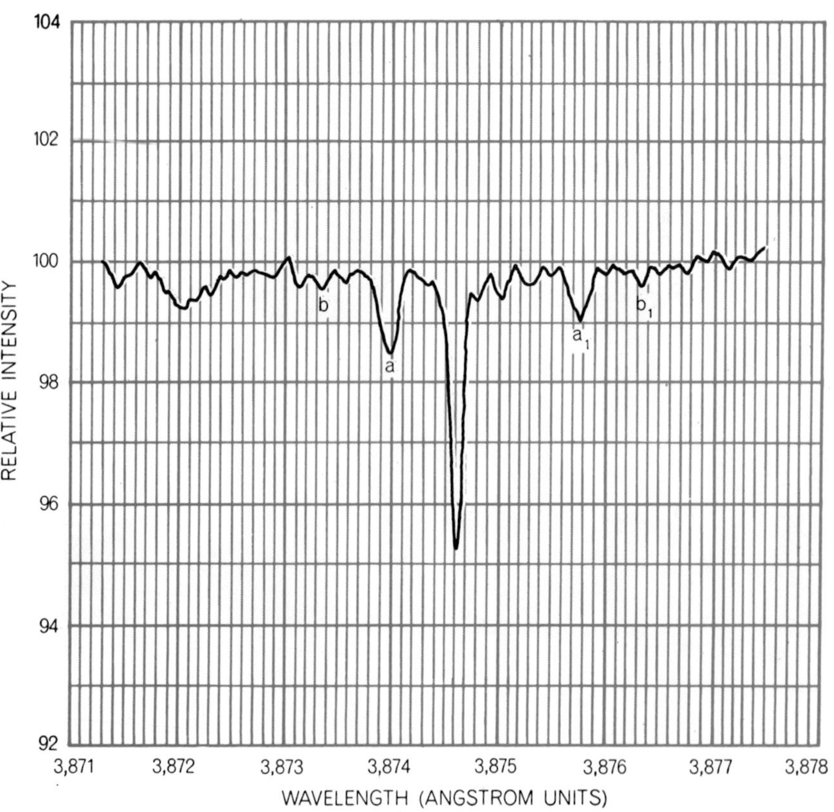

SPECTRUM of light from the star Zeta Ophiuchi was made by John F. Clauser of the Institute for Space Studies by summing the densitometer traces of a number of star plates. The deep trough marks the absorption line characteristic of the cyanogen ground state. Two dips (a, a_1) mark the two lines of the first excited state. Dips b and b_1, characteristic of the second excited state, may, if better resolved, provide a measurement at 1.3 millimeters.

absorption line and no excited-state line, we shall have to conclude that there is no cyanogen excitation in that cloud and therefore that the primeval fireball does not exist.

If one assumes that the universe in fact is isotropic, and if this newly discovered radiation in fact is the primeval fireball, the radiation should be isotropic. (Even the first assumption—that the universe is isotropic—should not be regarded as a self-evident principle. It is comforting to state assumptions as principles, but one must recognize the kinship between an assumption of isotropy and the old assumption that the earth is at the center of the universe. Both assumptions fit the poor observational data available at the time—and also the philosophical tenets of the day.) For the first time we now have a precise observational "handle" on the shape of the universe, and one of our current experiments at Princeton is aimed at making use of that handle.

We point the horn of our 3.2-centimeter radiometer toward the south at an angle of 45 degrees from the zenith so that it is directed approximately parallel to the plane of the earth's equator [*see top illustration on opposite page*]. As the earth rotates, the radiometer scans around this plane once a day. We cannot simply look at the record for daily variations and attribute them to some anisotropy in the primeval radiation; there are inevitably large daily effects due to solar heating, atmospheric changes and other phenomena. We correct for these variations by deflecting the antenna beam in the direction of the pole star every 15 minutes. Since that is a fixed point in the sky, it serves as a reference to which we can compare the reading along the equatorial plane. Keeping the apparatus running for many months further reduces the daily variations. Since any irregularity in the radiation would be fixed in relation to the stars, and therefore would traverse the antenna beam at different times of the day during different seasons of the year, we partly average out effects that have a period of one solar day. After about a year the experiment shows no differences between equatorial and polar radiation intensities greater than about .015 degree, which is to say it reveals no anisotropy greater than about ±.5 percent [*see bottom illustration on opposite page*].

Whatever the final explanation for what we now believe to be fireball radiation, it must account for this remarkable isotropy. The source cannot be our own galaxy; the solar system is off to one side of the galaxy, and the radiation would have to be more intense in the direction of the main body of the galaxy. If the source were in the solar system itself, there would be recurring variations each solar day, but there are no such effects. It seems clear that the radiation must be extragalactic.

If the earth is moving in relation to the local frame of reference defined by the average motion of the primeval fireball radiation, the radiation should seem a little hotter when we observe in the direction of the earth's motion and a little colder when we look "backward." One cannot be sure what the total velocity of the earth should be in relation to this standard, but we do know the earth is moving around the center of our galaxy at 200 kilometers per second. If we suppose the center of the galaxy is at rest in this frame of reference, the radiation would appear to be .07 percent hotter (or more intense) than average in the direction of the earth's motion (toward the constellation Cygnus) and .07 percent colder than average 180 degrees away. Since our instrument scans in the plane of the Equator, however, we should not observe this full effect but rather a variation of about .04 percent from the mean [*see bottom illustration on page 337*]. This is about half of the upper limit (roughly .1 percent) that we have been able to set so far for an anisotropy that has a period of 24 hours. (The radiation would appear hottest and coldest at 12-hour intervals.) We are now trying to improve the observations to a point at which we can actually see this effect of the earth's "absolute" motion through space.

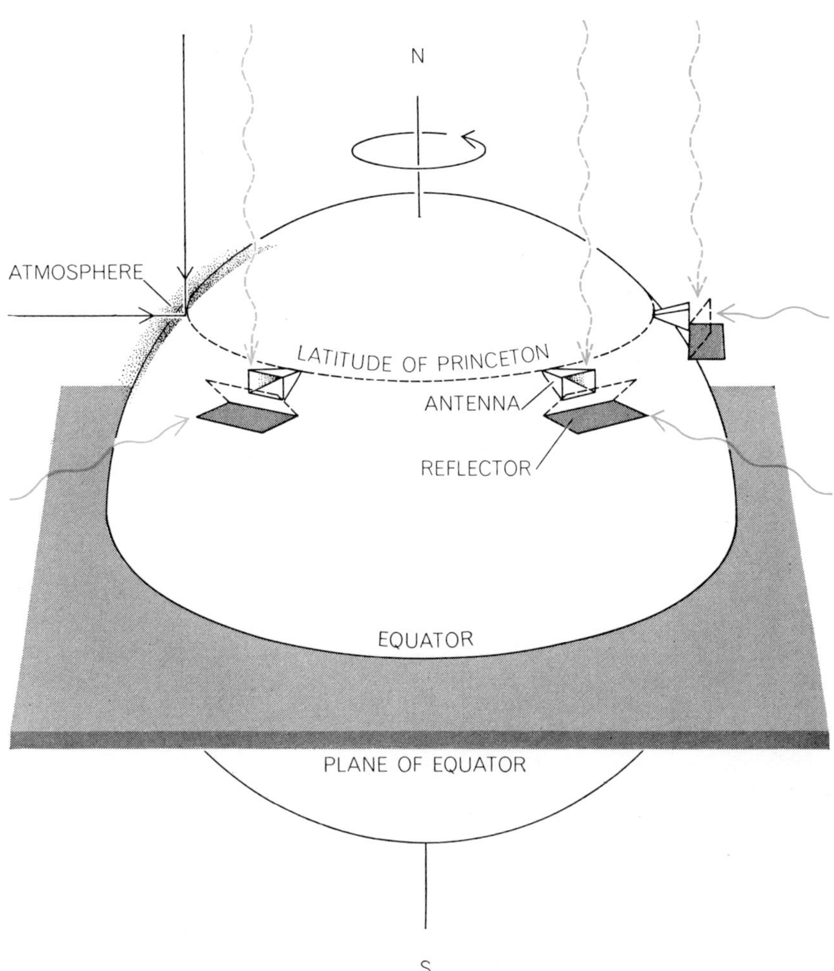

ISOTROPY is measured by pointing the horn antenna in the direction of the plane of the earth's equator and periodically raising a reflector that deflects radiation from the direction of the pole star into the horn. As the earth turns the radiometer measures the difference between the equatorial and the polar (or constant) radiation intensity. The length of the path through the atmosphere is the same in both directions (*left*), eliminating one source of error.

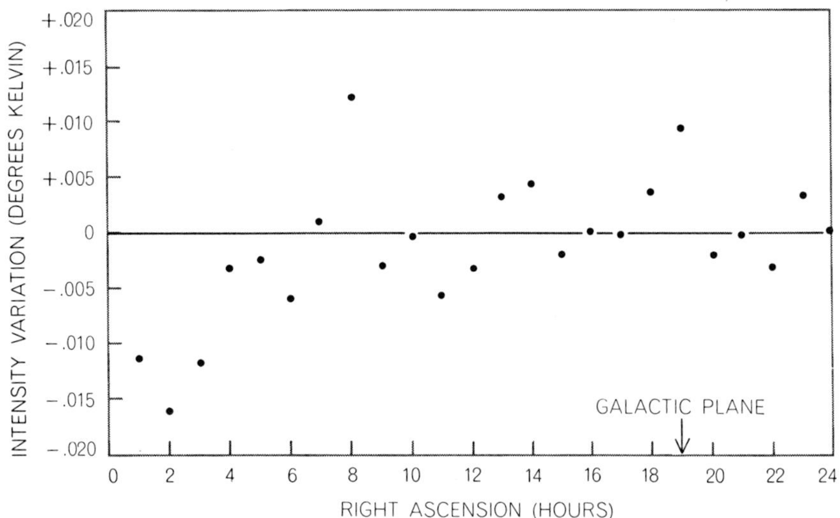

RESULTS of isotropy experiment, after about a year, indicate no anisotropy greater than about ±.5 percent. Each point represents the average difference between the equatorial radiation temperature in a given direction and the constant polar radiation temperature; the probable error in each point is about ±.003 degree. The scatter appears to be primarily random, although the dip at two hours may be real; more measurements are required to be sure.

Unequivocal proof of the phenomenon of a primeval fireball would seem to rule out a number of competing cosmologies. The steady-state theory would be ruled out because its universe was never in a dense state and therefore could not have manufactured black-body radiation. The fireball also creates severe difficulties for any cosmology that includes a visible edge to the matter-filled part of space, for example the cosmology of Oskar Klein and Hannes Alfvén [see "Antimatter and Cosmology," by Hannes Alfvén; SCI. AMER. Offprint 311]. If there is a visible boundary, then any radiation produced in the early days of the universe must long since have left the universe. Proponents of a visible-edge cosmology must therefore find a contemporary source for the radiation we attribute to a fireball. Unless the earth is right at the center of the universe—something most people would be reluc-

PRINCETON GROUP'S first fireball observations were made with an earlier version of the radiometer, here shown in position on the roof of the geology building. The slanted panels around the horn are wire-mesh screens that help to keep out ground radiation.

HORN ANTENNA of the Bell Telephone Laboratories receiver at Holmdel, N.J., was originally designed to collect signals reflected from Echo satellites. This was the antenna with which Arno A. Penzias and Robert W. Wilson first detected the fireball radiation.

tant to suppose—no contemporary source within the universe could produce the isotropic radiation we observe.

Our discussion of cosmology so far has been merely descriptive, but if cosmology is to be a respectable science it must attempt numerical confrontations between theory and observation. Such a confrontation is provided by what cosmologists are now calling the "helium problem." There is a theoretical connection between the temperature of the primeval fireball and the amount of helium in the matter that came out of the big bang and eventually condensed into galaxies. It is worth examining as an example of the development of cosmological ideas and of the way in which a single observational result can prompt new theoretical work that in turn calls for new observations.

The story begins in about 1930 with the pioneering work of R. C. Tolman on thermodynamics and thermal radiation in an expanding universe. In 1938 C. F. von Weizsäcker tried theoretically to produce the heavy elements by "cooking" hydrogen in an early "superstar" stage of the universe, which later exploded into the expanding universe. George Gamow pointed out in 1948, however, that according to general relativity the universe could not have existed in a static, high-temperature state. He proposed instead that the elements were largely formed—and also that black-body radiation was emitted—during the early and very rapid expansion of the universe. Later calculations showed that although helium would have been produced in such a stage, it was impossible to account for the formation of heavier elements. An improved theory of element formation in stars finally eclipsed theories of element formation in the big bang itself, and the idea of thermal radiation in a big bang dropped out of sight. It is remarkable, however, that this theory, as developed by Gamow, Ralph Alpher, Robert Herman and others, implied that the present temperature of the fireball would be about equal to the observed value of three degrees K.

Dicke arrived at the idea of a primeval fireball from a different direction. In the summer of 1964 he was considering not the origin of the elements but the origin of the universe. It is difficult to explain the apparently spontaneous creation of matter that is called for if one associates the beginning of the expansion of the universe with its actual origin. Dicke therefore preferred an oscillating model, in which the present expansion of the universe is considered to have been preceded by a collapsing phase.

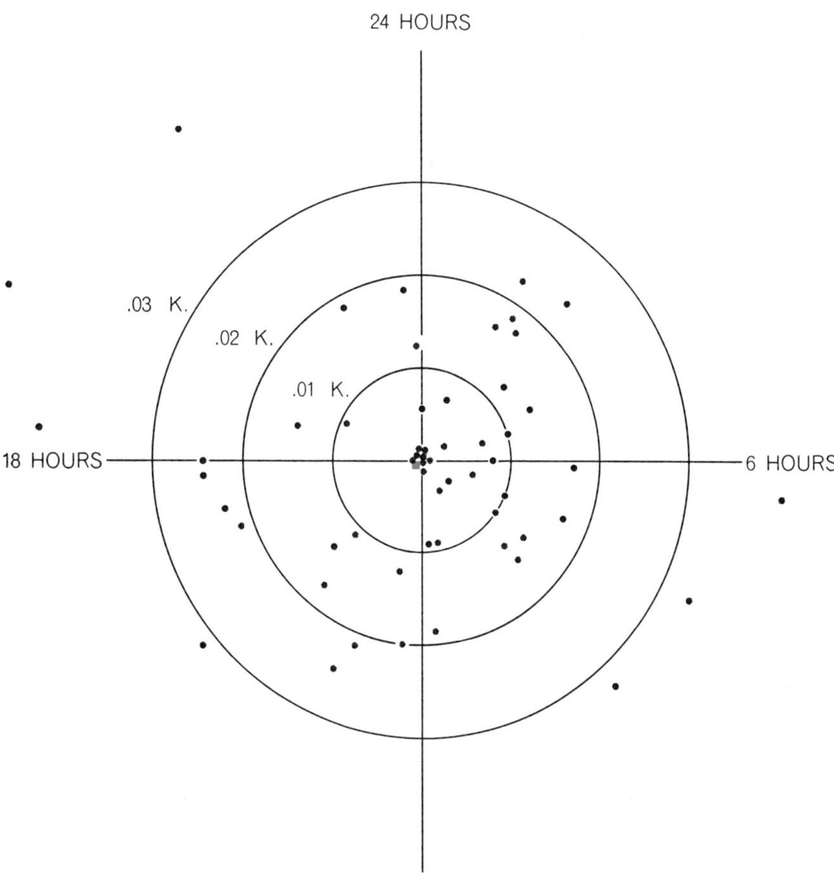

ISOTROPY DATA were analyzed in an effort to bring out any 24-hour periodicity. Each point gives the magnitude and direction of the maximum difference between equatorial and polar radiation temperatures on a day's run. (Solid dots are full runs, gray dots runs that were incomplete and were given half-weight.) The vector sum of all points yielded a maximum difference of about .001 degree K., an anisotropy of .03 percent (*colored square*).

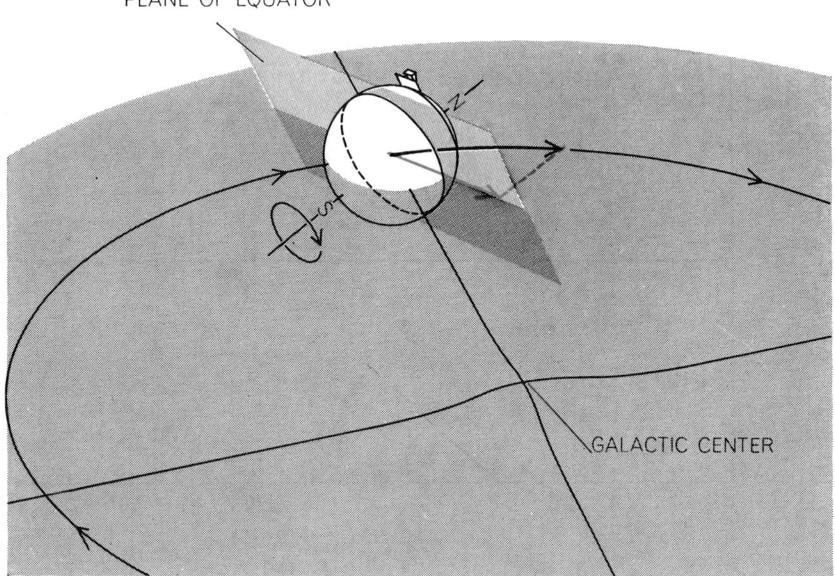

EARTH'S MOTION around the galaxy should be detectable as an apparent increase in radiation intensity in the direction of motion and a decrease in the opposite direction (see text). The direction of motion is not the same as the direction (in the plane of the Equator) in which the radiation is being observed, however. The observed effects should therefore be proportional to the equatorial projection (*colored arrow*) of the velocity of the earth.

The contraction of the universe would have heated up its contents, producing thermal radiation when the universe became dense enough; the temperature would have risen to at least 10 billion degrees, at which point complex nuclei would have evaporated, yielding pure hydrogen. Such a process could account for the elimination of heavy elements—the "ashes" from the hydrogen burned in stars in the previous cycle. Dicke, in other words, introduced the fireball to eliminate the heavy elements rather than to produce them.

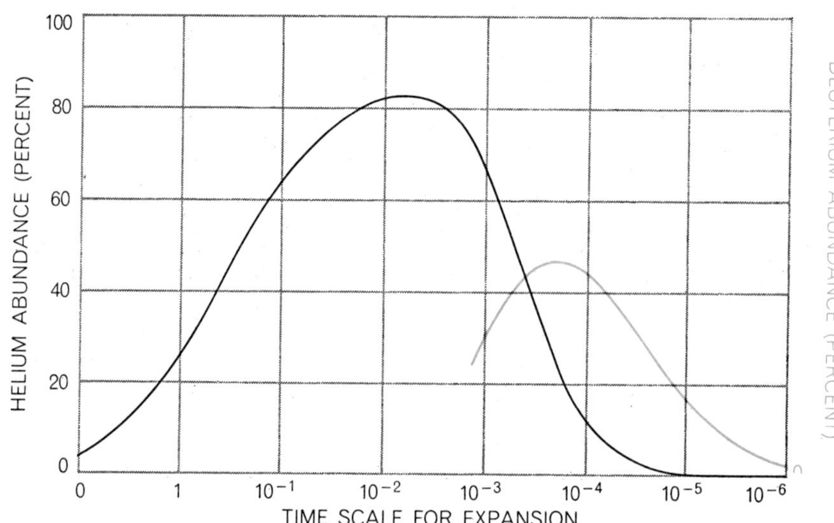

PRIMORDIAL ABUNDANCE of helium (*black*) and deuterium (*color*) is plotted for various assumptions about the rate of expansion of the universe. Unity on the horizontal scale corresponds to the time scale for expansion predicted by the general theory of relativity.

Perhaps because we are by training physicists rather than cosmologists, our group at Princeton was unaware for some time that there was a history of theoretical work on a primeval fireball and element production in the big bang. Having begun with the fireball idea, we reached the problem of element formation by a roundabout route. As our search for the fireball was getting under way we were concerned about how we would interpret the results we hoped to obtain. (One particularly wants to know this ahead of time when an experiment is highly speculative, as this one certainly was before the Bell Laboratories results became known.) We were anxious to establish some connection between a possible primeval fireball and some other observable quantity, so we asked ourselves what physical processes might be appropriate to the conditions encountered as the expansion of the universe is traced back in time to ever higher densities and temperatures.

We found that in the early stages of expansion conditions would have been right for the conversion of significant amounts of hydrogen to helium. The fractional amount of hydrogen that would have been converted to this primeval helium depends on two observable quantities: the present mean mass density in the universe and the present temperature of the primeval fireball. Given the present mass density, the primeval helium abundance would be lower for higher values of the present fireball temperature. This is because the helium would have formed at a certain temperature (about a billion degrees) and the amount of helium that formed would depend on the mass density at the epoch in which it formed. The higher the present fireball temperature, the less the radiation can have cooled since the epoch of helium production; the correspondingly lesser expansion of matter means a larger mass density at the helium epoch, and therefore more helium production.

We decided that if the present fireball temperature were 10 degrees K. or more, the primeval helium abundance could be well below the observed helium abundance in the sun. This seemed desirable because the sun also contains heavier elements thought to have been produced in earlier generations of stars, and we assumed that these earlier generations would also have produced substantial amounts of helium. It turned out, of course, that the fireball temperature is certainly not 10 degrees but rather three degrees. This pushes the calculated primeval helium abundance up to the range of 27 to 30 percent by mass, or just about the observed abundance of helium in the sun. That seemed surprisingly high.

Before considering the observational evidence that might confirm or rule out this theoretical finding, we should examine the assumptions that underlie the calculation. A number of variables from nuclear physics are involved, many of them actually measured and the rest derived from theory in which there is a good deal of confidence because it works well in conventional applications. Another basic ingredient in the calculation is the gravity theory, which determines how fast the universe would have expanded through the period of helium production. Here there are grounds for suspicion, because the conventional gravity theory—the general theory of relativity—has not been tested by a wide range of observations. The primordial helium abundance can be computed for various assumptions about the rate of expansion of the universe [*see illustration above*]. If the universe actually expanded just slightly faster than general relativity predicts, an unacceptably high amount of helium would have been produced. If the expansion rate were increased by a factor of about 10,000, the helium production would be acceptable but there would be too much deuterium. To avoid this requires an expansion rate so great that there would have been relatively little primeval helium. The principal competitor of general relativity, a generalization developed by Carl Brans and Dicke, predicts a faster expansion that might carry over into this area of negligible helium. If the rate of expansion could be shown to be somewhat slower, this difficult situation could be eased, but no one has suggested a reasonably attractive way to do this.

For the moment we are content to frame this question: Was the initial helium abundance in our galaxy very low or was it about equal to the solar value? The choice between these two clear-cut alternatives depends on the helium content of the oldest stars in the galaxy. Unfortunately these stars closely guard the secret of their helium abundance. They are small stars and generally have cool surfaces in which the spectral lines of helium are not seen, and so one cannot use the spectroscopic techniques that have been satisfactory for more massive stars with hotter surfaces. As for these massive stars, their lifetimes are relatively short, and the ones that formed early in the history of the galaxy have already burned out. The answer to the helium problem will be hard to obtain, but it will eventually add a fascinating piece of information to observational cosmology.

32

Cosmology before and after Quasars

DENNIS SCIAMA

September 1967

A review of *The Measure of the Universe,* by J. D. North. Oxford University Press, 1965.

I have often wondered what it must have been like to be a nuclear physicist in the early 1930's, particularly in 1932—that *annus mirabilis* which saw the discovery of the neutron and the positron and the first splitting of the nucleus by artificially accelerated particles. Now I think I know. As a cosmologist I have seen in the 1960's a similar stream of discoveries following one on another at an almost indecent rate. The years 1963 to 1965 stand out, beginning with the discovery of quasars, followed by the measurement of the fantastic red shifts possessed by some of them and culminating in what is perhaps the greatest discovery of them all: the cosmic black-body radiation. I should say at once that the evidence is not yet decisive that any of these discoveries has cosmological significance, but it is good enough to have reduced most cosmologists, who are traditionally starved for basic observations, to a state of bewildered euphoria.

These reflections are prompted by the publication of an interesting book called *The Measure of the Universe,* written by J. D. North, an Oxford philosopher. It is a history of modern cosmology that ends just before the new period begins. It barely mentions quasars and does not mention the cosmic black-body radiation at all. This is no criticism, because the book was written too early for it to have done so. It was therefore written at the right time to take stock of the first great period in cosmology. That period, which had only the expansion of the universe to explain, we might justly call the geometrical period. Today we are well and truly launched into the astrophysical period.

To be fair to the early theorists, they did predict the expansion of the universe before it was discovered. By the early 1920's it was clear to Willem de Sitter, Alexander Friedmann and Hermann Weyl that Einstein's field equations of general relativity had as solutions homogeneous and isotropic model universes whose material substratum was in a state of expansion, the relative velocity at which two particles moved apart simply being proportional to their distance (except for refinements for very widely separated particles). Moreover, if the debatable cosmical constant was dropped from the field equations, as Einstein later urged, then *all* the homogeneous and isotropic solutions exhibited expansion (or contraction if one cared to reverse the sense of time). It was not until 1929 that the Hubble law, that the observed red shift is proportional to the distance of a galaxy (as estimated by various more or less dubious criteria), was first stated.

North's book gives a thorough account of this classical phase of theoretical cosmology. There were many controversies at the time about the properties of the various models. That phase is now over, and the correct results are enshrined in standard theory. There were also controversies of a different nature; not everyone accepted the view that general relativity was uniquely fitted to deal with the universe as a whole. Various "heretical" theories were proposed, notably by Sir Arthur Eddington, E. A. Milne, P. A. M. Dirac and Pascual Jordan, and they are described too. I deliberately mention separately the steady-state theory of Herman Bondi, Thomas Gold and Fred Hoyle, because I think it is fair to say that of all the heretical theories this is the one that has irritated and excited the most people, has provoked the most good astrophysics and has more or less survived to the present day.

I say "more or less" because one of the consequences of the new turn of events—of cosmology becoming astrophysical—is that if the red shifts of the quasars are cosmological in origin, and if the universe is filled with black-body radiation, then the chances of the steady-state theory surviving are very small indeed. I want to make clear why this is so, and to discuss what further information we can hope to extract from the new results and their likely future extensions. I must add that for me the loss of the steady-state theory has been a cause of great sadness. The steady-state theory has a sweep and beauty that for some unaccountable reason the architect of the universe appears to have overlooked. The universe in fact is a botched job, but I suppose we shall have to make the best of it.

One of the botches is the existence of a singularity, that is, a moment when the density of the universe was infinite. To be more precise, this is what general relativity requires for the homogeneous and isotropic models to which I have referred. It has sometimes been suggested that the singularity would go away as soon as one admitted that the real universe was neither exactly homogeneous nor exactly isotropic; in such circumstances the galaxies would not move quite radially, and so the matter they are made of would not all have emerged from exactly one point in the past. It has recently been shown by Stephen Hawking and others, however, that the orthodox theory of general relativity, without on the one hand a cosmical constant and on the other assumptions of exact symmetry, still requires the physical properties of the universe to have been singular at some time or times.

It was to avoid such an unpleasant singularity (and for other reasons too) that Bondi, Gold and Hoyle proposed in 1948 a deviation from orthodox general relativity that would allow the continual creation of matter at a rate just compensating for the expansion of the universe. The resulting mean density of the universe (and indeed all its other average properties) would then be independent of time, leading to a steady state that would automatically persist forever. It is this magnificent conception we must now reluctantly abandon.

The first evidence against the steady-state theory came from counts of celestial radio sources, conducted notably by Sir Martin Ryle and his colleagues in Britain, but also by B. Y. Mills and J. G. Bolton in Australia and by M. Ceccarelli in Italy. These counts showed that the number of faint radio sources was far too large compared with the number of bright sources to be compatible with the steady-state theory. This evidence has given rise to much controversy, mainly because the majority of sources concerned have not yet been identified optically. Accordingly inferences drawn from these counts have been surrounded by an aura of uncertainty. A straightforward interpretation, stressed by Ryle and studied in detail by William Davidson and by Malcolm Longair, requires that the radio sources exhibit intrinsic evolution. That is to say, the faint sources, which are mostly at great distances and so are now being seen as they were a long time in the past (because of the time their radio waves take to reach us), must have average properties different from the bright sources, which are mostly relatively near and so are being seen almost contemporaneously with ourselves. Such evolution is of course incompatible with a steady-state universe, but it would be expected in a universe evolving from a dense state to a dilute one, a universe to which one can attach the concept of an age.

Needless to say, there have been several implausible attempts to evade Ryle's argument. My own attempt has turned out to be correct, but not in the way I intended. I proposed before the discovery of quasars that the radio sources in Ryle's catalogue consisted of two different populations. One population was to be the radio galaxies that had already been identified optically and were well known. For the second population I proposed the existence of radio stars in our galaxy, whose distribution between bright and faint sources would explain the anomalous counts but would have nothing to do with cosmology. It has turned out that a second population of starlike (that is, unresolvable) radio sources does exist. Moreover, they are just the ones responsible for the excess of faint sources (as has been shown independently by Philippe Véron and Longair). But these quasi-stellar radio sources, or quasars, have large red shifts and are therefore not the objects I had in mind.

It is true that a few physicists and astronomers (James Terrell, Geoffrey and Margaret Burbidge, Hoyle) hold, with differing degrees of assurance, that these large red shifts are not cosmological in origin, and that the quasars are within, or relatively close to, our galaxy. This would be in tune with my proposal, but I find their arguments unconvincing. If the red shifts have a Doppler origin, that is, if the quasars are receding from us rapidly as a result of a local explosion, the question arises of why we do not see any blue shifts from quasars fleeing from neighboring galaxies toward us. Of course, if quasar emission is a sufficiently rare process, the nearest such galaxies would be too far away for their quasars to be visible, but then why should we be privileged to witness such a rare event so close to us? Clearly this is possible but unlikely.

On the other hand, if the red shifts do not have a Doppler origin but arise, say, from the Einstein gravitational effect, and if the sources are distributed uniformly in space with the ones observed so far quite close to us, then we would not expect the source counts to manifest an excess of faint objects. The relative number of bright and faint sources should be the same as if there were no red shift (that is, the same as for a uniform distribution of stationary sources) and this is not what is observed. We conclude that the red shifts are most probably cosmological in origin. On this basis Martin Rees and I have carried out an analysis of the red shifts of the quasars, and we find again that there are too many faint sources with large red shifts to be compatible with the steady-state theory.

In weighing the significance of what I have said so far it is important to understand how accidental it is that we should be able to observe such large red shifts so easily. Objects with these large red shifts are so distant that the different cosmological theories make substantially different predictions about them, but such objects are visible only because quasars happen to be a hundred times brighter than galaxies. In contrast, the existence of cosmic black-body radiation, which also serves to distinguish among different theories, is intimately bound up with the development of the universe itself.

The detection in 1965 of excess radiation at microwave frequencies (that is, at wavelengths of a few centimeters) and the evidence that it has a black-body spectrum (that is, is in thermal equilibrium characterized by a single temperature) has been described in "The Primeval Fireball," by P. J. E. Peebles and David T. Wilkinson; page 329. The temperature observed is about 3 degrees absolute. I should like to make the following comments on this result:

1. No plausible noncosmological explanation has yet been proposed (and not for want of trying).

2. A natural cosmological explanation does exist if the universe was once very dense.

3. A temperature significantly greater than 3 degrees would not be compatible with our general knowledge of radio astronomy and high-energy astrophysics.

I shall say no more about the first point, but I should like to discuss the second and third in a little more detail. As in the case of the expansion of the universe, the existence of cosmic black-body radiation was predicted before it was observed. Around 1950 it was proposed by George Gamow and his associates that the early, dense stages of the universe were very hot, a state of affairs often described as the "hot big bang." Their reason for making this proposal was that in such conditions thermonuclear reactions could occur at an appreciable rate, converting primordial hydrogen into helium and possibly heavier elements. By choosing the right early conditions Gamow was able to account approximately for the abundance of helium with respect to hydrogen that is observed today. This helium problem is actually in a very confused state at the moment, but the important point here is that if the early, dense stages were hot, unquestionably there was ample time for matter and radiation to come into thermal equilibrium. At that time, then, the radiation would have had a black-body spectrum. Moreover, at all times thereafter the spectrum would remain that of a black body, the radiation simply cooling down as the universe expanded. Gamow's original calculations of helium formation led him to predict for the present temperature of the black-body radiation a value of about

30 degrees absolute, but modern calculations are compatible with a lower temperature, in particular with a temperature of 3 degrees absolute.

As I have mentioned in my third point, we now know that a temperature as high as 30 degrees can be ruled out. Cosmic ray protons and electrons interacting with such radiation would produce effects that could be observed, and they are not. Three degrees is about the highest permitted temperature from this point of view, and 3 degrees is just what has been found. The connection between the microwave observations and Gamow's theory was made by Robert H. Dicke and his colleagues at Princeton University. In fact, they had the bad luck to be setting up apparatus to look for the excess radiation when it was discovered accidentally down the road by Arno A. Penzias and Robert W. Wilson of the Bell Telephone Laboratories.

Can the steady-state theory account for the excess radiation? It would be reasonable to propose that along with the newly created matter there comes into existence newly created radiation; indeed, some such effect would be expected as a result of the creation process itself. But why the observed spectrum should be that of a black body over a wide range of wavelengths is totally obscure. It is therefore critically important to establish without doubt that the actual spectrum is that of a black body. The present evidence is strong but not decisive. Further work is being done, and this point should be settled fairly soon.

There is one final property of the radiation that I want to discuss because it is in some ways its most exciting feature, and that is its degree of isotropy—its uniformity with respect to direction of arrival. Just a few months ago R. B. Partridge and Wilkinson announced that any anisotropy is less than a few tenths of a percent. I have heard cynical scientists comment that this result throws doubt on the whole phenomenon, on the grounds that noise generated internally by the observing instruments would be more "isotropic" than externally generated noise, even noise coming from a highly isotropic universe. To silence this cynicism it is necessary to show that the universe is likely to be isotropic to the required degree. A first step in this direction has recently been taken by Charles W. Misner, who has shown that for a certain class of model universes any initial anisotropy would be rapidly removed by a rather exotic form of viscosity involving the pairs of neutrinos that would be excited by the high temperatures then prevailing. Misner's program is to allow the universe to start out as irregularly as it wishes and then to show that all irregularities would be damped out by the action of accepted physical processes, except for those irregularities we actually observe (such as clusters of galaxies).

Another intriguing aspect of the isotropy measurements is that they can be used to determine our "absolute" velocity, that is, our velocity with respect to the distant matter that last effectively scattered the radiation. Because of the Doppler effect such a velocity would reveal itself by leading to a slightly higher temperature for the radiation ahead of us and a slightly lower temperature for the radiation behind us. The present limit on the anisotropy corresponds to a velocity limit of 300 or 400 kilometers per second. One contribution to the expected velocity comes from the sun's known rotation around our galaxy of about 250 kilometers per second. Even this, however, depends on the correctness of Mach's principle [see "Inertia," by Dennis Sciama; SCIENTIFIC AMERICAN, February, 1957]. According to Mach's principle, the local nonrotating frame of reference as determined dynamically coincides with the frame in which distant matter is not rotating. The well-known rotation of our galaxy, with which is associated the galaxy's dynamical flattening, would then be a rotation with respect to distant matter, and therefore to the effective sources of the black-body radiation.

To obtain our net motion relative to these sources, however, we must also allow for the peculiar motion of our galaxy in the local group of galaxies (which has been estimated to be about 100 kilometers per second) and a possible systematic motion of our galaxy in the local supercluster. This supercluster is believed by some astronomers (Vera Rubin, Gérard de Vaucouleurs, K. F. Ogorodnikov) to be a flattened system of galaxies rotating around the Virgo cluster. I have recently rediscussed our possible motion in the supercluster and arrived at a tentative rough estimate for our net motion through the black-body radiation of about 400 kilometers per second in the general direction of the center of our galaxy.

Future observations of the black-body radiation should be able to test this prediction, and in view of the uncertainty surrounding the notion of a local supercluster I would not be at all surprised to find that it is wrong. My point is simply that yet another new range of problems has been opened up for the cosmologist by the existence of the black-body radiation. We have come a long way in a few years from the geometrical considerations described in North's book, and we can rejoice. Cosmology has at last become a science.

33

The Origin of Galaxies

MARTIN REES AND JOSEPH SILK

June 1970

Perhaps the most startling discovery made in astronomy this century is that the universe is populated by billions of galaxies and that they are systematically receding from one another, like raisins in an expanding pudding. If galaxies had always moved with their present velocities, they would have been crowded on top of one another about 10 billion years ago. This simple calculation has led to the cosmological hypothesis that the world began with the explosion of a primordial atom containing all the matter in the universe. A quite different line of speculation argues that the universe has always looked as it does now, that new matter is continuously being created and that new galaxies are formed to replace those that disappear over the "horizon."

On either hypothesis it is still necessary to account for the formation of galaxies. Why does matter tend to aggregate in bundles of this particular size? Why do galaxies comprise a limited hierarchy of shapes? Why do spiral galaxies rotate like giant pinwheels? Astrophysicists are trying to answer these and similar questions from first principles. The goal is to explain as many aspects of the universe as one can without invoking special conditions at the time of origin. In most of what follows we shall assume a cosmological model in which the universe starts with a "big bang." When we have finished, the reader will see, however, that some form of continuous creation of matter may not be ruled out.

Before the invention of the telescope the unaided human eye could see between 5,000 and 10,000 stars, counting all those visible in different seasons. Even modest telescopes revealed millions of stars and in addition disclosed the existence of many diffuse patches of light, not at all like stars. These extragalactic "nebulas," many of them beautiful spirals, are seen in all directions and in great profusion. As early as the 18th century Sir William Herschel and Immanuel Kant suggested that these nebulas were actually "island universes," huge aggregations of stars lying far beyond the limits of the Milky Way.

The validity of this hypothesis was not confirmed until 1924, when the American astronomer Edwin P. Hubble succeeded in measuring the distances to a number of spiral nebulas. Several years earlier Henrietta S. Leavitt had shown that Cepheid variables, named for the prototype Delta Cephei, a variable star discovered in 1784, had light curves that could be correlated with their magnitude. The distances of a number of Cepheids were later determined by independent means, so that it became possible to use more distant Cepheids as "standard candles" to establish a distance-magnitude relation. Hubble looked for Cepheid variables in some of the nearer external galaxies and found them. From their period he was able to deduce their absolute luminosity, and from this he was able to estimate their distance. Hubble soon established that the nearest spiral nebulas (or galaxies) were vast systems of stars situated a million or more light-years outside our own galaxy.

Subsequently Hubble developed a scheme for classifying galaxies according to their morphology, ranging from systems that are amorphous, reddish and elliptical to systems that are highly flattened disks with a complex spiral structure containing many blue stars and lanes of gas and dust [*see illustration on page 344*]. The spiral galaxies themselves vary in appearance. At one extreme are those with large, bright nuclei and inconspicuous, tightly coiled spiral arms. At the other extreme are galaxies in which the nuclei are less dominant and the spiral arms are loosely wound and prominent. The elliptical galaxies also form a sequence, ranging from almost spherical systems to flattened ellipsoids. In addition there are highly irregular systems showing very little structure of any kind.

In all these sequences there is a parallel progression in certain characteristics of the galaxies. In general spirals are rich in gas and dust, contain many blue supergiant stars, are highly flattened and rotate appreciably. Ellipticals, by contrast, seem to possess little gas or dust, usually contain late-type dwarf stars and exhibit scant rotation.

The masses of galaxies are found by several methods. Galaxies are often gravitationally bound together in pairs. If the distance between them and their relative velocities are known, Kepler's law can be used to find their total mass.

CLUSTER OF GALAXIES in the constellation of Hercules demonstrates the inhomogeneity of the distribution of galaxies in the sky. About 350 million light-years away, this cluster contains about 100 members and is some five million light-years across. It was photographed with the 200-inch Hale telescope on Palomar Mountain. Some very rich clusters contain 1,000 members or more and vary from one million to 10 million light-years across. There is some evidence that such clusters are in turn grouped together into superclusters of perhaps 100 members, spread over 100 million light-years. On scales larger than this the universe appears to be uniform. The bright circular spots with the spikes radiating from them are nearby stars; the spikes are produced by reflections within the telescope.

Another method, used mostly for spirals that are viewed edge on or obliquely, is to determine the velocity of rotation by measuring the Doppler shift of spectral lines emitted by ionized gas in various parts of the disk. (The spectral lines of approaching gas will be shifted toward the blue end of the spectrum, those of retreating gas toward the red end.) One can plot a rotation curve showing how the velocity of rotation varies with the distance from the center of the galaxy. The mass can then be estimated from the requirement that the centrifugal and gravitational (centripetal) forces must be in balance. It turns out that the masses of galaxies are typically about 10^{11} (100 billion) times the mass of the sun. The range, however, is fairly broad: from about 10^8 solar masses for some nearby dwarf galaxies to 10^{12} solar masses for giant ellipticals in more remote regions of the universe. The diameter of the larger spirals, such as our own galaxy, is about 100,000 light-years.

Galaxies also differ widely in the ratio of mass to luminosity. Taking the mass-to-luminosity ratio of the sun as unity, one finds that for large spirals, such as our own galaxy, the ratio varies from one up to 10. In other words, some spirals emit only a tenth as much light per unit of mass as the sun does. Ellipticals commonly emit even less: only about a fiftieth as much light per unit of mass. (Thus their mass-luminosity ratio is 50.)

The distribution of galaxies in the sky is quite inhomogeneous. There are many small groups of galaxies, and here and there some rich clusters containing up to 1,000 members or more. Such systems vary from one million light-years across to 10 million. Our own galaxy is a member of the "local group," an association of about 20 galaxies, only one of which, the Andromeda galaxy, has a mass comparable to that of ours. The local group is about three million light-years in diameter. The Andromeda galaxy is some two million light-years away; the nearest large cluster of galaxies is in Virgo, about 30 million light-years distant.

Even such clusters do not seem to be randomly distributed in space. Some astronomers have argued that there is evidence that clusters are grouped into superclusters of perhaps 100 members, spread over 100 million light-years. The universe appears to be uniform on scales larger than this.

Establishing the distance of galaxies was only part of Hubble's achievement. Working with the 100-inch telescope on Mount Wilson, he showed from red-shift measurements that the galaxies are in recession. Hubble found, moreover, that the red shift of a galaxy is directly proportional to its distance, as judged by its apparent luminosity. The most distant galaxies known are in a faint cluster in the constellation Boötes; Rudolph Minkowski discovered that the wavelength of light coming from this cluster is stretched by 45 percent. The corresponding velocity of recession is nearly half the speed of light. Light originating from some of the brilliant starlike objects known as quasars is red-shifted more than 200 percent, but astronomers disagree whether or not this red shift is due to the cosmological expansion of the universe.

The light from Minkowski's cluster of galaxies set out toward us about five bil-

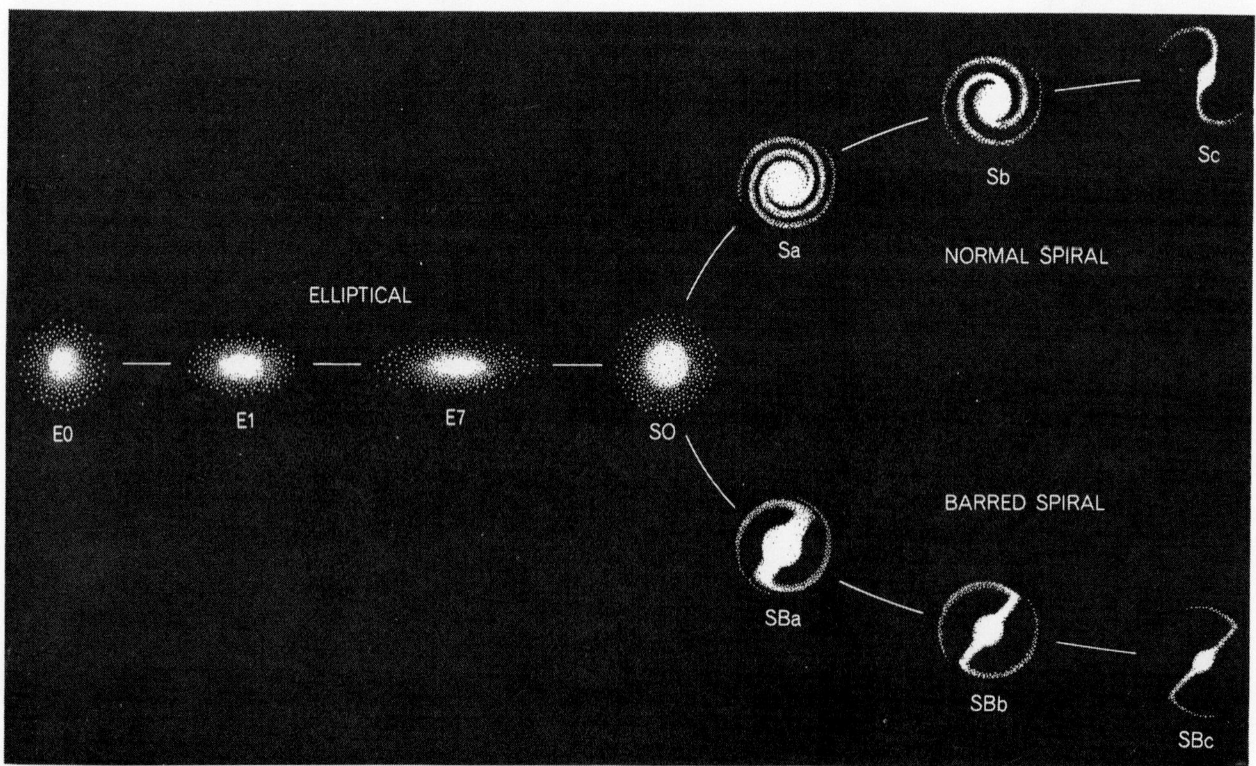

CLASSIFICATION SCHEME developed by Edwin P. Hubble in the early 1930's organizes galaxies according to their shape, ranging from amorphous elliptical systems containing many red stars and little gas and dust (*left*) to highly flattened spiral disks containing many blue stars and lanes of gas and dust (*right*). The elliptical galaxies range from almost spherical systems (designated E0) to highly flattened ellipsoids (E7). The spiral galaxies themselves form two sequences: normal spirals (*top right*) and barred spirals (*bottom right*). At one extreme in both sequences are galaxies with large bright nuclei and inconspicuous, tightly coiled spiral arms (Sa, SBa); at the other extreme are galaxies in which the nuclei are less dominant and the spiral arms are loosely wound and prominent (Sc, SBc). At the branching point of the diagram is a disklike form that resembles the spirals but lacks spiral arms (SO).

lion years ago, and so we can be sure that some galaxies are even older than that. On the other hand, as we have mentioned, all the galaxies must have been tightly packed together no more than 10 billion years ago, based on their present recession velocity. Estimates of the ages of stars suggest that our galaxy, and others like it, are unlikely to be much less than 10 billion years old. Hence we are presented with a remarkable coincidence: most galaxies appear to be about as old as the universe. This implies that galaxies must have formed when conditions in the universe were much different from those now prevailing.

It seems clear, then, that the formation of galaxies cannot be treated apart from cosmological considerations. The dynamics and structure of the universe in the large are beyond the scope of Newtonian physics; it is necessary to use Einstein's general theory of relativity. Because of the complexity of the theory, it is practicable to solve the equations only for cases having special symmetry. Until quite recently the only solutions for an expanding universe were those found in 1922 by the Russian mathematician Alexander A. Friedmann. In his idealized models matter is treated as a strictly uniform and homogeneous medium. The universe expands from a singular state of infinite density, with the rate of expansion decelerating as a consequence of the mutual gravitational attraction of its different parts. The universe may have enough energy to keep expanding indefinitely or the expansion may eventually cease and be followed by a general collapse back to a compressed state. Observations of the actual rate of expansion of the universe at different epochs, as determined by the red shift–luminosity relation of the most distant galaxies, fail to tell us unambiguously whether the expansion will finally stop and be reversed or whether it will continue indefinitely.

The clumping of matter into stars, galaxies and clusters of galaxies in the real universe might seem to make Friedmann's models, based on perfect homogeneity, empty exercises. In actuality the "graininess" we observe in the universe is on such a small scale that Friedmann's solutions remain valid. The reason is that the gravitational influence of local irregularities is swamped by that of more distant matter.

Perhaps the most convincing evidence in support of Friedmann's simple description of the universe was supplied

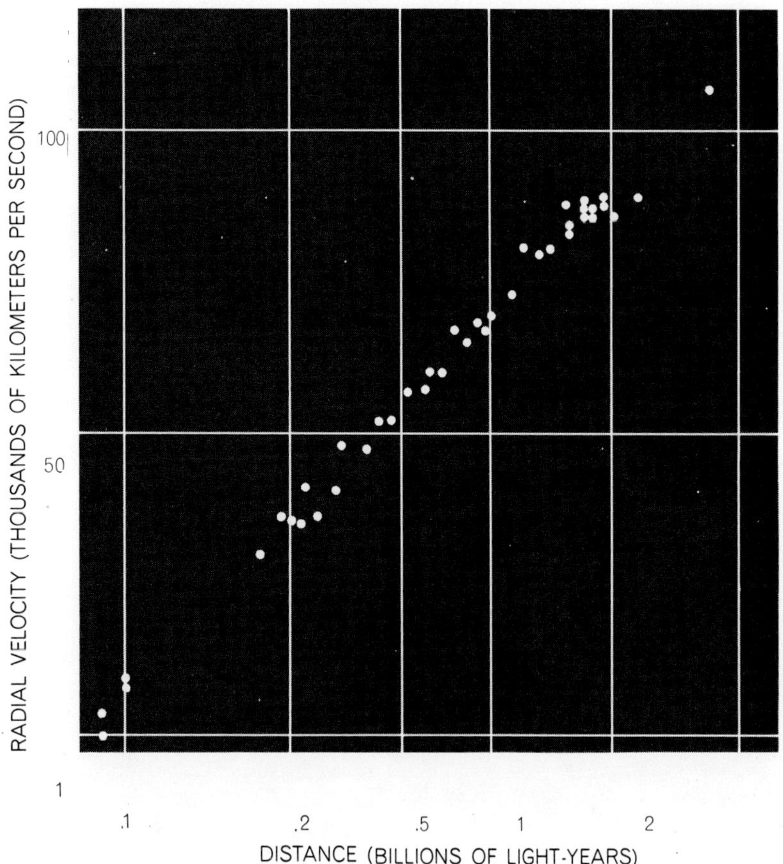

RECESSION VELOCITY OF A GALAXY is obtained by measuring the amount by which the radiation it emits is shifted to the red end of the spectrum. The velocity is directly proportional to the galaxy's distance, as judged by its apparent luminosity. In this diagram, adapted from a recent study by Allan R. Sandage of the Hale Observatories, the ratio of recession velocity to distance is shown for brightest galaxy in each of 41 clusters of galaxies.

in 1965 by the discovery that space is pervaded by a background radiation that peaks at the microwave wavelength of about two millimeters, corresponding to the radiation emitted by a black body at an absolute temperature of three degrees (three degrees Kelvin). This radiation could be the remnant "whisper" from the big bang of creation. The remarkable isotropy, or nondirectionality, of this radiation is impressive evidence for the isotropy of the universe.

The radiation was discovered independently and almost simultaneously at the Bell Telephone Laboratories and at Princeton University [see "The Primeval Fireball," by P. J. E. Peebles and David T. Wilkinson; page 329 in this volume]. The radiation has the spectrum characteristic of radiation that has attained thermal equilibrium with its surroundings as a result of repeated absorption and reemission, and it is generally interpreted as being a relic of a time when the entire universe was hot, dense and opaque. The radiation would have cooled and shifted toward longer wavelengths in the course of the universal expansion but would have retained a thermal spectrum. It thus constitutes remarkably direct evidence for the hot-big-bang model of the universe first examined in detail by George Gamow in 1940.

Assuming the general validity of the Friedmann model for the early stages of the universe, it seems clear that the material destined to condense into galaxies cannot always have been in discrete lumps but may have existed merely as slight enhancements above the mean density. There will be a tendency for the larger irregularities to be amplified simply because, on sufficiently large scales, gravitational forces predominate over pressure forces that tend to oppose collapse. This phenomenon, known as gravitational instability, was recognized by Newton, who, in a letter to Richard Bentley, the Master of Trinity College, wrote:

"It seems to me, that if the matter of our sun and planets, and all the matter of the Universe, were evenly scattered

SPIRAL GALAXY M 101 in Ursa Major is representative of the Sc type, which is characterized by a relatively inconspicuous nucleus and prominent, loosely wound spiral arms. Our own Milky Way galaxy is either of this type or of the slightly less open Sb type.

BARRED SPIRAL GALAXY NGC 1300 in Eridanus is classified SBb, which means that it is an intermediate type on the barred-spiral branch of the Hubble sequence. All the photographs shown on these two pages were made with the 200-inch Hale telescope.

through all the heavens, and every particle had an innate gravity towards all the rest, and the whole space throughout which this matter was scattered, was finite, the matter on the outside of this space would by its gravity tend towards all the matter on the inside, and by consequence fall down into the middle of the whole space, and there compose one great spherical mass. But if the matter were evenly disposed throughout an infinite space, it could never convene into one mass; but some of it would convene into one mass and some into another, so as to make an infinite number of great masses, scattered great distances from one to another throughout all that infinite space. And thus might the sun and fixed stars be formed, supposing the matter were of a lucid nature."

Newton envisaged a static universe, but the same qualitative picture occurs in an expanding Friedmann universe, as was shown by the Russian physicist Eugene Lifshitz in 1946.

Because of the atomic nature of matter the early universe could never have been completely smooth. It would obviously be gratifying if the inevitable random irregularities in the initial distribution of atoms sufficed ultimately to produce the bound systems of stars we see throughout the universe today. Unfortunately this type of statistical fluctuation fails by many orders of magnitude to account for the observed degree of structure in the universe. Moreover, it remained a puzzle why agglomerations of a certain mass, notably galaxies, should be so plentiful. It appeared necessary to postulate initial fluctuations in a seemingly *ad hoc* manner, and nothing had really been explained; "things are as they are because they were as they were."

Only in the past two or three years has it been realized that the background radiation acts as a gigantic homogenizer on certain preferred scales. To understand just how this works we must look more closely at Gamow's model of the universe. In the early stages, when the universe consisted of a primordial fireball, no structures such as galaxies or stars could have existed in anything like their present form. All space would have been filled with radiation (photons) and hot gas, consisting of the nuclei of hydrogen and helium and the accompanying electrons. The photons would be repeatedly scattered from the electrons. For at least the first 100,000 years of its history (beginning roughly 10 seconds after its emergence

SPHERICAL GALAXY, classified as type E0 in Hubble's scheme, is a member of the Virgo cluster of galaxies. A representative of the most massive type of galaxy, this system, designated M 87, contains about 30 times as many stars as a spiral system such as our own does.

ELLIPTICAL GALAXY in the constellation of Cassiopeia is a member of the local group of galaxies. Designated NGC 147, it is an E4, or intermediate, type of elliptical galaxy. Because of its comparative proximity to our system it can be resolved into individual stars.

from the initial singularity) the universe can be pictured as a composite gas in which some of the "atoms" are particles and the rest are photons. For the universe as a whole there are now at least 10 million times more photons than particles. From thermodynamic considerations one can conclude that photons must also have greatly outnumbered particles in the fireball. For a gas in equilibrium each species of particle contributes to the total pressure in proportion to its number. This still holds (very nearly) for photons, so that the radiation would make an overwhelmingly dominant contribution to the pressure. (During the first 10 seconds, when the temperature exceeds a few billion degrees, the situation is less simple because pairs of photons can interact to form an electron and a positron.)

As the expansion proceeds and the density decreases, the photons lose energy, the temperature drops and the particles move less rapidly. A key stage is reached after about 10^5 years, when the fireball has cooled to 3,000 degrees. The electrons are then moving so slowly that virtually all are captured by nuclei and retained in bound orbits. In this condition they can no longer scatter photons and the universe becomes transparent. Inasmuch as the background temperature today is only about three degrees absolute, one can conclude that the universe has expanded by a red-shift factor of 1,000 since the scattering stopped. (Wavelength is inversely proportional to temperature.)

The microwave background photons have probably propagated freely since the universe became transparent and therefore they should carry information about a "surface of last scattering" at a red shift of more than 1,000. Compare this with the red-shift factor of about one-half for the most distant galaxy known! Because these photons have been traveling unimpeded since long before galaxies existed, they should provide us with remarkably direct evidence of physical conditions in the early universe.

Let us return now to the epoch of the primordial fireball and ask: How were inhomogeneities in the fireball affected by the presence of the intense radiation field? Radiation would inhibit the process of gravitational collapse. Under radiation pressure nonuniformities in the fireball would take the form of oscillations, pressure waves or turbulence. These disturbances, in turn, will be dissipated by viscosity and the development of shock waves. Some wavelengths will be attenuated more severely than others, so that inhomogeneities of favored size will be preserved whereas those less favored will tend to be destroyed. The aim of recent work has been to determine what scales of perturbation are most likely to survive the various damping processes until the scattering of photons comes to an end. Any perturbation whose survival and growth is specially favored should eventually dominate, almost irrespective of how nonuniformities were initially distributed in the primordial fireball. An encouraging result that has already emerged from these studies is that 10^{12} solar masses, roughly the mass of a large galaxy, is one such preferred scale [*see illustration on page 350*].

After the electrons in the initial plasma have been bound into atoms, radiation no longer affects the distribution of mass. At this point the surviving perturbations are free to amplify gravitationally. (It should be noted, however, that on small scales—less than 10^6 solar masses—the kinetic energy of atoms exerts a pressure of its own that inhibits gravitational collapse.) The first generation of bound systems will therefore condense from whatever scale of fluctuations had the largest amplitude at the time of decoupling, that is, when the fireball ceased to be a plasma of electrons and other particles.

At what stage did protogalaxies stop expanding and separate out from the rest of the universe? We might guess that this happened when the mean density was comparable to the present density in the outlying parts of galaxies. In 1962 Olin J. Eggen, Donald Lynden-Bell and Allan R. Sandage of the Hale Observatories investigated the likely early history of our own galaxy by studying

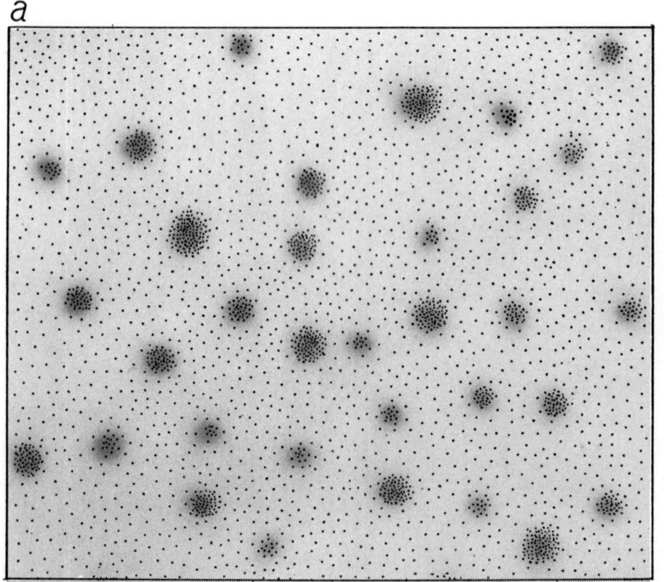

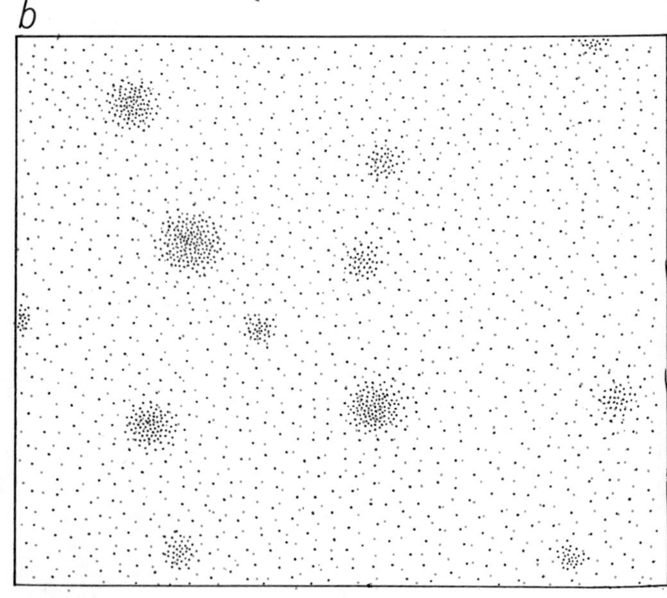

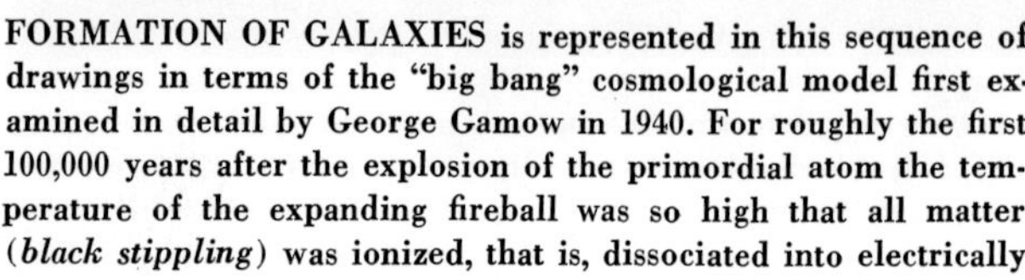

FORMATION OF GALAXIES is represented in this sequence of drawings in terms of the "big bang" cosmological model first examined in detail by George Gamow in 1940. For roughly the first 100,000 years after the explosion of the primordial atom the temperature of the expanding fireball was so high that all matter (*black stippling*) was ionized, that is, dissociated into electrically charged particles (*a*). In this situation photons of radiation could not travel very far without being scattered by the free electrons; as a result the universe during this period was effectively opaque (*light shade of color*). Nevertheless, slight random enhancements in the density of matter above the mean density presumably took place, usually accompanied by corresponding enhancements in the

very old stars in the galactic halo. These stars probably formed while the galaxy was collapsing to its present disklike shape (and before the birth of the stars in the Milky Way), and their orbits indicate that our galaxy attained a maximum radius of about 100,000 light-years. One can then tentatively estimate that galaxies such as our own formed when the universe was 1,000 times denser than it is now, about half a billion years after the expansion began.

Extrapolating backward in time, we find that the protogalaxies would have taken the form of nonuniformities roughly 1 percent denser than the average density of the universe at the decoupling epoch. It is an attractive possibility that these are the dominant surviving irregularities, all smaller scales having been smoothed out during the fireball phase. There are, however, some types of fluctuation that are not eradicated in the fireball, so that smaller gas clouds may have formed first and later collided and agglomerated into galaxies. Robert H. Dicke and P. J. E. Peebles of Princeton have suggested that globular clusters—compact groups of about 10^5 or 10^6 stars that orbit around galaxies—may represent that small fraction of clouds which managed to avoid collisions, fragmented into stars and survived. Clusters of galaxies would have evolved from initial irregularities of smaller amplitude but larger scale than those destined to form single galaxies.

The only contribution of cosmologists to date toward explaining galaxy formation has been to calculate what scales of perturbation are most likely to survive or amplify in the fireball, thereby reducing the need to build these preferred scales into the initial conditions. This removes one element of arbitrariness in the initial conditions prescribed for the universe. There still remains, however, the task of explaining both the origin of the nonuniformity of the universe on all scales except the very largest, and the apparent uniformity encountered on the largest scales.

In fact, the Friedmann models may not provide an adequate description of the fireball when large inhomogeneities are present. It would be conceptually attractive if there were processes that could transform an initially chaotic universe into one that displayed the large-scale uniformity of a Friedmann model. An encouraging step toward this goal has been taken by Charles W. Misner of the University of Maryland, who has considered a "mix master" universe, which expands anisotropically in such a way that all parts of the universe are causally related very early in its history. At the outset matter would be so densely packed that even neutrinos would interact with other particles at a significant rate. Acting like a blender, the neutrinos would destroy the original anisotropy of the fireball by the time it had cooled to about 20 billion degrees. Thereafter the expansion would mimic a homogeneous Friedmann model.

Several types of observation may help to test this general picture of galaxy formation. The fluctuations that develop into galaxies and clusters would give rise to random motions on the surface of last scattering. As a result the microwave background photons would not all have been red-shifted by exactly the same amount; in some directions they might have been scattered off material with a random velocity toward us, whereas in other directions the last-scattering surface may have been receding from us. As a consequence the microwave temperature would be slightly nonuniform over the sky. Edward R. Conklin and R. N. Bracewell of Stanford University, Arno A. Penzias, Johann B. Schraml and Robert W. Wilson at the Kitt Peak National Observatory and Yuri N. Parijsky of the Pulkovo Observatory can detect temperature fluctuations as small as a tenth of a percent on angular scales of a few minutes of arc, but so far they have found no positive effect. This technique, however, has the potentiality of detecting embryonic galaxies or clusters of galaxies when they were merely small enhancements above the mean gas density.

There are reasons to expect galaxies that have just condensed to be brighter than typical galaxies at the present epoch. The energy released by the collapse of the protogalaxy would probably have been radiated away by hot gas be-

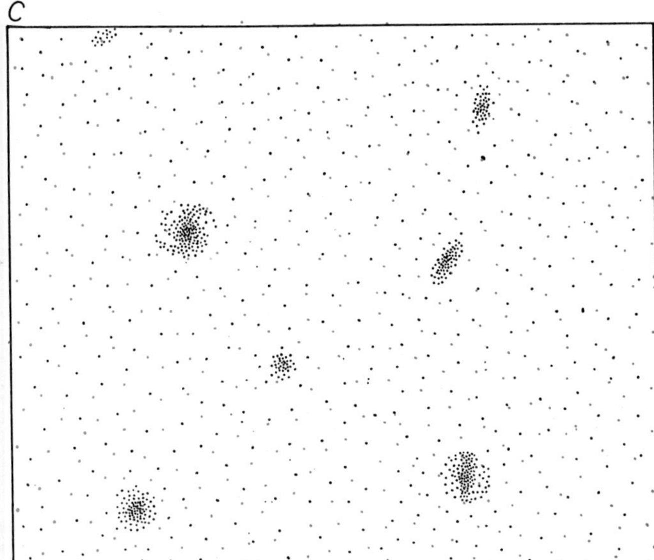

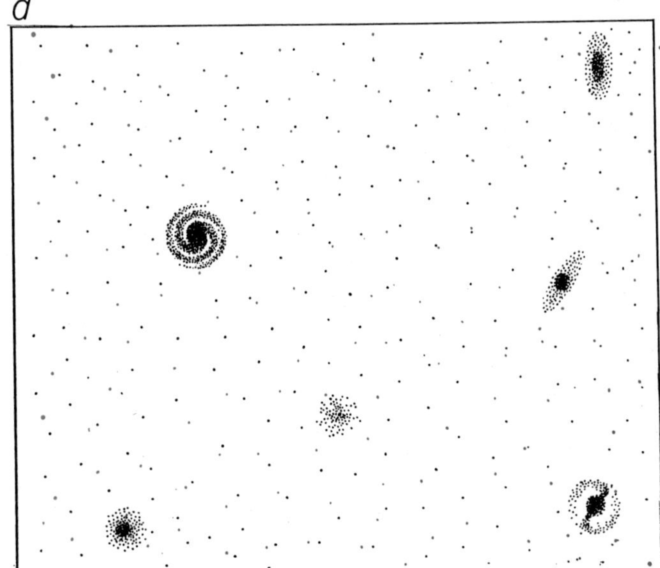

radiation density (adiabatic fluctuations). In such regions (*dark shade of color*) the radiation tended to damp fluctuations that would lead to further enhancements of matter if they were below a certain critical size (about 10^{11} solar masses). After about 100,000 years, when expanding fireball had cooled to about 3,000 degrees Kelvin, the negatively charged electrons were moving slowly enough to be captured by protons and retained in bound orbits, forming hydrogen atoms. In this condition electrons are much less effective in scattering photons and universe thus became transparent (*b*). Expansion and cooling of fireball continued and matter was progressively concentrated by gravitational forces, first into protogalaxies (*c*) and eventually into galactic types seen today (*d*).

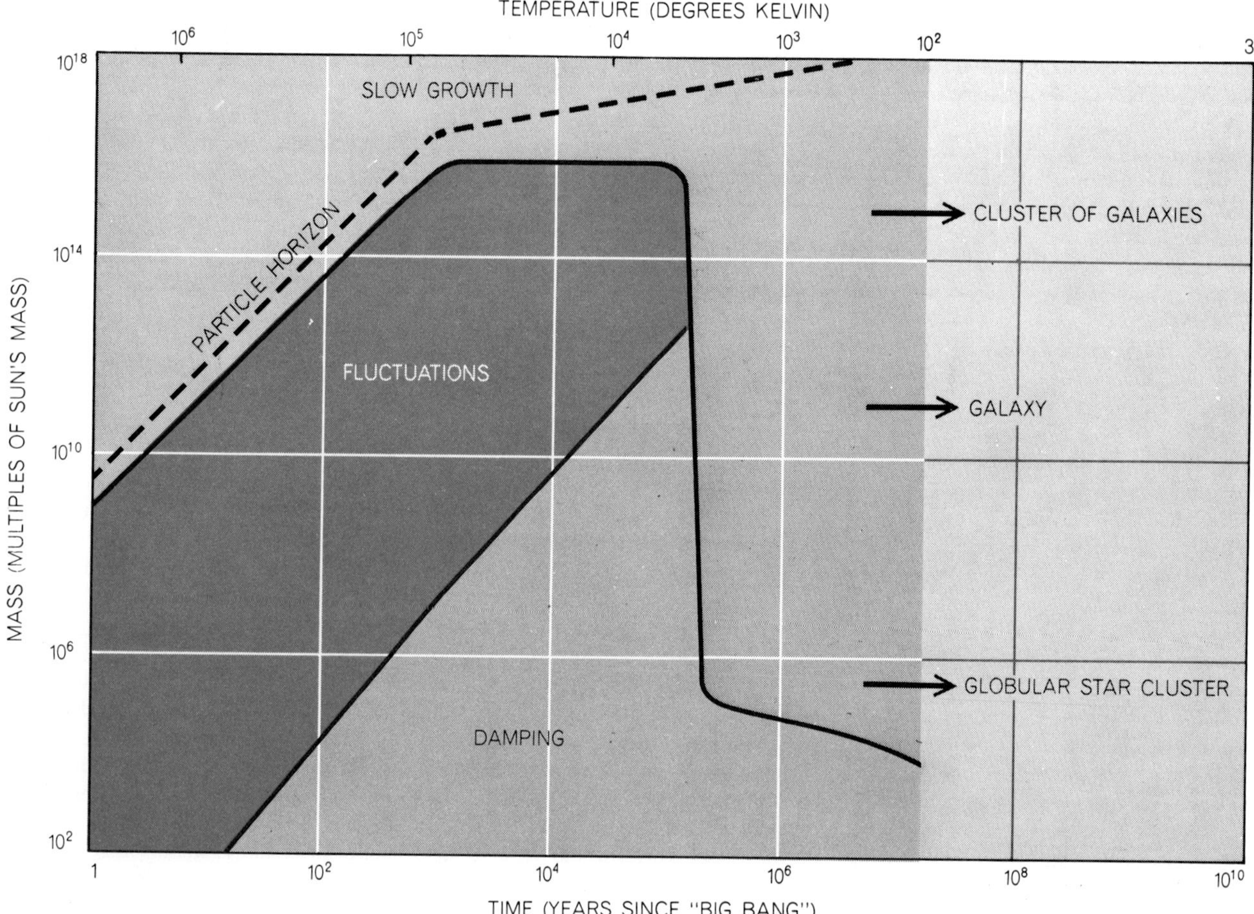

ANOTHER REPRESENTATION of the formation of galaxies is a graph relating the mass of a system to the age and ambient temperature of the universe. Any density enhancement that reaches a minimum value of some 10^{15} solar masses when the universe is about 1,000 years old (the epoch at which the density of matter first equals the mass density of the fireball radiation) has enough gravitational force to overwhelm the effects of radiation pressure. Such an enhancement thereupon enters on a lifetime of slow growth, culminating in a large cluster of galaxies. In an intermediate range (between 10^{11} and 10^{15} solar masses at decoupling) fluctuations in density persist until the decoupling stage is reached and radiation pressure ceases to interact effectively with matter; surviving density enhancements in this range become individual galaxies. Below a certain threshold (10^{11} solar masses at decoupling) radiation pressure damps out most density enhancements. Within this range, however, some density enhancements not accompanied by increases in radiation pressure (isothermal fluctuations) may survive to form globular star clusters, ranging from 10^5 to 10^6 solar masses. "Particle horizon" is boundary of the observable universe, where objects would be receding at speed of light.

fore most of the stars formed. Moreover, the first generation of stars would tend to be heavier and more luminous in relation to their mass than the stellar populations in present-day galaxies. Although most of this energy would be radiated in the ultraviolet, it would be received in the near infrared, owing to the red shift. Robert Bruce Partridge and Peebles at Princeton have suggested that it might be feasible to detect such young galaxies even though these may now have red shifts of about 10.

We are plainly still far from understanding even the broad outlines of the processes whereby the observed aggregations of matter in the universe came into being. We are even further from understanding the detailed morphology of the bewildering variety of different types of galaxies. For example, we have not yet discussed the possible origins of the angular momentum or magnetic fields of galaxies. Peebles has argued that the rotation of galaxies may be induced by tidal interactions soon after formation. Other authors, notably Leonid Ozernoi of the P. N. Lebedev Physical Institute in Moscow, have considered galactic rotation to be of primordial origin. One remarkable feature of the primordial fireball is that it can store rotation in the form of "photon whirls"; subsequently this stored rotation could be transferred to matter whirls.

Galactic magnetic fields may be produced after the formation of the galaxy by a mechanism of the dynamo type. Alternatively, magnetic fields may very well be of primordial origin. Edward R. Harrison of the University of Massachusetts has pointed out that the shear between the photon gas and the matter gas in the fireball could have generated a small magnetic field; if primordial photon whirls are assumed to be present, this mechanism leads to the production of a "seed" magnetic field many orders of magnitude below the value of the magnetic field observed in the spiral arms of our own galaxy. Harrison argues that rapid rotation of the protogalaxy may have subsequently produced sufficient winding of the primordial magnetic field to enhance it by the dynamo mechanism to the field currently observed. Primordial fields alone, he feels, would be insufficient to account for the observed galactic fields. The amount of rotation and the strength of the magnetic field in the protogalaxy probably help to determine whether it will evolve into an elliptical galaxy or into a spiral.

Galaxies are observed to possess random velocities with respect to the cosmic expansion. It is a curious coincidence that the rotational velocity of galaxies is of just the same order of magnitude—hundreds of kilometers per second—as these random motions. Perhaps this is simply a consequence of the primordial turbulence, which may have been the source of all structure in the universe.

Present data on the sizes of clusters of galaxies, and on possible "superclusters," are too sparse to enable us to assess the validity of theories that predict the mass spectrum of condensations. Moreover, our knowledge of the masses of galaxies is bedeviled by selection effects. Large and bright galaxies can be seen out to great distances, but small and intrinsically faint ones would only be noticed if they were comparatively close to us. Such objects may therefore occur much more frequently than is believed. A more drastic possibility is that most of the material in the universe may be in some nonluminous form. Evidence for the existence of such material comes from studies of the stability of clusters of galaxies.

This basic problem was first discussed in 1933 by Fritz Zwicky of the California Institute of Technology. For example, if one estimates the mass required to make the Virgo cluster a gravitationally bound system, one finds that the total observed mass in the member galaxies falls short by a factor of 50 or more. One possible way around this paradox is to assume that the Virgo system may be exploding, as the Soviet astrophysicist V. A. Ambartsumian has suggested. Perhaps even more puzzling is the apparent deficiency in mass of the Coma cluster. This system is so spherically symmetric and centrally condensed that astronomers believe it must be a stable system. Yet the observed mass, predominantly in elliptical galaxies, falls short of the mass required for stability by perhaps a factor of five, even if one assumes that the mass-luminosity ratio for ellipticals is around 50.

Similar results have been found for other clusters. Some astronomers have attempted to explain this problem by arguing that nonluminous matter is present in sufficient quantity to stabilize these systems. This material probably cannot all be in gaseous form; neutral hydrogen or ionized hydrogen, whether uniformly distributed or in clouds, ought to be observable either by radio or by X-ray observations.

Alternatively, the "missing mass" may be in the form of "dead," or burned-out, galaxies. An even more intriguing possibility is that concealed within the clusters are many objects that have undergone catastrophic gravitational collapse, as predicted by the general theory of relativity. The gravitational field around such objects would be so strong that no radiation could escape from them; only their gravitational influence could be detected by a distant observer.

Other arguments that indicate the apparent youthfulness of some galaxies stem from observations of clusters of galaxies. To be stable, one such chain of galaxies would require a mass-luminosity ratio of more than 5,000, or 100 times as much mass as the cluster seems to possess. One seems forced to the conclusion that here are newly formed galaxies, born within the past 100 million years. Zwicky has discovered an entire class of compact galaxies whose surface brightness resembles that found only in the nuclei of ordinary galaxies. Even more baffling is the discovery that some quasars emit as much radiation as 1,000 galaxies, the energy apparently coming from a colossal explosive event in a region less than 1 percent the size of the solar system. Seyfert galaxies display the same energetic phenomenon on a somewhat reduced scale.

Ambartsumian has long maintained that galactic nuclei are sources of matter and that indeed the galaxies themselves emerge out of dense primordial nuclei. In recent years Halton C. Arp of the Hale Observatories and Erik B. Holmberg of the University of Uppsala have found evidence that small galaxies may even have been ejected from larger galaxies. These phenomena certainly suggest that violent events, involving perhaps the birth of galaxies, are continually taking place in the nuclei of existing galaxies. One is reminded of Sir James Jeans's prescient conjecture, written in 1929, that "the centers of the nebulae are of the nature of 'singular points,' at which matter is poured into our Universe from some other, and entirely extraneous, spatial dimension, so that, to a denizen of our Universe, they appear as points at which matter is being continually created."

Further progress in this field must await fuller information on the distribution, masses and velocities of galaxies. Moreover, satellite observations in infrared, ultraviolet and X-ray wavelengths may soon reveal completely new and unsuspected types of objects, and should in any case give us confidence that we have a fairly complete inventory of the contents of the universe. We shall then be better able to relate theoretical abstractions to the universe in which we dwell.

CHAIN OF GALAXIES VV 172 was photographed by Halton C. Arp with the 200-inch Hale telescope. Four of the galaxies are 600 million light-years away; the fifth appears to be twice as distant. Conceivably it has been ejected from the cluster at high velocity.

EXPLODING GALAXY NGC 1275 was recently photographed in red light by C. Roger Lynds with the 84-inch telescope at Kitt Peak. The radiating filaments of gas, reminiscent of the Crab nebula, were not visible in earlier photographs taken in white light.

Biographical Notes and Bibliographies

I
The Earth and Moon

1. The Confirmation of Continental Drift

The Author

PATRICK M. HURLEY is professor of geology of the Massachusetts Institute of Technology. Born in Hong Kong, he lived there until he was nine years old. After being graduated from the University of British Columbia in 1934 with a degree of mining engineering he spent three years mining gold in British Columbia. He obtained a Ph.D. from M.I.T. in 1940 and has been on the faculty there since 1946. He serves as a consultant to industry and government on mineral resources and development. Hurley writes that in addition to his studies of continental drift his recent work "has been on the absolute abundance of minor elements in the earth and on the rate of separation of the crust from the earth's interior."

Bibliography

CONTINENTAL DRIFT. Edited by S. K. Runcorn. Academic Press, 1962.

THE ORIGIN OF CONTINENTS AND OCEANS. Alfred Wegener. Methuen & Co. Ltd., 1924.

SPREADING OF THE OCEAN FLOOR: NEW EVIDENCE. F. J. Vine in *Science,* Vol. 154, No. 3755, pages 1405–1415; 16 December 1966.

A SYMPOSIUM ON CONTINENTAL DRIFT. ORGANIZED FOR THE ROYAL SOCIETY. P. M. S. Blackett, F.R.S., Sir Edward Bullard, F.R.S., and S. K. Runcorn in *Philosophical Transactions of the Royal Society of London,* Series A, Vol. 258, pages vii–322; The Royal Society, 1965.

2. The Lunar Orbiter Missions to the Moon

The Authors

ELLIS LEVIN, DONALD D. VIELE and LOWELL B. ELDRENKAMP are members of engineering management in the Space Division of the Boeing Company. They worked closely in the design of the Lunar Orbiter missions: Levin as system engineering manager and as the project engineer for mission design and flight data; Viele as a supervisor in system engineering and as manager of mission design and analysis, and Eldrenkamp as supervisor of trajectory and guidance analysis. Levin was graduated from Tulane University in 1942 with a bachelor's degree in electrical engineering. Viele received a similar degree from the University of Colorado in 1950. Eldrenkamp was graduated from the University of Illinois in 1953 with bachelor's and master's degrees in aeronautical engineering. Each of them has been with Boeing since soon after graduation.

Bibliography

LUNAR ORBITER I, PHOTOGRAPHIC MISSION SUMMARY: NASA CR 782. The Boeing Company. Clearinghouse for Federal Scientific Information, April 1967.

LUNAR ORBITER II, PHOTOGRAPHIC MISSION SUMMARY: NASA CR 883. The Boeing Company. Clearinghouse for Federal Scientific and Technical Information, October 1967.

THE PRELIMINARY RESULTS OF LUNAR ORBITER I. L. R. Scherer and C. H. Nelson in *17th International Astronautical Congress, Madrid, 1966, Vol. I: Spacecraft Systems,* edited by Michal Lunc, P. Contensou, G. N. Duboshin and W. F. Hilton. Dunod Editeur, Gordon and Breach Science Publishers, Inc., and TWN Polish Scientific Publisher, 1967.

More recent studies of the moon are reported in:

APOLLO ASTRONAUTS PHOTOGRAPH THE MOON. *Sky and Telescope,* Vol. 37, pages 136–146; March 1969.

MAN ON THE MOON. Special issue of *Science Journal,* Vol. 5, No. 5; May 1969.

MARE ORIENTALE AND ITS INTRIGUING BASIN. William K. Hartman and Francis G. Yale in *Sky and Telescope,* Vol. 37, pages 4–7; January 1969.

PHOTOGRAPHIC RESULTS OF THE LUNAR ORBITER PROGRAM. William E. Brunk in *Highlights of Astronomy,* edited by Luboš Perek, pages 471–523. D. Reidel Publishing Company, 1968.

REPORTS ON THE APOLLO 11 LUNAR SAMPLES. Special issue of *Science,* Vol. 167, No. 3918; 30 January 1970.

II
The Planetary System

3. Radar Observations of the Planets

The Author

IRWIN I. SHAPIRO is professor of geophysics and physics at the Massachusetts Institute of Technology and a staff member of M.I.T.'s Lincoln Laboratory. After graduation from Cornell University in 1950 he began work at Harvard University, where he obtained a master's degree in 1951 and a Ph.D. in physics four years later. In 1954 he joined the Lincoln Laboratory. In addition to the subject on which he writes his interests include nuclear physics, statistical mechanics and the dynamics of satellites.

Bibliography

FOURTH TEST OF GENERAL RELATIVITY: PRELIMINARY RESULTS. Irwin I. Shapiro, Gordon H. Pettengill, Michael E. Ash, Melvin L. Stone, William B. Smith, Richard P. Ingalls and Richard A. Brockelman in *Physical Review Letters,* Vol. 20, No. 22, pages 1265–1269; 27 May 1968.

RADAR ASTRONOMY. Edited by John V. Evans and Tor Hagfors. McGraw-Hill Book Company, 1968.

RADAR ASTRONOMY. Gordon H. Pettengill and Irwin I. Shapiro in *Annual Review of Astronomy and Astrophysics,* Vol. 3, pages 377–410; 1965.

SPIN-ORBIT COUPLING IN THE SOLAR SYSTEM, II: THE RESONANT ROTATION OF VENUS. Peter Goldreich and Stanton Peale in *The Astronomical Journal,* Vol. 72, No. 1350, pages 662–668; June 1967.

THEORY OF THE AXIAL ROTATIONS OF MERCURY AND VENUS. E. Bellomo, G. Colombo and I. I. Shapiro in *Mantles of the Earth and Terrestrial Planets,* edited by S. K. Runcorn. Interscience Publishers, 1967.

4. The Atmospheres of Mars and Venus

The Author

VON R. ESHLEMAN is professor of electrical engineering at Stanford University and codirector of the Center for Radar Astronomy, a joint group of research workers from the university and the Stanford Research Institute. He attended George Washington University and received advanced degrees in electrical engineering at Stanford in 1950 and 1952. Since then he has been at Stanford as a researcher and teacher, heading a number of projects related to radio propagation and radar studies of the upper atmosphere, the sun, the moon, the planets and interplanetary space. He is a member of the Lunar and Planetary Missions Board of the National Aeronautics and Space Administration and a consultant to other Government and industrial groups.

Bibliography

THE ATMOSPHERE OF VENUS: PAPERS FROM THE SECOND ARIZONA CONFERENCE ON PLANETARY ATMOSPHERES. *Journal of the Atmospheric Sciences,* Vol. 25, No. 4; July 1968.

THE ATMOSPHERES OF VENUS AND MARS. Edited by John C. Brandt and Michael B. McElroy. Gordon and Breach, 1968.

LIFE DETECTION BY ATMOSPHERIC ANALYSIS. Dian R. Hitchcock and James E. Lovelock in *Icarus,* Vol. 7, No. 2, pages 149–159; March 1967.

MARINER V. *Science,* Vol. 158, No. 3809, pages 1665–1690; 29 December 1967.

MARINER IV MEASUREMENTS NEAR MARS: INITIAL RESULTS. *Science,* Vol. 149, No. 3689, pages 1226–1248, 10 September 1965.

THE SURFACE OF MARS. Robert B. Leighton in *Scientific American,* Vol. 222, No. 5, pages 26–41; May 1970.

VENUS: LOWER ATMOSPHERE NOT MEASURED. Von R. Eshleman, Gunnar Fjeldbo, John D. Anderson, Arvydas Kliore and Rolf B. Dyce in *Science,* Vol. 162, No. 3854, pages 661–665; 8 November 1968.

5. The Discovery of Icarus

The Author

ROBERT S. RICHARDSON, who tells the story of Icarus, has begun a career as a free-lance writer after 33 years as an astronomer and writer. Born in Kokomo, Ind., he received a bachelor's degree at the University of California at Los Angeles and a Ph.D. in astronomy at the University of California at Berkeley. He was on the staff of the Mount Wilson Observatory from 1931 to 1958 and served as associate director of the Griffith Observatory and Planetarium in Los Angeles from 1958 to 1964. Having begun to write articles and short stories in 1939, he found that writing was taking more and more of his time. So, he says, he "decided to take Nietzsche's advice to 'live dangerously' " and resigned to devote all his time to writing. He feels that he is in a particularly good position to assist in improving "communication between scientists and humanists."

Bibliography

BETWEEN THE PLANETS. Fletcher G. Watson. Harvard University Press, 1956.

EARTH SATELLITES AND RELATED ORBIT AND PERTURBATION THEORY. Samuel Herrick in *Space Technology,* edited by Howard Seifert. John Wiley & Sons, Inc., 1959.

ICARUS AND THE SPACE AGE. Maud W. Makemson in Leaflet No. 397, Astronomical Society of the Pacific, July 1962.

ICARUS AND THE VARIATION OF PARAMETERS. Samuel Herrick in *The Astronomical Journal,* Vol. 58, No. 6, pages 156–164; August 1953.

ICARUS FLIES PAST THE EARTH. *Sky and Telescope,* Vol. 36, pages 75–77; August 1968.

POPULAR ASTRONOMY. Simon Newcomb. Harper & Bros., 1878.

6. Jupiter's Great Red Spot

The Author

RAYMOND HIDE recently assumed the position of director of the geophysical fluid dynamics laboratory of the British Meteorological Office. He accepted this new responsibility after several years as professor of geophysics and physics at the Massachusetts Institute of Technology. Hide was born in England and received a bachelor's degree in physics at the University of Manchester and a Ph.D. in geophysics from the University of Cambridge. He has held appointments at the University of Chicago and, in England, at the Atomic Energy Research Establishment in Harwell and at King's College of the University of Durham (now part of the University of Newcastle upon Tyne).

Bibliography

THE ATMOSPHERE OF JUPITER. Tobias Owen in *Science,* Vol. 167, pages 1675–1681, 27 March 1970.

AN EXPERIMENTAL STUDY OF "TAYLOR COLUMNS." R. Hide and A. Ibbetson in *Icarus: International Journal of the Solar System,* Vol. 5, No. 3, pages 279–290; May 1966.

THE DISCOVERY OF JUPITER'S RED SPOT. Clark R. Chapman in *Sky and Telescope,* Vol. 35, pages 276–278, May 1968.

JUPITER'S GREAT RED SPOT. Wendell C. DeMarcus and Rupert Wildt in *Nature,* Vol. 209, No. 5018, page 62; 1 January 1966.

MOTIONS OF PLANETARY ATMOSPHERES. Richard Goody in *Annual Review of Astronomy and Astrophysics;* Vol. 7, pages 303–352; 1969.

ORIGIN OF JUPITER'S GREAT RED SPOT. R. Hide in *Nature,* Vol. 190, No. 4779, pages 895–896; 3 June 1961.

THE PLANET JUPITER. Bertrand M. Peek. Faber and Faber, 1958.

THE STRUCTURE AND COMPOSITION OF JUPITER AND SATURN. P. J. E. Peebles in *Astrophysical Journal,* Vol. 140, No. 1, pages 328–347; 1 July 1964.

7. The Solar System beyond Neptune

The Author

OWEN GINGERICH, the editor of this volume, is an astrophysicist at the Smithsonian Astrophysical Observatory and Professor of Astronomy and of the History of Science at Harvard University. He wrote this article while he was a graduate student at Harvard. "I have been in love with the stars as long as I can remember," he says. Born in Iowa in 1930, he was a life member of the American Association of Variable Star Observers by the time he was 15. "In college I completed my own eight-inch telescope, with which I independently found Honda's comet many weeks after it was known to the rest of the astronomical world!," wrote Gingerich in 1958 to the *Scientific American* editors, little realizing that within a decade he would become director of the international astronomical telegram bureau that names new comets, and that one of his first responsibilities would be to announce another Comet Honda. Gingerich continued, "One summer I worked for Harlow Shapley at the Harvard College Observatory; this experience led me to professional astronomy. In 1955 I was a member of the Harvard eclipse expedition to Ceylon, and for three years following I was director of the American University Observatory in Beirut."

Bibliography

THE FORMATION OF THE PLANETS. Gerard P. Kuiper in *Journal of the Royal Astronomical Society of Canada,* Vol. 50, pages 57–68, 105–121, 158–176; 1956.

THE GEORGE DARWIN LECTURE: THE MOTIONS OF THE OUTER PLANETS. Dirk Brouwer in *Monthly Notices of the Royal Astronomical Society,* Vol. 115, No. 3, pages 221–235; 6 April 1955.

THE HISTORY OF THE SOLAR SYSTEM. Fred L. Whipple in *Proceedings of the National Academy of Sciences,* Vol. 52, pages 565–594; August 1964.

THE ORIGIN OF THE SATELLITES AND THE TROJANS. Gerard P. Kuiper in *Vistas in Astronomy,* Vol. 2, pages 1,631–1,666; 1956.

ORIGIN OF THE SOLAR SYSTEM. Edited by Robert Jastrow and A. G. W. Cameron. Academic Press, 1963.

III
The Sun

8. The Solar Chromosphere

The Author

R. GRANT ATHAY, who leads off this section, is a member of the faculty in the department of physics and astrophysics at the University of Colorado and a member of the Senior Research Staff at the High Altitude Observatory in Boulder, Colo. Born in Smithfield, Utah, in 1923, Athay was trained as a meteorologist during World War II and served for several years as a weather forecaster for the Army Air Force. Having studied physics and radio engineering at Utah State Agricultural College, where he acquired a B.S. in 1947, Athay saw in the field of astrophysics a chance to combine his different scientific interests in a single area of research. He joined the High Altitude Observatory in 1950, received a Ph.D. in astrophysics from the University of Utah in 1953 and became a member of the Senior Research Staff that same year. From 1955 to 1956 he was research associate at the Harvard College Observatory. Athay has worked at the University of Colorado since 1956.

Bibliography

THE CHROMOSPHERE-CORONA TRANSITION REGION. G. W. Pneuman in *Sky and Telescope,* Vol. 39, pages 148–151; March 1970.

HOT SPOTS IN THE ATMOSPHERE OF THE SUN. Harold Zirin in *Scientific American,* Vol. 199, pages 34–41; August 1958.

OUR SUN. D. H. Menzel. Harvard University Press, 1959.

PHYSICS OF THE SOLAR CHROMOSPHERE. R. Grant Athay and Richard N. Thomas. Interscience Publishers, Inc., 1961.

THE SOLAR ATMOSPHERE. Harold Zirin. Blaisdell, 1956.

THE SUN. Edited by Gerard P. Kuiper. The University of Chicago Press, 1953.

THE SUN AND ITS INFLUENCE. M. A. Ellison. Routledge and Kegan Paul, Ltd., 1955.

9. The Solar Wind

The Author

E. N. PARKER is professor of physics at the University of Chicago; he is also a member of the staff of the university's Enrico Fermi Institute for Nuclear Studies. A graduate of Michigan State University, Parker received a Ph.D. in theoretical physics from the California Institute of Technology in 1951. He taught mathematics and astronomy at the University of Utah from 1951 to 1955, when he joined the staff of the Fermi Institute.

Bibliography

INTRODUCTION TO THE SOLAR WIND. John C. Brandt. W. H. Freeman & Company, 1970.

DIRECT OBSERVATIONS OF SOLAR-WIND PARTICLES. A. J. Hundhausen in *Space Science Reviews,* Vol. 8, pages 690–749; 1968.

DYNAMICAL PROPERTIES OF STELLAR CORONAS AND STELLAR WINDS: I. INTEGRATION OF THE MOMENTUM EQUATION; II. INTEGRATION OF THE HEAT-FLOW EQUATION. E. N. Parker in *The Astrophysical Journal,* Vol. 139, No. 1, pages 72–122; January 1964.

DYNAMICS OF THE INTERPLANETARY GAS AND MAGNETIC FIELDS. E. N. Parker in *The Astrophysical Journal,* Vol. 128, No. 3, pages 664–676; November 1958.

INTERPLANETARY DYNAMICAL PROCESSES. E. N. Parker. Interscience Publishers, 1963.

THE INTERPLANETARY MAGNETIC FIELD, SOLAR ORIGIN AND TERRESTRIAL EFFECTS. John M. Wilcox in *Space Science Reviews,* Vol. 8, pages 258–328; 1968.

NOTES ON THE SOLAR CORONA AND THE TERRESTRIAL IONOSPHERE. Sydney Chapman in *Smithsonian Contributions to Astrophysics,* Vol. 2, No. 1, pages 1–12; 1957.

10. Neutrinos from the Sun

The Author

JOHN N. BAHCALL is associate professor of theoretical physics at C.I.T. He was graduated from the University of California at Berkeley in 1956 and received his Ph.D. from Harvard University in 1961. He was a research fellow at Indiana University for two years before going to Cal. Tech. His wife, Neta, whom he married in Israel in 1966, received her Ph.D. in physics from the University of Tel Aviv. Bahcall writes: "I have had many able collaborators in the theoretical work described in this article. Everyone agrees, however, that my wife is the prettiest."

Bibliography

PRESENT STATUS OF THE THEORETICAL PREDICTIONS FOR THE ^{37}CL SOLAR-NEUTRINO EXPERIMENT. John N. Bahcall, Neta A. Bahcall and Giora Shaviv in *Physical Review Letters,* Vol. 20, No. 21, pages 1209–1212; 20 May 1968.

SEARCH FOR NEUTRINOS FROM THE SUN. Raymond Davis, Jr., Don S. Harmer and Kenneth C. Hoffman in *Physical Review Letters,* Vol. 20, No. 21, pages 1205–1209; 20 May 1968.

SOLAR NEUTRINOS. John N. Bahcall in *Physical Review Letters,* Vol. 17, No. 7, pages 398–401; 15 August 1966.

SOLAR NEUTRINOS, I: THEORETICAL. John N. Bahcall in *Physical Review Letters,* Vol. 12, No. 11, pages 300–302; 16 March 1964.

SOLAR NEUTRINOS, II: EXPERIMENTAL. Raymond Davis, Jr., in *Physical Review Letters,* Vol. 12, No 11, pages 303–305; 16 March 1964.

STRUCTURE AND EVOLUTION OF STARS. Martin Schwarzschild. Princeton University Press, 1958.

IV
Stellar Evolution

11. Life Outside the Solar System

The Author

SU-SHU HUANG has been an astronomer at the Dearborn Observatory of Northwestern University since 1965. Earlier he was a physicist at the National Aeronautics and Space Administration's Goddard Space Flight Center in Washington. He was born in Changshu in the Chinese province of Kiangsi in 1915. In high school, he says, he found physics and mathematics the easiest subjects "because they do not demand a good memory, which I lack." He settled on physics because he was more interested in natural phenomena than in abstract thought. He studied at the National Chekiang and Tsing Hua universities and came to the U. S. in 1947 on a fellowship given by the Nationalist Chinese Government. Huang earned his Ph.D. in astrophysics at the University of Chicago in 1949. A Guggenheim fellowship took him to the University of California, where he did research work for eight years. He has published papers in atomic physics, radiative transfer, stellar spectroscopy, celestial mechanics and applied mathematical problems. He states that his interest in the problem of life outside the solar system was aroused through reading various *Scientific American* articles concerning the origin and processes of life.

Bibliography

THE ORIGIN OF BINARY STARS. Su-Shu Huang in *Sky and Telescope,* Vol. 34, pages 368–370; December 1967.

OCCURRENCE OF LIFE IN THE UNIVERSE. Su-Shu Huang in *The American Scientist,* Vol. 47, No. 3, pages 397–402; Autumn 1959.

INTELLIGENT LIFE IN THE UNIVERSE. I. S. Shklovskii and Carl Sagan. Holden-Day, 1966.

THE ORIGIN OF LIFE ON EARTH. A. I. Oparin. Oliver & Boyd, 1957.

THE PROBLEM OF LIFE IN THE UNIVERSE AND THE MODE OF STAR FORMATION. Su-Shu Huang in *Publications of the Astronomical Society of the Pacific,* Vol. 71, No. 422, pages 421–424; October 1959.

12. The Crab Nebula

The Author

JAN H. OORT is professor of astronomy at the University of Leiden in the Netherlands and until recently he served as director of the observatory there.

He entered the University of Groningen in 1917 to study physics or astronomy, "two subjects by which I had been fascinated during my high-school years. There I soon came under the inspiring teaching of J. C. Kapteyn, who was one of the great pioneers of galactic research. This determined my further scientific life. From the beginning I was particularly attracted by problems of stellar dynamics, both for our own galactic system and for other systems, but always only insofar as the problems were directly connected with observations. This included, of course, the study of the interstellar medium, and now also includes radio astronomy. Besides, I worked on the Crab Nebula and on the origin of comets."

Immediately after he had graduated from Groningen, Oort was for two years a research assistant at the Yale Observatory. Since 1924, except for a three-year interruption during the war, he has been at Leiden. In 1935 he became general secretary of the International Astronomical Union, a post he filled for 13 years, and from 1958 to 1961 he served as president of the Union.

Bibliography

THE CRAB NEBULA. Chapter 3 in Supernovae by I. S. Shklovsky, pages 206–356; John Wiley and Sons, 1968.

POLARIZATION AND COMPOSITION OF THE CRAB NEBULA. J. H. Oort and Th. Walraven in *Bulletin of the Astronomical Institutes of the Netherlands,* Vol. 12, No. 462, pages 285–308; 5 May 1956.

THE POLARIZATION OF THE CRAB NEBULA ON PLATES TAKEN WITH THE 200-INCH TELESCOPE. W. Baade in *Bulletin of the Astronomical Institutes of the Netherlands,* Vol. 12, No. 462, page 312; 5 May 1956.

13. The Youngest Stars

The Author

GEORGE H. HERBIG is an astronomer at the Lick Observatory, where he has been a member of the staff since 1949. Herbig was graduated from U.C.L.A. in 1943 and received a Ph.D. from the University of California at Berkeley in 1948. He is a member of the International Astronomical Union and is also a member of the National Academy of Sciences. Considered the leading authority on T Tauri stars, Herbig is an active observer and specialist on pre-main sequence stars, including the so-called Herbig-Haro objects. These are spherical clouds of glowing interstellar material that may be in the process of star-formation.

Bibliography

EVOLUTION OF PROTOSTARS. Chushiro Hayashi in *Annual Review of Astronomy and Astrophysics,* Vol. 4, pages 171–192; 1966.

INFANT STARS. V. C. Reddish in *Science Journal,* Vol. 4, No. 7, pages 30–35; July 1968.

INFRARED PHOTOMETRY OF T TAURI STARS AND RELATED OBJECTS. Eugenio E. Mendoza V in *The Astrophysical Journal,* Vol. 143, No. 3, pages 1010–1014; March 1966.

IONISATION DANS LES OBJECTS DE HERBIG-HARO. Christian Magnan and Evry Schatzman in *Comptes Rendus des Séances de l'Académie des Sciences de l'Institut de France,* Vol. 260, No. 24, pages 6289–6291; Group 3, 14 June 1965.

THE PROPERTIES AND PROBLEMS OF T TAURI STARS AND RELATED OBJECTS. G. H. Herbig in *Advances in Astronomy and Astrophysics,* Vol. 1, pages 47–103. Academic Press, 1962.

T TAURI VARIABLE STARS. Alfred H. Joy in *The Astrophysical Journal,* Vol. 102, No. 2, pages 168–195; September 1945.

14. Dying Stars

The Author

JESSE L. GREENSTEIN is an astronomer at the Mount Wilson and Palomar Observatories and, in addition, heads the astronomy department at the California Institute of Technology. A New Yorker, he graduated from Harvard College in 1929, received his M.A. at Harvard in 1930, then rode out four depression years as an operator in real estate and investments. In 1934 he returned to Harvard for his Ph.D. Subsequently he joined the staff of the University of Chicago's Yerkes Observatory, first as a National Research Fellow, then as associate professor. Greenstein has worked at Mount Wilson and Palomar and Cal Tech since 1949. A spectroscopist, he has been interested in interstellar matter, in the chemical abundances in stars and the relation to nuclear processes, and in the ages and evolution of stars.

Bibliography

AN INTRODUCTION TO THE STUDY OF STELLAR STRUCTURE. S. Chandrasekhar. The University of Chicago Press, 1939.

A NIGHT AT THE OBSERVATORY. Henry S. F. Cooper in *Horizon,* Vol. 9, No. 3, pages 108–116; Summer, 1967. Reprinted in *Project Physics Reader 2;* Holt, Rinehart and Winston, 1969.

THE SPECTRA OF THE WHITE DWARFS. Jesse L. Greenstein in *Encyclopedia of Physics.* Springer-Verlag, 1958.

WHITE DWARFS. Evry Schatzman. Interscience, 1958.

V
The Milky Way

15. Stellar Populations

The Authors

MARGARET BURBIDGE and GEOFFREY BURBIDGE are a husband-and-wife team in science—she being an astronomer, he an astrophysicist. The Burbidges, who are English, met 10 years ago in a University of London lecture hall. At that time she was employed in the University of London Observatory, while he was a physicist. Their courtship resulted in his conversion to astrophysics and also in their marriage. In 1951, when Geoffrey Burbidge came to the Harvard College Observatory as an Agassiz Fellow, his wife became a fellow of the University of Chicago's Yerkes Observatory at Williams Bay, Wis. (the Burbidges commuted between the two places). In 1953 the Burbidges returned to England, where he worked for two years in the Cavendish Laboratory of the University of Cambridge. After returning to the United States in 1955 they worked at Mount Wilson and Palomar Observatories and at the University of Chicago; presently each of the Burbidges holds a professorship at the University of California at San Diego.

Bibliography

FRONTIERS OF ASTRONOMY. Fred Hoyle. William Heinemann, Ltd., 1955.

THE RESOLUTION OF MESSIER 32, NGC 205, AND THE CENTRAL REGION OF THE ANDROMEDA NEBULA. W. Baade in *Astrophysical Journal,* Vol. 100, No. 2, pages 137–146; September 1944.

STRUCTURE AND EVOLUTION OF THE STARS. Martin Schwarzschild. Princeton University Press 1958.

16. The Arms of the Galaxy

The Author

BART J. BOK is professor of astronomy at the Steward Observatory of the University of Arizona. A native of Hoorn in the Netherlands, he was educated at the universities of Leiden and Groningen. He came to Harvard University in 1929 as a fellow in astronomy and joined the faculty there in 1933. When he left in 1957 to go to Australia, he was Robert Wheeler Willson Professor of Astronomy. He served as director of the Mount Stromlo Observatory in Australia and professor of astronomy at the Australian National University until his return to the United States in 1966.

His major interest has always been the structure of the Milky Way. The center of our galaxy is best seen from the Southern Hemisphere, and Bok was in charge of installing the Baker-Schmidt telescope at Harvard's Boyden Station in South Africa.

Bibliography

THE DISTRIBUTION OF THE STARS IN SPACE. Bart J. Bok. The University of Chicago Press, 1937.

THE LARGE-SCALE STRUCTURE OF THE GALAXY. F. J. Kerr, J. V. Hindman and Martha Stahr Carpenter in *Nature,* Vol. 180, No. 4,588, pages 677–679; 5 October 1957.

RADIO ASTRONOMICAL STUDIES OF THE GALACTIC SYSTEM. Jan H. Oort in *Galaxies and the Universe,* edited by Lodewijk Woltjer, pages 1–32. Columbia University Press, 1968.

SPIRAL STRUCTURE IN GALAXIES. C. C. Lin in *Galaxies and the Universe,* edited by Lodewijk Woltjer, pages 33–51. Columbia University Press, 1968.

THE SPIRAL STRUCTURE OF OUR GALAXY. Bart J. Bok in *Sky and Telescope,* Vol. 38, pages 392–395; December 1969, and Vol. 39, pages 21–25, January 1970.

THE SPIRAL STRUCTURE OF THE GALAXY. Bart J. Bok in *American Scientist,* Vol. 55 pages 375–399; December 1967.

17. Interstellar Grains

The Author

J.MAYO GREENBERG, the author of this article, is professor of physics and astronomy at Rensselaer Polytechnic Institute. He has been at Rensselaer since 1952, when he joined the department of physics as assistant professor. Before that he had been a physicist with the National Advisory Committee for Aeronautics from 1944 to 1946, obtained a Ph.D. from Johns Hopkins University in 1948, spent three years teaching at the University of Delaware and worked for a year as research associate at the Institute of Fluid Dynamics and Applied Mathematics of the University of Maryland.

Bibliography

NEBULAE AND INTERSTELLAR MATTER. Edited by B. Middlehurst and L. H. Aller. University of Chicago Press 1968.

INTERSTELLAR DUST. Beverly T. Lynds and N. C. Wickramisinghe in *Annual Review of Astronomy and Astrophysics,* Vol. 6, pages 215–248; 1968.

INTERSTELLAR GRAINS. J. Mayo Greenberg in *Annual Review of Astronomy and Astrophysics,* Vol. 1, pages 267–290; 1963.

INTERSTELLAR MATTER IN GALAXIES. Edited by L. Woltjer. W. A. Benjamin, Inc., 1962.

THE INTERSTELLAR MEDIUM. S. A. Kaplan and S. B. Pikelner. Harvard University Press, 1970.

LIGHT SCATTERING BY SMALL PARTICLES. H. C. van de Hulst. John Wiley & Sons, 1957.

18. Hydroxyl Radicals in Space

The Author

BRIAN ROBINSON, whose article ends this section, is senior research scientist at the Radiophysics Laboratory of the Commonwealth Scientific and Industrial Research Organisation in Sydney. Born in Melbourne, he was graduated from the University of Sydney in 1952 and obtained a master's degree there a year later. After working briefly with the radio astronomy group at the C.S.I.R.O. he won a Royal Society scholarship to the University of Cambridge, where he received a Ph.D. in 1958. From 1958 to 1961 he was with the Netherlands Foundation for Radio Astronomy, leaving it to take up his present post. He writes that his research interests are "the study of neutral hydrogen in external galaxies and in intergalactic space; development of high-sensitivity receivers for radio astronomy and detection of new spectral lines at radio wavelengths."

Bibliography

RADIO OBSERVATIONS OF OH IN THE INTERSTELLAR MEDIUM. S. Weinreb, A. H. Barrett, M. L. Meeks and J. C. Henry in *Nature,* Vol. 200, No. 4909, pages 829–831; 30 November 1963.

THE INTERSTELLAR HYDROXYL RADIO EMISSION. Nannielou H. Dieter, Harold Weaver and David R. W. Williams in *Sky and Telescope,* Vol. 31, No. 3, pages 132–136; March 1966.

INTERSTELLAR MASERS. Dale F. Dickinson, Marvin M. Litvak and Benjamin Zuckerman in *Sky and Telescope,* Vol. 39, 4–7; January 1970.

OH MOLECULES IN THE INTERSTELLAR MEDIUM. B. J. Robinson and R. X. McGee in *Annual Review of Astronomy and Astrophysics,* Vol. 5, pages 183–212; 1967.

RADIO OBSERVATIONS OF INTERSTELLAR HYDROXYL RADICALS. Alan H. Barrett in *Science,* Vol. 157, No. 3791, pages 881–889; 25 August 1967.

RADIO SIGNALS FROM HYDROXYL RADICALS. Alan H. Barrett in *Scientific American,* Vol. 219, pages 36–44; December 1968.

VI
Galaxies

19. Pulsating Stars and Cosmic Distances

The Author

ROBERT P. KRAFT is professor of astronomy at the University of California at Santa Cruz. After receiving B.S. and M.S. degrees in mathematics from the University of Washington, Kraft joined the faculty of Whittier College in California. "There are two kinds of astronomers," he remarks: "those who are born to it—who want to do astronomy from the time they are 10 years old and grind their first mirror—and those who come into it later on from some other field, as in my own case. What prompted me was an interest that developed during my stay at Whittier in the philosophy of the physical sciences. This led me to turn from pure mathematics to a truly empirical science, in which theory really meets the test of experience." Aided by National Science Foundation fellowships, Kraft earned a Ph.D. in astronomy from the University of California in 1955 and went on to postdoctoral research at Mount Wilson and Palomar. He wrote "Pulsating and Cosmic Distances" while working as assistant professor of astronomy at the Yerkes Observatory of the University of Chicago. The following year he returned to Mount Wilson and Palomar where he served on the staff for several years.

Bibliography

EXTRA-GALACTIC DISTANCE SCALE. Sidney van den Bergh in *Nature,* Vol. 225, pages 503–505; 7 February 1970.

THE GALAXIES OF THE LOCAL GROUP. Sidney van den Bergh in *Journal of the Royal Astronomical Society of Canada,* Vol. 62, pages 145–180, 219–256; 1968.

THE PERIOD LUMINOSITY RELATION: A HISTORICAL REVIEW. J. D. Fernie in *Publications of the Astronomical Society of the Pacific,* Vol. 81, 707–731; December 1969.

THE PERIOD-LUMINOSITY RELATION OF THE CEPHEIDS. W. Baade in *Publications of the Astronomical Society of the Pacific,* Vol. 68, pages 5–16; February 1956.

20. Hydrogen in Galaxies

The Author

MORTON S. ROBERTS is a staff scientist at the National Radio Astronomy Observatory. Roberts received a B.A. in 1948 from Pomona College, where a job at the college observatory eventually led him to do graduate work in astronomy. From 1949 to 1952 Roberts taught physics at Occidental College (acquiring an M.Sc. from the California Institute of Technology in 1950). He worked for a year as a physicist in underwater ordnance for the Navy Department and then went to the University of California, which awarded him a Ph.D. in 1958. Roberts joined the Harvard Observatory in 1960 as research associate and lecturer in astronomy, a post he held until going to Green Bank, West Virginia in 1964. His radio astronomy research on the hydrogen content of galaxies is the subject of a film made by the American Astronomical Society.

Bibliography

COSMIC RADIO WAVES AND THEIR INTERPRETATION. J. L. Pawsey and E. R. Hill in *Reports on Progress in Physics,* Vol. 24, pages 69–115; 1961.

THE NEUTRAL HYDROGEN CONTENT OF LATE-TYPE SPIRAL GALAXIES. Morton S. Roberts in *The Astronomical Journal,* Vol. 67, No. 7, pages 437–446; September 1962.

RADIO ASTRONOMICAL STUDIES OF THE GALACTIC SYSTEM. Jan H. Oort in *Galaxies and the Universe,* edited by Lodewijk Woltjer, pages 1–32. Columbia University Press, 1968.

21. The Evolution of Galaxies

The Author

HALTON C. ARP took his present job as an astronomer on the staff of the Mount Wilson and Palomar Observatories in 1957. He received an A.B. from Harvard University in 1949 and a Ph.D. from the California Institute of Technology in 1953, was a Carnegie Fellow at the Mount Wilson and Palomar Observatories until 1955 and a research associate at Indiana University until 1957. His research interests include the discovery and analysis of peculiar galaxies, and the relation of the peculiar galaxies to the quasars.

Bibliography

THE GALACTIC SYSTEM AS A SPIRAL NEBULA. J. H. Oort, F. T. Kerr and C. Westerhout in *Monthly Notices of the Royal Astronomical Society,* Vol. 118, No. 4, pages 379–389; 1958.

GENERAL PHYSICAL PROPERTIES OF EXTERNAL GALAXIES. G. de Vaucouleurs in *Handbuch der Physik,* Vol. 53, pages 311–372; 1959.

THE HUBBLE ATLAS OF GALAXIES. Allan Sandage. *Carnegie Institution of Washington, Publication No. 618;* 1961.

ON THE ORIGIN OF ARMS IN SPIRAL GALAXIES. Halton C. Arp in *Sky and Telescope,* Vol. 38, pages 385–387; December 1969.

22. Exploding Galaxies

The Author

ALLAN R. SANDAGE, who also wrote article 28, is an astronomer at the Mount Wilson and Palomar Observatories. A graduate of the University of Illinois, Sandage received a Ph.D. in astronomy from the California Institute of Technology in 1953. He has worked at Mount Wilson and Palomar since 1952. In 1960 he was awarded the Helen Warner prize of the American Astronomical Society and in 1963 he won the Eddington Medal of the Royal Astronomical Society. In 1960 Sandage and Thomas A. Matthews were the first to isolate the quasars. (For further details of Sandage's earlier career, see the biographical note for article 28, "The Red Shift.")

Bibliography

EVIDENCE FOR AN EXPLOSION IN THE CENTER OF THE GALAXY M82. C. R. Lynds and A. R. Sandage in *The Astrophysical Journal,* Vol. 137, No. 4, pages 1005–1021; May 1963.

EVIDENCE FOR THE OCCURRENCE OF VIOLENT EVENTS IN THE NUCLEI OF GALAXIES. G. R. Burbidge, E. M. Burbidge and A. R. Sandage in *Reviews of Modern Physics,* Vol. 35, No. 4, pages 947–972; October 1963.

THE EXPLODING GALAXY M82: EVIDENCE FOR THE EXISTENCE OF A LARGE-SCALE MAGNETIC FIELD. Allan R.

Sandage and William C. Miller in *Science,* Vol. 144, No. 3617, pages 382–388; April 1964.

RADIO-GALAXIES AND QUASARS, II. A. Sandage in *Highlights of Astronomy,* edited by Luboš Perek, pages 45–70. D. Reidel Publishing Company, 1968.

23. Seyfert Galaxies

The Author

RAY J. WEYMANN is an astronomer at Steward Observatory of the University of Arizona. Weymann was awarded his bachelor's degree in astronomy from the California Institute of Technology in 1956 and his Ph.D. from Princeton University in 1959. He then returned to Cal Tech as a postdoctoral fellow for two years before joining the staff of Steward Observatory, where he has remained since except for the academic year 1963–1964, when he was at the University of California at Los Angeles. Weymann's interests other than Seyfert galaxies include mass loss from stars, destruction of the element lithium in the sun, the primordial radiation field and intergalactic matter.

Bibliography

NUCLEAR EMISSION IN SPIRAL NEBULAE. Carl K. Seyfert in *The Astrophysical Journal,* Vol. 97, No. 1, pages 28–40; January 1943.

PROCEEDINGS OF THE CONFERENCE ON SEYFERT GALAXIES AND RELATED OBJECTS. A. G. Pacholczyk and R. J. Weymann in *The Astronomical Journal,* Vol. 73, No. 9, pages 836–943; November 1968.

QUASISTELLAR OBJECTS AND SEYFERT GALAXIES. Stirling A. Colgate in *Physics Today,* Vol. 22, pages 27–35; January 1969.

VII
The New Astronomy

24. X-ray Astronomy

The Author

HERBERT FRIEDMAN is superintendent of the Atmosphere and Astrophysics Division and chief scientist at the E. O. Hulburt Center for Space Research of the Naval Research Laboratory. Friedmann received a B.A. from Brooklyn College in 1936 and a Ph.D. from Johns Hopkins University in 1940. After teaching physics for a year at Johns Hopkins, he joined the staff of the Metallurgy Division at the Naval Research Laboratory. In 1942 he assumed supervision of the Electron Optics Branch and in 1958 he was appointed to his present post. Friedman conducted his first experiments in rocket astronomy with a captured V-2 rocket in 1949. By 1964 he had participated in more than 100 rocket experiments and several satellite launchings. These experiments have traced the effect of solar-cycle variations on X-ray and ultraviolet radiations from the sun, produced the first astronomical photographs made in X-ray and ultraviolet wavelengths, discovered the hydrogen corona around the earth and discovered X-ray radiations from the Crab Nebula and from the spherical galaxy M87. Friedman was awarded the 1964 Eddington Medal of the Royal Astronomical Society of London.

Bibliography

COSMIC X-RAY SOURCES, GALACTIC AND EXTRAGALACTIC. E. T. Byram, T. A. Chubb and H. Friedman in *Science,* Vol. 152, pages 66–71; 1 April 1966.

EXPERIMENTAL X-RAY ASTRONOMY. Bruno Rossi in *Perspectives in Modern Physics,* edited by R. E. Marshak, pages 383–411. Interscience Publishers, 1966.

X-RAY STARS. Riccardo Giacconi in *Scientific American,* Vol. 217, pages 36–46; December 1967.

THE IDENTIFICATION OF THE X-RAY SOURCE IN SCORPIUS. Herbert Gursky in *Sky and Telescope,* Vol. 32, pages 252–255; November 1966.

OBSERVATION OF X-RAY SOURCES OUTSIDE THE SOLAR SYSTEM. Riccardo Giacconi and Herbert Gursky in *Space Science Reviews,* Vol. 4, No. 2, pages 151–175; March 1965.

OBSERVATIONAL TECHNIQUES IN X-RAY ASTRONOMY. R. Giacconi, H. Gursky and L. P. Van Speybroeck in *Annual Reviews of Astronomy and Astrophysics,* Vol. 6, pages 373–416; 1968.

SOME RECENT RESULTS OF X-RAY ASTRONOMY. Bruno Rossi in *Galaxies and the Universe,* edited by Lodewijk Woltjer, pages 53–73. Columbia University Press, 1968.

25. The Infrared Sky

The Author

G. NEUGEBAUER and ROBERT B. LEIGHTON are coauthors who are professors of physics at the California Institute of Technology. Neugebauer was graduated from Cornell University and obtained his Ph.D. from Cal Tech in 1960. After two years as an Army officer, during which time he served at the Jet Propulsion Laboratory, he returned to Cal Tech to pursue work in infrared astronomy. He participated in an infrared radiometric experiment flown past Venus on the spacecraft *Mariner II* and prepared a similar experiment for the two Mariner spacecraft that investigated Mars in 1969. Leighton was graduated from Cal Tech in 1941 and received his Ph.D. there in 1947. He worked on cosmic ray physics until about 1960, when his research interests turned toward solar physics and astrophysics. He participated in the television experiment carried out by *Mariner IV* and worked on a more comprehensive television experiment that was included in the Mars spacecraft in 1969. Neugebauer and Leighton cooperated in the construction and operation of an infrared telescope starting in 1963 and are now planning a larger instrument.

Bibliography

DISCOVERY OF AN INFRARED NEBULA IN ORION. D. E. Kleinmann and F. J. Low in *The Astrophysical Journal,* Vol. 149, No. 1, Part 2, pages L1–L4; July 1967.

FURTHER OBSERVATIONS OF EXTREMELY COOL STARS. B. T. Ulrich, G. Neugebauer, D. McCammon, R. B. Leighton, E. E. Hughes and E. Becklin in *The Astrophysical Journal,* Vol. 146, No. 1, pages 288–290; October 1966.

INFRARED ASTRONOMY. A. G. W. Cameron and P. J. Brancazio. Gordon & Breach, Science Publishers, Inc., 1968.

INFRARED PHOTOMETRY OF T TAURI STARS AND RELATED OBJECTS. Eugenio E. Mendoza V in *The Astrophysical Journal,* Vol. 143, No. 3, pages 1010–1014; March 1966.

OBSERVATIONS OF AN INFRARED STAR IN THE ORION NEBULA. E. E. Becklin and G. Neugebauer in *The Astrophysical Journal,* Vol. 147, No. 2, pages 799–802; February 1967.

OBSERVATIONS OF EXTREMELY COOL STARS. G. Neugebauer, D. E. Martz and R. B. Leighton in *The Astrophysical Journal,* Vol. 142, No. 1, pages 399–401; 1 July 1965.

26. Ultraviolet Astronomy

The Author

LEO GOLDBERG is Higgins Professor of Astronomy and chairman of the department of astronomy at Harvard University and director of the Harvard College Observatory. Professor Goldberg was a student at Harvard from 1930 to 1938, receiving his bachelor's degree in 1934, his master's degree in 1937 and his Ph.D. in 1938. For 19 years before returning to Harvard in 1960 he was at the University of Michigan, serving from 1946 to 1960 as director of its observatory and chairman of the department of astronomy. He describes "the application of atomic physics to the interpretation of astrophysical problems" as his "deepest interest." He also writes: "I have served on almost every conceivable committee in universities, in professional societies and in Government agencies. Only a very few of these have been genuinely satisfying."

Bibliography

ASTRONOMICAL RESEARCH WITH THE LARGE SPACE TELESCOPE. Lyman Spitzer, Jr., in *Science,* Vol. 161, No. 3838, pages 225–229; 19 July 1968.

THE SUN. Leo Goldberg and Edward R. Dyer, Jr., in *Science in Space,* edited by Lloyd V. Berkner and Hugh Odishaw. McGraw-Hill Book Company, 1961.

ULTRAVIOLET AND X RAYS FROM THE SUN. Leo Goldberg in *Annual Review of Astronomy and Astrophysics,* Vol. 5, pages 279–324; 1967.

ULTRAVIOLET ASTRONOMY. R. Wilson and A. Boksenberg in *Annual Review of Astronomy and Astrophysics,* Vol. 7, pages 421–472; 1969.

ULTRAVIOLET SOLAR IMAGES FROM SPACE. Leo Goldberg, Robert W. Noyes, William H. Parkinson, Edmond M. Reeves and George L. Withbroe in *Science,* Vol. 162, No. 3849, pages 95–99; 4 October 1968.

27. Pulsars

The Author

ANTONY HEWISH is a member of the group that discovered the first pulsar during observations made at the Mullard Radio Astronomy Observatory of the University of Cambridge. In addition to his work at the observatory, Hewish is a fellow of Churchill College and a university lecturer in physics. He entered the University of Cambridge as a student in 1942 but soon left for military service at the Royal Aircraft Establishment in Farnborough. There, he writes, he "met Ryle," who is now Sir Martin Ryle, professor of radio astronomy at Cambridge. "Returned to Cambridge 1946," Hewish continues, "graduated 1948 and joined Ryle's research group at Cavendish Laboratory. A natural choice—brand-new subject, prior acquaintance with Ryle, etc. Awarded Ph.D. 1952 and elected research fellow, Gonville and Caius College. Elected Fellow of Royal Society 1968." Describing his work, he says: "Besides pulsars, which came my way rather gratuitously, I am involved in investigations of interplanetary space using the scintillation of quasars. Starting from my early days as a research student, when I hacked through piles of brass tubing to make dipoles for Ryle, I have spent a good deal of my time designing feed arrays for the radio telescopes used in the 3*C* and 4*C* surveys at Cambridge."

Bibliography

THE ASTOUNDING PULSARS. Frank D. Drake in *Science Year,* pages 36–51; 1969.

THE 1968 TEXAS SYMPOSIUM: PULSARS. Louis C. Green in *Sky and Telescope,* Vol. 37, pages 214–218; April 1969.

THE PULSARS. F. Graham Smith in *Science Journal,* Vol. 5, No. 6, pages 32–39; June 1969.

PULSATING STARS, a *Nature* Reprint. Plenum Press, 1968.

PULSATING STARS 2, a *Nature* Reprint. Plenum Press, 1969.

STRUCTURE AND EVOLUTION OF THE STARS. Martin Schwarzschild. Princeton University Press, 1958.

VIII
Cosmology

28. The Red-Shift

The Author

ALLAN R. SANDAGE is astronomer at the Mount Wilson and Palomar Observatories. He attended Miami University in Ohio, transferred "by the grace of Uncle Sam" to the Navy in 1945 and completed his undergraduate work at the University of Illinois in 1948. He acquired his doctorate in astronomy at the California Institute of Technology in 1953, having meanwhile joined the staff of Mount Wilson and Palomar. In 1956 Sandage wrote to the editors of *SCIENTIFIC AMERICAN*, "As I recall, it was *Buck Rogers in the 25th Century* that steered me into astronomy at the unknowing age of 10. This unfortunate interest (from our neighbors' point of view) took the form of dragging a telescope, tables and other observing paraphernalia into our back yard at three in the morning to look at meteor showers and the like. For some reason the neighbors failed to understand the importance of such operations, and I failed to understand their need for sleep. Today, 20 years later, I sit sleepily at work on Palomar Mountain and wonder how young boys of 10 can take sleep so casually. From two to four A.M. *all* professional astronomers are sleepy. If questioned in this interval, most of them would express serious doubt as to whether astronomy was worth it all. (The doubts always disappear in the morning.) From 1951 to 1953 I was an assistant to the late Edwin P. Hubble. The associations with Hubble, Milton L. Humason and Walter Baade have been the high points in my scientific career."

Bibliography

COSMOLOGY: A SEARCH FOR TWO NUMBERS. Allan R. Sandage in *Physics Today*, Vol. 23, No. 2, pages 34–41; February 1970.

THE EXPANSION OF THE UNIVERSE. Paul Couderc. Faber & Faber, Ltd., 1952.

THE REALM OF THE NEBULAE. Edwin Hubble. Yale University Press, 1936.

RED-SHIFTS AND MAGNITUDES OF EXTRA-GALACTIC NEBULAE. M. L. Humason, N. U. Mayall and A. R. Sandage in *The Astronomical Journal*, Vol. 61, No. 3, pages 97–162; April 1956.

THE TIME SCALE FOR CREATION. Allan R. Sandage in *Galaxies and the Universe*, edited by Lodewijk Woltjer, pages 75–112. Columbia University Press, 1968.

29. The Evolutionary Universe

The Author

GEORGE GAMOW was professor of physics at the University of Colorado at the time of his death in 1968; he had just accepted that post in 1956 when this article was published. He was born in Odessa, Russia, in 1904, and studied nuclear physics at the University of Leningrad, where he received his doctoral degree in 1928; at the University of Copenhagen under Niels Bohr; and at the University of Cambridge under Ernest Rutherford. In 1934 Gamow emigrated to the U. S., where he served for over 20 years as professor of physics at George Washington University. During this period Gamow found his interest turning from the atomic nucleus to astrophysics, the theory of the expanding universe and later to fundamental problems of biology, including molecular genetics and the synthesis of proteins. In 1938 he began to write and illustrate a series of popular articles for the British magazine *Discovery* concerning the macro- and microcosmic adventures of "Mr. Tompkins," whose exploits have since appeared in several books and 19 languages.

A quixotic streak, which has enlivened his many popular books and articles, sometimes extends to his most serious scientific publications. On one occasion when he and Ralph A. Alpher were preparing a paper, they invited Hans Bethe of Cornell University to collaborate with them. The paper, which happened to concern the beginning of the universe, was therefore most appropriately authored by Alpher, Bethe and Gamow.

Bibliography

THE AGE OF CREATION. Allan R. Sandage in *Science Year*, pages 56–69; 1968.

THE CREATION OF THE UNIVERSE. George Gamow. The Viking Press, 1952.

FACT AND THEORY IN COSMOLOGY. G. C. McVittie. Eyre and Spottiswoode, 1961.

THE MYSTERY OF THE EXPANDING UNIVERSE. W. B. Bonner. Macmillan, 1964.

THE PRIMEVAL ATOM. Georges Lemaître. D. Van Nostrand Company, Inc., 1950.

RELATIVISTIC COSMOLOGY. Wolfgang Rindler in *Physics Today*, Vol. 20, No. 11, pages 23–31; November 1967.

30. The Problem of the Quasi-stellar Objects

The Authors

FRED HOYLE and GEOFFREY BURBIDGE have worked together on many problems in astrophysics. Hoyle is Plumian Professor of Astronomy and Experimental Philosophy at St. John's College in the University of Cambridge. He is also a member of the staff of the Mount Wilson and Palomar Observatories. He was born in Yorkshire in 1914; by the time he was six, he had taught himself the multiplication tables up to 12 times 12. Not all of his early experience with mathematics was so prodigious. One of his first teachers slapped him for miscounting the number of petals on a flower; his sense of justice was so outraged that he refused to return to the school. When he was 13 his parents bought him a three-inch telescope and allowed him to sit up all night with it. In 1939 he won a prize fellowship at St. John's. The next year he joined an Admiralty research group, but continued to devote much of his spare time to astronomy. After the war he returned to Cambridge. He is known not only as an expositor of science but also as a novelist and playwright.

Burbidge is professor of physics at the University of California at San Diego. His principal field is astrophysics, to which he turned from meson physics when he met and married an astronomer who was working at the University of London Observatory. (For other details of his career see the biographical note for article 15, "Stellar Populations.")

Bibliography

THE EXISTENCE OF A MAJOR NEW CONSTITUENT OF THE UNIVERSE: THE QUASI-STELLAR GALAXIES. Allan Sandage in *The Astrophysical Journal,* Vol. 141, No. 4, pages 1560–1578; 15 May 1965.

ON THE INTERPRETATION OF THE LINE SPECTRA OF QUASI-STELLAR OBJECTS. G. R. Burbidge, M. Burbidge, F. Hoyle and C. R. Lynds in *Nature,* Vol. 210, No. 5038, pages 774–778; 21 May 1966.

QUASARS. Harlan J. Smith in *Applied Optics,* Vol. 5, pages 1701–1719; November 1966.

QUASI-STELLAR OBJECTS. Geoffrey Burbidge and Margaret Burbidge. W. H. Freeman and Company, 1967.

QUASI-STELLAR OBJECTS. Maarten Schmidt in *Science Journal,* Vol. 2, pages 77–83; October 1966.

QUASISTELLAR OBJECTS. Maarten Schmidt in *Annual Review of Astronomy and Astrophysics,* Vol. 7, pages 527–552; 1969.

RELATION BETWEEN THE RED-SHIFTS OF QUASI-STELLAR OBJECTS AND THEIR RADIO AND OPTICAL MAGNITUDES. F. Hoyle and G. R. Burbidge in *Nature,* Vol. 210, No. 5043, pages 1346–1347; 25 June 1966.

31. The Primeval Fireball

The Authors

P. J. E. PEEBLES and DAVID T. WILKINSON both are professors of physics at Princeton University. Peebles was graduated from the University of Manitoba and obtained a Ph.D. at Princeton in 1961. Wilkinson received his bachelor's, master's and doctor's degrees from the University of Michigan. At Princeton their major concern is achieving a better understanding of gravity. They write: "Since gravity is such a miserably weak force in the laboratory, we have turned more and more to clues that might be uncovered on the much larger scale provided by geophysics, astrophysics and even cosmology." In their collaboration Peebles is the theorist, Wilkinson the experimentalist. They write that for outside activities Peebles "raises tropical fish and weird plants," Wilkinson "is active in the Unitarian Church and is a jazz fan" and "both enjoy beachcombing along the New Jersey shoreline."

Bibliography

COSMIC BLACK-BODY RADIATION. R. H. Dicke, P. J. E. Peebles, P. G. Roll and D. T. Wilkinson in *The Astrophysical Journal,* Vol. 142, No. 1, pages 414–419; 1 July 1965.

FROM GENESIS TO DOOMSDAY. E. L. Schücking and Brenda Biram in *Science Year,* pages 60–73; 1967.

A MEASUREMENT OF EXCESS ANTENNA TEMPERATURE AT 4080 MC/S. A. A. Penzias and R. W. Wilson in *The Astrophysical Journal,* Vol. 142, No. 1, pages 419–421; 1 July 1965.

THE PRIMEVAL FIREBALL TODAY. R. B. Partridge in *American Scientist,* Vol. 57, pages 37–74; Spring 1969.

32. Cosmology before and after Quasars

The Author

DENNIS W. SCIAMA holds a Peterhouse Fellowship and Lectureship in the Department of Applied Mathematics and Theoretical Physics, Cambridge University, where he has built up a group of graduate and post-doctoral students working on general relativity, cosmology, and astrophysics. Born in Manchester, England, in 1926, he was a student of the great theoretical physicist P. A. M. Dirac, and received his Ph.D. degree from Cambridge University in 1952. In that same year he obtained a Research Fellowship at Trinity College, Cambridge, and since then has been a member of the Institute for Advanced Study at Princeton and an Agassiz Fellow at Harvard University.

Bibliography

COSMOLOGY AFTER HALF A CENTURY. W. H. McCrea in *Science,* Vol. 160, pages 1295–1299; 21 June 1968.

THE COUNTS OF RADIO SOURCES. M. Ryle in *Annual Review of Astronomy and Astrophysics,* Vol. 6, pages 249–266; 1968.

THE EVOLUTION OF THE UNIVERSE. Hong-Yee Chiu in *Science Journal,* Vol. 4, No. 8, pages 33–38; August 1968.

THE PHYSICAL FOUNDATIONS OF GENERAL RELATIVITY. D. W. Sciama. Doubleday Anchor, 1969.

THE STRUCTURE OF THE UNIVERSE. E. L. Schatzman. World University Library, McGraw-Hill, 1968.

33. The Origin of Galaxies

The Authors

MARTIN J. REES and JOSEPH SILK, who provide the last selection, are respectively at the Institute of Theoretical Astronomy of the University of Cambridge and the Princeton University Observatory. Rees obtained his master's degree in mathematics and his Ph.D. in astrophysics from Cambridge. His interests include cosmology, diffuse matter in space and theoretical radio astronomy. Silk was a Cambridge undergraduate and obtained his Ph.D. at Harvard University. In the summer of 1970, he joined the department of astronomy of the University of California at Berkeley.

Bibliography

THE BLACK-BODY RADIATION CONTENT OF THE UNIVERSE AND THE FORMATION OF GALAXIES. P. J. E. Peebles in *The Astrophysical Journal,* Vol. 142, No. 4, pages 1317–1326; 15 November 1965.

THE CASE FOR A HIERARCHICAL COSMOLOGY. G. de Vaucouleurs in *Science,* Vol. 167, pages 1203–1213; 27 February 1970.

COSMIC BLACK-BODY RADIATION AND GALAXY FORMATION. Joseph Silk in *The Astrophysical Journal,* Vol. 151, No. 1, Part 2, pages 459–471; February 1968.

THE FORMATION AND EARLY DYNAMICAL HISTORY OF GALAXIES. G. B. Field in *Stars and Stellar Systems, Vol. IX: Galaxies and the Universe,* edited by A. Sandage and M. Sandage. The University of Chicago Press, in press.

THE FORMATION OF STARS AND GALAXIES: UNIFIED HYPOTHESES. David Layzer in *Annual Review of Astronomy and Astrophysics: Vol. II,* edited by Leo Goldberg, Armin J. Deutsch and David Layzer. Annual Reviews, Inc., 1964.

SOME CURRENT PROBLEMS IN GALAXY FORMATION. M. J. Rees in *Italian Physical Society Proceedings of the International School of Physics Enrico Fermi, Course 47: Relativity and Cosmology,* edited by R. K. Sachs. Academic Press, in press.

Index

NOTE: Page numbers in **boldface** indicate illustrations.